The La

W9-BLR-160

The sun is just a star.

Nuclear reactions make energy.

Earth Moon

(not to scale)

First hominids

One-inch line

Goal line

Ten thousand years ago, on the 0.0026 inch line, humans begin building cities and modern civilization begins.

50

40

30

20

10

Formation of the sun and planets from a cloud of interstellar gas and dust

Life begins in Earth's oceans.

Cambrian explosion 540 million years ago: Life in Earth's oceans becomes complex.

Life first emerges onto the land.

Age of Dinosaurs

TODAY

40

30

20

10

Over billions of years, generation after generation of stars have lived and died, cooking the hydrogen and helium of the big bang into the atoms of which you are made. Study the last inch of the time line to see the rise of human ancestors and the origin of civilization. Only in the last flicker of a moment on the time line have astronomers begun to understand the story.

Flash Reference: H–R Diagram

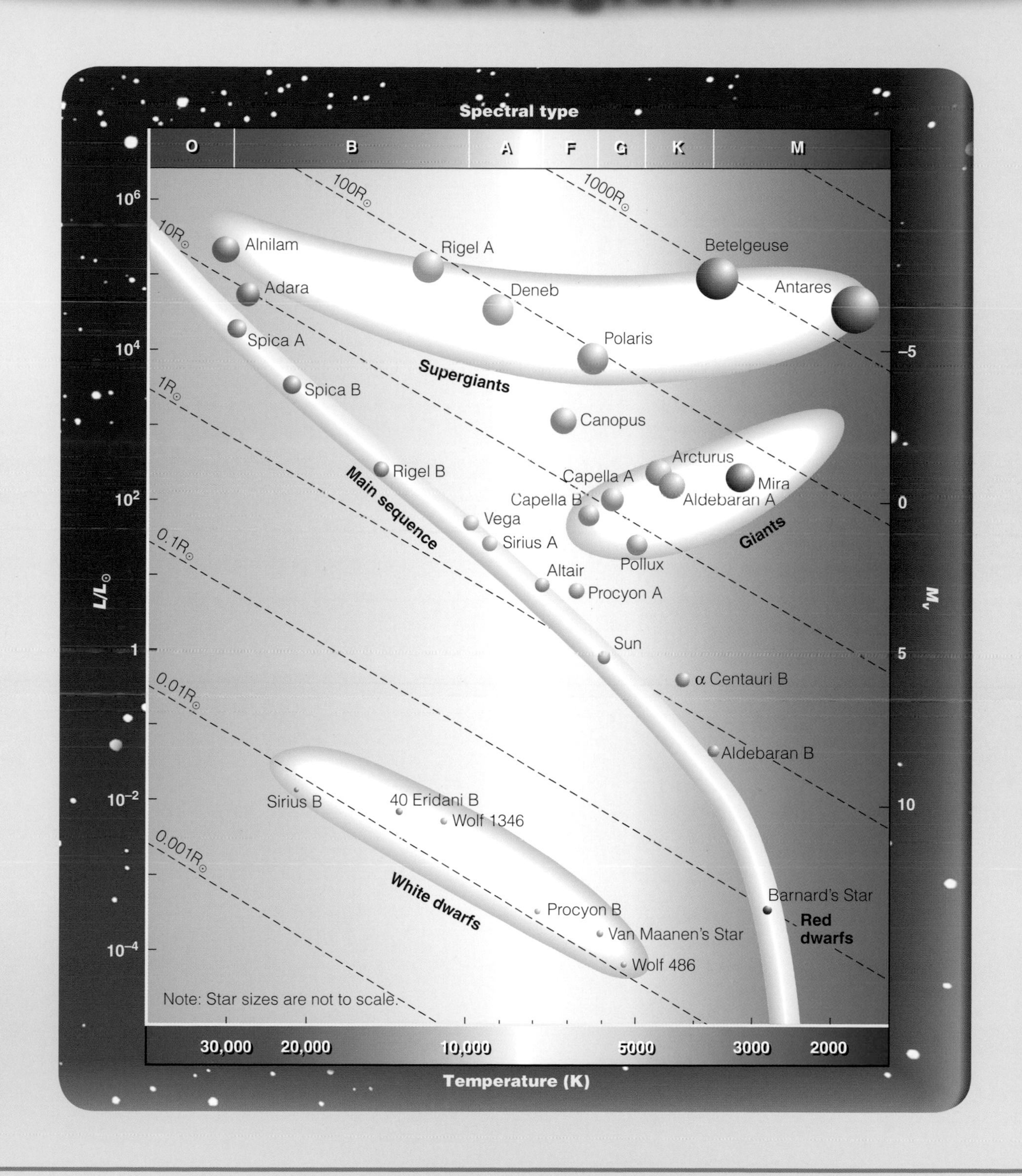

The H–R diagram is the key to understanding stars, their birth, their long lives, and their eventual deaths. Luminosity (*$L/L_\odot$*) refers to the total amount of energy that a star emits in terms of the sun's luminosity, and the temperature refers to the temperature of its surface. Together, the temperature and luminosity of a star locate it on the H–R diagram and tell astronomers its radius, its family relationships with other stars, and a great deal about its history and fate.

Flash Reference: Comparative Planetology

The terrestrial or Earthlike planets lie very close to the sun, and their orbits are hardly visible in a diagram that includes the outer planets.

Mercury, Venus, Earth and its moon, and Mars are small worlds made of rock and metal with relatively thin, or no, atmospheres.

The outer worlds of our solar system orbit far from the sun. Jupiter, Saturn, Uranus, and Neptune are Jovian or Jupiter-like planets much bigger than Earth. They contain large amounts of low-density gases.

The Terrestrial Worlds

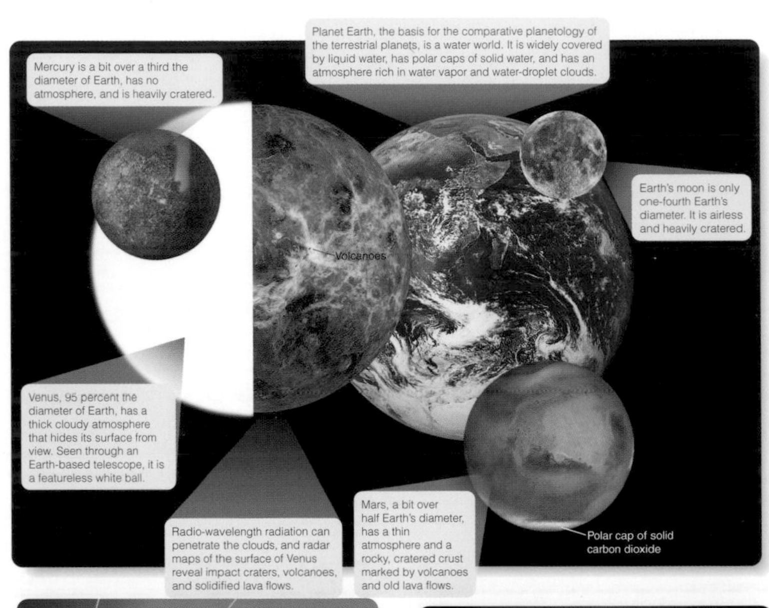

Mercury is a bit over a third the diameter of Earth, has no atmosphere, and is heavily cratered.
Planet Earth, the basis for the comparative planetology of the terrestrial planets, is a water world. It is widely covered by liquid water, has polar caps of solid water, and has an atmosphere rich in water vapor and water-droplet clouds.
Earth's moon is only one-fourth Earth's diameter. It is airless and heavily cratered.
Volcanoes
Venus, 95 percent the diameter of Earth, has a thick cloudy atmosphere that hides its surface from view. Seen through an Earth-based telescope, it is a featureless white ball.
Radio-wavelength radiation can penetrate the clouds, and radar maps of the surface of Venus reveal impact craters, volcanoes, and solidified lava flows.
Mars, a bit over half Earth's diameter, has a thin atmosphere and a rocky, cratered crust marked by volcanoes and old lava flows.
Polar cap of solid carbon dioxide

Planetary Orbits

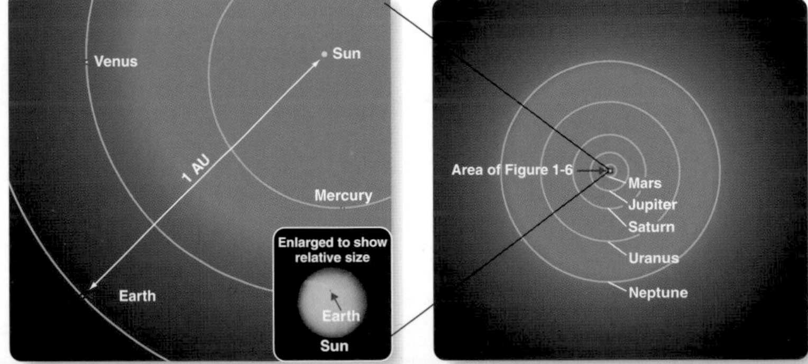

Venus
Sun
1 AU
Mercury
Earth
Enlarged to show relative size
Earth
Sun
Area of Figure 1-6
Mars
Jupiter
Saturn
Uranus
Neptune

The Outer Worlds

Jupiter, more than 11 times Earth's diameter, is the largest planet in our solar system.
The cloud belts and zones on Saturn are less distinct than those on Jupiter.
Uranus is four times Earth's diameter.
Shadow of one of Jupiter's many moons
Earth is the largest of the terrestrial worlds, but it is small compared with the Jovian planets.
Neptune, like Uranus, is generally blue because of small amounts of methane in its hydrogen-rich atmosphere.

• See page 444 for the terrestrial planets. See page 5 for the two orbital diagrams. See page 512 for the outer worlds.

Flash Reference: Arrows

This book is designed to use arrows to alert you to important concepts in diagrams and graphs. Some arrows point things out, but others represent motion, force, or even the flow of light. Look at arrows in the book carefully and use this Flash Reference card to catch all of the arrow clues.

Point at things:

You are here

Force:

Process flow: **Measurement:**

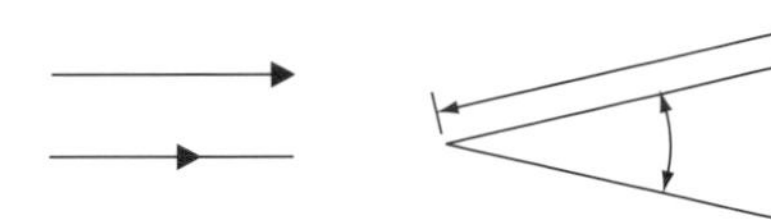

Direction: **Radio waves, infrared, photons:**

Motion:

Rotation 2-D

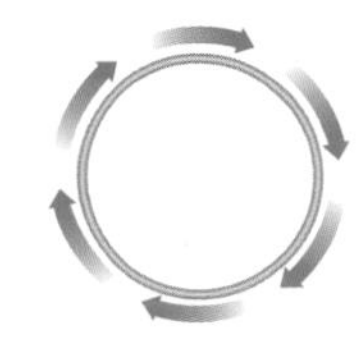

Rotation 3-D

Linear

Light flow:
Updated arrow style

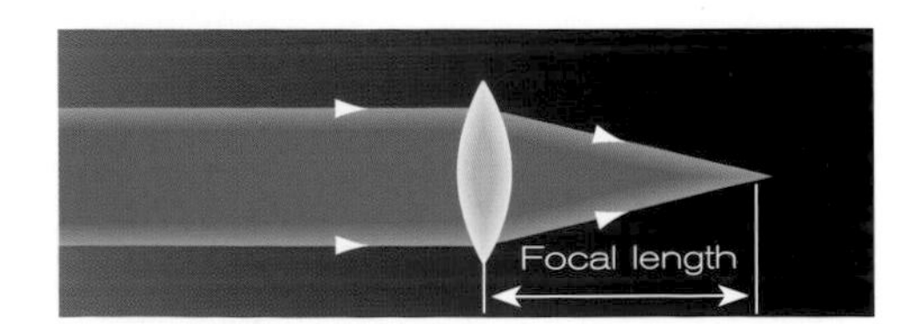

Focal length

Foundations of Astronomy

About the Author

Mike Seeds is Professor of Astronomy at Franklin and Marshall College, where he has taught astronomy since 1970. His research interests have focused on peculiar variable stars and the automation of astronomical telescopes. He extended his research by serving as Principal Astronomer in charge of the Phoenix 10, the first fully robotic telescope, located in southern Arizona. In 1989, he received the Christian R. and Mary F. Lindback Award for Distinguished Teaching. In addition to teaching, writing, and research, Mike has published educational systems for use in computer-smart classrooms. His interest in the history of astronomy led him to offer upper-level courses "Archaeoastronomy" and "Changing Concepts of the Universe," a history of cosmology from ancient times to Newton. He has also published educational software for preliterate toddlers. Mike was Senior Consultant in the creation of the 20-episode telecourse to accompany this text.

He is the author of *Horizons: Exploring the Universe,* Tenth Edition (2007), and *Astronomy: The Solar System and Beyond,* Fifth Edition (2006), published by Brooks/Cole.

About the Cover

Pictured on the cover is the Hale Telescope inside its dome on Palomar Mountain in San Diego County, California. The dome appears transparent in this time exposure, showing the 200-inch telescope inside. (© Roger Ressmeyer/CORBIS)

10 TENTH EDITION

Foundations of Astronomy

Michael A. Seeds

Joseph R. Grundy Observatory
Franklin and Marshall College

BROOKS/COLE

Australia • Brazil • Canada •Mexico • Singapore • Spain
United Kingdom •United States

For Emery and Helen Seeds

***Foundations of Astronomy,* Tenth Edition**
Michael A. Seeds

Astronomy Editor: *Chris Hall*
Development Editor: *Rebecca Heider*
Assistant Editor: *Sylvia Krick*
Editorial Assistants: *Shawn Vasquez, Stefanie Chase*
Technology Project Manager: *Sam Subity*
Marketing Manager: *Mark Santee*
Marketing Assistant: *Elizabeth Wong*
Marketing Communications Manager: *Darlene Amidon-Brent*
Project Manager, Editorial Production: *Hal Humphrey*
Art Director: *Vernon Boes*
Print Buyer: *Karen Hunt*
Permissions Editor: *Bob Kauser*
Production Service: *Graphic World Publishing Services*
Text Designer: *Liz Harasymczuk*
Photo Researcher: *Kathleen Olson*
Cover Designer: *Irene Morris*
Cover Printer: *Transcontinental Interglobe*
Compositor: *Graphic World Inc.*
Printer: *Transcontinental Interglobe*
Cover Image: © *Roger Ressmeyer/CORBIS.* Insets (top to bottom): (Pluto and Its Moons) NASA, ESA, H. Weaver (JHU/APL), A. Stern (SwRI), and the HST Pluto Companion Search Team; (Dark Matter Ring in Galaxy Cluster CI 0024+17) NASA, ESA, M.J. Jee and H. Ford (Johns Hopkins University); (Artist's Concept, Black Hole Digesting Remnants of a Star) NASA/JPL-Caltech.

© 2008, 2007 Thomson Brooks/Cole, a part of The Thomson Corporation. Thomson, the Star logo, and Brooks/Cole are trademarks used herein under license.

ALL RIGHTS RESERVED. No part of this work covered by the copyright hereon may be reproduced or used in any form or by any means—graphic, electronic, or mechanical, including photocopying, recording, taping, Web distribution, information storage and retrieval systems, or in any other manner—without the written permission of the publisher.

Printed in Canada
1 2 3 4 5 6 7 11 10 09 08 07

Library of Congress Control Number: 2007935690

ISBN-13: 978-0-495-38724-4
ISBN-10: 0-495-38724-X

Thomson Higher Education
10 Davis Drive
Belmont, CA 94002-3098
USA

For more information about our products, contact us at:
Thomson Learning Academic Resource Center
1-800-423-0563

For permission to use material from this text or product, submit a request online at **http://www.thomsonrights.com.**
Any additional questions about permissions can be submitted by e-mail to **thomsonrights@thomson.com.**

ExamView® and *ExamView Pro*® are registered trademarks of FSCreations, Inc. Windows is a registered trademark of the Microsoft Corporation used herein under license. Macintosh and Power Macintosh are registered trademarks of Apple Computer, Inc. Used herein under license.

© 2008 Thomson Learning, Inc. All Rights Reserved. Thomson Learning WebTutor™ is a trademark of Thomson Learning, Inc.

Brief Contents

Part 1: Exploring the Sky

CHAPTER 1 WHAT ARE WE? HOW DO WE KNOW? 1
CHAPTER 2 THE SKY 12
CHAPTER 3 CYCLES OF THE MOON 33
CHAPTER 4 THE ORIGIN OF MODERN ASTRONOMY 52
CHAPTER 5 GRAVITY 78
CHAPTER 6 LIGHT AND TELESCOPES 102

Part 2: The Stars

CHAPTER 7 ATOMS AND STARLIGHT 127
CHAPTER 8 THE SUN 149
CHAPTER 9 THE FAMILY OF STARS 174
CHAPTER 10 THE INTERSTELLAR MEDIUM 202
CHAPTER 11 THE FORMATION OF STARS 220
CHAPTER 12 STELLAR EVOLUTION 241
CHAPTER 13 THE DEATHS OF STARS 264
CHAPTER 14 NEUTRON STARS AND BLACK HOLES 287

Part 3: The Universe

CHAPTER 15 THE MILKY WAY GALAXY 314
CHAPTER 16 GALAXIES 342
CHAPTER 17 GALAXIES WITH ACTIVE NUCLEI 367
CHAPTER 18 COSMOLOGY IN THE 21ST CENTURY 389

Part 4: The Solar System

CHAPTER 19 THE ORIGIN OF THE SOLAR SYSTEM 416
CHAPTER 20 EARTH: THE STANDARD OF COMPARATIVE PLANETOLOGY 443
CHAPTER 21 THE MOON AND MERCURY: COMPARING AIRLESS WORLDS 460
CHAPTER 22 COMPARATIVE PLANETOLOGY OF VENUS AND MARS 482
CHAPTER 23 COMPARATIVE PLANETOLOGY OF JUPITER AND SATURN 510
CHAPTER 24 URANUS, NEPTUNE, AND THE DWARF PLANETS 543
CHAPTER 25 METEORITES, ASTEROIDS, AND COMETS 568

Part 5: Life

CHAPTER 26 LIFE ON OTHER WORLDS 598

Contents

Part 1: Exploring the Sky

Chapter 1 | What Are We? How Do We Know? 1

1-1 WHY STUDY ASTRONOMY? 2
1-2 WHERE ARE WE? 2
1-3 WHEN IS NOW? 8

Chapter 2 | The Sky 12

2-1 THE STARS 13
2-2 THE SKY AND ITS MOTION 17
2-3 THE CYCLES OF THE SUN 22
2-4 ASTRONOMICAL INFLUENCES ON EARTH'S CLIMATE 27

Chapter 3 | Cycles of the Moon 33

3-1 THE CHANGEABLE MOON 34
3-2 LUNAR ECLIPSES 35
3-3 SOLAR ECLIPSES 40
3-4 PREDICTING ECLIPSES 46

Chapter 4 | The Origin of Modern Astronomy 52

4-1 THE ROOTS OF ASTRONOMY 53
4-2 THE COPERNICAN REVOLUTION 59
4-3 THE PUZZLE OF PLANETARY MOTION 68
4-4 MODERN ASTRONOMY 74

Chapter 5 | Gravity 78

5-1 GALILEO AND NEWTON 79
5-2 ORBITAL MOTION AND TIDES 86
5-3 EINSTEIN AND RELATIVITY 94

Chapter 6 | Light and Telescopes 102

6-1 RADIATION: INFORMATION FROM SPACE 103
6-2 OPTICAL TELESCOPES 106
6-3 SPECIAL INSTRUMENTS 115
6-4 RADIO TELESCOPES 117
6-5 ASTRONOMY FROM SPACE 120

How Do We Know?

1-1 The Scientific Method 3
1-2 Scientific Arguments 8
2-1 Scientific Models 18
2-2 Pseudoscience 27
2-3 Evidence as the Foundation of Science 29
3-1 Scientific Imagination 41
4-1 Scientific Revolutions 64
5-1 Hypothesis, Theory, and Law 83
5-2 Cause and Effect 85
5-3 Testing a Theory by Prediction 93
6-1 Resolution and Precision 109

Concept Art Portfolios

The Sky Around You 20–21
The Cycle of the Seasons 24–25
The Phases of the Moon 36–37
The Ancient Universe 60–61
Orbiting Earth 88–89
Modern Astronomical Telescopes 112–113
The Great Observatories in Space 122–123

Focus on Fundamentals 1 | Mass 84
Focus on Fundamentals 2 | Energy 91

Part 2: The Stars

Chapter 7 | Atoms and Starlight 127

7-1 ATOMS 128

7-2 THE INTERACTION OF LIGHT AND MATTER 131

7-3 STELLAR SPECTRA 135

Chapter 8 | The Sun 149

8-1 THE SOLAR ATMOSPHERE 150

8-2 NUCLEAR FUSION IN THE SUN 156

8-3 SOLAR ACTIVITY 161

Chapter 9 | The Family of Stars 174

9-1 MEASURING THE DISTANCES TO STARS 175

9-2 INTRINSIC BRIGHTNESS 178

9-3 THE DIAMETERS OF STARS 180

9-4 THE MASSES OF STARS 187

9-5 A SURVEY OF THE STARS 194

Chapter 10 | The Interstellar Medium 202

10-1 VISIBLE-WAVELENGTH OBSERVATIONS 203

10-2 LONG- AND SHORT-WAVELENGTH OBSERVATIONS 209

10-3 A MODEL OF THE INTERSTELLAR MEDIUM 214

Chapter 11 | The Formation of Stars 220

11-1 MAKING STARS FROM THE INTERSTELLAR MEDIUM 221

11-2 THE SOURCE OF STELLAR ENERGY 230

11-3 STELLAR STRUCTURE 232

11-4 THE ORION NEBULA 235

Chapter 12 | Stellar Evolution 241

12-1 MAIN-SEQUENCE STARS 242

12-2 POST-MAIN-SEQUENCE EVOLUTION 249

12-3 EVIDENCE OF EVOLUTION: STAR CLUSTERS 255

12-4 EVIDENCE OF EVOLUTION: VARIABLE STARS 258

Chapter 13 | The Deaths of Stars 264

13-1 LOWER-MAIN-SEQUENCE STARS 265

13-2 THE EVOLUTION OF BINARY STARS 271

13-3 THE DEATHS OF MASSIVE STARS 275

Chapter 14 | Neutron Stars and Black Holes 287

14-1 NEUTRON STARS 288

14-2 BLACK HOLES 299

14-3 COMPACT OBJECTS WITH DISKS AND JETS 306

How Do We Know?

7-1 Quantum Mechanics 130

8-1 Scientific Confidence 160

8-2 Confirmation and Consolidation 166

9-1 Chains of Inference 189

9-2 Basic Scientific Data 195

10-1 Separating Facts from Theories 215

11-1 Theories and Proof 226

12-1 Mathematical Models 244

13-1 Toward Ultimate Causes 270

14-1 Checks on Fraud in Science 305

Concept Art Portfolios

Atomic Spectra 136–137

Sunspots and the Sunspot Cycle 162–163

Magnetic Solar Phenomena 168–169

The Family of Stars 196–197

Three Kinds of Nebulae 204–205

Observational Evidence of Star Formation 228–229

Star Formation in the Orion Nebula 236–237

Star Cluster H–R Diagrams 256–257

The Formation of Planetary Nebulae 268–269

The Lighthouse Model of a Pulsar 292–293

Focus on Fundamentals 3 | Temperature, Heat, and Thermal Energy 133

Focus on Fundamentals 4 | Density 145

Focus on Fundamentals 5 | Pressure 209

Celestial Profile 1 | The Sun 151

Part 3: The Universe

Chapter 15 | The Milky Way Galaxy 314

15-1 THE NATURE OF THE MILKY WAY GALAXY 315

15-2 THE ORIGIN OF THE MILKY WAY GALAXY 323

15-3 SPIRAL ARMS 329

15-4 THE NUCLEUS 338

Chapter 16 | Galaxies 342

16-1 THE FAMILY OF GALAXIES 343

16-2 MEASURING THE PROPERTIES OF GALAXIES 348

16-3 THE EVOLUTION OF GALAXIES 356

Chapter 17 | Galaxies with Active Nuclei 367

17-1 ACTIVE GALACTIC NUCLEI 368

17-2 QUASARS 378

Chapter 18 | Cosmology in the 21st Century 389

18-1 INTRODUCTION TO THE UNIVERSE 390

18-2 THE SHAPE OF SPACE AND TIME 400

18-3 21st-CENTURY COSMOLOGY 404

How Do We Know?

15-1 **Calibration 318**

15-2 **Nature as Processes 325**

16-1 **Classification in Science 345**

17-1 **Statistical Evidence 370**

18-1 **Reasoning by Analogy 392**

18-2 **Science: A System of Knowledge 400**

Concept Art Portfolios

Sagittarius A 336–337*

Galaxy Classification 346–347

Interacting Galaxies 360–361

Cosmic Jets and Radio Lobes 372–373

Part 4: The Solar System

Chapter 19 | The Origin of the Solar System 416

- 19-1 THEORIES OF EARTH'S ORIGIN 417
- 19-2 A SURVEY OF THE SOLAR SYSTEM 419
- 19-3 THE STORY OF PLANET BUILDING 426
- 19-4 PLANETS ORBITING OTHER STARS 434

Chapter 20 | Earth: The Standard of Comparative Planetology 443

- 20-1 A TRAVEL GUIDE TO THE TERRESTRIAL PLANETS 444
- 20-2 THE EARLY HISTORY OF EARTH 446
- 20-3 THE SOLID EARTH 447
- 20-4 EARTH'S ATMOSPHERE 454

Chapter 21 | The Moon and Mercury: Comparing Airless Worlds 460

- 21-1 THE MOON 461
- 21-2 MERCURY 473

Chapter 22 | Comparative Planetology of Venus and Mars 482

- 22-1 VENUS 483
- 22-2 MARS 494
- 22-3 THE MOONS OF MARS 505

Chapter 23 | Comparative Planetology of Jupiter and Saturn 510

- 23-1 A TRAVEL GUIDE TO THE OUTER PLANETS 511
- 23-2 JUPITER 512
- 23-3 JUPITER'S FAMILY OF MOONS 521
- 23-4 SATURN 528
- 23-5 SATURN'S MOONS 534

Chapter 24 | Uranus, Neptune, and the Dwarf Planets 543

- 24-1 URANUS 544
- 24-2 NEPTUNE 557
- 24-3 THE DWARF PLANETS 562

Chapter 25 | Meteorites, Asteroids, and Comets 568

- 25-1 METEORITES 569
- 25-2 ASTEROIDS 577
- 25-3 COMETS 584
- 25-4 IMPACTS ON EARTH 592

How Do We Know?

- 19-1 Evolution and Catastrophe 420
- 19-2 Courteous Skeptics 438
- 20-1 Studying an Unseen World 448
- 21-1 How Hypotheses and Theories Unify the Details 466
- 22-1 Data Manipulation 487
- 23-1 Science, Technology, and Engineering 517
- 23-2 Who Pays for Science? 534
- 24-1 Scientific Discoveries 546
- 25-1 Selection Effects 575

Concept Art Portfolios

- *Terrestrial and Jovian Planets* 422–423
- *The Active Earth* 452–453
- *Impact Cratering* 464–465
- *Volcanoes* 490–491
- *Jupiter's Atmosphere* 518–519
- *The Ice Rings of Saturn* 532–533
- *The Rings of Uranus and Neptune* 552–553
- *Observations of Asteroids* 578–579
- *Comet Observations* 586–587

- Celestial Profile 2 | Earth 445
- Celestial Profile 3 | The Moon 461
- Celestial Profile 4 | Mercury 475
- Celestial Profile 5 | Venus 483
- Celestial Profile 6 | Mars 495
- Celestial Profile 7 | Jupiter 513
- Celestial Profile 8 | Saturn 529
- Celestial Profile 9 | Uranus 547
- Celestial Profile 10 | Neptune 559

Part 5: Life

Chapter 26 | Life on Other Worlds 598

26-1 THE NATURE OF LIFE 599

26-2 THE ORIGIN OF LIFE 602

26-3 COMMUNICATION WITH DISTANT CIVILIZATIONS 609

How Do We Know?

26-1 **Judging Evidence 610**

Concept Art Portfolios

DNA: The Code of Life 600–601

AFTERWORD 615

APPENDIX A UNITS AND ASTRONOMICAL DATA 617

APPENDIX B OBSERVING THE SKY 627

GLOSSARY 640

ANSWERS TO EVEN-NUMBERED PROBLEMS 650

INDEX 651

A Note to the Student

From Mike Seeds

Hi,
I'm really glad you are taking an astronomy course. You are going to see some amazing things from the icy rings of Saturn to monster black holes. Our universe is so beautiful, it is sad to think that not everyone gets to take an astronomy course.

Two Goals

You will meet a lot of new ideas in this course, but there are two things I hope you find especially satisfying. This astronomy course will help you answer two important questions:

- What are we?
- How do we know?

By "What are we?" I mean, where do we fit in to the history of the universe? The atoms you are made of had their first birthday in the big bang when the universe began, but those atoms have been cooked and remade inside stars and now they are inside you. Where will they be in a billion years? Astronomy is the only course on campus than can tell you that story, and it is a story that everyone should know.

By "How do we know?" I mean how does science work? How can anyone know there was a big bang? In today's world, you need to think carefully about the things so-called experts say. Scientists have a special way of knowing based on evidence. Scientific knowledge isn't just opinion or policy or marketing or public relations. It is humanities' best understanding of nature. To understand the world around you, to evaluate the conflicting opinions that bombard you, to protect yourself and your family in a rapidly changing world, you should understand how science works.

These two questions are the message of astronomy and they are just for you. You need to know the answers to these questions so you can appreciate how wonderful the universe is and how special you are.

Expect to Be Astonished

One reason astronomy is exciting is that astronomers discover new things every day. Astronomers expect to be astonished. You can share in the excitement because I've worked hard to include the newest images, the newest discoveries, and the newest insights that will take you, in an introductory course, to the frontier of human knowledge. You'll see new evidence of ancient oceans and lakes on Mars and erupting geysers on Saturn's moon Enceladus. You'll visit the moon Titan, where it rains liquid methane, and visit the newly recognized dwarf planets so far from the sun that most gases freeze solid. You'll see stars die in violent explosions, and you'll share the struggle to understand new evidence that the expansion of the universe is speeding up. Huge telescopes in space and on remote mountaintops provide a daily dose of excitement that goes far beyond sensationalism. These new discoveries in astronomy are exciting because they are about us. They tell us more and more about what we are.

As you read this book, notice that it is not organized as lists of facts for you to memorize. That could make even astronomy boring. Rather, this book is organized to show you how scientists use evidence and theory to create logical arguments that show how nature works. Look at the list of special features that follows this note. Those features were carefully designed to help you understand astronomy as evidence and theory. Once you see science as logical arguments, you hold the key to the universe.

Do Not Be Humble

As a teacher, my quest is simple. I want you to understand your place in the universe—not just your location in space, but your location in the unfolding history of the physical universe. Not only do I want you to know where you are and what you are in the universe, but I want you to understand how scientists know. By the end of this book, I want you to know that the universe is very big but that it is described by a small set of rules and that we humans have found a way to figure out the rules—a method called science.

Do not be humble. Astronomy tells us that the universe is vast and powerful, but it also tells us that we are astonishing creatures. We humans are the parts of the universe that think. You are a small creature, but remember that it is your human brain that is capable of understanding the depth and beauty of the cosmos.

To appreciate your role in this beautiful universe, you must learn more than just the facts of astronomy. You must understand what we are and how we know. Every page of this book reflects that ideal.

Mike Seeds
mike.seeds@fandm.edu

Key Content and Pedagogical Changes to the Tenth Edition

- Every chapter has been reorganized to focus on the two main themes of the book. The What Are We? boxes at the end of each chapter provide a personal link between the student's life and the astronomy of the chapter, including the origin of the elements, the future of exploration in the solar system, and the astronomically short span of human civilization.
- The How Do We Know? boxes help students understand how science works and how scientists think about nature. They range from the scientific method, to the meaning of proof, and the way science is funded.
- Every chapter has been revised to place the "new terms" in context rather than present them as a vocabulary list. New terms are boldface where they first appear in each chapter and reappear in context as boldface terms in each chapter summary. New terms appear as boldface in Concept Art Portfolios and are previewed in italics as the portfolios are introduced.
- Guideposts have been rewritten to open each chapter with a short list of questions that focus the student's reading on the main objectives of the chapter.
- Every chapter summary has been revised to include the focus questions from the Guidepost and the boldfaced new terms to help the student review.
- The book is fully updated to include all of the newest discoveries in astronomy, including images of methane lakes on Titan and the most distant quasars. The controversy over the status of Pluto illuminates the role of the Kuiper Belt and the dwarf planets in the formation of the solar system and planet migration.

Special Features

- **What Are We?** Each chapter ends with a short essay that will help you understand your own role in the astronomy you have just learned.
- **How Do We Know?** commentaries appear in every chapter and will help you see how science works. They will point out where scientists use statistical evidence, why they think with analogies, and how they build confidence in theories.
- **Special two-page art spreads** provide an opportunity for you to create your own understanding and share in the satisfaction that scientists feel as they uncover the secrets of nature.
- **Guided discovery figures** illustrate important ideas visually and guide you to understand relationships and contrasts interactively.
- **Focus on Fundamentals** will help you understand five concepts from physics that are critical to understanding modern astronomy.
- **Guideposts** on the opening page of each chapter help you see the organization of the book by focusing on a small number of questions to be answered as you read the chapter.
- **Scientific Arguments** at the end of each text section are carefully designed questions to help you review and synthesize concepts from the section. A short answer follows to show how scientists construct scientific arguments from observations, evidence, theories, and natural laws that lead to a conclusion. A further question then gives you a chance to construct your own argument on a related issue.
- **End-of-Chapter Review Questions** are designed to help you review and test your understanding of the material.
- **End-of-Chapter Discussion Questions** go beyond the text and invite you to think critically and creatively about scientific questions. You can think about these questions yourself or discuss them in class.
- **Virtual Astronomy Laboratories.** This set of 20 online labs is free with a passcode included with every new copy of this textbook. The labs cover topics from helioseismology to dark matter and allow you to submit your results electronically to your instructor or print them out to hand in. The first page of each chapter in this textbook notes which labs correlate to that chapter.
- **ThomsonNOW.** Take charge of your learning with the first assessment-centered student learning tool for astronomy. Access ThomsonNOW free via the Web with the access code card bound into this book and begin to maximize your study time with a host of interactive tutorials and quizzes that help you focus on what you need to learn to master astronomy.
- **TheSky *Student Edition CD-ROM.*** With this CD-ROM, a personal computer becomes a powerful personal planetarium. Loaded with data on 118,000 stars and 13,000 deep-sky objects with images, it allows you to view the universe at any point in time from 4000 years ago to 8000 years in the future, to see the sky in motion, to view constellations, to print star charts, and much more.

Acknowledgments

I started writing astronomy textbooks in 1973, and over the years I have had the guidance of a great many people who care about astronomy and teaching.

I would like to thank all of the students and teachers who have responded so enthusiastically to *Foundations of Astronomy.* Their comments and suggestions have been very helpful in shaping this book.

Many observatories, research institutes, laboratories, and individual astronomers have supplied figures and diagrams for this edition. They are listed on the credits page, and I would like to thank them specifically for their generosity.

Writing about every branch of astronomy is a daunting task, and I could not do it without helpful contributions from experts in various fields. The textbook reviewers listed below provided insights into the newest research and current understanding. I especially want to thank Dana Backman, George Jacoby, Victoria Kaspi, Jackie Milingo, and William Keel, for their helpful guidance on technical issues.

Certain unique diagrams in Chapters 11, 12, 13, and 14 are based on figures I designed for my article "Stellar Evolution," which appeared in *Astronomy,* February 1979.

I am happy to acknowledge the use of images and data from a number of important programs. In preparing materials for this book I used NASA's Sky View facility located at NASA Goddard Space Flight Center. I have used atlas images and mosaics obtained as part of the Two Micron All Sky Survey (2MASS), a joint project of the University of Massachusetts and the Infrared Processing and Analysis Center/California Institute of Technology, funded by the National Aeronautics and Space Administration and the National Science Foundation. A number of solar images are used by the courtesy of the SOHO consortium, a project of international cooperation between ESA and NASA.

I would like to thank my daughter, Kathryn Coolidge, for her word-by-word, comma-by-comma assistance with the writing in this new edition. Her work has made the book much more readable.

It is always a pleasure to work with the Brooks/Cole team. Special thanks go to all of the people who have contributed to this project, including Hal Humphrey, Carol O'Connell, Kathleen Olson, Sam Subity, and Sylvia Krick.

I have enjoyed working with Margaret Pinette of Heckman & Pinette, and I want to thank my developmental editor Rebecca Heider for her detailed guidance in this new edition and her patience with my efforts. I would especially like to thank my editor Chris Hall for her understanding and help on this project.

Most of all, I would like to thank my wife, Janet, and my daughter, Kate, for putting up with "the books." They know all too well that textbooks are made of time.

Mike Seeds

Reviewers

John J. Cowan, University of Oklahoma
Andrew Cumming, McGill University
Joshua P. Emery, NASA Ames Research Center
Jonathan Fortney, NASA Ames Research Center
Jennifer Heldmann, Santa Clara University
Chris Littler, North Texas University

The longest journey begins with a single step.

LAO TSE

YOU ARE ABOUT to embark on a voyage out to the end of the universe, past the moon, sun, and other planets, past the stars you see in the evening sky, and past billions more that can be seen only with the aid of the largest telescopes. You will journey through great whirlpools of stars to the most distant galaxies visible from Earth—and then you will continue on, looking for the structure of the universe itself.

Knowing where you are in space and time is part of the story of astronomy. You will learn how Earth circles the sun, how the sun circles the galaxy and how our galaxy drifts through space with billions of other galaxies. You will learn how stars are born and how they die. But more importantly you will learn about the natural processes that connect you with the stars and galaxies that fill the universe. Astronomy can help you not only see where you are in the universe but understand what you are.

1-1 Why Study Astronomy?

YOUR EXPLORATION OF THE UNIVERSE will help you answer two fundamental questions:

- What are we?
- How do we know?

As you study stars and galaxies, you will be learning about your place in the cycles of the universe.

What are we? That is the first organizing theme of this book. Astronomy is important to you because it will tell you what you are. Notice that the question is not, "*Who* are we?" If you want to know who we are, you may want to talk to a sociologist, theologian, paleontologist, artist, or poet. "*What* are we?" is a fundamentally different question.

By "What are we?" I mean, "Where do we fit into the history of the universe?" For example, the atoms in your body had their first birthday in the big bang when the universe began. Those atoms have been cooked and remade inside stars, and now after billions of years, they are inside you. Where will they be in another billion years? This is a story everyone should know, and astronomy is the only course on campus that can tell you that story.

Every chapter in this book ends with a short segment entitled **What Are We?** This summary shows how the astronomy presented in the chapter relates to your role in the story of the universe. If you know astronomy, you know what you are.

How do we know? That is the second organizing theme of this book. You should ask that question over and over, not only as you study astronomy but whenever you encounter statements by so-called experts in any field. Should you follow a diet recommended by a TV star? Should you vote for a candidate who warns of an energy crisis? To understand the world around you, to make wise decisions for yourself, for your family, and for your nation, you need to understand how science works.

You can use astronomy as a case study in science. In every chapter of this book, you will find short essays entitled **How Do We Know?** They are designed to help you think not about *what* is known but about *how* it is known. That is, they will explain different aspects of scientific reasoning and in that way help you understand how scientists know about the natural world

Over the last four centuries, scientists have developed a way to understand nature that is called the **scientific method (How Do We Know? 1-1)**. You will see this process applied over and over as you read about exploding stars, colliding galaxies, and whirling planets. The universe is very big, but it is described by a small set of rules, and we humans have found a way to figure out the rules—a method called *science*.

1-2 Where Are We?

ASTRONOMY DISCUSSES BIG THINGS and huge distances, and it is sometimes hard to find your place in the universe. Where are we? To find yourself among the stars and to grasp the relative sizes of things, you can take a cosmic zoom, a ride out through space to preview the kinds of objects that fill the universe.

You can begin with something familiar. ■ Figure 1-1 shows a region about 52 feet across occupied by a human being, a sidewalk, and a few trees—all objects whose size you can understand.

■ **Figure 1-1**

(M. Seeds)

How Do We Know? 1-1

The Scientific Method

How does science work? The **scientific method** is the process by which scientists form theories and test them against evidence gathered by experiment or observation. If a theory is contradicted, it must be revised or discarded. If a theory is confirmed, it must be tested further. The scientific method is a way of testing and refining ideas to create improved descriptions of how nature works.

For example, Gregor Mendel (1822–1884) was an Austrian abbot who liked plants. He formed a theory that offspring usually inherited traits from their parents not as a smooth blend as most scientists of the time believed but according to strict mathematical rules. Mendel cultivated and tested over 28,000 pea plants, noting which produced smooth peas and which wrinkled peas and how that trait was inherited by successive generations. His study of pea plants and other plants confirmed his theory and allowed him to expand it into a series of laws of inheritance. Although the importance of his work was not recognized in his lifetime, it was combined with the recognition of chromosomes in 1915, and Mendel is now called the father of modern genetics.

Scientists rarely think of the scientific method. It is such an ingrained way of knowing about nature that scientists use it almost automatically, forming, testing, revising, and discarding theories almost minute by minute. Sometimes, however, a scientist will devise a theory that is so important that he or she will spend years devising an experiment and gathering the data to test the idea.

The scientific method is not a mechanical way of grinding facts into understanding. It takes insight and ingenuity to form a good theory and to devise a way to test the theory. Rather, the scientific method is a way of knowing how the universe works.

Whether peas are wrinkled or smooth is an inherited trait. (Inspirestock/jupiterimages)

Each successive picture in this cosmic zoom will show you a region of the universe that is 100 times wider than the preceding picture. That is, each step will widen your **field of view,** the region you can see in the image, by a factor of 100.

Widening your field of view by a factor of 100 allows you to see an area 1 mile in diameter (■ Figure 1-2). People, trees, and sidewalks are now too small to see, but now you can see a college campus and the surrounding streets and houses. The dimensions of houses and streets are familiar. This is the world you know, and you can relate such objects to the scale of your body.

The photo in Figure 1-2 is 1.609 kilometers (1 mile) in diameter. A kilometer (abbreviated km) is a bit under two-thirds of a mile—a short walk across a neighborhood. Even though you started your adventure using feet and miles, in your study of astronomy you should use the metric system of units. Not only is it used by all scientists around the world, but it makes calculations much easier. If you are not already familiar with the metric system, or if you need a review, study Appendix A before reading on.

The view in ■ Figure 1-3 spans 160 km (100 mi). In this infrared photo, green foliage shows up as various shades of red. The college campus is now invisible, and the patches of gray are small cities. Wilmington, Delaware, is visible at the lower right. At this scale, you can see the natural features of Earth's surface. The Allegheny Mountains of southern Pennsylvania cross the image in the upper left, and the Susquehanna River flows southeast into Chesapeake Bay. What look like white bumps are a few puffs of clouds.

Because Figure 1-3 is an infrared photograph, healthy green leaves and crops show up as red. Human eyes are sensitive only to a narrow range of colors. As you explore the universe, you will learn to use a wide range of other "colors," from X rays to radio waves, to reveal sights invisible to unaided human eyes.

■ **Figure 1-2**

(USGS)

At the next step in your journey, you can see your entire planet (■ Figure 1-4), which is 12,756 km in diameter. The photo shows most of the daylight side of the planet. The blurriness at the extreme right is the sunset line. Earth rotates on its axis once a day, exposing half of its surface to daylight at any particular moment. The rotation of Earth carries you eastward, and as you cross the sunset line into darkness, you see the sun set in the west. It is the rotation of the planet that causes the cycle of day and night. This is a good example of how a photo can give you visual clues to understanding a concept. Special questions called **Learning to Look** at the end of each chapter give you a chance to use your own imagination to connect images with the theories that describe astronomical objects.

Enlarge your field of view by a factor of 100, and you see a region 1,600,000 km wide (■ Figure 1-5). Earth is the small blue dot in the center, and the moon, whose diameter is only one-fourth that of Earth, is an even smaller dot along its orbit 380,000 km from Earth.

■ **Figure 1-4**

(NASA)

These numbers are so large that it is inconvenient to write them out. Astronomy is sometimes known as the science of big numbers, and you will use numbers much larger than these to discuss the universe. Rather than writing out these numbers as in the previous paragraph, it is convenient to write them in **scientific notation.** This is nothing more than a simple way to write very big or very small numbers without writing lots of zeros. In scientific notation, you would write 380,000 as 3.8×10^5. If you are not familiar with scientific notation, read the section on powers of 10 notation in the Appendix. The universe is too big to discuss without using scientific notation.

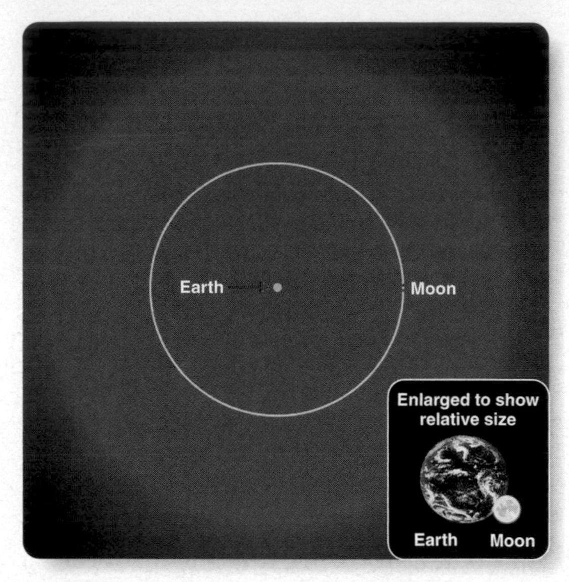

■ **Figure 1-5**

(NASA)

■ **Figure 1-3**

(NASA infrared photograph)

When you once again enlarge your field of view by a factor of 100 (■ Figure 1-6), Earth, the moon, and the moon's orbit all lie in the small red box at lower left. But now you can see the sun and two other planets that are part of our solar system. Our **solar system** consists of the sun, its family of planets, and some smaller bodies such as moons and comets.

Like Earth, Venus and Mercury are **planets,** small, spherical, nonluminous bodies that orbit a star and shine by reflected light. Venus is about the size of Earth, and Mercury is a bit larger than Earth's moon. On this diagram, they are both too small to be seen as anything but tiny dots. The sun is a **star,** a self-luminous ball of hot gas that generates its own energy. Even though the sun is 109 times larger in diameter than Earth (inset), it is nothing more than a dot in this diagram.

This diagram represents an area with a diameter of 1.6×10^8 km. One way astronomers deal with large numbers is to use larger units of measurement. The average distance from Earth to the sun is a unit of distance called the **astronomical unit (AU),** a distance of 1.5×10^{11} m. Using this unit, you can say that the average distance from Venus to the sun is about 0.7 AU. The average distance from Mercury to the sun is about 0.39 AU.

The orbits of the planets are not perfect circles, and this is particularly apparent for Mercury. Its orbit carries it as close to the sun as 0.307 AU and as far away as 0.467 AU. You can see the variation in the distance from Mercury to the sun in Figure

1-6. Earth's orbit is more circular, and its distance from the sun varies by only a few percent.

Enlarge your field of view again, and you can see the entire solar system (■ Figure 1-7). The details of the preceding figure are now lost in the red square at the center of this diagram. You see only the brighter, more widely separated objects. The sun, Mercury, Venus, and Earth lie so close together that you cannot see them separately them at this scale. Mars, the next outward planet, lies only 1.5 AU from the sun. In contrast, Jupiter, Saturn, Uranus, and Neptune are so far from the sun that they are easy to place in this diagram. These are cold worlds far from the sun's warmth. Light from the sun reaches Earth in only 8 minutes, but it takes over 4 hours to reach Neptune.

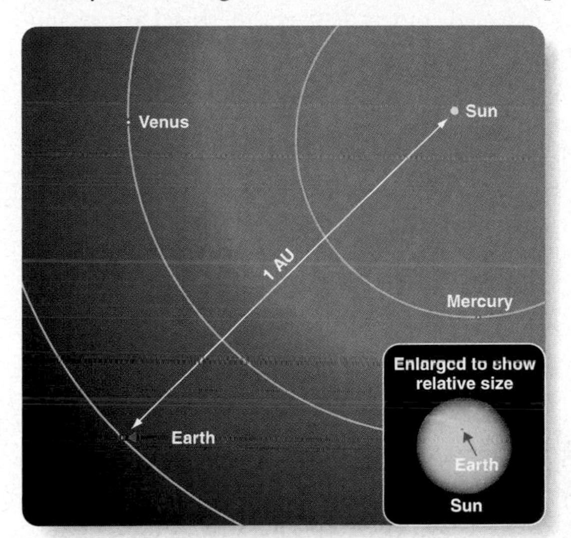

■ **Figure 1-6**

(NOAO)

When you again enlarge your field of view by a factor of 100, the solar system vanishes (■ Figure 1-8). The sun is only a point of light, and all the planets and their orbits are now crowded into the small red square at the center. The planets are too small and reflect too little light to be visible so near the brilliance of the sun.

Nor are any stars visible except for the sun. The sun is a fairly typical star, and it seems to be located in a fairly average neighborhood in the universe. Although there are many billions of stars like the sun, none are close enough to be visible in this diagram, which shows an area only 11,000 AU in diameter. The stars are typically separated by distances about 10 times larger than the distance represented by the diameter of this diagram.

In ■ Figure 1-9, your field of view has expanded to a diameter of a bit over 1 million AU. The sun is at the center, and you can see a few of the nearest stars. These stars are so distant that it is not reasonable to give their distances in astronomical units. To express distances so large, astronomers define a new unit of distance, the light-year. The diameter of your field of view in Figure 1-9 is 17 ly. One **light-year (ly)** is the distance that light travels in one year, roughly 10^{13} km or 63,000 AU. It is a **Common Misconception** that a light-year is a unit of time. The next time you hear someone say, "It will take me take light-years to finish my history paper," you can tell that person that a light-year is a distance, not a time.

Another **Common Misconception** is that stars look like disks when seen through a telescope. Although stars are roughly the same size as the sun, they are so far away that astronomers cannot see them as anything but points of light. Even the closest star to the sun—Alpha Centauri, only 4.2 ly from Earth—looks like a point of light through any telescope. Any planets that might circle stars are much too small, too faint, and too close to the glare of their star to be visible directly. Astronomers have used indirect methods to detect over 200 planets orbiting other stars, but you can't see them by just looking through a telescope.

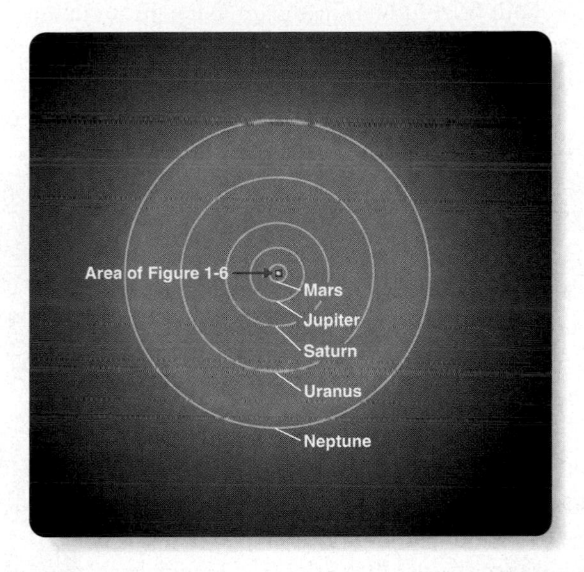

■ **Figure 1-7**

In Figure 1-9, the sizes of the dots represent not the sizes of the stars but their brightness. This is the custom in astronomical diagrams, and it is also how star images are recorded on photographs. Bright stars make larger spots on a photograph than faint stars, so the size of

■ **Figure 1-8**

a star image in a photograph tells you not how big the star is but only how bright it looks.

In ■ Figure 1-10, you expand your field of view by another factor of 100, and the sun and its neighboring stars vanish into the background of thousands of other stars. The field of view is now 1700 ly in diameter. Of course, no one has ever journeyed thousands of light-years from Earth to look back and photograph the solar neighborhood, so this is a representative photograph of the sky. The sun is a relatively faint star that would not be easily located in a photo at this scale.

If you expand your field of view by a factor of 100, you see our galaxy, a disk of stars about 80,000 ly in diameter (■ Figure 1-11). A **galaxy** is a great cloud of stars, gas, and dust bound together by the combined gravity of all the matter. Galaxies range from 1500 to over 300,000 ly in diameter and can contain over 100 billion stars. In the night sky, you see our galaxy as a great, cloudy wheel of stars ringing the sky. This band of stars is known as the **Milky Way,** and our galaxy is called the **Milky Way Galaxy.**

Of course, no one can journey far enough into space to look back and photograph our home galaxy. Using evidence and theory as guides, astronomers can imagine what the Milky Way looks like, and then artists can use those scientific conceptions to create a painting. Many images in this book are artist's renderings of objects and events that are too big or too dim to see clearly, or objects that emit energy your eyes cannot detect, or processes that happen too slowly or too rapidly for humans to sense. As you explore, notice how astronomers use their scientific imagination and astronomical art to accurately depict cosmic events.

■ **Figure 1-9**

Figure 1-11 shows an artist's conception of the Milky Way. Our sun would be invisible in such an image, but if you could see it, you would find it in the disk of the galaxy about two-thirds of the way out from the center.

Our galaxy, like many others, has graceful **spiral arms** winding outward through the disk. You will discover that stars are born in great clouds of gas and dust as they cross through the spiral arms.

Ours is a fairly large galaxy. Only a century ago astronomers thought it was the entire universe—an island cloud of stars in an otherwise empty vastness. Now they know that our galaxy is not unique; it is only one of many billions of galaxies scattered throughout the universe.

When you expand your field of view by another factor of 100, our galaxy appears as a tiny luminous speck surrounded by other specks (■ Figure 1-12). This diagram includes a region 17 million ly in diameter, and each of the dots represents a galaxy. Notice that our galaxy is part of a cluster of a few dozen galaxies. Galaxies are commonly grouped together in such clusters. Some of these galaxies have beautiful spiral patterns like our own galaxy, but others do not. Some are strangely distorted. One of the mysteries of modern astronomy is what produces these differences among the galaxies.

■ **Figure 1-10**

This ▪ box represents the relative size of the previous frame. (NOAO)

Now is a chance for you to correct a **Common Misconception**. People often say "galaxy" when they mean "solar system," and they sometimes confuse those terms with "universe." Your cosmic zoom has shown you the difference. The solar system is the sun and its planets. The galaxy contains billions of stars and whatever planets orbit around them. The universe includes ev-

■ **Figure 1-11**

(© Mark Garlick/space-art.com)

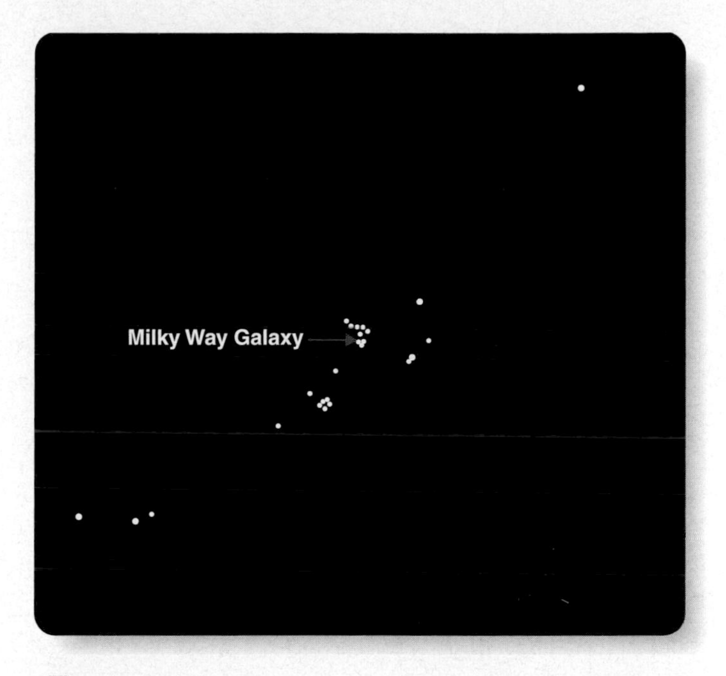

■ **Figure 1-12**

erything, all of the galaxies, stars, and planets, including our galaxy and our solar system.

If you again expand your field of view, you can see that the clusters of galaxies are connected in a vast network (■ Figure 1-13). Clusters are grouped into superclusters—clusters of clusters—and the superclusters are linked to form long filaments and walls outlining voids that seem nearly empty of galaxies. These appear to be the largest structures in the universe. Were you to expand your field of view another time, you would probably see a uniform fog of filaments and voids. When you puzzle over the origin of these structures, you are at the frontier of human knowledge.

Astronomers say the entire universe began in an event called the big bang, but how could anyone know there was a big bang? Astronomers say amazing things, but you should not believe astronomers or anyone else just because they claim to be experts. You have a right to see the evidence. Scientists are accustomed to organizing evidence and theory in logical arguments. **How Do We Know? 1-2** expands on the ways scientists organize their ideas in logical arguments. Throughout this book, chapter sections end with short reviews called **Scientific Arguments.** These feature a review question, which is then analyzed in a scientific argument. A second question gives you a chance to build your own scientific argument. Use these Scientific Arguments to review chapter material but also to practice thinking like a scientist. Once you see science as logical arguments, you hold the key to the universe.

◄ SCIENTIFIC ARGUMENT ►

Why can't astronomers see planets orbiting other stars?

The planets of our solar system shine by reflected sunlight, and planets orbiting other stars must also shine by reflecting light from their star. That means planets can't be very bright, certainly not as bright as stars. Also, planets orbit close to their stars, so their images are lost in the glare from the star. Astronomers have used indirect methods to detect planets orbiting other stars, but they are not easily visible.

Now construct an argument of your own. **Why do astronomers create new units of measurement such as astronomical units and light-years?**

◄ ►

■ **Figure 1-13**

This box ■ represents the relative size of the previous frame. (Detail from galaxy map from M. Seldner, B. L. Siebers, E. J. Groth, and P. J. E. Peebles, *Astronomical Journal 82* [1977].)

How Do We Know? | 1-2

Scientific Arguments

How is a scientific argument different from an advertisement? Advertisements sometimes make claims that sound scientific, but advertisements are fundamentally different from scientific arguments. An advertisement is designed to convince you to buy a product. "Our shampoo promises 85% shinier hair." The statement may sound like science, but it doesn't provide all of the evidence. It also doesn't mention any negatives, like waxy build-up. An advertiser's only goal is a sale, so they don't provide everything you might need to know to make a wise decision.

Scientists construct arguments because they want to test their own ideas and give an accurate explanation of some aspect of nature. For example, in the 1960s, biologist E. O. Wilson presented a scientific argument to show that ants communicate by smell. The argument included a description of his careful observations and the ingenious experiments he had conducted to test his theory. He also considered other evidence and other theories for ant communication. Scientists can include any evidence or theory that supports their claim, but they must observe one fundamental rule of science: They must be totally honest—they must include all of the evidence and all of the theories.

Scientists publish their work in scientific arguments, but they also think in scientific arguments. If, in thinking through his research, Wilson had found a contradiction, he would have known he was on the wrong track. That is why scientific arguments must be complete and honest. Scientists who ignore inconvenient evidence or brush aside other theories are only fooling themselves.

A good scientific argument gives you all the information you need to decide for yourself whether the argument is correct. Wilson's study of ant communication is now widely understood and is being applied to other fields such as telecommunications networks and pest control.

Scientists have discovered that ants communicate with a large vocabulary of smells. (Eye of Science/Photo Researchers, Inc.)

1-3 When Is Now?

ONCE YOU HAVE AN IDEA where you are in space, you need to know where you are in time. The stars have shone for billions of years before the first human looked up and wondered what they were. Fitting yourself into the universe means finding yourself in cosmic history as well as cosmic space. To get a good sense of your place in time, all you need is a long red ribbon.

Imagine stretching a ribbon from goal line to goal line down the center of a football field as shown on the inside front cover of this book. Imagine that one end of the ribbon is *now* and that the other end represents *the big bang*—the beginning of the universe. In Chapter 18 you will see evidence that the universe is about 14 billion years old. Then the long red ribbon represents 14 billion years, the entire history of the universe.

Imagine beginning at the goal line labeled *big bang.* You could replay the entire history of the universe by walking along your ribbon toward the goal line labeled *now.* Observations tell astronomers that the big bang filled the entire universe with hot, dense gas, but that gas cooled rapidly, and the universe went dark. All that happened in the first half inch on the ribbon. There was no light for the first 400 million years until gravity was able to pull the gas together to form the first stars. That seems like a lot of years, but if you stick a little flag beside the ribbon to mark the birth of the first stars it would be not quite 3 yards from the goal line where the universe began.

You would go only about 5 yards before galaxies formed in large numbers. Our home galaxy would be one of those taking shape. By the time you crossed the 50-yard line, the universe would be full of galaxies, but the sun and Earth would not have formed yet. You would have to walk past the 50-yard line down to the 35-yard line before you could finally stick a flag to mark the formation of the sun and planets—our solar system.

You would have to carry your flags a few yards further to the 29-yard line to mark the appearance of the first life on Earth, but you would still be marking the origin only of microscopic creatures in the oceans. You would have to walk all the way to the 3-yard line before you could mark the emergence of life on land, and your dinosaur flag would go just inside the 2-yard line. Dinosaurs would go extinct as you passed the one-half-yard line.

What about people? You could put a little flag for the first humanlike creatures only about three-quarters of an inch from the goal line labeled *now.* Civilization, the building of cities, began about 10,000 years ago. You have to try to fit that flag in only 0.0026 inches from the goal line. That's half the thickness of a sheet of paper. Compare the history of human civilization with the history of the universe. Every war you have ever heard of, every person whose name is recorded, every building ever built from Stonehenge to the building you are in right now fits into that 0.0026 inches.

Humanity is very new to the universe. Our civilization on Earth has existed for only a flicker of an eye blink in the history of the universe. As you will discover in the chapters that follow, only in the last hundred years or so have astronomers began to understand where we are in space and in time.

◄ SCIENTIFIC ARGUMENT ►

What produced the first light after the big bang cooled?

The hot gas of the big bang emitted light, but it cooled quickly and the universe faded into darkness. Gravity pulled the gas together to form the first stars about 400 million years after the universe began. Those stars produced the first light, and stars have been producing light ever since.

Are you wondering how astronomers know about the first stars? The rest of this book will tell you not only what astronomers know, but how they know it and how they use the scientific method to test theories against evidence and understand nature.

Now construct your own argument. **Why do scientists say that humanity is a recent development in the history of the universe?**

◄ ►

What Are We?

Finding Your Astronomical Perspective

Astronomy will give you perspective on what it means to be here on Earth. This chapter used astronomy to locate you in space and time. Once you realize how vast our universe is, people on the other side of Earth seem like neighbors. And in the entire history of the universe, the human story is only the blink of an eye. This may seem humbling at first, but you can be proud of how much we humans have understood in such a short time.

Not only does astronomy locate you in space and time, it places you in the physical processes that govern the universe. Gravity and atoms work together to make stars, light the universe, generate energy, and create the chemical elements in your body. Astronomy locates you in that cosmic process.

Although you are very small and your kind have existed in the universe for only a short time, you are an important part of something very large and very beautiful.

Study and Review

Summary

1-1 | Why Study Astronomy?

Why should you study astronomy?

- Although astronomy seems to be about stars and planets, it describes the universe in which you live, so it is really about you.
- Knowing where you are among the planets, stars, and galaxies is the first step to knowing what you are.
- When you locate yourself in time, you discover that human civilization emerged very recently on this planet.
- To understand any science, you need to understand the **scientific method** by which scientists test theories against evidence to understand how nature works.

How do scientists know about nature?

- Science is a way of knowing how nature works.
- Scientists expect statements to be supported by evidence compared with theory in logical scientific arguments.

1-2 | Where Are We?

Where are you in the universe?

- You surveyed the universe by taking a cosmic zoom in which each **field of view** was 100 times wider than the previous field of view.
- Astronomers use the metric system because it simplifies calculations and **scientific notation** for very large or very small numbers.
- You live on a **planet,** Earth, which orbits our **star,** the sun, once a year. As Earth rotates once a day, you see the sun rise and set.
- The moon is only one-fourth the diameter of Earth, but the sun is 109 times larger in diameter than Earth—a typical size for a star.
- The **solar system** includes the sun at the center and all of the planets that orbit around it—Mercury, Venus, Mars, Jupiter, Saturn, Uranus, and Neptune.
- The **astronomical unit (AU)** is the average distance from Earth to the sun. Mars, for example, orbits 1.5 AU from the sun. The **light-year (ly)** is the distance light can travel in one year. The nearest star is 4.2 ly from the sun.
- Many stars seem to have planets, but such small, distant worlds are difficult to detect. Only a few hundred have been found so far, but planets seem to be common, so you can probably trust that there are lots of planets in the universe including some like Earth.
- The **Milky Way,** the hazy band of light that encircles the sky is the **Milky Way Galaxy** seen from inside. The sun is just one out of the billions of stars that fill the Milky Way Galaxy.
- **Galaxies** contain many billions of stars. Our galaxy is about 80,000 ly in diameter and contains over 100 billion stars.
- Some galaxies, including our own, have graceful **spiral arms** bright with stars, but some galaxies are plain clouds of stars.
- Our galaxy is just one of billions of galaxies that fill the universe in great clusters, clouds, filaments, and walls.
- The largest things in the universe are the vast filaments and walls containing many clusters of galaxies.

1-3 | When Is Now?

How does human history fit on the time scale of the universe?

- The universe began about 14 billion years ago in an event called the big bang, which filled the universe with hot gas.
- The hot gas cooled, the first galaxies began to form, and stars began to shine only about 400 million years after the big bang.
- The solar system formed and the sun began to shine about 4.6 billion years ago.
- Life began in Earth's oceans soon after Earth formed but did not emerge onto land until only 400 million years ago. Dinosaurs evolved not long ago and went extinct only 65 million years ago.
- Humans appeared on Earth only about 3 million years ago, and human civilizations appeared only about 10,000 years ago.

Review Questions

ThomsonNOW™ Assess your understanding of this chapter's topics with additional quizzing and animations at www.thomsonedu.com.

WebAssign The problems from this chapter may be assigned online in WebAssign.

1. In what way is astronomy about you?
2. What is the largest dimension you have personal knowledge of? Have you run a mile? Hiked 10 miles? Run a marathon?
3. What is the difference between our solar system, our galaxy, and the universe?
4. Why are light-years more convenient than miles, kilometers, or astronomical units for measuring certain distances?
5. Why is it difficult to detect planets orbiting other stars?
6. What does the size of the star image in a photograph tell you?
7. What is the difference between the Milky Way and the Milky Way Galaxy?
8. What are the largest known structures in the universe?
9. **How Do We Know?** How do scientists use the scientific method to test their ideas?
10. **How Do We Know?** What are the distinguishing characteristics of a scientific argument?

Discussion Questions

1. Do you think you have a right to know the astronomy described in this chapter? Do you think you have a duty to know it? Can you think of ways this knowledge helps you enjoy a richer life and be a better citizen?
2. How is a statement in a political campaign speech different from a statement in a scientific argument? Find examples in newspapers, magazines, and this book.

Problems

1. The diameter of Earth is 7928 miles. What is its diameter in inches? In yards? If the diameter of Earth is expressed as 12,756 km, what is its diameter in meters? In centimeters?
2. If a mile equals 1.609 km and the moon is 2160 miles in diameter, what is its diameter in kilometers?
3. One astronomical unit is about 1.5×10^8 km. Explain why this is the same as 150×10^6 km.
4. Venus orbits 0.7 AU from the sun. What is that distance in kilometers?
5. Light from the sun takes 8 minutes to reach Earth. How long does it take to reach Mars?
6. The sun is almost 400 times farther from Earth than is the moon. How long does light from the moon take to reach Earth?
7. If the speed of light is 3×10^5 km/s, how many kilometers are in a light-year? How many meters?
8. How long does it take light to cross the diameter of our Milky Way Galaxy?
9. The nearest galaxy to our own is about 2 million light-years away. How many meters is that?
10. How many galaxies like our own would it take laid edge-to-edge to reach the nearest galaxy? (*Hint:* See Problem 9.)

Learning to Look

1. In Figure 1-4, the division between daylight and darkness is at the right on the globe of Earth. How do you know this is the sunset line and not the sunrise line?
2. Look at Figure 1-6. How can you tell that Mercury follows an elliptical orbit?
3. Of the objects listed here, which would be contained inside the object shown in the photograph at the right? Which would contain the object in the photo? Stars, planets, galaxy clusters, filaments, spiral arms

Bill Schoening/NOAO/AURA/NSF

4. In the photograph shown here, which stars are brightest, and which are faintest? How can you tell? Why can't you tell which stars in this photograph are biggest or which have planets?

NOAO

■ **Figure 2-2**

(a) In antiquity, constellation boundaries were poorly defined, as shown on this map by the curving dotted lines that separate Pegasus from Andromeda. (From Duncan Bradford, *Wonders of the Heavens,* Boston: John B. Russell, 1837) (b) Modern constellation boundaries are precisely defined by international agreement.

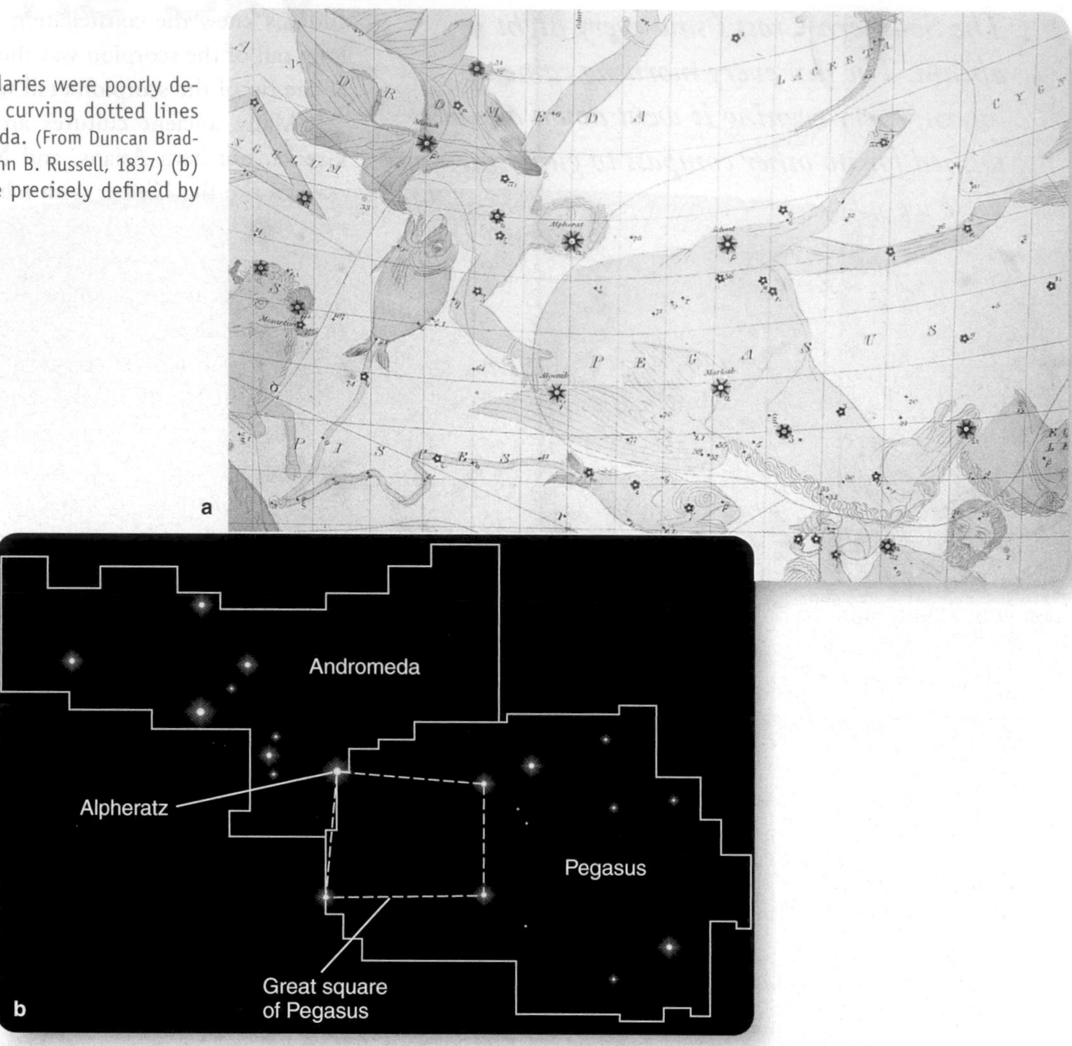

In addition to the 88 official constellations, the sky contains a number of less formally defined groupings called **asterisms.** The Big Dipper, for example, is a well-known asterism that is part of the constellation Ursa Major (the Great Bear). Another asterism is the Great Square of Pegasus (Figure 2-2b), which includes three stars from Pegasus and one from Andromeda. The star charts at the end of this book will introduce you to the brighter constellations and asterisms.

Although constellations and asterisms are named based on their appearance in the sky, it is important to remember that most of these groups are made up of stars that are not physically associated with one another. Some stars may be many times farther away than others and moving through space in different directions. The only thing they have in common is that they lie in approximately the same direction from Earth (■ Figure 2-3).

The Names of the Stars

In addition to naming groups of stars, ancient astronomers named the brighter stars, and modern astronomers still use many of those names. The constellation names come from Greek translated into Latin—the language of science from the fall of Rome to the 19th century—but most star names come from ancient Arabic, though they have been much altered by the passing centuries. The name of Betelgeuse, the bright red star in Orion, for example, comes from the Arabic *yad al jawza,* meaning "armpit of Jawza [Orion]." Names such as Sirius (the Scorched One), Capella (the Little She Goat), and Aldebaran (the Follower of the Pleiades) are beautiful additions to the mythology of the sky.

Giving the stars individual names is not very helpful because you can see thousands of stars, and their names do not help you locate the star in the sky. In which constellation is Antares, for example? In 1603, Bavarian lawyer Johann Bayer published an atlas of the sky called *Uranometria* in which he assigned lowercase Greek letters to the brighter stars of each constellation. In many constellations, the letters follow the order of brightness, but in some constellations, by tradition, mistake, or the personal preference of early chartmakers, there are exceptions. Astronomers have used those Greek letters ever since. (See the Appendix table with the Greek alphabet.) In this way, the brightest star is usually designated α (alpha), the second-brightest β (beta), and so on (■ Figure 2-4). To identify a star in this way, give the Greek letter followed by the Latin possessive form of the constellation name, such as α Scorpii (sometimes written alpha Scorpii) for Antares. That designation tells you that Antares is in the constellation Scorpius and that it is probably the brightest star in the constellation.

Favorite Stars

■ Figure 2-5 identifies eight bright stars that you can adopt as Favorite Stars. Getting to know a few bright stars as individuals

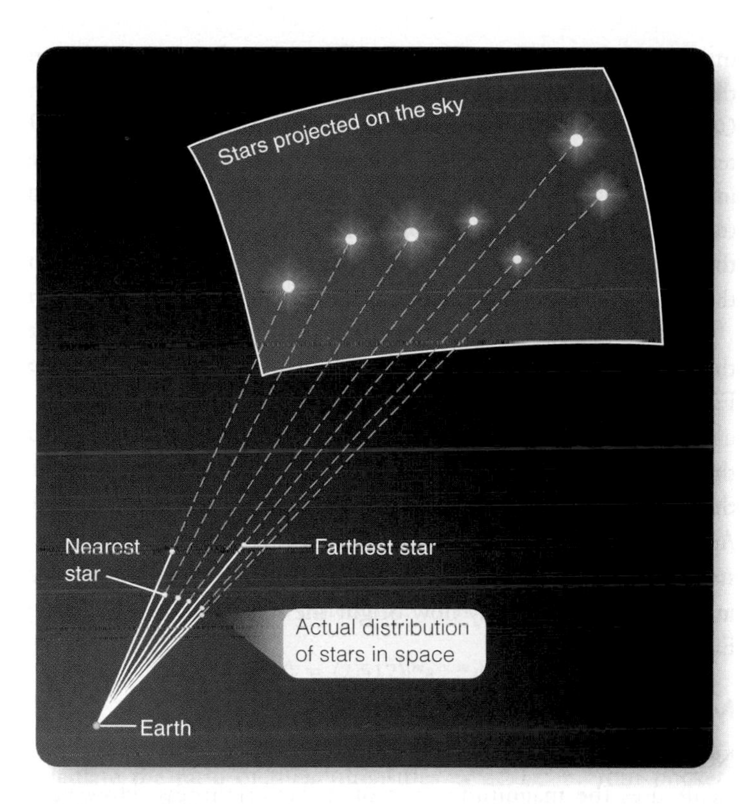

Figure 2-3

You see the Big Dipper in the sky because you are looking through a group of stars scattered through space at different distances from Earth. You see them as if they were projected on a screen, where they form the shape of the Dipper.

will give you a personal connection to the sky and make glancing at the night sky much like encountering old friends. You may want to add more Favorite Stars to your list, but you will certainly find that these eight stars have interesting personalities. When you see Betelgeuse, for example, you will think of it as not just a bright point of light but as an aging, cool, red star over 800 times larger in diameter than the sun.

You can use the star charts at the end of this book to help locate these Favorite Stars. You can see Polaris year round, but Sirius, Betelgeuse, Rigel, and Aldebaran are in the winter sky. Spica is a summer star, and Vega is visible evenings in late summer or fall. Alpha Centauri is a special star, but you will have to travel as far south as southern Florida to glimpse it above the southern horizon.

Knowing that Favorite Star Beteleguse is α Orionis tells you that it is probably a bright star, but to be precise, you need an accurate way of referring to the brightness of stars. For that you must consult one of the first great astronomers.

The Brightness of Stars

Astronomers measure the brightness of stars using the **magnitude scale,** a system that first appeared in the writings of the ancient astronomer Claudius Ptolemy about 140 AD. The system may have originated earlier than Ptolemy, and most astronomers attribute it to the Greek astronomer Hipparchus (160-127 BC).

The ancient astronomers divided the stars into six magnitudes. The brightest were called first-magnitude stars and those that were slightly fainter, second-magnitude. The scale continued downward to sixth-magnitude stars, the faintest visible to the human eye. Thus, the larger the magnitude number, the fainter the star. This may seem awkward at first, but it makes sense if you think of the bright stars as first-class stars and the faintest stars visible as sixth-class stars.

Hipparchus is believed to have compiled the first star catalog, and he may have used the magnitude system in that catalog. Almost 300 years later Ptolemy used the magnitude system in his

Figure 2-4

Stars in a constellation can be identified by Greek letters and by names derived from Arabic. The spikes on the star images in the photograph were produced by the optics in the camera. (William Hartmann)

How Do We Know? 2-1

Scientific Models

How can Tinkertoys help explain genetics? A scientific model is a carefully devised conception of how something works, a framework that helps scientists think about some aspect of nature, just as the celestial sphere helps astronomers think about the motions of the sky.

Chemists, for example, use colored balls to represent atoms and sticks to represent the bonds between them, kind of like Tinkertoys. Using these molecular models, chemists can see the three-dimensional shape of molecules and understand how the atoms interconnect. The molecular model of DNA proposed by Watson and Crick in 1953 led to our modern understanding of the mechanisms of genetics. You have probably seen elaborate ball-and-stick models of DNA, but does the molecule really look like Tinkertoys? No, but the model is both simple enough and accurate enough to help scientists think about their theories.

A scientific model is not a statement of truth; it does not have to be precisely true to be useful. In an idealized model, some complex aspects of nature can be simplified or omitted. The ball-and-stick model of a molecule doesn't show the relative strength of the chemical bonds, for instance. A model gives scientists a way to think about some aspect of nature but need not be true in every detail.

When you use a scientific model, it is important to remember the limitations of that model. If you begin to think of a model as true, it can be misleading instead of helpful. The celestial sphere, for instance, can help you think about the sky, but you must remember that it is only a model. The universe is much larger and much more interesting than this ancient scientific model of the heavens.

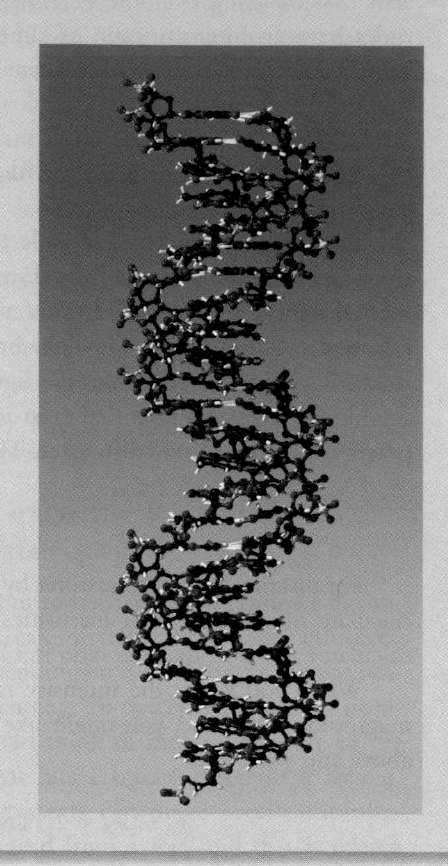

Balls represent atoms and rods represent chemical bonds in this model of a DNA molecule. (Digital Vision/Getty Images)

The Concept Art Portfolio **The Sky Around You** on pages 20–21 takes you on an illustrated tour of the sky. Throughout this book, these two-page art spreads introduce new concepts and new terms through photos and diagrams. These concepts and new terms are not discussed elsewhere, so examine the art spreads carefully. Notice that **The Sky Around You** introduces you to three important principles and 16 new terms that will help you understand the sky:

1. The sky appears to rotate westward around Earth each day, but that is a consequence of the eastward rotation of Earth. That produces day and night. Notice how reference points on the *celestial sphere* such as the *zenith, nadir, horizon, celestial equator* and *celestial poles* define the four directions, *north point, south point, east point,* and *west point.*
2. Astronomers measure *angular distance* across the sky as angles and express them as *degrees, minutes,* and *seconds of arc.*
3. What you can see of the sky depends on where you are on Earth. If you lived in Australia, you would see many constellations and asterisms invisible from North America, but you would never see the Big Dipper. How many *circumpolar constellations* you see depends on where you are. Remember your Favorite Star Alpha Centauri? It is in the southern sky and isn't visible from most of the United States. You could just glimpse it above the southern horizon if you were in Miami, but you could see it easily from Australia.

Pay special attention to the new terms on pages 20-21. You need to know these terms to describe the sky and its motions, but don't fall into the trap of just memorizing new terms. The goal of science is to understand nature, not just to memorize definitions. Study the diagrams and see how the geometry of the celestial sphere and its motions produce the sky you see above you.

This is a good time to eliminate a couple **Common Misconceptions**. Lots of people, without thinking about it much, assume that the stars are not in the sky during the daytime. You can see that the stars are there day and night; they are just invisible during the day because the sky is lit up by sunlight. Also, many people insist that Favorite Star Polaris is the brightest star in the sky. You can see that Polaris is important because of its position, not because of its brightness.

In addition to the obvious daily motion of the sky, Earth's daily rotation conceals a very slow celestial motion that can be detected only over centuries.

Precession

Over 2000 years ago, Hipparchus compared a few of his star positions with those recorded nearly two centuries before and realized that the celestial poles and equator were slowly moving across the sky. Later astronomers understood that this motion is caused by a slow drift in the direction of Earth's rotational axis.

If you have ever played with a gyroscope or top, you have seen how the spinning mass resists any change made to the direction of its axis of rotation. The more massive the top and the more rapidly it spins, the more difficult it is to change the orientation of its axis of rotation. But you probably recall that tops wobble. The axis of even the most rapidly spinning top sweeps around in a conical motion. Physicists understand how the weight of the top tends to make it tip over, and this combined with its rapid rotation makes its axis sweep around in that conical motion called **precession** (■ Figure 2-7a).

Earth spins like a giant top, and it does not spin upright in its orbit; it is inclined 23.5° from vertical. At present, Earth's axis of rotation happens to be pointed toward a spot near the star Polaris, the North Star. The axis would not wander at all if Earth were a perfect sphere. However, because of its rotation, Earth has a slight bulge around its middle, and the gravity of the sun and moon pull on this bulge, tending to twist Earth upright in its orbit. The combination of these forces with Earth's rotation causes Earth's axis to precess in a conical motion, taking about 26,000 years for one cycle (Figure 2-7b).

Because the celestial poles and equator are defined by Earth's rotational axis, precession moves these reference marks on the sky. You would notice no change at all from night to night or year to year, but precise measurements reveal the precessional motion of the celestial poles and equator.

Over centuries, precession has dramatic effects. Egyptian records show that 4800 years ago the north celestial pole was pointed to a spot on the sky near the star Thuban (α Draconis). The pole is now approaching Polaris and will be closest to it in about 2100. In about 12,000 years, the pole will have moved to within 5° of Favorite Star Vega (α Lyrae). Figure 2-7c shows the path followed by the north celestial pole. You will discover in later chapters that precession is common among rotating celestial bodies.

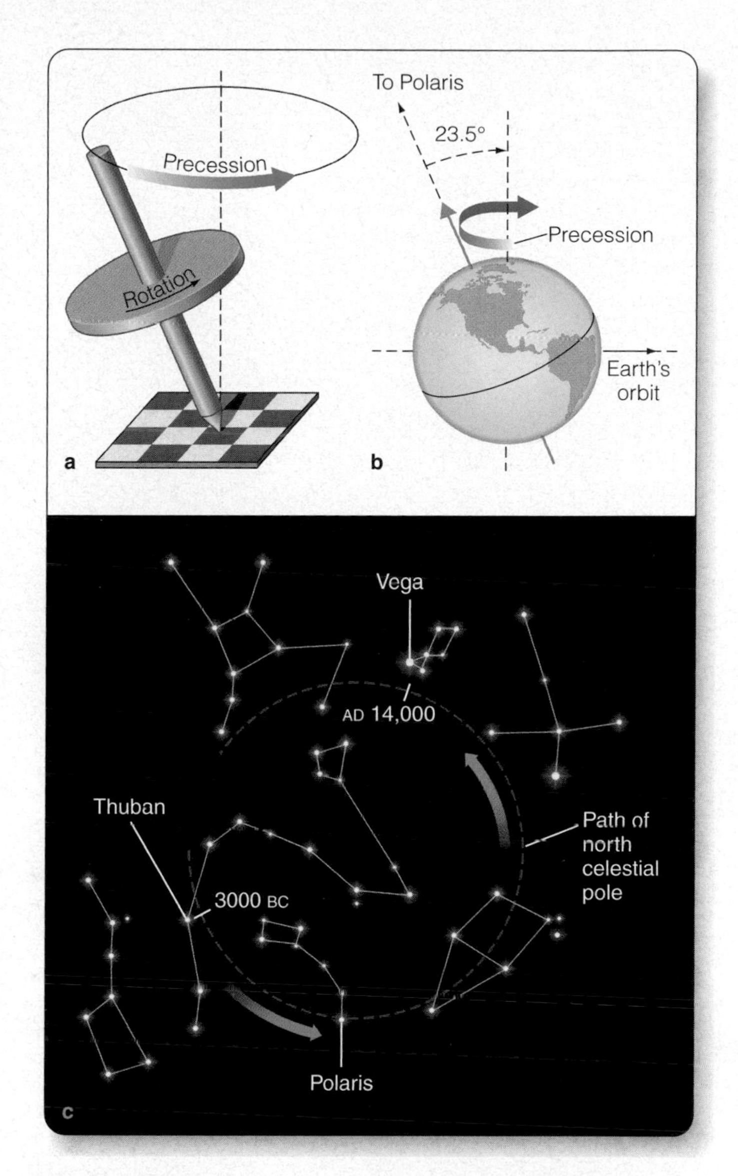

■ Figure 2-7

Precession. (a) A spinning top precesses in a conical motion around the perpendicular to the floor because its weight tends to make it fall over. (b) Earth precesses around the perpendicular to its orbit because the gravity of the sun and moon tend to twist it upright. (c) Precession causes the north celestial pole to drift among the stars, completing a circle in 26,000 years.

◄ SCIENTIFIC ARGUMENT ►

Does everyone see the same circumpolar constellations?

Here you must use your imagination and build your argument with great care. You can use the celestial sphere as a convenient model of the sky. A circumpolar constellation is one that does not set or rise. Which constellations are circumpolar depends on your latitude. If you live on Earth's equator, you see all the constellations rising and setting, so there are no circumpolar constellations at all. If you live at Earth's North Pole, all the constellations north of the celestial equator never set, and all the constellations south of the celestial equator never rise. In that case, every constellation is circumpolar. At intermediate latitudes, the circumpolar regions are caps on the sky whose angular radius equals the latitude of the observer. If you live in Iceland, the caps are very large, and if you live in Egypt, near the equator, the caps are much smaller.

Locate Ursa Major and Orion on the star charts at the end of this book. For people in Canada, Ursa Major is circumpolar, but people in Mexico see most of this constellation slip below the horizon. From much of the United States, some of the stars of Ursa Major set, and some do not. In contrast, Orion rises and sets as seen from nearly everywhere on Earth. Explorers at Earth's poles, however, never see Orion rise or set.

Now use the argument you have just built. **How would you improve the definition of a circumpolar constellation to clarify the status of Ursa Major? Would your definition help in the case of Orion?**

◄ ►

The Sky Around You

1 The eastward rotation of Earth causes the sun, moon, and stars to move westward in the sky as if the **celestial sphere** were rotating westward around Earth. From any location on Earth you see only half of the celestial sphere, the half above the **horizon.** The **zenith** marks the top of the sky above your head, and the **nadir** marks the bottom of the sky directly under your feet. The drawing at right shows the view for an observer in North America. An observer in South America would have a dramatically different horizon, zenith, and nadir.

The apparent pivot points are the **north celestial pole** and the **south celestial pole** located directly above Earth's north and south poles. Halfway between the celestial poles lies the **celestial equator.** Earth's rotation defines the directions you use every day. The **north point** and **south point** are the points on the horizon closest to the celestial poles. The **east point** and the **west point** lie halfway between the north and south points. The celestial equator always touches the horizon at the east and west points.

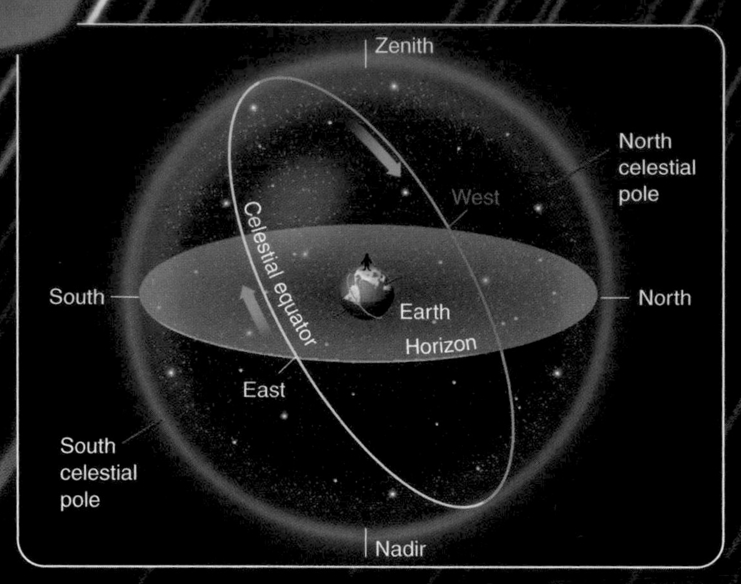

ThomsonNOW™

Sign in at www.thomsonedu.com and go to ThomsonNOW to see Active Figure "Celestial Sphere." Notice how each location on Earth has its unique horizon.

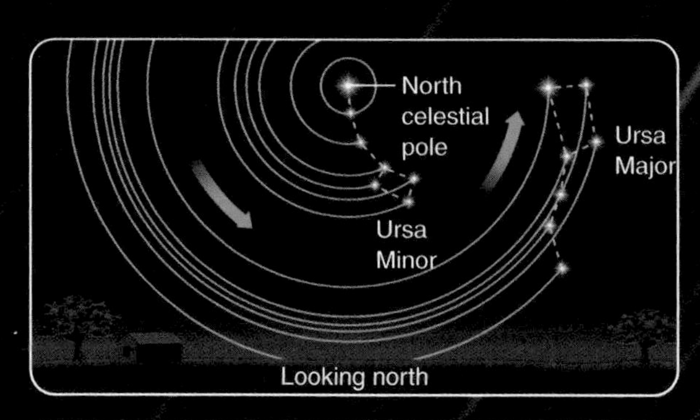

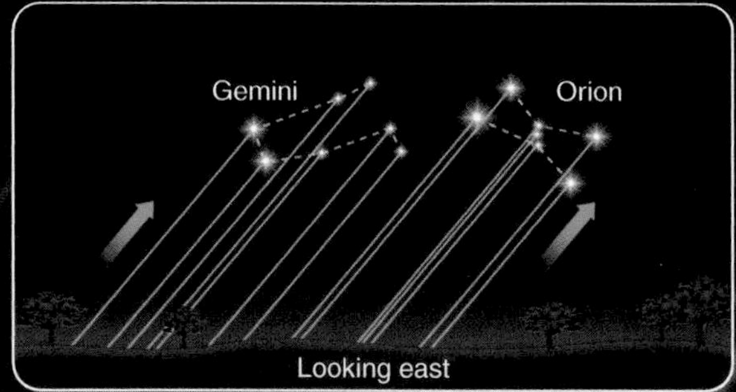

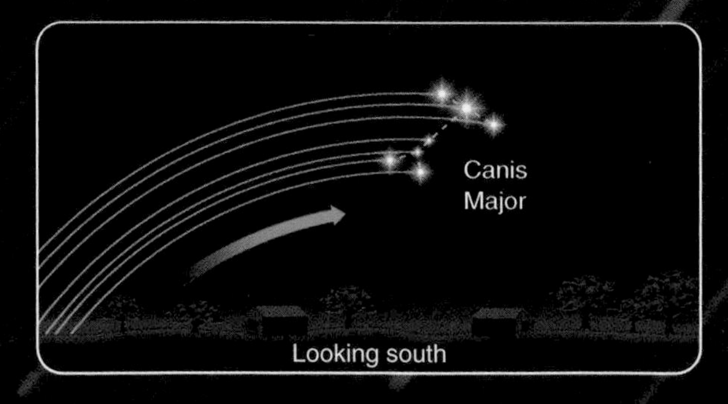

ThomsonNOW™

Sign in at www.thomsonedu.com and go to ThomsonNOW to see Active Figure "Rotation of the Sky." Look in different directions and compare the motions of the stars.

1a This time exposure of about 30 minutes shows stars as streaks, called star trails, rising behind an observatory dome. The camera was facing northeast to take this photo. The motion you see in the sky depends on which direction you look, as shown at right. Looking north, you see the star Polaris, the North Star, located near the north celestial pole. As the sky appears to rotate westward, Polaris hardly moves, but other stars circle the celestial pole. Looking south from a location in North America, you can see stars circling the south celestial pole, which is invisible below the southern horizon.

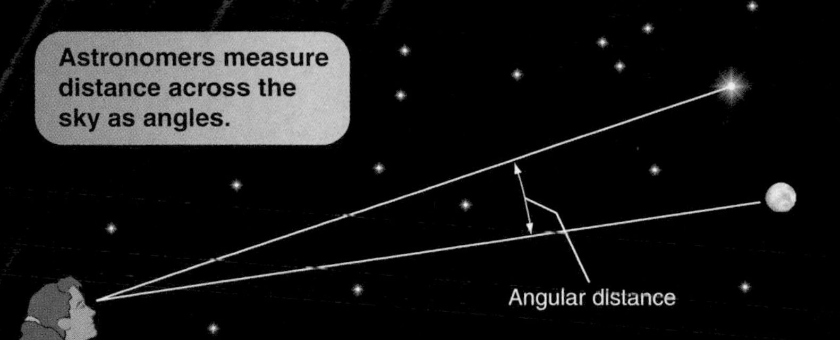

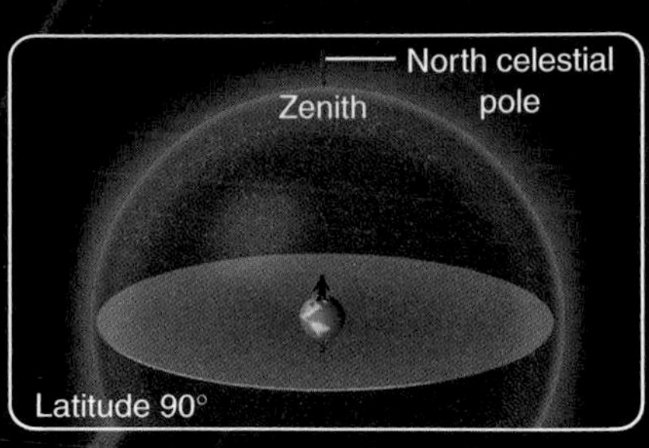

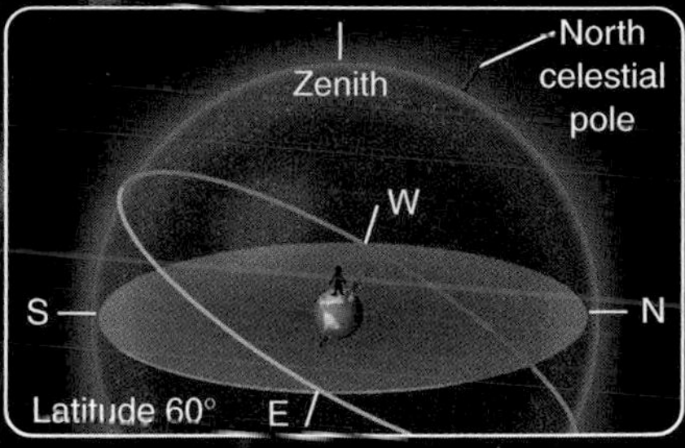

2 Astronomers might say, "The star was only 2 degrees from the moon." Of course, the stars are much farther away than the moon, but when you think of the celestial sphere, you can measure distances *on the sky* as **angular distances** in degrees, minutes of arc, and seconds of arc. A **minute of arc** is 1/60th of a degree, and a **second of arc** is 1/60th of a minute of arc. Then the **angular diameter** of an object is the angular distance from one edge to the other. The sun and moon are each about half a degree in diameter, and the bowl of the Big Dipper is about 10° wide.

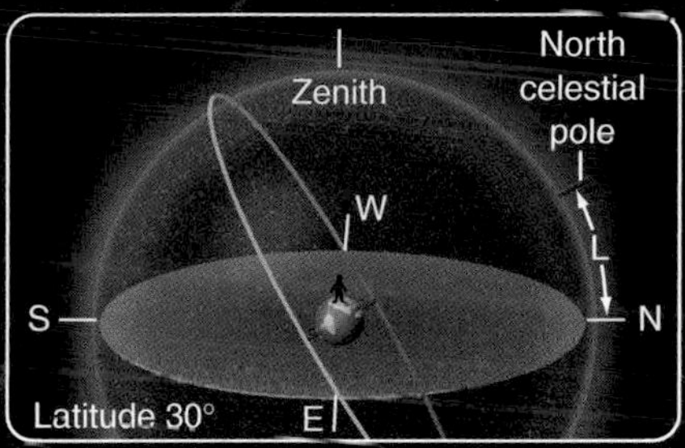

3 What you see in the sky depends on your latitude as shown at right. Imagine that you begin a journey in the ice and snow at Earth's North Pole with the north celestial pole directly overhead. As you walk southward, the celestial pole moves toward the horizon, and you can see further into the southern sky. The angular distance from the horizon to the north celestial pole always equals your latitude (L)—the basis for celestial navigation. As you cross Earth's equator, the celestial equator would pass through your zenith, and the north celestial pole would sink below your northern horizon.

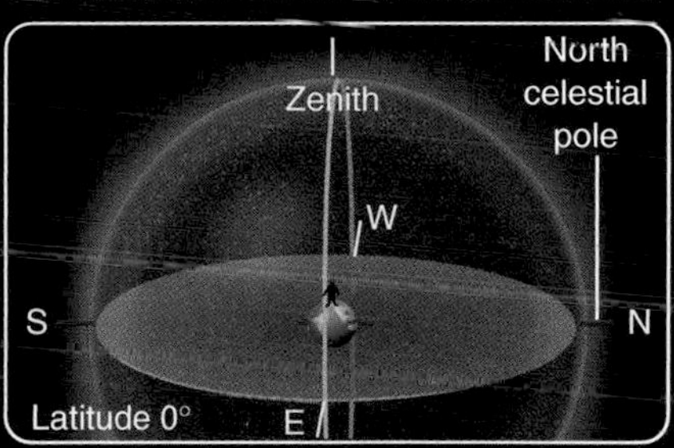

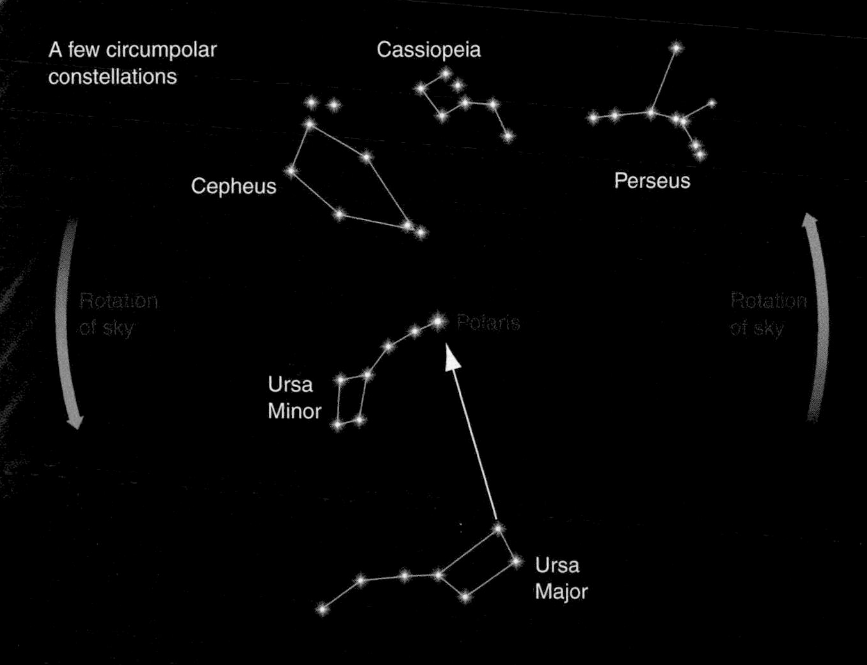

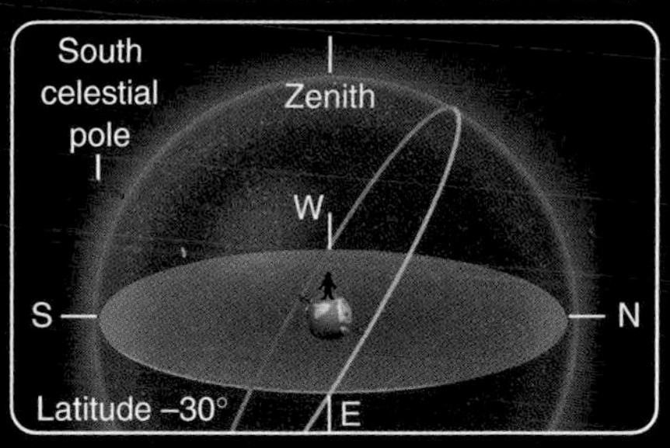

3a **Circumpolar constellations** are those that never rise or set. From mid-northern latitudes, as shown at left, you see a number of familiar constellations circling Polaris and never dipping below the horizon. As the sky rotates, the pointer stars at the front of the Big Dipper always point toward Polaris. Circumpolar constellations near the south celestial pole never rise as seen from mid-northern latitudes. From a high latitude such as Norway, you would have more circumpolar constellations, and from Quito, Ecuador, located on Earth's equator, you would have no circumpolar constellations at all.

ThomsonNOW™

Sign in at www.thomsonedu.com and go to ThomsonNOW to see Active Figure "Constellations from Different Latitudes."

2-3 The Cycles of the Sun

The MOTIONS OF EARTH produce dramatic cycles in the sky. **Rotation** is the turning of a body about an axis through its center. Thus Earth rotates on its axis once a day. **Revolution** is the motion of a body around a point located outside the body, and Earth revolves around the sun once a year.

As you saw in the previous section, the cycle of day and night is caused by the rotation of Earth on its axis, and that means that the time of day is different at different locations around the world. You can see this if you watch international live news coverage on TV; it may be lunchtime where you are, but it can already be dark in the Middle East. In ■ Figure 2-8 you can see that four people in different places on Earth have different times of day.

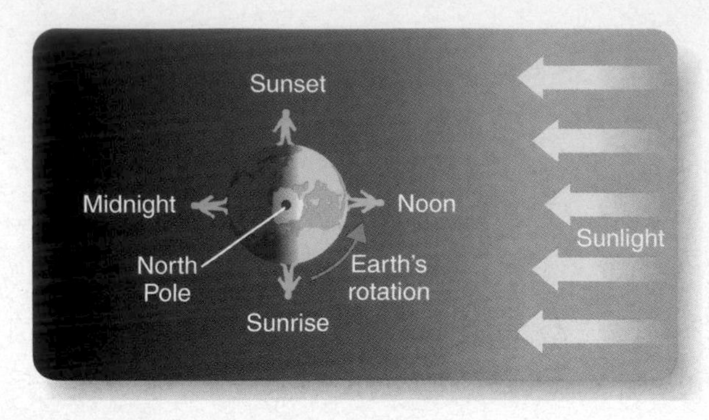

■ Figure 2-8

Looking down on Earth from above the North Pole shows how the time of day or night depends on your location on Earth.

Earth's rotation takes a day and causes day and night. Earth's revolution around the sun takes a year and produces the cycle of the seasons. To understand that motion, you must imagine that you can make the sun fainter.

The Annual Motion of the Sun

Even in the daytime, the sky is filled with stars, but the glare of sunlight fills Earth's atmosphere with scattered light, so that you can see only the brilliant sun. If the sun were fainter, you could see the stars and sun at the same time, and at dawn you would be able to see the sun rise surrounded by stars. Earth's rotation causes both the sun and stars to move westward across the sky during the day, but if you watched carefully, you would notice that the stars in the background were moving westward slightly faster than the sun. That is, while the sky whirls around you, the sun drifts slowly eastward against the background of stars.

That eastward motion of the sun is caused by Earth's motion along its orbit as it revolves around the sun. You can see that in ■ Figure 2-9. As Earth moves along its circular orbit (blue), the sun appears to drift slowly eastward in the sky. In January, you

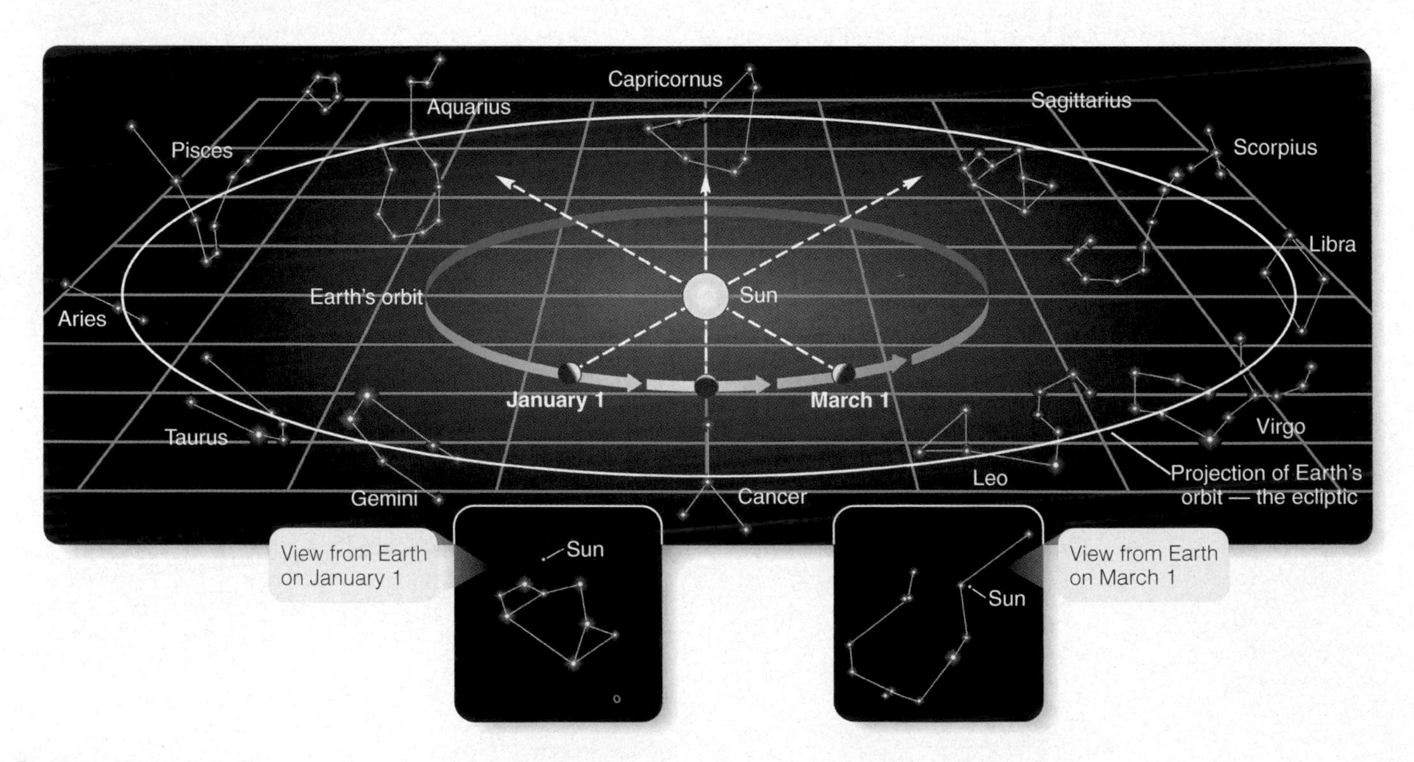

■ Active Figure 2-9

Earth's motion around the sun makes the sun appear to move against the background of the stars. Earth's circular orbit is thus projected on the sky as the circular path of the sun, the ecliptic. If you could see the stars in the daytime, you would notice the sun crossing in front of the distant constellations as Earth moves along its orbit.

would see the sun crossing in front of the constellation Sagittarius. By March 1 the sun would be drifting slowly across Aquarius.

Although people often say the sun is "in Sagittarius" or "in Aquarius," it isn't quite correct to say that the sun is "in" a constellation. The sun is only 1 AU away, and the stars visible in the sky are at least a million times farther away. Nevertheless, in March of each year, you can see the sun against the background of the stars in Aquarius, and people like to say, "The sun is in Aquarius."

If you continue watching the sun against the background of stars throughout the year, you could plot its path as a line on a star chart. After one full year, you would see the sun begin to retrace its path as it continued its annual cycle of motion around the sky. This line, the apparent path of the sun in its yearly motion around the sky, is called the **ecliptic.** Another way to define the ecliptic is to say it is the projection of Earth's orbit on the sky. If the sky were a great screen illuminated by the sun at the center, then the shadow cast by Earth's orbit would be the ecliptic. Yet a third way to define the ecliptic is to imagine extending the plane of Earth's orbit out to touch the celestial sphere; the intersection is the ecliptic. These three definitions of the ecliptic are equivalent, and it is worth considering them all because the ecliptic is one of the most important reference lines on the sky.

Earth circles the sun in 365.26 days, and consequently the sun appears to circle the sky in the same period. That means the sun, traveling 360° around the ecliptic in 365.26 days, travels about 1° eastward each day, about twice its angular diameter. You don't notice this motion because you can't see the stars in the daytime, but the motion of the sun has an important consequence that you do notice—the seasons.

The Seasons

The seasons are caused by a simple fact: Earth does not rotate upright in its orbit. Its axis of rotation is tipped 23.5° from the perpendicular to its orbit. Study **The Cycle of the Seasons** on pages 24–25 and notice that the art introduces you to two important principles and six new terms:

1. The seasons are not caused by any variation in the distance from Earth to the sun. That is a very **Common Misconception.** Earth's orbit is nearly circular, so it is always about the same distance from the sun. Rather the seasons are marked by the north-south motion of the sun as it circles the ecliptic. Notice how the two *equinoxes* and the two *solstices* mark the beginning of the seasons. Further, notice the minor effect of Earth's slightly elliptical orbit as marked by the two terms *perihelion* and *aphelion.*

2. The seasons are caused by the changes in solar energy that Earth's northern and southern hemispheres receive at different times of the year. Because of circulation patterns in Earth's atmosphere, the northern and southern hemispheres are mostly isolated from each other and exchange little heat. When one hemisphere receives more solar energy than the other, it grows rapidly warmer.

Now that you know the seasons well, you can alert your friends to a **Common Misconception** that is among the silliest misunderstandings in science. For some reason, many people believe you can stand an egg on end on the day of the vernal equinox. Radio and TV announcers love to talk about it, but it just isn't true. You can stand a raw egg on end on any day of the year if you have steady hands. (*Hint:* It helps to shake the egg really hard to break the yolk inside so it can settle to the bottom.)

In ancient times, the cycle of the seasons and the solstices and equinoxes were celebrated with rituals and festivals. Shakespeare's play *A Midsummer Night's Dream* describes the enchantment of the summer solstice night. (In Shakespeare's time, the equinoxes and solstices were taken to mark the midpoint of the seasons.) Many North American Indians marked the summer solstice with ceremonies and dances. Early church officials placed Christmas day in late December to coincide with an earlier pagan celebration of the winter solstice.

The Moving Planets

The planets of our solar system shine by reflected sunlight, and Mercury, Venus, Mars, Jupiter, and Saturn are all visible to the naked eye. Uranus is usually too faint to be seen, and Neptune is never bright enough. Like the sun, the planets move generally eastward along the ecliptic. In fact, the word *planet* comes from a Greek word meaning "wanderer."

The planets with orbits outside Earth's can move completely around the sky. Mars takes a bit less than 2 years, but Saturn, farther from the sun, takes nearly 30 years. Venus and Mercury can never move far from the sun because their orbits are inside Earth's orbit. They sometimes appear near the western horizon just after sunset or near the eastern horizon just before sunrise. Venus is easier to locate because its larger orbit carries it higher above the horizon than Mercury (■ Figure 2-10). Mercury is hard to see against the sun's glare and is often hidden in the clouds and haze near the horizon. At certain times when it is farthest from the sun, however, Mercury shines brightly, and you might be able to find it near the horizon in the evening or morning sky. (See the Appendix for the best times to observe Venus and Mercury.)

By tradition, any planet visible in the evening sky is called an **evening star,** although planets are not stars. Any planet visible in the sky shortly before sunrise is called a **morning star.** Perhaps the most beautiful is Venus, which can become as bright as minus fourth magnitude.

Seen from Earth, the planets move gradually eastward along the ecliptic, but, because their orbits are tipped slightly, they don't follow the ecliptic exactly. Also, each travels at its own pace and seems to speed up and slow down at various times. To the ancients, this complex motion reflected the moods of the sky

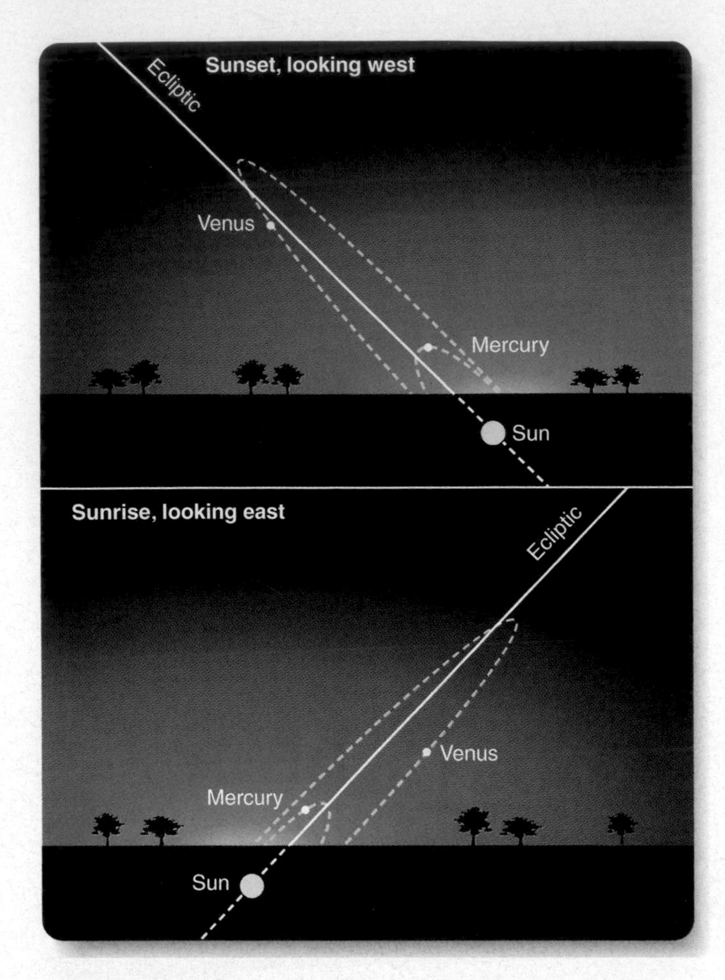

■ **Figure 2-10**

Mercury and Venus follow orbits that keep them near the sun, and they are visible only soon after sunset or before sunrise. Venus takes 584 days to move from morning sky to evening sky and back again, but Mercury zips around in only 116 days.

gods, and gave rise to astrology, the superstitious belief that motions in the sky influence human events.

Ancient astrologers defined the **zodiac,** a band 18° wide centered on the ecliptic, as the highway the planets follow. They divided this band into 12 segments named for the constellations along the ecliptic—the signs of the zodiac. A **horoscope** shows the location of the sun, moon, and planets among the zodiacal signs with respect to the horizon at the moment of a person's birth as seen from that longitude and latitude. As you can see, horoscopes are quite specific, and horoscopes published in newspapers and tabloids can't have been calculated accurately for individual readers.

Astrology believers argue that a person's personality, life history, and fate are revealed in his or her horoscope, but the evidence contradicts this. Astrology has been tested many times over the centuries, and no correlation has ever been found. Believers, however, don't give up on it, perhaps because it comforts us to believe that our sweetheart has rejected us because of the motions of the planets rather than to admit that we behaved badly on our last date. Astrology is a poor basis for life decisions because astrology is a **pseudoscience** that depends on blind belief and not a science that depends on evidence **(How Do We Know? 2-2).**

One reason astronomers object to astrology is that it has no link to the physical world. For example, precession has moved the constellations so that they no longer match the zodiacal signs. Whatever sign you were "born under," the sun was probably in the previous zodiacal constellation. In fact, if you were born on or between November 30 and December 17, the sun was passing through a corner of the nonzodiacal constellation Ophiuchus, and you have no zodiacal sign.* Furthermore, as astronomers like to point out, there is no mechanism by which the planets could influence us. The gravitational influence of a doctor who is delivering a baby is many times more powerful than the gravitational influence of the planets.

Even though astrology is not scientifically valid, it does demonstrate the long-standing fascination humans have had with the stars. Although astrology is not related to science and the physical world at all, astrology makes sense when you think of the world as the ancients did. They believed in a mystical world in which the moods of sky gods altered events on Earth.

Modern science left astrology behind centuries ago, but astrology survives as a fascinating part of human history—an early attempt to understand the meaning of the sky.

◄ SCIENTIFIC ARGUMENT ►

If Earth had a significantly elliptical orbit, how would its seasons be different?

Sometimes as you review your understanding it is helpful to build a scientific argument with one factor slightly exaggerated. Suppose Earth had an orbit so elliptical that Earth's distance from the sun changed significantly. In July, at perihelion, Earth would be closer to the sun, and the entire surface of Earth would be a bit warmer. In July, it would be summer in the northern hemisphere and winter in the southern hemisphere, and both would be warmer than they are now. It could be a dreadfully hot summer in Canada, and southern Argentina could have a mild winter. Six months later, at aphelion, Earth would be a bit farther from the sun, and if that occurred in January, winter in northern latitudes could be frigid. Argentina, in the southern hemisphere, could be experiencing an unusually cool summer.

Of course, this doesn't happen. Earth's orbit is nearly circular, and the seasons are caused not by a variation in the distance of Earth from the sun but by the inclination of Earth in its orbit.

Nevertheless, Earth's orbit is slightly elliptical. Earth passes perihelion about January 3 and aphelion about July 5. Although Earth's oceans tend to store heat and reduce the importance of this effect, this very slight variation in distance does affect the seasons. Now use your scientific argument to analyze the seasons. **Does the elliptical shape of Earth's orbit make your winters slightly warmer or cooler?**

◄ ►

*The author of this book was born on December 14 and thus has no astrological sign. An astronomer friend claims that the author must therefore have no personality.

How Do We Know? 2-2

Pseudoscience

Do pyramids have healing powers? Astronomers have a low opinion of beliefs such as pyramid power and astrology, not so much because they are groundless but because they *pretend* to be sciences. They are pseudosciences, from the Greek *pseudo,* meaning false. Now that you know the traits of a scientific argument, you should be able to identify a pseudoscientific claim.

A pseudoscience is a set of beliefs that appear to be based on scientific ideas but that fail to obey the most basic rules of science. For example, in the 1970s a claim was made that pyramidal shapes focus cosmic forces on anything underneath and might even have healing properties. For example, it was claimed that a pyramid made of paper, plastic, or other materials would preserve fruit, sharpen razor blades, and do other miraculous things. Many books promoted the idea of the special power of pyramids, and this idea led to a popular fad.

A key characteristic of science is that its claims can be tested and verified. In this case, simple experiments showed that any shape, not only pyramids, protects a piece of fruit from airborne spores and allows it to dry without rotting. Likewise, any shape allows oxidation to improve the cutting edge of a razor blade. Because experimental evidence contradicted the claim and because supporters of the theory declined to abandon or revise their claims, you can recognize pyramid power as a pseudoscience. Disregard of contradictory evidence and alternate theories is a sure sign of a pseudoscience.

Pseudoscientific claims can be self-fulfilling. For example, some believers in pyramid power slept under pyramidal tents to improve their rest. Although there is no logical mechanism by which such a tent could affect a sleeper, because people wanted and expected the claim to be true they reported that they slept more soundly. Vague claims based on personal testimony that cannot be tested are another sign of a pseudoscience.

Why do people continue to believe in pseudosciences despite contradictory evidence?

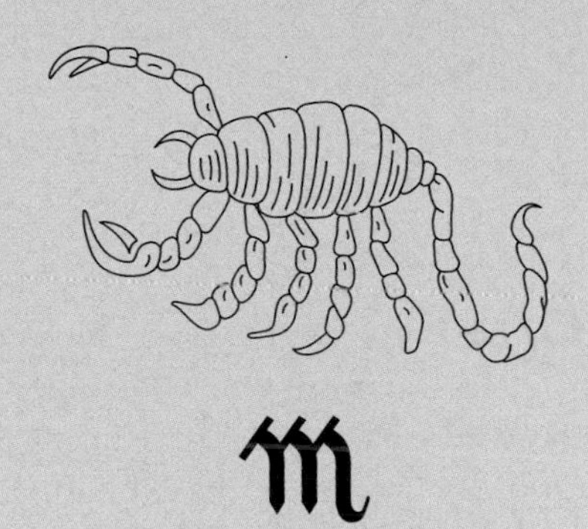

Astrology may be the oldest pseudoscience.

Many pseudosciences appeal to our need to understand and control the world around us. Many pseudoscientific claims involve medical cures, ranging from using magnetic bracelets and crystals to focus mystical power to astonishingly expensive, illegal, and dangerous treatments for cancer. Logic is a stranger to pseudoscience, but human fears and needs are not.

2-4 Astronomical Influences on Earth's Climate

WEATHER IS WHAT HAPPENS TODAY; climate is the average of what happens over decades. Earth has gone through past episodes, called ice ages, when the worldwide climate was cooler and dryer and thick layers of ice covered northern latitudes.

The earliest known ice age occurred about 570 million years ago and the next about 280 million years ago. The most recent ice age began only about 3 million years ago and is still going on. You are living in one of the periodic episodes when the glaciers melt and Earth grows slightly warmer. The current warm period began about 20,000 years ago.

Ice ages seem to occur with a period of roughly 250 million years, and cycles of glaciation within ice ages occur with a period of about 40,000 years. These cyclic changes have an astronomical origin.

The Hypothesis

Sometimes a theory or hypothesis is proposed long before scientists can find the critical evidence to test it. That happened in 1920 when Yugoslavian meteorologist Milutin Milankovitch proposed what became known as the **Milankovitch hypothesis**—that changes in the shape of Earth's orbit, in precession, and in inclination affect Earth's climate and trigger ice ages. You can examine each of these three motions in turn.

First, astronomers know that the elliptical shape of Earth's orbit varies slightly over a period of about 100,000 years. At present, Earth's orbit carries it 1.7 percent closer than average to the sun during northern hemisphere winters and 1.7 percent farther away in northern hemisphere summers. This makes the northern climate very slightly less extreme, and that is critical—most of the landmass where ice can accumulate is in the northern hemisphere. If Earth's orbit became more elliptical, northern summers might be too cool to melt all of the snow and ice from the previous winter. That would allow glaciers to grow larger.

A second factor is also at work. Precession causes Earth's axis to sweep around a cone with a period of about 26,000 years, and that changes the location of the seasons around Earth's orbit. Northern summers now occur when Earth is 1.7 percent farther from the sun, but in 13,000 years northern summers will occur on the other side of Earth's orbit where Earth is slightly closer to the sun. Northern summers will be warmer, which could melt all of the previous winter's snow and ice and reduce the growth of glaciers.

The third factor is the inclination of Earth's equator to its orbit. Currently at 23.5°, this angle varies from 22° to 24° with a period of roughly 41,000 years. When the inclination is greater, seasons are more severe.

In 1920, Milankovitch proposed that these three factors cycled against each other to produce complex periodic variations in Earth's climate and the advance and retreat of glaciers ■ Figure 2-11a). But no evidence was available to test the theory in 1920, and scientists treated it with skepticism. Many thought it was laughable.

The Evidence

By the middle 1970s, Earth scientists could collect the data that Milankovitch needed. Oceanographers could drill deep into the seafloor and collect samples, and geologists could determine the age of the samples from the natural radioactive atoms they contained. From all this, scientists constructed a history of ocean temperatures that convincingly matched the predictions of the Milankovitch hypothesis (Figure 2-11b).

The evidence seemed very strong, and, by the 1980s, the Milankovitch hypothesis was widely discussed as the leading hypothesis. But science follows a mostly unstated set of rules that holds that a hypothesis must be tested over and over against all available evidence **(How Do We Know? 2-3)**. In 1988, scientists discovered contradictory evidence.

A water-filled crack in Nevada called Devil's Hole contains deposits of the mineral calcite. Diving with scuba gear, scientists drilled out samples of the calcite and analyzed the oxygen atoms found there. For 500,000 years, layers of calcite have built up in Devil's Hole, recording in their oxygen atoms the temperature of the atmosphere when rain fell there. Finding the ages of the mineral samples was difficult, but the results seemed to show that the previous ice age ended thousands of years too early to have been caused by Earth's motions.

These contradictory findings are irritating because we all prefer certainty, but such circumstances are common in science. The disagreement between ocean floor samples and Devil's Hole samples triggered a scramble to understand the problem. Were the ages of one or the other set of samples wrong? Were the ancient temperatures wrong? Or were scientists misunderstanding the significance of the evidence?

In 1997, a new study of the ages of the samples confirmed that those from the ocean floor are correctly dated. This seems to give scientists renewed confidence in the Milankovitch hypoth-

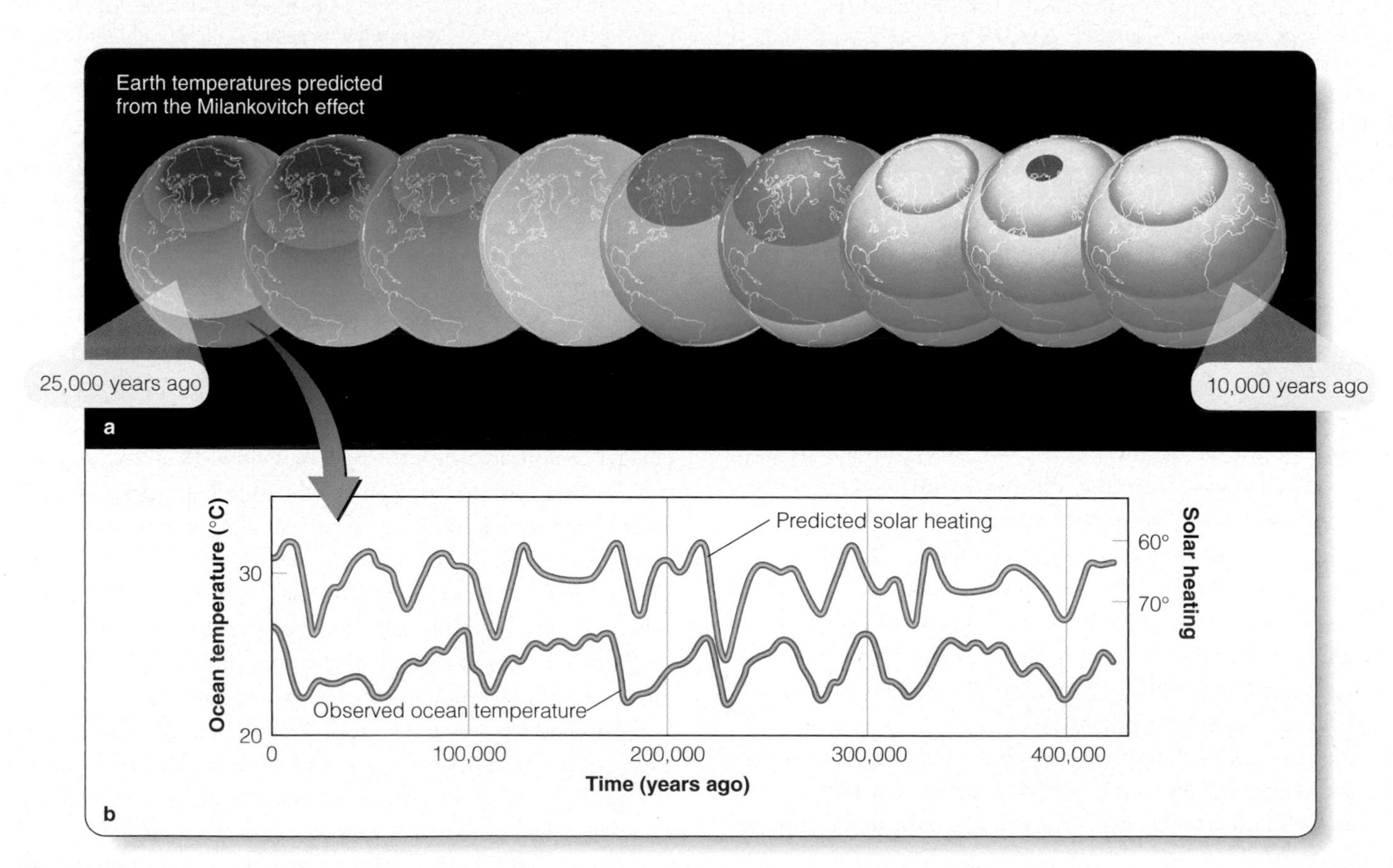

■ **Figure 2-11**

(a) Mathematical models of the Milankovich effect can be used to predict temperatures on Earth over time. In these Earth globes, cool temperatures are represented by violet and blue and warm temperatures by yellow and red. These globes show the warming that occurred beginning 25,000 years ago, which ended the last ice age. (Courtesy Arizona State University, Computer Science and Geography Departments) (b) Over the last 400,000 years, changes in ocean temperatures measured from fossils found in sediment layers from the seabed match calculated changes in solar heating. (Adapted from Cesare Emiliani)

How Do We Know? 2-3

Evidence as the Foundation of Science

How is scientific knowledge more than opinion or speculation? From colliding galaxies to the inner workings of atoms, scientists love to speculate and devise theories, but all scientific knowledge is ultimately based on evidence from observations and experiments. Evidence is reality, and scientists constantly check their ideas against reality.

When you think of evidence, you probably think of criminal investigations in which detectives collect fingerprints and eyewitness accounts. In court, that evidence is used to try to understand the crime, but there is a key difference in how lawyers and scientists use evidence. A defense attorney can call a witness and intentionally fail to ask a question that would reveal evidence harmful to the defendant. In contrast, the scientist must be objective and not ignore any known evidence. The attorney is presenting only one side of the case, but the scientist is searching for the truth. In a sense, the scientist must deal with the evidence as both the prosecution and the defense.

It is a characteristic of scientific knowledge that it is supported by evidence. A scientific statement is more than an opinion or a speculation because it has been tested objectively against reality.

As you read about any science, look for the evidence in the form of observations and experiments. Every theory or conclusion should have supporting evidence. If you can find and understand the evidence, the science will make sense. All scientists, from astronomers to zoologists, demand evidence. You should, too.

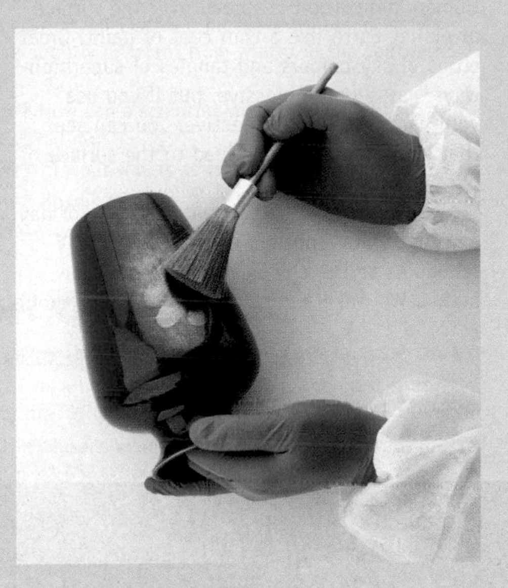

Fingerprints are evidence to past events. (Dorling Kindersley/Getty Images)

esis. But the same study found that the ages of the Devil's Hole samples are also correct. Evidently the temperatures at Devil's Hole record local climate changes in the region that became the southwestern United States. The ocean floor samples record global climate changes, and they fit well with the Milankovitch hypothesis. In this way, the Milankovitch hypothesis, though widely accepted today, is still being tested as scientists try to understand the world we live on.

◄ SCIENTIFIC ARGUMENT ►

How do precession and the shape of Earth's orbit interact to affect Earth's climate?

Here, as in an earlier section of this chapter, exaggeration is a useful analytical tool in your argument. If you exaggerate the variation in the shape of Earth's orbit, you can see dramatically the influence of precession. At present, Earth reaches perihelion during winter in the northern hemisphere and aphelion during summer. The variation in distance is only about 1.7 percent, and that difference doesn't cause much change in the severity of the seasons. But if Earth's orbit were much more elliptical, then winter in the northern hemisphere would be much warmer, and summer would be much cooler.

Now you can see the importance of precession. As Earth's axis precesses, it points gradually in different directions, and the seasons occur at different places in Earth's orbit. In 13,000 years, northern winter will occur at aphelion, and, if Earth's orbit were highly elliptical, northern winter would be terrible. Similarly, northern summer would occur at perihelion, and the heat would be awful. Such extremes might deposit large amounts of ice in the winter but then melt it away in the hot summer, thus preventing the accumulation of glaciers.

Continue this analysis by exaggeration in your own scientific argument. **What effect would precession have if Earth's orbit were more circular?**

◄ ►

Even a man who is pure in heart
And says his prayers by night
May become a wolf when the wolfbane
blooms
And the moon shines full and bright.

PROVERB FROM OLD WOLFMAN MOVIES

DID ANYONE EVER WARN YOU, "Don't stare at the moon—you'll go crazy"?* For centuries, the superstitious have associated the moon with insanity. The word *lunatic* comes from a time when even doctors thought that the insane were "moonstruck." It is a **Common Misconception** that people tend to act crazy at full moon. Actual statistical studies of records from schools, prisons, hospitals, police departments, and so on show that it isn't true. There are always a few people who misbehave; the moon has nothing to do with it. The moon is so bright and its cycles through the sky are so dramatic people simply *expect* it to influence them in some way.

In fact, the moon produces some of the most beautiful and exciting phenomena visible to the naked eye. Not only does it cycle through its phases, but it occasionally produces dramatic eclipses of the sun and moon.

3-1 The Changeable Moon

STARTING THIS EVENING, begin looking for the moon in the sky. As you watch for the moon on successive evenings, you will see it following its orbit around Earth and cycling through its phases as it has done for billions of years.

The Motion of the Moon

If you watch the moon night after night, you will notice two things about its motion. First, you will see it moving eastward against the background of stars; second, you will notice that the markings on its face don't change. These two observations will help you understand the motion of the moon and the origin of its phases.

The moon moves rapidly among the constellations. If you watch the moon for just an hour, you can see it move eastward against the background of stars by slightly more than its angular diameter. In the previous chapter, you learned that the moon is about 0.5° in angular diameter, so it moves eastward a bit more than 0.5° per hour. In 24 hours, it moves 13°. Each night when you look at the moon, you see it about 13° eastward of its location the night before. This eastward movement is the result of the motion of the moon along its orbit around Earth.

The Cycle of Phases

The changing shape of the moon as it orbits Earth is one of the most easily observed phenomena in astronomy. Everyone has noticed the full moon rising dramatically in the evening sky. Study **The Phases of the Moon** on pages 36–37 and notice that it introduces three important concepts and two new terms:

1. The moon always keeps the same side facing Earth. "The man in the moon" is produced by the familiar features on the moon's near side, but you never see the far side of the moon.
2. The changing shape of the moon as it passes through its cycle of phases is produced by sunlight illuminating different parts of the side of the moon you can see.
3. Notice the difference between the orbital period of the moon around Earth *(sidereal period),* and the length of the lunar phase cycle *(synodic period).* That difference is a good illustration of how your view from Earth is produced by the combined motions of Earth and other heavenly bodies such as the sun and moon.

You can make a moon-phase dial from the middle diagram on page 36. Cover the lower half of the moon's orbit with a sheet of paper, aligning the edge of the paper to pass through the word *Full* at the left and the word *New* at the right. Push a pin through the edge of the paper at Earth's North Pole to make a pivot, and under the word *Full* write on the paper *Eastern Horizon.* Under the word *New* write *Western Horizon.* The paper now represents the horizon you see facing south.

You can set your moon-phase dial for a given time by rotating the diagram behind the horizon paper. Set the dial to sunset by turning the diagram until the human figure labeled *sunset* is standing at the top of the Earth globe; the dial shows, for example, that the full moon at sunset would be at the eastern horizon.

The lunar cycle of phases have produced a **Common Misconception** about the moon. You may hear people mention "the dark side of the moon," but you will be able to assure them that there is no dark side. Any location on the moon is sunlit for two weeks and is in darkness for two weeks as the moon rotates in sunlight. The phases of the moon are dramatic and beautiful (■ Figure 3-1). Watch for the moon and enjoy its cycle of phases. It is one of the perks of living on this planet.

*When I was very small, my grandmother told me if I gazed at the moon, I might go crazy. But it was too beautiful, and I ignored her warning. I secretly watched the moon from my window, became fascinated by the sky, and grew up to be an astronomer.

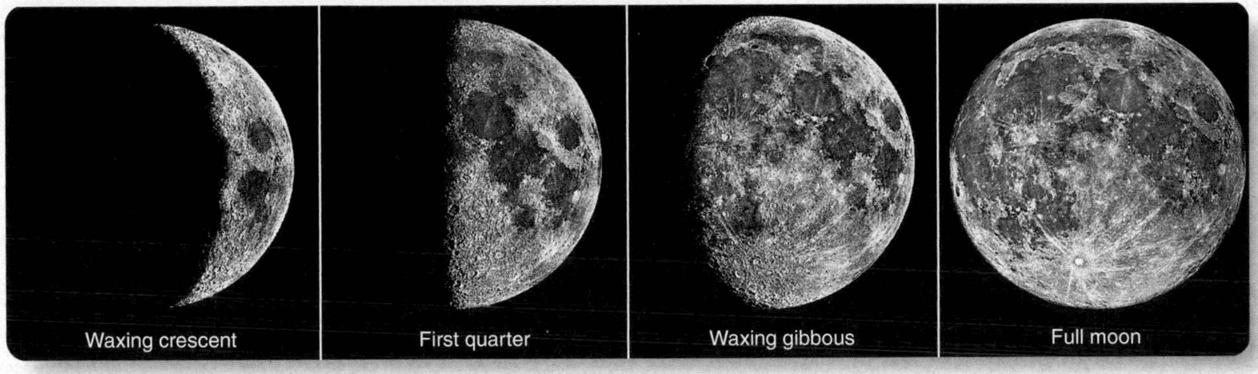

■ Figure 3-1

In this sequence of the waxing moon, you see the same face of the moon, the same mountains, craters, and plains, but the changing direction of sunlight produces the lunar phases. (© UC Regents/Lick Observatory)

ThomsonNOW Sign in at www.thomsonedu.com and go to ThomsonNOW to see Astronomy Exercises "Phases of the Moon" and "Moon Calender."

◄ SCIENTIFIC ARGUMENT ►

Why is the moon sometimes visible in the daytime?

Lots of people are surprised when they notice the moon in the daytime sky, but you won't be because you can explain it with a simple scientific argument involving the geometry of the moon's motion. The full moon rises at sunset and sets at sunrise, so it is visible in the night sky but never in the daytime sky. At other phases, it is possible to see the moon in the daytime. For example, when the moon is a waxing gibbous moon, it rises a few hours before sunset. If you look in the right spot, you can see it in the late afternoon in the southeast sky. It looks pale and washed out because of sunlight illuminating Earth's atmosphere, but it is quite visible once you notice it. You can also locate the waning gibbous moon in the morning sky. It sets an hour or two after sunrise, so you would look for it in the southwestern sky in the morning.

A simple scientific argument analyzing the motion of the moon can explain a lot about what you see and what you don't see. **Why is it extremely difficult to see the crescent moon in the daytime?**

◄ ►

3-2 Lunar Eclipses

In cultures all around the world, the sky is a symbol of order and power, and the moon is the regular counter of the passing days. So it is not surprising that people are startled and sometimes worried when they see the moon grow dark and angry red. To understand these events, you must begin with Earth's shadow.

Earth's Shadow

Earth's shadow consists of two parts. The **umbra** is the region of total shadow. If you were floating in space in the umbra of Earth's shadow, you would see no portion of the sun; it would be completely hidden behind Earth. However, if you moved into the **penumbra,** you would be in partial shadow and would see part of the sun peeking around Earth's edge. In the penumbra, the sunlight is dimmed but not extinguished.

You can make a model of Earth's shadow by pressing a map tack into the eraser of a pencil and holding the tack between a lightbulb a few feet away and a white cardboard screen (■ Figure 3-2). The lightbulb represents the sun, and the map tack represents Earth. If you hold the screen close to the tack, you will see that the umbra is nearly as large as the tack and that the penumbra is only slightly larger. However, if you move the screen away from the tack, the umbra shrinks, and the penumbra expands. Beyond a certain point, the shadow has no dark core at all, indicating that the screen is beyond the end of the umbra.

The umbra of Earth's shadow is over three times longer than the distance to the moon and points directly away from the sun. A giant screen placed in the shadow at the average distance of the moon would reveal a dark umbral shadow about 2.5 times the diameter of the moon. The faint outer edges of the penumbra would mark a circle about 4.6 times the diameter of the moon. Consequently, when the moon's orbit carries it through the umbra, it has plenty of room to become completely immersed in shadow.

Total Lunar Eclipses

A **lunar eclipse** occurs when the moon passes through Earth's shadow and grows dark. If the moon passes through the umbra and no part of the moon remains outside the umbra in the partial sunlight of the penumbra, the eclipse is a **total lunar eclipse.**

■ Figure 3-3 illustrates the stages of a total lunar eclipse. As the moon begins to enter the penumbra, it is only slightly dimmed, and a casual observer may not notice anything odd. After an hour, the moon is deeper in the penumbra and dimmer;

3-3 Solar Eclipses

FOR EONS, CULTURES WORLDWIDE HAVE UNDERSTOOD that the sun is the source of life, so you can imagine the panic people felt at the fearsome sight of the sun gradually disappearing in the middle of the day. Many imagined that the sun was being devoured by a monster (■ Figure 3-5). Modern scientists must use their imaginations to visualize how nature works, but with a key difference **(How Do We Know? 3-1)**. You can take comfort that today's astronomers explain solar eclipses without imagining celestial monsters.

A **solar eclipse** occurs when the moon moves between Earth and the sun. If the moon covers the disk of the sun completely, you see a spectacular **total solar eclipse** (■ Figure 3-6). If, from your location, the moon covers only part of the sun, you see a less dramatic **partial solar eclipse**. During a particular solar eclipse, people in one place on Earth may see a total eclipse, while people only a few hundred kilometers away see a partial eclipse.

The Angular Diameter of the Sun and Moon

Solar eclipses are spectacular because Earth's moon happens to have nearly the same angular diameter as the sun, so it can cover the sun almost exactly. You learned about the angular diameter of an object in Chapter 2; now you need to think carefully about how the size and distance of an object like the moon determine its angular diameter. This is the key to understanding solar eclipses.

Linear diameter is simply the distance between an object's opposite sides. You use linear diameter when you order a 16-inch pizza—the pizza is 16 inches across. The linear diameter of the moon is 3476 km.

The angular diameter of an object is the angle formed by lines extending toward you from opposite sides of the object and meeting at your eye (■ Figure 3-7). Clearly, the farther away an object is, the smaller its angular diameter.

To find the angular diameter of the moon, you need the **small-angle formula.** It gives you a way to figure out the angular diameter of any object, whether it is a pizza, the moon, or a galaxy. In the small-angle formula, you should always express angular diameter in seconds of arc* and always use the same units for distance and linear diameter:

$$\frac{\text{angular diameter}}{206{,}265''} = \frac{\text{linear diameter}}{\text{distance}}$$

You can use this formula to find any one of these three quantities if you know the other two; in this case, you are interested in finding the angular diameter of the moon.

The moon has a linear diameter of 3476 km and a distance from Earth of about 384,000 km. What is its angular diameter? The moon's linear diameter and distance are both given in the same units, kilometers, so you can put them directly into the small-angle formula:

*The number 206,265" is the number of seconds of arc in a radian. When you divide by 206,265", you convert the angle from seconds of arc to radians.

■ Figure 3-5

(a) A 12th-century Mayan symbol believed to represent a solar eclipse. The black-and-white sun symbol hangs from a rectangular sky symbol, and a voracious serpent approaches from below. (b) The Chinese representation of a solar eclipse shows a monster usually described as a dragon flying in front of the sun. (From the collection of Yerkes Observatory) (c) This wall carving from the ruins of a temple in Vijayanagaara in southern India symbolizes a solar eclipse as two snakes approach the disk of the sun. (T. Scott Smith)

How Do We Know? | 3-1

Scientific Imagination

Is an atom more like a plum pudding or a tiny solar system? Good scientists are invariably creative people with strong imaginations who can look at raw data about some invisible aspect of nature such as an atom and construct mental pictures as diverse as a plum pudding or a solar system. These scientists share the same human impulse to understand nature that drove ancient cultures to imagine eclipses as serpents devouring the sun.

As the 20th century began, physicists were busy trying to imagine what an atom was like. No one can see an atom, but English physicist J. J. Thomson used what he knew from his experiments and his powerful imagination to create an image of what an atom might be like. He suggested that an atom was a ball of positively charged material with negatively charged electrons distributed throughout like plums in a plum pudding.

The key difference between using a plum pudding to represent the atom and a hungry serpent to represent an eclipse is that the plum pudding model was based on experimental data and could be tested against new evidence. As it turned out, Thomson's student, Ernest Rutherford, performed ingenious new experiments and proposed a better representation of the atom. He imagined a tiny positively charged nucleus surrounded by negatively charged electrons.

Like other scientific models, scientific images simplify a complex reality. Today we know that electrons don't really orbit the atomic nucleus like planets. Despite its limitations, Rutherford's model of the atom is so useful and compelling that it has become a universally recognized symbol for atomic energy.

Ancient cultures pictured the sun being devoured by a serpent. Thomson, Rutherford, and scientists like them used their scientific imaginations to visualize natural processes. The critical difference is that scientific imagination is continually tested against evidence and is revised when necessary.

$$\frac{\text{angular diameter}}{206{,}265''} = \frac{3476 \text{ km}}{384{,}000 \text{ km}}$$

To solve for angular diameter, you can multiply both sides by 206,265 and find that the angular diameter is 1870 seconds of arc. If you divide by 60, you get 31 minutes of arc or, dividing by 60 again, about 0.5°. The moon's orbit is slightly elliptical, so it can sometimes look a bit larger or smaller, but its angular diameter is always close to 0.5°. It is a **Common Misconception** that the moon is larger when it is on the horizon. Certainly the rising full moon looks big when you see it on the horizon, but that is an optical illusion. In reality, the moon is the same size on the horizon as when it is high overhead.

You can repeat this small-angle calculation for the angular diameter of the sun. The sun is 1.39×10^6 km in linear diameter and 1.50×10^8 km from Earth. If you put these numbers into the small-angle formula, you will discover that the sun has an angular diameter of 1900 seconds of arc, which is 32 minutes of arc or about 0.5°. Earth's orbit is slightly elliptical, and consequently the sun can sometimes look slightly larger or smaller, but it, like the moon, is always close to 0.5° in angular diameter.

By fantastic good luck, you live on a planet with a moon that is almost exactly the same angular diameter as your sun. When

■ Figure 3-6

Solar eclipses are dramatic. In June 2001, an automatic camera in southern Africa snapped pictures every 5 minutes as the afternoon sun sank lower in the sky. From upper right to lower left, you can see the moon crossing the disk of the sun. A longer exposure was needed to record the total phase of the eclipse. (©2001 F. Espenak, www.MrEclipse.com)

at the sun in the example at lower left, and no eclipses are possible. At lower right, the line of nodes points toward the sun, and the shadows produce eclipses.

The shadows of Earth and moon, seen from space, are very long and thin, as shown in the lower part of Figure 3-14. It is easy for them to miss their mark at new moon or full moon and fail to produce an eclipse. Only during an eclipse season, when the line of nodes points toward the sun, do the long, skinny shadows produce eclipses.

If you watched for years from your point of view in space, you would see the orbit of the moon precess like a hubcap spinning on the ground. This precession is caused mostly by the gravitational influence of the sun, and it makes the line of nodes rotate once every 18.6 years. People back on Earth see the nodes slipping westward along the ecliptic 19.4° per year. This means that, according to the calendar, the eclipse seasons begin about 19 days earlier every year (■ Figure 3-15). So the sun does not need a full year to go from a node all the way around the ecliptic and back to that same node. The node is slipping westward to meet the sun, and the sun will cross the node after only 346.6 days (an **eclipse year**). This means the eclipses gradually move around the year occurring about 19 days earlier each year. If you see an eclipse in late December one year, you will see eclipses in early December the next year, and so on.

The cyclic pattern of eclipses shown in Figure 3-15 should give you another clue as to how to predict eclipses. Eclipses follow a pattern, and if you were an ancient astronomer and understood the pattern, you could predict eclipses without ever knowing what the moon was or how an orbit works.

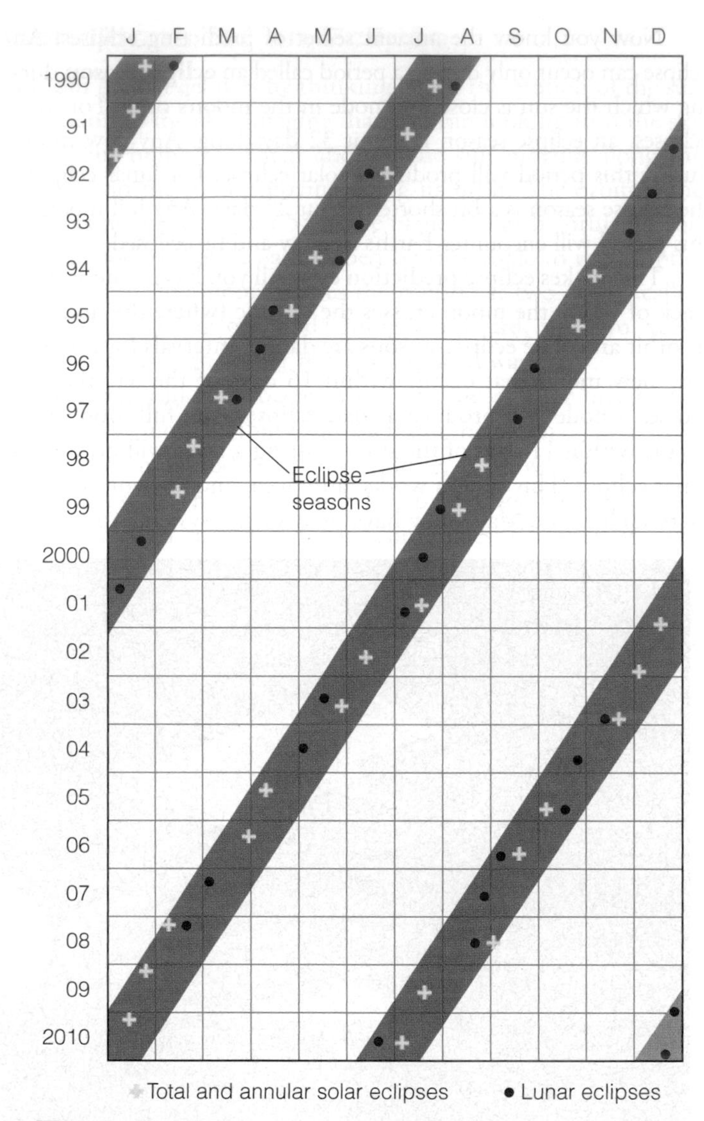

■ Figure 3-15

A calendar of eclipse seasons. Each year the eclipse seasons begin about 19 days earlier. Any new moon or full moon that occurs during an eclipse season results in an eclipse. Only total and annular eclipses are shown here.

The Saros Cycle

Ancient astronomers could predict eclipses in an approximate way using the eclipse seasons, but they could have been much more accurate if they had recognized that eclipses occur following certain patterns. The most important of these is the **saros cycle** (sometimes referred to simply as the saros). After one saros cycle of 18 years 11⅓ days, the pattern of eclipses repeats. In fact, *saros* comes from a Greek word that means "repetition."

The eclipses repeat because, after one saros cycle, the moon and the nodes of its orbit return to the same place with respect to the sun. One saros contains 6585.321 days, which is equal to 223 lunar months. Therefore, after one saros cycle, the moon is back to the same phase it had when the cycle began. But one saros is also equal to 19 eclipse years. After one saros cycle, the sun has returned to the same place it occupied with respect to the nodes of the moon's orbit when the cycle began. If an eclipse occurs on a given day, then 18 years 11⅓ days later the sun, the moon, and the nodes of the moon's orbit return to nearly the same relationship, and the eclipse occurs all over again.

Although the eclipse repeats almost exactly, it is not visible from the same place on Earth. The saros cycle is one-third of a day longer than 18 years 11 days. When the eclipse happens again, Earth will have rotated one-third of a turn farther east, and the eclipse will occur one-third of the way westward around Earth (■ Figure 3-16). That means that after three saros cycles—a period of 54 years 1 month—the same eclipse occurs in the same part of Earth.

One of the most famous predictors of eclipses was the Greek philosopher Thales of Miletus (about 640–546 BC), who supposedly learned of the saros cycle from the Chaldeans, who had discovered it. No one knows which eclipse Thales predicted, but some scholars suspect the eclipse of May 28, 585 BC. In any case, the eclipse occurred at the height of a battle between the Lydians and the Medes, and the mysterious darkness in midafternoon so startled the two factions that they concluded a truce.

Although there are historical reasons to doubt that Thales actually predicted the eclipse, the important point is that he

What Are We?

Special Effects

The moon is a companion in our daily lives, our history, and our mythology. It makes a dramatic sight as is moves through the sky, cycling through a sequence of phases that has repeated for billions of years. The moon has been humanity's timekeeper. Moses, Jesus, and Muhammad saw the same moon that you see counting out the days, weeks, and months.

The moon is part of our human heritage. Have you enjoyed watching a beautiful moonrise? Have you skied by moonlight? Famous paintings, poems, plays, and music celebrate the beauty of the moon. Lunar and solar eclipses add a hint of mystery and spectacle. Earth would be a poorer planet if it had no moon.

We live on a planet with astonishing special effects. The phases of the moon and lunar and solar eclipses punctuate our lives and our history, giving rise to great legends and ancient superstitions. The moon plays out the grand mechanisms of our solar system on the wide screen of our sky and reminds us that we are riding a planet as it whirls through space.

could have done it. If he had had records of past eclipses of the sun visible from the area, he could have discovered that they tended to recur with a period of 54 years 1 month (three saros cycles). Indeed, he could have predicted the eclipse without ever understanding what the sun and moon were or how they moved.

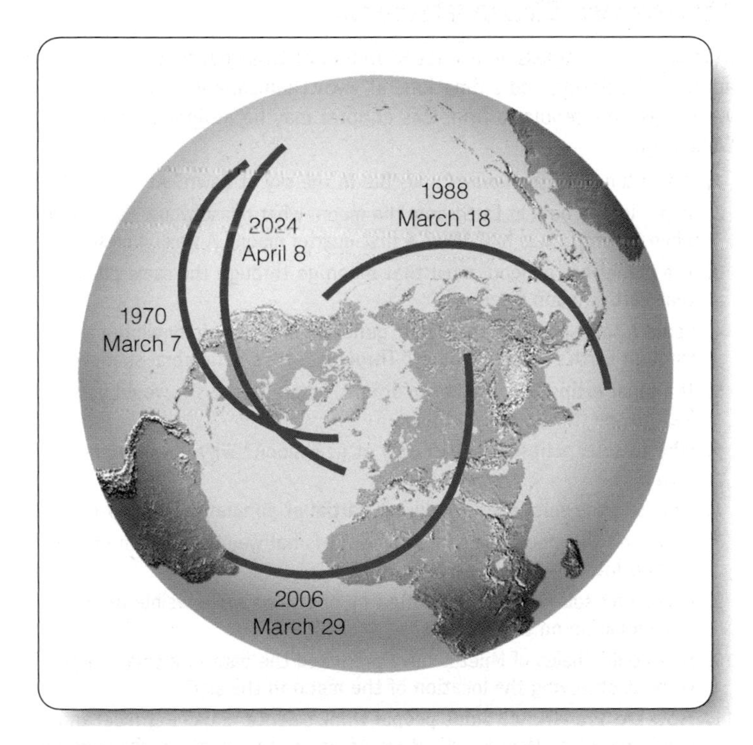

Figure 3-16

The saros cycle at work. The total solar eclipse of March 7, 1970, recurred after 18 years 11⅓ days over the Pacific Ocean. After another interval of 18 years 11⅓ days, the same eclipse was visible from Asia and Africa. After a similar interval, the eclipse will again be visible from the United States.

◄ SCIENTIFIC ARGUMENT ►

Why can't two successive full moons be totally eclipsed?

Most people suppose that eclipses occur at random or in some pattern so complex you need a big computer to make predictions. You know the geometry is fairly simple, so you can build a simple argument to analyze this question. Remember that a lunar eclipse can happen only when the sun is near one node and the moon crosses Earth's shadow at the other node.

Now you can apply what you know about the moon's phases. An eclipse season for a total lunar eclipse is only 22 days long, but the moon takes 29.5 days to go from one full moon to the next. If one full moon is totally eclipsed, the next full moon 29.5 days later will occur long after the end of the eclipse season, and there can be no second eclipse.

Now use your knowledge of the cycles of the sun and moon to revise your argument. **How can the sun be eclipsed by two successive new moons?**

◄ ►

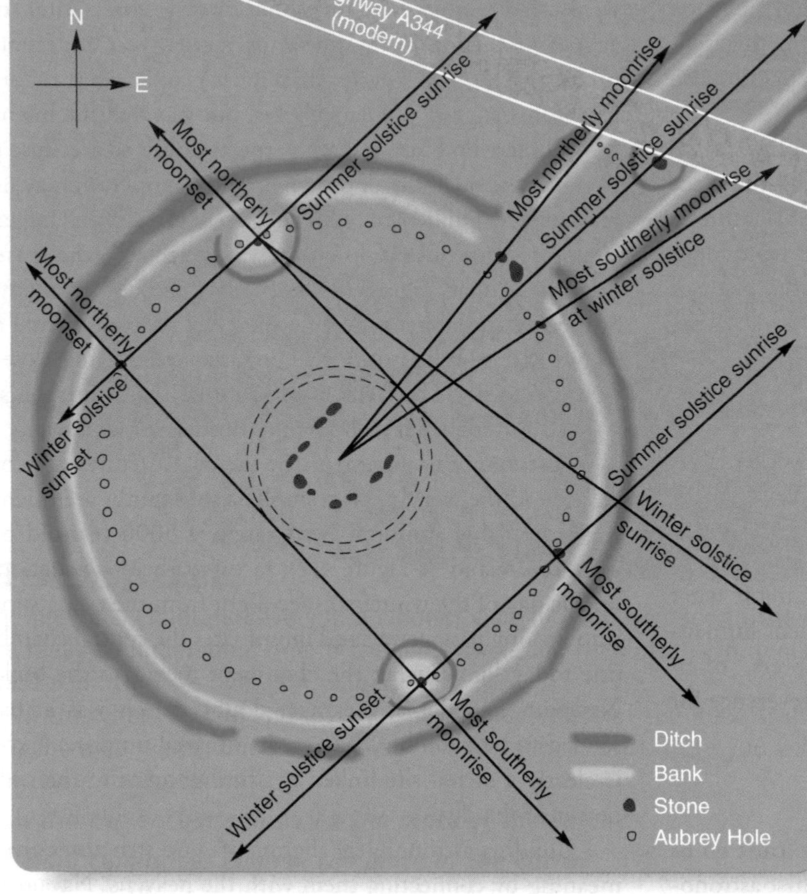

■ Figure 4-1

The central horseshoe of upright stones is only the most obvious part of Stonehenge. The best-known astronomical alignment at Stonehenge is the summer solstice sun rising over the Heelstone. Although a number of astronomical alignments have been found at Stonehenge, experts debate their significance. (Photo: Jamie Backman)

■ Figure 4-2

Newgrange was built on a small hill in Ireland about 3200 BC. A long passageway extends from the entryway back to the center of the mound, and sunlight shines down the passageway into the central chamber at dawn on the day of the winter solstice. Other passage graves have similar alignments, but their purpose is unknown. (Newgrange: Benelux Press/Index Stock Imagery)

High on Fajada Butte,
the Sun Dagger is off
limits to visitors.
The spiral pattern is the
size of a dinner plate.

■ Figure 4-3

In the ancient Native American settlement known as Chaco Canyon, New Mexico, sunlight shines between two slabs of stone high on the side of 440-foot-high Fajada Butte to form a dagger of light on the cliff face. About noon on the day of the summer solstice, the dagger of light slices through the center of a spiral pecked into the sandstone. (NPS Chaco Culture National Historic Park)

An American site in New Mexico is known as the Sun Dagger, but there is no surviving mythology to help you understand it. At noon on the day of the summer solstice, a narrow dagger of sunlight shines across the center of a spiral carved on a cliff face high above the desert floor (■ Figure 4-3). The purpose of the Sun Dagger is open to debate, but similar examples have been found throughout the American Southwest. It may have been more a symbolic and ceremonial marker than a precise calendrical indicator. In any case, it is just one of the many astronomical alignments that ancient people built into their structures to link themselves with the sky.

Some scholars are looking not at ancient structures but at small artifacts from thousands of years ago. Scratches on certain bone and stone implements follow a pattern that may record the phases of the moon (■ Figure 4-4). Some scientists contend that humanity's first attempts at writing were stimulated by a desire to record and predict lunar phases.

The study of archaeoastronomy is uncovering the earliest roots of astronomy and simultaneously revealing some of the first human efforts at systematic inquiry. The most important lesson of archaeoastronomy is that humans don't have to be technologically sophisticated to admire and study the universe. Trying to understand the sky is a universal part of human nature.

One thing about archaeoastronomy is especially sad. Although the methods of archaeoastronomy can show how ancient people observed the sky, their thoughts about their universe are in many cases lost. Many cultures had no written language. In other cases, the written record has been lost. Dozens, perhaps hundreds, of beautiful Mayan manuscripts, for instance, were burned by Spanish missionaries who believed that the books were the work of Satan. Only four of these books have survived, and all four contain astronomical references. One contains sophisticated tables that allowed the Maya to predict the motion of Venus and eclipses of the moon. No one will ever know what was burnt.

The fate of the Mayan books illustrates one reason why histories of astronomy usually begin with the Greeks. Some of their writing has survived, and you can discover what they thought about the shape and motion of the heavens.

The Astronomy of Greece

The names of the people who built Stonehenge will never be known, but the names of the greatest Greek philosophers have been famous for thousands of years, and their ideas shaped the development of astronomy long after their deaths.

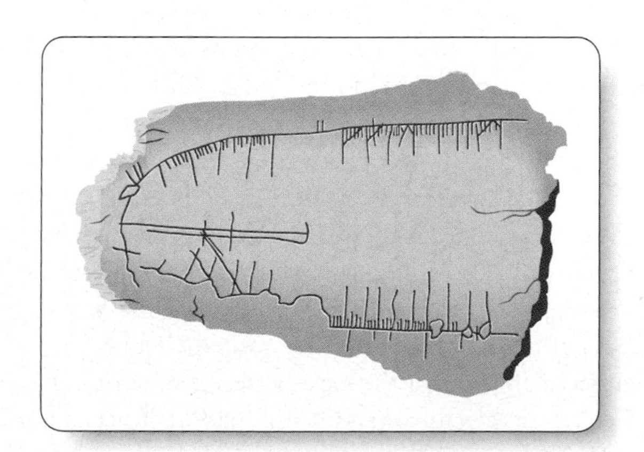

■ Figure 4-4

A fragment of a 27,000-year-old mammoth tusk found at Gontzi in Ukraine contains scribe marks on its edge, simplified in this drawing. These markings have been interpreted as a record of four cycles of lunar phases. Although controversial, such finds suggest that some of the first human attempts at recording events in written form were stimulated by astronomical phenomena.

Greek astronomy was derived from Babylon and Egypt, but the Greek philosophers took a new approach. Rather than relying on religion and astrology, the Greeks proposed a rational universe whose secrets could be understood through logic and reason.

This new attitude toward the heavens was a first step toward modern science, and it was made possible by two early Greek philosophers. Thales of Miletus (c. 624–547 BC) lived and worked in what is now Turkey. He taught that the universe is rational and that the human mind can understand why the universe works the way it does. This view contrasts sharply with that of earlier cultures, which believed that the ultimate causes of things are mysteries beyond human understanding. To Thales and his followers, the mysteries of the universe are mysteries because they are unknown, not because they are unknowable.

The other philosopher who made the new scientific attitude possible was Pythagoras (c. 570–500 BC). He and his students noticed that many things in nature seem to be governed by geometrical or mathematical relations. Musical pitch, for example, is related in a regular way to the lengths of plucked strings. This led Pythagoras to propose that all nature was underlain by musical principles, by which he meant mathematics. One result of this philosophy was the later belief that the harmony of the celestial movements produced actual music, the music of the spheres. But, at a deeper level, the teachings of Pythagoras made Greek astronomers look at the universe in a new way. Thales said that the universe could be understood, and Pythagoras said that the underlying rules were mathematical.

In trying to understand the universe, Greek astronomers did something that Babylonian astronomers had never done—they tried to construct descriptions based on geometrical forms. Anaximander (c. 611–546 BC) described a universe made up of wheels filled with fire: The sun and moon are holes in the wheels through which the flames can be seen. Philolaus (fifth century BC) argued that Earth moves in a circular path around a central fire (not the sun), which is always hidden behind a counterearth located between the fire and Earth. This, by the way, was the first theory to suppose that Earth is in motion.

Plato (428–347 BC) was not an astronomer, but his teachings influenced astronomy for 2000 years. Plato argued that the reality humans see is only a distorted shadow of a perfect, ideal form. If human observations are distorted, then observation can be misleading, and the best path to truth, said Plato, is through pure thought on the ideal forms that underlie nature.

Plato also argued that the most perfect geometrical form was the sphere, and therefore, he said, the perfect heavens must be made up of spheres rotating at constant rates and carrying objects around in circles. Consequently, later astronomers tried to describe the motions of the heavens by imagining multiple, rotating spheres. This became known as the principle of **uniform circular motion.**

Eudoxus of Cnidus (409–356 BC), a student of Plato, combined a system of 27 nested spheres rotating at different rates about different axes to produce a mathematical description of the motions of the universe (■ Figure 4-5).

At the time of the Greek philosophers, it was common to refer to systems such as that of Eudoxus as descriptions of the world, where the word *world* included not only Earth but all of the heavenly spheres. The reality of these spheres was open to debate. Some thought of the spheres as nothing more than mathematical ideas that described motion in the world model, while others began to think of the spheres as real objects made of perfect celestial material. Aristotle, for example, seems to have thought of the spheres as real.

Aristotle and the Nature of Earth

Aristotle (384–322 BC), one of Plato's students, made his own unique contributions to philosophy, history, politics, ethics, poetry, drama, and other subjects (■ Figure 4-6). Because of his sensitivity and insight, he became the greatest authority of antiquity, and his astronomical model was accepted with minor variations for almost 2000 years.

Much of what Aristotle wrote about scientific subjects was wrong, but that is not surprising. The scientific method, depending on evidence and hypothesis, had not yet been invented. Aristotle, like other philosophers of his time, attempted to understand their world by reasoning logically and carefully from first principles. A first principle is something that is obviously true. The perfection of the heavens was, for Aristotle, a first principle. Once a principle is recognized as true, whatever can be logically derived from it must also be true.

■ Figure 4-5

The spheres of Eudoxus explain the motions in the heavens by means of nested spheres rotating about various axes at different rates. Earth is located at the center.

■ Figure 4-6

Aristotle, honored on this Greek stamp, wrote on such a wide variety of subjects and with such deep insight that he became the great authority on all matters of learning. His opinions on the nature of Earth and the sky were widely accepted for almost two millennia.

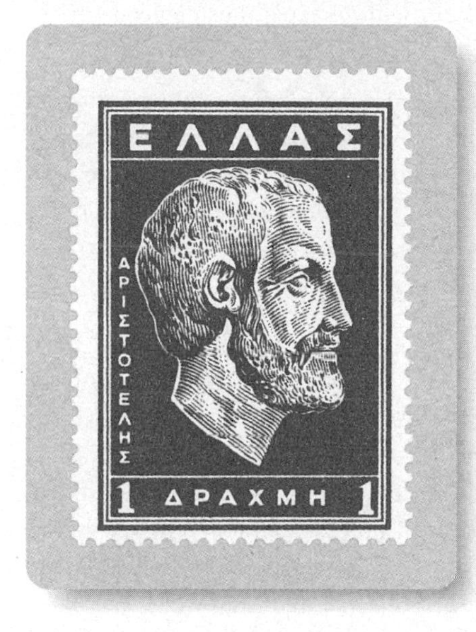

Aristotle believed that the universe was divided into two parts-Earth, imperfect and changeable; and the heavens, perfect and unchanging. Like most of his predecessors, he believed that Earth was the center of the universe, so his model is called a **geocentric universe.** The heavens surrounded Earth, and he devised 55 crystalline spheres turning at different rates and at different angles to carry the sun, moon, and planets across the sky. The lowest sphere, that of the moon, marked the boundary between the changeable imperfect region of Earth and the unchanging perfection of the celestial realm above the moon.

Because he believed Earth to be immobile, Aristotle had to make this entire nest of spheres whirl westward around Earth every 24 hours to produce day and night, but the spheres had to move more slowly with respect to one another to produce the motions of the sun, moon, and planets against the background of the stars. Because his model was geocentric, he taught that Earth could be the only center of motion. All of his whirling spheres had to be centered on Earth. Like most other Greek philosophers, Aristotle viewed the universe as a perfect heavenly machine that was not many times larger than Earth itself.

About a century after Aristotle, the Alexandrian philosopher Aristarchus proposed a theory that Earth rotated on its axis and revolved around the sun. This theory is, of course, correct, but most of the writings of Aristarchus were lost, and his theory was not well known. Later astronomers rejected any suggestion that Earth could move, because it conflicted with the teachings of the great philosopher Aristotle.

Aristotle had taught that Earth had to be a sphere because it always casts a round shadow during lunar eclipses, but he could only estimate its size. About 200 BC, Eratosthenes, working in the great library in Alexandria, found a way to calculate Earth's radius. He learned from travelers that the city of Syene (Aswan) in southern Egypt contained a well into which sunlight shone vertically on the day of the summer solstice. This told him that the sun was at the zenith at Syene; but, on that same day in Alexandria, he noted that the sun was 1/50 of the circumference of the sky (about 7°) south of the zenith.

Because sunlight comes from such a great distance, its rays arrive at Earth traveling almost parallel. That allowed Eratosthenes to use simple geometry to find that the distance from Alexandria to Syene was 1/50 of Earth's circumference (■ Figure 4-7).

To find Earth's circumference, Eratosthenes had to know the distance from Alexandria to Syene. Travelers told him it took 50 days to cover the distance, and he knew that a camel can travel about 100 stadia per day. That meant the total distance was about 5000 stadia. If 5000 stadia is 1/50 of Earth's circumference, then Earth must be 250,000 stadia in circumference, and, dividing by 2π, Eratosthenes found Earth's radius to be 40,000 stadia.

How accurate was Eratosthenes? The stadium (singular of *stadia*) had different lengths in ancient times. If you assume 6 stadia to the kilometer, then Eratosthenes's result was too big by only 4 percent. If he used the Olympic stadium, his result was 14 percent too big. In any case, this was a much better measure-

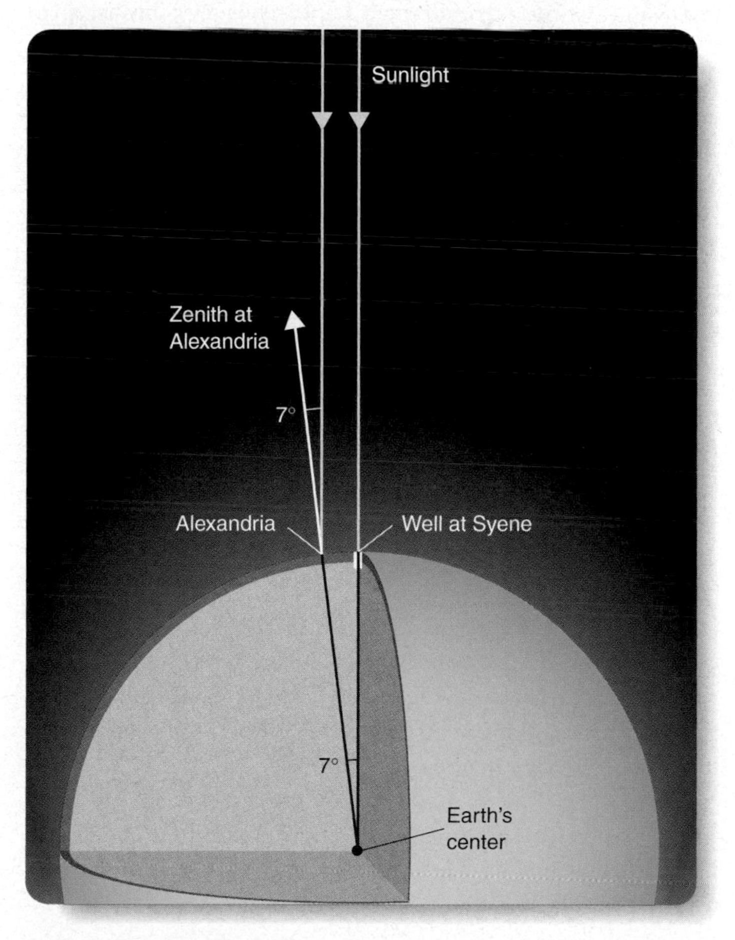

■ Active Figure 4-7

On the day of the summer solstice, sunlight fell to the bottom of a well at Syene, but the sun was about 1/50 of a circle (7°) south of the zenith at Alexandria. From this, Eratosthenes was able to calculate Earth's radius.

ment of Earth's radius than Aristotle's estimate, which was much too small, about 40 percent of the true radius.

You might think this is just a disagreement between two ancient philosophers, but it is related to a **Common Misconception**. Christopher Columbus did not have to convince Queen Isabella that the world was round. At the time of Columbus, all educated people knew the world was round and not flat, but they weren't sure how big it was. Columbus, like many others, adopted Aristotle's diameter for Earth, so he thought Earth was small enough that he could sail west and reach Japan and the Spice Islands of the East Indies. If he had accepted Eratosthenes' diameter, Columbus would never have risked the voyage. As it turned out, he and his crew were lucky that North America was in the way. If there had been open ocean all the way to Japan, they would have starved to death long before they reached land.

Aristotle, Aristarchus, and Eratosthenes were philosophers, but the next person you need to meet was a real astronomer who observed the sky in detail. Little is known about him. Hipparchus lived during the second century BC, about two centuries after Aristotle. He is usually credited with the invention of trigonometry, he compiled the first star catalog, and he discovered precession (Chapter 2). Although Hipparchus probably accepted the idea of the rotating celestial spheres, the important point was that the sphere carries its planet around in a circle. Hipparchus described the motion of the sun, moon and planets as following circular paths with Earth near, but not at, their centers. These off-center circles are now known as **eccentrics.** Hipparchus recognized that he could produce the same motion by having the sun, for instance, travel around a small circle that followed a larger circle around Earth. The compounded circular motion that he devised became the key element in the masterpiece of the last great astronomer of classical times, Claudius Ptolemy.

ThomsonNOW™ Sign in at www.thomsonedu.com and go to ThomsonNOW to see Astronomy Exercise "Eratosthenes' Calculations." Try his experiment on worlds of different diameters.

The Ptolemaic Universe

Claudius Ptolemaeus was one of the great astronomer-mathematicians of antiquity. His nationality and birth date are unknown, but he lived and worked in the Greek settlement at Alexandria in what is now Egypt about AD 140. He ensured the continued acceptance of Aristotle's universe by transforming it into a sophisticated mathematical model.

When you read **The Ancient Universe** on pages 60-61, you will encounter three important ideas and five new terms that show how first principles influenced early descriptions of the universe and its motions:

1. Ancient philosophers and astronomers accepted as first principles that the heavens were geocentric with Earth located at the center and sun, moon, and planets moving in uniform circular motion. It seemed clear to them that Earth was not moving because they saw no *parallax* in the positions of the stars.

2. Notice how the observed motion of the planets, the evidence, did not fit the theory very well. The *retrograde motion* of the planets was very difficult to explain using geocentrism and uniform circular motion.

3. Also, notice how Ptolemy attempted to explain the motion of the planets by devising a small circle, an *epicycle,* rotating along the edge of a larger circle; the *deferent,* which enclosed a slightly off-center Earth; and the *equant,* from which the center of the epicycle appeared to move at a constant rate. That meant the speed of the planets had to vary slightly as they circled Earth.

Ptolemy lived roughly five centuries after Aristotle, and although Ptolemy believed in the Aristotelian universe, he was interested in a different problem—the motion of the planets. He was a brilliant mathematician, and he was mainly interested in creating a mathematical description of the motions he saw in the heavens. For him, first principles took second place to mathematical precision.

Aristotle's universe, as embodied in the mathematics of Ptolemy, dominated ancient astronomy, but it was wrong. The planets don't follow circles at uniform speeds. At first, the Ptolemaic system predicted the positions of the planets well; but, as centuries passed, errors accumulated. If your watch gains only one second a year, it will keep time well for many years, but the error will gradually become noticeable. So, too, did the errors in the Ptolemaic system gradually accumulate as the centuries passed, but, because of the deep respect people had for the writings of Aristotle, the Ptolemaic system was not abandoned. Islamic and later European astronomers tried to update the system, computing new constants and adjusting epicycles. In the middle of the 13th century, a team of astronomers supported by King Alfonso X of Castile studied the *Almagest* for 10 years. Although they did not revise the theory very much, they simplified the calculation of the positions of the planets using the Ptolemaic system and published the result as *The Alfonsine Tables,* the last great attempt to make the Ptolemaic system of practical use.

ThomsonNOW™ Sign in at www.thomsonedu.com and go to ThomsonNOW to see Astronomy Exercise "Parallax I." Notice that the parallax angle depends on the length of the baseline.

◄ SCIENTIFIC ARGUMENT ►

How did the astronomy of Hipparchus and Ptolemy violate the principles of the early Greek philosophers Plato and Aristotle?

Today, scientific arguments depend on evidence and theory, but in classical times, they started from first principles. Hipparchus and Ptolemy lived very late in the history of classical astronomy, and they concentrated more on the mathematical problems and less on philosophical principles. They replaced the perfect spheres of Plato with nested circles in the form of epicycles and deferents. Earth was moved slightly away from the center of the deferent, so their models of the universe were not exactly geocentric, and the epicycles moved uniformly only as seen from the equant. The celestial motions were no longer precisely uniform, and the principles of geocentrism and uniform circular motion were weakened.

The work of Hipparchus and Ptolemy led eventually to a new understanding of the heavens, but first astronomers had to abandon uniform circular motion. Construct a scientific argument in the classical style based on first principles to answer the following: **Why did Plato argue for uniform circular motion?**

◄ ►

4-2 The Copernican Revolution

YOU WOULD NOT EXPECT NICOLAUS COPERNICUS (■ Figure 4-8) to have triggered an earthshaking revision in human thought. He was born in 1473 to a merchant family in Poland. Orphaned at the age of 10, he was raised by his uncle, an important bishop, who sent him to the University of Cracow and then to the best universities in Italy. Although he studied law and medicine and pursued a lifelong career as an important administrator in the Church, his real passion was astronomy.

Copernicus the Revolutionary

If you had been in astronomy class with Copernicus, you would have studied the Ptolemaic universe. The central location of Earth was widely accepted, and everyone knew that the heavens moved by the combination of uniform circular motions. For most scholars, questioning these principles was not an option because, over the course of centuries, Aristotle's proposed geometry of the heavens had become linked with the teachings of the Christian Church. According to the Aristotelian universe, the most perfect region was in the heavens, and the most imperfect at Earth's center. This classical geocentric universe matched the commonly held Christian geometry of heaven and hell, and anyone who criticized the Ptolemaic model was questioning Aristotle's geometry and indirectly challenging belief in heaven and hell.

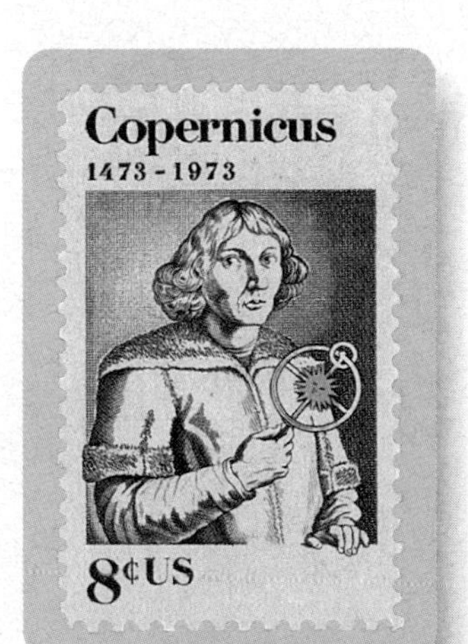

■ Figure 4-8

Copernicus proposed that the sun and not Earth was the center of the universe. Notice the heliocentric model on this stamp issued in 1973 to commemorate the 500th anniversary of his birth.

Copernicus studied the Ptolemaic universe and probably found it difficult at first to consider alternatives. Throughout his life, he was associated with the Church. His uncle was an important bishop in Poland, and through his uncle's influence, Copernicus became a canon at the cathedral in Frauenberg at the unusually young age of 24. (A canon was not a priest but a Church administrator.) This gave Copernicus an income, although he remained at the universities in Italy. When he did leave the university life, he joined his uncle and served as secretary and personal physician until his uncle died in 1512. At that point, Copernicus moved to quarters adjoining the cathedral in Frauenburg, where he served as canon for the rest of his life.

His close connection with the Church notwithstanding, Copernicus began to consider an alternative to the Ptolemaic universe, probably while he was still at university. Sometime before 1514, he wrote an essay proposing a **heliocentric universe** in which the sun, not Earth, was the center of the universe, and in which Earth rotated on its axis and revolved around the sun. He distributed this commentary in handwritten form, without a title and in some cases anonymously, to friends and astronomical correspondents. He may have been cautious out of modesty, out of respect for the Church, or out of fear that his revolutionary ideas would be attacked unfairly. Although his essay discusses every major aspect of his later work, it does not include observations and calculations to add support. His ideas needed further work, and he began gathering data and making detailed calculations in order to publish a book that would demonstrate the truth of his revolutionary ideas.

De Revolutionibus

Copernicus worked on his book *De Revolutionibus Orbium Coelestium (On the Revolutions of the Celestial Spheres)* over a period of many years. Although he essentially finished by about 1530, he hesitated to publish it, even though other astronomers knew of his theories and Church officials concerned about the reform of the calendar sought his advice and looked forward to the publication of his book.

One reason he hesitated was that the idea of a heliocentric universe would be highly controversial. This was a time of rebellion in the Church—Martin Luther was speaking harshly about fundamental church teachings, and others, both scholars and scoundrels, were questioning the authority of the Church. Even matters as abstract as astronomy could stir controversy. Remember that Earth's place in astronomical theory was linked to the geometry of heaven and hell, so moving Earth from its central place was a controversial and perhaps heretical idea.

Another reason Copernicus may have hesitated to publish was that his work was incomplete. His model could not accurately predict planetary positions. Copernicus was clearly concerned about how his ideas would be received, but in 1540 he allowed the visiting astronomer Joachim Rheticus (1514–1576) to publish an account of the Copernican universe in Rheticus's book *Prima Narratio (First Narrative)*. In 1542, Copernicus sent the manuscript for *De Revolutionibus* off to be printed (■ Figure 4-9a). He died in the spring of 1543 before the printing was completed.

The most important idea in *De Revolutionibus* was the placement of the sun at the center of the universe. That single innovation had an astonishing consequence—the retrograde motion of the planets was immediately explained in a straightforward way without the large epicycles that Ptolemy used. In the Copernican system, Earth moves faster along its orbit than the planets that lie further from the sun. Consequently, Earth periodically overtakes and passes these planets. Imagine that you are in a race car, driving rapidly along the inside lane of a circular racetrack. As you pass slower cars driving in the outer lanes, they fall behind, and if you did not know you were moving, it would seem that the cars in the outer lanes occasionally slowed to a stop and then backed up. The same thing happens as Earth passes a planet such as Mars. Although Mars moves steadily along its orbit, as seen from Earth it appears to slow to a stop and move westward (retrograde) as Earth passes it (■ Figure 4-10). Because the planetary orbits do not lie in precisely the same plane, a planet does not resume its eastward motion in precisely the same path it followed earlier. Consequently, it describes a loop whose shape depends on the angle between the orbital planes.

Copernicus could explain retrograde motion without epicycles, and that was impressive. The Copernican system was elegant and simple compared with the whirling epicycles and off-center equants of the Ptolemaic system. However, *De Revolutionibus* failed to disprove the geocentric model for one critical reason—the Copernican theory could not predict the positions of the

NICOLAI COPERNICI

net, in quo terram cum orbe lunari tanquam epicyclo contineri diximus. Quinto loco Venus nono menſe reducitur. Sextum denique locum Mercurius tenet, octuaginta dierum ſpacio circū currens. In medio uero omnium reſidet Sol. Quis enim in hoc

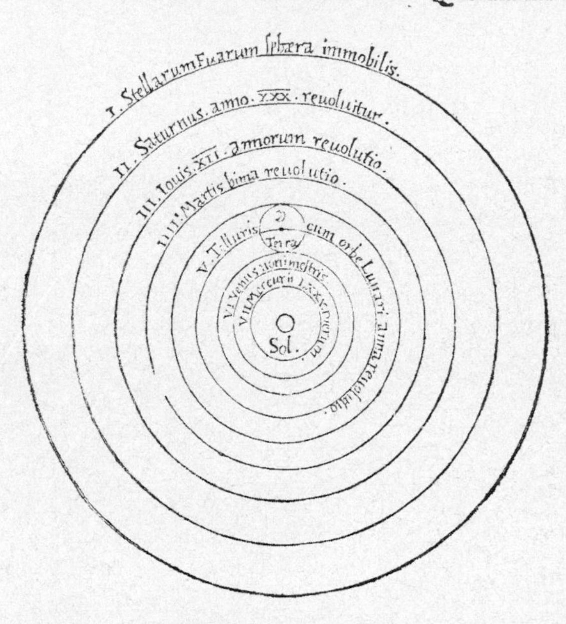

pulcherrimo templo lampadem hanc in alio uel meliori loco poneret, quàm unde totum ſimul poſsit illuminare? Siquidem non inepte quidam lucernam mundi, alij mentem, alij rectorem uocant. Trimegiſtus uiſibilem Deum, Sophoclis Electra intuentē omnia. Ita profecto tanquam in ſolio regali Sol reſidens circum agentem gubernat Aſtrorum familiam. Tellus quoque minime fraudatur lunari miniſterio, ſed ut Ariſtoteles de animalibus ait, maximam Luna cum terra cognationē habet. Cōcipit interea à Sole terra, & impregnatur annno partu. Inuenimus igitur ſub hac

a

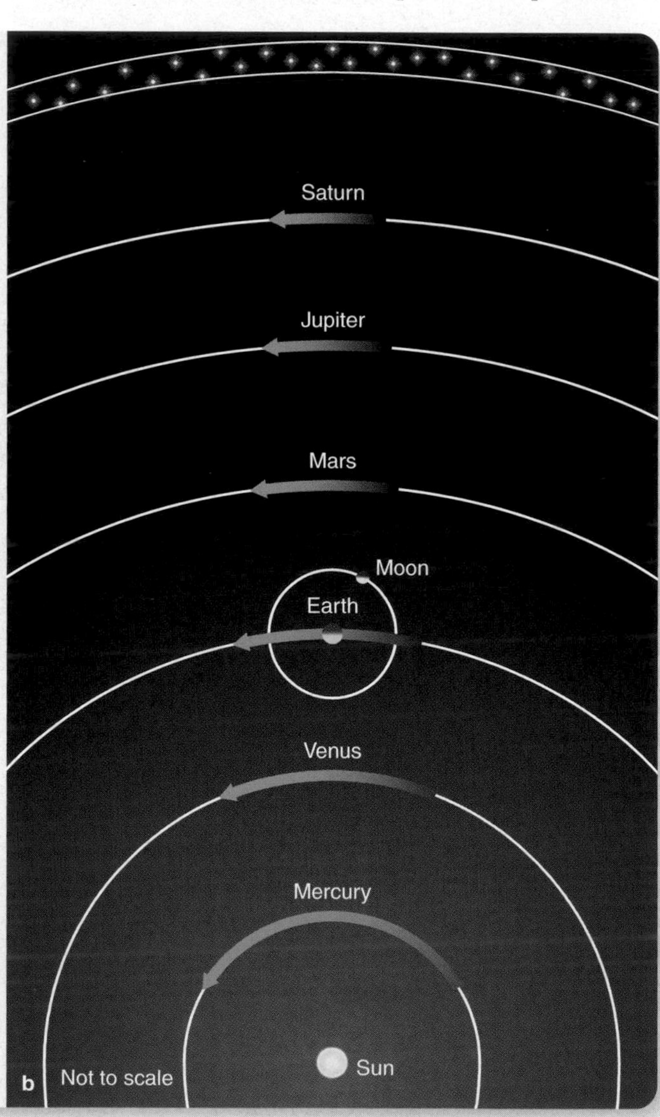

■ Figure 4-9

(a) The Copernican universe as reproduced in *De Revolutionibus*. Earth and all the known planets revolve in separate circular orbits about the sun (Sol) at the center. The outermost sphere carries the immobile stars of the celestial sphere. Notice the orbit of the moon around Earth (Terra). (Yerkes Observatory) (b) The model was elegant not only in its arrangement of the planets but also in their motions. Orbital speed (blue arrows) decreased from Mercury, the fastest, to Saturn, the slowest. Compare the elegance of this model with the complexity of the Ptolemaic model on page 61.

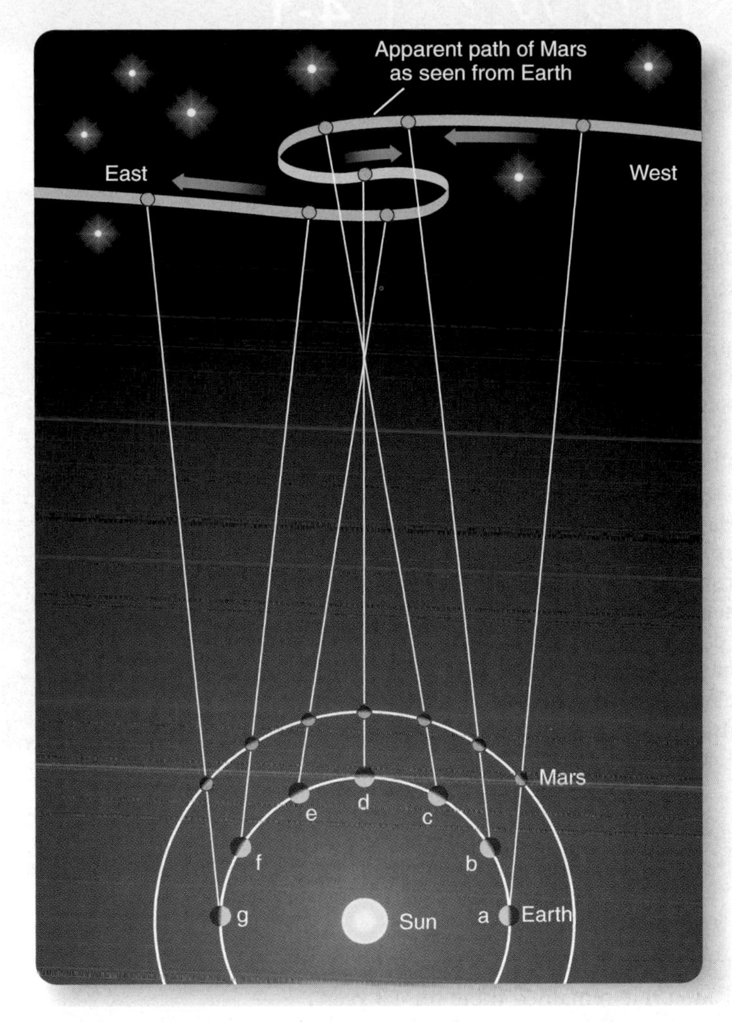

Figure 4-10

The Copernican explanation of retrograde motion: As Earth overtakes Mars (a-c), Mars appears to slow its eastward motion. As Earth passes Mars (d), Mars appears to move westward. As Earth draws ahead of Mars (e-g), Mars resumes its eastward motion against the background stars. Compare with the illustration of retrograde motion on page 60. The positions of Earth and Mars are shown at equal intervals of one month.

planets any more accurately than the Ptolemaic system could. To understand why it failed this critical test, you must understand Copernicus and his world.

Although Copernicus proposed a revolutionary idea in making the planetary system heliocentric, he was a classical astronomer with tremendous respect for the old concept of uniform circular motion. In fact, Copernicus objected strongly to Ptolemy's use of the equant. It seemed arbitrary to Copernicus, a direct violation of the elegance of Aristotle's philosophy of the heavens. Copernicus called equants "monstrous" in that they violated both geocentrism and uniform circular motion. In devising his model, Copernicus returned to a strong belief in uniform circular motion. Although he did not need epicycles to explain retrograde motion, Copernicus discovered that the sun, moon, and planets suffered other smaller variations in their motions that he could not explain with uniform circular motion centered on the sun. Today astronomers recognize that those variations are typical of objects following elliptical orbits, but Copernicus held firmly with uniform circular motion, so he had to introduce small epicycles to reproduce these minor variations in the motions of the sun, moon, and planets.

Because Copernicus imposed uniform circular motion on his model, it could not accurately predict the motions of the planets. *The Prutenic Tables* (1551) were based on the Copernican model, and they were not significantly more accurate than *The Alfonsine Tables* (1251), which were based on Ptolemy's model. Both could be in error by as much as 2°, four times the angular diameter of the full moon.

If the Copernican model was no better than the Ptolemaic model at making predictions, then why would anyone support it? The most important factor may be the elegance of the idea. Placing the sun at the center of the universe produced a symmetry among the motions of the planets that was pleasing to the eye and to the intellect (Figure 4-9b). All of the planets moved in the same direction at speeds that were simply related to their distance from the sun. In the Ptolemaic model, Venus and Mercury were treated differently from the other planets; their epicycles had to remain centered on the Earth-sun line. In the Copernican model, all planets were treated the same. The model may have eventually won support for its elegance more than its accuracy.

The Copernican *model* was inaccurate. It relies on uniform circular motion and consequently does not precisely describe the motions of the planets. But the Copernican *hypothesis* that the universe is heliocentric was correct. (Keep in mind that, given how little astronomers of the time knew of other stars and galaxies, saying that the planets circle the sun, not Earth, meant that the universe as they knew it was heliocentric.) Despite his flawed model, Copernicus's hypothesis was a groundbreaking moment in the history of astronomy.

The most astonishing consequence of the Copernican hypothesis was not what it said about the sun but what it said about Earth. By placing the sun at the center, Copernicus made Earth move around the sun just as the other planets did, and that made Earth a planet. By revealing that Earth is a planet, Copernicus revolutionized humanity's view of its place in the universe and triggered a controversy that would eventually bring Galileo before the Inquisition.

Although astronomers throughout Europe read and admired *De Revolutionibus,* most did not immediately accept the Copernican hypothesis. The mathematics was elegant, and the astronomical observations and calculations were of tremendous value. Yet most astronomers found it hard to believe, at first, that the sun actually was the center of the planetary system and that Earth moved. The gradual acceptance of the Copernican hypothesis has been named the Copernican Revolution because it involved not just the adoption of a new idea but a total revolution in the way astronomers thought about the place of the Earth **(How Do We Know? 4-1)**.

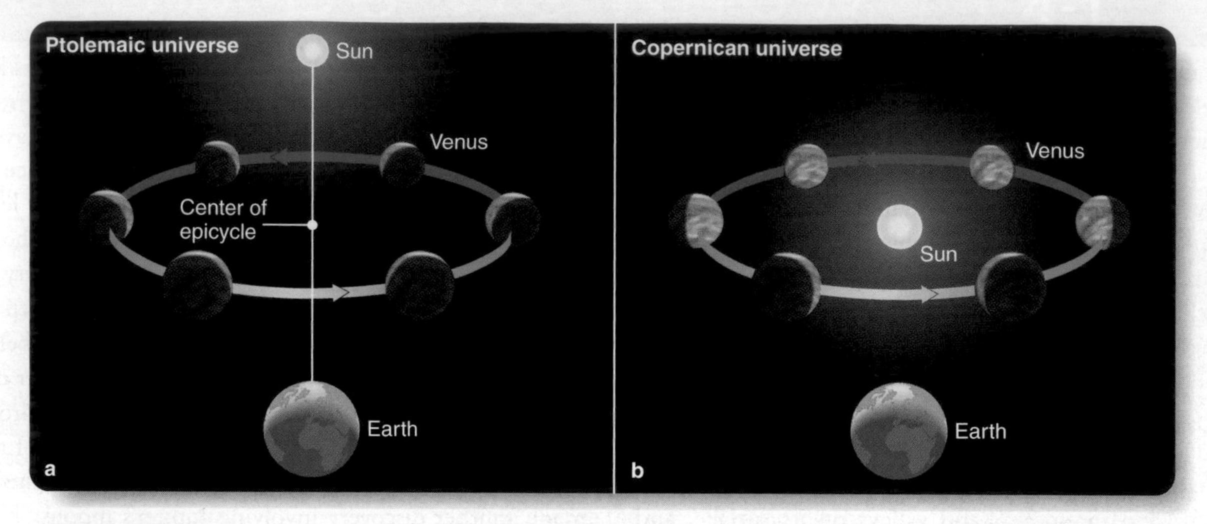

■ Figure 4-13

(a) If Venus moved in an epicycle centered on the Earth-sun line (see page 61), it would always appear as a crescent. (b) Galileo's telescope showed that Venus goes through a full set of phases, proving that it must orbit the sun.

■ Figure 4-14

Galileo's telescope made him famous, and he demonstrated his telescope and discussed his observations with powerful people. Some thought the telescope was the work of the devil and would deceive anyone who looked. In any case, Galileo's discoveries produced intense and, in some cases, angry debate. (Yerkes Observatory)

Pope Paul V decided to end the disruption, so when Galileo visited Rome in 1616 Cardinal Bellarmine interviewed him privately and ordered him to cease debate. There is some controversy today about the nature of Galileo's instructions, but he did not pursue astronomy for some years after the interview. Books relevant to Copernicanism were banned, including *De Revolutionibus,* although owners were allowed to keep their books if they changed certain phrases to make it clear that the central place of the sun was only a theory and not a fact.

In 1621 Pope Paul V died, and his successor, Pope Gregory XV, died in 1623. The next pope was Galileo's friend Cardinal Barberini, who took the name Urban VIII. Galileo rushed to Rome hoping to have the prohibition of 1616 lifted; and, although the new pope did not revoke the orders, he did encourage Galileo. When he got back home, Galileo began to write his great defense of the Copernican model, finally completing it on December 24, 1629. After some delay, the book was approved by both the local censor in Florence and the head censor of the Vatican in Rome. It was printed in February 1632.

Called *Dialogo Dei Due Massimi Sistemi (Dialogue Concerning the Two Chief World Systems),* it confronts the ancient astronomy of Aristotle and Ptolemy with the Copernican model and with telescopic observations as evidence. Galileo wrote the book as a debate among three friends. Salviati, a swift-tongued defender of Copernicus, dominates the book; Sagredo is intelligent but largely uninformed. Simplicio is the dismal defender of Ptolemy. In fact, he does not seem very bright.

The publication of *Dialogo* created a storm of controversy, and it was sold out by August 1632, when the Inquisition ordered sales stopped. The book was a clear defense of Copernicus, and, either intentionally or unintentionally, Galileo exposed the pope's authority to ridicule. Urban VIII was fond of arguing that, as God was omnipotent, he could construct the universe in any form while making it appear to humans to have a different form, and thus its true nature could not be deduced by mere observation. Galileo placed the pope's argument in the mouth of Simplicio, and Galileo's enemies showed the passage to the pope as an example of Galileo's disrespect. The pope thereupon ordered Galileo to face the Inquisition.

The Trial of Galileo

The trial of Galileo was one of the turning points in the history of science and human learning, but historians still argue about what happened and why. The trial involved the highest religious principles and the lowest of behind-the-scenes political maneuvering. One thing you can be sure of: The trial changed the way humanity thought about the world and marked the beginning of modern science as a way to understand nature.

Galileo was interrogated by the Inquisition four times and was threatened with torture. He must have thought often of a member of the Dominican order, Giordano Bruno, who was tried, condemned, and burned at the stake in Rome in 1600. One of Bruno's offenses had been Copernicanism. But Galileo's trial did not center on his belief in Copernicanism. After all, *Dialogo* had been approved by two censors. Rather, the trial centered on the instructions given Galileo in 1616. From his file in the Vatican, his accusers produced a record of the meeting between Galileo and Cardinal Bellarmine that included the statement that Galileo was "not to hold, teach, or defend in any way" the principles of Copernicus. Some historians believe that this document, which was signed neither by Galileo nor by Bellarmine nor by a legal secretary, was a forgery. Others suspect it may be a draft that was never used. In any case, it is possible that Galileo's true instructions were much less restrictive. But Bellarmine was dead and could not testify at Galileo's trial.

The Inquisition condemned him not for heresy but for disobeying the orders given him in 1616. On June 22, 1633, at the age of 70, kneeling before the Inquisition, Galileo read a recantation admitting his errors. Tradition has it that as he rose he whispered, *"E pur si muove"* ("Still it moves"), referring to Earth. Although he was sentenced to life imprisonment, he was actually confined at his villa for the next 10 years, perhaps through the intervention of the pope. He died there on January 8, 1642, 99 years after the death of Copernicus.

Galileo was not condemned for heresy, nor was the Inquisition interested when he tried to defend Copernicanism. He was tried and condemned on a charge you might call a technicality. Then why is his trial so important that historians have studied it for almost four centuries? Why have some of the world's greatest authors, including Bertolt Brecht, written about Galileo's trial? Why in 1979 did Pope John Paul II create a commission to reexamine the case against Galileo?

To understand the trial, you must recognize that it was the result of a conflict between two ways of understanding the universe. Plato had argued that observation was deceptive and that the only way to find truth was through pure thought. Since the Middle Ages, scholars had taught that the only path to true understanding was through religious faith. St. Augustine (AD 354-430) wrote *"Credo ut intelligame,"* which can be translated as, "Believe in order to understand." But Galileo and other scientists of the Renaissance used their own observations to try to understand the universe; and when their observations contradicted Scripture, they assumed their observations of reality were correct and that Scripture was not being correctly understood (■ Figure 4-15). Galileo paraphrased Cardinal Baronius in saying, "The Bible tells us how to go to heaven, not how the heavens go."

Galileo's discoveries produced intense, and in some cases angry, debate. Various passages of Scripture seemed to contradict observation. For example, Joshua is said to have commanded the sun to stand still, not Earth to stop rotating (Joshua 10:12–13). In response to such passages, Galileo argued that you should "read the book of nature"—that is, you should observe the universe with your own eyes. This ultimate reliance on evidence is a distinguishing characteristic of science.

The trial of Galileo was not about the place of the Earth. It was not about Copernicanism. It wasn't really about the instructions Galileo received in 1616. It was about the birth of modern science as a rational way to understand our universe. The commission appointed by John Paul II in 1979, reporting its conclusions in October 1992, said of Galileo's inquisitors, "This subjective error of judgment, so clear to us today, led them to a disciplinary measure from which Galileo 'had much to suffer.'"

■ Figure 4-15

Although he did not invent it, Galileo will always be associated with the telescope because it was the source of the observational evidence he used to try to understand the universe. By depending on evidence instead of first principles, Galileo led the way to the invention of modern science as a way to know about the natural world.

Mundi (The Harmony of the World), in which he returned to the cosmic mysteries of *Mysterium Cosmographicum.* The main thing of note in *Harmonice Mundi* is his discovery that the radii of the planetary orbits are related to the planets' orbital periods. That and his two previous discoveries are now recognized as Kepler's three laws of planetary motion.

Kepler's Three Laws of Planetary Motion

Although Kepler dabbled in the philosophical arguments of the day, he was a mathematician, and his triumph was the solution of the problem of the motion of the planets. The key to his solution was the ellipse.

An **ellipse** is a figure that is drawn around two points called the foci of the ellipse in such a way that the distance from one focus to any point on the ellipse and back to the other focus equals a constant. You can easily draw ellipses with two thumbtacks and a loop of string. Press the thumbtacks into a board, loop the string about the tacks, and place a pencil in the loop. If you keep the string taut as you move the pencil, it traces out an ellipse (■ Figure 4-20a).

The geometry of an ellipse is described by two simple numbers. The **semimajor axis, *a*,** is half of the longest diameter. The **eccentricity** of an ellipse, ***e*,** is the distance from either focus to the center of the ellipse divided by the semimajor axis. If you want to draw a circle with the string and tacks as shown in Figure 4-20a, you would move the two thumbtacks together, which shows that a circle is really just an ellipse with eccentricity equal to zero. As you move the thumbtacks farther apart, the ellipse becomes flatter, and the eccentricity moves closer to 1.

■ Table 4-1 | Kepler's Laws of Planetary Motion

I. The orbits of the planets are ellipses with the sun at one focus.
II. A line from a planet to the sun sweeps over equal areas in equal intervals of time.
III. A planet's orbital period squared is proportional to its average distance from the sun cubed:

$$P^2_{yr} = a^3_{AU}$$

Kepler used ellipses to describe the motion of the planets. His three fundamental rules have been tested and confirmed so many times that astronomers now refer to them as laws. They are commonly called Kepler's laws of planetary motion (■ Table 4-1).

Kepler's first law states that the orbits of the planets around the sun are ellipses with the sun at one focus. Thanks to the precision of Tycho's observations and the sophistication of Kepler's mathematics, Kepler was able to recognize the elliptical shape of the orbits, even though they are nearly circular. Of the planets in our solar system, Mercury has the most elliptical orbit, but even it deviates only slightly from a circle (■ Figure 4-21a).

Kepler's second law states that a line from the planet to the sun sweeps over equal areas in equal intervals of time. This means that when the planet is closer to the sun and the line connecting it to the sun is shorter, the planet moves more rapidly to sweep over the same area that is swept over when the planet is farther from the sun. So the planet in Figure 4-21b would move from point *A* to point *B* in one month, sweeping over the area shown. But when the planet is farther from the sun, one month's motion would be shorter, from *A′* to *B′*.

The time that a planet takes to travel around the sun once is its orbital period, *P,* and its average distance from the sun equals the semimajor axis of its orbit, *a.* Kepler's third law says these two quantities are related: The orbital period squared is proportional to the semimajor axis cubed. If you measure *P* in years and *a* in astronomical units, you can summarize the third law as:

$$P^2_{yr} = a^3_{AU}$$

The subscripts are reminders that you must express the period in years (yr) and the semimajor axis in astronomical units (AU).

■ Figure 4-20

The geometry of elliptical orbits: Drawing an ellipse with two tacks and a loop of string is easy. The semimajor axis, *a*, is half of the longest diameter. The sun lies at one of the foci of the elliptical orbit of a planet.

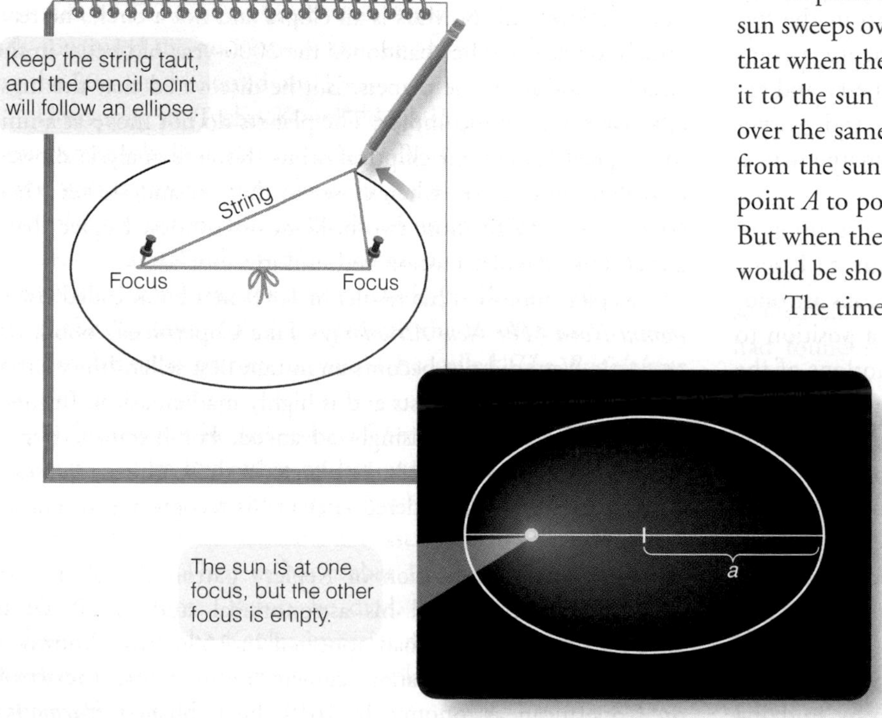

Keep the string taut, and the pencil point will follow an ellipse.
String
Focus
Focus
The sun is at one focus, but the other focus is empty.
a

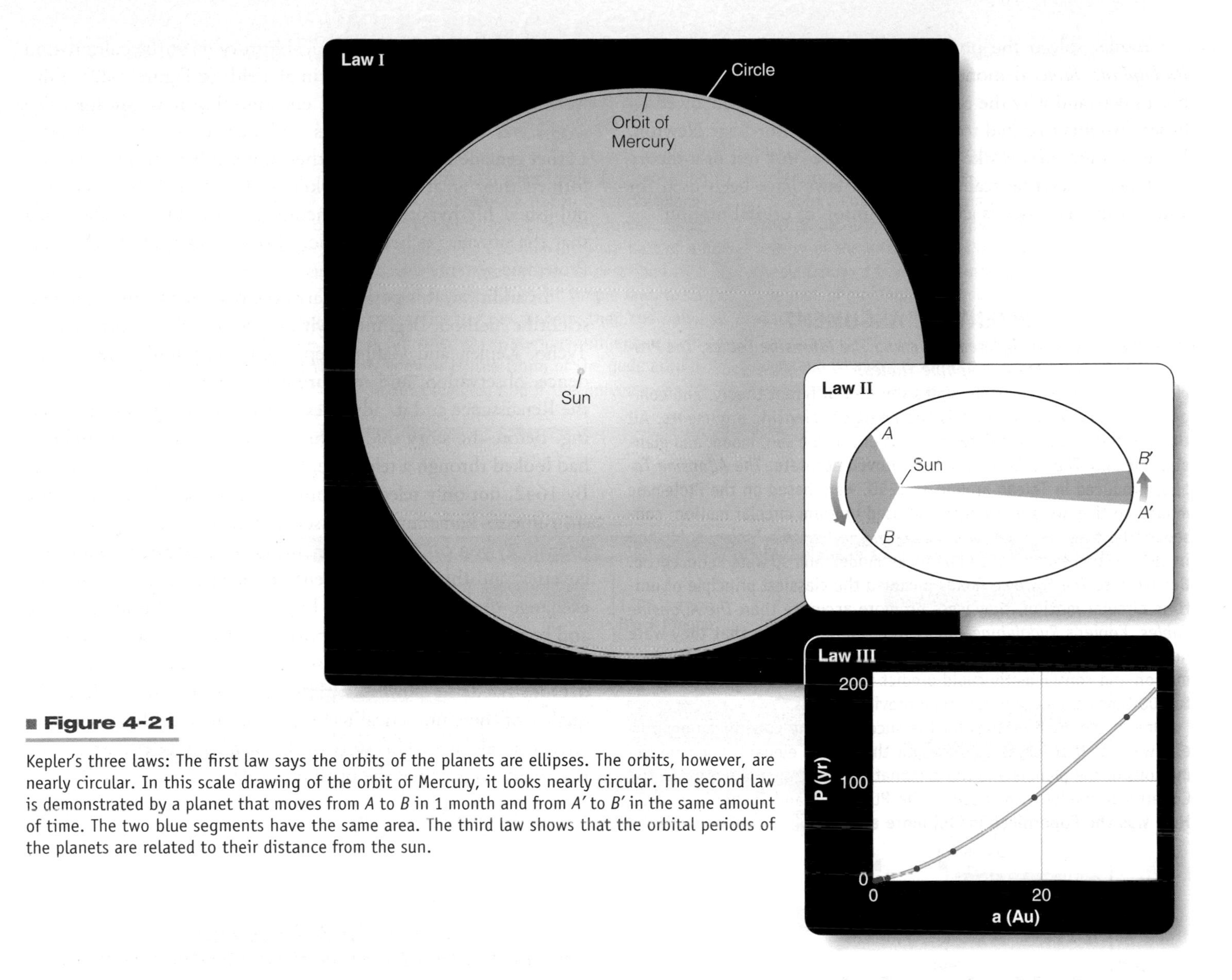

■ Figure 4-21

Kepler's three laws: The first law says the orbits of the planets are ellipses. The orbits, however, are nearly circular. In this scale drawing of the orbit of Mercury, it looks nearly circular. The second law is demonstrated by a planet that moves from *A* to *B* in 1 month and from *A′* to *B′* in the same amount of time. The two blue segments have the same area. The third law shows that the orbital periods of the planets are related to their distance from the sun.

You can use Kepler's third law to make simple calculations. For example, Jupiter's average distance from the sun is 5.20 AU. What is its orbital period? If a equals 5.20, then a^3 equals 140.6. The orbital period must be the square root of 140.6, which equals about 11.8 years.

You should note that Kepler's three laws are empirical. That is, they describe a phenomenon without explaining why it occurs. Kepler derived them from Tycho's extensive observations, not from any fundamental assumption or theory. In fact, Kepler never knew what held the planets in their orbits or why they continued to move around the sun. His books are a fascinating blend of careful observation, mathematical analysis, and mystical theory.

The Rudolphine Tables

In spite of Kepler's recurrent involvement with astrology and numerology, he continued to work on *The Rudolphine Tables.* At last, in 1627, they were ready, and he financed their printing himself, dedicating them to the memory of Tycho Brahe. In fact, Tycho's name appears in larger type on the title page than Kepler's own. This is especially surprising when you recall that Kepler based the tables on the heliocentric model of Copernicus and his own elliptical orbits and not on the Tychonic system. The reason for Kepler's evident deference was Tycho's family, still powerful and still intent on protecting the memory of Tycho. They even demanded a share of the profits and the right to censor the book before publication, though they changed nothing but a few words on the title page and added an elaborate dedication to the emperor.

The Rudolphine Tables were Kepler's masterpiece. They could predict the positions of the planets 10 to 100 times more accurately than previous tables. Kepler's tables were the precise model of planetary motion that Copernicus had sought but failed to find. The accuracy of *The Rudolphine Tables* was strong evidence that both Kepler's model for planetary motion and the Copernican hypothesis for the place of the Earth were correct. Copernicus would have been pleased.

5 Gravity

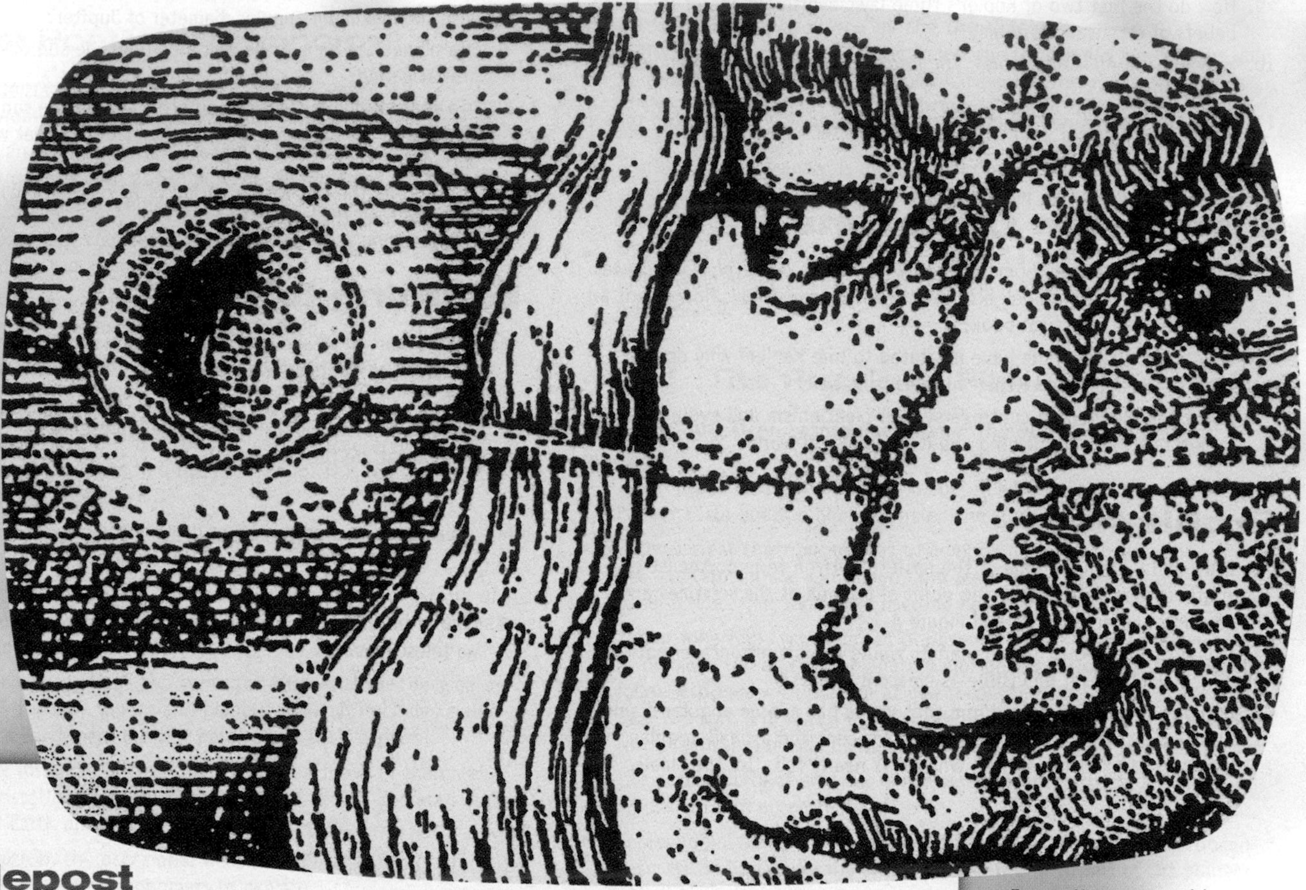

Isaac Newton could explain orbital motion using gravity.

Guidepost

If only Renaissance astronomers had understood gravity, they wouldn't have had so much trouble describing the motion of the planets, but that insight didn't appear until three decades after the trial of Galileo. Isaac Newton, starting from the work of Galileo, devised a way to explain motion and gravity, and that allowed astronomers to understand orbital motion and tides. Then, in the early 20th century, Albert Einstein found an even better way to describe motion and gravity.

This chapter is about gravity, the master of the universe. Here you will find answers to five essential questions:

- **What happens when an object falls?**
- **How did Newton discover gravity?**
- **How does gravity explain orbital motion?**
- **How does gravity explain the tides?**
- **How did Einstein better describe motion and gravity?**

Gravity rules. The moon orbiting Earth, matter falling into black holes, and the overall structure of the universe are dominated by gravity. As you study gravity, you will see science in action and find answers to three important questions:

- How Do We Know? **What are the differences among a hypothesis, a theory, and a law?**
- How Do We Know? **Why is the principle of cause and effect so important to scientists?**
- How Do We Know? **How are a theory's predictions useful in science?**

The rest of this book will tell the story of matter and gravity. The universe is a swirling waltz of matter dancing to the music of gravity, and you are along for the ride.

Nature and Nature's laws lay hid in night:
God said, "Let Newton be!" and all was light.

ALEXANDER POPE

ISN'T IT WEIRD that Isaac Newton is said to have "discovered" gravity in the late 17th century—as if people didn't have gravity before that, as if they floated around holding onto tree branches? Of course, everyone experienced gravity without noticing it. Newton realized that a force had to exist that made things fall, and that realization changed the way people thought about nature (■ Figure 5-1).

5-1 Galileo and Newton

ISAAC NEWTON WAS BORN in Woolsthorpe, England, on December 25, 1642, and on January 4, 1643. This was not a biological anomaly but a calendrical quirk. Most of Europe, following the lead of the Catholic countries, had adopted the Gregorian calendar, but Protestant England continued to use the Julian calendar. So December 25 in England was January 4 in Europe. If you take the English date, then Newton was born in the same year that Galileo Galilei died.

Newton went on to become one of the greatest scientists who ever lived, but even he admitted the debt he owed to those who had studied nature before him. He said, "If I have seen farther than other men, it is because I stood on the shoulders of giants." One of those giants was Galileo. Although Galileo is remembered as the defender of Copernicanism, he was also a talented scientist who studied the motions of falling bodies, and that was the key that led Newton to gravity and an explanation of planetary motion.

Johannes Kepler discovered three laws of planetary motion, but he never understood why the planets move along their orbits. He thought they might be pulled along by magnetic forces from the sun, and he considered and dismissed the idea that the planets were pushed along their orbits by angels.

Newton refined Kepler's model of planetary motion but did not perfect it. In science, a model is an intellectual conception of

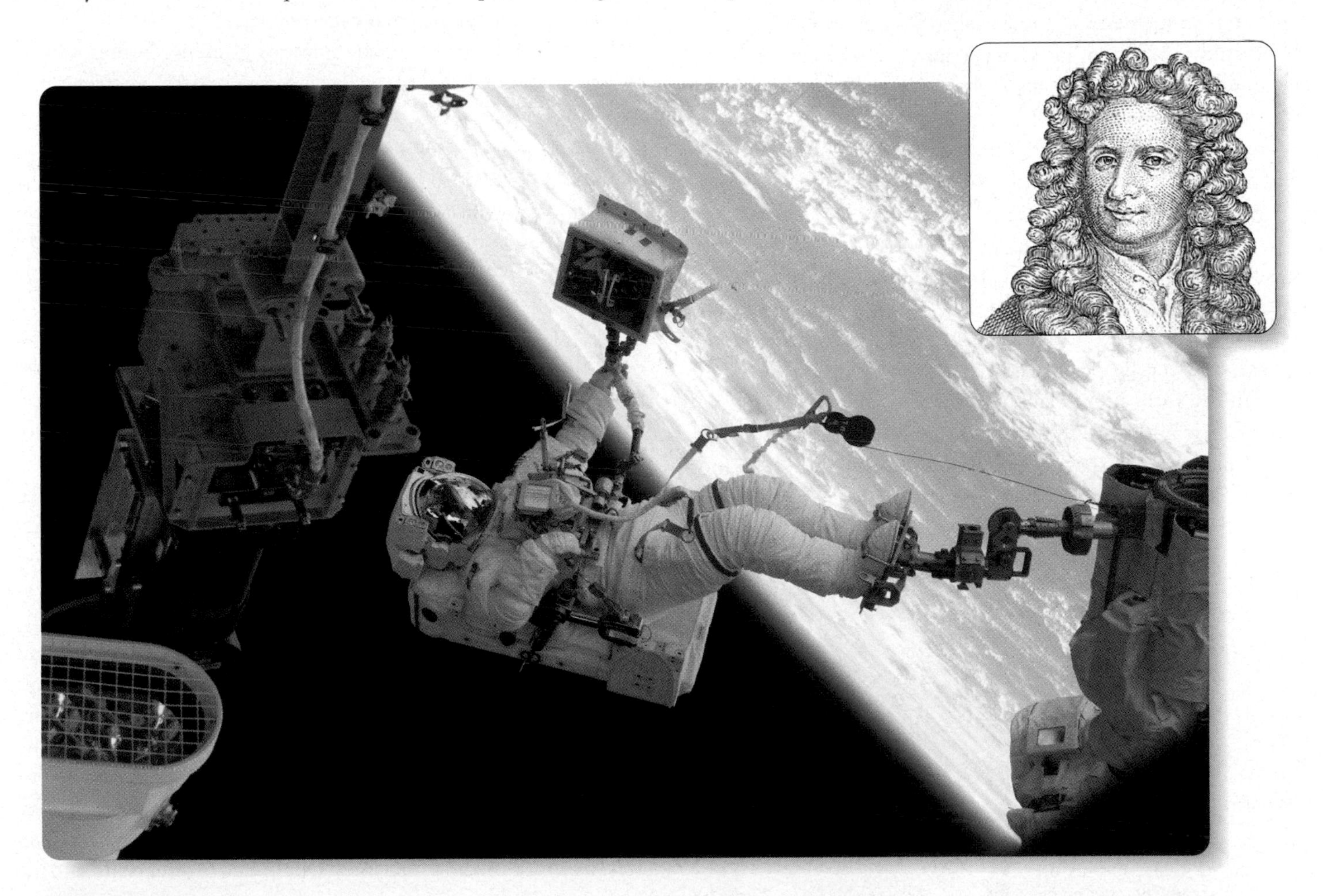

■ **Figure 5-1**

Space stations and astronauts, as well as planets, moons, stars, and galaxies, follow paths called orbits that are described by three simple laws of motion and a theory of gravity first understood by Isaac Newton (1642–1727). Newtonian physics is adequate to send astronauts to the moon and analyze the rotation of the largest galaxies. (NASA/JSC)

how nature works (see How Do We Know? 2-1). No model is perfect. Kepler's model was better than Aristotle's, but Newton improved Kepler's model by expanding it into a general theory of motion and gravity. In fact, most scientists now refer to Newton's *law* of gravity. Whether it is called a model, a theory, or a law, it is not perfect. Newton never understood what gravity was. It was as mysterious as an angel pushing the moon inward toward Earth instead of forward along the moon's orbit.

To understand science, you must understand the nature of scientific descriptions. The scientist studies nature by either creating new theories or refining old theories. Yet a theory can never be perfect, because it can never represent the universe in all its intricacies. Instead, a theory must be a limited description of a single phenomenon, such as orbital motion. It is fitting that Newton's discoveries all began with Kepler's fellow Copernican, Galileo.

Galileo and Motion

Even before Galileo built his first telescope, he had begun studying the motion of freely moving bodies (■ Figure 5-2). After the Inquisition condemned and imprisoned him in 1633, he continued his study of motion. He seems to have realized that he would have to understand motion before he could truly understand the Copernican system. Galileo's ability to set aside the authority of the ancients and think for himself allowed him to formulate principles that later led Newton to the laws of motion and the theory of gravity.

Aristotle's ideas on motion still held sway in Galileo's time. Aristotle said that the world is made up of four classical elements: earth, water, air, and fire, each located in its proper place. The proper place for earth (meaning soil and rock) is the center of the universe, and the proper place of water is just above earth. Air and then fire form higher layers, and above them lies the realm of the planets and stars. (You can see the four layers of the classical elements in the diagram at the top of page 60.) The four elements were believed to have a natural tendency to move toward their proper place in the cosmos. Things made up mostly of air or fire—smoke, for instance—tend to move upward. Things composed mostly of earth and water—wood, rock, flesh, bone, and so on—tend to move downward. According to Aristotle, objects fall downward because they are moving toward their proper place.*

Aristotle called these motions **natural motions** to distinguish them from **violent motions** produced, for instance, when you push on an object and make it move other than toward its proper place. According to Aristotle, such motions stop as soon as the force is removed. To explain how an arrow could continue to move upward even after it had left the bowstring, he said currents in the air around the arrow carried it forward even though the bowstring was no longer pushing it.

In Galileo's time and for the two preceding millennia, scholars had commonly tried to resolve problems of science by referring to authority. To analyze the flight of a cannonball, for instance, they would turn to the writings of Aristotle and other classical philosophers and try to deduce what those philosophers would have said on the subject. This generated a great deal of discussion but little real progress. Galileo broke with this tradition and conducted his own experiments.

He began by studying the motions of falling bodies, but he quickly discovered that the velocities were so great and the times so short that he could not measure them accurately. Consequently, he began using polished bronze balls rolling down gently sloping inclines. In that instance, the velocity is lower and the time longer. Using an ingenious water clock, he was able to measure the time the balls took to roll given distances down the incline, and he correctly recognized that these times are proportional to the times taken by falling bodies.

He found that falling bodies do not fall at constant rates, as Aristotle had said, but are accelerated. That is, they move faster with each passing second. Near Earth's surface, a falling object will have a velocity of 9.8 m/s (32 ft/s) at the end of 1 second, 19.6 m/s (64 ft/s) after 2 seconds, 29.4 m/s (96 ft/s) after 3 seconds, and so on. Each passing second adds 9.8 m/s (32 ft/s) to the object's velocity (■ Figure 5-3). In modern terms, this steady increase in the velocity of a falling body by 9.8 m/s each second (usually written 9.8 m/s^2) is called the **acceleration of gravity** at Earth's surface.

Galileo also discovered that the acceleration does not depend on the weight of the object. This, too, is contrary to the teachings of Aristotle, who believed that heavy objects, containing more earth and water, fall with higher velocity. Galileo found that the acceleration of a falling body is the same whether it is heavy or light. According to some accounts, he demonstrated this by dropping balls of iron and wood from the top of the Leaning Tower of Pisa to show that they would fall together and hit the

■ **Figure 5-2**

Although Galileo is often associated with the telescope, as on this Italian stamp, he also made systematic studies of the motion of falling bodies and made discoveries that led to the law of inertia.

*This is one reason why Aristotle had to have a geocentric universe. If Earth's center had not also been the center of the cosmos, his explanation of gravity would not have worked.

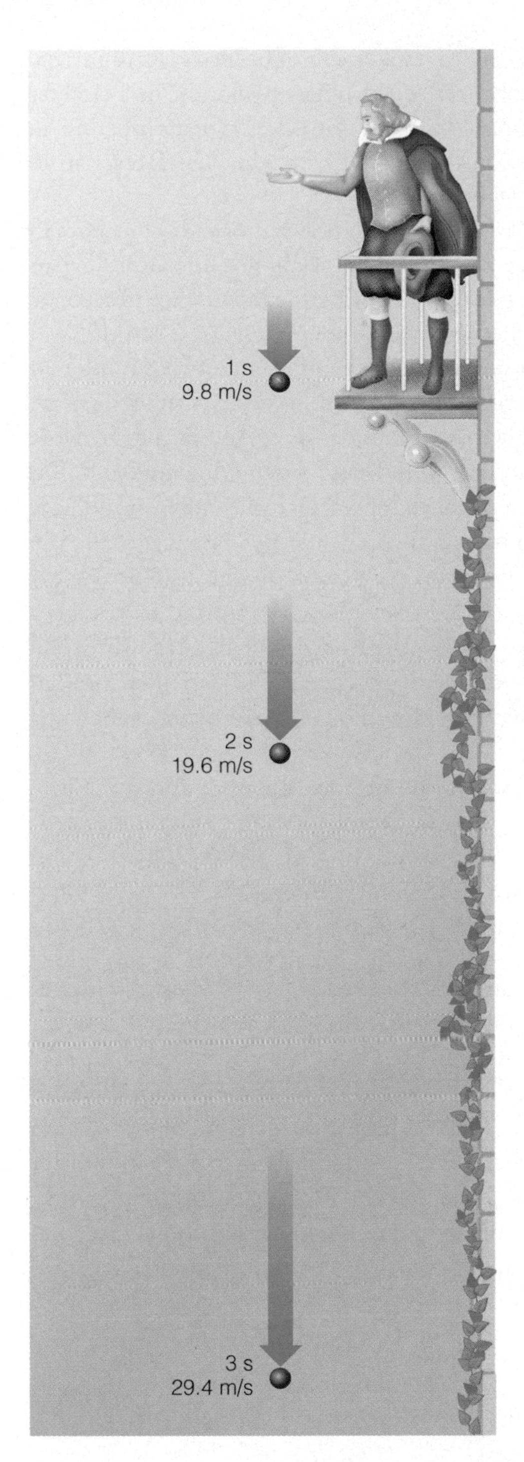
1 s
9.8 m/s
2 s
19.6 m/s
3 s
29.4 m/s

■ Figure 5-3

Galileo found that a falling object is accelerated downward. Each second, its velocity increases by 9.8 m/s (32 ft/s).

Air resistance would have slowed the wooden ball more and ruined Galileo's demonstration.
a
On the airless moon, there is no air resistance to slow the feather.
Hammer
Feather
b

■ Figure 5-4

(a) According to tradition, Galileo demonstrated that the acceleration of a falling body is independent of its weight by dropping balls of iron and wood from the Leaning Tower of Pisa. In fact, air resistance would have confused the result. (b) In a historic television broadcast from the moon on August 2, 1971, David Scott dropped a hammer and a feather at the same instant. They fell with the same acceleration and hit the surface together. (NASA)

ground at the same time (■ Figure 5-4a). In fact, he probably didn't perform this experiment. It would not have been conclusive anyway because of air resistance. More than 300 years later, Apollo 15 astronaut David Scott, standing on the airless moon, demonstrated Galileo's discovery by dropping a feather and a steel geologist's hammer. They fell at the same rate and hit the lunar surface at the same time (Figure 5-4b).

Having described natural motion, Galileo turned his attention to violent motion—that is, motion directed other than toward an object's proper place in the cosmos. He pointed out that an object rolling down an incline is accelerated and that an object rolling up the same incline is decelerated. If the incline were perfectly horizontal and frictionless, he reasoned, there could be no acceleration or deceleration to change the object's velocity, and, in the absence of friction, the object would continue to move forever. In his own words, "any velocity once imparted to a moving body will be rigidly maintained as long as the external causes of acceleration or retardation are removed."

Remember how Aristotle said that motion must be sustained by a force? Remove the force and the motion stops. No, said Galileo. Once begun, motion continues until something changes it. In fact, Galileo's statement is a perfectly valid summary of the law of inertia, which became Newton's first law of motion.

Galileo published his work on motion in 1638, two years after he had become entirely blind and only four years before his death. The book was called *Mathematical Discourses and Demonstrations Concerning Two New Sciences, Relating to Mechanics and to Local Motion.* It is known today as *Two New Sciences.*

The book is a brilliant achievement for a number of reasons. To understand motion, Galileo had to abandon the authority of the ancients, devise his own experiments, and draw his own conclusions. In a sense, this was the first example of experimental science. But Galileo also had to generalize his experiments to discover how nature worked. Though his apparatus was finite and his results skewed by friction, he was able to imagine an infinite, frictionless plane on which a body moves at constant velocity. In his workshop, the law of inertia was obscure, but in his imagination it was clear and precise.

ThomsonNOW Sign in at www.thomsonedu.com and go to ThomsonNOW to see the Astronomy Exercise "Falling Bodies."

Newton and the Laws Of Motion

Newton's three laws of motion (■ Table 5-1) are critical to understanding gravity and orbital motion. They are general laws of nature **(How Do We Know? 5-1)** and apply to any moving object, from an automobile driving along a highway to galaxies colliding with each other.

■ Table 5-1 | Newton's Three Laws of Motion

I. A body continues at rest or in uniform motion in a straight line unless acted upon by some force.
II. The acceleration of a body is inversely proportional to its mass, directly proportional to the force, and in the same direction as the force.
III. To every action, there is an equal and opposite reaction.

The first law is really a restatement of Galileo's law of inertia. An object continues at rest or in uniform motion in a straight line unless acted upon by some force. Astronauts drifting in space will travel at constant rates in straight lines forever if no forces act on them (■ Figure 5-5a).

Newton's first law also explains how a projectile continues to move after all forces have been removed—for instance, how an arrow continues to move after leaving the bowstring. The object continues to move because it has momentum. You can think of an object's **momentum** as a measure of its amount of motion.

An object's momentum is equal to its velocity times its mass. A paper clip tossed across a room has low velocity and therefore little momentum, and you could easily catch it in your hand. But the same paper clip fired at the speed of a rifle bullet would have tremendous momentum, and you would not dare try to catch it. Momentum also depends on the mass of an object **(Focus on Fundamentals 1)**. Now imagine that, instead of tossing a paper clip, someone tosses you a bowling ball. A bowling ball contains much more mass than a paper clip and therefore has much greater momentum, even though it is moving at the same velocity.

Newton's second law of motion discusses forces. Where Galileo spoke only of accelerations, Newton saw that an acceleration is the result of a force acting on a mass (Figure 5-5b). Newton's second law is commonly written as

$$F = ma$$

As always, you must define terms carefully when you look at an equation. An **acceleration** is a change in velocity, and a **velocity** is a directed speed. Most people use the words *speed* and *velocity* interchangeably, but they mean two different things. Speed is a rate of motion and does not have any direction associated with it, but velocity does. If you drive a car in a circle at 55 mph, your speed is constant, but your velocity is changing because your direction of motion is changing. An object experiences an acceleration if its speed changes or if its direction of motion changes.

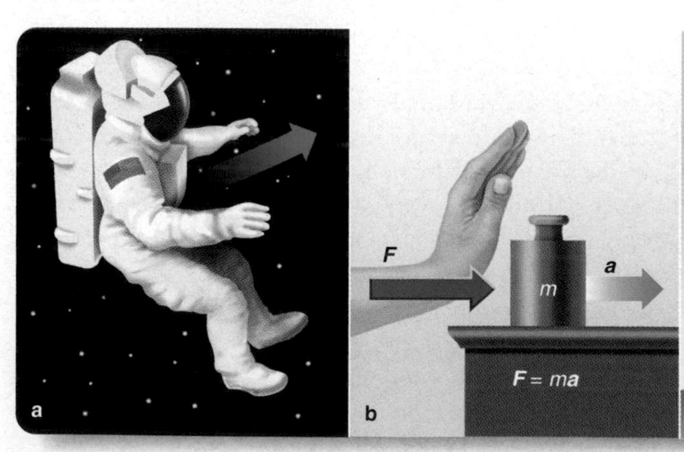

■ Figure 5-5

Newton's three laws of motion.

How Do We Know? 5-1

Hypothesis, Theory, and Law

How does sour milk help explain the spread of disease? Scientists study nature by devising new hypotheses and then developing those ideas into theories and laws that describe how nature works. A good example is the connection between sour milk and the spread of disease.

A scientist's first step in solving a natural mystery is to propose a reasonable explanation based on what is known so far. This proposal, called a **hypothesis,** is a single assertion or conjecture that must then be tested through observation and experimentation. Since the time of Aristotle, scientists believed that food spoils as a result of the spontaneous generation of life—maggots appeared out of rotting meat and mold out of drying bread. French chemist Louis Pasteur (1822–1895) hypothesized that microorganisms were not spontaneously generated but were carried through the air on particles of dust. To test his hypothesis, he sealed a nutrient broth in glass, completely protecting it from the dust particles in the air; no mold grew, effectively disproving spontaneous generation. Although others had argued against spontaneous generation before Pasteur, it was Pasteur's meticulous testing of his hypothesis through experimentation that finally convinced the scientific community.

The process of testing and confirming a hypothesis is only one step of the scientific process, although for some scientists this step constitutes their life's work. A **theory** generalizes the specific results of well-confirmed hypotheses to give a broader description of nature, which can be applied to a wide variety of circumstances. For instance, Pasteur's specific hypothesis about mold growing in broth contributed to a broader theory that disease is caused by microorganisms and that some can travel through the air. This theory, called the germ theory of disease, is a cornerstone of modern medicine.

Sometimes when a theory has been refined, tested, and confirmed so often that scientists have great confidence in it, it is called a **natural law.** Natural laws are the most fundamental principles of scientific knowledge. Newton's laws of motion are good examples.

Scientists generally have more confidence in a theory than in a hypothesis and the most confidence in a natural law. However, there is no precise distinction between a theory and a law, and use of these terms is sometimes a matter of tradition. For instance, some textbooks refer to the Copernican "theory" of heliocentrism, but it had not been well tested and is more rightly called the Copernican hypothesis. At the other extreme, Darwin's "theory" of evolution, containing many hypotheses that have been tested and confirmed over and over for nearly 150 years, might more rightly be called a natural law.

A fossil of a 500-million-year-old trilobite: Darwin's theory of evolution has been tested successfully many times, but by custom it is called a theory and not a law. (From the collection of John Coolidge III)

Every automobile has three accelerators—the gas pedal, the brake pedal, and the steering wheel. All three change the car's velocity.

In a way, the second law is just common sense; you experience it every day. The acceleration of a body is proportional to the force applied to it. If you push gently against a grocery cart, you expect a small acceleration. The second law of motion also says that the acceleration depends on the mass of the body. If your grocery cart were filled with bricks and you pushed it gently, you would expect very little result. If it were full of inflated balloons, however, it would move easily in response to a gentle push. Finally, the second law says that the resulting acceleration is in the direction of the force. This is also what you would expect. If you push on a cart that is not moving, you expect it to begin moving in the direction you push.

The second law of motion is important because it establishes a precise relationship between cause and effect (**How Do We Know? 5-2**). Objects do not just move. They accelerate due to the action of a force. Moving objects do not just stop. They decelerate due to a force. Also, moving objects don't just change direction for no reason. Any change in direction is a change in velocity and requires the presence of a force. Aristotle said that objects move because they have a tendency to move. Newton said that objects move due to a specific cause, a force.

Newton's third law of motion specifies that for every action there is an equal and opposite reaction. In other words, forces must occur in pairs directed in opposite directions. For example, if you stand on a skateboard and jump forward, the skateboard will shoot away backward. As you jump, your feet exert a force against the skateboard, which accelerates it toward the rear. But forces must occur in pairs, so the skateboard must exert an equal but opposite force on your feet that accelerates your body forward (Figure 5-5c).

Mass

One of the most fundamental parameters in science is **mass,** a measure of the amount of matter in an object. A bowling ball, for example, contains a large amount of matter and so is more massive than a child's rubber ball of the same size.

Mass is not the same as weight. Your weight is the force that Earth's gravity exerts on the mass of your body. Because gravity pulls you downward, you press against the bathroom scale, and you can measure your weight. Floating in space, you would have no weight at all; a bathroom scale would be useless. But your body would still contain the same amount of matter, so you would still have mass.

Sports analogies illustrate the importance of mass in dramatic ways. A bowling ball, for example, must be massive in order to have a large effect on the pins it strikes. Imagine trying to knock down all the pins with a balloon instead of a bowling ball. In space, where the bowling ball would be weightless, a bowling ball would still have more effect on the pins than a balloon. On the other hand, runners want track shoes that have low mass and thus are easy to move. Imagine trying to run a 100-meter dash wearing track shoes that were as massive as bowling balls. They would be very hard to move, and it would be difficult to accelerate away from the starting blocks. The shot put takes muscle because the shot is massive, not because it is heavy. Imagine throwing the shot in space where it would have no weight. It would still be massive, and it would take great effort to start it moving.

Mass is a unique measure of the amount of material in an object. Using the metric system (Appendix A), mass is measured in kilograms.

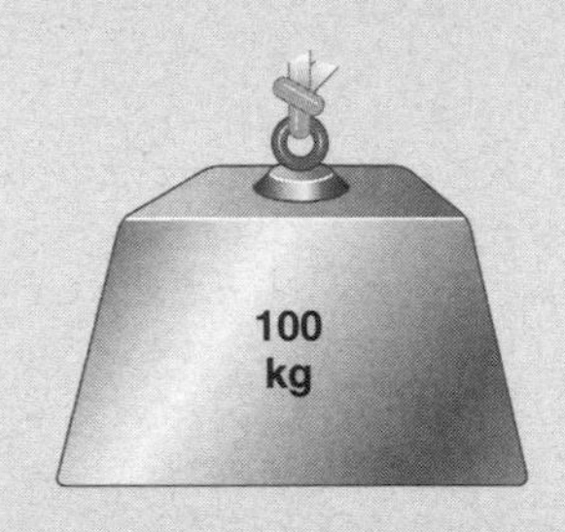

Mass is not the same as weight.

Mutual Gravitation

The three laws of motion led Newton to consider the force that causes objects to fall. The first and second laws tell you that falling bodies accelerate downward because some force must be pulling downward on them. Newton wondered what that force could be.

Newton was also aware that some force has to act on the moon. The moon follows a curved path around Earth, and motion along a curved path is accelerated motion. The second law says that an acceleration requires a force, so a force must be making the moon follow that curved path.

Newton wondered if the force that holds the moon in its orbit could be the same force that causes apples to fall—gravity. He was aware that gravity extends at least as high as the tops of mountains, but he did not know if it could extend all the way to the moon. He believed that it could, but he thought it would be weaker at greater distances, and he guessed that its strength would decrease as the square of the distance increased.

This relationship, the **inverse square law,** was familiar to Newton from his work on optics, where it applied to the intensity of light. A screen set up 1 meter from a candle flame receives a certain amount of light on each square meter. However, if that screen is moved to a distance of 2 meters, the light that originally illuminated 1 square meter must cover 4 square meters ■ Figure 5-6). Consequently, the intensity of the light is inversely proportional to the square of the distance to the screen.

Newton made two assumptions that enabled him to predict the strength of Earth's gravity at the distance of the moon. He assumed that the strength of gravity follows the inverse square law and that the critical distance is not the distance from Earth's surface but the distance from Earth's center. Because the moon is about 60 Earth radii away, Earth's gravity at the distance of the moon should be about 60^2 times less than at Earth's surface. Instead of being 9.8 m/s^2 at Earth's surface, it should be about 0.0027 m/s^2 at the distance of the moon.

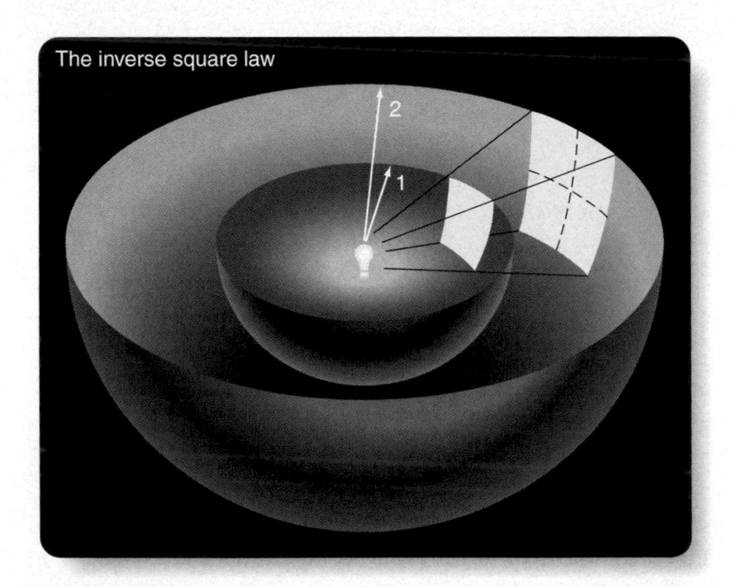

■ **Figure 5-6**

As light radiates away from a source, it spreads out and becomes less intense. Here the light falling on one square meter on the inner sphere must cover four square meters on a sphere twice as big. This shows how the intensity of light is inversely proportional to the square of the distance.

How Do We Know? 5-2

Cause and Effect

Why are cause and effect so important to scientists? One of the most often used and least often stated principles of science is cause and effect. Modern scientists all believe that events have causes, but ancient philosophers such as Aristotle argued that objects moved because of tendencies. They said that earth and water, and objects made mostly of earth and water, had a natural tendency to move toward the center of the universe. This natural motion had no cause but was inherent in the nature of the objects. Newton's second law of motion *($F = ma$)* was the first clear statement of the principle of cause and effect. If an object (of mass *m*) changes its motion (*a* in the equation), then it must be acted on by a force (*F* in the equation). Any effect *(a)* must be the result of a cause *(F)*.

The principle of cause and effect goes far beyond motion. It gives scientists confidence that every effect has a cause. The struggle against disease is an example. Cholera is a horrible disease that can kill its victims in hours. Long ago it was probably blamed on bad magic or the will of the gods, and only two centuries ago it was blamed on "bad air." When an epidemic of cholera struck England in 1854, Dr. John Snow carefully mapped cases in London showing that the victims had drunk water from a small number of wells contaminated by sewage. In 1876, the German Dr. Robert Koch traced cholera to an even more specific cause when he identified the microscopic bacillus that causes the disease. Step by step, scientists tracked down the cause of cholera.

If the universe did not depend on cause and effect, then you could never expect to understand how nature works. Newton's second law of motion was arguably the first clear statement that the behavior of the universe depends rationally on causes.

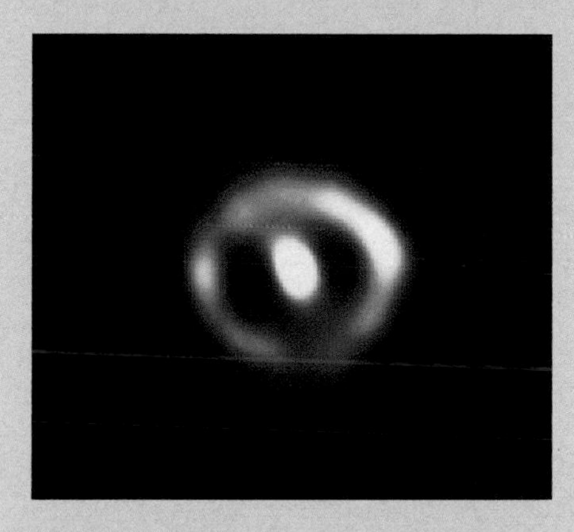

Cause and effect: Why did this star explode in 1992? There must have been a cause. (ESA/STScI and NASA)

Now, Newton wondered, could this acceleration keep the moon in orbit? He knew the moon's distance and its orbital period, so he could calculate the actual acceleration needed to keep it in its curved path. The answer is 0.0027 m/s^2, as his calculations predicted. The moon is held in its orbit by gravity, and gravity obeys the inverse square law.

Newton's third law says that forces always occur in pairs, and this leads to the conclusion that gravity is mutual. If Earth pulls on the moon, then the moon must pull on Earth. Gravitation is a general property of the universe. The sun, the planets, and all their moons must also attract each other by mutual gravitation. In fact, every particle of mass in the universe must attract every other particle, which is why Newtonian gravity is often called universal mutual gravitation.

Clearly the force of gravity depends on mass. Your body is made of matter, and you have your own personal gravitational field. But your gravity is weak and does not attract personal satellites orbiting around you. Larger masses have stronger gravity. From an analysis of the third law of motion, Newton realized that the mass that resists acceleration in the first law must be the same as the mass associated with gravity. Newton performed precise experiments with pendulums and confirmed this equivalence between the mass that resists acceleration and the mass that causes gravity.

From this, combined with the inverse square law, he was able to write the famous formula for the gravitational force between two masses, *M* and *m:*

$$F = -\frac{GMm}{r^2}$$

The constant *G* is the gravitational constant; it is the constant that connects mass to gravity. In the equation, *r* is the distance between the masses. The negative sign means that the force is attractive, pulling the masses together and making *r* decrease. In plain English, Newton's law of gravitation says: The force of gravity between two masses *M* and *m* is proportional to the product of the masses and inversely proportional to the square of the distance between them.

Newton's description of gravity was a difficult idea for physicists of his time to accept because it is an example of action at a distance. Earth and moon exert forces on each other even though there is no physical connection between them. Modern scientists resolve this problem by referring to gravity as a **field.** Earth's presence produces a gravitational field directed toward Earth's center. The strength of the field decreases according to the inverse square law. Any particle of mass in that field experiences a force that depends on the mass of the particle and the strength of the field at the particle's location. The resulting force is directed toward the center of the field.

The field is an elegant way to describe gravity, but it still does not say what gravity is. Later in this chapter, when you learn about Einstein's theory of curved space-time, you will get a better idea of what gravity really is.

◄ SCIENTIFIC ARGUMENT ►

What do the words universal and mutual mean when you say "universal mutual gravitation"?

Newton argued that the force that makes an apple accelerate downward is the same as the force that accelerates the moon and holds it in its orbit. You can learn more by thinking about Newton's third law of motion, which says that forces always occur in pairs. If Earth attracts the moon, then the moon must attract Earth. That is, gravitation is *mutual* between any two objects.

Furthermore, if Earth's gravity attracts the apple and the moon, then it must attract the sun, and the third law says that the sun must attract Earth. But if the sun attracts Earth, then it must also attract the other planets and even distant stars, which, in turn, must attract the sun and each other. Step by step, Newton's third law of motion leads to the conclusion that gravitation must apply to all masses in the universe. That is, gravitation must be *universal.*

Aristotle explained gravity in a totally different way. **Could Aristotle's explanation of a falling apple on Earth account for a hammer falling on the surface of the moon?**

◄ ►

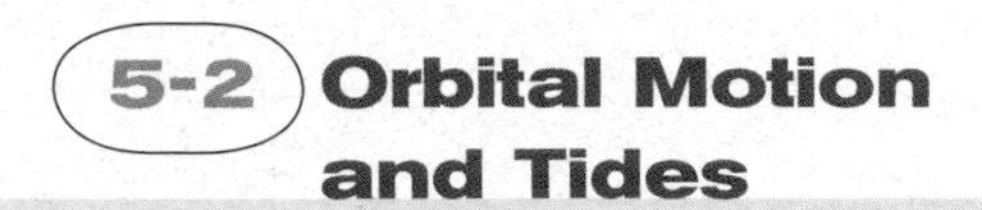

5-2 Orbital Motion and Tides

ORBITAL MOTION AND TIDES are two different kinds of gravitational phenomena. As you think about the orbital motion of the moon and planets, you need to think about how gravity pulls on an object. When you think about tides, you must think about how gravity pulls on different parts of an object. Analyzing these two kinds of phenomena will give you a deeper insight into how gravity works.

Orbits

Newton was the first person to realize that objects in orbit are falling. You can explore Newton's insight by analyzing the motion of objects orbiting Earth.

Carefully read **Orbiting Earth** on pages 88–89 and notice three important concepts and six new terms that will help you discuss orbital motion:

1. An object orbiting Earth is actually falling (being accelerated) toward Earth's center. The object continuously misses Earth because of its orbital velocity. To follow a circular orbit, the object must move at *circular velocity*, and at the right distance from Earth it could be a very useful *geosynchronous satellite.*

2. Also, notice that objects orbiting each other actually revolve around their *center of mass.*

3. Finally, notice the difference between *closed orbits* and *open orbits*. If you want to leave Earth never to return, you must accelerate your spaceship at least to *escape velocity* so it will follow an open orbit.

ThomsonNOW Sign in at **www.thomsonedu.com** and go to ThomsonNOW to see the Astronomy Exercise "Orbital Motion." Experiment with an object in orbit.

Orbital Velocity

If you were about to ride a rocket into orbit, you would have to answer a critical question. "How fast must I go to stay in orbit?" An object's circular velocity is the lateral velocity it must have to remain in a circular orbit. If you assume that the mass of your spaceship is small compared with the mass of Earth, then the circular velocity is:

$$V_c = \sqrt{\frac{GM}{r}}$$

In this formula, M is the mass of the central body (Earth in this case) in kilograms, r is the radius of the orbit in meters, and G is the gravitational constant, 6.67×10^{-11} m^3/s^{2}kg. This formula is all you need to calculate how fast an object must travel to stay in a circular orbit.

For example, how fast does the moon travel in its orbit? Earth's mass is 5.98×10^{24} kg, and the radius of the moon's orbit is 3.84×10^8 m. Then the moon's velocity is:

$$V_c = \sqrt{\frac{6.67 \times 10^{-11} \times 5.98 \times 10^{24}}{3.84 \times 10^8}} = \sqrt{\frac{39.9 \times 10^{13}}{3.84 \times 10^8}}$$
$$= \sqrt{1.04 \times 10^6} = 1020 \text{ m/s} = 1.02 \text{ km/s}$$

This calculation shows that the moon travels 1.02 km along its orbit each second. That is the circular velocity at the distance of the moon.

A satellite just above Earth's atmosphere is only about 200 km above Earth's surface, or 6578 km from Earth's center, so Earth's gravity is much stronger, and the satellite must travel much faster to stay in a circular orbit. You can use the formula above to find that the circular velocity just above Earth's atmosphere is about 7790 m/s, or 7.79 km/s. This is about 17,400 miles per hour, which shows why putting satellites into Earth orbit takes such large rockets. Not only must the rocket lift the satellite above Earth's atmosphere, but the rocket must then tip over and accelerate the satellite to circular velocity.

A **Common Misconception** holds that there is no gravity in space. You can see that space is filled with gravitational forces from Earth, the sun, and all other objects in the universe. An astronaut who appears weightless in space is actually falling along an orbit at the urging of the combined gravitational fields in the vicinity. Just above Earth's atmosphere, the orbital motion of the astronaut is dominated by Earth's gravity.

Calculating Escape Velocity

If you launch a rocket upward, it will consume its fuel in a few moments and reach its maximum speed. From that point on, it will coast upward. How fast must a rocket travel to coast away from Earth and escape? Of course, no matter how far it travels, it can never escape from Earth's gravity. The effects of Earth's gravity extend to infinity. It is possible, however, for a rocket to travel so fast initially that gravity can never slow it to a stop. Then the rocket could leave Earth.

Escape velocity is the velocity required to escape from the surface of an astronomical body. Here you are interested in escaping from Earth or a planet; later chapters will consider the escape velocity from stars, galaxies, and even a black hole.

The escape velocity, V_e, is given by a simple formula:

$$V_e = \sqrt{\frac{2GM}{r}}$$

Here G is the gravitational constant 6.67×10^{-11} m^3/s^2kg, M is the mass of the astronomical body in kilograms, and r is its radius in meters. (Notice that this formula is very similar to the formula for circular velocity.)

You can find the escape velocity from Earth by looking up its mass, 5.98×10^{24} kg, and its radius, 6.38×10^6 m. Then the escape velocity is:

$$V_e = \sqrt{\frac{2 \times 6.67 \times 10^{-11} \times 5.98 \times 10^{24}}{6.38 \times 10^6}} = \sqrt{\frac{7.98 \times 10^{14}}{6.38 \times 10^6}}$$

$$= \sqrt{1.25 \times 10^8} = 11{,}200 \text{ m/s} = 11.2 \text{ km/s}$$

This is equal to about 25,000 miles per hour.

Notice from the formula that the escape velocity from a body depends on both its mass and radius. A massive body might have a low escape velocity if it has a very large radius. You will meet such objects in the discussion of giant stars. On the other hand, a rather low-mass body could have a very large escape velocity if it had a very small radius, a condition you will meet in the discussion of black holes.

Circular velocity and escape velocity are two aspects of Newton's laws of gravity and motion. Once Newton understood gravity and motion, he could do what Kepler had failed to do—he could explain why the planets obey Kepler's laws of planetary motion.

ThomsonNOW www.thomsonedu.com and go to ThomsonNOW to see the Astronomy Exercise "Escape Velocity."

Kepler's Laws Reexamined

Now that you understand Newton's laws, gravity, and orbital motion, you can understand Kepler's laws of planetary motion in a new way.

Kepler's first law says that the orbits of the planets are ellipses with the sun at one focus. The orbits of the planets are ellipses because gravity follows the inverse square law. In one of his most famous problems, Newton proved that if a planet moves in a closed orbit under the influence of an attractive force that follows the inverse square law, then the planet must follow an elliptical path.

Even though Kepler correctly identified the shape of the planets' orbits, he still wondered why the planets keep moving along these orbits, and now you know the answer. They move because there is nothing to slow them down. Newton's first law says that a body in motion stays in motion unless acted on by some force. The gravity of the sun accelerates the planets inward toward the sun and holds them in their orbits, but it doesn't pull backward on the planets, so they don't slow to a stop. In the absence of friction, they must continue to move.

Kepler's second law says that a planet moves faster when it is near the sun and slower when it is farther away. Once again, Newton's discoveries explain why. Imagine you are in an elliptical orbit around the sun. As you round the most distant part of the ellipse, aphelion, you begin to move back closer to the sun, and the sun's gravity pulls you slightly forward in your orbit. You pick up speed as you fall closer to the sun, so, of course, you go faster as you approach the sun. As you round the closest point to the sun, perihelion, you begin to move away from the sun, and the sun's gravity pulls slightly backward on you, slowing you down as you climb away from the sun. So Kepler's second law makes sense when you analyze it in terms of forces and motions.

Physicists explain Kepler's second law in a slightly more elegant way. Earlier you saw that a body moving on a frictionless surface will continue to move in a straight line until it is acted on by some force; that is, the object has momentum. In a similar way, an object set rotating on a frictionless surface will continue rotating until something acts to speed it up or slow it down. Such an object has **angular momentum,** a measure of the rotation of the body about some point. A planet circling the sun has a given amount of angular momentum; and, with no outside influences to alter its motion, it must conserve its angular momentum. That is, its angular momentum must remain constant. Mathematically, a planet's angular momentum is the product of its mass, velocity, and distance from the sun. This explains why a planet must speed up as it comes closer to the sun along an elliptical orbit. Because its angular momentum is conserved, as its distance from the sun decreases, its velocity must increase. Conversely, the planet's velocity must decrease as its distance from the sun increases.

The conservation of angular momentum is actually a common human experience. Skaters spinning slowly can draw their arms and legs closer to their axis of rotation and, through conservation of angular momentum, spin faster (■ Figure 5-7). To slow their rotation, they can extend their arms again. Similarly, divers can spin rapidly in the tuck position and then slow their rotation by stretching into the extended position.

Kepler's third law is also explained by a conservation law, but in this case it is the law of conservation of energy (**Focus on Fundamentals 2**). A planet orbiting the sun has a specific amount of energy that depends only on its average distance from the sun. That energy can be divided between energy of motion and energy stored in the gravitational attraction between the

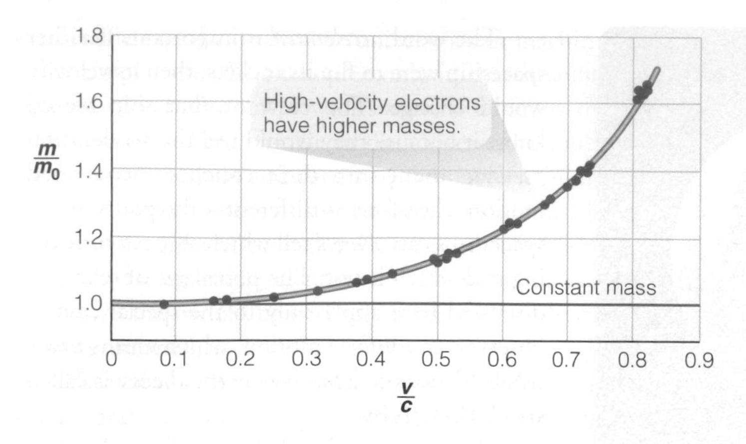

Figure 5-12

The observed mass of moving electrons depends on their velocity. As the ratio of their velocity to the velocity of light, v/c, gets larger, the mass of the electrons in terms of their mass at rest, m/m_0, increases. Such relativistic effects are quite evident in particle accelerators, which accelerate atomic particles to very high velocities.

The constant c is the speed of light, and m_0 is the mass of the particle when it is at rest. This simple formula shows that mass and energy are related, and you will see in later chapters how nature can convert one into the other inside stars.

For example, suppose that you convert 1 kg of matter into energy. The velocity of light as 3×10^8 m/s, so your result is 9×10^{16} joules (J) (approximately equal to a 20-megaton nuclear bomb). (Recall that a joule is a unit of energy roughly equivalent to the energy given up when an apple falls from a table to the floor.) This simple calculation shows that the energy equivalent of even a small mass is very large.

Other relativistic effects include the slowing of moving clocks and the shrinkage of lengths measured in the direction of motion. A detailed discussion of the major consequences of the special theory of relativity is beyond the scope of this book, but you can have confidence that these strange effects have been confirmed many times in experiments. Rather than pursue those details, you can consider Einstein's second advance, his general theory.

The General Theory of Relativity

In 1916, Einstein published a more general version of the theory of relativity that dealt with accelerated as well as uniform motion. This **general theory of relativity** contained a new description of gravity.

Einstein began by thinking about observers in accelerated motion. Imagine an observer sitting in a windowless spaceship. Such an observer cannot distinguish between the force of gravity and the inertial forces produced by the acceleration of the spaceship (■ Figure 5-13). This led Einstein to conclude that gravity

Figure 5-13

(a) An observer in a closed spaceship on the surface of a planet feels gravity. (b) In space, with the rockets smoothly firing and accelerating the spaceship, the observer feels inertial forces that are equivalent to gravitational forces.

and acceleration are related, a conclusion now known as the equivalence principle:

> **Equivalence principle:** Observers cannot distinguish locally between inertial forces due to acceleration and uniform gravitational forces due to the presence of a massive body.

This should not surprise you. Earlier in this chapter you read that Newton concluded that the mass that resists acceleration is the same as the mass that exerts gravitational forces. He even performed an elegant experiment with pendulums to test the equivalence of the mass related to motion and the mass related to gravity.

The importance of the general theory of relativity lies in its description of gravity. Einstein concluded that gravity, inertia, and acceleration are all associated with the way space is related to time in what is now referred to as space-time. This relation is often referred to as curvature, and a one-line description of general relativity explains a gravitational field as a curved region of space-time:

> **Gravity according to general relativity:** Mass tells space-time how to curve, and the curvature of space-time (gravity) tells mass how to accelerate.

So you feel gravity because Earth's mass causes a curvature of space-time. The mass of your body responds to that curvature by accelerating toward Earth's center. According to general relativity, all masses cause curvature, and the larger the mass, the more severe the curvature. That's gravity.

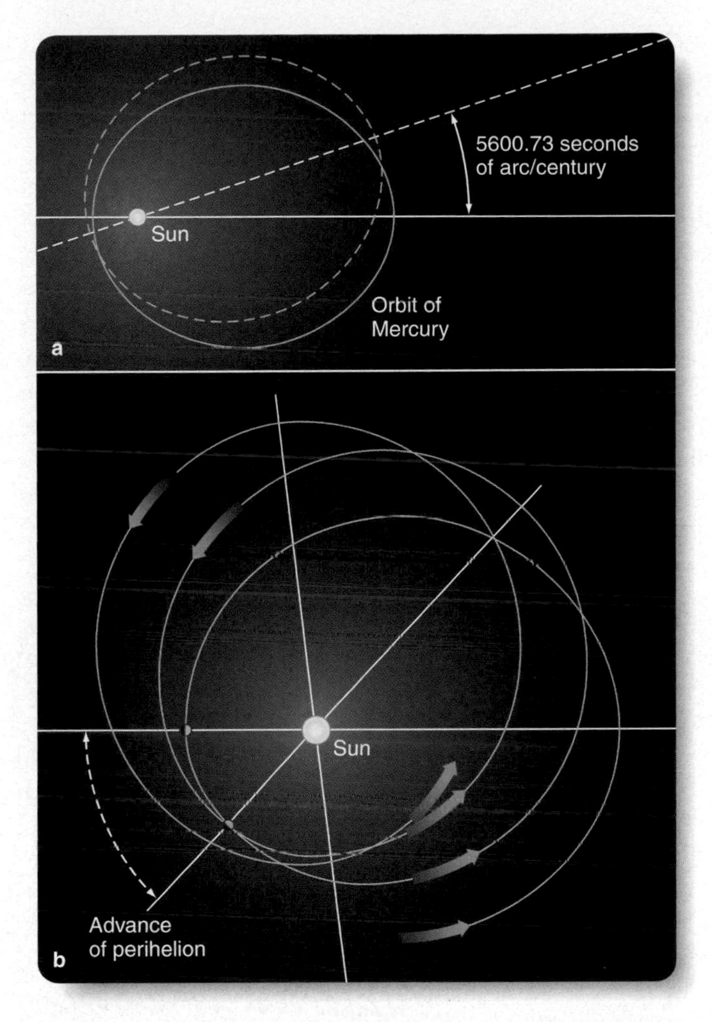

■ Figure 5-14

(a) Mercury's orbit precesses 43.11 seconds of arc per century faster than predicted by Newton's laws. (b) Even when you ignore the influences of the other planets, Mercury's orbit is not a perfect ellipse. Curved space-time near the sun distorts the orbit from an ellipse into a rosette. The advance of Mercury's perihelion is exaggerated about a million times in this figure.

Confirmation of the Curvature of Space-Time

Einstein's general theory of relativity has been confirmed by a number of experiments, but two are worth mentioning here because they were among the first tests of the theory. One involves Mercury's orbit, and the other involves eclipses of the sun.

Johannes Kepler understood that the orbit of Mercury is elliptical, but only since 1859 have astronomers known that the long axis of the orbit sweeps around the sun in a motion called precession (■ Figure 5-14). The total observed precession is 5600.73 seconds of arc per century (as seen from Earth), which equals about 1.5° per century. This precession is produced by the gravitation of Venus, Earth, and the other planets. However, when astronomers used Newton's description of gravity to account for the gravitational influence of all of the planets, they calculated that the precession should amount to only 5557.62 seconds of arc per century. So Mercury's orbit is advancing 43.11 seconds of arc per century faster than Newton's law predicts.

This is a tiny effect. Each time Mercury returns to perihelion, its closest point to the sun, it is about 29 km (18 mi) past the position predicted by Newton's laws. This is such a small distance compared with the planet's diameter of 4850 km that it could never have been detected had it not been cumulative. Each orbit, Mercury gains 29 km, and in a century it gains over 12,000 km—more than twice its own diameter. This tiny effect, called the advance of perihelion of Mercury's orbit, accumulated into a serious discrepancy in the Newtonian description of the universe.

The advance of perihelion of Mercury's orbit was one of the first problems to which Einstein applied the principles of general relativity. First he calculated how much the sun's mass curves space-time in the region of Mercury's orbit, and then he calculated how Mercury moves through the space-time. The theory predicted that the curved space-time should cause Mercury's orbit to advance by 43.03 seconds of arc per century, well within the observational accuracy of the excess (Figure 5-13b).

Einstein was elated with this result, and he would be even happier with modern studies that have shown that Mercury,

Venus, Earth, and even Icarus, an asteroid that comes close to the sun, have orbits observed to be slipping forward due to the curvature of space-time near the sun (■ Table 5-2). This same effect has been detected in pairs of stars that orbit each other.

A second test was directly related to the motion of light through the curved space-time near the sun. The equations of general relativity predicted that light would be deflected by curved space-time, just as a rolling golf ball is deflected by undulations in a putting green. Einstein predicted that starlight grazing the sun's surface would be deflected by 1.75 seconds of arc (■ Figure 5-15). Starlight passing near the sun is normally lost in the sun's glare, but during a total solar eclipse stars beyond the sun could be seen. As soon as Einstein published his theory, astronomers rushed to observe such stars and test the curvature of space-time.

■ Table 5-2 | Precession in Excess of Newtonian Physics

Planet	Observed Excess Precession (Sec of arc per century)	Relativistic Prediction (Sec of arc per century)
Mercury	43.11 ± 0.45	43.03
Venus	8.4 ± 0.48	8.6
Earth	5.0 ± 1.2	3.8
Icarus	9.8 ± 0.8	10.3

The first solar eclipse following Einstein's announcement in 1916 was June 8, 1918. It was cloudy at some observing sites, and results from other sites were inconclusive. The next occurred on May 29, 1919, only months after the end of World War I, and was visible from Africa and South America. British teams went to both Brazil and Príncipe, an island off the coast of Africa. Months before the eclipse, they photographed that part of the sky where the sun would be located during the eclipse and measured the positions of the stars on the photographic plates. Then, during the eclipse, they photographed the same star field with the eclipsed sun located in the middle. After measuring the plates, they found slight changes in the positions of the stars. During the eclipse, the positions of the stars on the plates were shifted outward, away from the sun (■ Figure 5-16). If a star had been located at the edge of the solar disk, it would have been shifted outward by about 1.8 seconds of arc. This represents good agreement with the theory's prediction.

This test has been repeated at many total solar eclipses since 1919, with similar results. The most accurate results were obtained in 1973 when a Texas-Princeton team measured a deflection of 1.66 ± 0.18 seconds of arc—good agreement with Einstein's theory.

The general theory of relativity is critically important in modern astronomy. You will meet it again in the discussion of black holes, distant galaxies, and the big bang universe. The theory revolutionized modern physics by providing a theory of gravity based on the geometry of curved space-time. Thus, Galileo's inertia and Newton's mutual gravitation are shown to be not just descriptive rules but fundamental properties of space and time.

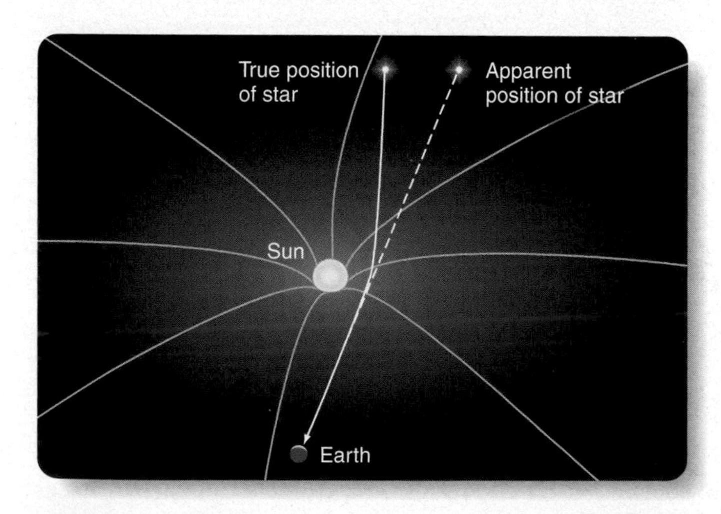

■ Figure 5-15

Like a depression in a putting green, the curved space-time near the sun deflects light from distant stars and makes them appear to lie slightly farther from the sun than their true positions.

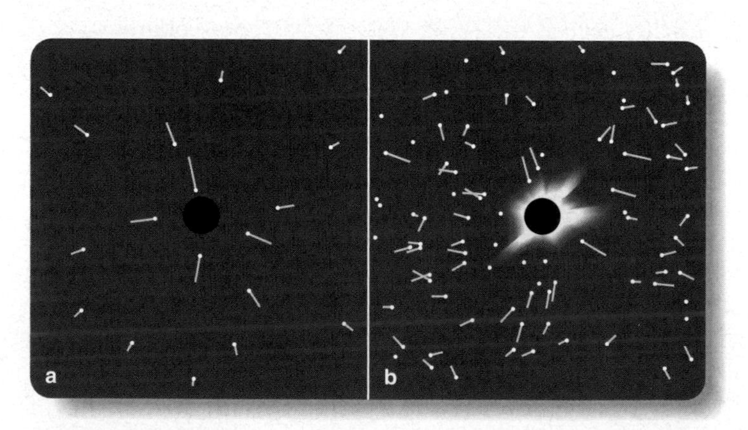

■ Figure 5-16

(a) Schematic drawing of the deflection of starlight by the sun's gravity. Dots show the true positions of the stars as photographed months before the eclipse. Lines point toward the positions of the stars during the eclipse. (b) Actual data from the eclipse of 1922. Random uncertainties of observation cause some scatter in the data, but in general the stars appear to move away from the sun by 1.77 seconds of arc at the edge of the sun's disk. The deflection of stars is magnified by a factor of 2300 in both (a) and (b).

What Are We?

The Falling Universe

Everything in the universe is falling. The moon is falling around Earth. Earth is falling along its orbit around the sun, and the sun and every other star in our galaxy are falling along their orbits around the center of our galaxy. Stars in other galaxies are falling around the center of those galaxies, and every galaxy in the universe is falling as it feels the gravitational tugs of every bit of matter that exists.

Newton's explanation of gravity as a force between two unconnected masses was action at a distance, and it offended many of the scientists of his time. They thought Newton's gravity seemed like magic. Einstein explained that gravity is a curvature of space-time and that every mass accelerates according to the curvature it feels around it. That's not action at a distance, and it can give you a new insight into how the universe works.

The mass of every atom in the universe contributes to the curvature, creating a universe filled with three-dimensional hills and valleys of curved space-time. You and your world, your sun, your galaxy, and every other object in the universe are falling through space guided by the curvature of space-time.

◄ SCIENTIFIC ARGUMENT ►

What does the equivalence principle tell you?

The equivalence principle says that there is no observation you can make inside a closed spaceship to distinguish between uniform acceleration and gravitation. Of course, you could open a window and look outside, but then you would no longer be in a closed spaceship. As long as you make no outside observations, you can't tell whether your spaceship is firing its rockets and accelerating through space or resting on the surface of a planet where gravity gives you weight.

Einstein took the equivalence principle to mean that gravity and acceleration through space-time are somehow related. The general theory of relativity gives that relationship mathematical form and shows that gravity is really a distortion in space-time that physicists refer to as curvature. Consequently, you can say "mass tells space-time how to curve, and space-time tells mass how to move." The equivalence principle led Einstein to an explanation for gravity.

Einstein began his work by thinking carefully about common things such as what you feel when you are moving uniformly or accelerating. This led him to deep insights now called postulates. Special relativity sprang from two postulates. **Why does the second postulate have to be true if the first postulate is true?**

◄ ►

Study and Review

Summary

5-1 | Galileo and Newton

What happens when an object falls?

- Aristotle argued that the universe was composed of four elements, earth at the center, with water, air and fire in layers above. **Natural motion** occurred when a displaced object returned to its natural place. **Violent motion** was motion other than natural motion and had to be sustained by a force.
- Galileo found that a falling object is accelerated; that is, it falls faster and faster with each passing second. The rate at which it accelerates, termed the **acceleration of gravity**, is 9.8 m/s^2 (32 ft/s^2) at Earth's surface and does not depend on the weight of the object, contrary to what Aristotle said.
- According to tradition, Galileo demonstrated this by dropping balls of iron and wood from the Leaning Tower of Pisa to show that they would fall together. Air resistance would have ruined the experiment, but a feather and a hammer dropped on the airless moon by an astronaut did fall together.
- Galileo stated the law of inertia. In the absence of friction, a moving body on a horizontal plane will continue moving forever.

How did Newton discover gravity?

- The first of Newton's three laws of motion was based on Galileo's law of inertia. A body continues at rest or in uniform motion in a straight line unless it is acted on by some force.
- **Momentum** is the tendency of a moving body to continue moving.
- **Mass** is the amount of matter in a body.
- Newton's second law says that a change in motion, an **acceleration** (a change in velocity), must be caused by a force. A **velocity** is a directed speed so a change in speed or direction is an acceleration.
- Newton's third law says that forces occur in pairs acting in opposite directions.
- A **hypothesis** is a single statement about nature subject to testing. A **theory** is usually a more elaborate system of rules and principles that has been tested and widely applied. A **natural law** is a theory that has been so thoroughly tested scientists have great confidence in it.

whereas the wavelength of ocean waves might be a hundred meters or more. There is no restriction on the wavelength of electromagnetic radiation. Wavelengths can range from smaller than the diameter of an atom to larger than that of Earth.

Whereas radio waves have wavelengths that can be measured in millimeters or kilometers, the wavelength of light is so short that you will need more convenient units. This book uses **nanometers (nm)** because this unit is consistent with the International System of units. One nanometer is 10^{-9} meter, and visible light has wavelengths that range from about 400 nm to about 700 nm. Another unit that astronomers commonly use, and a unit that you will see in many references on astronomy, is the **Angstrom (Å).** One Angstrom is 10^{-10} meter, and visible light has wavelengths between 4000 Å and 7000 Å. Astronomers may also use centimeters, millimeters, or micrometers (microns), depending on their field of specialization. No matter which unit is used to describe the wavelength, all electromagnetic radiation is the same phenomenon.

Wavelength is related to **frequency,** the number of cycles that pass in one second. Short-wavelength radiation has a high frequency; long-wavelength radiation has a low frequency. To understand this, imagine watching an electromagnetic wave race past while you count its peaks (■ Figure 6-2). If the wavelength is short, you will count many peaks in one second; if the wavelength is long, you will count few peaks per second. The dials on radios are marked in frequency, but they could just as easily be marked in wavelength. Because all electromagnetic radiation travels at the speed of light, the relation between wavelength and frequency is a simple one:

$$\lambda = \frac{c}{f}$$

That is, the wavelength equals the speed of light c divided by the frequency f. Notice that the larger (higher) the frequency, the smaller (shorter) the wavelength. In most cases, astronomers use wavelength rather than frequency.

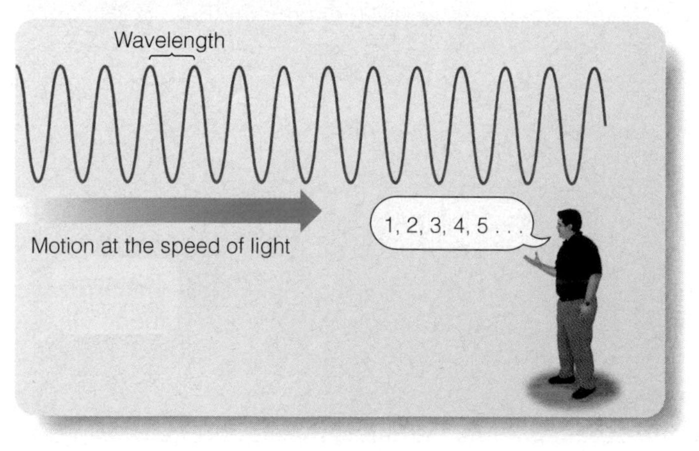

■ **Figure 6-2**

All electromagnetic waves travel at the speed of light. The wavelength is the distance between successive peaks. The frequency of the wave is the number of peaks that pass you in one second.

What exactly is electromagnetic radiation? Is it a particle or a wave? Throughout his life, Newton believed that light was made up of particles, but modern physicists now recognize that light can behave as both particle and wave. The modern model of light is more complete than Newton's, and it refers to "a particle of light" as a **photon.** You can recognize its dual nature by thinking of it as a bundle of waves. Because the bundle contains waves, a photon has a wavelength. Because the waves are bundled, the photon has a specific amount of energy.

The energy of a photon depends on its wavelength. The shorter the wavelength, the more energy the photon carries; the longer the wavelength, the less energy it contains. This is easy to remember because short wavelengths have high frequencies, and you would naturally expect rapid fluctuations to be more energetic. A simple formula expresses the relationship between energy and wavelength:

$$E = \frac{hc}{\lambda}$$

Here h is Planck's constant (6.6262×10^{-34} joule s), c is the speed of light (3×10^8 m/s), and λ is the wavelength in meters. A photon of visible light carries a very small amount of energy, but a photon with a very short wavelength can carry much more.

The Electromagnetic Spectrum

A spectrum is an array of electromagnetic radiation displayed in order of wavelength. You are most familiar with the spectrum of visible light, which you see in rainbows. The colors of the visible spectrum differ in wavelength, with red having the longest wavelength and violet the shortest. The visible spectrum is shown at the top of ■ Figure 6-3.

The average wavelength of visible light is about 0.00005 cm. You could put 50 light waves end to end across the thickness of a sheet of household plastic wrap. Measured in nanometers, the wavelength of visible light ranges from about 400 to 700 nm. Just as you sense the wavelength of sound as pitch, you sense the wavelength of light as color. Light near the short-wavelength end of the visible spectrum (400 nm) looks violet to your eyes, and light near the long-wavelength end (700 nm) looks red.

Figure 6-3 shows that the visible spectrum makes up only a small part of the entire electromagnetic spectrum. Beyond the red end of the visible spectrum lies **infrared radiation,** where wavelengths range from 700 nm to about 0.1 cm. Your eyes are not sensitive to this radiation, but your skin senses it as heat. For example, a "heat lamp" warms you by giving off infrared radiation.

Beyond the infrared part of the electromagnetic spectrum lie radio waves. Microwaves have wavelengths of a millimeter to a

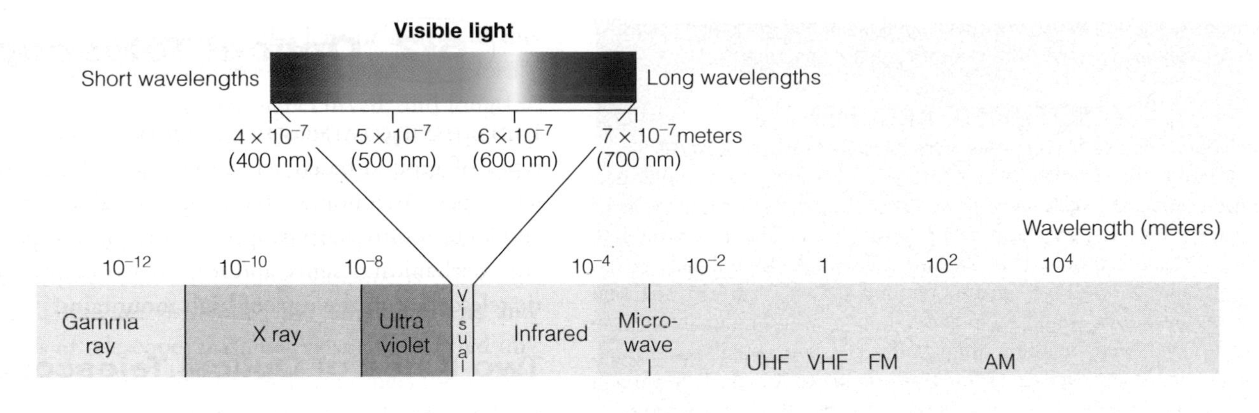

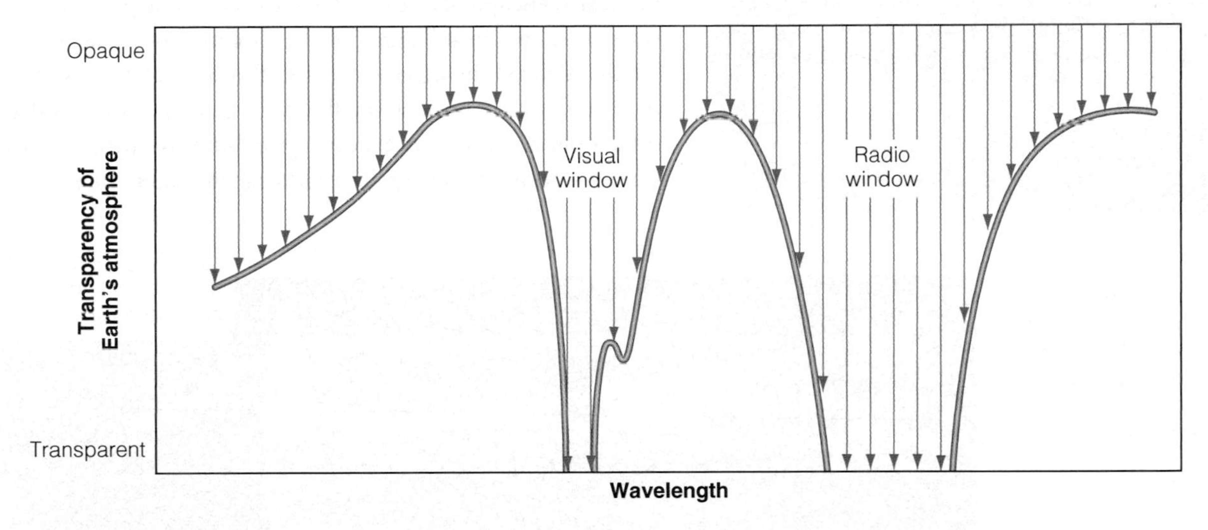

Figure 6-3

The spectrum of visible light, extending from red to violet, is only part of the electromagnetic spectrum. Most radiation is absorbed in Earth's atmosphere, and only radiation in the visual window and the radio window can reach Earth's surface.

few centimeters and are used for radar and long-distance telephone communication. Longer wavelengths are used for UHF and VHF television transmissions. FM, military, governmental, and ham radio signals have wavelengths up to a few meters, and AM radio waves can have wavelengths of kilometers.

The boundaries between the wavelength ranges are not sharp. Long-wavelength infrared radiation blends smoothly into the shortest microwave radio waves. Similarly, there is no natural division between the short-wavelength infrared and the long-wavelength part of the visible spectrum.

Look once again at the electromagnetic spectrum in Figure 6-3 and notice that electromagnetic waves shorter than violet are called **ultraviolet.** Electromagnetic waves even shorter are called X rays, and the shortest are gamma rays. Again, the boundaries between these wavelength ranges are not clearly defined.

Remember the formula for the energy of a photon? Extremely short wavelength photons such as X rays and gamma rays have high energies and can be dangerous. Even ultraviolet photons have enough energy to do harm. Small doses of ultraviolet produce a suntan and larger doses sunburn and skin cancers. Contrast this to the lower-energy infrared photons. Individually they have too little energy to affect skin pigment, a fact that explains why you can't get a tan from a heat lamp. Only by concentrating many low-energy photons in a small area, as in a microwave oven, can you transfer significant amounts of energy.

Astronomers are interested in electromagnetic radiation because it carries clues to the nature of stars, planets, and other celestial objects. Earth's atmosphere is opaque to most electromagnetic radiation, as shown by the graph at the bottom of Figure 6-3. Gamma rays, X rays, and some radio waves are absorbed high in Earth's atmosphere, and a layer of ozone (O_3) at an altitude of about 30 km absorbs ultraviolet radiation. Water vapor in the lower atmosphere absorbs the longer-wavelength infrared radiation. Only visible light, some shorter-wavelength infrared, and some radio waves reach Earth's surface through two wavelength regions called **atmospheric windows.** Obviously, if you wish to study the sky from Earth's surface, you must look out through one of these windows.

magnifying power of a telescope, or its ability to make the image bigger, is actually the least significant of the three powers. Because the amount of detail you can see is limited by the seeing conditions and the resolving power, very high magnification does not necessarily show more detail. Also, you can change the magnification by changing the eyepiece, but you cannot alter the telescope's light-gathering power or resolving power without changing the diameter of the objective lens or mirror, and that would be so expensive that you might as well build a whole new telescope.

You can calculate the magnification of a telescope by dividing the focal length of the objective by the focal length of the eyepiece:

$$M = \frac{F_o}{F_e}$$

For example, if a telescope has an objective with a focal length of 80 cm and you use an eyepiece whose focal length is 0.5 cm, the magnification is 80/0.5, or 160 times.

Notice that the two most important powers of the telescope, light-gathering power and resolving power, depend on the diameter of the telescope. This explains why astronomers refer to telescopes by diameter and not by magnification. Astronomers will refer to a telescope as an 8-meter telescope or a 10-meter telescope, but they would never identify a telescope as a 200-power telescope.

The search for light-gathering power and high resolution explains why nearly all major observatories are located far from big cities and usually on high mountains. Astronomers avoid cities because **light pollution,** the brightening of the night sky by light scattered from artificial outdoor lighting, can make it impossible to see faint objects (■ Figure 6-10). In fact, many residents of cities are unfamiliar with the beauty of the night sky because they can see only the brightest stars. Even far from cities, nature's own light pollution, the moon, is so bright it drowns out fainter objects, and astronomers are often unable to observe on the nights near full moon when faint objects cannot be detected even with the largest telescopes on high mountains.

Astronomers prefer to place their telescopes on carefully selected high mountains. The air there is thin and more transparent. The air is very dry at high altitudes and is more transparent to infrared radiation. Most important, for the best seeing, astronomers select mountains where the air flows smoothly and is not turbulent. Building an observatory on top of a high mountain far from civilization is difficult and expensive, as you can imagine from the photo in Figure 6-10, but the dark sky and steady seeing make it worth the effort.

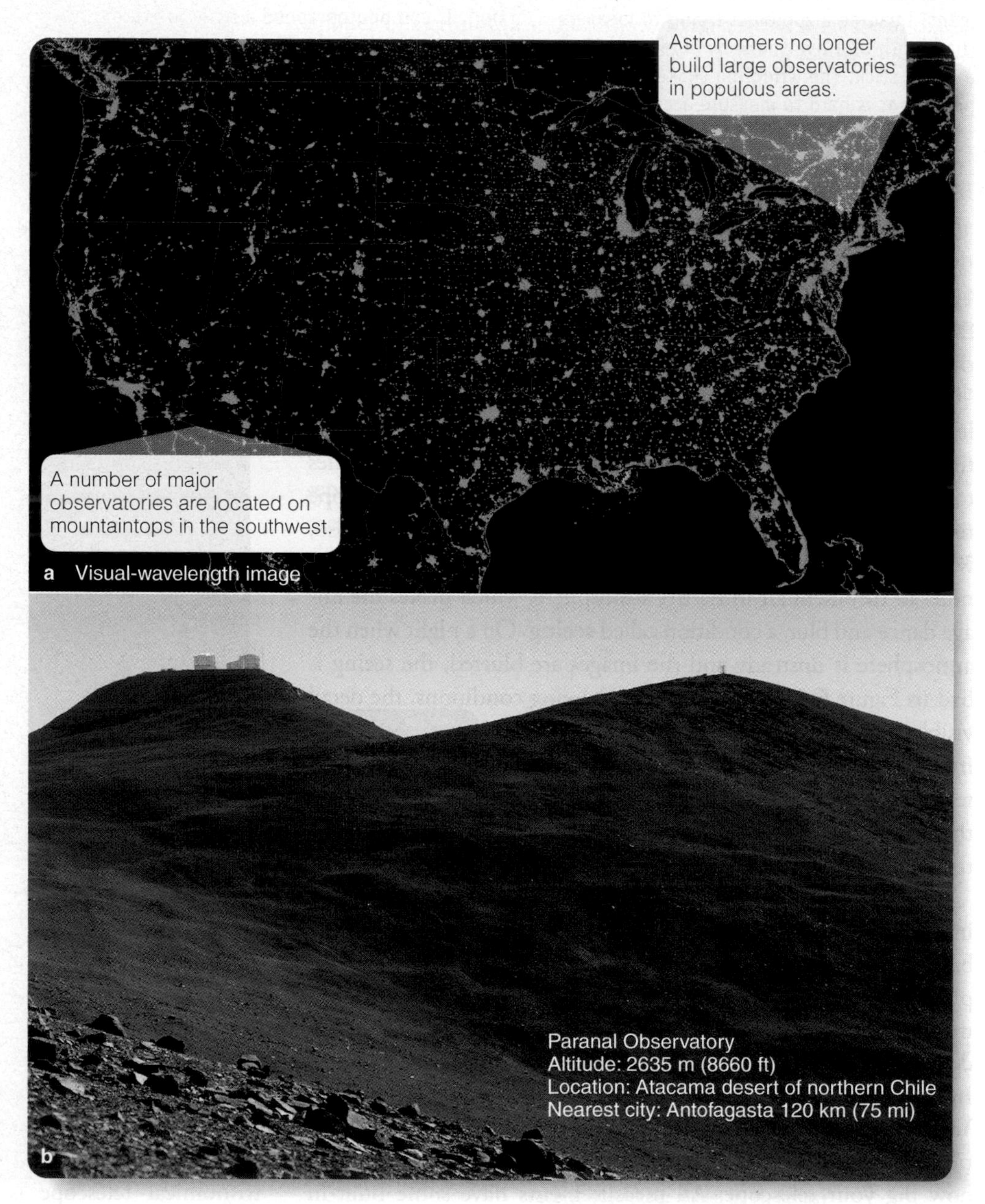

Astronomers no longer build large observatories in populous areas.
A number of major observatories are located on mountaintops in the southwest.
a Visual-wavelength image
Paranal Observatory
Altitude: 2635 m (8660 ft)
Location: Atacama desert of northern Chile
Nearest city: Antofagasta 120 km (75 mi)
b

■ Figure 6-10

(a) This satellite view of the continental United States at night shows the light pollution and energy waste produced by outdoor lighting. Observatories cannot be located near large cities. (NOAA) *(b)* The domes of four giant telescopes are visible at upper left at Paranal Observatory, built by the European Southern Observatory. The Atacama Desert is believed to be the driest place on Earth. (ESO)

ThomsonNOW Sign in at www.thomsonedu.com and go to ThomsonNOW to see Astronomy Exercises "Telescopes and Resolution I," "Telescopes and Resolution II," and "Particulate, Heat, and Light Pollution."

Buying a Telescope

Thinking about how to shop for a new telescope will not only help you if you decide to buy one but will also illustrate some important points about astronomical telescopes.

Assuming you have a fixed budget, you should buy the highest-quality optics and the largest-diameter telescope you can afford. Of the two things that limit what you see, only optical quality is under your control. You can't make the atmosphere less turbulent, but you should buy good optics. If you buy a telescope from a toy store and it has plastic lenses, you shouldn't expect to see very much. Also, you want to maximize the light-gathering power of your telescope, so you want to purchase the largest-diameter telescope you can afford. Given a fixed budget, that means you should buy a reflecting telescope rather than a refracting telescope. Not only will you get more diameter per dollar, but your telescope will not suffer from chromatic aberration.

You can safely ignore magnification. Department stores and camera shops may advertise telescopes by quoting their magnification, but it is not an important number. What you can see is fixed by light-gathering power, optical quality, and Earth's atmosphere. Besides, you can change the magnification by changing eyepieces.

Other things being equal, you should choose a telescope with a solid mounting that will hold the telescope steady and allow it to point at objects easily. Computer-controlled pointing systems are available for a price on many small telescopes. A good telescope on a poor mounting is almost useless.

You might be buying a telescope to put in your backyard, but you must think about the same issues astronomers consider when they design giant telescopes to go on mountaintops. Designing new, giant telescopes has led astronomers to solve some traditional problems in new ways, as you will see in the next section.

New-Generation Telescopes

For most of the 20th century, astronomers faced a serious limitation on the size of astronomical telescopes. Traditional telescope mirrors were made thick to avoid sagging that would distort the reflecting surface, but those thick mirrors were heavy. The 5-m (200-in.) mirror on Mount Palomar weighs 14.5 tons. These traditional telescopes were big, heavy, and expensive.

Modern astronomers have solved these problems in a number of ways. Read **Modern Astronomical Telescopes** on pages 112–113 and notice four important points about telescope design and 11 new terms that describe astronomical telescopes and their operation:

1. Traditional telescopes use large, solid, heavy mirrors to focus starlight to a *prime focus*, or, by using a *secondary mirror*, to a *Cassegrain focus*. Some small telescopes have a *Newtonian focus* or a *Schmidt-Cassebrain focus*.
2. Telescopes must have a *sidereal drive* to follow the stars; an *equatorial mounting* with easy motion around a *polar axis* is the traditional way to provide that motion. Today, astronomers can build simpler, lighter-weight telescopes on *alt-azimuth mountings* and depend on computers to move the telescope and follow the westward motion of the stars as Earth rotates.
3. *Active optics,* computer control of the shape of telescope mirrors, allows the use of thin, lightweight mirrors—either "floppy" mirrors or segmented mirrors. Lowering the weight of the mirror lowers the weight of the rest of the telescope and makes it stronger and less expensive. Also, thin mirrors cool faster at nightfall and produce better images.
4. Astronomers use *adaptive optics,* high-speed computers rapidly adjusting the shape of telescope mirrors, to reduce seeing distortion caused by Earth's atmosphere. Only a few decades ago, many astronomers argued that it wasn't worth building more large telescopes on Earth's surface because of the limitations set by seeing. Now adaptive optics can cancel out part of the seeing distortion, and a number of new giant telescopes have been built, with more in development.

Did you notice in the concept art spread that astronomical telescopes must be aligned with the north celestial pole? Polaris, the North Star, is one of your Favorite Stars in the list in Chapter 1. It marks the location of the north celestial pole. Equatorial mountings have an axis that points toward Polaris, and alt-azimuth telescopes are run by computers, which align their motion with Polaris. Even telescopes in the Southern Hemisphere, where the north celestial pole lies below the horizon, must tip their hats toward Polaris. That's one reason Polaris deserves to be one of your Favorite Stars; whenever you notice Polaris in the night sky, think of all the astronomical telescopes in backyards and observatories all over the world that bow toward Polaris.

High-speed computers have allowed astronomers to build new, giant telescopes with unique designs. A few are shown in ■ Figure 6-11. The European Southern Observatory has built the Very Large Telescope (VLT) high in the remote Andes Mountains of northern Chile. The VLT consists of four telescopes with computer-controlled mirrors 8.2 m in diameter and only 17.5 cm (6.9 in.) thick. The four telescopes can work singly or can combine their light to work as one large telescope. Italian and American astronomers have built the Large Binocular Telescope, which carries a pair of 8.4-m mirrors on a single mounting. Other giant telescopes are being planned with segmented mirrors or with multiple mirrors such as the Giant Magellan Telescope planned to carry seven thin mirrors on a single mounting.

Modern Astronomical Telescopes

1 The traditional telescopes described on this page are limited by complexity, weight, and Earth's atmosphere. Modern solutions are shown on the opposite page.

In larger telescopes the light can be focused to a **prime focus** position high in the telescope tube as shown at the right. Although it is a good place to image faint objects, the prime focus is inconvenient for large instruments. A **secondary mirror** can reflect the light through a hole in the primary mirror to a **Cassegrain focus.** This focal arrangement may be the most common form of astronomical telescope.

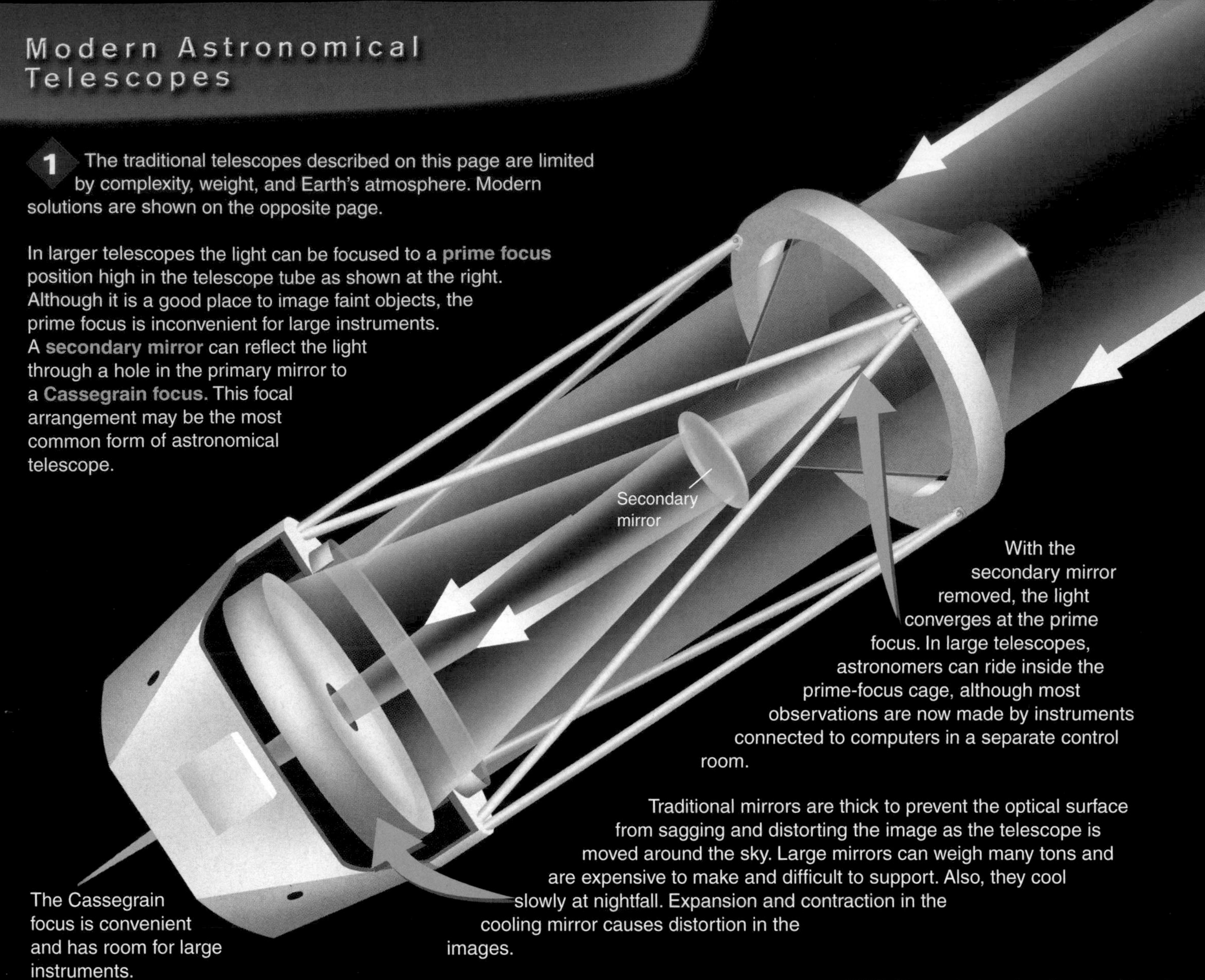

With the secondary mirror removed, the light converges at the prime focus. In large telescopes, astronomers can ride inside the prime-focus cage, although most observations are now made by instruments connected to computers in a separate control room.

Traditional mirrors are thick to prevent the optical surface from sagging and distorting the image as the telescope is moved around the sky. Large mirrors can weigh many tons and are expensive to make and difficult to support. Also, they cool slowly at nightfall. Expansion and contraction in the cooling mirror causes distortion in the images.

The Cassegrain focus is convenient and has room for large instruments.

1a Smaller telescopes are often found with a **Newtonian focus,** the arrangement that Isaac Newton used in his first reflecting telescope. The Newtonian focus is inconvenient for large telescopes as shown at right.

Newtonian focus

Thin correcting lens

Schmidt-Cassegrain telescope

1b Many small telescopes such as the one on your left use a **Schmidt-Cassegrain focus.** A thin correcting plate improves the image but is too slightly curved to introduce serious chromatic aberration.

1c Shown below, the 4-meter Mayall Telescope at Kitt Peak National Observatory in Arizona can be used at either the prime focus or the Cassegrain focus. Note the human figure at lower right.

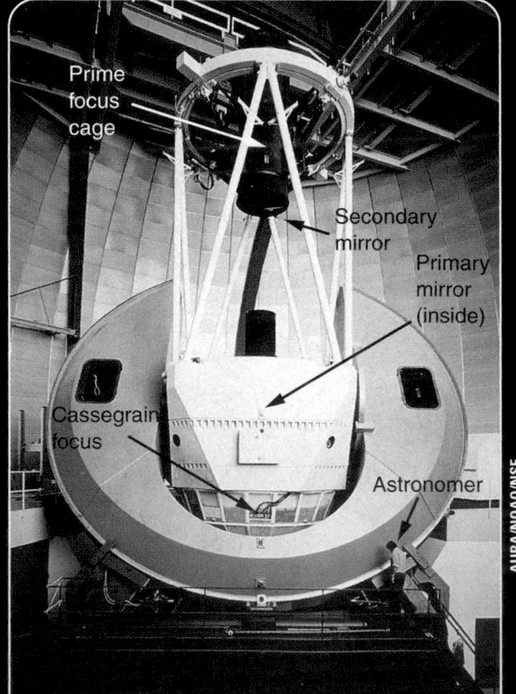

Equatorial mounting

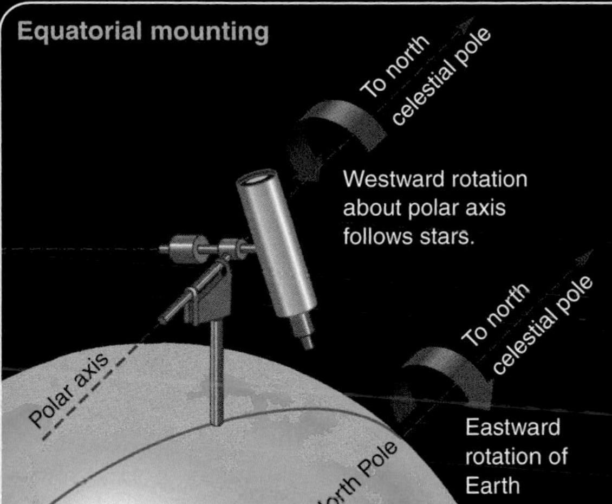

Alt-azimuth mounting

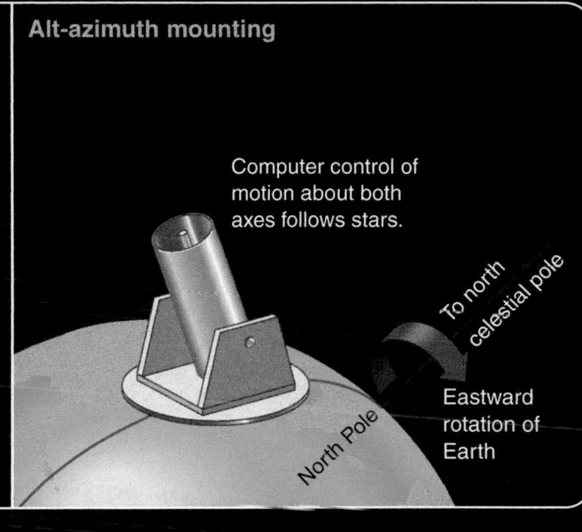

2 Telescope mountings must contain a **sidereal drive** to move smoothly westward and counter the eastward rotation of Earth. The traditional **equatorial mounting** (far left) has a **polar axis** parallel to Earth's axis, but the modern **alt-azimuth mounting** (near left) moves like a cannon — up and down and left to right. Such mountings are simpler to build but need computer control to follow the stars.

3 Unlike traditional thick mirrors, thin mirrors, sometimes called floppy mirrors as shown at right, weigh less and require less massive support structures. Also, they cool rapidly at nightfall and there is less distortion from uneven expansion and contraction.

Floppy mirror

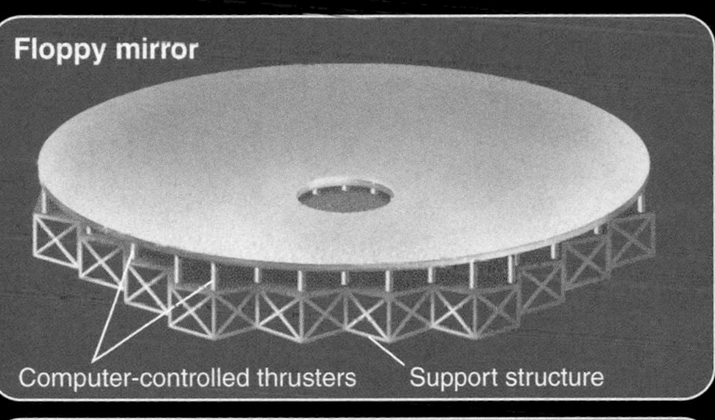

3a Grinding a large mirror may remove tons of glass and take months, but new techniques speed the process. Some large mirrors are cast in a rotating oven that causes the molten glass to flow to form a concave upper surface. Grinding and polishing such a preformed mirror is much less time consuming.

3b Mirrors made of segments are economical because the segments can be made separately. The resulting mirror weighs less and cools rapidly. See image at right.

Segmented mirror

Computer-controlled thrusters

Support structure

3c Both floppy mirrors and segmented mirrors sag under their own weight. Their optical shape must be controlled by computer-driven thrusters under the mirror in what is called **active optics**.

3d Large telescopes with segmented mirrors have been very successful, and that has led astronomers to propose huge telescopes.

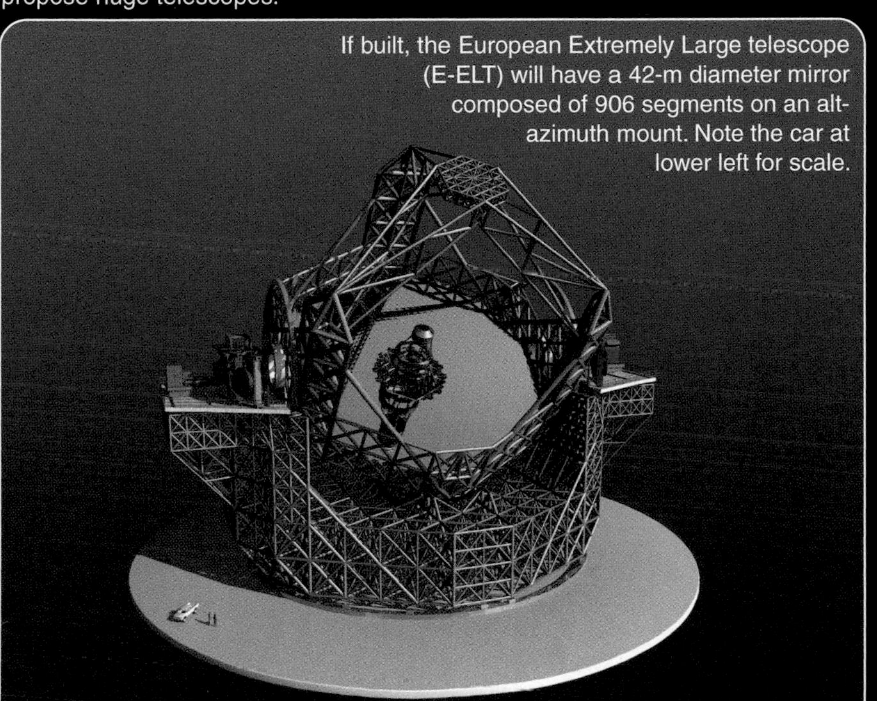

If built, the European Extremely Large telescope (E-ELT) will have a 42-m diameter mirror composed of 906 segments on an alt-azimuth mount. Note the car at lower left for scale.

Adaptive optics in telescopes

Adaptive optics off

Object appears to be a single star.

Adaptive optics on

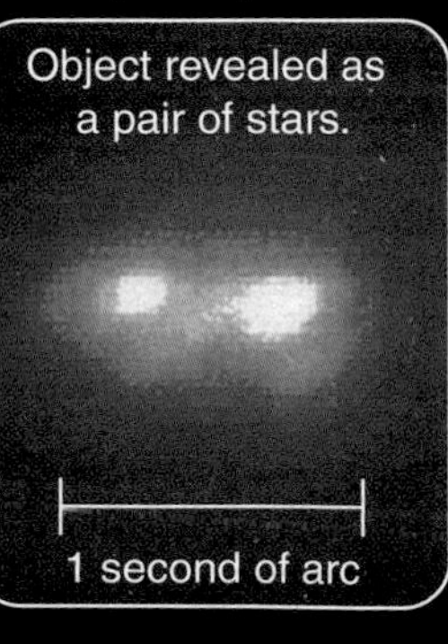

Paul Kalas

4 **Adaptive optics** uses high-speed computers to monitor the image distortion caused by Earth's atmosphere and adjust the optics many times a second to compensate. This can reduce the blurring due to seeing and dramatically improve image quality in Earth-based telescopes.

Large Binocular Telescope
The mirrors in the VLT telescopes are each 8.2 m in diameter.
Note the human figure for scale in this computer graphic visualization.

■ **Figure 6-11**

The four telescopes of the VLT are housed in separate domes at Paranal Observatory in Chile (Figure 6-10). The Large Binocular Telescope (LBT) carries two 8.4-m mirrors that combine their light. The entire building rotates as the telescope moves. The proposed Giant Magellan Telescope will have the resolving power of a telescope 24.5 meters in diameter when it is finished about 2016. (VLT: ESO; LBT: Large Binocular Telescope Project and European Industrial Engineer; GMT: ESO)

High-speed computers have improved astronomical telescopes in another way that might surprise you. Computer control and data handling have made possible huge surveys of the sky in which millions of objects are observed. The Sloan Digital Sky Survey, for example, is mapping the sky, measuring the position and brightness of 100 million stars and galaxies at a number of wavelengths. The Two-Micron All Sky Survey (2MASS) has mapped the entire sky at three wavelengths in the infrared. Other surveys are being made at many other wavelengths. Every night large telescopes scan the sky, and billions of bytes of data are compiled automatically in immense sky atlases. Astronomers will study those data banks for decades to come.

The days when astronomers worked beside their telescopes through long, dark, cold nights are nearly gone. The complexity and sophistication of telescopes require a battery of computers, and almost all research telescopes are run from control rooms that astronomers call "warm rooms." Astronomers don't need to be kept warm, but computers demand comfortable working conditions (■ Figure 6-12).

Interferometry

One of the reasons astronomers build big telescopes is to increase resolving power, and astronomers have been able to achieve very high resolution by connecting multiple telescopes together to work as if they were a single telescope. This method of synthesizing a larger telescope is known as **interferometry** (■ Figure 6-13).

To work as an interferometer, the separate telescopes must combine their light through a network of mirrors, and the path that each light beam travels must be controlled so that it does not vary more than some small fraction of the wavelength. Turbulence in Earth's atmosphere constantly distorts the light, and high-speed computers must continuously adjust the light paths. Recall that the wavelength of light is very short, roughly 0.00005 cm, so building optical interferometers is one of the most difficult technical problems that astronomers face. Infrared- and radio-wavelength interferometers are slightly easier to build because the wavelengths are longer. In fact, as you will discover later in this chap-

■ **Figure 6-12**

In the control room of the 4-meter telescope atop Kitt Peak National Observatory, the telescope operator at left manages the operation and safety of the telescope. The astronomer at right operates the instruments, records data, and makes decisions on the observing program. Astronomers work through the night controlling the computers that control the telescope and its instruments. (NOAO/AURA/NSF)

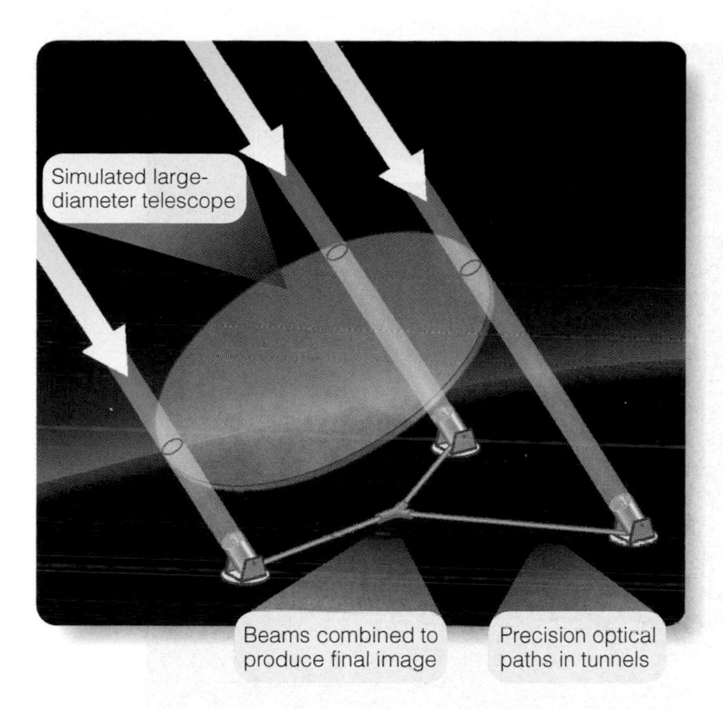

■ Figure 6-13

In an astronomical interferometer, smaller telescopes can combine their light through specially designed optical tunnels to simulate a larger telescope with a resolution set by the separation of the smaller telescopes.

ter, the first astronomical interferometers worked at radio wavelengths.

The VLT shown in Figure 6-11 consists of four 8.2-m telescopes that can operate separately, but they can be linked together through underground tunnels with three 1.8-m telescopes on the same mountaintop. The resulting optical interferometer provides the resolution of a telescope 200 m in diameter. Other telescopes can work as interferometers. The two Keck 10-m telescopes can be used as an interferometer. The CHARA array on Mt. Wilson combines six 1-m telescopes to create the equivalent of a telescope one-fifth of a mile in diameter. The Large Binocular Telescope shown in Figure 6-11 can be used as an interferometer.

Although turbulence in Earth's atmosphere can be partially averaged out in an interferometer, plans are being made to put interferometers in space to avoid atmospheric turbulence altogether. The Space Interferometry Mission, for example, will work at visual wavelengths and study everything from the cores of erupting galaxies to planets orbiting nearby stars.

◄ SCIENTIFIC ARGUMENT ►

Why do astronomers build observatories at the tops of mountains?

To develop this argument you need to think about the powers of a telescope. Astronomers have joked that the hardest part of building a new observatory is constructing the road to the top of the mountain. It certainly isn't easy to build a large, delicate telescope at the top of a high mountain, but it is worth the effort. A telescope on top of a high mountain is above the thickest part of Earth's atmosphere. There is less air to dim the light, and there is less water vapor to absorb infrared radiation. Even more important, the thin air on a mountaintop causes less disturbance to the image, and consequently the seeing is better. A large telescope on Earth's surface has a resolving power much better than the distortion caused by Earth's atmosphere. So, it is limited by seeing, not by its own diffraction. It really is worth the trouble to build telescopes atop high mountains.

Astronomers not only build telescopes on mountaintops, they also build gigantic telescopes many meters in diameter. Revise your argument to focus on telescope design. **What are the problems and advantages in building such giant telescopes?**

◄ ►

6-3 Special Instruments

JUST LOOKING THROUGH A TELESCOPE doesn't tell you much. A star looks like a point of light. A planet looks like a little disk. A galaxy looks like a hazy patch. To use an astronomical telescope to learn about the universe, you must be able to analyze the light the telescope gathers. Special instruments attached to the telescope make that possible.

Imaging Systems

The original imaging device in astronomy was the photographic plate. It could record faint objects in long time exposures and could be stored for later analysis. But photographic plates have been almost entirely replaced in astronomy by electronic imaging systems.

Most modern astronomers use **charge-coupled devices (CCDs)** to record images. A CCD is a specialized computer chip containing roughly a million microscopic light detectors arranged in an array about the size of a postage stamp. These devices can be used like small photographic plates, but they have dramatic advantages. They can detect both bright and faint objects in a single exposure, are much more sensitive than photographic plates, and can be read directly into computer memory for later analysis. Although CCDs for astronomy are extremely sensitive and therefore expensive, less sophisticated CCDs are used in video and digital cameras.

The image from a CCD is stored as numbers in computer memory, so it is easy to manipulate the image to bring out details that would not otherwise be visible. For example, astronomical images are often reproduced as negatives with the sky white and the stars dark. This makes the faint parts of the image easier to see (■ Figure 6-14). Astronomers also manipulate images to produce **false-color images** in which the colors represent different levels of intensity and are not related to the true colors of the object. You can see an example in Figure 6-14. In fact, false-color images are common in many fields such as medicine and meteorology.

Measurements of intensity and color were made in the past using a photometer, a highly sensitive light meter attached to a

Advantages of a Radio Telescope

Building large radio telescopes in isolated locations is expensive, but three factors make it all worthwhile. First, and most important, a radio telescope can reveal clouds of cool hydrogen in space. These hydrogen clouds are important because, for one thing, they are the places where stars are born. Also, 90 percent of the atoms in the universe are hydrogen, so it is important to be able to map the hydrogen. Large clouds of cool hydrogen are completely invisible to normal telescopes because they produce no visible light of their own and reflect too little to be detected on photographs. However, cool hydrogen emits a radio signal at the specific wavelength of 21 cm. (You will see how the hydrogen produces this radiation in the discussion of the gas clouds in space in Chapter 10.) The only way astronomers can detect these clouds of gas is with a radio telescope that receives the 21-cm radiation, so that is one reason that radio telescopes are important.

The second reason is related to dust in space. Astronomers observing at visual wavelengths can't see through the dusty clouds in space. Light waves are short, and they interact with tiny dust grains floating in space; as a result, the light is scattered and never gets through the dust to reach optical telescopes on Earth. However, radio signals have wavelengths much longer than the diameters of dust grains, and radio waves from far across the galaxy pass unhindered through the dust, giving radio astronomers an unobscured view.

Finally, radio telescopes are important because they can detect objects that are more luminous at radio wavelengths than at visible wavelengths. This includes everything from cold clouds of gas that give birth to stars to intensely hot gas expelled by gas orbiting black holes. Some of the most violent events in the universe are detectable at radio wavelengths.

Figure 6-19

(a) The largest steerable radio telescope in the world is the GBT located in Green Bank, West Virginia. With a diameter of 100 m, it stands higher than the Statue of Liberty. (Mike Bailey: NRAO/AUII) (b) The 300-m (1000-ft) radio telescope in Arecibo, Puerto Rico, hangs from cables over a mountain valley. The Arecibo Observatory is part of the National Astronomy and Ionosphere Foundation operated by Cornell University and the National Science Foundation. (David Parker/SPL/Photo Researchers, Inc.)

◄ SCIENTIFIC ARGUMENT ►

Why do optical astronomers build big telescopes, while radio astronomers build groups of widely separated smaller telescopes?

Once again you can learn a lot by building a scientific argument based on comparison. Optical astronomers build large telescopes to maximize light-gathering power, but the problem for radio telescopes is resolving power. Because radio waves are so much longer than light waves, a single radio telescope can't resolve details in the sky much smaller than the moon. By linking radio telescopes that are many kilometers apart, radio astronomers build a radio interferometer that can simulate a radio telescope kilometers in diameter and thus increase the resolving power.

The difference between the wavelengths of light and radio waves makes a big difference in building the best telescopes. Keep that difference in mind as you build a new argument: **Why don't radio astronomers want to build their telescopes on mountaintops as optical astronomers do?**

◄ ►

6-5 Astronomy from Space

YOU HAVE LEARNED about the observations that ground-based telescopes can make through the two atmospheric windows in the visible and radio parts of the electromagnetic spectrum. Most of the rest of the electromagnetic radiation—infrared, ultraviolet, X ray, and gamma ray—never reaches Earth's surface; it is absorbed high in Earth's atmosphere. To observe at these wavelengths, telescopes must fly above the atmosphere in high-flying aircraft, rockets, balloons, and satellites. The only exceptions are observations that can be made in the near-infrared and the near-ultraviolet.

The Ends of the Visual Spectrum

Astronomers can observe in the near-infrared just beyond the red end of the visible spectrum. You can't see this light, but some of it leaks through the atmosphere in narrow, partially open atmospheric windows scattered from 1200 nm to about 40,000 nm. Infrared astronomers usually measure wavelength in micrometers (10^{-6} meters), so they refer to this wavelength range as 1.2 to 40 micrometers (or microns for short). In this range, much of the radiation is absorbed by water vapor, but carbon dioxide and oxygen molecules also absorb infrared. As you saw earlier in this chapter, it is an advantage to place telescopes on mountaintops where the air is thin and dry. For example, a number of important infrared telescopes observe from the 4150-m (13,600-ft) summit of Mauna Kea in Hawaii. At this altitude, the telescopes are above much of the water vapor in Earth's atmosphere (■ Figure 6-20).

The far-infrared range, which includes wavelengths longer than 40 micrometers, carries clues to the nature of comets, planets, forming stars, and other cool objects, but these wavelengths are absorbed high in Earth's atmosphere—much higher than mountaintops. Infrared telescopes have flown to high altitudes under balloons and in airplanes. NASA is now building the Stratospheric Observatory for Infrared Astronomy (SOFIA), a Boeing 747 that will carry a 2.5-m telescope, control systems, and a team of technicians and astronomers to the fringes of the atmosphere. Once at that altitude, they can open a door above the telescope and make infrared observations for hours as the plane flies a precisely calculated path. You can see the door in the photo in Figure 6-20.

To reduce internal noise, the light-sensitive detectors in astronomical telescopes are cooled to very low temperatures, usually with liquid nitrogen, as shown in Figure 6-20. This is especially necessary for a telescope observing at infrared wavelengths, and, to observe at the longest infrared wavelengths, astronomers must cool the entire telescope. Infrared radiation is emitted by heated objects, and if the telescope is warm it will emit many times more infrared radiation than that coming from a distant object. Imagine trying to look for rabbits at night through binoculars that are themselves glowing.

At the short wavelength end of the spectrum, astronomers can observe in the near-ultraviolet. Your eyes don't detect this radiation, but it can be recorded by photographic plates and CCDs. Wavelengths shorter than about 290 nm, the far-ultraviolet, are completely absorbed by the ozone layer extending from 20 km to about 40 km above Earth's surface. No mountaintop is that high, and no airplane can fly to such an altitude. To observe in the far-ultraviolet or beyond at X-ray or gamma-

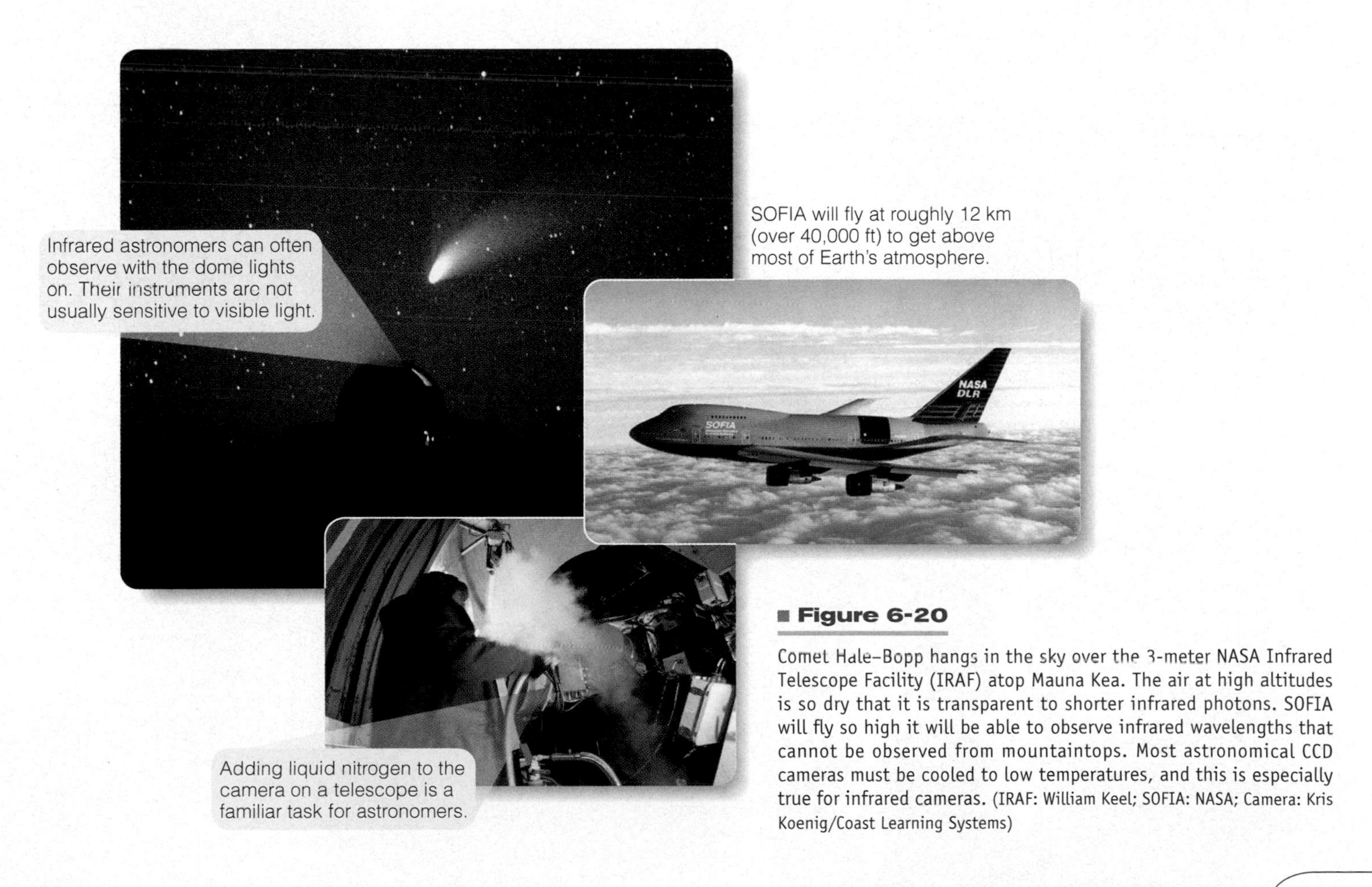

Infrared astronomers can often observe with the dome lights on. Their instruments are not usually sensitive to visible light.
SOFIA will fly at roughly 12 km (over 40,000 ft) to get above most of Earth's atmosphere.
NASA DLR
SOFIA
Adding liquid nitrogen to the camera on a telescope is a familiar task for astronomers.

■ **Figure 6-20**

Comet Hale–Bopp hangs in the sky over the 3-meter NASA Infrared Telescope Facility (IRAF) atop Mauna Kea. The air at high altitudes is so dry that it is transparent to shorter infrared photons. SOFIA will fly so high it will be able to observe infrared wavelengths that cannot be observed from mountaintops. Most astronomical CCD cameras must be cooled to low temperatures, and this is especially true for infrared cameras. (IRAF: William Keel; SOFIA: NASA; Camera: Kris Koenig/Coast Learning Systems)

Awake! for Morning in the Bowl of Night
Has flung the Stone that puts the Stars to
Flight:
And Lo! the Hunter of the East has caught
The Sultan's Turret in a Noose of Light.

THE RUBÁIYÁT OF OMAR KHAYYÁM,
TRANS. EDWARD FITZGERALD

THE UNIVERSE IS FILLED with fabulously beautiful clouds of glowing gas illuminated by brilliant stars, but it is all hopelessly beyond reach. No laboratory jar on Earth holds a sample labeled "star stuff," and no space probe has ever visited the inside of a star. The stars are far away, and the only information you can obtain about them comes hidden in starlight (■ Figure 7-1).

Earthbound humans knew almost nothing about stars until the early 19th century, when the Munich optician Joseph von Fraunhofer studied the solar spectrum and found it interrupted by some 600 dark lines. As scientists realized that the lines were related to the various atoms in the sun and found that stellar spectra had similar patterns of lines, the door to an understanding of stars finally opened.

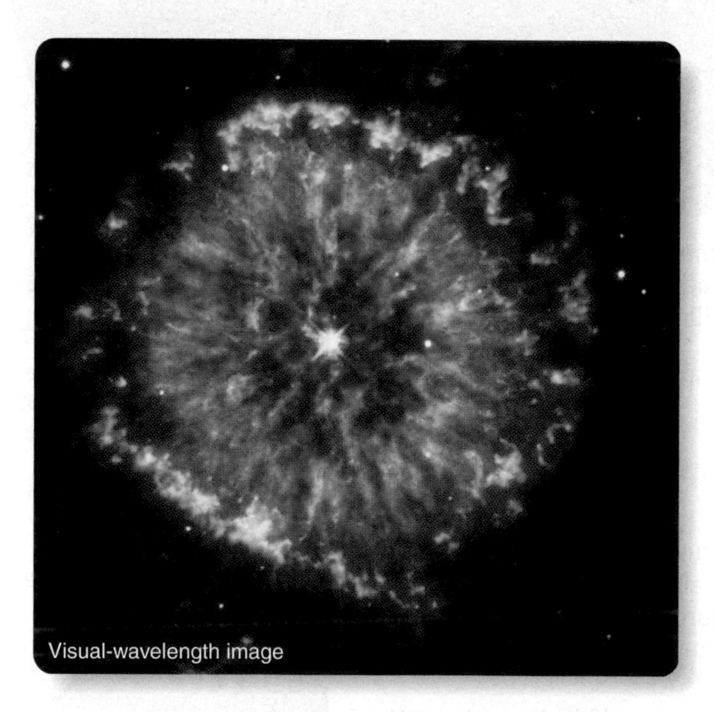

■ Figure 7-1

What's going on here? The sky is filled with beautiful and mysterious objects that lie far beyond your reach—in the case of the nebula NGC 6751, about 6500 ly beyond your reach. The only way to understand such objects is by analyzing their light. Such an analysis reveals that this object is a dying star surrounded by the expanding shell of gas it ejected a few thousand years ago. You will learn more about this phenomenon in Chapter 13. (NASA Hubble Heritage Team/STScI/AURA)

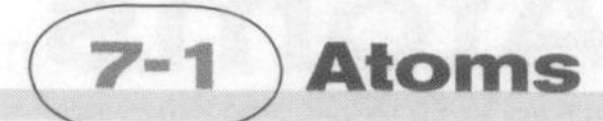

7-1 Atoms

STARS ARE GREAT BALLS OF HOT GAS, and the atoms in the surface layers of stars leave their marks on the light the stars emit. By understanding what atoms are and how they interact with light, you can decode the spectra of the stars and learn their secrets.

A Model Atom

To think about atoms and how they interact with light, you need a working model of an atom. In Chapter 2, you used a working model of the sky, the celestial sphere. You identified and named the important parts and described how they were located and how they interacted. In this chapter, you will begin your study of atoms by creating a model of an atom.

Your model atom contains a positively charged **nucleus** at the center, which consists of two kinds of particles. **Protons** carry a positive electrical charge, and **neutrons** have no charge, leaving the nucleus with a net positive charge.

The nucleus in this model atom is surrounded by a whirling cloud of orbiting **electrons,** low-mass particles with negative charges. In a normal atom, the number of electrons equals the number of protons, and the positive and negative charges balance to produce a neutral atom. Because protons and neutrons each have a mass 1836 times greater than that of an electron, most of the mass of an atom lies in the nucleus. The hydrogen atom is the simplest of all atoms. The nucleus is a single proton orbited by a single electron, with a total mass of only 1.67×10^{-27} kg, about a trillionth of a trillionth of a gram.

An atom is mostly empty space. To see this, imagine constructing a simple scale model. The nucleus of a hydrogen atom is a proton with a diameter of about 0.0000016 nm, or 1.6×10^{-15} m. If you multiply this by one trillion (10^{12}), you can represent the nucleus of your model atom with something about 0.16 cm in diameter—a grape seed would do. The region of a hydrogen atom that contains the whirling electron has a diameter of about 0.4 nm, or 4×10^{-10} m. Multiplying by a trillion increases the diameter to about 400 m, or about 4.5 football fields laid end to end (■ Figure 7-2). When you imagine a grape seed in the midst of a sphere 4.5 football fields in diameter, you can see that an atom is mostly empty space.

Now you can understand a **Common Misconception.** Most people, without thinking about it much, imagine that matter is solid, but you have seen that atoms are mostly empty space. The chair you sit on, the floor you walk on, are mostly not there. In Chapter 14, you will see what happens when stars collapse and most of the empty space gets squeezed out of the atoms.

Different Kinds of Atoms

There are over a hundred chemical elements. Which element an atom is depends only on the number of protons in the nucleus. For example, carbon has six protons in its nucleus. An atom with

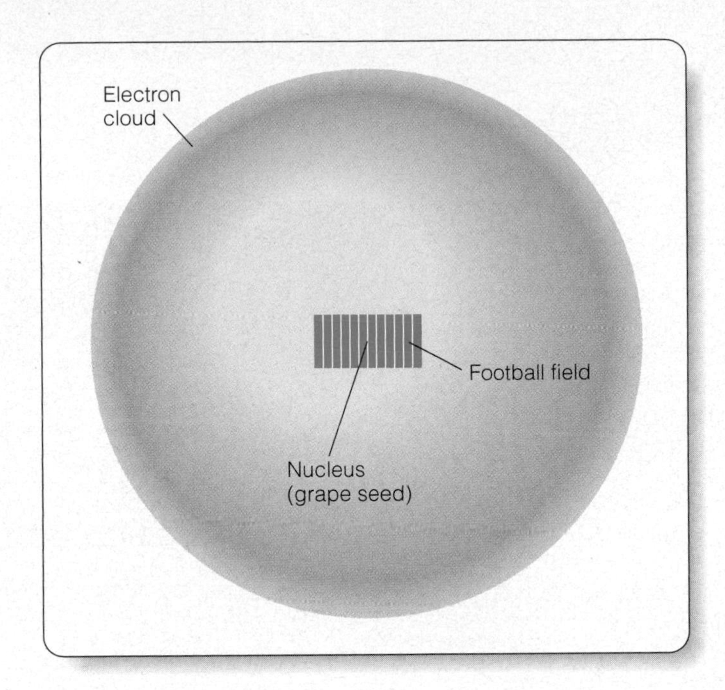

■ **Figure 7-2**

Magnifying a hydrogen atom by 10^{12} makes the nucleus the size of a grape seed and the diameter of the electron cloud about 4.5 times longer than a football field. The electron itself is still too small to see.

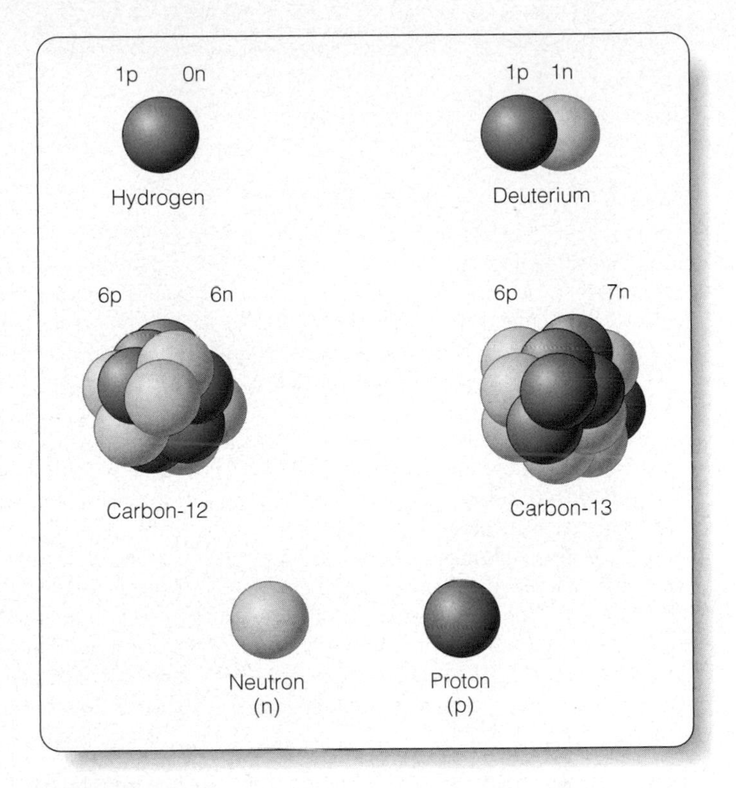

■ **Figure 7-3**

Some common isotopes. A rare isotope of hydrogen, deuterium, contains a proton and a neutron in its nucleus. Two isotopes of carbon are carbon-12 and carbon-13.

one more proton than this is nitrogen, and an atom with one fewer proton is boron.

Although the number of protons in an atom of a given element is fixed, the number of neutrons is less restricted. For instance, if you added a neutron to a carbon nucleus, you would still have carbon, but it would be slightly heavier than normal carbon. Atoms that have the same number of protons but a different number of neutrons are **isotopes.** Carbon has two stable isotopes. One contains six protons and six neutrons for a total of 12 particles and is thus called carbon-12. Carbon-13 has six protons and seven neutrons in its nucleus (■ Figure 7-3).

Protons and neutrons are bound tightly into the nucleus, but the electrons are held loosely in the electron cloud. Running a comb through your hair creates a static charge by removing a few electrons from their atoms. This process is called **ionization,** and an atom that has lost one or more electrons is an **ion.** A neutral carbon atom has six electrons to balance the positive charge of the six protons in its nucleus. If you ionize the atom by removing one or more electrons, the atom is left with a net positive charge. Under some circumstances, an atom may capture one or more extra electrons, giving it more negative charges than positive. Such a negatively charged atom is also considered an ion.

Atoms that collide may form bonds with each other by exchanging or sharing electrons. Two or more atoms bonded together form a **molecule.** Atoms do collide in stars, but the high temperatures cause violent collisions that are unfavorable for chemical bonding. Only in the coolest stars are the collisions gentle enough to permit the formation of chemical bonds. You will see later that the presence of molecules such as titanium oxide (TiO) in a star is a clue that the star is very cool. In later chapters, you will see that molecules can form in cool gas clouds in space and in the atmospheres of planets.

Electron Shells

So far you have been thinking of the cloud of the whirling electrons in a general way, but now it is time to be more specific as to how the electrons behave within the cloud.

Electrons are bound to the atom by the attraction between their negative charge and the positive charge on the nucleus. This attraction is known as the **Coulomb force,** after the French physicist Charles-Augustin de Coulomb (1736–1806). To ionize an atom, you need a certain amount of energy to pull an electron away from the nucleus. This energy is the electron's **binding energy,** the energy that holds it to the atom.

The size of an electron's orbit is related to the energy that binds it to the atom. If an electron orbits close to the nucleus, it is tightly bound, and a large amount of energy is needed to pull it away. Consequently, its binding energy is large. An electron orbiting farther from the nucleus is held more loosely, and less energy is needed to pull it away. That means it has less binding energy.

Nature permits atoms only certain amounts (quanta) of binding energy, and the laws that describe how atoms behave are

How Do We Know? | 7-1

Quantum Mechanics

What are atoms made of? You can see objects such as stars, planets, aircraft carriers, and hummingbirds, but you can't see individual atoms. As scientists apply the principle of cause and effect, they study the natural effects they can see and work backward to find the causes. Invariably that quest for causes leads back to the invisible world of atoms.

Quantum mechanics is the set of rules that describe how atoms and subatomic particles behave. On the atomic scale, particles behave in ways that seem unfamiliar. One of the principles of quantum mechanics specifies that you cannot know simultaneously the exact location and motion of a particle. This is why physicists prefer to describe the electrons in an atom as if they were a cloud of negative charge surrounding the nucleus rather than small particles following orbits.

This raises some serious questions about reality. Is an electron really a particle at all? If you can't know simultaneously the position and motion of a specific particle, how can you know how it will react to a collision with a photon or another particle? The answer is that you can't know, and that seems to violate the principle of cause and effect.

Modern physicists are trying to understand the nature of the particles that make up atoms. Are protons and neutrons made up of even smaller particles? What are these ultimate particles made of? The world you experience is shaped and animated by subatomic particles whose true nature lies at one of the most exciting frontiers of science.

The world you see, including these neon signs, is animated by the properties of atoms and subatomic particles. (Jeff Greenberg/PhotoEdit)

called the laws of **quantum mechanics (How Do We Know? 7-1).** Much of this discussion of atoms is based on the laws of quantum mechanics.

Because atoms can have only certain amounts of binding energy, your model atom can have orbits of only certain sizes, called **permitted orbits.** These are like steps in a staircase: you can stand on the number-one step or the number-two step, but not on the number-one-and-one-quarter step. The electron can occupy any permitted orbit but not orbits in between.

The arrangement of permitted orbits depends primarily on the charge of the nucleus, which in turn depends on the number of protons. Consequently, each kind of element has its own pattern of permitted orbits (■ Figure 7-4). Isotopes of the same elements have nearly the same pattern because they have the same number of protons. However, ionized atoms have orbital patterns that differ from their un-ionized forms. Thus the arrangement of permitted orbits differs for every kind of atom and ion.

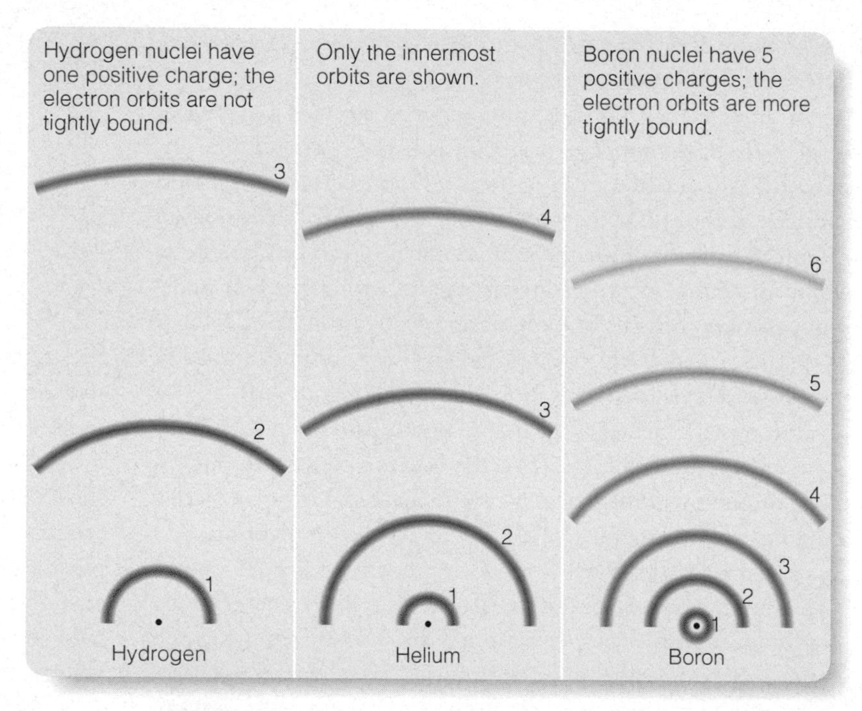

Hydrogen nuclei have one positive charge; the electron orbits are not tightly bound.
Only the innermost orbits are shown.
Boron nuclei have 5 positive charges; the electron orbits are more tightly bound.
3
2
1
Hydrogen
4
3
2
1
Helium
6
5
4
3
2
1
Boron

■ Figure 7-4

The electron in an atom may occupy only certain permitted orbits. Because different elements have different charges on their nuclei, the elements have different, unique patterns of permitted orbits.

◄ SCIENTIFIC ARGUMENT ►

How many hydrogen atoms would it take to cross the head of a pin? This is not a frivolous question. In answering it, you will discover how small atoms really are, and you will see how powerful physics and mathematics can be as a way to understand nature. Many scientific arguments are convincing because they have the precision of mathematics. To begin, assume that the head of a pin is about 1 mm in diameter. That is 0.001 m. The size of a hydrogen atom is represented by the diameter of the electron cloud, roughly 0.4 nm. Because 1 nm equals 10^{-9} m, you can multiply and discover that 0.4 nm equals 4×10^{-10} m. To find out how many atoms would stretch 0.001 m, you can divide the diameter of the pinhead by the diameter of an atom. That is, divide 0.001 m by 4×10^{-10} m, and you get 2.5×10^{6}. It would take 2.5 million hydrogen atoms lined up side by side to cross the head of a pin.

Now you can see how tiny an atom is and also how powerful a bit of physics and mathematics can be. It reveals a view of nature beyond the capability of your eyes. Now build an argument using another bit of arithmetic: **How many hydrogen atoms would you need to add up to the mass of a paper clip (1 g)?**

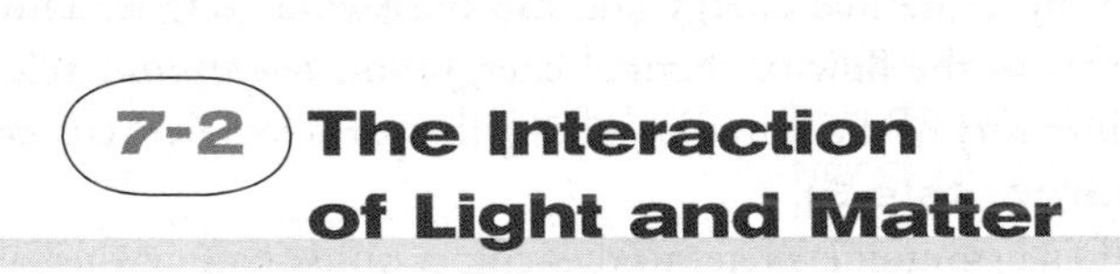

7-2 The Interaction of Light and Matter

IF LIGHT DID NOT INTERACT with matter, you would not be able to see these words. In fact, you would not exist, because, among other problems, photosynthesis would be impossible, and there would be no grass, wheat, bread, beef, cheeseburgers, or any other kind of food. The interaction of light and matter makes your life possible, and it also makes it possible for you to understand the universe.

You should begin your study of light and matter by considering the hydrogen atom. As you read earlier, hydrogen is both simple and common. Roughly 90 percent of all atoms in the universe are hydrogen.

The Excitation of Atoms

Each orbit in an atom represents a specific amount of binding energy, so physicists commonly refer to the orbits as **energy levels.** Using this terminology, you can say that an electron in its smallest and most tightly bound orbit is in its lowest permitted energy level. You could move the electron from one energy level to another by supplying enough energy to make up the difference between the two energy levels. It would be like moving a flowerpot from a low shelf to a high shelf; the greater the distance between the shelves, the more energy you would need to raise the pot. The amount of energy needed to move the electron is the energy difference between the two energy levels.

If you move the electron from a low energy level to a higher energy level, you can call the atom an **excited atom.** That is, you have added energy to the atom in moving its electron. An atom can become excited by collision. If two atoms collide, one or both may have electrons knocked into a higher energy level. This happens very commonly in hot gas, where the atoms move rapidly and collide often.

Another way an atom can get the energy that moves an electron to a higher energy level is to absorb a photon. Only a photon with exactly the right amount of energy can move the electron from one level to another. If the photon has too much or too little energy, the atom cannot absorb it. Because the energy of a photon depends on its wavelength, only photons of certain wavelengths can be absorbed by a given kind of atom. ■ Figure 7-5 shows the lowest four energy levels of the hydrogen atom, along with three photons the atom could absorb. The longest-wavelength photon has only enough energy to excite the electron to the second energy level, but the shorter-wavelength photons can excite the electron to higher levels. A photon with too much or too little energy cannot be absorbed. Because the hydrogen atom has many more energy levels than shown in Figure 7-5, it can absorb photons of many different wavelengths.

Atoms, like humans, cannot exist in an excited state forever. An excited atom is unstable and must eventually (usually within 10^{-6} to 10^{-9} seconds) give up the energy it has absorbed and return its electron to a lower energy level. The lowest energy level an electron can occupy is called the **ground state.**

When an electron drops from a higher to a lower energy level, it moves from a loosely bound level to one more tightly bound. The atom then has a surplus of energy—the energy difference between the levels—that it can emit as a photon. Study

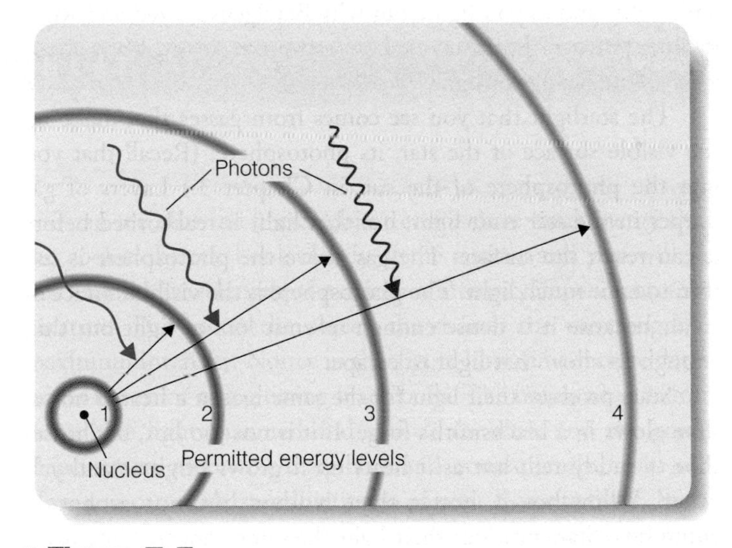

■ **Figure 7-5**

A hydrogen atom can absorb only those photons that move the atom's electron to one of the higher-energy orbits. Here three different photons are shown along with the change they would produce if they were absorbed.

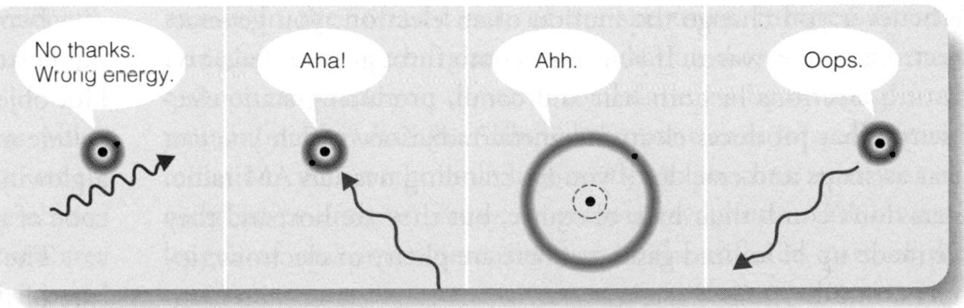

■ **Figure 7-6**

An atom can absorb a photon only if the photon has the correct amount of energy. The excited atom is unstable and within a fraction of a second returns to a lower energy level, reradiating the photon in a random direction.

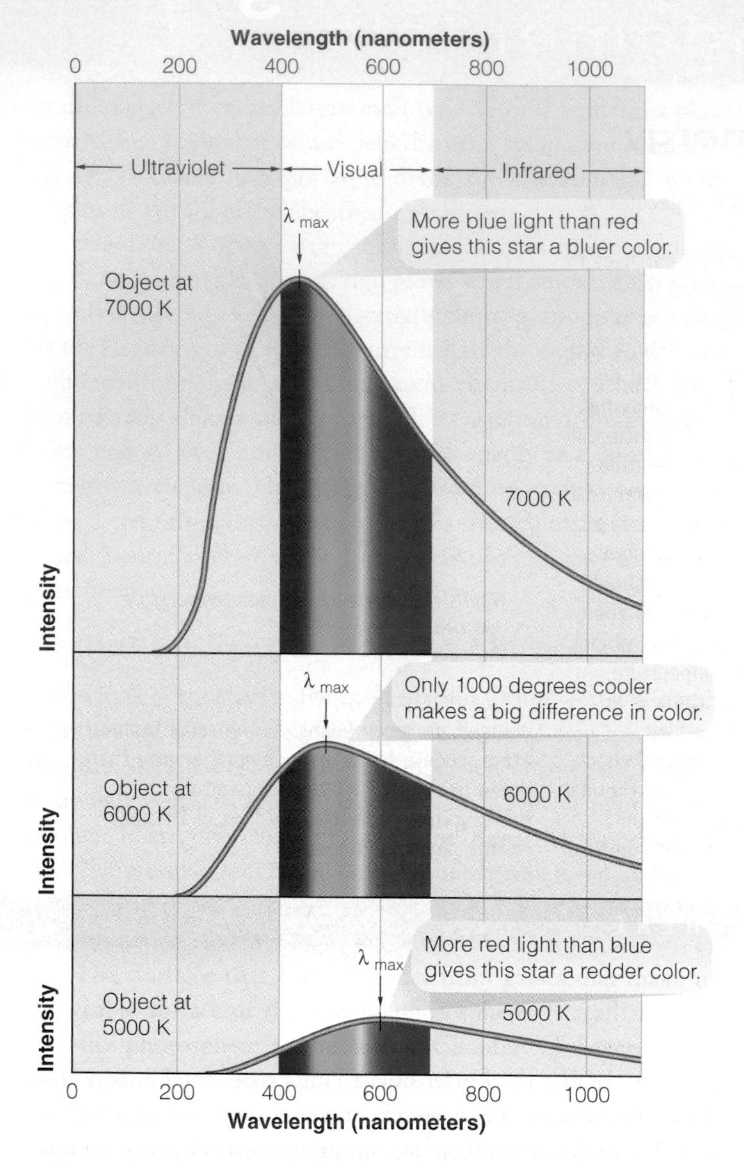

Figure 7-7

Black body radiation from three bodies at different temperatures demonstrates that a hot body radiates more total energy and that the wavelength of maximum intensity is shorter for hotter objects. The hotter object here will look blue to your eyes, while the cooler object will look red.

Two Radiation Laws

The two features of black body radiation that you have just considered can be given precise mathematical form, and they have proven so dependable, they are known as laws. One law is related to energy and one to color.

As you saw in the previous section, a hot object emits more black body radiation than a cool object. That is, it emits more energy. Recall from Chapter 5 that energy is expressed in units called joules (J); 1 joule is about the energy of an apple falling from a table to the floor. The total radiation given off by 1 square meter of the surface of the object in joules per second equals a constant number, represented by σ, times the temperature raised to the fourth power.* This relationship is called the Stefan–Boltzmann law:

$$E = \sigma T^4 \text{ (J/s/m}^2\text{)}$$

How does this help you understand stars? Suppose a star the same size as the sun had a surface temperature that was twice as hot as the sun's surface. Then each square meter of that star would radiate not twice as much energy but 2^4, or 16, times as much energy. From this law you can see that a small difference in temperature can produce a very large difference in the amount of energy a star's surface emits.

The second radiation law is related to the color of stars. In the previous section, you saw that hot stars look blue and cool stars look red. Wien's law tells you that the wavelength at which a star radiates the most energy, its wavelength of maximum intensity (λ_{max}), depends only on the star's temperature:

$$\lambda_{max} = \frac{3{,}000{,}000}{T}$$

That is, the wavelength of maximum radiation in nanometers equals 3 million divided by the temperature on the Kelvin scale.

This law is a powerful tool in astronomy, because it means you can relate the temperature of a star to its wavelength of maximum intensity. For example, you might find a star emitting light with a wavelength of maximum intensity of 1000 nm—in the near-infrared. Then the surface temperature of the star must be 3000 K. Later you will meet stars much hotter than the sun; such stars radiate most of their energy at very short wavelengths. The hottest stars, for instance, radiate most of their energy in the ultraviolet.

ThomsonNOW Sign in at www.thomsonedu.com and go to ThomsonNOW to see Astronomy Exercises "Black Body" and "Stefan–Boltzmann Law."

◄ SCIENTIFIC ARGUMENT ►

The infrared radiation coming out of your ear can tell a doctor your temperature. How does that work?

You know two radiation laws, so your argument must use the right one. Doctors and nurses use a handheld device to measure body temperature by observing the infrared radiation emerging from a patient's ear. You might suspect the device depends on the Stefan–Boltzmann law and measures the intensity of the infrared radiation. A person with a fever will emit more energy than a healthy person. However, a healthy person with a large ear canal would emit more than a person with a small ear canal, so measuring intensity would not be accurate. The device actually depends on Wien's law in that it measures the "color" of the infrared radiation. A patient with a fever will emit at a slightly shorter wavelength of maximum intensity, and the infrared radiation emerging from his or her ear will be a tiny bit "bluer" than normal.

*For the sake of completeness, you can note that the constant σ equals 5.67×10^{-8} J/m^{2}s degree^4.

Astronomers can measure the temperatures of stars the same way. Adapt your argument for stars. **Use Figure 7-7 to explain how the colors of stars reveal their temperatures.**

◄ ►

7-3 Stellar Spectra

SCIENCE IS A WAY OF UNDERSTANDING nature, and the spectrum of a star tells you a great deal about such things as temperature, motion, and composition. In later chapters, you will use spectra to study galaxies and planets, but you can begin with the spectra of stars, including that of the sun.

The Formation of a Spectrum

The spectrum of a star is formed as light passes outward through the gases near its surface. Read **Atomic Spectra** on pages 136–137 and notice that it describes three important properties of spectra and defines 12 new terms that will help you discuss astronomical spectra:

1. There are three kinds of spectra: *continuous spectra*, *absorption* or *dark-line spectra* with *absorption lines*, and *emission* or *bright-line spectra* with *emission lines*. These spectra are described by *Kirchhoff's laws*. When you see one of these types of spectra, you can recognize the kind of matter that emitted the light.

2. Photons are emitted or absorbed when an electron in an atom makes a *transistion* from one energy level to another. The wavelengths of the photons depend on the energy difference between the two levels. Hydrogen atoms can produce many spectral lines in series such as the *Lyman*, *Balmer*, and *Paschen* series. Only three lines in the Balmer series are visible to human eyes. The emitted photons coming from a hot cloud of hydrogen gas have the same wavelengths as the photons absorbed by hydrogen atoms in the gases of a star.

3. Most modern astronomy books display spectra as graphs of intensity versus wavelength. Be sure you see the connection between dark absorption lines and dips in the graphed spectrum.

Whatever kind of spectrum astronomers look at, the most common spectral lines are the Balmer lines of hydrogen. In the next section, you will see how Balmer lines can tell you the temperature of a star's surface.

ThomsonNOW Sign in at www.thomsonedu.com and go to ThomsonNOW to see Astronomy Exercise "Emission and Absorption Spectra."

The Balmer Thermometer

You can use the Balmer absorption lines as a thermometer to find the temperatures of stars. From the discussion of black body radiation, you already know how to estimate temperature from color—red stars are cool, and blue stars are hot. You can estimate temperature from color, but the Balmer lines give you much greater accuracy.

Recall that astronomers use the Kelvin temperature scale when referring to stellar temperatures. These temperatures range from 40,000 K to 2000 K and refer to the temperature of the surface of the star. The centers of stars are much hotter—millions of degrees—but the colors and spectra of stars tell you about the surface. That's where the light comes from.

The Balmer thermometer works because the Balmer absorption lines are produced only by atoms whose electrons are in the second energy level. If the star is cool, there are few violent collisions between atoms to excite the electrons, and most atoms have their electrons in the ground state. Electrons in the ground state can't absorb photons in the Balmer series. As a result, you should expect to find weak Balmer absorption lines in the spectra of cool stars.

In the surface layers of hotter stars, on the other hand, there are many violent collisions between atoms. These collisions can excite electrons to high energy levels or ionize some atoms by knocking the electron out of the atoms. Consequently, few hydrogen atoms have their electron in the second orbit to form Balmer absorption lines, and you should expect hot stars, like cool stars, to have weak Balmer absorption lines.

At an intermediate temperature, roughly 10,000 K, the collisions are just right to excite large numbers of electrons into the second energy level. The gas absorbs Balmer wavelength photons very well and produces strong Balmer lines.

To summarize, the strength of the Balmer lines depends on the temperature of the star's surface layers. Both hot and cool stars have weak Balmer lines, but medium-temperature stars have strong Balmer lines.

Theoretical calculations can predict just how strong the Balmer lines should be for stars of various temperatures. Such calculations are the key to finding temperatures from stellar spectra. The curve in ■ Figure 7-8a shows the strength of the Balmer lines for various stellar temperatures. You could use this as a temperature indicator, except that the curve gives two possible answers. A star with Balmer lines of a certain strength might have either of two temperatures, one high and one low. How do you know which is right? You must examine other spectral lines to choose the correct temperature.

You have seen how the strength of the Balmer lines depends on temperature. Temperature has a similar effect on the spectral lines of other elements, but the temperature at which the lines reach their maximum strength differs for each element (Figure 7-8b). If you add a number of chemical elements to your graph,

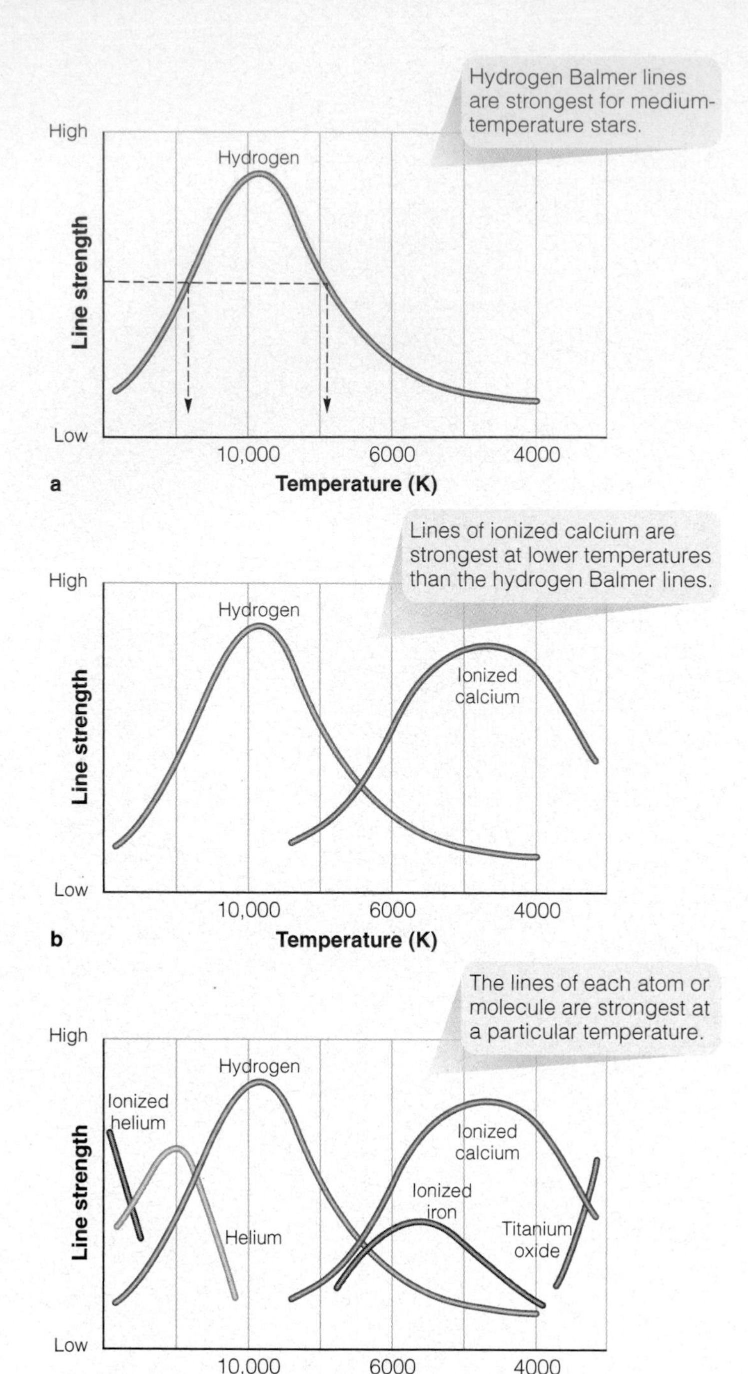

■ **Figure 7-8**

The strength of spectral lines can tell you the temperature of a star. (a) Balmer hydrogen lines alone are not enough because they give two answers. Balmer lines of a certain strength could be produced by a hotter star or a cooler star. (b) Adding another atom to the diagram helps, and (c) adding many atoms and molecules to the diagram creates a precise aid to find the temperatures of stars.

you get a powerful aid for finding the stars' temperatures (Figure 7-8c).

Now you can determine a star's temperature by comparing the strengths of its spectral lines with your graph. For instance, if you recorded the spectrum of a star and found medium-strength Balmer lines and strong helium lines, you could conclude it had a temperature of about 20,000 K. But if the star had weak hydrogen lines and strong lines of ionized iron, you would assign it a temperature of about 5800 K, similar to that of the sun.

The spectra of stars cooler than about 3000 K contain dark bands produced by molecules such as titanium oxide (TiO). Because of their structure, molecules can absorb photons at many wavelengths, producing numerous, closely spaced spectral lines that blend together to form bands. These molecular bands appear only in the spectra of the coolest stars because, as mentioned before, molecules in cool stars are not subject to the violent collisions that would break them up in hotter stars. Consequently, the presence of dark bands in a star's spectrum indicates that the star is very cool.

From stellar spectra, astronomers have found that the hottest stars have surface temperatures above 40,000 K and the coolest about 2000 K. Compare these with the surface temperature of the sun, about 5800 K.

Spectral Classification

You have seen that the strengths of spectral lines depend on the surface temperature of the star. From this you can conclude that all stars of a given temperature should have similar spectra. If you learn to recognize the pattern of spectral lines produced by a 6000 K star, for instance, you need not use Figure 7-8c every time you see that kind of spectrum. You can save time by classifying stellar spectra rather than analyzing each one individually.

The first widely used classification system was devised by astronomers at Harvard during the 1890s and 1900s. One of the astronomers, Annie J. Cannon, personally inspected and classified the spectra of over 250,000 stars. The spectra were first classified into groups labeled A through Q, but some groups were later dropped, merged with others, or reordered. The final classification includes the seven major **spectral classes,** or **types,** still used today: O, B, A, F, G, K, M.*

This sequence of spectral types, called the **spectral sequence,** is important because it is a temperature sequence. The O stars are the hottest, and the temperature decreases along the sequence to the M stars, the coolest. For maximum precision, astronomers divide each spectral class into 10 subclasses. For example, spectral class A consists of the subclasses A0, A1, A2, . . . A8, A9. Next come F0, F1, F2, and so on. This finer division gives a star's temperature to an accuracy within about 5 percent. The sun, for example, is not just a G star, but a G2 star.

■ Table 7-1 breaks down some of the information in Figure 7-8c and presents it in tabular form according to spectral class.

*Generations of astronomy students have remembered the spectral sequence using the mnemonic "Oh, Be A Fine Girl (Guy), Kiss Me." More recent suggestions from students include, "Oh Boy, An F Grade Kills Me," and "Only Bad Astronomers Forget Generally Known Mnemonics."

Table 7-1 | Spectral Classes

Spectral Class	Approximate Temperature (K)	Hydrogen Balmer Lines	Other Spectral Features	Naked-Eye Example
O	40,000	Weak	Ionized helium	Meissa (O8)
B	20,000	Medium	Neutral helium	Achernar (B3)
A	10,000	Strong	Ionized calcium weak	Sirius (A1)
F	7500	Medium	Ionized calcium weak	Canopus (F0)
G	5500	Weak	Ionized calcium medium	Sun (G2)
K	4500	Very weak	Ionized calcium strong	Arcturus (K2)
M	3000	Very weak	TiO strong	Betelgeuse (M2)

For example, if a star has weak Balmer lines and lines of ionized helium, it must be an O star.

Thirteen stellar spectra are arranged in ■ Figure 7-9 from the hottest at the top to the coolest at the bottom. You can easily see how the strength of spectral lines depends on temperature. The Balmer lines are strongest in A stars, where the temperature is moderate but still high enough to excite the electrons in hydrogen atoms to the second energy level, where they can absorb Balmer wavelength photons. In the hotter stars (O and B), the Balmer lines are weak because the higher temperature excites the electrons to energy levels above the second or ionizes the atoms. The Balmer lines in cooler stars (F through M) are also weak but for a different reason. The lower temperature cannot excite many electrons to the second energy level, so few hydrogen atoms are capable of absorbing Balmer wavelength photons.

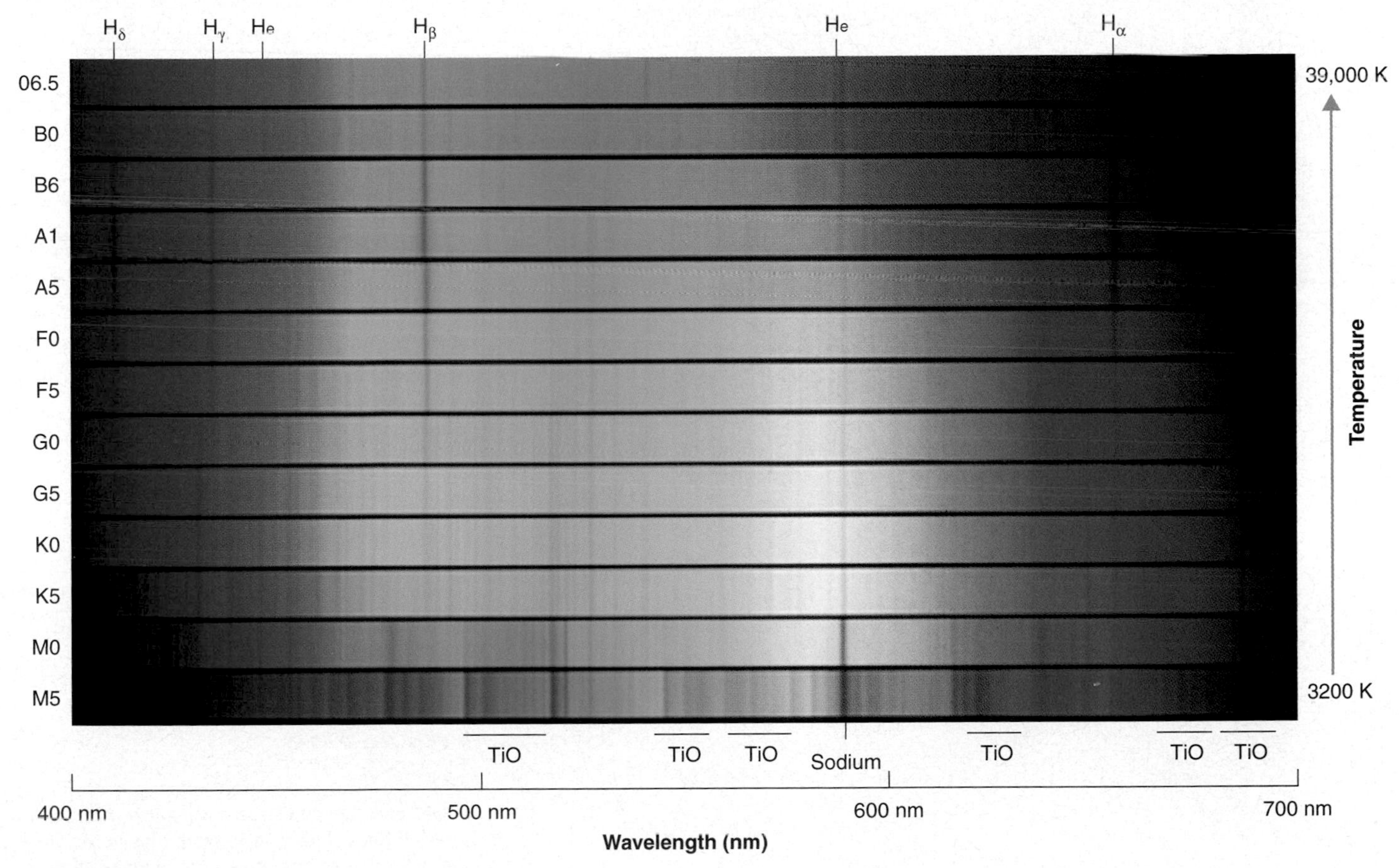

■ Figure 7-9

These spectra show stars from hot O stars at the top to cool M stars at the bottom. The Balmer lines of hydrogen are strongest about A0, but the two closely spaced lines of sodium in the yellow are strongest for very cool stars. Helium lines appear only in the spectra of the hottest stars. Notice that the helium line visible in the top spectrum has nearly but not exactly the same wavelength as the sodium lines visible in cooler stars. Bands produced by the molecule titanium oxide are strong in the spectra of the coolest stars. (AURA/NOAO/NSF)

Although these spectra are attractive, astronomers rarely work with spectra as color images. Rather, they display spectra as graphs of intensity versus wavelength that show dark absorption lines as dips in the graph (■ Figure 7-10). Such graphs allow more detailed analysis than photographs. Notice, for example, that the overall curves are similar to black body curves. The wavelength of maximum intensity is in the infrared for the coolest stars and in the ultraviolet for the hottest stars. Look carefully at these graphs, and you can see that helium is visible only in the spectra of the hottest classes, and titanium oxide bands only in the coolest. Two lines of ionized calcium increase in strength from A to K and then decrease from K to M. Because the strength of these spectral lines depends on temperature, it requires only a few moments to study a star's spectrum and determine its temperature.

Now you can learn something new about your Favorite Stars. Sirius, brilliant in the winter sky, is an A1 star; and Vega, bright overhead in the summer sky, is an A0 star. They have nearly the same temperature and color, and both have strong Balmer lines in their spectra. The bright red star in Orion is Betelgeuse, a cool M2 star, but blue-white Rigel is a hot B8 star. Polaris, the North Star, is an F8 star a bit hotter than our sun, and Alpha Centauri, the closest star to the sun, seems to be a G2 star just like the sun.

The study of spectral types is a century old, but astronomers continue to discover new types of stars. The **L dwarfs,** found in 1998, are cooler and fainter than M stars. The spectra of L dwarfs show that they are clearly a different type of star. The spectra of M stars contain bands produced by metal oxides such as titanium oxide (TiO), but L dwarf spectra contain bands produced by molecules such as iron hydride (FeH). The **T dwarfs,** discovered in 2000, are even cooler and fainter than L dwarfs. Their spectra show absorption by methane (CH_4) and water vapor (■ Figure 7-11). The development of giant telescopes and highly sensitive infrared cameras and spectrographs is allowing astronomers to find and study these coolest of stars.

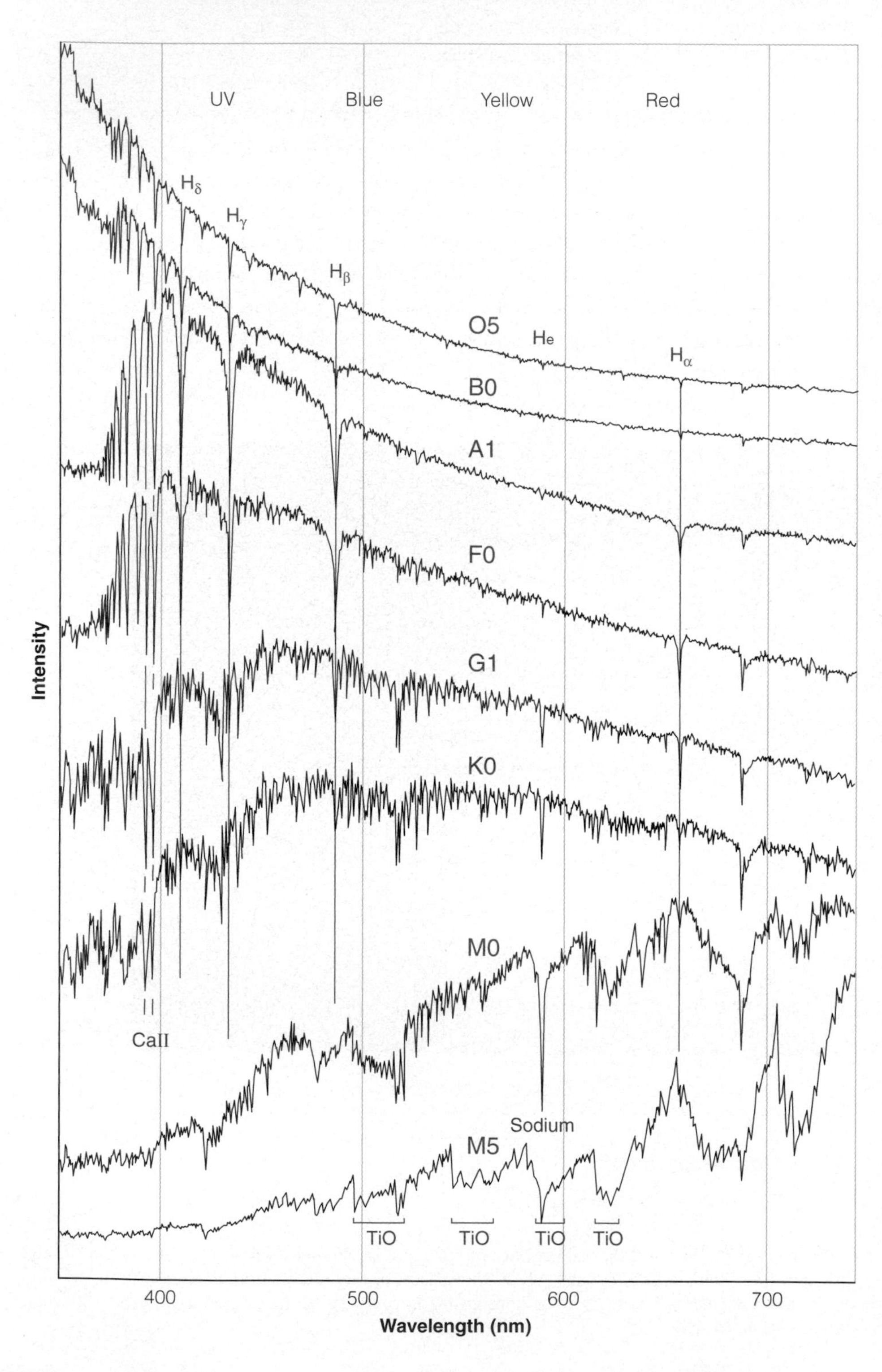

■ Figure 7-10

Modern digital spectra are often represented as graphs of intensity versus wavelength with dark absorption lines appearing as sharp dips in the curves. The hottest stars are at the top and the coolest at the bottom. Hydrogen Balmer lines are strongest at about A0, while lines of ionized calcium (CaII) are strong in K stars. Titanium oxide (TiO) bands are strongest in the coolest stars. Compare these spectra with Figures 7-8c and 7-9. (Courtesy NOAO, G. Jacoby, D. Hunter, and C. Christian)

The Composition of the Stars

It seems as though it should be easy to find the composition of the sun and stars just by looking at their spectra, but this is actually a

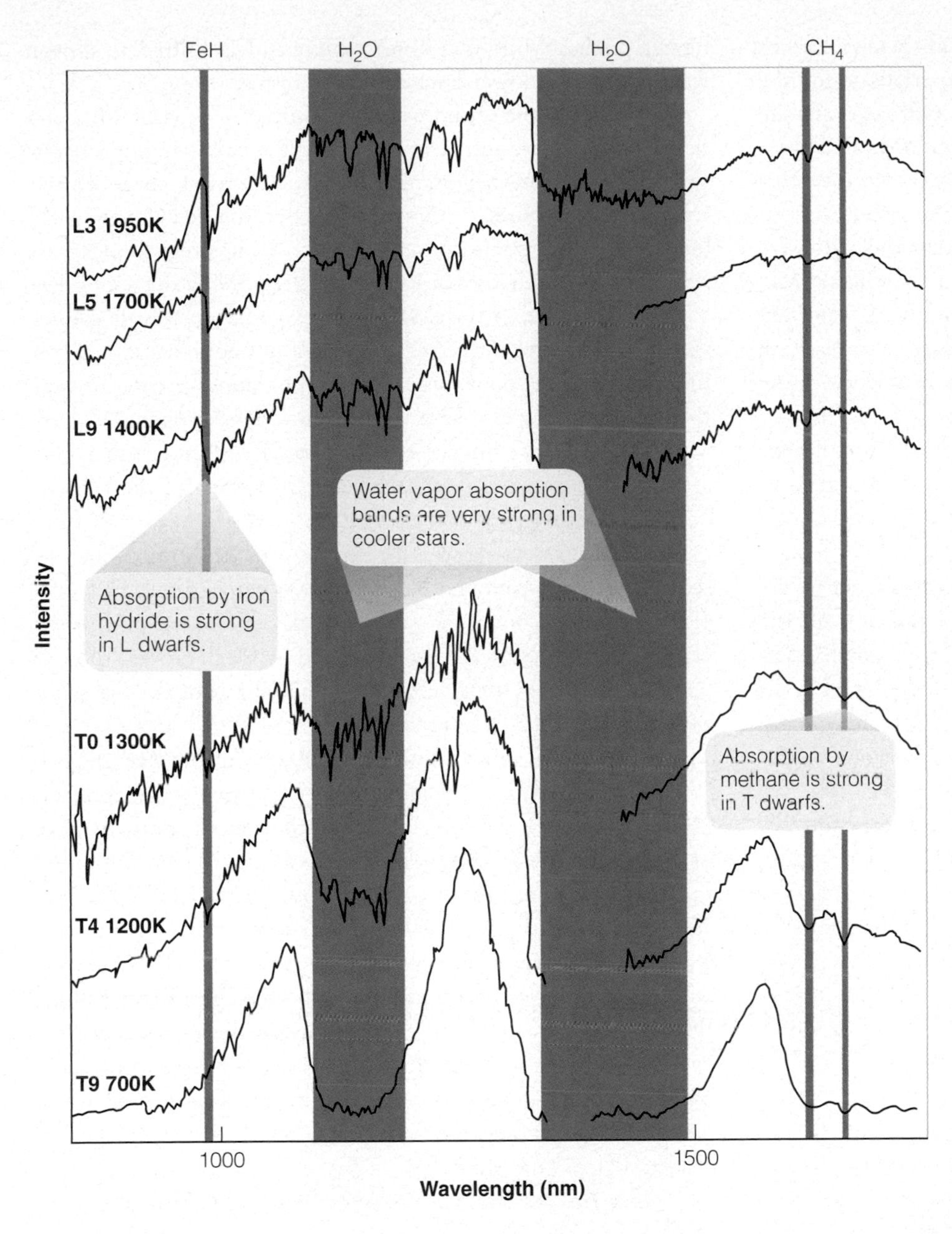
FeH
H₂O
H₂O
CH₄
L3 1950K
L5 1700K
L9 1400K
T0 1300K
T4 1200K
T9 700K
Water vapor absorption bands are very strong in cooler stars.
Absorption by iron hydride is strong in L dwarfs.
Absorption by methane is strong in T dwarfs.
Intensity
1000
1500
Wavelength (nm)

■ Figure 7-11

These six infrared spectra show the dramatic differences between L dwarfs and T dwarfs. Spectra of M stars show titanium oxide bands (TiO), but L and T dwarfs are so cool that TiO molecules do not form. Other molecules such as iron hydride (FeH), water (H_2O), and methane (CH_4) can form in these very cool stars. (Adapted from Thomas R. Geballe, Gemini Observatory, from a graph that originally appeared in *Sky and Telescope Magazine,* February 2005, p. 37.)

difficult problem that wasn't well understood until the 1920s. The story is worth telling, not only because it is the story of an important American astronomer who never got proper credit, but also because it illustrates the temperature dependence of spectral features. The story begins in England.

As a child in England, Cecilia Payne (1900–1979) excelled in classics, languages, mathematics, and literature, but her first love was astronomy. After finishing Newnham College in Cambridge, she left England, sensing that there were no opportunities there for a woman of science. In 1922, Payne arrived at Harvard, where she eventually earned her Ph.D., although her degree was awarded by Radcliffe because Harvard did not then admit women.

In her thesis, Payne attempted to relate the strength of the absorption lines in stellar spectra to the physical conditions in the atmospheres of the stars. This was not easy because a given spectral line can be weak because the atom is rare or because the temperature is too high or too low for it to absorb efficiently. If you see sodium lines in a star's spectrum, you can be sure that the star contains sodium atoms, but if you see no sodium lines, you must consider the possibility that sodium is present but the star is too hot or too cool for the atom to produce spectral lines.

Payne's problem was to untangle these two factors and find both the true temperatures of the stars and the true abundance of the atoms in their atmospheres. Recent advances in atomic physics gave her the theoretical tools she needed. About the time Payne left Newnham College, Indian physicist Meghnad Saha published his work on the ionization of atoms. Drawing from such theoretical work, Payne was able to show that over 90 percent of the atoms in stars (including the sun) are hydrogen and most of the rest helium (■ Table 7-2). The heavier atoms like calcium, sodium, and iron seem more abundant only because they are better at absorbing photons at the temperatures of stars.

■ Table 7-2 | The Most Abundant Elements in the Sun

Element	Percentage by Number of Atoms	Percentage by Mass
Hydrogen	91.0	70.9
Helium	8.9	27.4
Carbon	0.03	0.3
Nitrogen	0.008	0.1
Oxygen	0.07	0.8
Neon	0.01	0.2
Magnesium	0.003	0.06
Silicon	0.003	0.07
Sulfur	0.002	0.04
Iron	0.003	0.1

At the time, astronomers found it hard to believe Payne's abundances of hydrogen and helium. They especially found the abundance of helium unacceptable. After all, hydrogen lines are at least visible in most stellar spectra, but helium lines are almost invisible in the spectra of all but the hottest stars. Nearly all astronomers assumed that the stars had roughly the same composition as Earth's surface; that is, they believed that the stars were composed mainly of heavier atoms such as carbon, silicon, iron, and aluminum. Even the most eminent astronomers dismissed Payne's result as illusory. Faced with this pressure and realizing the limited opportunities available to women in science in the 1920s, Payne could not press her discovery.

By 1929, astronomers generally understood the importance of temperature on measurements of composition derived from stellar spectra. At that point, they recognized that stars are mostly hydrogen and helium, but Payne received no credit.

Payne worked for many years as a staff astronomer at the Harvard College Observatory with no formal position on the faculty. She married Russian astronomer Sergei Gaposchkin in 1934 and was afterward known as Cecilia Payne-Gaposchkin. In 1956, when Harvard accepted women to its faculty, she was appointed a full professor and chair of the Harvard astronomy department.

Cecilia Payne-Gaposchkin's work on the chemical composition of the stars illustrates the importance of fully understanding the interaction between light and matter. It was her detailed understanding of the physics that led her to the correct composition. As you turn your attention to other information that can be derived from stellar spectra, you will again discover the importance of understanding light.

ThomsonNOW™ Sign in at www.thomsonedu.com and go to ThomsonNOW to see Astronomy Exercise "Stellar Atomic Absorption Lines."

The Doppler Effect

Surprisingly, one of the pieces of information hidden in a spectrum is the velocity of the light source. Astronomers can measure the wavelengths of lines in a star's spectrum and find the velocity of the star. The **Doppler effect** is an apparent change in the wavelength of radiation caused by the motion of the source.

When astronomers talk about the Doppler effect, they are talking about a shift in the wavelength of electromagnetic radiation. But the Doppler shift can occur in all forms of wave phenomena, including sound waves, so you probably hear the Doppler effect every day without noticing.

The pitch of a sound is determined by its wavelength. Sounds with long wavelengths have low pitches, and sounds with short wavelengths have higher pitches. You hear a Doppler shift every time a car or truck passes you and the pitch of its engine noise drops. Its sound is shifted to shorter wavelengths and higher pitches while it is approaching and is shifted to longer wavelengths and lower pitches after it passes.

To see why the sound waves are shifted in wavelength, consider a fire truck approaching you with a bell clanging once a second. When the bell clangs, the sound travels ahead of the truck to reach your ears. One second later, the bell clangs again, but, during that one second, the fire truck has moved closer to you, so the bell is closer at its second clang. Now the sound has a shorter distance to travel and reaches your ears a little sooner than it would have if the fire truck were not approaching. If you timed the clangs, you would find that the clangs were slightly less than one second apart. After the fire truck passes you and is moving away, you hear the clangs sounding slightly more than one second apart, because now each successive clang of the bell occurs farther from you.

■ Figure 7-12a shows a fire truck moving toward one observer and away from another observer. The position of the bell at each clang is shown by a small black bell. The sound of the clangs spreading outward is represented by black circles. You can see how the clangs are squeezed together ahead of the fire truck and stretched apart behind.

Now you can substitute a source of light for the clanging bell (Figure 7-12b). Imagine the light source emitting waves continuously as it approaches you. Each time the source emits the peak of a wave, it will be slightly closer to you than when it emitted the peak of the previous wave. From your vantage point, the successive peaks of the wave will seem closer together in the same way that the clangs of the bell seemed closer together. The light will appear to have a shorter wavelength, making it slightly bluer. Because the light is shifted slightly toward the blue end of the spectrum, this is called a **blueshift.** After the light source has passed you and is moving away, the peaks of successive waves seem farther apart, so the light has a longer wavelength and is redder. This is a **redshift.**

The terms *redshift* and *blueshift* are used to refer to any range of wavelengths. The light does not actually have to be red or blue, and the terms apply equally to wavelengths in other parts of the electromagnetic spectrum such as X rays and radio waves. *Red* and *blue* refer to the direction of the shift, not to actual color.

The amount of change in wavelength, and thus the magnitude of the Doppler shift, depends on the velocity of the source. A moving car has a smaller Doppler shift than a jet plane, and a slow moving star has a smaller Doppler shift than one that is moving more quickly. The next section will show how astronomers can convert Doppler shifts into velocities.

Police measure Doppler shifts of passing cars using radar guns, and astronomers measure the Doppler shifts of lines in a star's spectrum. If a star is moving toward Earth, it has a blueshift, and each of its spectral lines is shifted very slightly to shorter wavelengths. If the star is moving away from Earth, it is redshifted, and each of its spectral lines is shifted very slightly toward longer wavelengths. The shifts are much too small to change the color of a star, but they are easily detected in spectra.

When you think about the Doppler effect, it is important to remember two points. Earth itself moves, so a measurement of a Doppler shift really measures the relative motion between Earth and the star. Figure 7-12c shows the Doppler effect in two spectra of the star Arcturus. Lines in the top spectrum are slightly blueshifted because the spectrum was recorded when Earth, in the course of its orbit, was moving toward Arcturus. Lines in the bottom spectrum are redshifted because it was recorded six months later, when Earth was moving away from Arcturus.

The second point to remember is that the Doppler shift is sensitive only to the part of the velocity directed away from you or toward you. This is the **radial velocity (V_r).** You cannot use the Doppler effect to detect any part of the velocity that is perpendicular to your line of sight. A star moving to the left, for example, would have no blueshift or redshift because its distance from Earth would not be decreasing or increasing. This is why police using radar guns park right next to the highway. They want to measure your full velocity as you drive down the highway, not just part of your velocity. This is shown in ■ Figure 7-13.

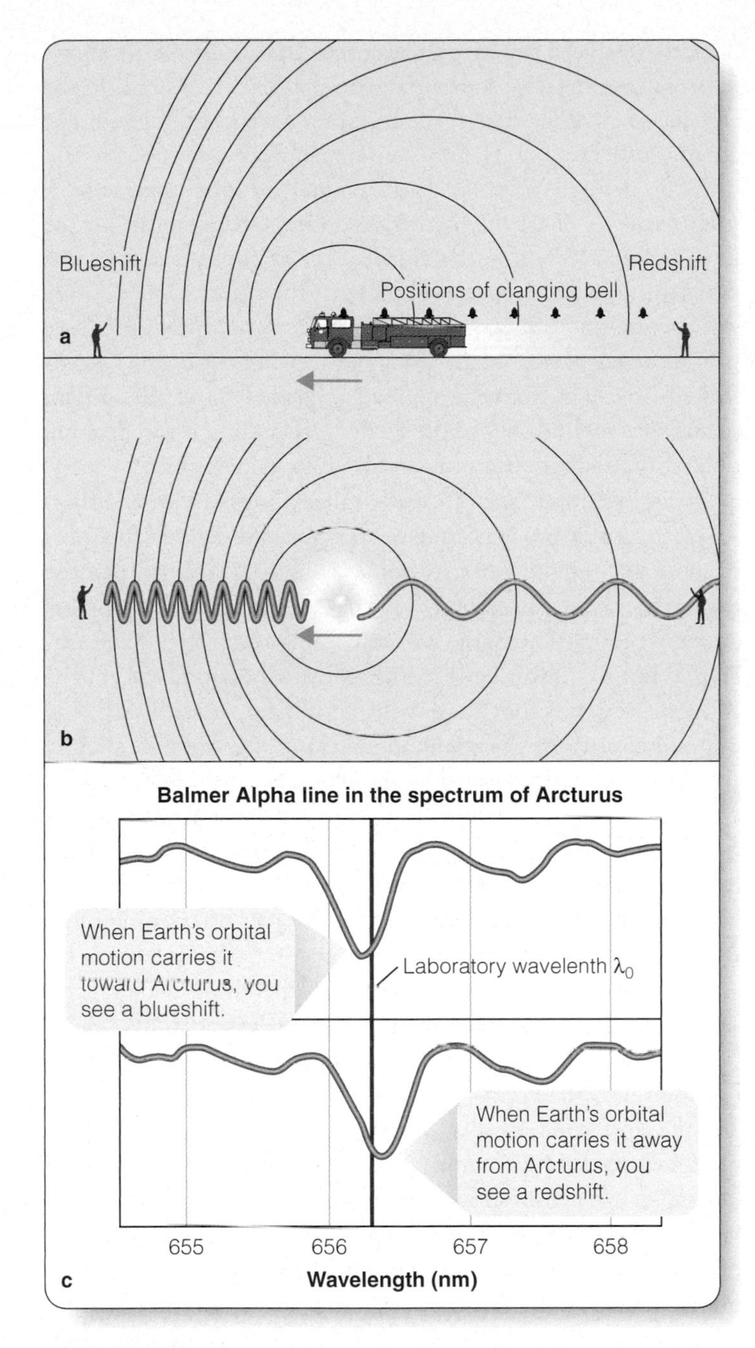

■ Figure 7-12

The Doppler effect. (a) The clanging bell on a moving fire truck produces sounds that move outward (black circles). An observer ahead of the truck hears the clangs closer together, while an observer behind the truck hears them farther apart. (b) A moving source of light emits waves that move outward (black circles). An observer in front of the light source observes a shorter wavelength (a blueshift), and an observer behind the light source observes a longer wavelength (a redshift). (c) Absorption lines in the spectrum of the bright star Arcturus are shifted to the blue in winter, when Earth's orbital motion carries it toward the star, and to the red in summer when Earth moves away from the star.

Calculating the Doppler Velocity

It is easy to calculate the radial velocity of an object from its Doppler shift. The formula is a simple proportion relating the radial velocity V_r divided by the speed of light c to the change in wavelength, λ, divided by the unshifted wavelength, λ_0:

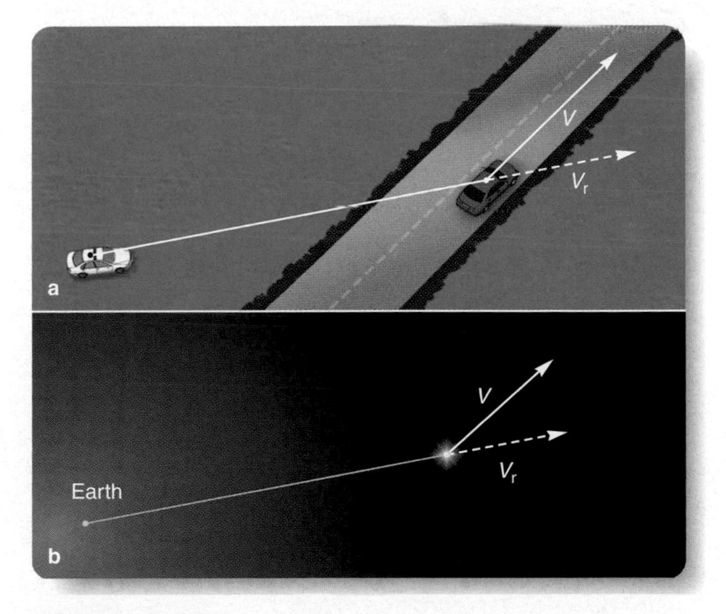

■ Figure 7-13

(a) Police radar can measure only the radial part of your velocity *(V_r)* as you drive down the highway, not your true velocity along the pavement *(V)*. That is why police using radar never park far from the highway. (b) From Earth, astronomers can use the Doppler effect to measure the radial velocity *(V_r)* of a star, but they cannot measure its true velocity, *V*, through space.

$$\frac{V_r}{c} = \frac{\Delta\lambda}{\lambda_0}$$

For example, suppose you observed a line in a star's spectrum with a wavelength of 600.1 nm. Laboratory measurements show that the line should have a wavelength of 600 nm. That is, its unshifted wavelength is 600 nm. What is the star's radial velocity? First note that the change in wavelength is 0.1 nm:

$$\frac{V_r}{c} = \frac{0.1}{600} = 0.000167$$

Multiplying by the speed of light, 3×10^5 km/s, gives the radial velocity, 50 km/s. Because the wavelength is shifted to the red (lengthened), the star must be receding.

Now that you understand the Doppler shift you can understand a final illustration of the information hidden in stellar spectra. Even the shapes of the spectral lines can reveal secrets about the stars.

ThomsonNOW Sign in at www.thomsonedu.com and go to ThomsonNOW to see Astronomy Exercise "Doppler Shift."

The Shapes of Spectral Lines

When astronomers refer to the shape of a spectral line, they mean the variation of intensity across the line. An absorption line, for instance, is darkest in the center and brighter to each side. Two examples are shown in ■ Figure 7-14.

The exact shape of a line can reveal a great deal about a star, but the most important characteristic is the width of the line. Spectral lines are not perfectly narrow; if they were, they would be undetectable. They have a natural width because nature allows an atom some leeway in the energy it may absorb or emit. In the absence of all other effects, spectral lines have a natural width of about 0.001 to 0.00001 nm—very narrow indeed.

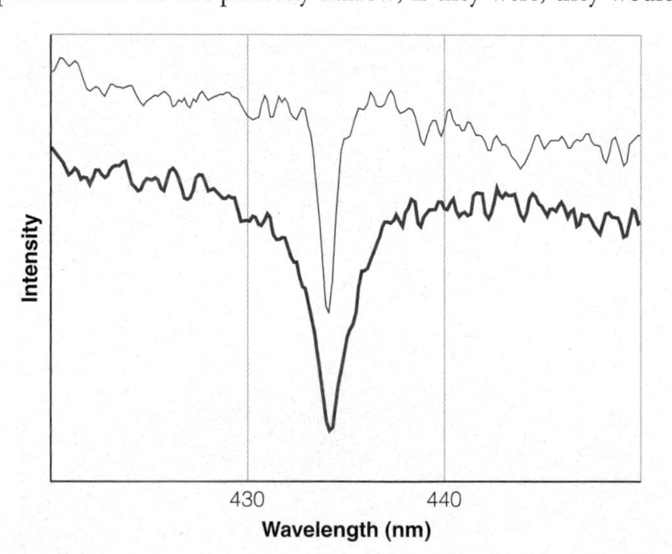

■ Figure 7-14

Here two dark absorption lines are magnified from the spectra of two A1 stars. The upper line is quite narrow, but the bottom line is much broader. Because the two stars have the same spectral type, they must have the same temperature. The stars differ not in temperature but in gas density. The star with the narrow spectral lines has a very low-density atmosphere. Precise observations of the shapes of spectral lines can reveal a great deal about stars. (Courtesy NOAO, G. Jacoby, D. Hunter, and C. Christian)

The natural widths of spectral lines are not important in most branches of astronomy because other effects smear out the lines and make them much broader. For example, if a star spins rapidly, the Doppler effect will broaden the spectral lines. As the star rotates, one side will recede from Earth, and the other side will approach Earth. Light from the receding side will be redshifted, and light from the approaching side will be blueshifted, so any spectral lines will be broadened. Astronomers can measure a star's rotation rate from the width of its spectral lines.

Another important process is called **Doppler broadening.** To consider this process, imagine that you photograph the spectrum of a jar full of hydrogen atoms (■ Figure 7-15). Because the gas has some thermal energy (it is not at absolute zero), the gas atoms are in motion. Some will be coming toward your spectrograph, and some will be receding. Most, of course, will not be traveling very fast, but some will be moving very quickly. The photons emitted by the atoms approaching you will have slightly shorter wavelengths because of the Doppler effect, and photons emitted by atoms receding from you will have slightly longer wavelengths. Thus, the Doppler shifts due to the motions of the individual atoms will smear the spectral line out and make it broader. This summary describes the Doppler broadening of an emission line, but the effect is the same for absorption lines.

The extent of Doppler broadening depends on the temperature of the gas. If the gas is cold, the atoms travel at low velocities, and the Doppler shifts are small (Figure 7-15a). If the gas is hot, however, the atoms travel faster, Doppler shifts are larger, and the lines will be wider (Figure 7-15b). Sometimes astronomers will estimate the temperature of a cloud of gas in space by looking at the widths of its spectral lines.

Another form of broadening, **collisional broadening,** is caused by collisions between atoms, and consequently it depends on the density of the gas. **Density** refers to the amount of matter per unit volume in a body **(Focus on Fundamentals 4).** Densities in astronomy cover an enormous range, from one atom per cubic centimeter in space to millions of tons of atoms per cubic centimeter inside dead stars. Clearly, such densities affect the way atoms collide with one another and how they absorb and emit photons.

Collisional broadening spreads out spectral lines when the atoms absorb or emit photons while they are colliding with other atoms, ions, or electrons. The collisions disturb the energy levels in the atoms, making it possible for the atoms to absorb a slightly wider range of wavelengths. Because of this, the spectral lines are wider. Because atoms in a dense gas collide more often than atoms in a low-density gas, collisional broadening depends on the density of the gas. Temperature is also an important factor. Atoms in a hot gas travel faster and collide more often and more

Density

You are about as dense as an average star. What does that mean? As you study astronomy, you will use the term *density* often, so you should be sure to understand this fundamental concept. Density is a measure of the amount of matter in a given volume. Density is expressed as mass per volume, such as grams per cubic centimeter. The density of water, for example, is about 1 g/cm^3, and you are almost as dense as water.

To get a feel for density, imagine holding a brick in one hand and a similar- sized block of Styrofoam in the other hand. You can easily tell that the brick contains more matter than the Styrofoam block, even though both are the same size. The brick weighs more than the Styrofoam, but it isn't really the weight that you should consider. Rather, you should think about the mass of the two objects. In space, where they have no weight, the brick and the Styrofoam would still have mass, and you could tell just by moving them around that the brick contains more mass than the Styrofoam. For example, imagine tapping each object gently against your ear. The massive brick would be easy to distinguish from the low-mass Styrofoam block, even in weightlessness.

Density is a fundamental idea in science because it is a general property of materials. Metals tend to be dense; lead, for example, has a density of about 7 g/cm^3. Rock, in contrast, has a density of 3 to 4 g/cm^3. Water and ice have densities of about 1 g/cm^3. If you knew that a small moon orbiting Saturn had a density of 1.5 g/cm^3, you could immediately draw some conclusions about what kinds of materials the little moon might be made of—ice and a little rock, but not much metal. The density of an object is a basic clue to its composition.

Astronomical bodies can have dramatically different densities. The gas in a nebula can have a very low density, but the same kind of gas in a star can have a much higher density. The sun, for example, has an average density of about 1 g/cm^3, about the same as your body. As you study astronomical objects, pay special attention to their densities.

A brick would be dense even in space where it had no weight.

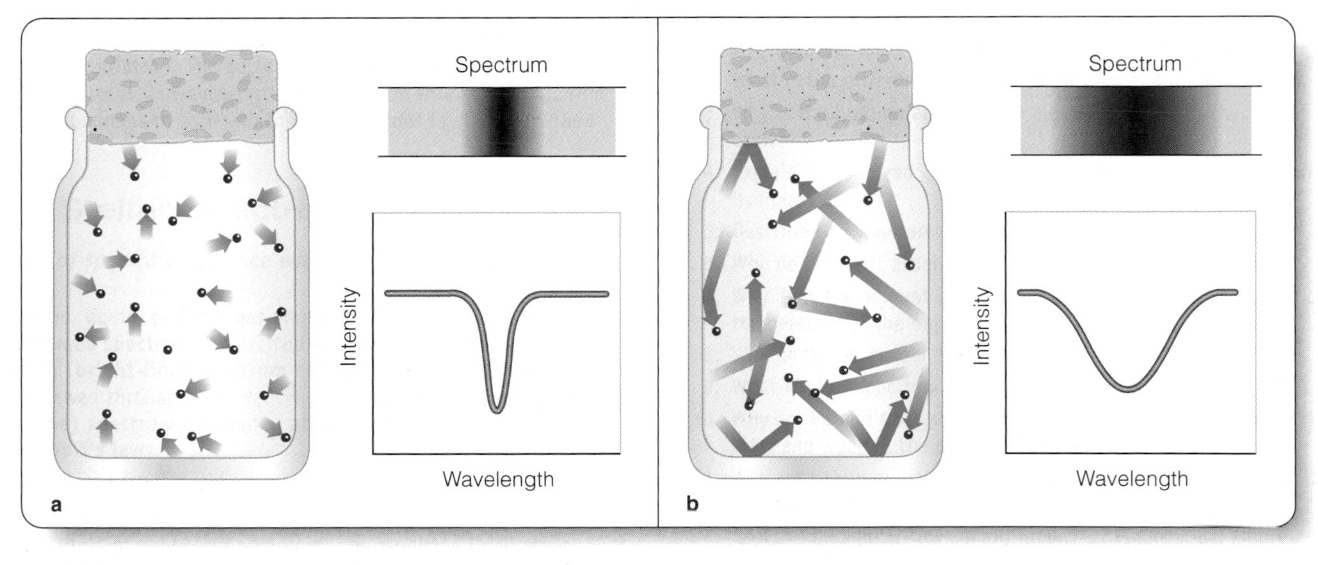

Figure 7-15

Doppler broadening. The atoms of a gas are in constant motion. Photons emitted by atoms moving toward the observer will have slightly shorter wavelengths, and those emitted by atoms moving away will have slightly longer wavelengths. This broadens the spectral line. If the gas is cool (a), the atoms do not move very fast, the Doppler shifts are small, and the line is narrow. If the gas is hot (b), the atoms move faster, the Doppler shifts are larger, and the line is broader.

All cannot live on the piazza, but everyone may enjoy the sun.

ITALIAN PROVERB

A WIT ONCE REMARKED that solar astronomers would know a lot more about the sun if it were farther away, and that contains a grain of truth; the sun is only a humdrum star, and although there are billions like it in the sky, the sun is the only one close enough to show surface detail such as swirling currents of gas and arching bridges of magnetic force. All of that detail makes the sun a challenge to understand.

Life on Earth depends on the sun. Without it, Earth would be a frozen ball of rock and ice. Yet the sun is a surprisingly simple object. It is 109 times Earth's diameter and about 333,000 times more massive. That means it is only slightly denser than water (**Celestial Profile 1**). From its center to its surface, it is a hot gas held together by its own gravity.

8-1 The Solar Atmosphere

THE SUN'S ATMOSPHERE is made up of three layers: the photosphere, the chromosphere, and the corona. (You met these terms when you studied solar eclipses in Chapter 3.) The visible surface is the photosphere, with the transparent gases of the chromosphere lying just above the photosphere. The thin gases of the corona extend far above the chromosphere and are visible only during total solar eclipses (■ Figure 8-1).

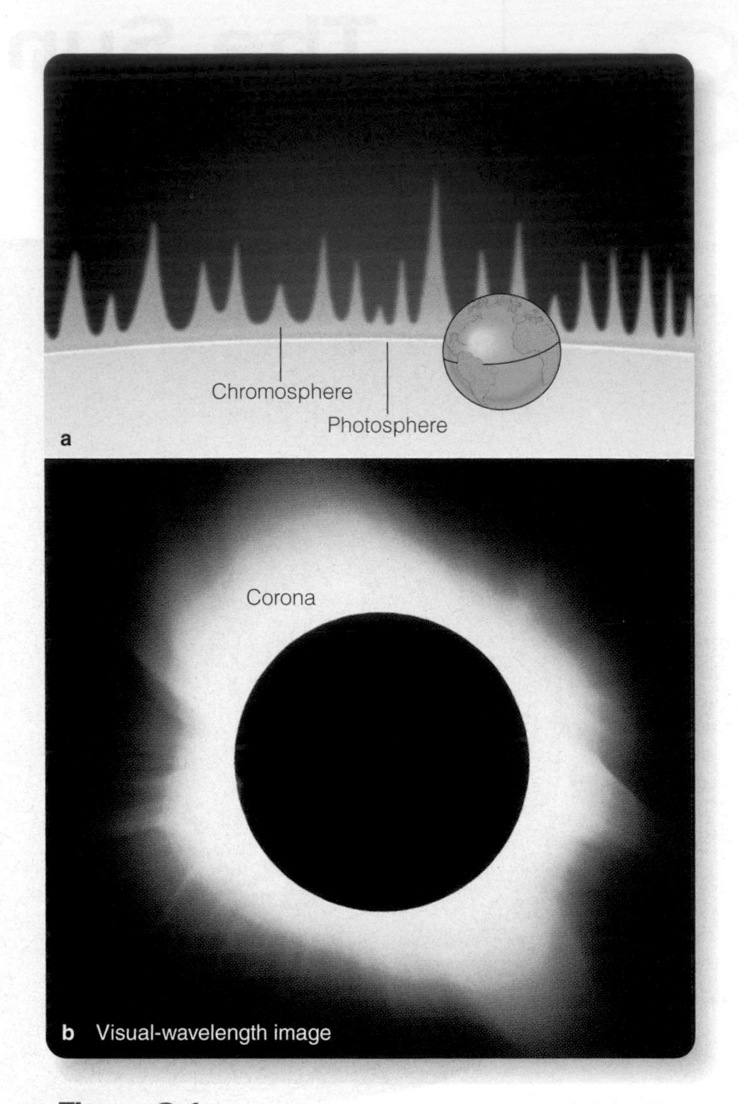

■ **Figure 8-1**

(a) A cross section at the edge of the sun shows the relative thickness of the photosphere and chromosphere. Earth is shown for scale. On this scale, the disk of the sun would be more than 1.5 m (5 ft) in diameter. (b) The corona extends from the top of the chromosphere to great height above the photosphere. (b) This photograph, made during a total solar eclipse, shows only the inner part of the corona. (Daniel Good)

The Photosphere

When the sun is dimmed at sunset and is safe to observe with your unprotected eye, the visible surface looks like a smooth, featureless layer of gas. Dark **sunspots** come and go on the sun's surface, but only very rarely is one large enough to see with the unaided eye at sunset. The photosphere is the thin layer of gas from which Earth receives most of the sun's light.

The photosphere is less than 500 km deep and has an average temperature of about 5800 K. Although the photosphere appears to be substantial, it is really a very-low-density gas. Even in the deepest and densest layers visible, the photosphere is 3400 times less dense than the air you breathe. To find gases as dense as the air you breathe, you would have to descend about 70,000 km below the photosphere, about 10 percent of the way to the sun's center. With fantastically efficient insulation to protect you from the heat, you could fly a spaceship right through the photosphere.

Below the photosphere, the gas is denser and hotter and therefore radiates plenty of light, but that light cannot escape from the sun because of the outer layers of gas. So you cannot detect light from these deeper layers. Above the photosphere, the gas is less dense and so is unable to radiate much light. The photosphere is the layer in the sun's atmosphere that is dense enough to emit plenty of light but not so dense that the light can't escape.

If the sun magically shrank to the size of a bowling ball, the photosphere would be no thicker than a layer of tissue paper wrapped around the ball. One reason the photosphere is so shallow is related to the hydrogen atom. Because the temperature of the photosphere is high enough to ionize some atoms, there are a large number of free electrons in the gas. Neutral hydrogen atoms can add an extra electron and become an H-minus (H^-) ion, but this extra electron is held so loosely that almost any photon has enough energy to free it. In the process, of course, the photon is absorbed. That makes H-minus ions very good absorbers of photons and makes the gas of the photosphere opaque. Light from below cannot escape easily, and you can't see very deeply into the hot fog—the photosphere.

The spectrum of the sun is an absorption spectrum, and that can tell you a great deal about the photosphere. You know from Kirchhoff's third law that an absorption spectrum is produced when a source of a continuous spectrum is viewed through a gas. In the case of the photosphere, the deeper layers are dense enough to produce a continuous spectrum, but atoms in the photosphere absorb photons of specific wavelengths, producing the absorption lines you see.

In good photographs, the photosphere has a mottled appearance because it is made up of dark-edged regions called granules, and the visual pattern they produce is called **granulation** (■ Figure 8-2a). Each granule is about the size of Texas and lasts for only 10 to 20 minutes before fading away. Faded granules are continuously replaced by new granules. Spectra of these granules

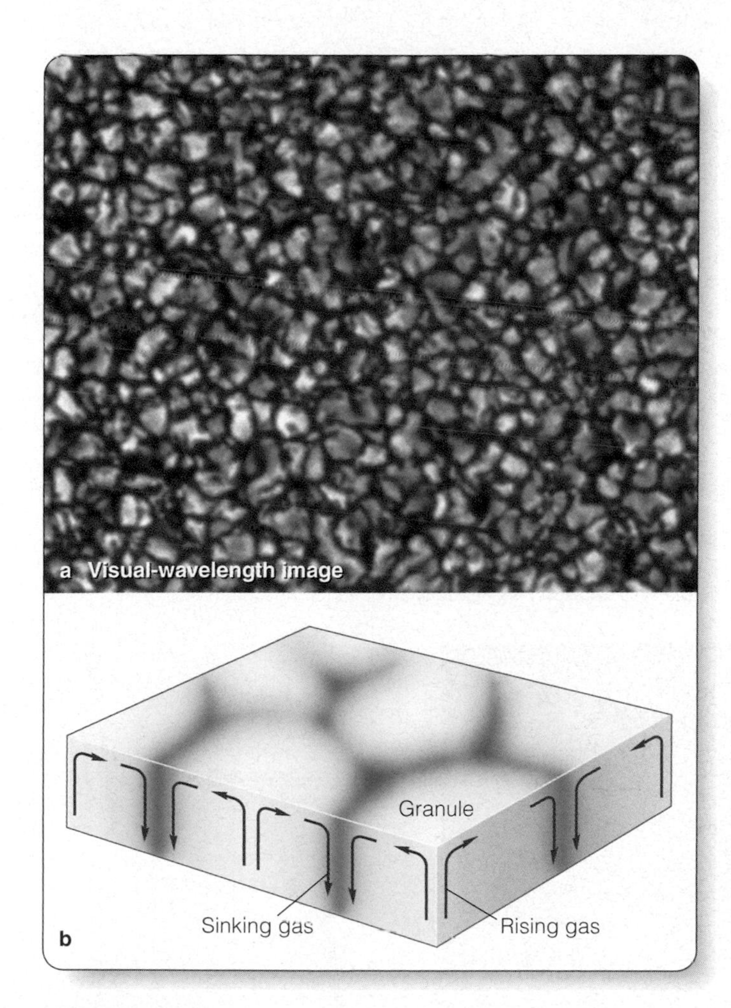

■ Figure 8-2

(a) This ultra-high-resolution image of the photosphere shows granulation. The largest granules here are about the size of Texas. (P. N. Brandt, G. Scharmer, G. W. Simon, Swedish Vacuum Solar Telescope) (b) Granulation is the tops of rising convection currents just below the photosphere. Heat flows upward as rising currents of hot gas and downward as sinking currents of cool gas.

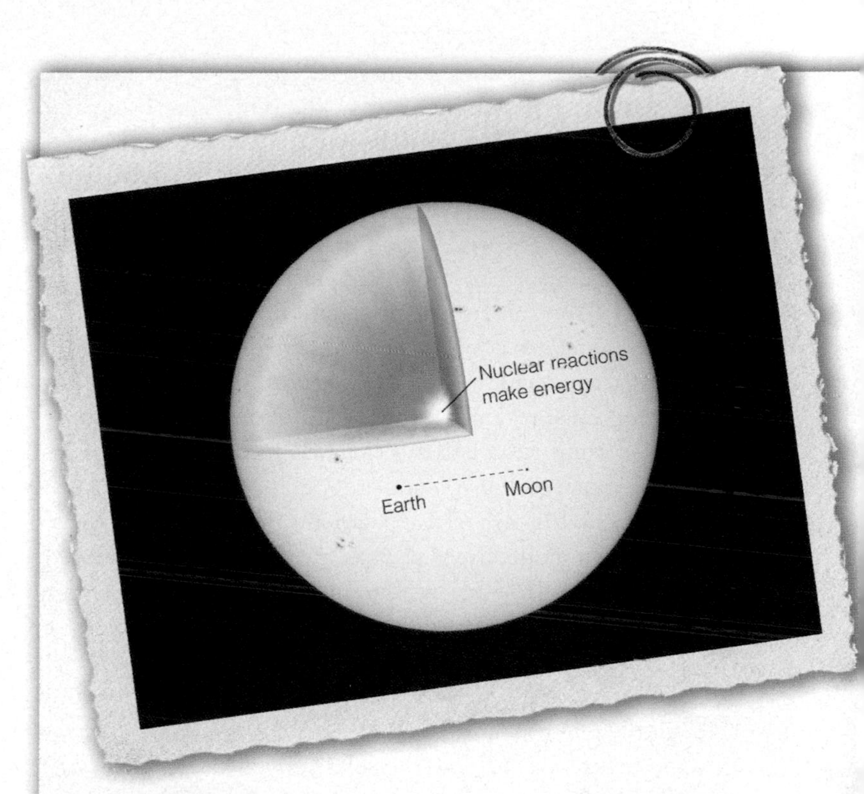

This visual wavelength image of the sun shows a few sunspots and is cut away to show the location of energy generation at the sun's center. The Earth–moon system is shown for scale. (Dan Good)

Celestial Profile 1: The Sun

From Earth:

Average distance from Earth	1.00 AU (1.495979×10^8 km)
Maximum distance from Earth	1.0167 AU (1.5210×10^8 km)
Minimum distance from Earth	0.9833 AU (1.4710×10^8 km)
Average angular diameter	0.53° (32 minutes of arc)
Period of rotation	25.38 days at equator
Apparent visual magnitude	−26.74

Characteristics:

Radius	6.9599×10^5 km
Mass	1.989×10^{30} kg
Average density	1.409 g/cm^3
Escape velocity at surface	617.7 km/s
Luminosity	3.826×10^{26} J/s
Surface temperature	5800 K
Central temperature	15×10^6 K
Spectral type	G2 V
Absolute visual magnitude	4.83

Personality Point:

In Greek mythology, the sun was carried across the sky in a golden chariot pulled by powerful horses and guided by the sun-god, Helios. When Phaeton, son of Helios, drove the chariot one day, he lost control of the horses, and Earth was nearly set ablaze before Zeus smote Phaeton from the sky. Even in classical times, people understood that life on Earth depends critically on the sun.

show that the centers are a few hundred degrees hotter than the edges, and Doppler shifts reveal that the centers are rising and the edges are sinking at speeds of about 0.4 km/s.

From this evidence, astronomers recognize granulation as the surface effects of convection just below the photosphere. **Convection** occurs when hot fluid rises and cool fluid sinks, as when, for example, a convection current of hot gas rises above a candle flame. You can create convection in a liquid by adding a bit of cool nondairy creamer to an unstirred cup of hot coffee. The cool creamer sinks, warms, rises, cools, sinks again, and so on, creating small regions on the surface of the coffee that mark the tops of convection currents. Viewed from above, these regions look much like solar granules.

In the sun, the tops of rising currents of hot gas are brighter than their surroundings. As the gas cools slightly, it is pushed aside by rising gas from below. The cooler gas, sinking at the edge of the granules, is slightly dimmer. Consequently, granules have bright centers and dimmer edges (Figure 8-2b). The presence of granulation is clear evidence that energy is flowing upward through the photosphere.

Spectroscopic studies of the solar surface have revealed another kind of granulation. **Supergranules** are regions about 30,000 km in diameter (about 2.3 times Earth's diameter) and include about 300 granules. These supergranules are regions of very slowly rising currents that last a day or two. They may be the surface traces of larger currents of rising gas deeper under the photosphere.

The edge, or **limb,** of the solar disk is dimmer than the center (see the figure in Celestial Profile 1). This **limb darkening** is caused by the absorption of light in the photosphere. When you look at the center of the solar disk, you are looking directly down into the sun, and you see deep, hot, bright layers in the photosphere. But when you look near the limb of the solar disk, you are looking at a steep angle to the surface and cannot see as deeply. The photons you see come from shallower, cooler, dimmer layers in the photosphere. Limb darkening proves that the temperature in the photosphere increases with depth, yet another confirmation that energy is flowing up from below.

The Chromosphere

Above the photosphere lies the chromosphere. Solar astronomers define the lower edge of the chromosphere as lying just above the visible surface of the sun with its upper regions blending gradually with the corona. You can think of the chromosphere as an irregular layer with an average depth of less than Earth's diameter (see Figure 8-1). Because the chromosphere is roughly 1000 times fainter than the photosphere, you can see it with your unaided eyes only during a total solar eclipse when the moon covers the brilliant photosphere. Then, the chromosphere flashes into view as a thin line of pink just above the photosphere. The word *chromosphere* comes from the Greek word *chroma,* meaning "color." The pink color is produced by the combined light of three bright emission lines—the red, blue, and violet Balmer lines of hydrogen.

Astronomers know a great deal about the chromosphere from its spectrum. The chromosphere produces an emission spectrum, so Kirchhoff's second law tells you the chromosphere must be an excited, low-density gas. Its density is about 10^8 times less than the air you breathe.

Spectra reveal that atoms in the lower chromosphere are ionized, and atoms in the higher layers of the chromosphere are even more highly ionized. That is, they have lost more electrons. From the ionization of the gas, astronomers can find the temperature in different parts of the chromosphere. Just above the photosphere the temperature falls to a minimum of about 4500 K and then rises rapidly (■ Figure 8-3). The region where the temperature increases fastest is called the **transition region** because it makes the transition from the lower temperatures of the photosphere and chromosphere to the extremely high temperatures of the corona. What heats the chromosphere so hot? You will discover an important clue when you study the sun's corona in the next section.

Solar astronomers can take advantage of some elegant physics to study the chromosphere. The gases of the chromosphere are transparent to nearly all visible light, but atoms in the gas are very good at absorbing photons of specific wavelengths. This produces certain dark absorption lines in the spectrum of the

■ Figure 8-3

The chromosphere. If you could place thermometers in the sun's atmosphere, you would discover that the temperature increases from 5800 K at the photosphere to 10^6 K at the top of the chromosphere.

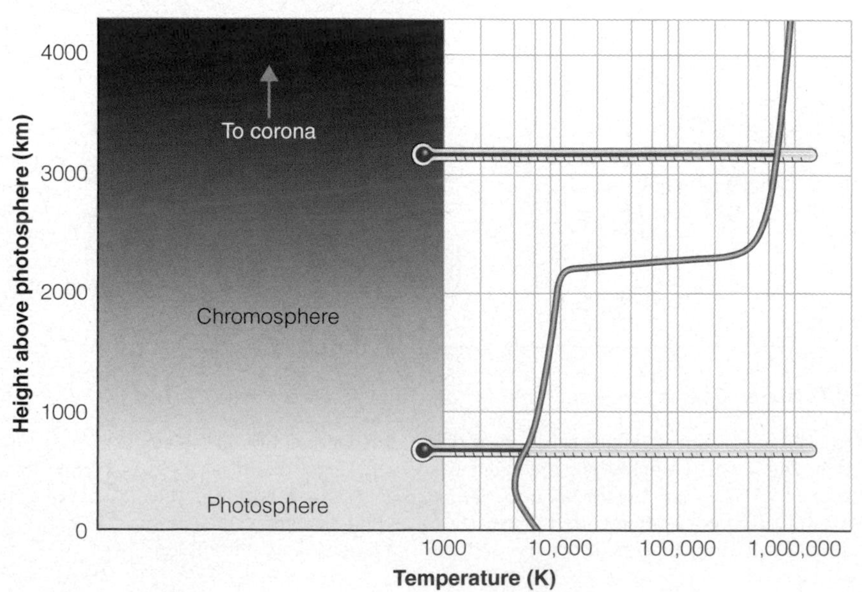

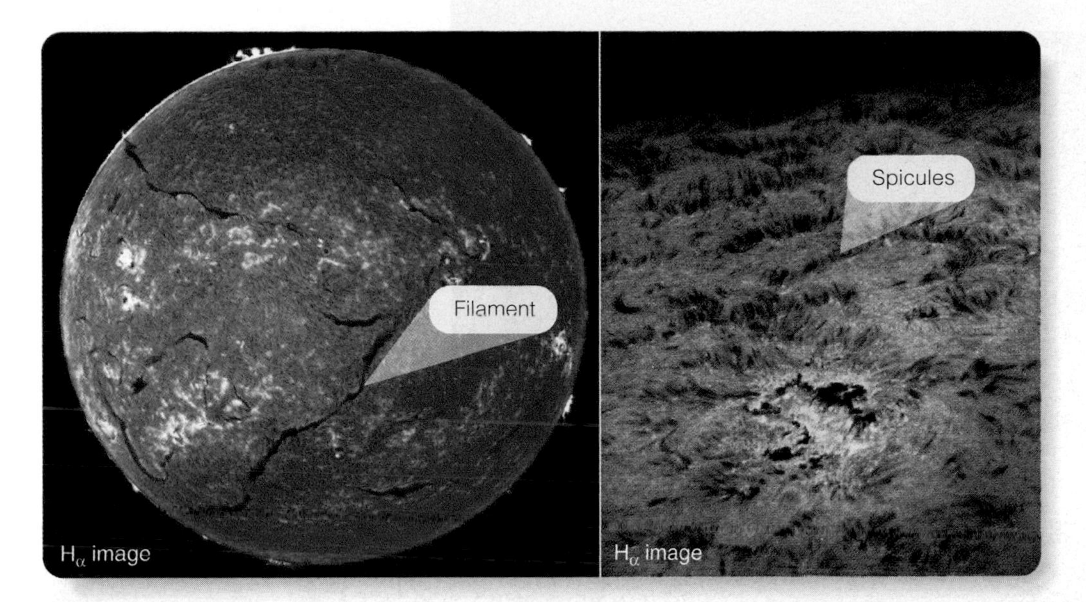

■ **Figure 8-4**

H_α filtergrams reveal complex structure in the chromosphere, including long, dark filaments and spicules springing from the edges of supergranules twice the diameter of Earth. (NOAA/SEL/USAF; © 1971 NOAO/NSO)

photosphere. A photon having one of those wavelengths that is emitted in a deeper layer is very unlikely to escape from the chromosphere without being absorbed. If a photon at one of these easily absorbed wavelengths reaches Earth, you can be sure it came from higher in the sun's atmosphere. A **filtergram** is a photograph made using light in one of those dark absorption lines. In this way filtergrams reveal detail in the upper layers of the chromosphere. In a similar way, an image recorded in the far-ultraviolet or in the X-ray part of the spectrum reveals other structures in the solar atmosphere.

■ Figure 8-4 shows a filtergram made at the wavelength of the H_α Balmer line. This image shows complex structure in the chromosphere including long, dark **filaments** silhouetted against the brighter surface. **Spicules** are flamelike jets of gas extending upward into the chromosphere and lasting 5 to 15 minutes. Seen at the limb of the sun's disk, these spicules blend together and look like flames covering a burning prairie (Figure 8-1a), but they are not flames at all. Spectra show that spicules are cooler gas from the lower chromosphere extending upward into hotter regions. Images of the chromosphere at the center of the solar disk show that spicules spring up around the edge of supergranules like weeds around flagstones (Figure 8-4b). Although spicules are not yet well understood, they are clearly driven by the outward flow of energy in the sun.

Spectroscopic analysis of the chromosphere alerts you that it is a low-density gas in constant motion where the temperature increases rapidly with height. Just above the chromosphere lies even hotter gas.

The Solar Corona

The outermost part of the sun's atmosphere is called the corona, after the Greek word for crown. The corona is so dim that it is not visible in Earth's daytime sky because of the glare of scattered light from the brilliant photosphere. During a total solar eclipse, however, when the moon covers the photosphere, you can see the innermost parts of the corona, as shown in Figure 8-1b. Observations made with specialized telescopes called **coronagraphs** on Earth or in space can block the light of the photosphere and image the corona out beyond 20 solar radii, almost 10 percent of the way to Earth. Such images show streamers in the corona that appear to follow lines of magnetic force (■ Figure 8-5).

The spectrum of the corona tells you a great deal about the coronal gases and simultaneously illustrates how astronomers analyze a spectrum. Some of the light from the outer corona produces a spectrum with absorption lines the same as the photosphere's spectrum. This light is just sunlight reflected from dust particles in the corona. In contrast, some of the light from the corona produces a continuous spectrum that lacks absorption lines. That happens when sunlight from the photosphere is scattered off free electrons in the ionized coronal gas. Because the coronal gas has a temperature over 1 million K, the electrons travel very fast, and the reflected photons suffer large, random Doppler shifts that smear out solar absorption lines to produce a continuous spectrum.

Superimposed on the corona's continuous spectrum are emission lines of highly ionized gases. In the lower corona, the atoms are not as highly ionized as they are at higher altitudes, and this tells you that the temperature of the corona rises with altitude. Just above the chromosphere, the temperature is about 500,000 K; but in the outer corona the temperature can be as high as 2 million K or more.

The corona is made up of exceedingly hot gas, but it is not very bright. Its density is very low, only 10^6 atoms/cm^3 in its lower regions. That is about a trillion times less dense than the air you breath. In its outer layers the corona contains only 1 to 10 atoms/cm^3, better than the best vacuum on Earth. Because of this low density, the hot gas does not emit much radiation.

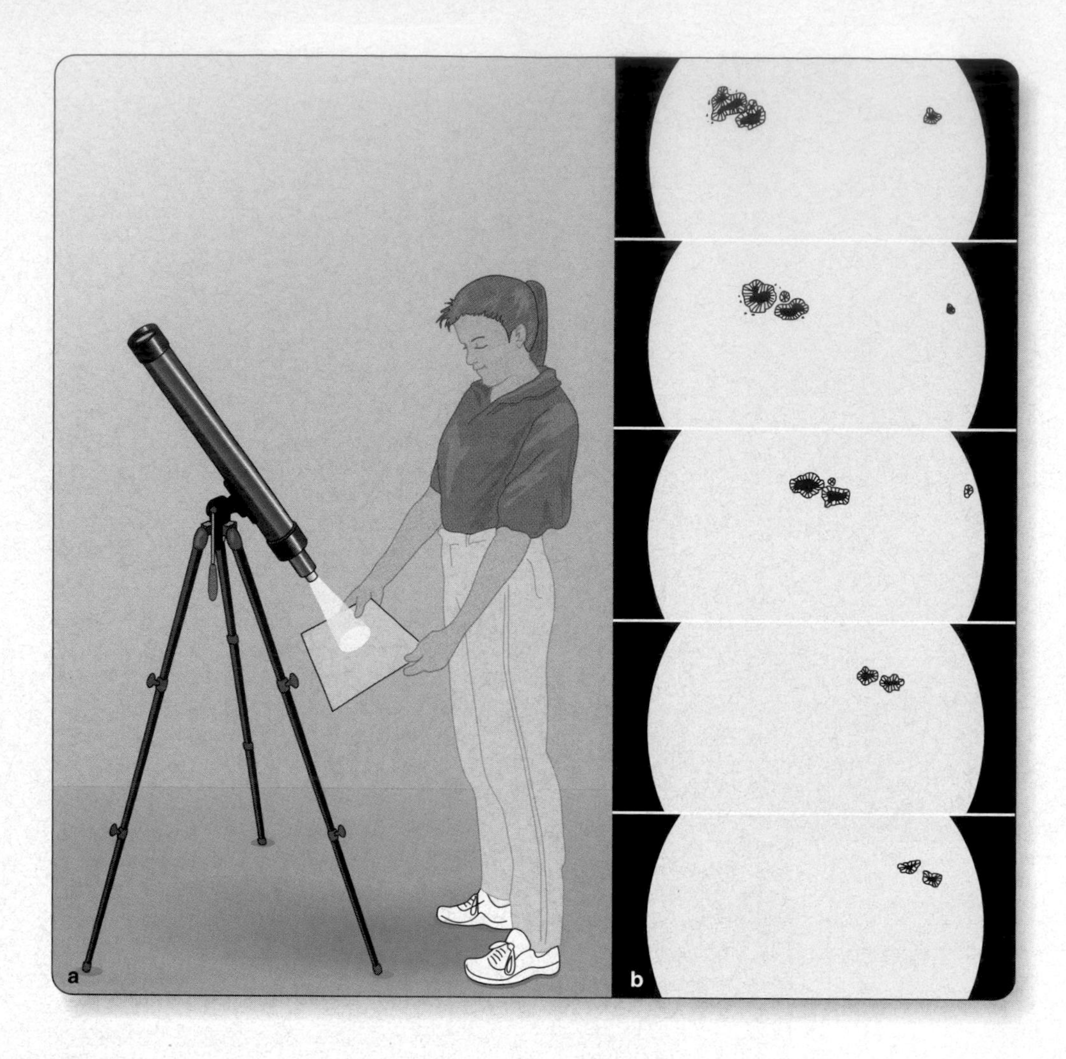

■ **Figure 8-12**

(a) Looking through a telescope at the sun is dangerous, but you can always view the sun safely with a small telescope by projecting its image on a white screen. (b) If you sketch the location and structure of sunspots on successive days, you will see the rotation of the sun and gradual changes in the size and structure of sunspots just as Galileo did in 1610.

The sunspot groups are merely the visible traces of magnetically active regions. But what causes this magnetic activity? The answer appears to be linked to the waxing and waning of the sun's magnetic field.

ThomsonNOW Sign in at www.thomsonedu.com and go to ThomsonNOW to see Astronomy Exercises "Zeeman Effect," "Sunspot Cycle I," and "Sunspot Cycle II."

The Sun's Magnetic Cycle

Sunspots are magnetic phenomena, so the 11-year cycle of sunspots must be caused by cyclical changes in the sun's magnetic field. To explore that idea, begin with the sun's rotation.

The sun does not rotate as a rigid body. It is a gas from its outermost layers down to its center, so some parts of the sun rotate faster than other parts. From the study of sunspots, astronomers can tell that the equatorial region of the photosphere rotates faster than do regions further from the equator (■ Figure 8-13a). At the equator, the photosphere rotates once every 25 days, but at latitude 45° one rotation takes 27.8 days. Furthermore, helioseismology can map the rotation rates throughout the interior (Figure 8-13b). Because different parts of the sun rotate at different rates, its motion is called **differential rotation,** and that motion is clearly linked to the magnetic cycle.

The sun's magnetic field appears to be powered by the energy flowing outward through the moving currents of gas. The gas is highly ionized, so it is a very good conductor of electricity. When an electrical conductor rotates rapidly and is stirred by convection, it can convert some of the energy flowing outward as convection into a magnetic field. This process is called the **dynamo effect,** and it is believed to produce Earth's magnetic field as well. Helioseismologists have found evidence that the sun's magnetic field is generated at the bottom of the convection zone deep under the photosphere. The details of this process are still poorly understood, but the sun's magnetic cycle is clearly related to the creation of its magnetic field.

Sunspots provide an insight into how the magnetic cycle works. Sunspots tend to occur in groups or pairs, and the magnetic field around the pair resembles the magnetic field around a bar magnet in that one end is magnetic north and the other end is magnetic south. At any one time, sunspot pairs south of the sun's equator have reversed polarity compared to those north of the sun's equator. ■ Figure 8-14 illustrates this by showing sunspot pairs south of the sun's equator with magnetic south poles leading and sunspots north of the sun's equator with magnetic north poles leading. At the end of an 11-year sunspot cycle, the new spots appear with reversed magnetic polarity.

This magnetic cycle is not fully understood, but the **Babcock model** (named for its inventor) explains the magnetic cycle as a progressive tangling of the solar magnetic field. Because the electrons in an ionized gas are free to move, the gas is a very good conductor of electricity, and any magnetic field in the gas is "frozen" into the gas. If the gas moves, the magnetic field must move with it. Differential rotation wraps the sun's magnetic field around the sun like a long string caught on a hubcap. Rising and sinking gas currents twist the field into ropelike tubes, which

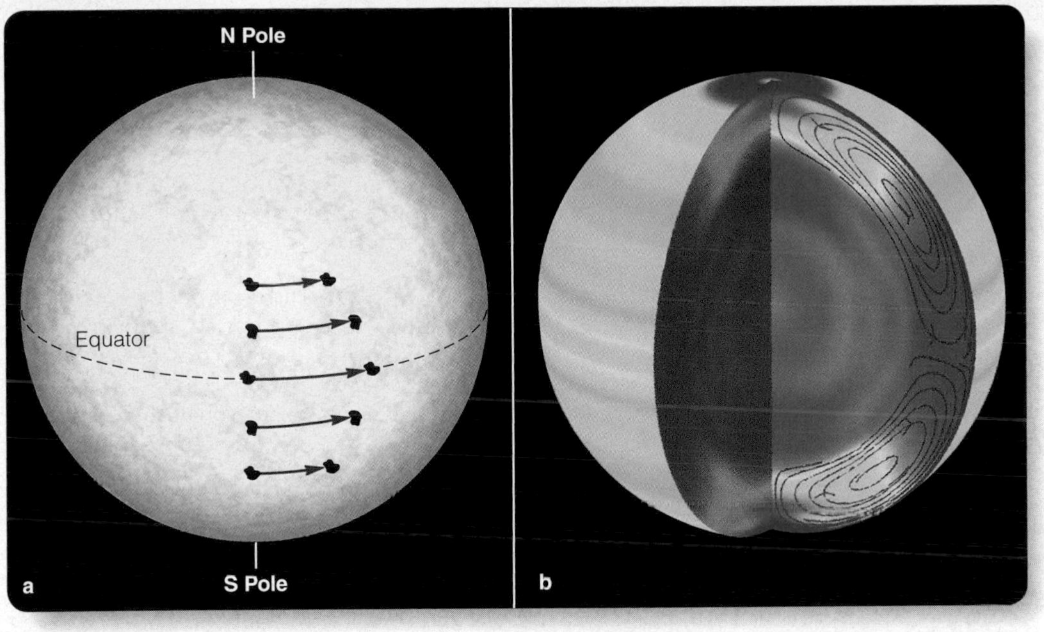

■ **Figure 8-13**

(a) In general, the photosphere of the sun rotates faster at the equator than at higher latitudes. If you started five sunspots in a row, they would not stay lined up as the sun rotates. (b) Detailed analysis of the sun's rotation from helioseismology reveals regions of slow rotation (blue) and rapid rotation (red). Such studies show that the interior of the sun rotates differentially and that currents similar to the trade winds in Earth's atmosphere flow through the sun. (NASA/SOI)

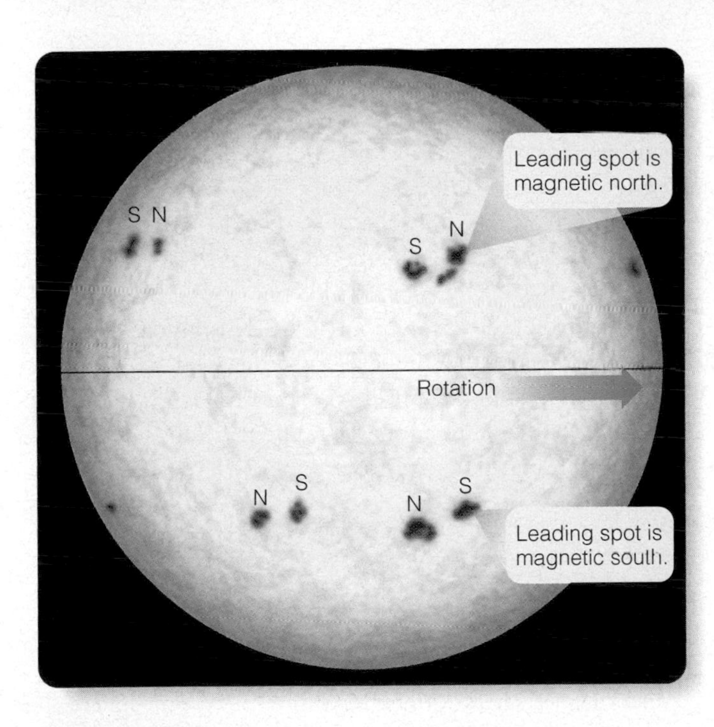

■ **Figure 8-14**

In sunspot groups, here simplified into pairs of major spots, the leading spot and the trailing spot have opposite magnetic polarity. Spot pairs in the southern hemisphere have reversed polarity from those in the northern hemisphere.

tend to float upward and burst through the sun's surface in great arches like magnetic rainbows. Sunspots occur at the two bases of an arch where the magnetic field emerges from below and then plunges back into the sun (■ Figure 8-15). Because the magnetic field points in opposite directions in the two spots, they have opposite magnetic polarity as in Figure 8-14.

The Babcock model explains the reversal of the sun's magnetic field from cycle to cycle. As the magnetic field becomes tangled, adjacent regions of the sun are dominated by magnetic fields that point in different directions. After about 11 years of tangling, the field becomes so complex that adjacent regions of the sun begin changing their magnetic field to agree with neighboring regions. The entire field quickly rearranges itself into a simpler pattern, and differential rotation begins winding it up to start a new cycle. But the newly organized field is reversed, and the next sunspot cycle begins with magnetic north replaced by magnetic south. Evidently the complete magnetic cycle is 22 years long, and the sunspot cycle is 11 years long.

This magnetic cycle may explain the Maunder butterfly diagram. As a sunspot cycle begins, the twisted tubes of magnetic force first begin to float upward and produce sunspot pairs farther north and south of the equator. Consequently the first sunspots in a cycle appear farther from the equator. Later in the cycle, when the field is more tightly wound, the tubes of magnetic force arch up through the surface closer to the sun's equator. As a result, the later sunspot pairs in a cycle appear closer to the equator.

Notice the power of a scientific model. The Babcock model may in fact be incorrect in some details, but it provides a framework on which to organize all of the complex solar activity. Even though the models of the sky in Chapter 2 and the atom in Chapter 7 were only partially correct, they served as organizing themes to guide your thinking. Similarly, although the precise details of the solar magnetic cycle are not yet understood, the Babcock model gives you a general picture of the behavior of the sun's magnetic field **(How Do We Know? 8-2)**.

ThomsonNOW Sign in at www.thomsonedu.com and go to ThomsonNOW to see Astronomy Exercise "Convection and Magnetic Fields."

How Do We Know? | 8-2

Confirmation and Consolidation

What do scientists do all day? The scientific method is sometimes portrayed as a kind of assembly line where scientists crank out new hypotheses and then test them through observation. In reality, scientists don't often generate entirely new hypotheses. It is rare that an astronomer makes an observation that disproves a long-held theory and triggers a revolution in science. Then what is the daily grind of science really about?

Many observations and experiments merely confirm already tested hypotheses. The biologist knows that all worker bees in a hive are sisters. All of the workers are female, and they all had the same mother, the queen bee. A biologist can study the DNA from many workers and confirm that hypothesis. By repeatedly confirming a hypothesis, scientists build confidence in the hypothesis and may be able to extend it. Do all of the workers in a hive have the same father, or did the queen mate with more than one male drone?

Another aspect of routine science is consolidation, the linking of a hypothesis to other well-studied phenomena. A biologist can study wasps from a single nest and discover that the wasps are not sisters. There is no queen wasp. Each female wasp lays her own eggs, and the wasps share the nest for convenience and protection. From her study of wasps, the biologist consolidates what she knows about bees and could conclude that bees and wasps have evolved in different ways.

The Babcock model of the solar magnetic cycle is an astronomical example of the scientific process. Solar astronomers know that the model explains some solar features but has shortcomings. Although most astronomers don't expect to discard the entire model, they work through confirmation and consolidation to better understand how the solar magnetic cycle works and how it is related to cycles in other stars.

Spots and Magnetic Cycles on Other Stars

The sun seems to be a representative star, so you should expect other stars to have similar cycles of starspots and magnetic fields. This is a difficult topic, because, except for the sun, the stars are so far away that no surface detail is visible. Some stars, however, vary in brightness in ways that suggest they are mottled by dark spots. As these stars rotate, their total brightnesses change slightly, depending on the number of spots on the side facing Earth. High-precision spectroscopic analysis has even allowed astronomers to map the locations of spots on the surfaces of certain stars (■ Figure 8-16a). Such results confirm that the sunspots you see on our sun are not unusual.

Certain features in stellar spectra are associated with magnetic fields. Regions of strong magnetic fields on the solar surface emit strongly at the central wavelengths of the two strongest lines of ionized calcium. This calcium emission appears in the spectra of other sunlike stars and suggests that these stars, too, have strong magnetic fields on their surfaces. In some cases, the strength of this calcium emission varies over periods of days or weeks and suggests that the stars have active regions and are rotating with periods similar to that of the sun. These stars presumably have starspots as well.

In 1966, astronomers began a long-term project that monitored the strength of this calcium emission in the spectra of certain stars similar to the sun. With temperatures ranging from 1000 K hotter than the sun to 3000 K cooler, these stars were considered most likely to have sunlike magnetic activity on their surfaces.

The observations show that the strength of the calcium emission varies over periods of years. The calcium emission averaged over the sun's disk varies with the sunspot cycle, and similar periodic variations can be seen in the spectra of some of the stars studied (Figure 8-16b). The star 107 Piscium, for instance, appears to have a starspot cycle lasting nine years. This kind of evidence suggests that stars like the sun have similar magnetic cycles.

These observations confirm that the sun is a typical star. Most other stars like our sun have magnetic fields and starspots and go through magnetic cycles.

ThomsonNOW™ Sign in at www.thomsonedu.com and go to ThomsonNOW to see Astronomy Exercise "Convection and Magnetic Fields."

Chromospheric and Coronal Activity

The solar magnetic fields extend high into the chromosphere and corona, where they produce beautiful and powerful phenomena. Read **Magnetic Solar Phenomena** on pages 168–169 and notice three important points and six new terms:

1. Solar activity is magnetic. The arched shapes of *prominences* are produced by magnetic fields. The filaments shown in Figure 8-4 are prominences seen from above.

2. Tremendous energy can be stored in arches of magnetic field, and when two arches encounter each other a *reconnection* can release powerful eruptions called *flares*. Although these eruptions occur far from Earth, they can affect us in dramatic ways, and *coronal mass ejections (CMEs)* can trigger communications blackouts and *auroras*.

The Solar Magnetic Cycle

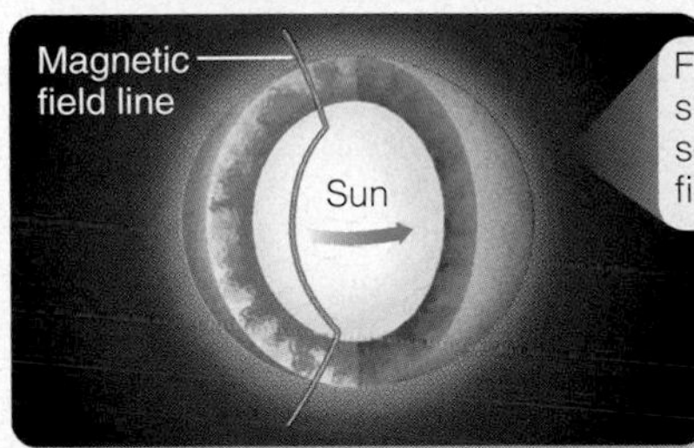

Magnetic field line
Sun
For simplicity, a single line of the solar magnetic field is shown.

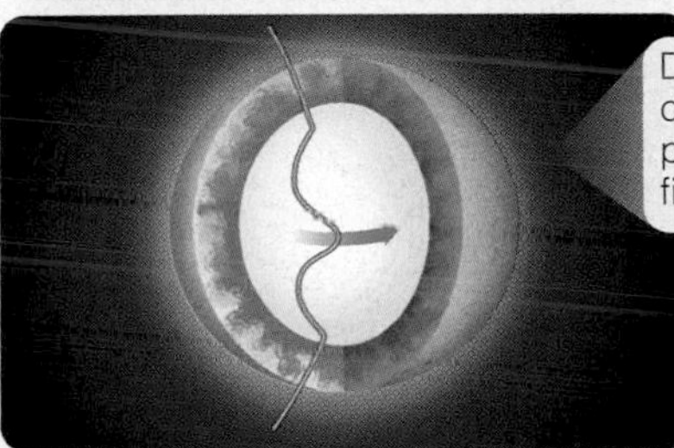

Differential rotation drags the equatorial part of the magnetic field ahead.

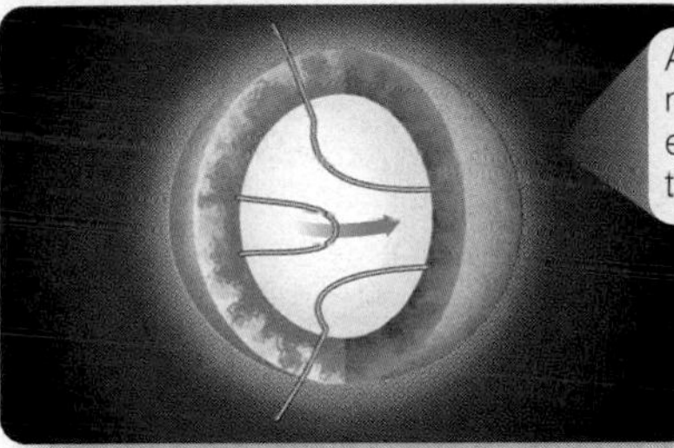

As the sun rotates, the magnetic field is eventually dragged all the way around.

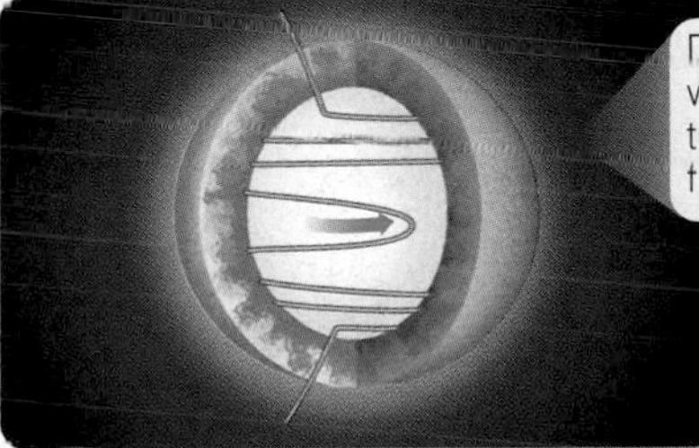

Differential rotation wraps the sun in many turns of its magnetic field.

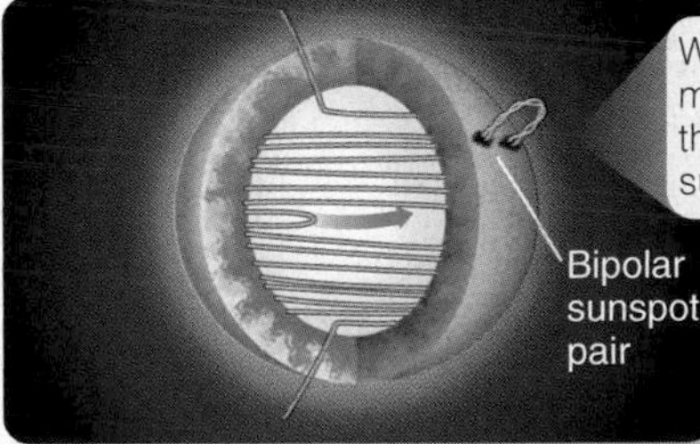

Where loops of tangled magnetic field rise through the surface, sunspots occur.
Bipolar sunspot pair

■ Figure 8-15

The Babcock model of the solar magnetic cycle explains the sunspot cycle as a consequence of the sun's differential rotation gradually winding up the magnetic field.

3 In some regions of the solar surface, the magnetic field does not loop back. High energy gas from these *coronal holes* flows outward and produces much of the solar wind.

You may have heard a **Common Misconception** that auroras are caused by sunlight reflected from ice at Earth's poles. It is fun to think about polar bears and icebergs, but the truth is more exciting. Auroras on Earth are caused by energy from the sun, so auroras are part of the sun's magnetic weather. Earth's weather is not magnetic because Earth's magnetic field is weak, and Earth's atmosphere is not ionized and so is free to move independent of the magnetic field. On the sun, however, the weather is a magnetic phenomenon of great power.

ThomsonNOW Sign in at www.thomsonedu.com and go to ThomsonNOW to see Astronomy Exercise "Auroras."

The Solar Constant

Even a small change in the sun's energy output could produce dramatic changes in Earth's climate. The continued existence of the human species depends on the constancy of the sun, but we humans know very little about the variation of the sun's energy output.

The energy production of the sun can be measured by adding up all of the energy falling on 1 square meter of Earth's surface during 1 second. Of course, some correction for the absorption of Earth's atmosphere is necessary, and you must count all wavelengths from X rays to radio waves. The result, which is called the **solar constant,** amounts to about 1360 joules per square meter per second. A change in the solar constant of only 1 percent could change Earth's average temperature by 1 to 2°C (about 1.8 to 3.6°F). For comparison, during the last ice age Earth's average temperature was about 5°C cooler than it is now.

Some of the best measurements of the solar constant were made by instruments aboard the Solar Maximum Mission satellite. These have shown variations in the energy received from the sun of about 0.1 percent that lasted for days or weeks. Superimposed on that random variation is a long-term decrease of about 0.018 percent per year that has been confirmed by observations made by sounding rockets, balloons, and satellites. This long-term decrease may be related to a cycle of activity on the sun with a period longer than the 22-year magnetic cycle.

Small, random fluctuations will not affect Earth's climate, but a long-term decrease over a decade or more could cause worldwide cooling. History contains some evidence that the solar constant may have varied in the past. As you saw on page 163, the "Little Ice Age" was a period of unusually cool weather in Europe and America that lasted from about 1500 to 1850.* The average temperature worldwide was about 1°C cooler than it is now. This period of cool weather corresponded to the Maunder minimum, a period of reduced solar activity—few sunspots, no auroral displays, and no solar coronas visible during solar eclipses.

In contrast, an earlier period called the Grand Maximum, lasting from about AD 1100 to about 1250, saw a warming

*Ironically, the Maunder minimum coincides with the reign of Louis XIV of France, the "Sun King."

Magnetic Solar Phenomena

1 Magnetic phenomena in the chromosphere and corona, like magnetic weather, result as constantly changing magnetic fields on the sun trap ionized gas to produce beautiful arches and powerful outbursts. Some of this solar activity can affect Earth's magnetic field and atmosphere.

This ultraviolet image of the solar surface was made by the NASA TRACE spacecraft. It shows hot gas trapped in magnetic arches extending above active regions. At visual wavelengths, you would see sunspot groups in these active regions.

The gas in prominences may be 60,000 to 80,000 K, quite cold compared with the low-density gas in the corona, which may be as hot as a million Kelvin.

Trace/NASA

1a A **prominence** is composed of ionized gas trapped in a magnetic arch rising up through the photosphere and chromosphere into the lower corona. Seen during total solar eclipses at the edge of the solar disk, prominences look pink because of the three Balmer emission lines. The image below shows the arch shape suggestive of magnetic fields. Seen from above against the sun's bright surface, prominences form dark filaments.

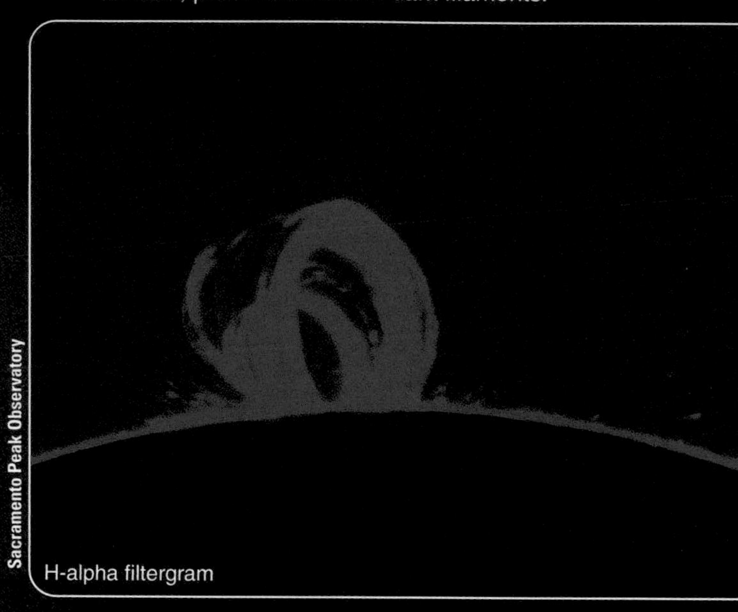
H-alpha filtergram

Sacramento Peak Observatory

1b Quiescent prominences may hang in the lower corona for many days, whereas eruptive prominences burst upward in hours. The eruptive prominence below is many Earth diameters long.

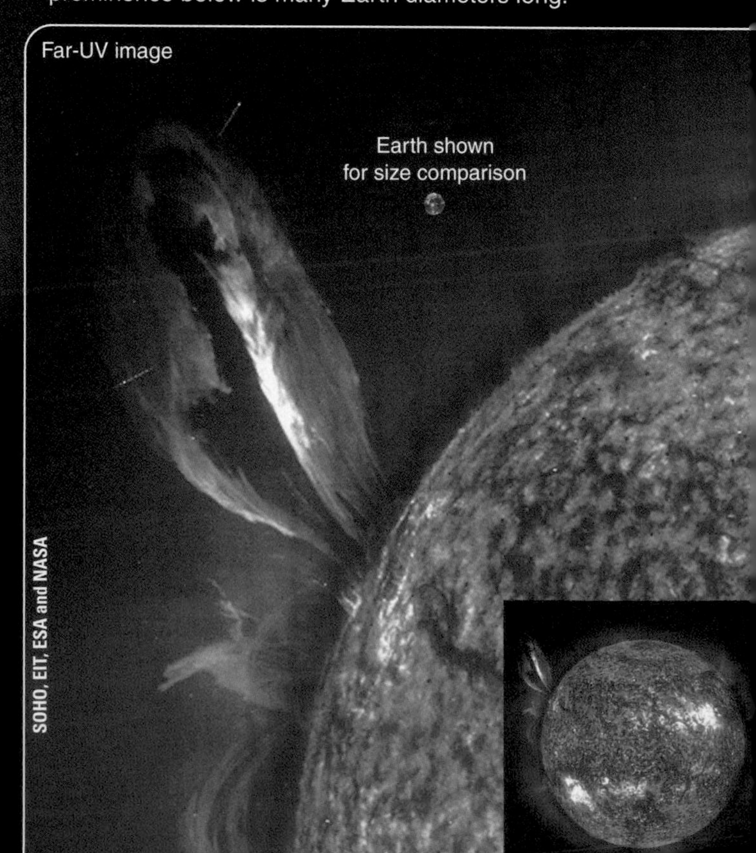

Far-UV image

SOHO, EIT, ESA and NASA

2 Solar **flares** rise to maximum in minutes and decay in an hour. They occur in active regions where oppositely directed magnetic fields meet and cancel each other out in what astronomers call **reconnections.** Energy stored in the magnetic fields is released as short-wavelength photons and as high-energy protons and electrons. X-ray and ultraviolet photons reach Earth in 8 minutes and increase ionization in our atmosphere, which can interfere with radio communications. Particles from flares reach Earth hours or days later as gusts in the solar wind, which can distort Earth's magnetic field and disrupt navigation systems. Solar flares can also cause surges in electrical power lines and damage to Earth satellites.

An ultraviolet image shows an active region experiencing a flare.

Far-UV image

NASA

2a At right, waves rush outward at 50 km/sec from the site of a solar flare 40,000 times stronger than the 1906 San Francisco earthquake. The biggest solar flares can be a billion times more powerful than a hydrogen bomb.

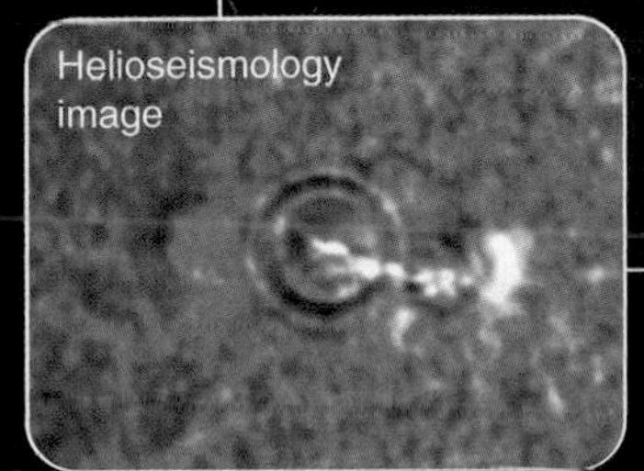

SOHO/MDI, ESA, and NASA

2b The solar wind, enhanced by eruptions on the sun, interacts with Earth's magnetic field and can create electrical currents up to a million megawatts. Those currents flowing down into a ring around Earth's magnetic poles excite atoms in Earth's upper atmosphere to emit photos as shown below. Seen from Earth's surface, the gas produces glowing clouds and curtains of **aurora.**

Auroras occur about 130 km above the Earth's surface.

North magnetic pole

Coronal mass ejection

Ring of aurora

2c Magnetic reconnections can release enough energy to blow large amounts of ionized gas outward from the corona in **coronal mass ejections (CMEs).** If a CME strikes Earth, it can produce especially violent disturbances in Earth's magnetic field.

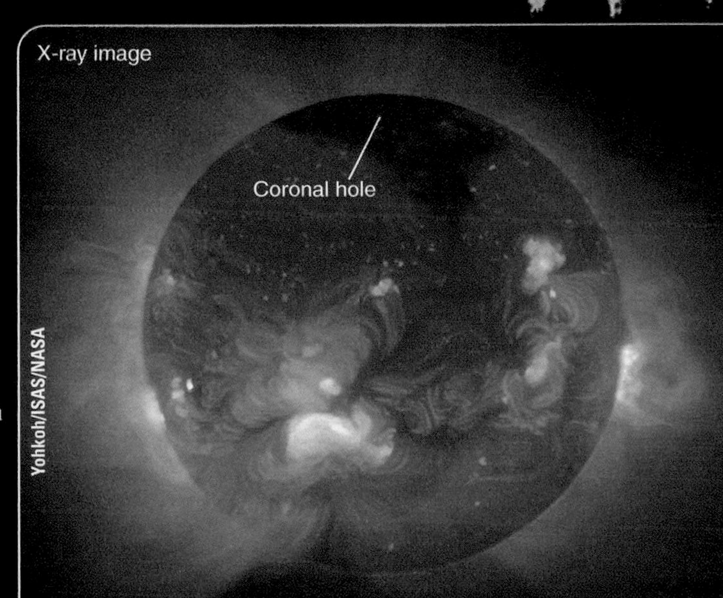

Yohkoh/ISAS/NASA

3 Much of the solar wind comes from **coronal holes,** where the magnetic field does not loop back into the sun. These open magnetic fields allow ionized gas in the corona to flow away as the solar wind. The dark area in this X-ray image at right is a coronal hole.

surements. No matter how it is done, the important point is that if you measure the baseline and the two angles, you can figure out the distance across the river.

The more distant an object is, the longer the baseline you must use to measure the distance to the object accurately. You could use a baseline 50 m long to find the distance across a river, but to measure the distance to a mountain on the horizon, you might need a baseline 1000 m long. Great distances require very long baselines.

The Astronomer's Method

To find the distance to a star, you must use an extremely long baseline; the diameter of Earth's orbit suffices for the nearest stars. If you take a photograph of a nearby star and then wait six months, Earth will have moved halfway around its orbit. You can then take another photograph of the star. This second photograph is taken on the other side of Earth's orbit, 2 AU (astronomical units) from the point where the first photograph was taken. So your baseline equals the diameter of Earth's orbit, or 2 AU.

You now have two photographs of the same part of the sky taken from slightly different locations in space. When you examine the photographs, you will discover that the nearby star is not in exactly the same place in the two photographs. This apparent shift in the position of the star is called *parallax* (■ Figure 9-3).

■ Figure 9-3

You can measure the parallax of a nearby star by photographing it from two points along Earth's orbit. For example, you might photograph it now and again 6 months from now. Half of the star's total change in position from one photograph to the other is its stellar parallax, *p*.

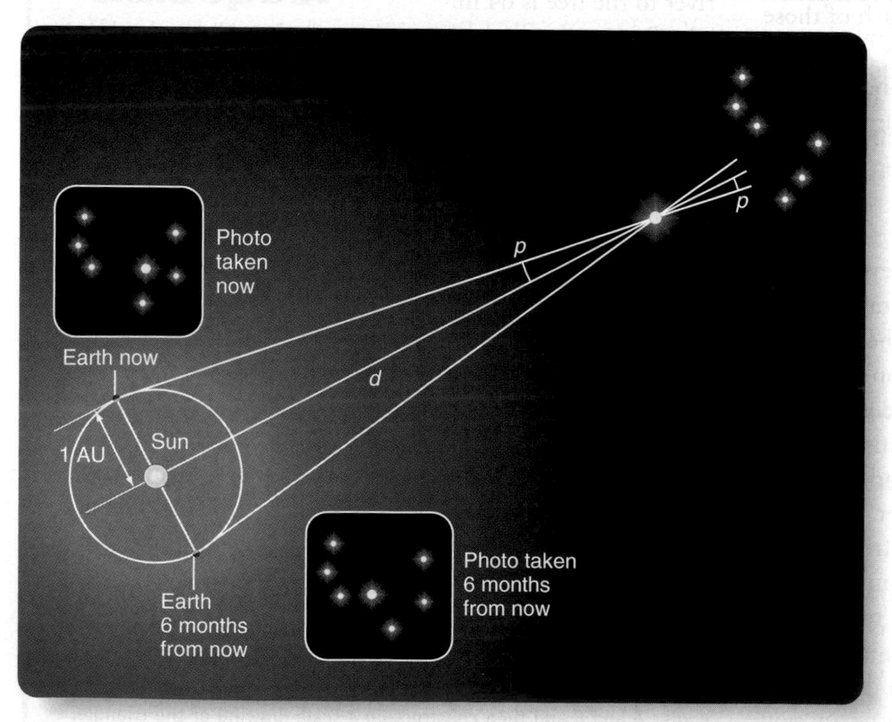

Parallax is the apparent change in the position of an object due to a change in the location of the observer. In Chapter 4 you saw an everyday example of parallax. Your thumb, held at arm's length, appears to shift position against a distant background when you look with first one eye and then with the other. In this case, the baseline is the distance between your eyes, and the parallax is the apparent movement of your thumb when you change eyes. The farther away you hold your thumb, the smaller the parallax.

Because the stars are so distant, their parallaxes are very small angles, usually expressed in seconds of arc. The quantity that astronomers call **stellar parallax** (p) is half the total shift of the star, as shown in Figure 9-3. Astronomers measure the parallax, and surveyors measure the angles at the ends of the baseline, but both measurements do the same thing—reveal the shape of the triangle and allow you to find the distance to the object in question.

Astronomers have defined a special unit of distance, the **parsec (pc)**,* for use in distance calculations. A parsec is defined as the distance to an imaginary star with a parallax of 1 second of arc. One parsec turns out to equal 206,265 AU, or 3.26 ly. This makes it very easy to calculate the distance to a star given the parallax. The formula is simply:

$$d = \frac{1}{p}$$

where the parallax, p, is measured in seconds of arc and the distance, d, is measured in parsecs. For example, one of your Favorite Stars, Sirius, has a parallax of 0.375 second of arc. Then the distance to Sirius in parsecs is 1 divided by 0.375, which equals 2.7 pc. If you want to convert to light-years, multiply 2.7 pc by 3.26, and you find that Sirius is 8.8 ly away.

Measuring the small angle p is very difficult. The nearest star to the sun is α Centauri, another Favorite Star. It has a parallax of only 0.76 second of arc and is only 1.3 pc (4.3 ly) distant. More distant stars have even smaller parallaxes. To see how small these angles are, hold a piece of paper edgewise at arm's length. The thickness of the paper covers an angle of about 30 seconds of arc. You can see that the parallax of a star, smaller than 1 second of arc, must be very difficult to measure accurately.

The blurring caused by Earth's atmosphere smears star images into tiny blobs of light no smaller than half a second of arc in diameter, and that makes it difficult to measure parallax from

*The parsec is used throughout astronomy because it simplifies the calculation of distance. However, there are instances when the light-year is also convenient. Consequently, the chapters that follow use either parsecs or light-years as convenience and custom dictate.

Earth's surface. Even when astronomers average together many observations, they cannot measure parallax with an uncertainty smaller than about 0.002 second of arc. If you measure a parallax of 0.02 second of arc, the uncertainty is about 10 percent. That means that astronomers observing from Earth's surface can't measure accurate parallaxes smaller than about 0.02 seconds of arc, which corresponds to a distance of 50 pc; consequently, ground-based parallax measurements are limited to only the closest stars. Since the first stellar parallax was measured in 1838, ground-based astronomers have been able to measure accurate parallaxes for only about 10,000 stars.

In 1989, the European Space Agency launched the satellite Hipparcos to measure stellar parallaxes from orbit above the blurring effects of Earth's atmosphere. The little satellite observed for four years, and the data were reduced by the most sophisticated computers to produce two parallax catalogs in 1997. One catalog contains 120,000 stars with parallaxes 20 times more accurate than ground-based measurements. The other catalog contains over a million stars with parallaxes as accurate as ground-based parallaxes. By producing such huge amounts of accurate data, Hipparcos has allowed astronomers new insights into the nature of stars.

Before dropping the subject of parallax measurement, you should learn about a related observation that reveals the motions of the stars through space.

ThomsonNOW Sign in at www.thomsonedu.com and go to ThomsonNOW to see Astronomy Exercises "Parallax I" and "Parallax II."

The Changing Shape of the Big Dipper

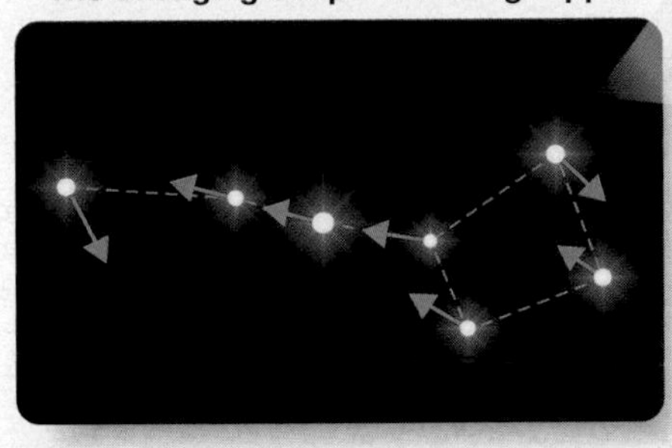

100,000 years ago the Big Dipper had a different shape.

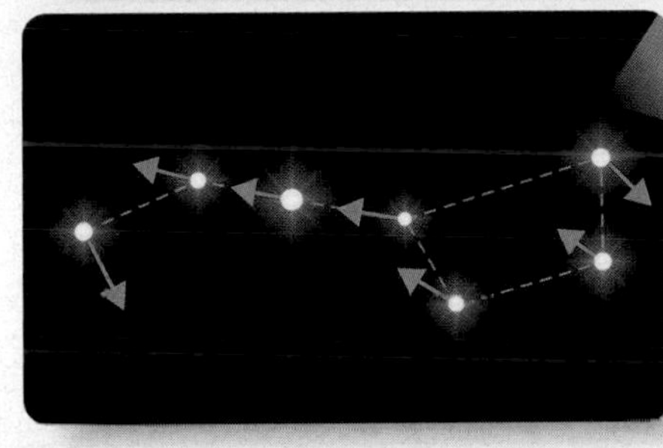

Proper motion is moving the stars of the Big Dipper across the sky.

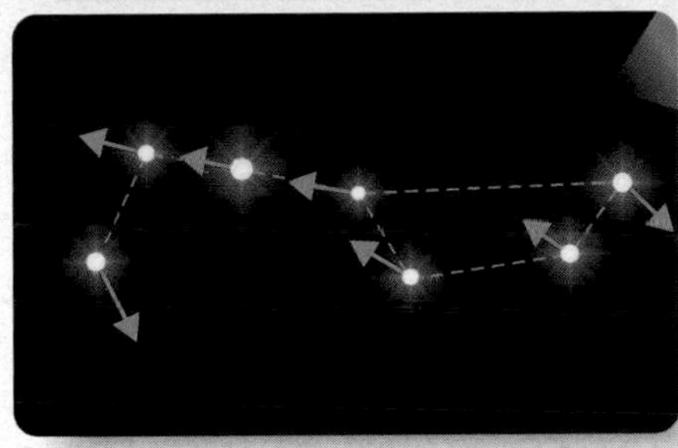

100,000 years in the future, the Big Dipper will have a distorted shape.

■ Figure 9-4

Proper motion refers to the slow movement of the stars across the sky.

Proper Motion

All the stars in the sky, including the sun, are moving along orbits around the center of our galaxy. That motion isn't obvious over periods of years, but over the centuries it can significantly distort the shape of constellations (■ Figure 9-4). Detecting this slow motion of the stars requires high-precision measurements.

If you photograph a small area of the sky on two dates separated by 10 years or more, you can notice that some of the stars in the photograph have moved very slightly against the background stars. This motion, expressed in units of seconds of arc per year, is the **proper motion** of the stars. As examples, consider two stars from your list of Favorite Stars. Vega is a bright blue-white star in the summer sky, and it has a proper motion of 0.327 second of arc per year. Rigel, a bright blue-white star in the winter sky, has a proper motion of only 0.002 second of arc per year. The two stars are nearly the same brightness in the sky and nearly the same temperature, but the proper motion of Rigel is over a hundred times less than that of Vega. What does that mean?

A star might have a small proper motion if it is moving almost directly toward or away from you; then its position on the sky would change very slowly, so it would have a small proper motion. That is unlikely, but it does happen. Another reason a star might have a small proper motion is that it could be quite far away from you. Then even if the star were moving rapidly through space, it would not have a large proper motion. That explains why Rigel, at a distance of 237 pc, has a smaller proper motion, and Vega, at a distance of only 7.7 pc, has a larger proper motion. By the way, this alerts you to something interesting. Although Rigel is 31 times further away than Vega, they have nearly the same brightness in the sky. Rigel must be emitting a lot more light than Vega.

Astronomers can use proper motion to look for nearby stars. If you see a star with a small (or zero) proper motion, it is probably a distant star, but a star with a large proper motion is probably quite close. You may have seen this effect if you watch birds. Distant geese move slowly across the sky, but a nearby bird flits quickly across your field of view. In this way, proper motions can give astronomers a way to locate nearby stars for further study.

◄ SCIENTIFIC ARGUMENT ►

Why are parallax measurements made from space better than parallax measurements made from Earth?

At first you might suppose that a satellite in orbit can measure the parallax of the stars better because the satellite is closer to the stars, but that will lead your argument astray. When you recall the immense distances to the stars, you can see that being in space doesn't really put the satellite significantly closer to the stars. Rather, the reason for increased accuracy is that the satellite is above Earth's atmosphere. When astronomers try to measure parallax, the turbulence in Earth's atmosphere blurs the star images and smears them out into blobs roughly 1 second of arc in diameter. It isn't possible to measure the position of these fuzzy images accurately. Astronomers can't measure parallax smaller than about 0.02 second of arc. A parallax of 0.02 second of arc corresponds to a distance of 50 pc, so ground-based astronomers can't measure parallax accurately beyond that distance.

A satellite in orbit, however, is above Earth's atmosphere, so the only blurring in the star images is that produced by diffraction in the optics. In other words, the star images are very sharp, and a satellite in orbit can measure the positions of stars and thus their parallaxes to high accuracy.

Now extend your argument one more step. **If a satellite can measure parallaxes as small as 0.001 second of arc, then how far are the most distant stars it can measure?**

◄ ►

9-2 Intrinsic Brightness

YOUR EYES TELL YOU that some stars look brighter than others, and in Chapter 2 you used the scale of apparent magnitudes to refer to stellar brightness. The faintest stars you can see with the naked eye are about sixth magnitude. Brighter stars have magnitudes represented by smaller numbers, and the brightest stars you see in the sky have negative magnitudes. Sirius, for example, has an apparent magnitude of -1.47.

The scale of apparent magnitudes only tells you how bright stars look, however, and you need to know their true, or intrinsic, brightness. *Intrinsic* means "belonging to the thing." When astronomers refer to the intrinsic brightness of a star, they mean a measure of the total amount of light the star emits. Apparent magnitudes can't tell you the intrinsic brightness of the stars, only how bright they look. An intrinsically very bright star would appear faint if it were far enough away. To find the true brightness of stars, you must correct the apparent magnitudes for the influence of distance.

Brightness and Distance

If you see lights at night, it is difficult to determine which are less powerful but nearby and which are highly luminous but farther away (■ Figure 9-5). You face the same problem when you look at stars, and to resolve that problem, you must think carefully about how brightness depends on distance.

When you look at a light, your eyes respond to the visual-wavelength energy falling on your eyes' retinas, telling you how bright the object looks. Recall from Chapter 2 that the light energy falling on one square meter in one second is called the flux. The light flux entering your eye is directly related to the intensity you perceive. The more flux entering your eye, the brighter the light looks.

If you placed a screen 1 meter square near a lightbulb, a certain amount of flux would fall on the screen. If you moved the screen twice as far from the bulb, the light that previously fell on the screen would spread out to cover an area four times larger, and the screen's surface would receive only one-fourth as much light per square meter. If you tripled the distance to the screen, its surface would receive only one-ninth as much light per square meter. In this way, the flux you receive from a light source is inversely proportional to the square of the distance to the source. This is known as the inverse square relation. (You first encountered the inverse square relation in Chapter 5, where it was applied to the strength of gravity. See Figure 5-6.)

Now you understand how the brightness of a star depends on its distance. If you knew the apparent magnitude of a star and its distance from Earth, you could use the inverse square law to correct for the effect of distance. Astronomers do that using a special kind of magnitude scale as described in the next section.

Absolute Visual Magnitude

If all the stars were the same distance away, you could compare one with another and decide which was emitting more light and which less. Of course, the stars are scattered at different distances, and you can't shove them around to line them up for comparison. But you can use what you have learned about the inverse square relation to calculate the brightness a star would

■ Active Figure 9-5

To judge the true brightness of a light source, you need to know how far away it is. With no clues to distance, the distant headlight on a truck might look as bright as the nearby headlight on a bicycle.

have at some standard distance. Astronomers take 10 pc as the standard distance and refer to the intrinsic brightness of the star as its **absolute visual magnitude** ***(M_v)***, the apparent visual magnitude the star would have it if were 10 pc away.

The symbol for absolute visual magnitude is an uppercase *M* with a subscript *v*. Recall from Chapter 2 that the symbol for apparent visual magnitude is a lowercase *m* with a subscript *v*. The subscript reminds you that the visual magnitude system is based only on the wavelengths of light human eyes can see. Other magnitude systems are based on other parts of the electromagnetic spectrum such as the infrared, ultraviolet, and so on.

The intrinsically brightest stars known have absolute visual magnitudes of about −8 and the faintest about +19. The sun has an absolute magnitude of 4.78. If the sun were only 10 pc away from Earth, it would look no brighter than the faintest star in the handle of the Little Dipper. Look at the list of Favorite Stars. The nearest star to the sun, alpha Centauri, is only 1.4 pc away, and its apparent magnitude is 0.0, indicating that it looks bright in the sky. However, its absolute magnitude is 4.39, about the same as the sun. Remember Vega and Rigel from the previous section? Vega has an absolute magnitude of 0.6, but Rigel has an absolute magnitude of –6.8. The two stars look the same in the sky, but Rigel is producing a lot more energy than Vega.

Calculating Absolute Visual Magnitude

How exactly do astronomers find the absolute visual magnitude of a star? This question leads to one of the most important formulas in astronomy, a formula that relates a star's magnitude and its distance.

The **magnitude–distance formula** relates the apparent magnitude m_v, the absolute magnitude M_v, and the distance *d* in parsecs:

$$m_v - M_v = -5 + 5 \log_{10}(d)$$

If you know any two of the parameters in this formula, you can easily calculate the third. If you want to find the absolute magnitude of a star, then you need to know its distance and apparent magnitude. Suppose a star has a distance of 50 pc and an apparent magnitude of 4.5. A calculator tells you that the log of 50 is 1.70, and $-5 + 5 \times 1.70$ equals 3.5, so you know that the absolute magnitude is 3.5 magnitudes brighter than the apparent magnitude. That means the absolute magnitude is 1.0 because 4.5 minus 3.5 is 1.0. (Remember that smaller numbers mean brighter magnitudes.) If this star were 10 pc away, it would look bright in the sky, a first-magnitude star.

Astronomers also use the magnitude–distance formula to calculate the distance to a star if the apparent and absolute magnitudes are known. For that purpose, it is handy to rewrite the formula in the following form:

$$d = 10^{(m_v - M_v + 5)/5}$$

If you knew that a star had an apparent magnitude of 7 and an absolute magnitude of 2, then m_v - M_v is 5 magnitudes, and the distance would be 10^2 or 100 parsecs.

The magnitude difference $m_v - M_v$ is known as the **distance modulus,** a measure of how far away the star is. The larger the distance modulus, the more distant the star. You could use the magnitude–distance formula to construct a table of distance and distance modulus (■ Table 9-1).

The magnitude–distance formula may seem awkward at first, but a calculator makes it easy to use. It is important because it performs a critical function in astronomy: It allows astronomers to convert observations of distance and apparent magnitude into absolute magnitude, a measure of the true brightness of the star. Once you know the absolute magnitude, you can go one step further and figure out the total amount of energy a star is radiating into space.

ThomsonNOW Sign in at www.thomsonedu.com and go to ThomsonNOW to see Astronomy Exercise "Apparent Brightness and Distance."

Luminosity

The second of your goals for this chapter is to find out how much energy stars emit. With the absolute magnitudes of the stars in hand, you can now compare other stars with the sun. That is easiest if you convert absolute magnitude into luminosity. The **luminosity** ***(L)*** of a star is the total amount of energy the star radiates in 1 second—not just visible light, but all wavelengths. To find a star's luminosity, you begin with its absolute visual magnitude, make a small correction, and compare the star with the sun.

■ **Table 9-1 | Distance Moduli**

m_v - M_v	*d* (pc)
0	10
1	16
2	25
3	40
4	63
5	100
6	160
7	250
8	400
9	630
10	1000
⋮	⋮
15	10,000
⋮	⋮
20	100,000
⋮	⋮

The correction you must make adjusts for the radiation emitted at wavelengths humans cannot see. Recall that absolute visual magnitude includes only visible light. The absolute visual magnitudes of hot stars and cool stars will underestimate their total luminosities because those stars radiate significant amounts of radiation in the ultraviolet or infrared parts of the spectrum. You can correct for the missing radiation because the amount of missing energy depends only on the star's temperature. For hot and cool stars, the correction can be large, but for medium-temperature stars like the sun, the correction is small. Adding the proper correction to the absolute visual magnitude changes it into the **absolute bolometric magnitude**—the absolute magnitude the star would have if you could see all wavelengths.

Once you know a star's absolute bolometric magnitude, you can find its luminosity by comparing it with the sun. The absolute bolometric magnitude of the sun is 4.7. For every magnitude a star is brighter than 4.7, it is 2.512 times more luminous than the sun. (Recall from Chapter 2 that a difference of 1 magnitude corresponds to an intensity ratio of 2.512.) That means that a star with an absolute bolometric magnitude of 2.7, which is 2 magnitudes brighter than the sun, must be 6.3 times more luminous (6.3 is approximately 2.512×2.512).

Favorite Star Aldebaran makes a convenient example. It has an absolute bolometric magnitude of -0.39. That makes it just a bit over 5 magnitudes brighter than the sun. A difference of 5 magnitudes is defined to be a factor of 100 in brightness, so the luminosity of Aldebaran is 100 times the sun's luminosity, or 100 $L_{\odot}$. Aldebaran is the red eye of Taurus the Bull; next time you see it in the winter sky, nudge your friends and say, "See that star? It emits just over 100 times more energy than the sun."

Remember Favorite Stars Vega and Rigel? Earlier in this chapter you noted that they look the same in the sky, but Rigel is much further away. If you analyze their brightnesses you will discover that Rigel is a hundred times more luminous than Vega.

The symbol $L_{\odot}$ represents the luminosity of the sun, a number astronomers can calculate in a direct way. Earth satellites can measure the total solar energy hitting 1 square meter in 1 second just above Earth's atmosphere (the solar constant defined in the previous chapter). The distance from Earth to the sun is known, so it is a simple matter to calculate how much energy the sun must radiate in all directions to provide Earth with the energy it receives per second (see Problem 9 at the end of this chapter). The measured luminosity of the sun is about 4×10^{26} joules/s. You can use that to convert luminosity in terms of the sun into actual joules per second. For example, if Aldebaran is 100 times the luminosity of the sun, it must be emitting a total luminosity of 4×10^{28} joules/s.

Finding the luminosity of a star is an important process, so it is worth a quick review: If you can measure the parallax of a star, you can find its distance, use its apparent visual magnitude to calculate its absolute visual magnitude, correct for the light you can't see to find the absolute bolometric magnitude, and then find the luminosity in terms of the sun. Then you can multiply by the sun's luminosity to find the luminosity of the star in joules per second.

Some stars are a million times more luminous than the sun, and some are almost a million times less luminous. Clearly, the family of stars is filled with interesting characters.

◄ SCIENTIFIC ARGUMENT ►

How can two stars look the same in the sky but have dramatically different luminosities?

You can answer this question by building a scientific argument that relates three factors: the appearance of a star, its true luminosity, and its distance. The further away a star is, the fainter it looks, and that is just the inverse square law. If two stars such as Vega and Rigel have the same apparent visual magnitude, then your eyes must be receiving the same amount of light from them. But Rigel is much more luminous than Vega, so it must be further away. Parallax observations from the Hipparcos satellite confirm that Rigel is 31 times further away.

Distance is often the key to understanding the brightness of stars, but temperature can also be important. Build a scientific argument to answer the following: **Why must astronomers make a correction in converting the absolute visual magnitude of very hot or very cool stars into luminosities?**

◄ ►

9-3 The Diameters of Stars

MANY PEOPLE ASSUME that you can look through a large astronomical telescope and see the stars as disks, but that is a **Common Misconception.** As you learned in Chapter 6, no telescope can resolve the disk of a star; the stars look like points of light, so finding the diameters of stars, your third goal in this chapter, takes a little ingenuity.

You already know how to find the temperatures and luminosities of stars, and that gives you a way to find their diameters. The relationship between temperature, luminosity, and diameter will allow you to sort the stars and will introduce you to the most important diagram in astronomy, where you will discover more family relations among the stars.

Luminosity, Radius, and Temperature

The luminosity and temperature of a star can tell you its diameter if you understand the two factors that affect a star's luminosity, surface area and temperature. You can eat dinner by candlelight because the candle flame has a small surface area, and, although it is very hot, it cannot radiate much heat; it has a low luminosity. However, if the candle flame were 12 ft tall, it would have a very large surface area from which to radiate, and although it might be no hotter than a normal candle flame, its luminosity would drive you from the table (■ Figure 9-6).

■ **Figure 9-6**

Molten lava pouring from a volcano is not as hot as a candle flame, but a lava flow has more surface area and radiates more energy than a candle flame. Approaching a lava flow without protective gear is dangerous. (Karafft/Photo Researchers. Inc.)

In a similar way, a hot star may not be very luminous if it has a small surface area. It could be highly luminous if it were larger. Even a cool star could be very luminous if it were very large and so had a large surface area from which to radiate. Both temperature and surface area help determine the luminosity of a star.

Stars are spheres, and the surface area of a sphere is $4\pi R^2$. If you express radius in meters, the area is the number of square meters on the surface of the star. Each square meter radiates like a black body, and you will remember from Chapter 7 that the total energy given off each second from each square meter is σT^4. So the total luminosity of the star is the surface area multiplied by the energy radiated per square meter:

$$L = 4\pi R^2 \sigma T^4$$

If you divide by the same quantities for the sun, you can cancel out the constants and get a simple formula for the luminosity of a star in terms of its radius and temperature:

$$\frac{L}{L_\odot} = \left(\frac{R}{R_\odot}\right)^2 \left(\frac{T}{T_\odot}\right)^4$$

Here the symbol $\odot$ stands for the sun, and the formula says that the luminosity of a star in terms of the sun equals the radius of the star in terms of the sun squared times the temperature of the star in terms of the sun raised to the fourth power.

Suppose a star is 10 times the sun's radius but only half as hot. How luminous would it be?

$$\frac{L}{L_\odot} = \left(\frac{10}{1}\right)^2 \left(\frac{1}{2}\right)^4 = \left(\frac{100}{1}\right)\left(\frac{1}{16}\right) = 6.25$$

The star would be 6.25 times more luminous than the sun.

How can you use this formula to find the diameters of the stars? If you see a cool star that is very luminous, you know it must be very large, and if you see a hot star that is not very luminous, you know it must be very small. Suppose that a star is 40 times the luminosity of the sun and twice as hot. If you put these numbers into the formula, you get:

$$\frac{40}{1} = \left(\frac{R}{R_\odot}\right)^2 \left(\frac{2}{1}\right)^4$$

Solving for the radius, you get:

$$\left(\frac{R}{R_\odot}\right)^2 = \frac{40}{2^4} = \frac{40}{16} = 2.5$$

So the radius is:

$$\frac{R}{R_\odot} = \sqrt{2.5} = 1.58$$

The star is 1.58 times larger in radius than the sun.

The H–R Diagram

Because a star's luminosity depends on its surface area and its temperature, you can use luminosity and temperature to sort the stars into groups. Astronomers use a special diagram for that sorting. The **Hertzsprung–Russell (H–R) diagram,** named after its originators, Ejnar Hertzsprung and Henry Norris Russell, is a graph that separates the effects of temperature and surface area on stellar luminosities and sorts the stars according to their sizes. Before you study the details of the H–R diagram (as it is often called), try looking at a similar diagram you might use to sort automobiles.

You could plot a diagram such as ■ Figure 9-7 to show engine power versus weight for various makes of cars. In so doing, you would find that, in general, the more a car weighs, the more power it has. Most cars fall somewhere along the sequence of cars running from heavy, high-powered cars to light, low-powered models. You might call this the main sequence of cars. But some cars have much more power than normal for their weight—the sport or racing models—and the economy models have less power than normal for cars of the same weight. Just as this diagram sorts cars into family groups, the H–R diagram sorts the stars into groups based on size.

The H–R diagram is a graph with luminosity on the vertical axis and temperature on the horizontal axis. A star is represented by a point on the graph that marks both the luminosity of the star and its temperature. The H–R diagram in ■ Figure 9-8 also contains a scale of spectral type across the top. Because a star's spectral type is determined by its temperature, you could use either spectral type or temperature on the horizontal axis.

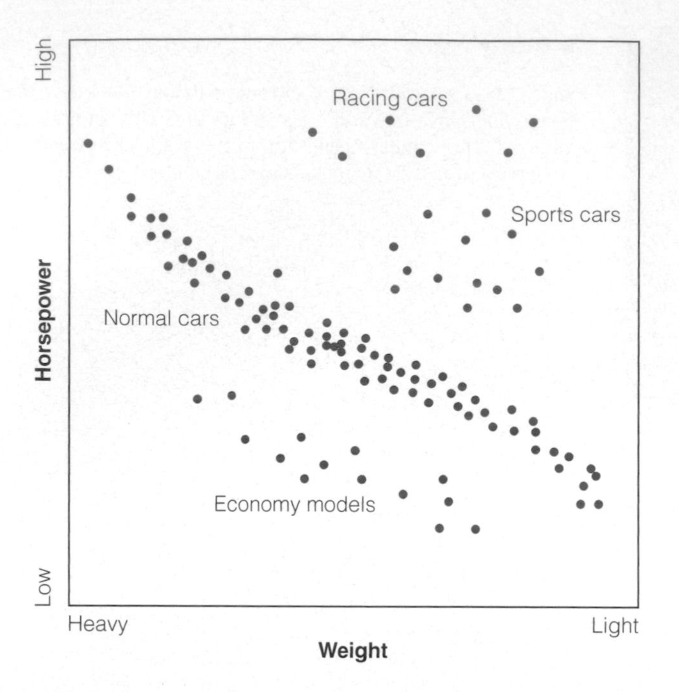

■ Figure 9-7

You could analyze automobiles by plotting their horsepower versus their weight and thus reveal relationships between various models. Most would lie somewhere along the main sequence of "normal" cars.

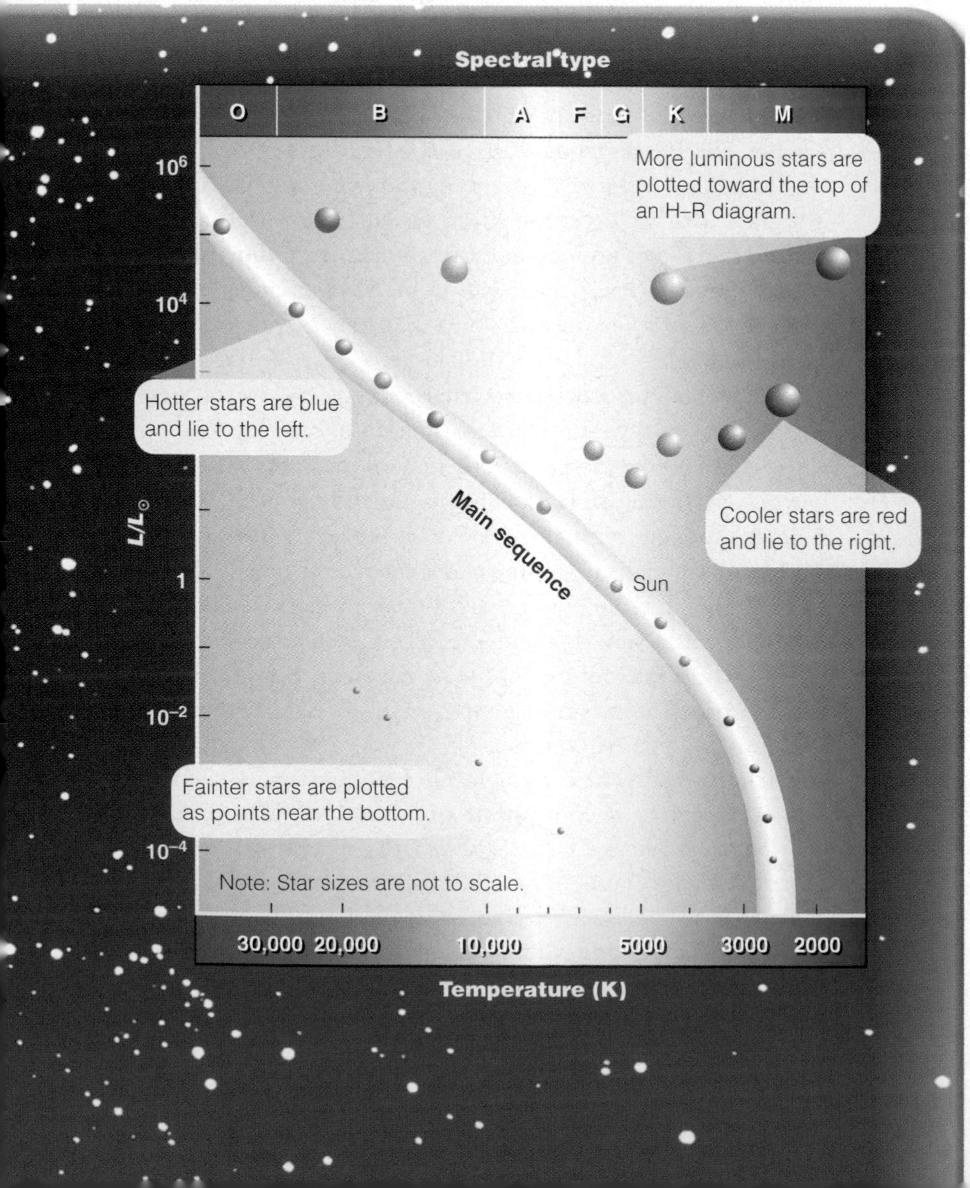

In an H–R diagram, the location of a point tells you a great deal about the star it represents. Points near the top of the diagram represent very luminous stars, and points near the bottom represent very low-luminosity stars. Also, points near the right edge of the diagram represent very cool stars, and points near the left edge of the diagram represent very hot stars. Notice in Figure 9-8 how the artist has used color to represent temperature. Cool stars are red, and hot stars are blue. You can see that dramatically in the photo of real stars in ■ Figure 9-9.

Astronomers use H–R diagrams so often that they usually skip the words "the point that represents the star." Rather they will say that a star is located in a certain place in the diagram. The location of a star in the H–R diagram has nothing to do with the location of the star in space. Furthermore, a star may move in the H–R diagram as it ages and its luminosity and temperature change, but such motion in the diagram has nothing to do with the star's motion in space.

Notice that the vertical axis of the H–R diagram is an exponential scale. That's a convenient way to graph data that covers a wide range. On a graph of body weight, you could plot the weight of a cat at the bottom, your own weight just above that, and above that the weight of an elephant and then a whale. You could get all of your data on one graph, but notice that an exponential scale compresses the high numbers and spreads out the low numbers. You can see that effect in the H-R diagram.

Giants, Supergiants, and Dwarfs

The H-R diagram reveals the family secrets of the stars. Look again at Figure 9-8 and notice the **main sequence,** a region of the H–R diagram running from upper left to lower right. It includes roughly 90 percent of all normal stars. As you might expect, the hot main-sequence stars are more luminous than the cool main-sequence stars. There are, however, stars that don't fall on the main sequence. That alerts you that temperature is not the only thing that determines the luminosity of a star. Size is important too.

For stars, luminosity, radius, and temperature have a precise mathematical relationship that can be used to draw lines of constant radius across an H–R diagram. ■ Figure 9-10 is an H–R diagram on which slanting dashed lines show the location of stars of

■ Figure 9-8

In an H–R diagram, a star is represented by a dot that shows the luminosity and temperature of the star. The background color in this diagram indicates the temperature of the stars. The sun is a yellow-white G2 star. Most stars fall along the main sequence running from hot luminous stars at upper left to cool low luminosity stars at lower right.

■ Figure 9-9

Notice the colors of the stars in the small star cluster M39. The brightest stars are either hot and blue or cool and red. Compare with the most luminous stars in Figure 9-8. (Heidi Schweiker/NAOA/AURA/NSF)

certain radii. For example, locate the dashed line labeled 1 $R_\odot$. That line passes through the point marked "Sun" and represents the location of any star whose radius equals that of the sun. Of course, the line slants down to the right because cooler stars are always fainter than hotter stars of the same size.

The H–R diagram reveals relationships within the family of stars. The stars called **giants** lie at the right above the main sequence. Although these stars are cool, they are luminous because they are 10 to 100 times larger than the sun. Look in Figure 9-10 and find the star Capella, composed of two stars. The stars have the same temperature as the sun, but they are much more luminous, so they must be larger. The **supergiants** are even more luminous and lie near the top of the H–R diagram; they are 10 to 1000 times the radius of the sun. Now you can understand why Rigel is so much more luminous than Vega. They have nearly the same temperature, but Rigel is a supergiant and has a much larger surface area from which to radiate. Another of your Favorite Stars is Betelgeuse in Orion, also a supergiant. If it magically replaced the sun at the center of our solar system, it would swallow up Mercury, Venus, Earth, and Mars. The largest stars known have radii of about 7 AU; if one of them replaced the sun, it would extend nearly to the orbit of Saturn.

At the bottom of the H–R diagram lie the economy models, stars that are very low in luminosity because they are very small. At the bottom end of the main sequence, the **red dwarfs** are not only small, they are also cool, and that means they can't radiate much energy; they have very low luminosities. In contrast, the **white dwarfs** lie in the lower left of the H–R diagram, and, although some are very hot, they are so small they can't be very bright. They are all roughly the size of Earth and can't radiate much energy from their small surface areas.

The H–R diagram shows that there is a great range in the sizes of stars. The largest are 100,000 times larger than the smallest. Notice that the size of the dots in the H–R diagrams here are only symbolic of the true sizes of the stars. If those dots were plotted in true size, your book would need to be as big as a billboard (■ Figure 9-11).

The distribution of stars in the H–R diagram according to size is a clue to how stars are born and how they die. You will follow those clues in the chapters that follow, but first you need to gather more information about the diverse family of stars.

ThomsonNOW Sign in at www.thomsonedu.com and go to ThomsonNOW to see Astronomy Exercise "Stefan–Boltzmann Law II."

Interferometric Observations of Diameter

Is there any way to check the diameters predicted by the H-R diagram? One way is to use interferometers such as the Center for High Angular Resolution Astronomy (CHARA) Array on Mount Wilson. There six 1-meter telescopes combine their light to produce the resolving power of a virtual telescope 330 meters in diameter. Such interferometers can resolve the diameters of large, bright, nearby stars.

You can see Favorite Star Vega high in the sky on summer evenings. It is an A0 star, and observations with the CHARA array confirm that it is about 2.5 times larger in diameter than the sun—about what you would expect from its location in the H-R diagram. The observations reveal, however, that it is spinning about twice a day compared to once a month for the sun and is flattened with its poles 2300 K hotter than its equator (■ Figure 9-12).

Observations with interferometers confirm that upper-main-sequence stars are larger than the sun. Altair (A7) is about 50 percent larger than the sun, and Regulus, a B7 star bright in the sky on spring evenings, is about four times larger than the sun. Achernar, a B3 star is still larger. All three of these stars are flattened by their rapid spin, as you can see in Figure 9-12.

Interferometric observations of giant and supergiant stars such as Betelgeuse (M2) show that they are indeed very large. Betelgeuse is at least 500 times the sun's diameter and can sometimes puff itself up to 800 times. In contrast, observations of red dwarfs confirm that although they are quite small, they are 15 to 20 percent bigger than expected. Evidently models of red dwarfs need further refining.

In short, interferometric observations confirm the sizes predicted by the H-R diagram. Stars really do range from roughly the size of Earth to hundreds of times bigger than the sun.

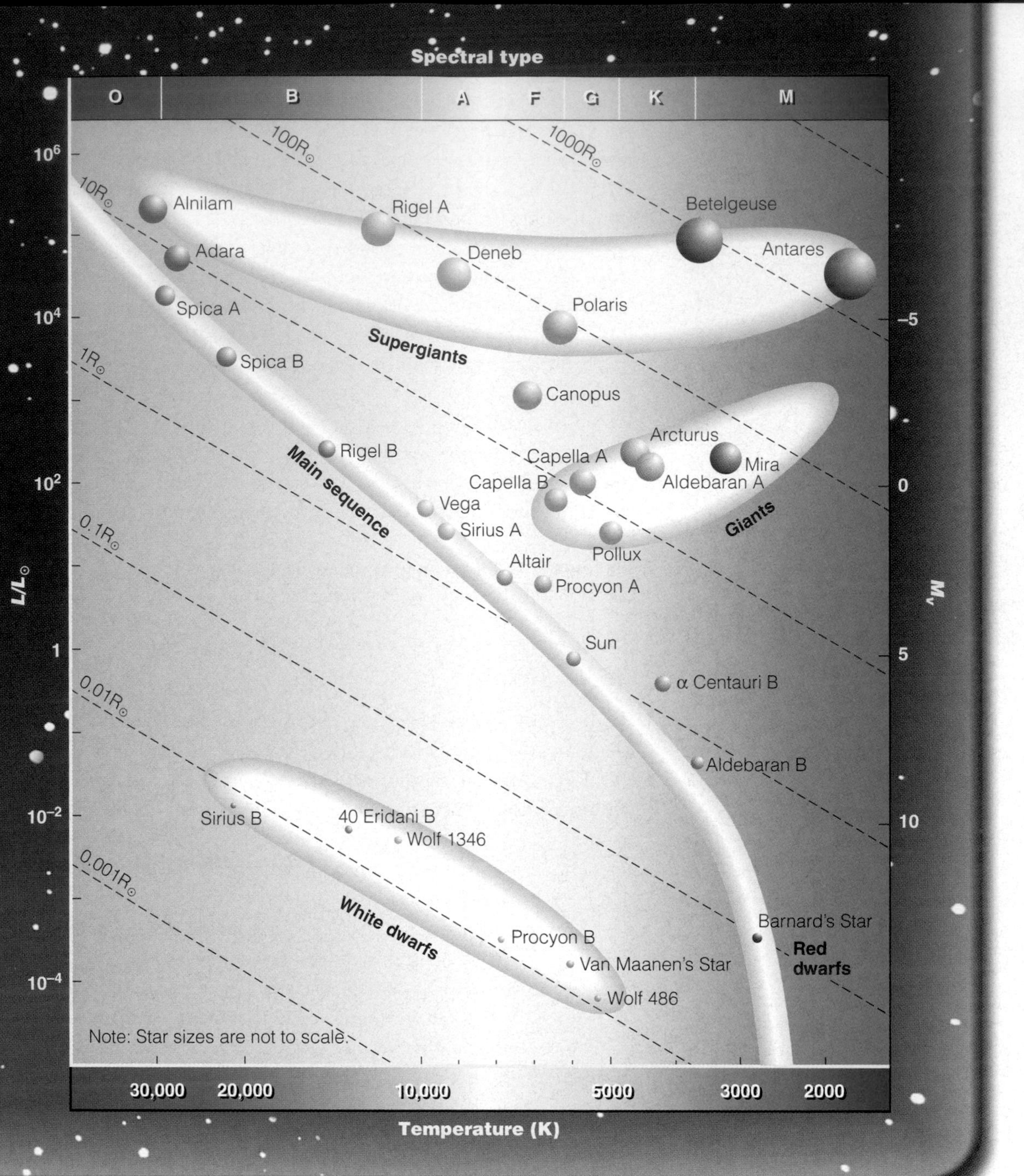

■ **Figure 9-10**

An H–R diagram showing the luminosity and temperature of many well-known stars. The dashed lines are lines of constant radius. The star sizes on this diagram are not to scale. (Individual stars that orbit each other are designated A and B, as in Spica A and Spica B.)

Luminosity Classification

You can tell from a star's spectrum whether it is a main-sequence star, a giant, or a supergiant. The clue is in the spectral lines.

Recall from Chapter 7 that collisional broadening can make spectral lines wider when the gas is dense and the atoms collide often. Main-sequence stars are relatively small and have dense atmospheres in which the gas atoms collide often and distort their electron energy levels. That makes the lines in the spectra of main-sequence stars broad. On the other hand, giant stars are larger, their atmospheres are less dense, and the atoms disturb one another relatively little. Spectra of giant stars have narrower spectral lines, and spectra of supergiants have very narrow lines (■ Figure 9-13).

That means you can look at a star's spectrum and tell roughly how big it is. These size divisions derived from spectra are called **luminosity classes** because the size of the star is the dominating factor in determining luminosity. The luminosity classes are represented by the Roman numerals I through V, with supergiants further subdivided into types Ia and Ib, as follows:

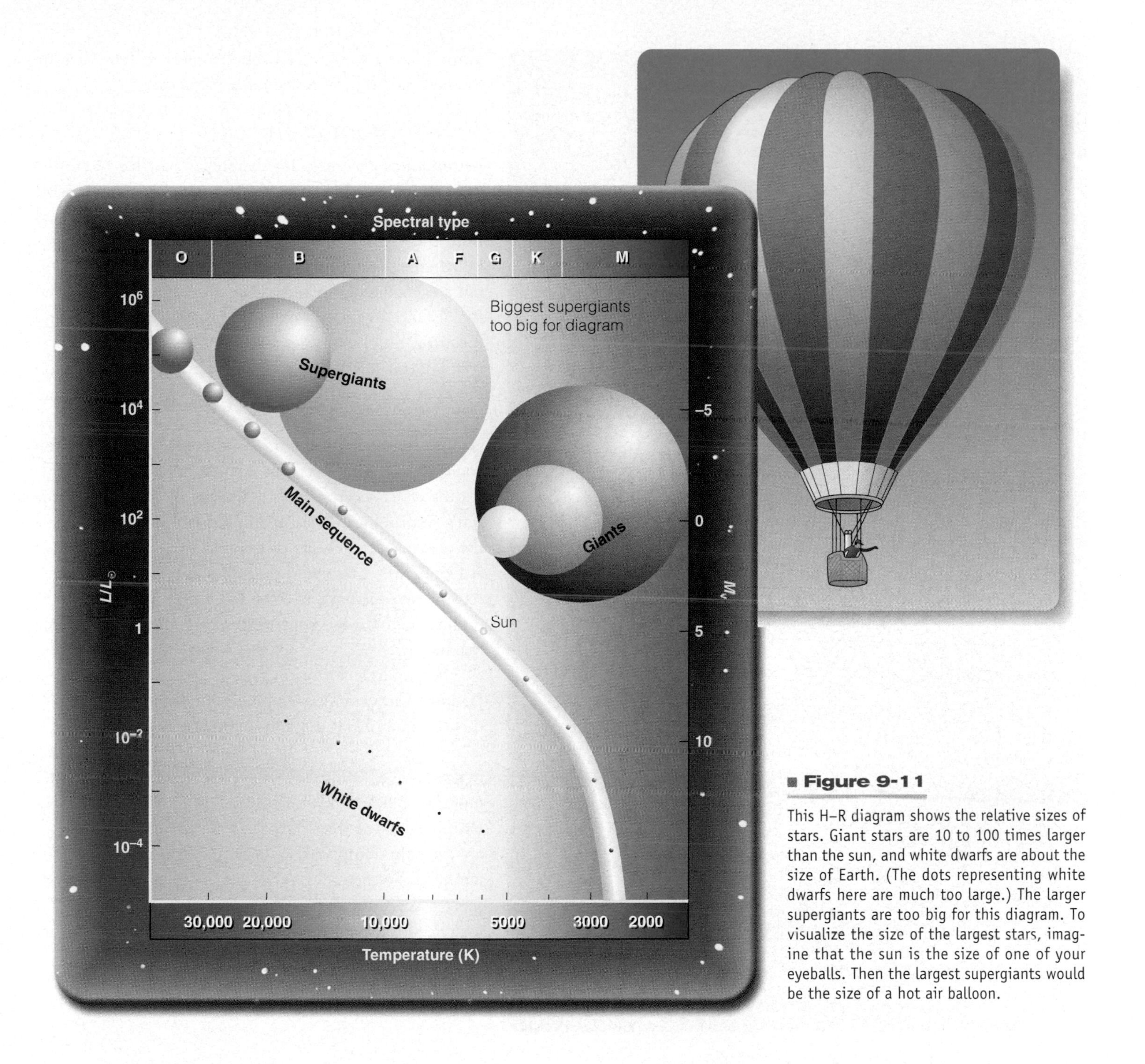

■ **Figure 9-11**

This H–R diagram shows the relative sizes of stars. Giant stars are 10 to 100 times larger than the sun, and white dwarfs are about the size of Earth. (The dots representing white dwarfs here are much too large.) The larger supergiants are too big for this diagram. To visualize the size of the largest stars, imagine that the sun is the size of one of your eyeballs. Then the largest supergiants would be the size of a hot air balloon.

Luminosity Classes

Ia Bright supergiant
Ib Supergiant
II Bright giant
III Giant
IV Subgiant
V Main-sequence star

You can distinguish between the bright supergiants (Ia) such as Rigel and the regular supergiants (Ib) such as Polaris, the North Star. The star Adhara is a bright giant (II), Aldebaran is a giant (III), and Altair is a subgiant (IV). The sun is a main-sequence star (V). The luminosity class is written after the spectral type, as in G2 V for the sun. (White dwarfs don't enter into this classification, because their spectra are peculiar.) Some of your Favorite Stars are quite different from the sun. Next time you look at the North Star, remind yourself that it is a supergiant.

If you plot the positions of the luminosity classes on the H–R diagram you get a figure like ■ Figure 9-14. Remember that these are rather broad classifications and that the lines on the

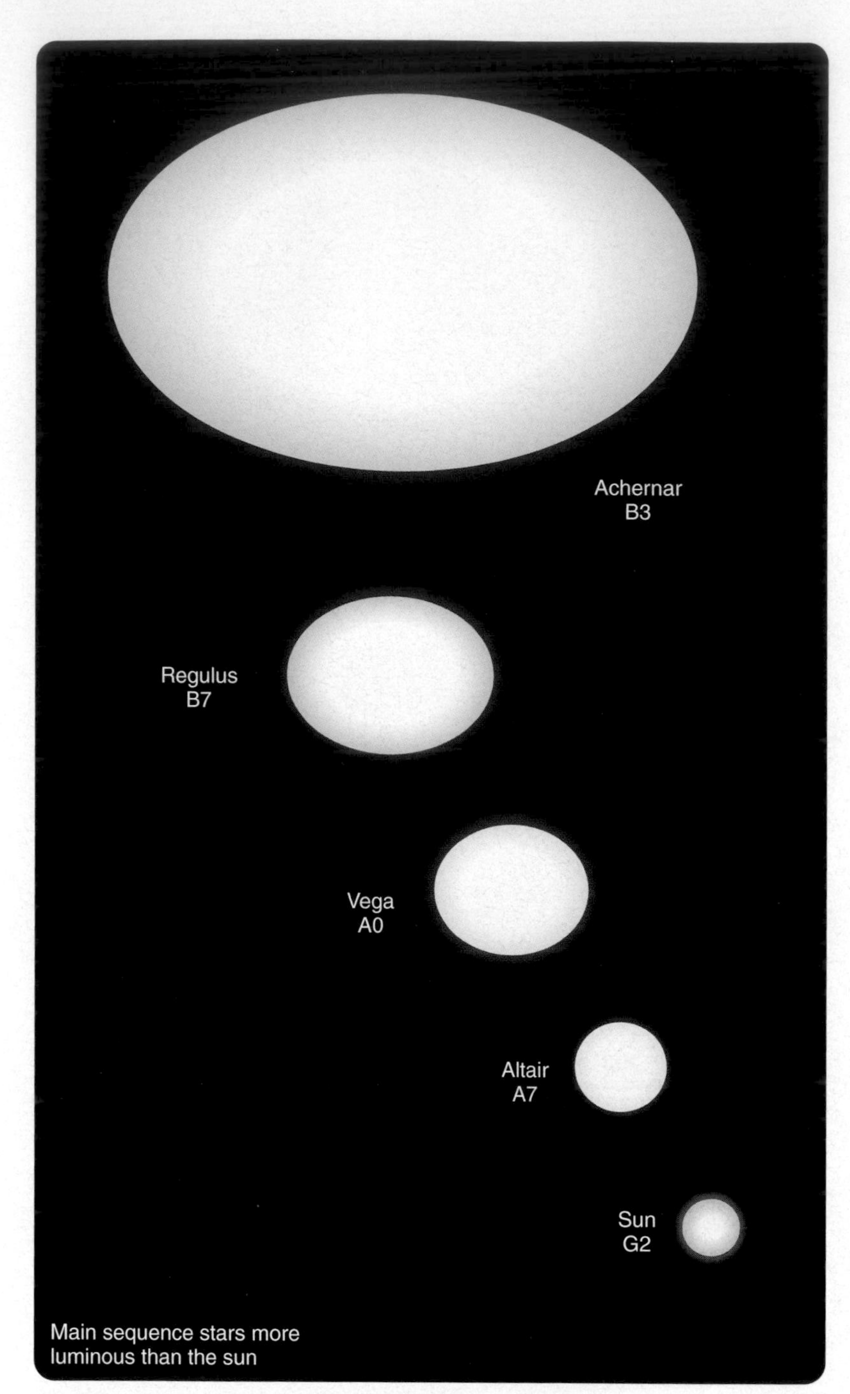

■ Figure 9-12

Observations with interferometers can resolve the size and shape of some nearby stars. The stars of the upper main sequence are indeed larger than the sun, as predicted by the H-R diagram. The examples shown here are flattened by rapid rotation, but most stars rotate slower and are more nearly spherical. On this scale, the supergiant Betelegeuse would have a diameter similar to that of a typical classroom.

diagram are only approximate. A star of luminosity class III may lie slightly above or below the line labeled III. Luminosity classification is subtle and not too accurate, but it is an important tool in modern astronomy. As you will see in the next section, luminosity classification, combined with the H–R diagram, gives astronomers a way to find the distances to stars that are too far away to have measurable parallaxes.

Spectroscopic Parallax

Astronomers can measure the stellar parallax of nearby stars, but most stars are too distant to have measurable parallaxes. The distances to these stars can be estimated from their spectra and apparent magnitude in a process called **spectroscopic parallax.** Spectroscopic parallax is not an actual measure of parallax but rather a way to find the distance to the star from its apparent magnitude and spectrum.

The method of spectroscopic parallax depends on the H–R diagram. If you recorded the spectrum of a star, you could determine its spectral class, and that would tell you its horizontal location in the H–R diagram. You could also determine its luminosity class by looking at the widths of its spectral lines, and that would tell you the star's vertical location in the diagram. Once you plotted the point that represents the star in an H–R diagram such as Figure 9-14, you could read off its absolute magnitude. As you have seen earlier in this chapter, you can find the distance to a star by comparing its apparent and absolute magnitudes.

For example, Favorite Star Betelgeuse is classified M2 Ia, and its apparent magnitude is about 0.5. You can plot this star in an H–R diagram such as that in Figure 9-14, where you would find that it should have an absolute magnitude of about −7.2. That means its distance modulus is 0.5 minus (−7.2), or about 7.7, and the distance (estimated from Table 9-1) is about 350 pc. Parallax from the Hipparcos satellite shows that the true distance to Betelgeuse is 520 pc, so the estimate from spectroscopic parallax is only approximate. An error of 1 magnitude changes the distance by a *factor* of 1.6. That's an error of 60 percent, so spectroscopic parallax isn't very accurate, but it can provide an estimate for stars so distant that parallax can't be measured.

◄ SCIENTIFIC ARGUMENT ►

What evidence can you give that giant stars really are bigger than the sun?

Scientific arguments are based on evidence, so you need to proceed step by step here. Stars exist that have the same spectral type as the sun but are clearly more luminous. Capella A, for example, is a G star with an absolute magnitude of 0. Because it is a G star, it must have about the same temperature as the sun, but its absolute magnitude is 5 magnitudes brighter than the sun's. A magnitude difference of 5 magnitudes corresponds to an intensity ratio of 100, so Capella A must be about 100 times more luminous than the sun. If it has the same surface temperature as the sun but is 100 times more luminous, then it must have a surface area 100 times greater than the sun's. Because the surface area of a sphere is proportional to the square of the radius,

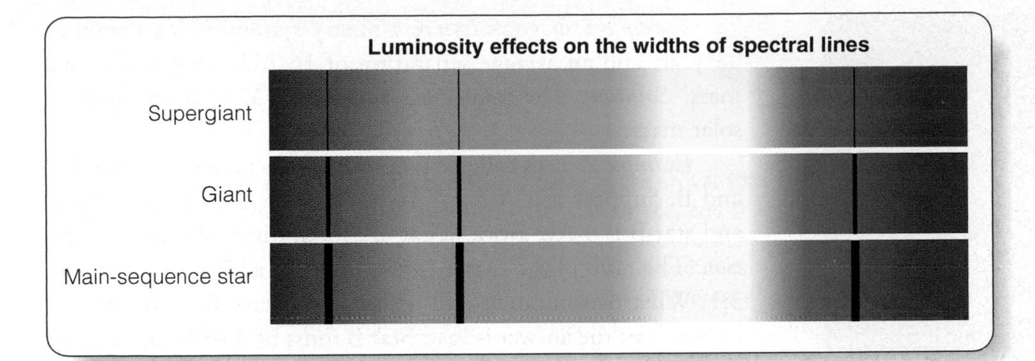

■ **Figure 9-13**

These schematic spectra show how the widths of spectral lines reveal a star's luminosity classification. Supergiants have very narrow spectral lines, and main-sequence stars have broad lines. In addition, certain spectral lines are more sensitive to this effect than others, so an experienced astronomer can inspect a star's spectrum and determine its luminosity classification.

Capella A must be 10 times larger in radius. That is clear observational evidence that Capella A is a giant star.

In Figure 9-10, you can see that Procyon B is a white dwarf slightly warmer than the sun but about 10,000 times less luminous than the sun. Build a scientific argument based on evidence to resolve this question. **Why do astronomers conclude that white dwarfs must be small stars?**

◄ ►

9-4 The Masses of Stars

YOUR FINAL GOAL is to find out how much matter stars contain, that is, to know their masses. Do they all contain about the same mass as our sun, or are some more massive and others less? Unfortunately, it's difficult to determine the mass of a star. Looking through a telescope at a star, you see only a point of light that tells you nothing about the star. Gravity is the key. Matter produces a gravitational field, and you can figure out how much matter a star contains if you watch an object move through the star's gravitational field. To find the masses of stars, you must study **binary stars,** pairs of stars that orbit each other.

Binary Stars in General

The key to finding the mass of a binary star is understanding orbital motion. Chapter 5 illustrated orbital motion with an imaginary cannonball fired from a high mountain (see page 88). If Earth's gravity didn't act on the cannonball, it would follow a straight-line path and leave Earth forever. Because Earth's gravity pulls it away from its straight-line path, the cannonball follows a curved path around Earth—an orbit. When two stars orbit each other, their mutual gravitation pulls them away from straight-line paths and makes them follow closed orbits around a point between the stars.

Each star in a binary system moves in its own orbit around the system's center of mass, the balance point of the system. If the stars were connected by a massless rod and placed in a uniform gravitational field such as that near Earth's surface, the system would balance at its center of mass like a child's seesaw (see page 89). If one star were more massive than its companion, then the massive star would be closer to the center of mass and would travel in a smaller orbit, while the lower-mass star would whip around in a larger orbit (■ Figure 9-15). The ratio of the masses of the stars M_A/M_B equals r_B/r_A, the inverse of the ratio of the radii of the orbits. If one star has an orbit twice as large as the other star's orbit, then it must be half as massive. Getting the ratio of the masses is easy, but that doesn't tell you the individual masses of the stars, which is what you really want to know. That takes further analysis.

■ **Figure 9-14**

The approximate location of the luminosity classes on the H–R diagram.

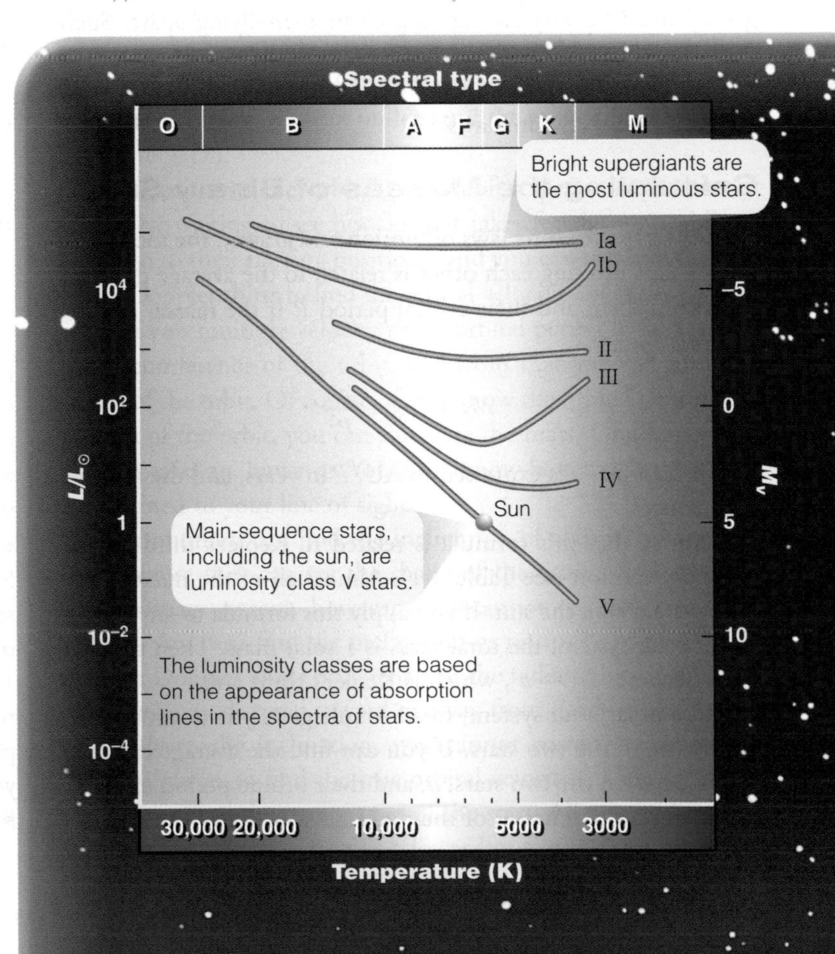

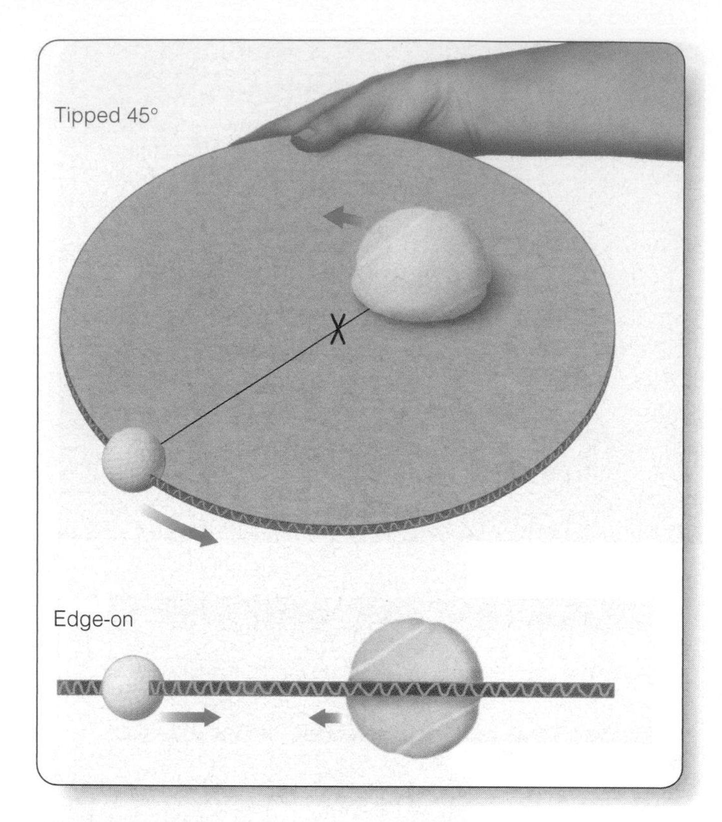

■ **Figure 9-20**

Imagine a model of a binary system with balls for stars and a disk of cardboard for the plane of the orbits. Only if you view the system edge-on do you see the stars cross in front of each other.

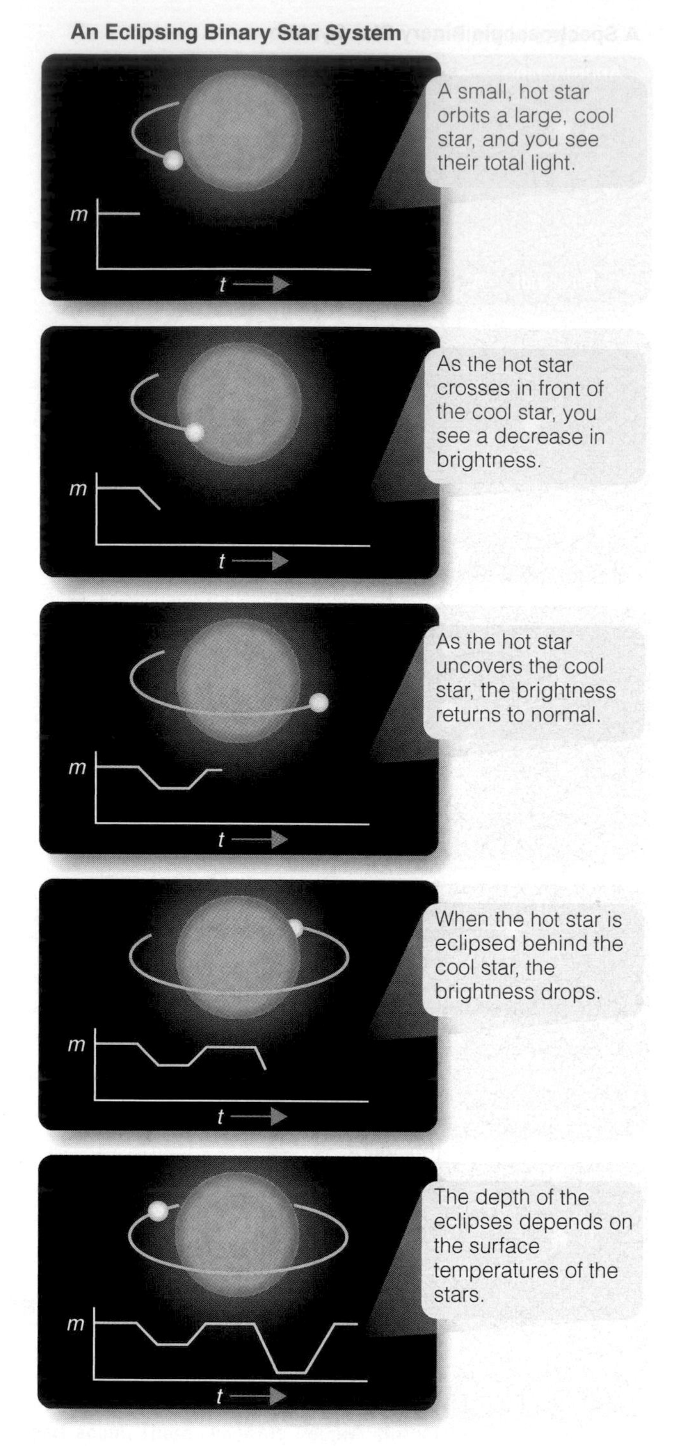

An Eclipsing Binary Star System
A small, hot star orbits a large, cool star, and you see their total light.
As the hot star crosses in front of the cool star, you see a decrease in brightness.
As the hot star uncovers the cool star, the brightness returns to normal.
When the hot star is eclipsed behind the cool star, the brightness drops.
The depth of the eclipses depends on the surface temperatures of the stars.
m
t

■ **Figure 9-21**

From Earth, an eclipsing binary looks like a single point of light, but changes in brightness reveal that two stars are eclipsing each other. Doppler shifts in the spectrum combined with the light curve, shown here as magnitude versus time, can reveal the size and mass of the individual stars.

because you don't have to untip the orbits; you know they are nearly edge-on, or there would not be eclipses. Then you can find the size of the orbits and the masses of the stars.

Earlier in this chapter, luminosity and temperature were used to calculate the radii of stars, but eclipsing binary systems provide a way to measure the sizes of stars directly. From the light curve you could tell how long it took for the small star to cross the large star. Multiplying this time interval by the orbital velocity of the small star would give you the diameter of the larger star. You could also determine the diameter of the small star by noting how long it took to disappear behind the edge of the large star. For example, if it took 300 seconds for the small star to disappear while traveling 500 km/s relative to the large star, then it would have to be 150,000 km in diameter.

Of course, there are complications due to the inclination and eccentricity of orbits, but often these effects can be taken into account, and astronomers can find not only the masses of the two stars but also their diameters.

Algol (alpha Persei) is one of the best-known eclipsing binaries, because its eclipses are visible to the naked eye. Normally, its magnitude is about 2.15, but its brightness drops to 3.4 in eclipses that occur every 68.8 hours. Although history books say an English astronomer "discovered" the variation of Algol in

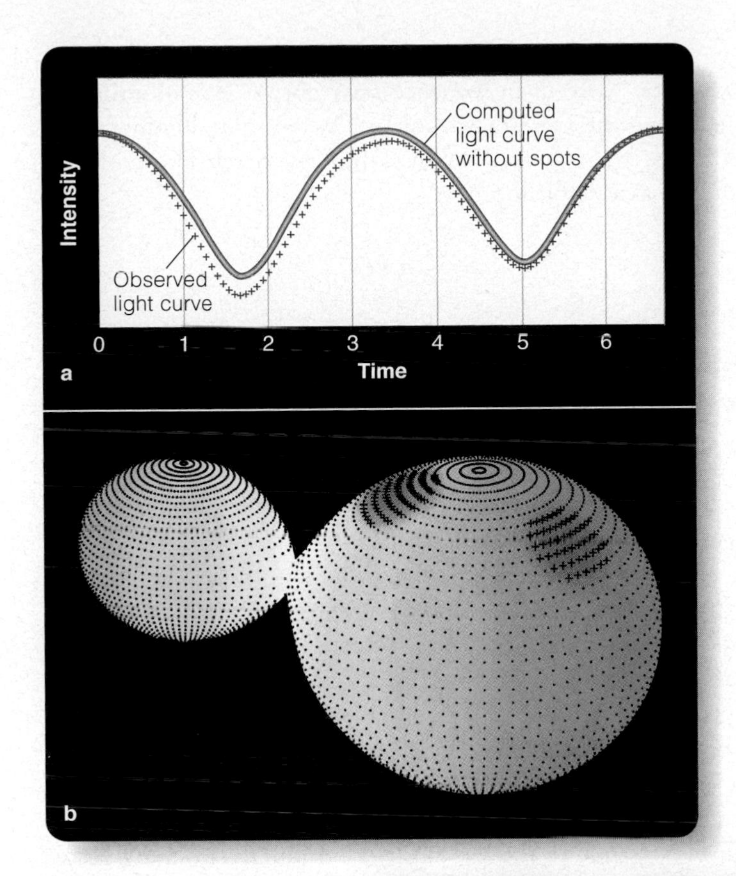

■ **Figure 9-22**

The observed light curve of the binary star VW Cephei (lower curve) shows that the two stars are so close together that their gravity distorts their shapes. Slight distortions in the light curve reveal the presence of dark spots at specific places on the star's surface. The upper curve shows what the light curve would look like if there were no spots. (Graphics created with Binary Maker 2.0)

1783, its periodic dimming was probably known to the ancients. *Algol* comes from the Arabic for "the demon's head," and it is associated in constellation mythology with the severed head of Medusa, the sight of whose serpentine locks turned mortals to stone (■ Figure 9-23). Indeed, in some accounts, Algol is the winking eye of the demon.

From the study of binary star systems, astronomers have found that the masses of stars range from 0.08 solar mass at the low end to as high as 150 solar masses at the high end. The most massive stars ever found in a binary system are a pair of stars with masses of 83 and 82 solar masses. A few other stars are believed to be more massive, but they do not lie in binary systems, so astronomers can only estimate their masses. Stars near the upper limit are very rare, and few are known, so this upper limit is uncertain.

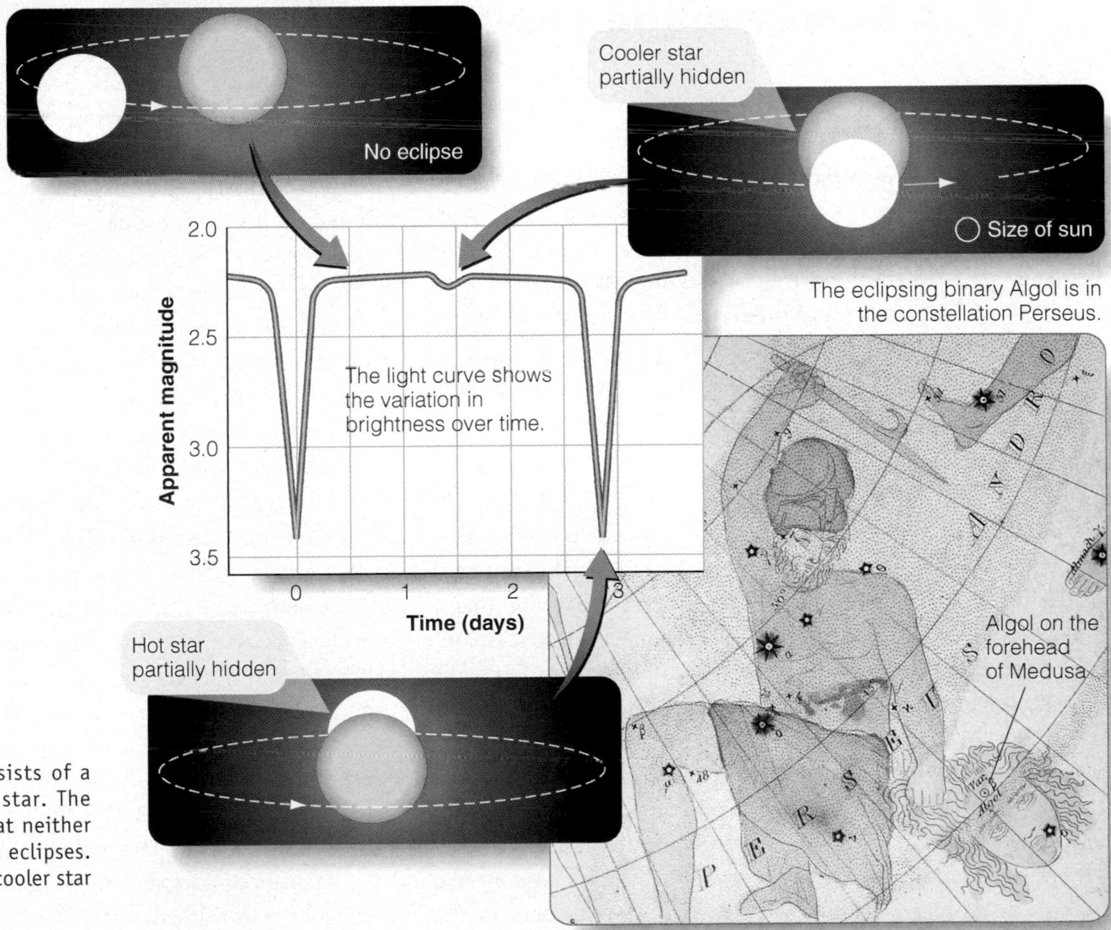

■ **Figure 9-23**

The eclipsing binary Algol consists of a hot B star and a cooler G or K star. The eclipses are partial, meaning that neither star is completely hidden during eclipses. The orbit here is drawn as if the cooler star were stationary.

ThomsonNOW Sign in at www.thomsonedu.com and go to ThomsonNOW to see Astronomy Exercise "Eclipsing Binaries."

◄ SCIENTIFIC ARGUMENT ►

When you look at the light curve for an eclipsing binary system with total eclipses, how can you tell which star is hotter?

This scientific argument brings together a number of things you have learned about light and binary stars. If you assume that the two stars in an eclipsing binary system are not the same size, then you can refer to them as the larger star and the smaller star. When the smaller star moves behind the larger star, you lose the light coming from the total area of the small star. And when the smaller star moves in front of the larger star, it blocks off light from the same amount of area on the larger star. In both cases, the same amount of area, the same number of square meters, is hidden from your sight. Then the amount of light lost during an eclipse depends only on the temperature of the hidden surface because temperature is what determines how much each square meter can radiate per second. When the surface of the hotter star is hidden, the brightness will fall dramatically, but when the surface of the cooler star is hidden, the brightness will not fall as much. So you can look at the light curve and point to the deeper of the two eclipses and say, "That is where the hotter star is behind the cooler star."

Now change the argument to consider the diameters of the stars. **How could you look at the light curve of an eclipsing binary with total eclipses and find the ratio of the diameters?**

◄ ►

9-5 A Survey of the Stars

YOU HAVE ACHIEVED the goals set for you at the start of this chapter. You know how to find the distances, luminosities, diameters, and masses of stars. Over the last two centuries, astronomers have collected a huge amount of data concerning stars **(How Do We Know? 9-2)**, and now you can put that data together to paint a family portrait of the stars; as in most family portraits, both similarities and differences are important clues to the history of the family. As you begin, you can ask a simple question: What is the average star like? Answering that question is both challenging and illuminating.

Mass, Luminosity, and Density

With enough data plotted in an H–R diagram, you can see patterns. If you label an H–R diagram with the masses of the plotted stars, as in ■ Figure 9-24, you will discover that the main-sequence stars are ordered by mass. The most massive main-sequence stars are the hot stars. As you run your eye down the main sequence, you find lower-mass stars; and the lowest-mass stars are the coolest, faintest main-sequence stars.

Stars that do not lie on the main sequence are not in order according to mass. Giant stars are a jumble of different masses, and supergiants, although they tend to be more massive than giants, are in no particular order in the H-R diagram. In contrast, all white dwarfs have about the same mass, somewhere in the narrow range of 0.5 to about 1 solar mass.

Because of the systematic ordering of mass along the main sequence, these main-sequence stars obey a **mass–luminosity relation**—the more massive a star is, the more luminous it is ■ Figure 9-25). In fact, the mass–luminosity relation can be expressed as a simple formula:

$$L = M^{3.5}$$

That is, a star's luminosity (in terms of the sun's luminosity) equals its mass (in solar masses) raised to the 3.5 power. For example, a star of 4 solar masses has a luminosity of approximately $4^{3.5}$, which equals $4 \times 4 \times 4 \times \sqrt{4}$. This equals 64×2, or 128. So a 4-solar-mass star will have a luminosity of about 128 times the luminosity of the sun. This is only an approximate equation, as shown by the red line in Figure 9-25.

Notice how large the range in luminosity is. The observed range of masses extends from about 0.08 solar mass to about 83 solar masses—a factor of about 1000. But the range of luminosities extends from about 10^{-6} to about 10^{6} solar luminosities—a factor of 10^{12}. Clearly, a small difference in mass causes a large difference in luminosity.

Although giants and supergiants do not follow the mass–luminosity relation very closely, and white dwarfs not at all, the link between mass and luminosity is critical in astronomy. In the next chapters, the mass–luminosity relation will help you understand how stars generate their energy.

Though mass alone does not reveal any pattern among giants, supergiants, and white dwarfs, density does. Once you know a star's mass and diameter, you can calculate its average density by dividing its mass by its volume. Stars are not uniform in density but are most dense at their centers and least dense near their surface. The center of the sun, for instance, is about 100 times as dense as water; its density near the visible surface is about 3400 times less dense than Earth's atmosphere at sea level. A star's average density is intermediate between its central and surface densities. The sun's average density is approximately 1 g/cm^3—about the density of water.

Main-sequence stars have average densities similar to the sun's, but giant stars, being large, have low average densities, ranging from 0.1 to 0.01 g/cm^3. The enormous supergiants have still lower densities, ranging from 0.001 to 0.000001 g/cm^3. These densities are thinner than the air you breathe; if you could insulate yourself from the heat, you could fly an airplane through these stars. Only near the center would you be in any danger, for there the material is very dense—about 3,000,000 g/cm^3.

The white dwarfs have masses about equal to the sun's but are very small, only about the size of Earth. That means the matter is compressed to densities of 3,000,000 g/cm^3 or more. On Earth, a teaspoonful of this material would weigh about 15 tons.

Density divides stars into three groups. Most stars are main-sequence stars with densities like the sun's. Giants and supergiants are very-low-density stars, and white dwarfs are high-density

How Do We Know? 9-2

Basic Scientific Data

How is scuba diving on an ancient shipwreck like observing a binary star? In a simple sense, science is the process by which scientists look at data and search for relationships. It sometimes requires large amounts of data to establish these relationships. For example, astronomers need to know the masses and luminosities of many stars before they can begin to understand the mass luminosity relationship.

Compiling basic data is one of the common forms of scientific work. It is the first step toward scientific analysis and understanding. An archeologist may spend months or even years diving to the floor of the Mediterranean Sea to study an ancient Greek shipwreck. She will carefully measure the position of every wooden timber and bronze fitting. She will photograph and recover everything from broken pottery to tools and weapons. The care with which she records data on the site pays off when she begins her analysis. For every hour the archaeologist spends recovering an object, she may spend days or weeks in her office, a library, or a museum identifying and understanding the object. Why was there a Phoenician hammer on a Greek ship? What does that reveal about the economy of ancient Greece?

Finding, identifying, and understanding that ancient hammer is only one bit of information, but the contributions of many scientists eventually build a picture of how ancient Greeks saw their world. Solving a single binary star system to find the masses of the stars does not tell an astronomer a great deal about nature. Over the years, however, many astronomers have added their results to the growing data file on stellar masses. Astronomers can now analyze that data to better understand how stars work.

Collecting mineral samples can be hard work, but it is also fun. Scientists sometimes collect large amounts of data because they enjoy the process. (M. A. Seeds)

stars. You will see in later chapters that these densities reflect different stages in the evolution of stars.

ThomsonNOW Sign in at www.thomsonedu.com and go to ThomsonNOW to see Astronomy Exercises "Mass–Luminosity Relation" and "H–R/Mass–Luminosity 3-D Graph."

Surveying the Stars

If you want to know what the average person thinks about a certain subject, you take a survey. If you want to know what the average star is like, you must survey the stars. Such surveys reveal important relationships among the family of stars.

Not many decades ago, surveying large numbers of stars was an exhausting task, but modern computers have changed that. Specially designed telescopes controlled by computers can make millions of observations per night, and high-speed computers can compile and analyze these vast surveys and create easy-to-use databases. Chapter 6 mentioned the Sloan Digital Sky Survey and the 2Mass infrared survey. Those and other surveys produce mountains of data that astronomers can mine, searching for relationships within the family of stars.

What could you learn about stars from a survey of the stars near the sun? Because the sun is thought to be in a typical place in the universe, such a survey could reveal general characteristics of the stars and might reveal unexpected processes in the formation and evolution of stars. Read **The Family of Stars** on pages 196–197 and notice three important points:

1. First, taking a survey is difficult because you must be sure you get an honest sample. If you don't survey enough stars or if you don't notice some kinds of stars, you can get biased results.

2. The second important point is that most stars are faint, and the most luminous stars are rare. The most common kinds of stars are the lower-main-sequence red dwarfs and the white dwarfs.

3. Finally, notice that what you see in the sky is deceptive. Stars near the sun are quite faint, but stars that are luminous, although they are rare, are easily visible even at great distances. Many of the brighter stars in the sky are highly luminous stars that you see even though they lie far away.

The night sky is a beautiful carpet of stars, but they are not all the same. Some are giants and supergiants, and some are dwarfs. The family of the stars is rich in its diversity.

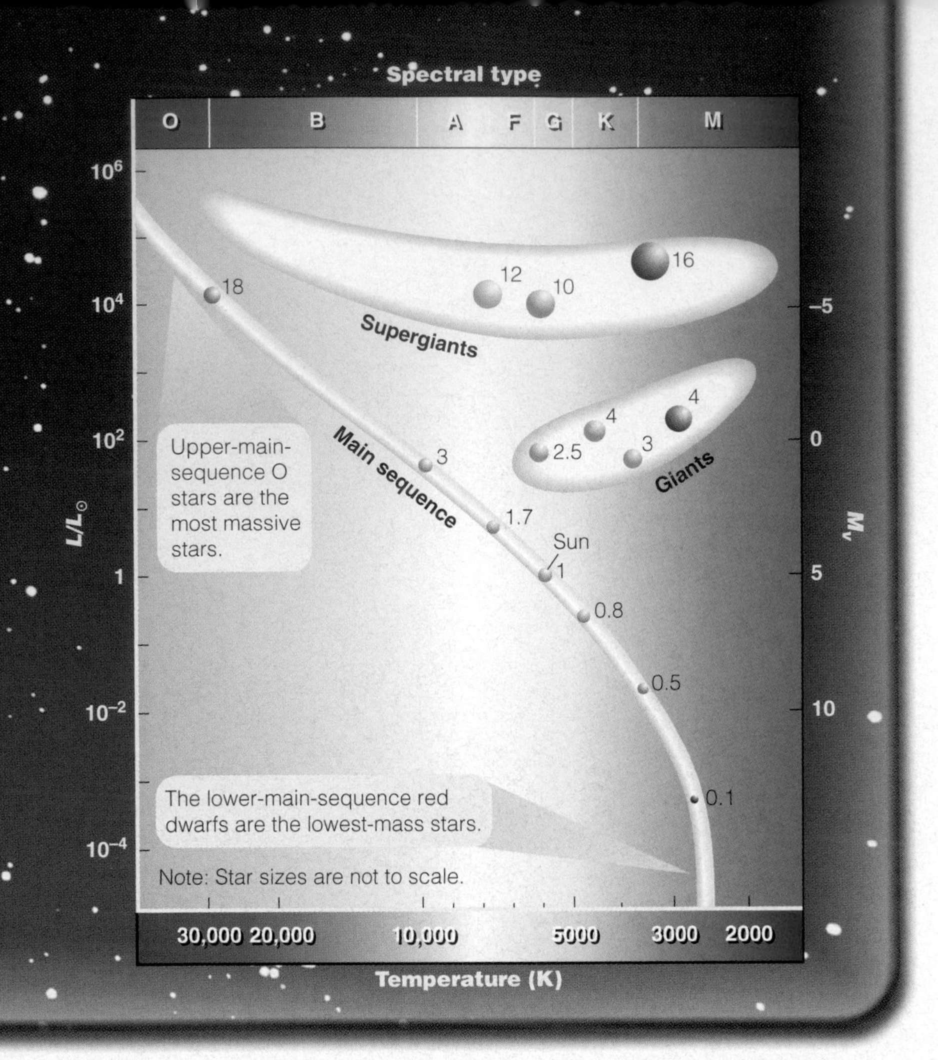

■ Figure 9-24

The masses of the plotted stars are labeled on this H–R diagram. Notice that the masses of main-sequence stars decrease from top to bottom but that masses of giants and supergiants are not arranged in any ordered pattern.

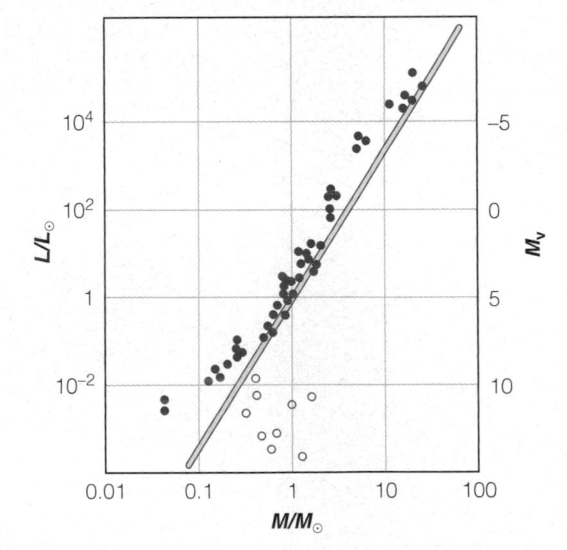

■ Figure 9-25

The mass–luminosity relation shows that the more massive a main-sequence star is, the more luminous it is. The open circles represent white dwarfs, which do not obey the relation. The red line represents the equation $L = M^{3.5}$.

◄ SCIENTIFIC ARGUMENT ►

What kind of stars do you see if you look at a few of the brightest stars in the sky?

This argument shows how careful you must be to interpret simple observations. When you look at the night sky, the brightest stars are mostly giants and supergiants. Most of the bright stars in Canis Major, for instance, are supergiants. Sirius, in Canis Major, is the brightest star in the sky, but it is a main-sequence star; it looks bright because it is very nearby, not because it is very luminous. In general, the supergiants and giants are so luminous that they stand out and look bright, even though they are not nearby. When you look at a bright star in the sky, you are probably looking at a highly luminous star—a supergiant or a giant. You can check the argument above by consulting the tables of the brightest and nearest stars in the Appendix.

Now revise your argument. **What kind of star do you see if you look at a few of the stars nearest to the sun?**

◄ ►

What Are We?

Medium Creatures

We humans are medium creatures, and we experience medium things. You can see trees and flowers and small insects, but you cannot see the beauty of the microscopic world without ingenious instruments and special methods. Similarly, you can sense the grandeur of a mountain range, but larger objects, such as stars, are too big to experience unless you use special methods to understand them. You must use your ingenuity and imagination to really understand such large objects. That is what science does for us. We live between the microscopic world and the astronomical world, and science enriches our lives by revealing the parts of the universe beyond our daily experience.

Humans have a natural drive to understand as well as experience. Now that you have experienced the stars as objects ranging from hot O stars to cool red dwarfs, it is natural for you to wonder why these stars are so different. As you explore that story in the following chapters, you will discover that the universe you live in is even more exciting and more beautiful than your medium senses suggest.

Study and Review

Summary

9-1 | Measuring the Distances to Stars

How far away are the stars?

- Your goal in this chapter was to characterize the stars by finding their distances, luminosities, diameters, and masses. You must find the distance first because finding luminosity, diameter, and mass depend on distance.
- Astronomers can measure the distance to nearer stars by observing their **stellar parallaxes.**
- Stellar distances are commonly expressed in **parsecs.** One parsec is 206,265 AU—the distance to an imaginary star whose parallax is 1 second of arc.
- Stellar parallaxes are very small angles and are difficult to measure from Earth's surface. A satellite orbiting above Earth's atmosphere has made high-precision parallax measurements of thousands of stars.
- **Proper motion,** the slow drift of stars across the sky, can be measured and can help astronomers find nearby stars by their rapid proper motions.

9-2 | Intrinsic Brightness

How much energy do stars make?

- Once you know the distance to a star, you can find its intrinsic brightness expressed as its **absolute visual magnitude** (M_v). The absolute magnitude of a star equals the apparent magnitude it would have if it were 10 pc away.
- The **magnitude-distance formula** relates apparent visual magnitude, absolute visual magnitude, and distance.
- The **distance modulus** ($m_v - M_v$) is simply related to the distance to the star.
- To find energy output of a star, astronomers must correct for the light at wavelengths that are not visible to convert absolute visual magnitude into **absolute bolometric magnitude,** which can be converted into **luminosity** (L), the total amount of energy the star radiates in a second.

9-3 | The Diameters of Stars

How big are stars?

- The **H–R (Hertzsprung-Russell) diagram** is a plot of luminosity versus surface temperature. It is an important graph in astronomy because it sorts the stars into categories by size.
- Roughly 90 percent of normal stars, including the sun, fall on the **main sequence,** with the more massive stars being hotter and more luminous. The **giants** and **supergiants,** however, are much larger and lie above the main sequence.
- **Red dwarfs** are small stars at the lower end of the main sequence, and the **white dwarfs** are also very small but fall below the main sequence. Some white dwarfs are very hot.
- Observations made with interferometers confirm the sizes of stars implied by the H-R diagram and reveal that rapidly rotating stars are slightly flattened.
- It is possible to assign stars to **luminosity classes** by the widths of their spectral lines. The large size of the giants and supergiants means their atmospheres have low densities and their spectra have sharper spectral lines than the spectra of main-sequence stars. Class V stars are main-sequence stars. Giant stars, class III, have sharper lines; and supergiants, class I, have extremely sharp spectral lines.
- **Spectroscopic parallax** allows astronomers to estimate the distance to stars too far away for direct parallax measurements. By classifying a star according to spectral type and luminosity class, an astronomer can estimate its absolute magnitude from the H-R diagram and then find its distance by comparing absolute and apparent magnitudes.

Now you can understand why these forbidden lines aren't visible in laboratories on Earth where extremely low density gas can't be produced. The atoms in a dense gas collide with each other so often that there isn't time for an electron in a metastable level to decay to a lower level. The atoms collide so often that such electrons get excited back up to higher levels before they can drop downward and emit a photon.

And that tells you something important about emission nebulae. In those nebulae, the gas has a very low density, and an atom could go for an hour or more between collisions, giving an electron stuck in a metastable level time to decay to a lower level and emit a photon at a so-called forbidden wavelength. Two good examples are the strong green lines at 495.9 nm and 500.7 nm produced by oxygen atoms that have lost two electrons. (Following the convention for naming ions, twice-ionized oxygen is OIII.) The oxygen ions can become excited by collision with a high-energy photon or a rapidly moving ion or electron, and the atom can emit various-wavelength photons as its electron cascades back down to lower energy levels. Some of those electrons get caught in metastable levels and eventually emit photos at forbidden wavelengths. The forbidden lines are clear evidence that nebulae have very low densities. This is a dramatic example of how astronomers can use knowledge of atomic physics to understand astronomical objects.

Nebulae are particularly beautiful manifestations of the interstellar medium, but there is much more to learn. Although you can't see most of the interstellar medium, the thin gas and dust still affect the light that passes through it.

Extinction and Reddening

The dust in space is called **interstellar dust.** Its presence is made dramatically evident by dark nebulae, but simple observations tell astronomers that interstellar dust is spread throughout space, making up roughly 1 percent of the mass of the interstellar medium.

One way astronomers know that dust is present in the interstellar medium is that it makes distant stars appear fainter than they would if space were perfectly transparent. This phenomenon is called **interstellar extinction,** and in the neighborhood of the sun it amounts to about 2 magnitudes per thousand parsecs. That is, a star 1000 pc from Earth will look about 2 magnitudes fainter than it would if space were perfectly empty. If it were 2000 pc away, it would look about 4 magnitudes dimmer, and so on. This is a dramatic effect, and it shows that the interstellar medium is not confined to a few nebulae scattered here and there. The so-called vacuum of space is filled with a low-density, dusty gas. The spaces between the stars are far from empty.

Another way dust reveals its presence is through the effect it has on the colors of stars. An O star should be blue, but some stars with the spectrum of an O star look much redder than they should. **Interstellar reddening** is produced by dust particles scattering light. As you saw in the case of the reflection nebulae, the dust particles are small, with diameters roughly equal to the wavelength of light, and they scatter shorter wavelengths better than longer wavelengths. As light from a distant star travels toward Earth it loses some of its shorter wavelength (blue) photons because of scattering, and consequently the star looks redder (■ Figure 10-2).

As discussed earlier, the scattering of blue light in Earth's atmosphere is what makes the sky blue, but it is also what makes distant city lights look yellow. If you view the lights of a city at night from a high-flying aircraft or a distant mountaintop, the lights will look yellow. As you descend toward the city, the lights will look less yellow. The light from the city is reddened by microscopic particles in the air. If the particles are especially dense, people call them smog.

Astronomers can measure the amount of reddening by comparing two stars of the same spectral type, one of which is dimmed more than the other. If you plot the difference in brightness between the two stars against wavelength, you get a curve that shows the reddening. That is, it shows how the starlight is dimmed at different wavelengths (■ Figure 10-3). In general, the light is dimmed in proportion to the reciprocal of the wavelength, a pattern typical of scattering from small dust particles. Laboratory measurements show that the high extinction at about 220 nm is caused by a form of carbon, evidence that some of the dust particles are carbon. Other evidence suggests that some grains contain silicates and metals and may have coatings of water ice, frozen ammonia, or carbon-based molecules.

Interstellar Absorption Lines

If you looked at the spectra of distant stars, you could see dramatic evidence of an interstellar medium. Of course, you would see spectral lines produced by the gas in the atmospheres of the stars, but you would also see sharp spectral lines produced by the gas in the interstellar medium. These **interstellar absorption lines** provide a new way to study the gas between the stars. Astronomers can recognize interstellar absorption lines in three ways: by their ionization, by their widths, and by their multiple components.

Some stellar spectra contain absorption lines that just don't belong because they represent the wrong ionization state. For example, if you looked at the spectrum of a very hot star such as an O star, you would expect to see no lines of once-ionized calcium (CaII) because that ion cannot exist at the high temperatures in the atmosphere of such a hot star. But many O-star spectra contain lines of CaII. These lines must have been produced not in the star but in the interstellar medium.

The widths of the interstellar lines also give away their identity. Recall from Chapter 7 that Doppler broadening and collisional broadening widen the spectral lines in a star's atmosphere. This blurring makes the lines in the spectrum of a main-sequence

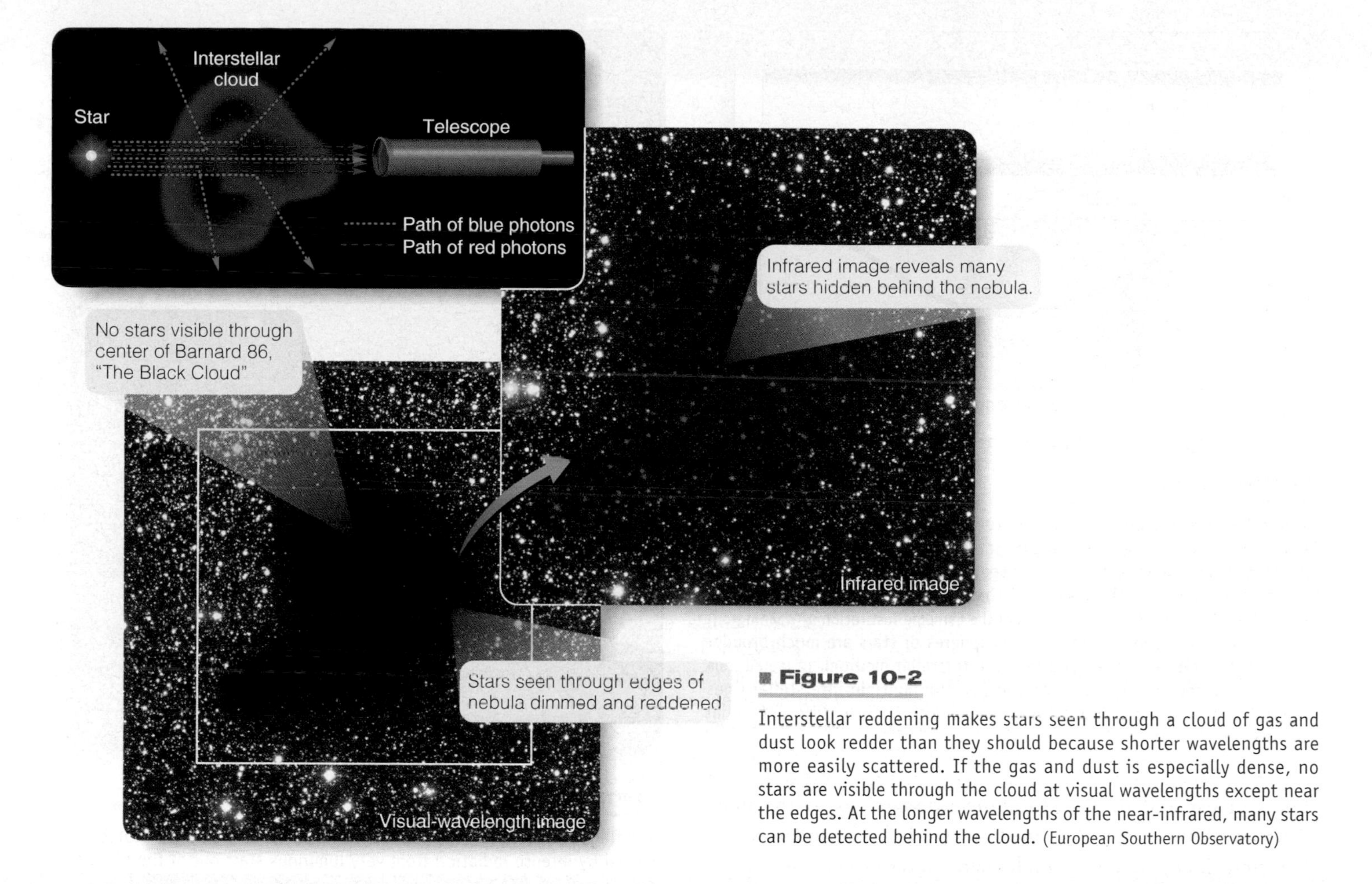

■ **Figure 10-2**

Interstellar reddening makes stars seen through a cloud of gas and dust look redder than they should because shorter wavelengths are more easily scattered. If the gas and dust is especially dense, no stars are visible through the cloud at visual wavelengths except near the edges. At the longer wavelengths of the near-infrared, many stars can be detected behind the cloud. (European Southern Observatory)

star quite broad, but even in the atmosphere of a giant or supergiant star, where the gas is less dense, the gas atoms move rapidly enough and collide often enough to broaden the spectral lines. Interstellar lines, on the other hand, are exceedingly sharp, confirming that the interstellar matter is extremely cold and has a low density. If it were hot, Doppler broadening would smear out the lines due to the motions of individual atoms. If the gas were dense, collisional broadening would produce wider lines.

Another revealing characteristic of the interstellar lines is that they are often split into two or more components. The multiple components have slightly different wavelengths and appear to have been produced when the light from the star passed through different clouds of gas on its way to Earth. Because separate clouds of gas have slightly different radial velocities, they produce absorption lines with slightly Doppler-shifted wavelengths. ■ Figure 10-4 illustrates all three of these characteristics of interstellar absorption lines—the wrong ionization, narrow widths, and multiple components.

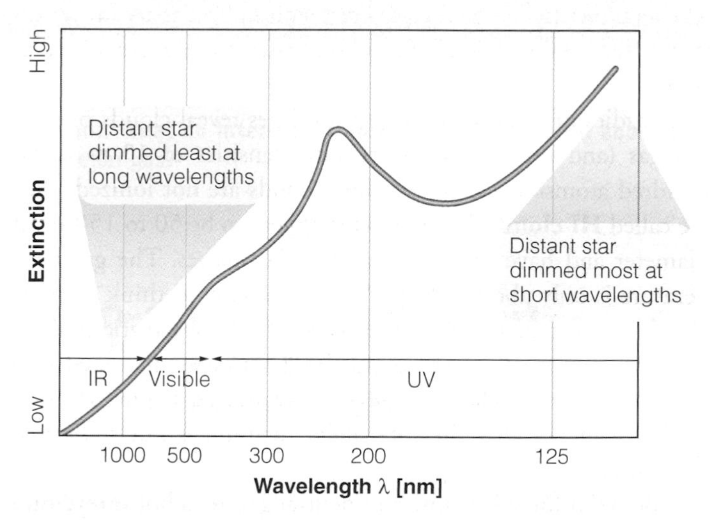

■ **Figure 10-3**

Interstellar extinction, the dimming of starlight by dust between the stars, depends strongly on wavelength. Infrared radiation is only slightly affected, but ultraviolet light is strongly scattered. The strong extinction at about 220 nm is caused by a certain form of carbon dust in the interstellar medium.

Clouds and What's in Between

Emission nebulae, reflection nebulae, and dark nebulae are only the most obvious parts of the interstellar medium. Space is filled with low-density gas and in some places with slightly denser

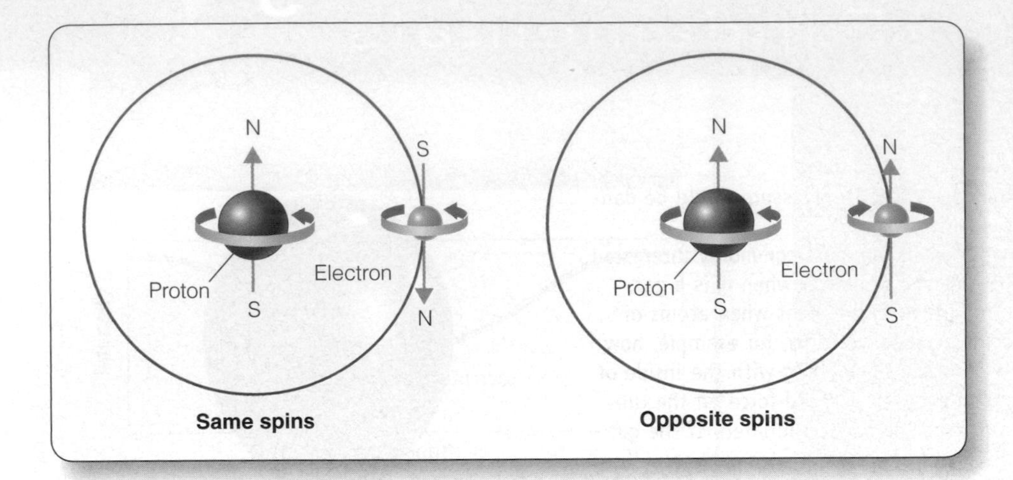

■ Figure 10-6

Both the proton and electron in a neutral hydrogen atom spin and consequently have small magnetic fields. When they spin in the same direction, their magnetic fields are reversed, and when they spin in opposite directions, their magnetic fields are aligned. As explained in the text, this allows cold, neutral hydrogen in space to emit radio photons with a wavelength of 21 cm.

The existence of the 21-cm radiation was predicted theoretically in the 1940s by H. C. van de Hulst, but it was not detected until 1951. You can understand how the theoretical prediction was made by thinking of the structure of a hydrogen atom. A hydrogen atom consists of a proton and an electron, and physicists know that both of these particles must spin perpetually like tiny tops. Because these particles have an electrostatic charge, their rapid spin has the same effect as the circulation of an electric current through a coil of wire: it creates a magnetic field. Because the charge on the proton is positive and the charge on the electron is negative, the two magnetic fields have opposite polarity when the particles spin in the same direction. If the particles spin in opposite directions, their magnetic fields are aligned.

You have probably played with small magnets and noticed that they repel each other in one orientation and attract each other if you turn one around. In the same way, the small magnetic fields produced by the spinning proton and electron can repel or attract each other, and that affects the binding energy that holds the electron to the proton in a hydrogen atom. In one orientation the electron is slightly less tightly bound, and in the other orientation it is slightly more tightly bound (■ Figure 10-6). The energy difference is very small, but it makes a big difference in astronomy.

Now you can follow van de Hulst's logic and predict the existence of 21-cm radiation. Because an electron in the ground state of a hydrogen atom could spin in either of two directions—the same way as the proton or the opposite way—the ground state must really be two energy levels separated by a tiny amount of energy—the difference between the two ways the electron could spin. If an electron is spinning such that it is in the higher of the two states, it can spontaneously flip over and spin in the other direction. Then the atom must drop to the lower of the two energy levels, and the excess energy is radiated away as a photon. The energy difference is small, so the photon has a long wavelength—21 cm (■ Figure 10-7).

The existence of 21-cm radiation was predicted theoretically, but it could not be detected in laboratory experiments. Of the two closely spaced energy levels, the upper one is metastable. Once an electron gets caught in the upper energy level, it will, on average, stay there for 11 million years before spontaneously dropping to the lower energy level and emitting a photon of 21-cm wavelength. The atoms in a gas in a laboratory jar will collide with each other millions of times a second, so none of those atoms can remain undisturbed long enough to produce a photon of 21-cm radiation. Atoms in space, however, collide much less often, so a few do manage to emit 21-cm radiation. That's why the

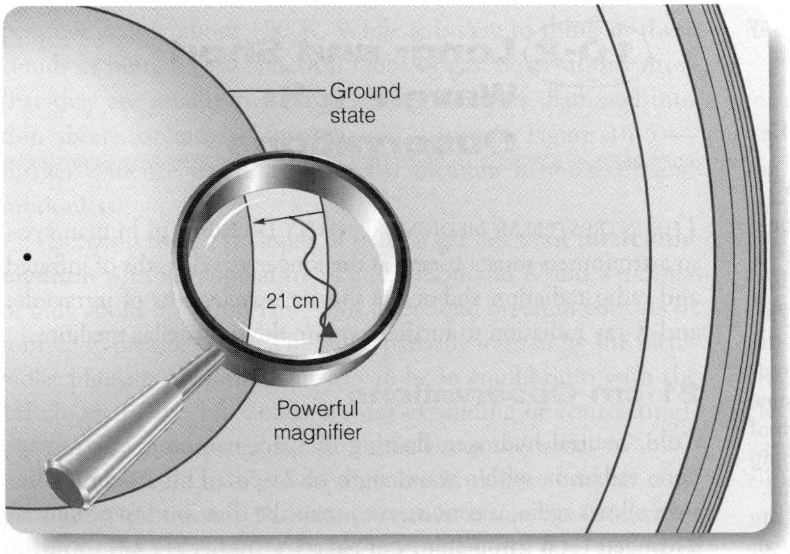

■ Figure 10-7

The lowest energy level of the hydrogen atom, the ground state, is actually two closely spaced energy levels that differ because the proton and electron spin. In this diagram, you would need a magnifying glass to distinguish the energy levels that make up the ground state. When an atom decays from the upper energy level to the lower energy level, it emits a photon with a wavelength of 21 cm.

existence of 21-cm radiation had to be confirmed observationally by radio astronomers who detected it coming from clouds of neutral hydrogen in space.

The 21-cm observations give astronomers a way to map the cold, neutral hydrogen that fills much of our galaxy. Astronomers can locate individual clouds and measure their motion by observing the Doppler shifts in the 21-cm radiation (■ Figure 10-8).

Radio observations of 21-cm radiation can map only neutral hydrogen. Ionized hydrogen lacks an electron, so it can't emit 21-cm radiation. Further, hydrogen atoms locked in molecules are also unable to emit 21-cm wavelength photons. Astronomers must find other ways to study these parts of the interstellar medium.

Molecules in Space

Radio telescopes can also detect radiation from various molecules in the interstellar medium. A molecule can store energy in a number of different ways. For example, it can rotate at different rates, or the atoms in a molecule can vibrate as if they were linked together by small springs. If a molecule suffers a collision or absorbs a photon, it can be excited to vibrate and rotate in some higher energy state. However, it will return to a lower energy state quickly and radiate the excess energy as a photon. These energy levels are closely spaced, so the emitted photons typically have low energies, and they fall in the radio or far-infrared part of the electromagnetic spectrum. Just as neutral hydrogen radiates at a specific wavelength of 21 cm, many natural molecules radiate at their own unique wavelengths.

Unfortunately for astronomers, molecules of hydrogen (H_2) are difficult to detect. In the far-ultraviolet, molecular hydrogen can be detected by the photons it absorbs, but these observations must be made by specialized telescopes in space. Mapping the location of molecular hydrogen would be easier if the molecules emitted radio-wavelength photons, but a molecule containing two identical atoms does not radiate in the radio part of the spectrum. However, clouds of gas dense enough to form molecular hydrogen also form tiny amounts of other molecules, and many of those molecules are good emitters of radio energy. A **molecular cloud** is a region of the interstellar medium that is dense enough to form molecules. Nearly 100 different molecules have been detected (■ Table 10-1). Some are quite complex, and it is not clear how they form. Most astronomers believe that the atoms meet and bond to form molecules on the surfaces of dust grains. Some of the molecules have not yet been synthesized on

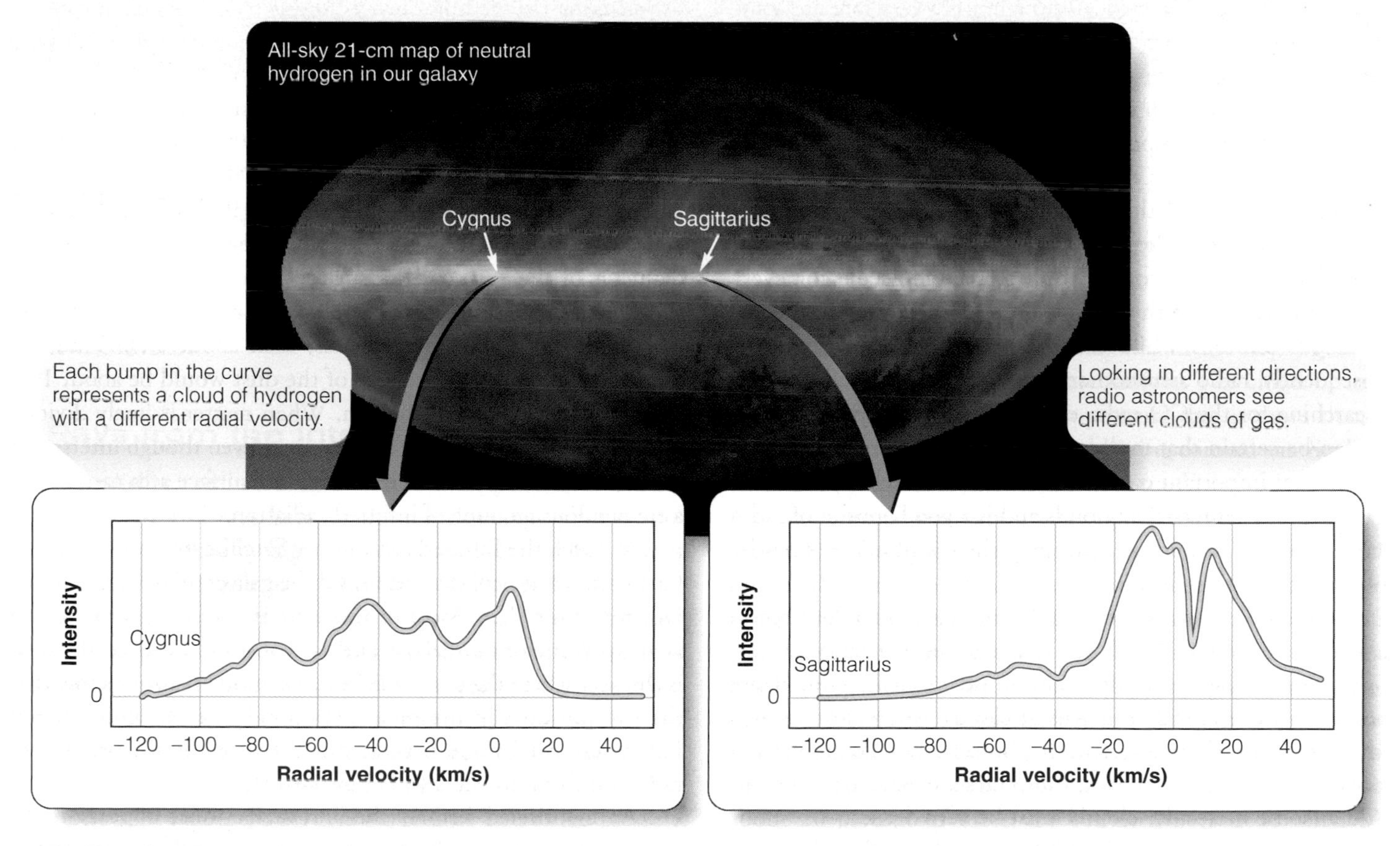

■ **Figure 10-8**

Most of the neutral hydrogen in the Milky Way Galaxy is in the plane of the disk-shaped galaxy—the bright band running from left to right. Notice how wispy and irregular the hydrogen is. (J. Dickey, UMn, F. Lockman, NRAO, SkyView) 21-cm radio observations reveal many clouds of neutral hydrogen orbiting the center of the galaxy. (Adapted from observations by Burton)

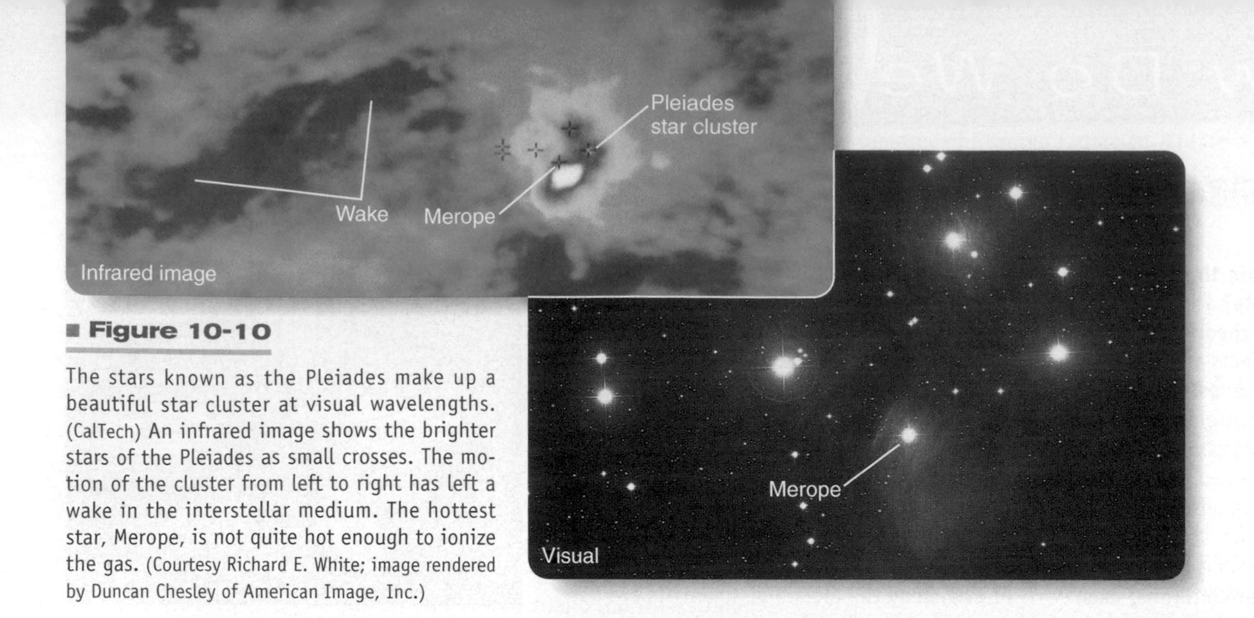

■ **Figure 10-10**

The stars known as the Pleiades make up a beautiful star cluster at visual wavelengths. (CalTech) An infrared image shows the brighter stars of the Pleiades as small crosses. The motion of the cluster from left to right has left a wake in the interstellar medium. The hottest star, Merope, is not quite hot enough to ionize the gas. (Courtesy Richard E. White; image rendered by Duncan Chesley of American Image, Inc.)

■ **Figure 10-11**

(a) The nebula N44C is part of a larger region where young, hot, massive stars have formed and where at least one supernova has blown a bubble filled with hot gas. (The Hubble Heritage Team STScI/ AURA/NASA) (b) This image of the entire N44 region combines a visual-wavelength image (white) with an X-ray image (red). X rays reveal regions of very-high-temperature gas that has been ejected from supernova explosions. (R. Chris Smith, CTIO and Y.-H. Chu, UICU)

The composition of the interstellar dust suggests that it is formed mostly in the atmospheres of cool stars. There the temperatures are low enough for some atoms to condense into specks of solid matter, much as soot can condense in a candle flame. The pressure of the starlight can push these dust specks out of the star where they replenish the interstellar medium. Other stars that eject mass into space, such as supernovae, probably also add to the supply of interstellar dust.

Supernova explosions keep the interstellar medium stirred and create the vast regions filled with coronal gas (■ Figure 10-11). Much of the motion in interstellar clouds and their twisted shapes is probably produced by these supernova explosions and the hot coronal gas they eject. It seems that our galaxy produces about two supernova explosions per century (although most are not visible from Earth because they are obscured by dust). In fact, the sun lies inside such a region of high-temperature, low-density gas called the local bubble or void. With a typical diameter of a few hundred parsecs, the local bubble may be a cavity in the interstellar medium inflated by a supernova explosion that occurred within the last million years or so.

What Are We?

The Cosmic Ocean

City dwellers see only the brightest stars scattered across the sky, but even those who live far from city lights see almost nothing in the night sky except the stars. It is easy to imagine that space is empty. Only a few hazy nebulae such as that in Orion's belt hint at the hidden richness of space.

We live in a universe filled with beauty beyond our senses. Special telescopes and cameras can capture photographs of glowing gas containing brilliant stars and dark clouds of gas twisting through space. In contrast, the cold clouds of neutral gas and the hot bubbles of coronal gas don't show up in most astronomical images, but they are part of the complex beauty of the interstellar medium. All around us, stars drift through an invisible ocean whose peaceful reaches are blasted here and there by storms of light and heat. Science enriches our lives by revealing the beauty we cannot sense.

■ Figure 10-12

The Trifid Nebula embodies an important part of the interstellar cycle. A dense cloud of gas has given birth to stars, and the hottest are ionizing the gas to produce the pink emission nebula, while dust scatters light to produce the blue reflection nebula. As the nebula is disrupted by the newborn stars, the gas and dust merge with the interstellar medium. (NOAO; Inset: NASA, ESA, and The Hubble Heritage Team, AURA/STScI)

The agitation caused by supernova explosions and the hot gas blowing away from the hottest stars combined with the natural gravitation of gas clouds and their orbital motions around the galaxy produces collisions between clouds that may gradually build more massive clouds. In the most massive clouds, the dust protects the interior from ultraviolet photons, and molecules can form. These giant molecular clouds eventually give birth to new stars, and the cycle begins all over again.

The Trifid Nebula (■ Figure 10-12) is a dramatic illustration of this cycle. Measuring over 12 pc in diameter, the nebula is illuminated by a number of hot, young stars that have apparently formed recently from the gas. Near the stars, the gas is ionized and glows as a pink-red HII region; but farther from the stars, where the ultraviolet radiation is weaker, the gas is not ionized. Nevertheless, dust in the nebula scatters blue light, so this un-ionized part of the nebula is visible as a blue reflection nebula. Dark lanes of obscuring dust cross the face of the nebula as if to remind Earth's astronomers again of the importance of dust in the interstellar cycle.

Star Formation in the Orion Nebula

1 The visible Orion Nebula shown below is a pocket of ionized gas on the near side of a vast, dusty molecular cloud that fills much of the southern part of the constellation Orion. The molecular cloud can be mapped by radio telescopes. To scale, the cloud would be many times larger than this page. As the stars of the Trapezium were born in the cloud, their radiation has ionized the gas and pushed it away. Where the expanding nebula pushes into the larger molecular cloud, it is compressing the gas (see diagram at right) and may be triggering the formation of the protostars that can be detected at infrared wavelengths within the molecular cloud.

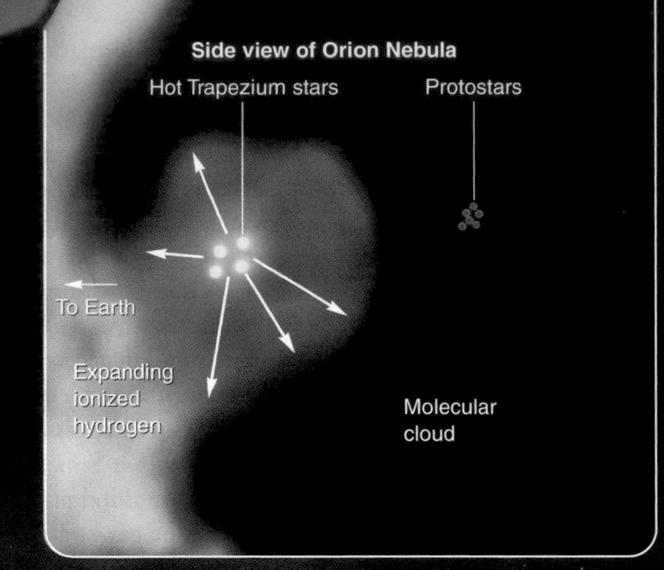

Hundreds of stars lie within the nebula, but only the four brightest, those in the Trapezium, are easy to see with a small telescope. A fifth star, at the narrow end of the Trapezium, may be visible on nights of good seeing.

The cluster of stars in the nebula is less than 2 million years old. This must mean the nebula is similarly young.

Trapezium

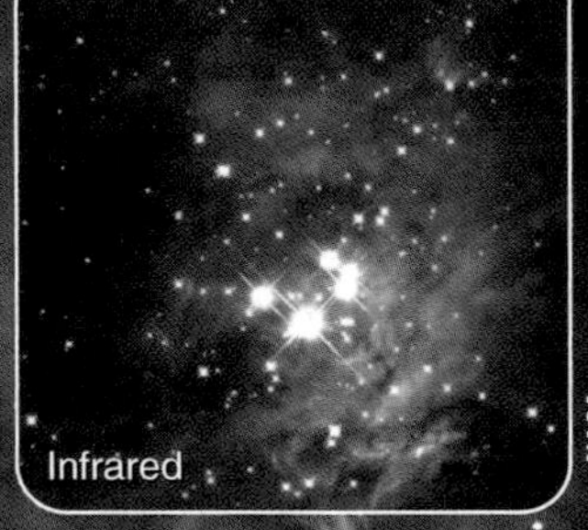

NASA

The near-infrared image above reveals about 50 low-mass, very cool stars that must have formed recently.

NASA/CXC/SAO

Roughly 1000 young stars with hot chromospheres appear in this X-ray image of the Orion Nebula.

Visual-wavelength image

Credit: NASA, ESA, M. Robberto, STScI and the Hubble Space Telescope Orion Treasury Project Team

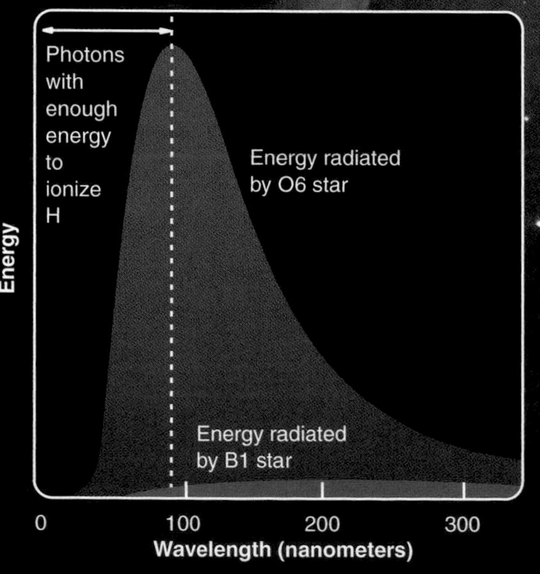

2 Of all the stars in the Orion Nebula, only one is hot enough to ionize the gas. Only photons with wavelengths shorter than 91.2 nm can ionize hydrogen. The second-hottest stars in the nebula are B1 stars, and they emit little of this ionizing radiation. The hottest star, however, is an O6 star 30 times the mass of the sun. At a temperature of 40,000 K, it emits plenty of photons with wavelengths short enough to ionize hydrogen. Remove that one star, and the nebula would turn off its emission.

What Are We?

The Cosmic Ocean

City dwellers see only the brightest stars scattered across the sky, but even those who live far from city lights see almost nothing in the night sky except the stars. It is easy to imagine that space is empty. Only a few hazy nebulae such as that in Orion's belt hint at the hidden richness of space.

We live in a universe filled with beauty beyond our senses. Special telescopes and cameras can capture photographs of glowing gas containing brilliant stars and dark clouds of gas twisting through space. In contrast, the cold clouds of neutral gas and the hot bubbles of coronal gas don't show up in most astronomical images, but they are part of the complex beauty of the interstellar medium. All around us, stars drift through an invisible ocean whose peaceful reaches are blasted here and there by storms of light and heat. Science enriches our lives by revealing the beauty we cannot sense.

■ Figure 10-12

The Trifid Nebula embodies an important part of the interstellar cycle. A dense cloud of gas has given birth to stars, and the hottest are ionizing the gas to produce the pink emission nebula, while dust scatters light to produce the blue reflection nebula. As the nebula is disrupted by the newborn stars, the gas and dust merge with the interstellar medium. (NOAO; Inset: NASA, ESA, and The Hubble Heritage Team, AURA/STScI)

The agitation caused by supernova explosions and the hot gas blowing away from the hottest stars combined with the natural gravitation of gas clouds and their orbital motions around the galaxy produces collisions between clouds that may gradually build more massive clouds. In the most massive clouds, the dust protects the interior from ultraviolet photons, and molecules can form. These giant molecular clouds eventually give birth to new stars, and the cycle begins all over again.

The Trifid Nebula (■ Figure 10-12) is a dramatic illustration of this cycle. Measuring over 12 pc in diameter, the nebula is illuminated by a number of hot, young stars that have apparently formed recently from the gas. Near the stars, the gas is ionized and glows as a pink-red HII region; but farther from the stars, where the ultraviolet radiation is weaker, the gas is not ionized. Nevertheless, dust in the nebula scatters blue light, so this unionized part of the nebula is visible as a blue reflection nebula. Dark lanes of obscuring dust cross the face of the nebula as if to remind Earth's astronomers again of the importance of dust in the interstellar cycle.

◄ SCIENTIFIC ARGUMENT ►

How can the coronal gas make up only a few percent of the mass but 20 percent of the volume of the interstellar medium?

Once again your argument needs to focus on some simple physics. The solution to this puzzle is the extremely low density of the coronal gas, roughly one particle for every thousand cubic centimeters. The coronal gas is the least dense part of the interstellar medium. It is a much better vacuum than any laboratory vacuum on Earth. It doesn't take much mass in coronal gas to occupy a large volume. In contrast, molecular clouds are the densest part, and they contain roughly a quarter of the mass, although they occupy only a small part of the volume.

The density and temperature of the different components of the interstellar medium determine how you could observe them. Create a new argument to consider observations. **What techniques would you use to observe the coronal gas and the molecular clouds?**

◄ ►

Study and Review

Summary

10-1 | Visible-Wavelength Observations

How do visual-wavelength observations show that space isn't empty?

- The **interstellar medium**, the gas and dust between the stars, is mostly concentrated near the plane of our Milky Way Galaxy.
- A **nebula** is a cloud of gas in space, and an **emission nebula** (also known as an **HII region)** is produced when ultraviolet radiation from nearby hot stars ionizes nearby gas, making it glow like a giant neon sign. The red, blue, and violet Balmer lines blend together to produce the characteristic pink-red color of ionized hydrogen.
- A **reflection nebula** is produced by gas and dust illuminated by a star that is not hot enough to ionize the gas. Rather, the dust scatters the starlight to produce a reflection of the stellar absorption spectrum. Because shorter-wavelength photons scatter more easily than longer-wavelength photons, reflection nebulae look blue. The daytime sky looks blue for the same reason.
- A **dark nebula** is a cloud of gas and dust that is visible because it blocks the light of distant stars. The irregular shapes of these dark nebulae reveal the turbulence in the interstellar medium.
- The gas in nebulae has a low density, and the atoms collide so rarely that an electron caught in a **metastable level** can remain there long enough to finally fall to a lower level. This produces so-called **forbidden lines** that are not seen in laboratory spectra on Earth where the atoms in the gas collide too often.
- **Interstellar dust** makes up roughly 1 percent of the mass of the interstellar medium.
- **Interstellar extinction**, or dimming, makes the distant stars look fainter than they should. **Interstellar reddening** makes distant stars appear too red because dust particles in the interstellar medium scatter blue light more easily than red light. The dependence of this extinction on wavelength shows that the scattering dust particles are very small. The dust is made of carbon, silicates, iron, and ice.
- **Interstellar absorption lines** in the spectra of distant stars are very narrow. The interstellar gas is cold and has a very low density, and this makes the interstellar lines much narrower than the spectral lines produced in stars. Multiple interstellar lines reveal that the light has passed through more than one interstellar cloud on its way to Earth.
- **HI clouds** of neutral hydrogen are separated by a hotter but lower-density gas called the **intercloud medium,** and the two appear to have similar **pressures** and are in equilibrium.

10-2 | Long- and Short-Wavelength Observations

How are observations at nonvisible wavelengths evidence of invisible matter between the stars?

- The **21-cm radiation** allows radio astronomers to map the distribution of neutral hydrogen gas in the interstellar medium.
- Radio telescopes tuned to other wavelengths have detected nearly 100 different molecules in the interstellar medium.
- Hydrogen molecules (H_2) do not radiate at radio wavelengths, but molecules such as CO and OH, which are present as impurities, do emit radio energy and allow astronomers to map cold, dense **molecular clouds.** The largest of these are called **giant molecular clouds.**
- Molecules can be broken up by short-wavelength photons, but the dust in molecular clouds scatters these photons and shields the molecules in the inner part of the cloud.
- Infrared observations, some made from above Earth's atmosphere, have mapped the distribution of the cold interstellar dust. Although it is very cold, the dust has a large surface area and emits significant amounts of infrared.
- X-ray observations have detected very hot **coronal gas** produced by supernova explosions. Far-ultraviolet observations show that the sun is located in a **local bubble** (or **local void**) of this coronal gas.

10-3 | A Model of the Interstellar Medium

How does the matter between the stars interact with the stars?

- The four main components of the interstellar medium are the small neutral HI clouds, the warm intercloud medium, coronal gas, and molecular clouds.

- Stars are born in the dense molecular clouds, and the energy from hot stars and supernova explosions causes currents in the interstellar medium and creates the coronal gas. The dust in the interstellar medium is formed in the atmospheres of cool stars and from gas ejected by supernova explosions.
- The collision of gas clouds builds massive clouds in which new stars are born.

Review Questions

ThomsonNOW Assess your understanding of this chapter's topics with additional quizzing and animations at http:www.thomsonedu.com

WebAssign The problems from this chapter may be assigned online in WebAssign.

1. What evidence can you cite that the spaces between the stars are not totally empty?
2. What evidence can you cite that the interstellar medium contains both gas and dust?
3. How do the spectra of HII regions differ from the spectra of reflection nebulae? Why?
4. Why are interstellar lines so narrow? Why do some spectral lines forbidden in spectra on Earth appear in spectra of interstellar clouds and nebulae? What does that tell you?
5. How is the blue color of a reflection nebula related to the blue color of the daytime sky?
6. Why do distant stars look redder than their spectral types suggest?
7. If starlight on its way to Earth passed through a cloud of interstellar gas that was hot instead of very cold, would you expect the interstellar absorption lines to be broader or narrower than usual? Why?
8. How can the HI clouds and the intercloud medium have similar pressures when their temperatures are so different?
9. Why can the 21-cm radio emission line of neutral hydrogen be observed in the interstellar medium but not in the laboratory?
10. What does the shape of the 21-cm radio emission line of neutral hydrogen tell you about the interstellar medium?
11. What produces the coronal gas?
12. **How Do We Know?** Why are scientists free to adjust their theories but not their facts?

Discussion Questions

1. When you see distant streetlights through smog, they look dimmer and redder than they do normally. But when you see the same streetlights through fog or falling snow, they look dimmer but not redder. Use your knowledge of the interstellar medium to discuss the relative sizes of the particles in smog, fog, and snowstorms compared to the wavelength of light.
2. If you could see a few stars through a dark nebula, how would you expect their spectra and colors to differ from similar stars just in front of the dark nebula?

Problems

1. A small nebula has a diameter of 20 seconds of arc and a distance of 1000 pc from Earth. What is the diameter of the nebula in parsecs? In meters?
2. The dust in a molecular cloud has a temperature of about 50 K. At what wavelength does it emit the maximum energy? (*Hint:* Consider black body radiation, Chapter 7.)
3. Extinction dims starlight by about 1.9 magnitudes per 1000 pc. What fraction of photons survives a trip of 1000 pc? (*Hint:* Consider the definition of the magnitude scale in Chapter 2.)
4. If the total extinction through a dark nebula is 10 magnitudes, what fraction of photons makes it through the cloud? (*Hint:* See Problem 3.)
5. The density of air in a child's balloon 20 cm in diameter is roughly the same as the density of air at sea level, 10^{19} particles/cm^3. To how large a diameter would you have to expand the balloon to make the gas inside the same density as the interstellar medium, about 1 particle/cm^3? (*Hint:* The volume of a sphere is $\frac{4}{3}\pi R^3$.)
6. If a giant molecular cloud has a diameter of 30 pc and drifts relative to neighboring clouds at 20 km/s, how long will it take to travel its own diameter?
7. An HI cloud is 4 pc in diameter and has a density of 100 hydrogen atoms/cm^3. What is its total mass in kilograms? (*Hints:* The volume of a sphere is $\frac{4}{3}\pi R^3$, and the mass of a hydrogen atom is 1.67×10^{-27} kg.)
8. Find the mass in kilograms of a giant molecular cloud that is 30 pc in diameter and has a density of 300 hydrogen molecules/cm^3. (*Hint:* See Problem 7.)
9. At what wavelength does the coronal gas radiate most strongly? (*Hint:* Consider black body radiation, Chapter 7.)

Learning to Look

1. The bright-blue star at lower left in Figure 10-1 is surrounded by a blue nebula. What can you conclude about the temperature of this star?
2. In Figure 10-9b, imagine that the Cygnus Loop was the same distance as the Cygnus Superbubble and compare their diameters.
3. The image here shows two nebulae, one pink in the background and one black in the foreground. What kind of nebulae are these?

NASA and the Hubble Heritage Team, STScI/AURA

11 The Formation of Stars

Nebula NGC 1333 is filled with newborn stars caught in the act of forming in this false-color infrared image. (NASA/JPL-Caltech/R. A. Gutermuth, Harvard-Smithsonian CfA)

Guidepost

The last chapter introduced you to the gas and dust between the stars. Here you will begin putting together observations and theories to understand how nature makes stars. That will answer four essential questions:

How are stars born?

How do stars make energy?

How do stars maintain their stability?

What evidence do astronomers have that theories of star formation are correct?

Astronomers have developed a number of theories that explain the birth of stars. Are they true? That raises one of the most important questions you will meet concerning science:

How Do We Know? **How certain can a theory be?**

As you learn how nature makes new stars you will see science in action as evidence and theory combine to produce real understanding.

Jim he allowed [the stars] was made, but I allowed they happened. Jim said the moon could'a laid them; well, that looked kind of reasonable, so I didn't say nothing against it, because I've seen a frog lay most as many, so of course it could be done.

MARK TWAIN, *THE ADVENTURES OF HUCKLEBERRY FINN*

THE STARS YOU SEE TONIGHT are the same stars your parents, grandparents, and great-grandparents saw. Stars change hardly at all in a human lifetime, but they are not eternal. Stars are born, and stars die. This chapter begins that story.

In this chapter, you will see how gravity creates stars from the thin gas of space and how nuclear reactions inside stars generate energy. You will see how the flow of that energy outward toward the stars' surfaces balances gravity and makes the stars stable. To understand that story you will plunge from the cold gas of the interstellar medium into the hot cores of the stars themselves.

11-1 Making Stars from the Interstellar Medium

It is a **Common Misconception** that the stars are eternal. The stars you see tonight are the stars your ancestors saw centuries ago, so it is reasonable to think the stars are unchanging. In fact, astronomers find strong evidence that stars age and die and that new stars are being born all around the sky.

Astronomers find hundreds of places in the sky where clouds of gas glow brightly because they are illuminated by massive, luminous, hot stars. Stars like these pour out such floods of energy they cannot live very long. For example, your Favorite Star Spica (α Virginis), a B1 main-sequence star, cannot last more than 10 million years and must have formed recently. Apparently these stars formed from the gas and dust around them (■ Figure 11-1).

The evidence shows that stars form from such clouds, much as raindrops condense from the water vapor in a thundercloud. Indeed, the giant molecular clouds discussed in the preceding chapter can give birth to entire clusters of new stars. But how can these large, low-density, cold clouds of gas become comparatively small, high-density, hot stars? Gravity is the key.

Star Birth in Giant Molecular Clouds

Giant molecular clouds are sites of active star formation, yet at first glance they are nothing like stars. With a typical diameter of 50 pc and a typical mass exceeding 10^5 solar masses, a giant molecular cloud is vastly larger than a star. Also, the gas in a giant molecular cloud is about 10^{20} times less dense than a star and has

Nebula N44
Roughly 40 young stars are inflating a bubble of hot gas inside the nebula from which they formed.
100 ly
Visual-wavelength image
Small dark nebulae may form stars.
NGC 3603
Aging supergiant has ejected ring of gas.
Newborn stars still in their birth nebulae
Visual-wavelength image

■ Figure 11-1

Nebula N44 has given birth to a large cluster of stars. (ESO) Various stages of star formation are evident in NGC3603, including a massive star ejecting gas as it approaches its end. (W. Brandner, JPL/IPAC, E.K. Grebel, Univ. of Washington, Y. Chu, Univ. of Illinois, and NASA)

temperatures of only a few degrees Kelvin. These clouds can form stars if gravity can force some small regions of the clouds to contract to high density and high temperature.

Both theory and observations suggest that many giant molecular clouds cannot begin the formation of stars spontaneously. At least four factors resist the contraction of a gas cloud, and gravity must overcome those four factors before star formation can begin.

First, thermal energy in the gas is present as motion among the atoms and molecules. Even at the very low temperature of 10 K, the average hydrogen molecule moves at about 0.35 km/s (almost 800 mph). This thermal motion would make the cloud drift apart if gravity were too weak to hold it together.

The interstellar medium is permeated by a magnetic field only about 0.0001 times as strong as that on Earth, but it can act like an internal spring and prevent the gas from contracting. Neutral atoms and molecules are unaffected by a magnetic field, but ions, having an electric charge, cannot move freely through a magnetic field. Although the gas in a molecular cloud is mostly neutral, there are some ions, and that means a magnetic field can exert a force on the gas. In some cases, the gas in giant molecular clouds can gradually recombine with free electrons and become less ionized. Neutral gas is free to "slip past" the magnetic field and contract. This gradual process has been observed inside isolated molecular clouds. In any case, gravity must overcome the interstellar magnetic field if it is to make the gas contract.

The third factor is rotation. Everything in the universe rotates. As a gas cloud begins to contract, it spins more and more rapidly as it conserves angular momentum, just as ice-skaters spin faster as they pull in their arms (Figure 5-7). This rotation can become so rapid that it resists further contraction of the cloud.

Turbulence in the interstellar medium is the fourth thing that could prevent a cloud from contracting. In the previous chapter, you learned that nebulae are often twisted and distorted by strong currents. This turbulence could make it difficult for a large molecular cloud to contract.

Given these four resistive factors, it seems surprising that any giant molecular clouds can contract at all, but radio observations show that at least some giant molecular clouds develop regions called dense cores that are only 0.1 pc in radius and that contain roughly 1 solar mass. A single giant molecular cloud may contain many of these dense cores and, as gas and dust fall into the dense cores, the cloud can give birth to star clusters containing hundreds of stars.

Both theory and observation suggest that many giant molecular clouds are triggered to form stars by a passing **shock wave,** the astronomical equivalent of a sonic boom (■ Figure 11-2). During such a triggering event, a few regions of the large cloud can be compressed to such high densities that the resistive factors can no longer oppose gravity, and star formation begins.

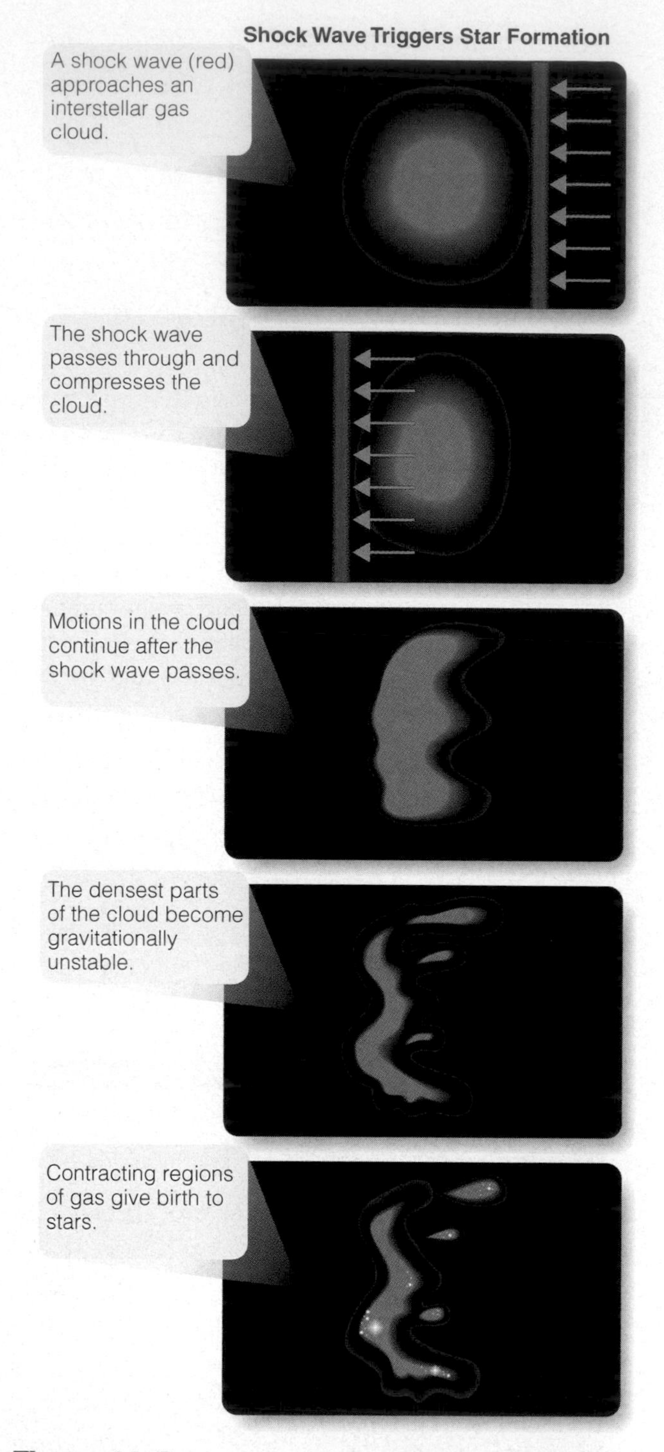

Shock Wave Triggers Star Formation
A shock wave (red) approaches an interstellar gas cloud.
The shock wave passes through and compresses the cloud.
Motions in the cloud continue after the shock wave passes.
The densest parts of the cloud become gravitationally unstable.
Contracting regions of gas give birth to stars.

■ **Figure 11-2**

In this summary of a computer model, an interstellar gas cloud is triggered into star formation by a passing shock wave. The events summarized here might span about 6 million years.

Shock waves are not uncommon in the interstellar medium. Supernova explosions (Chapter 13) can produce powerful shock waves that rush through the interstellar medium. Also, the ignition of very hot stars can ionize nearby gas and drive it away, producing a shock wave where it pushes into the colder, denser interstellar matter. A third trigger is the collision of molecular clouds. Because the clouds are large, they are likely to run into each other occasionally; and, because they contain magnetic fields, they cannot pass through each other. A collision between such clouds can compress parts of the clouds and trigger star formation. The fourth trigger is the spiral pattern of our Milky Way Galaxy (see Figure 1-11). One theory suggests that the spiral arms are shock waves that travel around the galaxy like the moving hands of a clock (Chapter 15). As a cloud passes through a spiral arm, the cloud could be compressed, and star formation could begin. Astronomers have found regions of star formation where these processes can be identified (■ Figure 11-3).

A single giant molecular cloud containing a million solar masses does not contract to form a single humongous star. The cloud fragments and the densest parts form a number of dense cores. Exactly why a cloud fragments isn't fully understood, but its rotation, magnetic field, and turbulence probably play important roles. Whatever the reason, a giant cloud of gas typically contracts to form a number of newborn stars.

Heating by Contraction

You can understand how low-density clouds of interstellar gas can fall together to become dense enough to make stars, but how can the cold gas become hot enough to become a star? The answer, once again, is gravity.

To see how gravity can heat the gas, shift your attention to a single dense core destined to become a single star. Once the small cloud of gas begins to contract, gravity draws the atoms toward the center. That means the atoms are falling, and, like all things that fall, they gather speed as they fall. In fact, astronomers refer to this early stage in the formation of a star as **free-fall contraction.** Whereas the atoms may have had low velocities to start with, by the time they have fallen most of the way to the center of the cloud, they are traveling at high velocities. Thermal energy is the agitation of the particles in a gas, so this increase in velocity is a step toward heating the gas. But you can't say that the gas is hot simply because all of the atoms are moving rapidly. The air in the cabin of a jet airplane is traveling rapidly, but it isn't hot because all of the atoms are moving in generally the same direction along with the plane. To convert the high velocity of the infalling atoms into thermal energy, their motion must become randomized, and that happens when the atoms begin to collide with one another as they fall into the central region of the cloud. The jumbled, random motion of the colliding atoms represents thermal energy, and the temperature of the gas increases.

This is an important principle in astronomy. Whenever a cloud of gas contracts, the atoms move downward in the gravitational field, pick up speed, and collide more rapidly, and the gas

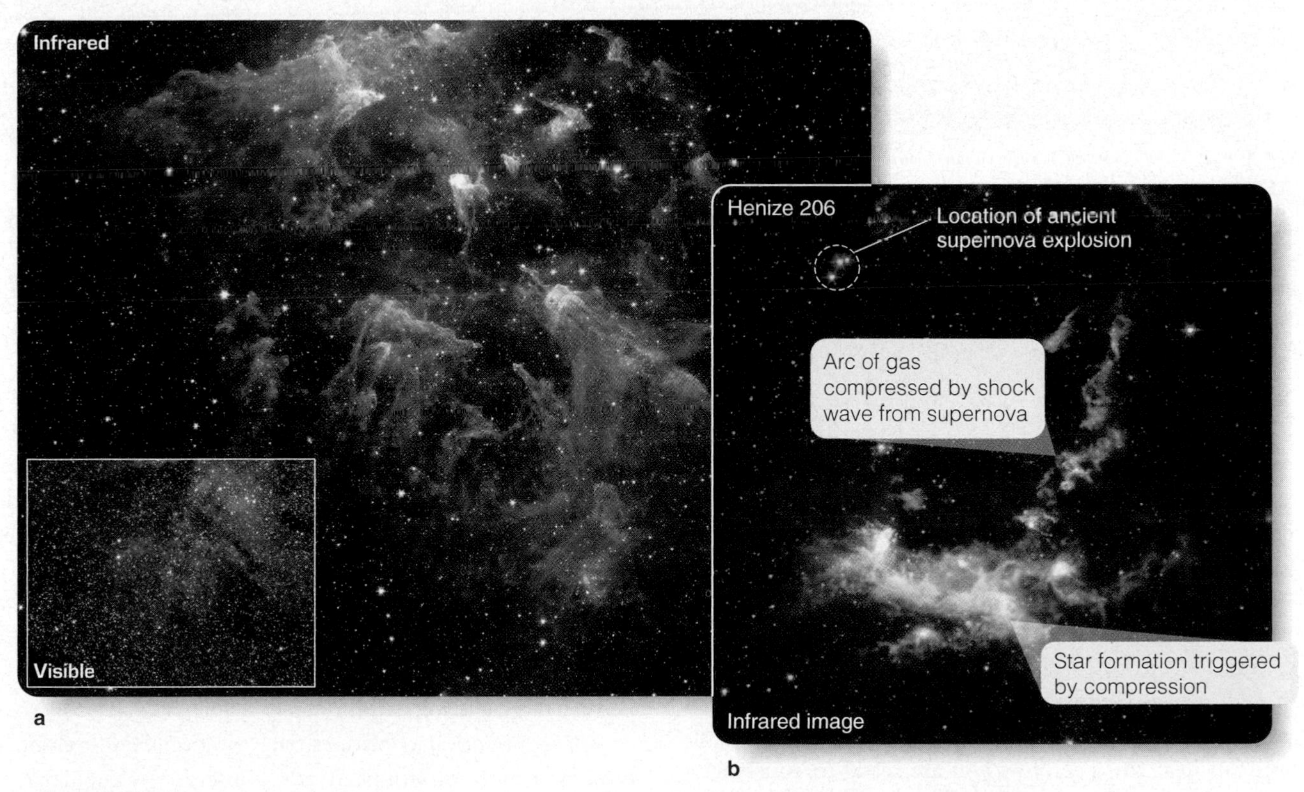

■ **Figure 11-3**

(a) Light and gas flowing away from the massive star Eta Carena out of the picture at the top are eroding this nebula and compressing parts of it. Forming stars are being exposed as the gas and dust is blown away. Little detail is detectable at visual wavelengths.(NASA/JPL-Calthech/N. Smith) (b) An expanding shockwave from a supernova explosion a few million years ago has compressed a nearby cloud of gas and triggered the birth of new stars. Most of the bright, young stars in this arc-shaped nebula are hidden deep in dust clouds and are not yet visible at visual wavelengths. (NASA/JPL-Caltech/N. Smith, Univ. of Colorado at Boulder [right] and V. Gorjian, NOAO [left])

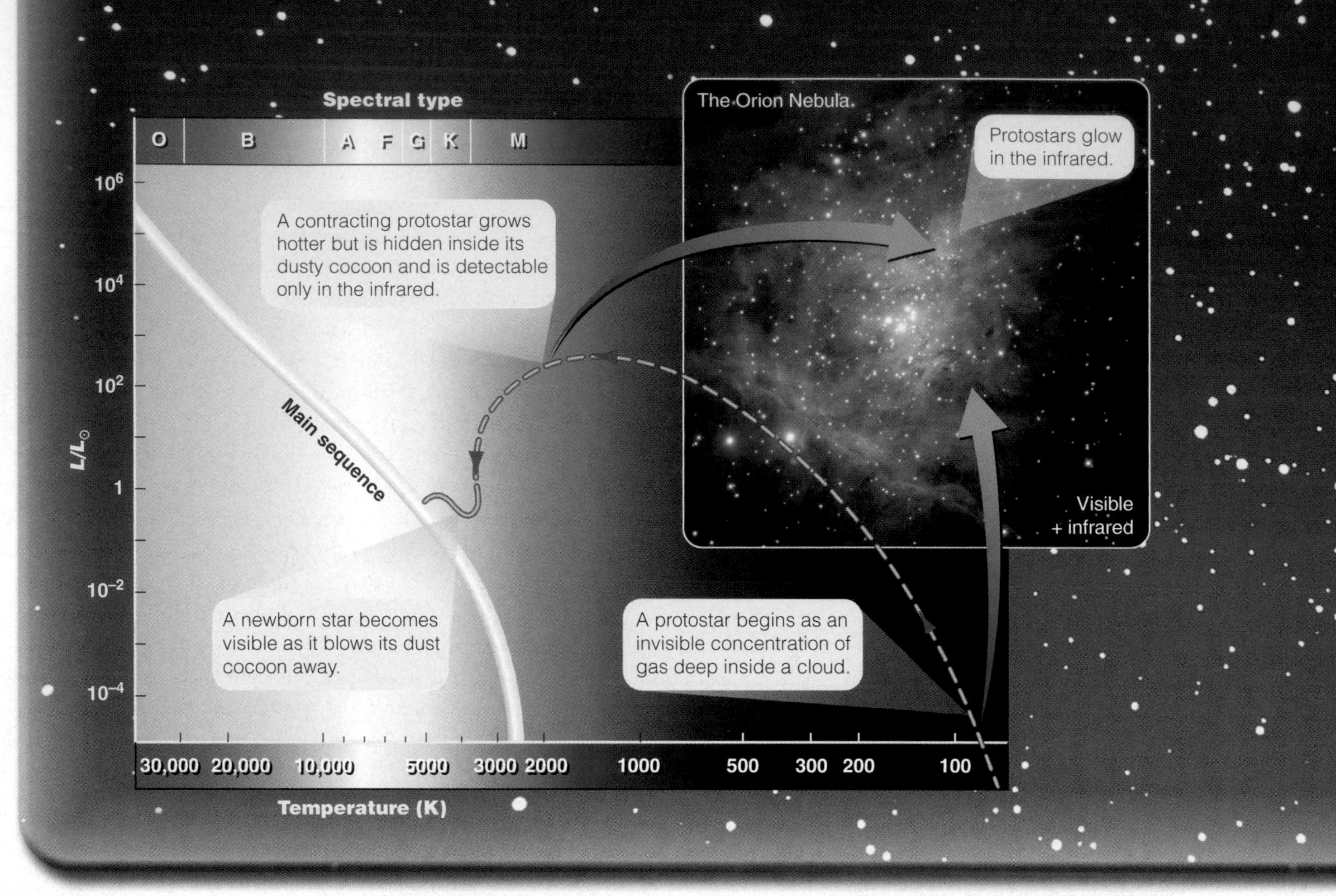
Spectral type
O
B
A
F
G
K
M
10^6
10^4
10^2
1
10^{-2}
10^{-4}
$L/L_\odot$
A contracting protostar grows hotter but is hidden inside its dusty cocoon and is detectable only in the infrared.
Main sequence
A newborn star becomes visible as it blows its dust cocoon away.
A protostar begins as an invisible concentration of gas deep inside a cloud.
The Orion Nebula.
Protostars glow in the infrared.
Visible + infrared
30,000 20,000 10,000 5000 3000 2000 1000 500 300 200 100
Temperature (K)

■ **Figure 11-4**

This H–R diagram has been extended to very low temperatures to show schematically the contraction of a dim, cool protostar. At visual wavelengths, protostars are invisible because they are deep inside dusty clouds of gas, but they are detectable at infrared wavelengths. The Orion Nebula contains both protostars and newborn stars that are just blowing their dust cocoons away. (ESO)

grows hotter. Astronomers express this by saying that gravitational energy is converted into thermal energy. Whenever a gas cloud expands, gas atoms move upward against gravity and lose speed, and the gas becomes cooler. Astronomers say that in this case thermal energy is converted into gravitational energy. This principle applies not only to clouds of interstellar gas but also to contracting and expanding stars, as you will see in the following chapters.

Your study of gas clouds has shown how the contraction of dense cores in giant molecular clouds can begin and how this contraction can heat the gas. Now you are ready to construct a detailed story of the transformation from gas cloud to star.

Protostars

To understand star formation farther, you must continue to follow the contraction of a dense core as matter falls in and heats up and the object begins to behave like a star. Although the term is used rather loosely by astronomers, a **protostar** will be defined here as a prestellar object that is hot enough to radiate infrared radiation but not hot enough to generate energy by nuclear fusion.

Early in its life, a protostar develops a higher-density region at the center and a low-density envelope. Mass continues to flow inward from the outer parts of the cloud. That is, the cloud contracts from the inside out, with the protostar taking shape deep inside an enveloping cloud of cold, dusty gas. These clouds have been called **cocoons** because they hide the forming protostar from view as it takes shape. Hidden within its cocoon of dusty gas, the protostar is not visible, but the cocoon absorbs the protostar's visible radiation and, growing warm, reradiates the energy as infrared radiation. If you could see the protostar, it would be very large and very luminous (■ Figure 11-4).

As contraction continues, the rotation of the cloud grows more and more pronounced as it conserves its angular momentum. The rapidly spinning core of the cloud must flatten into a spinning disk like a blob of pizza dough spun into the air. Gas that has lost its angular momentum through collisions can sink directly to the center of the cloud, where the protostar grows larger, surrounded by the disk. As more gas falls inward, it passes through the disk, giving up much of its angular momentum by

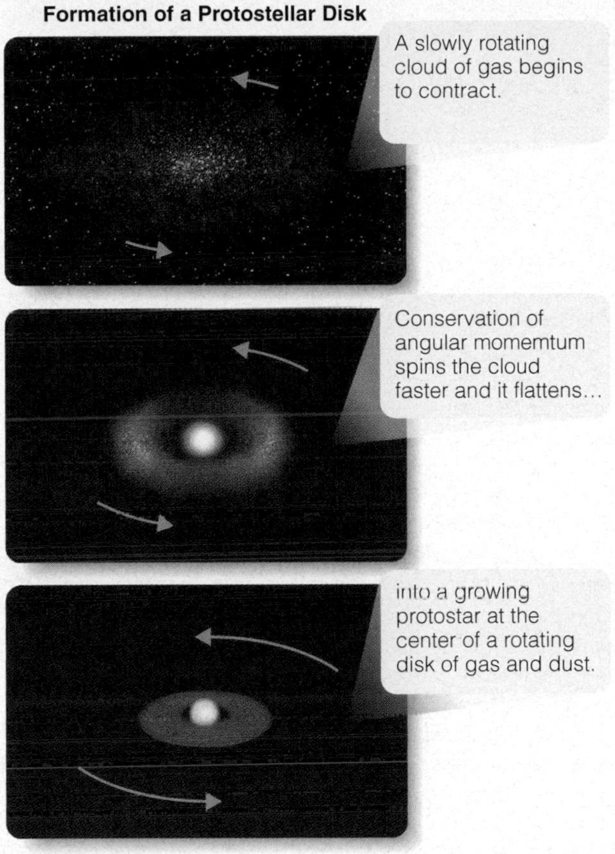

Formation of a Protostellar Disk

A slowly rotating cloud of gas begins to contract.

Conservation of angular momemtum spins the cloud faster and it flattens...

into a growing protostar at the center of a rotating disk of gas and dust.

■ **Figure 11-5**

The rotation of a contracting gas cloud forces it to flatten into a disk, and the protostar grows at the center.

collisions in the disk and possibly by interactions with the disk's magnetic field before it sinks into the protostar (■ Figure 11-5).

The disks that form around protostars are called **protostellar disks**, and they are important because astronomers conclude that planets form within these disks. Earth formed in a disk around the protosun 4.6 billion years ago. As you will see in Chapter 19, the evidence is very strong that planetary systems form in these protostellar disks.

When protostars become hot enough, they drive away the gas and dust of their cocoon and become visible. Just as the sun exhales a solar wind, stars exhale a **stellar wind,** and it can be quite strong for hot, young stars. Also, when photons encounter gas atoms or dust specks in space, the photons can exert **radiation pressure.** Stellar winds and radiation pressure combine to blow the cocoons apart.

The location in the H–R diagram where protostars first shed their cocoons and become visible is called the **birth line.** Once a star crosses the birth line, it continues to contract and move toward the main sequence with a speed that depends on its mass. More massive stars have stronger gravity and contract more rapidly (■ Figure 11-6). The sun took about 30 million years to reach the main sequence, but a 30-solar-mass star takes only 30,000 years. A 0.2-solar-mass star needs about 1 billion years to contract from a gas cloud to the main sequence.

The theory of star formation takes you into an unearthly realm filled with unfamiliar processes and objects. How can anyone really know how stars are born? The theory of star formation, like all scientific theories, can never be absolutely proven, although scientists can certainly build their confidence in it through testing and observation **(How Do We Know? 11-1).**

Evidence of Star Formation

In astronomy, evidence means observations. Consequently, astronomers must ask what observations confirm their theories of star formation. Unfortunately, a protostar is not easy to observe.

■ **Figure 11-6**

The more massive a protostar is, the faster it contracts. A 1-$M_\odot$ star requires 30 million years to reach the main sequence. (Recall that $M_\odot$ means "solar mass.") The dashed line is the birth line, where contracting protostars first become visible as they dissipate their surrounding clouds of gas and dust. Compare with Figure 11-4, which shows the evolution of a protostar of about 1 $M_\odot$ as a dashed line up to the birth line and as a solid line from the birth line to the main sequence. (Illustration design by author)

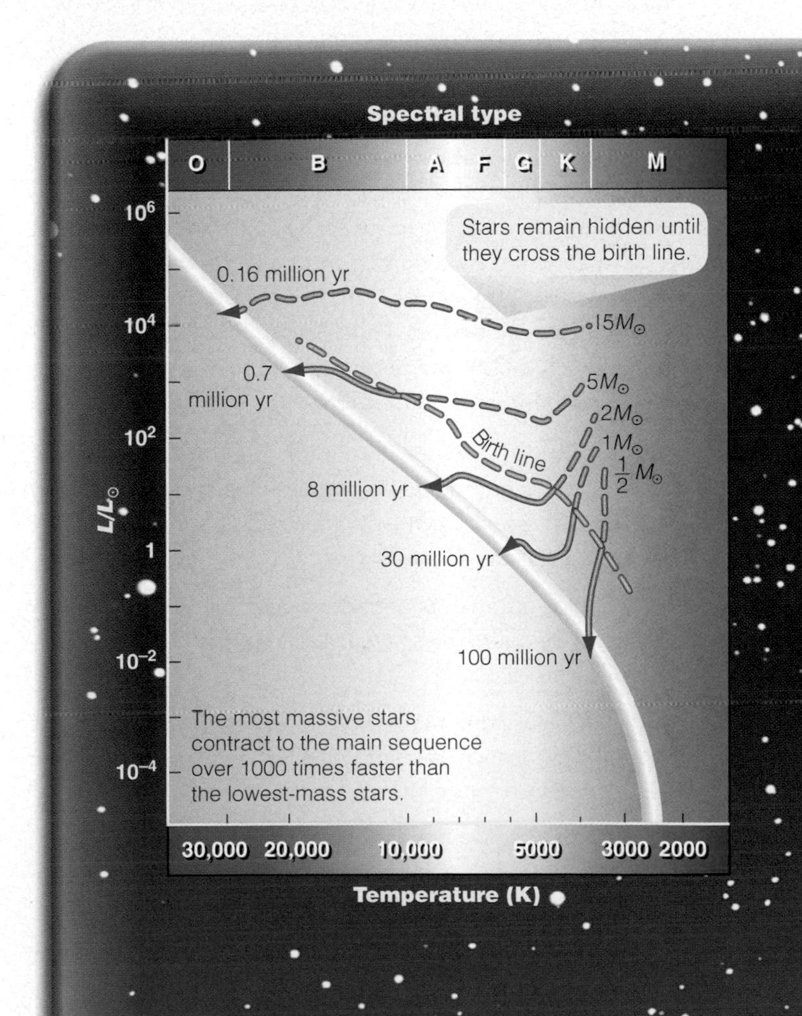

How Do We Know? | 11-1

Theories and Proof

How do astronomers know the sun isn't made of burning coal? People say dismissively of a theory they dislike, "That's only a theory," as if a theory were just a random guess. In fact, a theory can be a well-tested truth in which all scientists have great confidence. Yet no matter how many tests and experiments you conduct, you can never prove that any scientific theory is absolutely true. It is always possible that the next observation you make will disprove the theory.

There have always been theories about why the sun is hot. Some astronomers once thought the sun was a ball of burning coal, and over a century ago most astronomers accepted the theory that the sun was hot because gravity was making it contract. In the late 19th century, geologists showed that Earth was much older than the sun could be if it was powered by gravity, so the gravity theory had to be wrong. It wasn't until 1920 that another promising theory was proposed by Sir Arthur Eddington, who suggested the sun was powered somehow by the energy in atomic nuclei. In 1938 the German-American astrophysicist Hans Bethe showed how nuclear fusion could power the sun. He won the Nobel Prize in 1967.

No one will ever go to the center of the sun, so you can't *prove* the fusion theory is right. Many observations and model calculations support this theory, and in Chapter 8 you saw further evidence in the neutrinos that have been detected coming from the sun's core. Nevertheless there remains some tiny possibility that all the observations and models are misunderstood and that the theory will be overturned by some future discovery. Astronomers have tremendous confidence that the sun is powered by fusion and not gravity or coal, but a scientific theory can never be proven conclusively correct.

There is a great difference between a theory that is a far-fetched guess and a scientific theory that has undergone decades of testing and confirmation with observations, experiments, and models. But no theory can ever be proven absolutely true. It is up to you as a consumer of knowledge and a responsible citizen to distinguish between a flimsy guess and a well-tested theory that deserves to be treated like truth—at least pending further information.

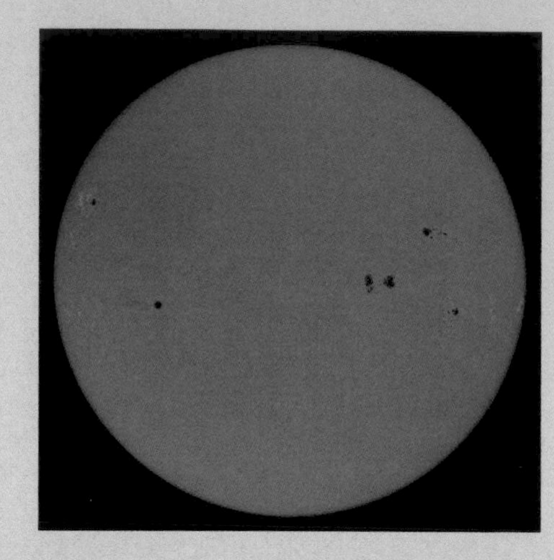

Technically it is still a theory, but astronomers have tremendous confidence that the sun gets its power from nuclear fusion and not from burning coal. (SOHO/MDI)

The protostar stage is less than 0.1 percent of a star's lifetime, and although that is a long time in human terms, you cannot expect to find many stars in the protostar stage. Furthermore, protostars form deep inside clouds of dusty gas that absorb any light the protostar might emit. Astronomers must depend on observations at infrared wavelengths to search for hidden protostars.

Read **Observational Evidence of Star Formation** on pages 228–229 and notice that it makes four important points and introduces three new terms:

1. You can be sure that star formation is going on right now because you can find regions containing stars so young they must have formed recently. *T Tauri* stars, for example, are still in the process of contracting.
2. Visual and infrared observations can reveal small dusty clouds of gas called *Bok globules* that seem to be in the process of forming stars.
3. In the H–R diagram, newborn stars lie between the birth line and the main sequence —just where you would expect to find stars that have recently blown away their dust cocoons.
4. Finally, notice how observations provide clues to the process by which stars form. Disks around protostars eject gas in *bipolar flows* that push into the surrounding interstellar medium and produce *Herbig–Haro objects.* Also, observations of small globules of gas and dust within larger nebulae show how star formation begins.

In some cases, these dark disks of gas and dust are clearly visible around newborn stars (■ Figure 11-7), but it isn't clear how the protostellar disk can produce jets. Certainly the contracting, spinning disk contains tremendous energy, and theorists suspect that magnetic fields become twisted tightly around the disk. Exactly how those fields squeeze hot gas out above and below the disk along the axis of rotation is not yet clear. But the detection of these jets was one of the first pieces of evidence that protostars are born at the centers of spinning disks.

Some evidence of star formation is not obvious. An **association** is a widely distributed star cluster that is not held together by its own gravity—its stars wander away as the association ages. It is not clear why some gas clouds give birth to compact star clusters held together by their own gravity and others give birth to larger associations not bound together by gravity. You can conclude that these associations must consist of young stars because the stars wander apart so quickly. The constellation Orion, a known region of star formation, is filled with T Tauri stars in a **T association.** T Tauri stars are relatively low-mass objects ranging from 0.75 to 3 solar masses, but the stars in **O associations,**

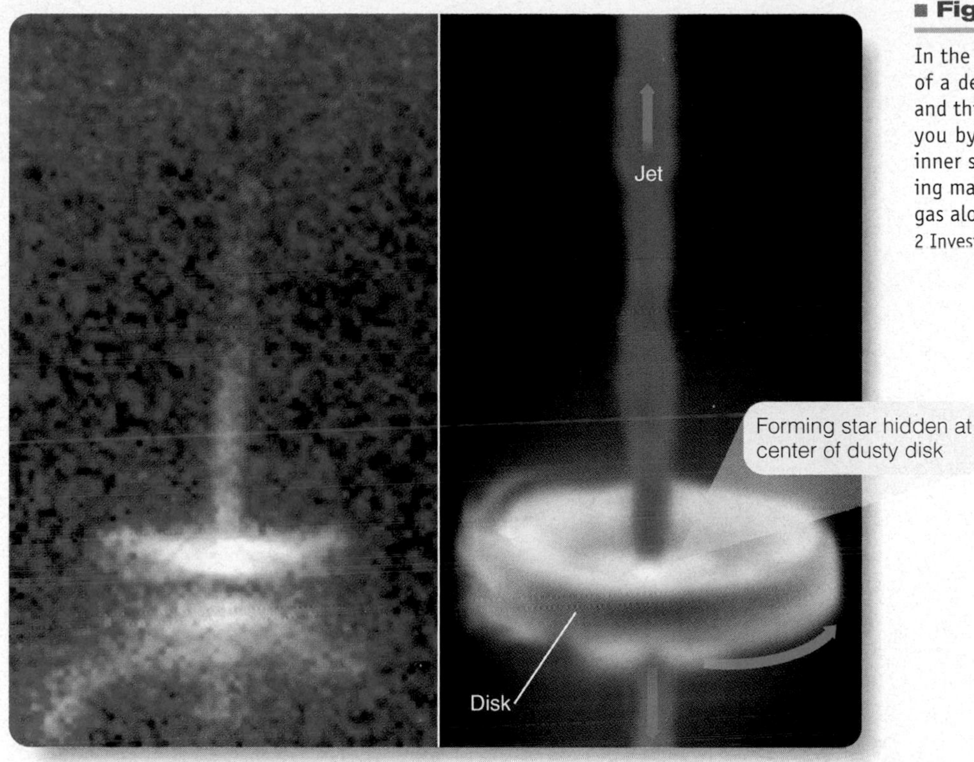

■ **Figure 11-7**

In the object HH30, a newly formed star lies at the center of a dense disk of dusty gas that is narrow near the star and thicker farther away. Although the star is hidden from you by the edge-on dusty disk, the star illuminates the inner surface of the disk. Interactions between the infalling material in the disk and the spinning star eject jets of gas along the axis of rotation. (C. Burrows, STScI & ESA, WFPC 2 Investigation Definition Team, NASA)

extended groups of O stars, are more massive. Associations are dramatic evidence of recent star formation.

Not only can astronomers locate evidence of star formation, but they have also found evidence that star formation can stimulate more star formation. If a gas cloud produces massive stars, those massive stars ionize the gas nearby and drive it away. Where the intense radiation and hot gas pushes into surrounding gas, it can compress the gas and trigger more star formation. One sign of this process is the presence of **star-formation pillars,** columns of gas that point back toward the young massive star. Such pillars are produced by denser regions that protect the gas behind them as the blast of intense radiation and hot gas flows past (■ Figure 11-8). You can recognize this process in the overall shape of the Eagle Nebula on page 229 as well as in the smaller pillars within the nebula. Another way massive stars trigger star formation is by exploding as a super-

■ **Figure 11-8**

The hot, massive stars in the star cluster 30 Doradus are pouring out intense ultraviolet radiation and powerful winds of hot gas that are pushing back and compressing the surrounding nebula. Slightly denser regions in the nebula protect the gas behind them to form star-formation pillars a few light-years long that point back at the star cluster. The dense blobs are being compressed and may form more stars within the next few million years. (N. Walborn and J. Maíz Apellániz, STScI, R. Barbá, La Plata Observatory, and NASA)

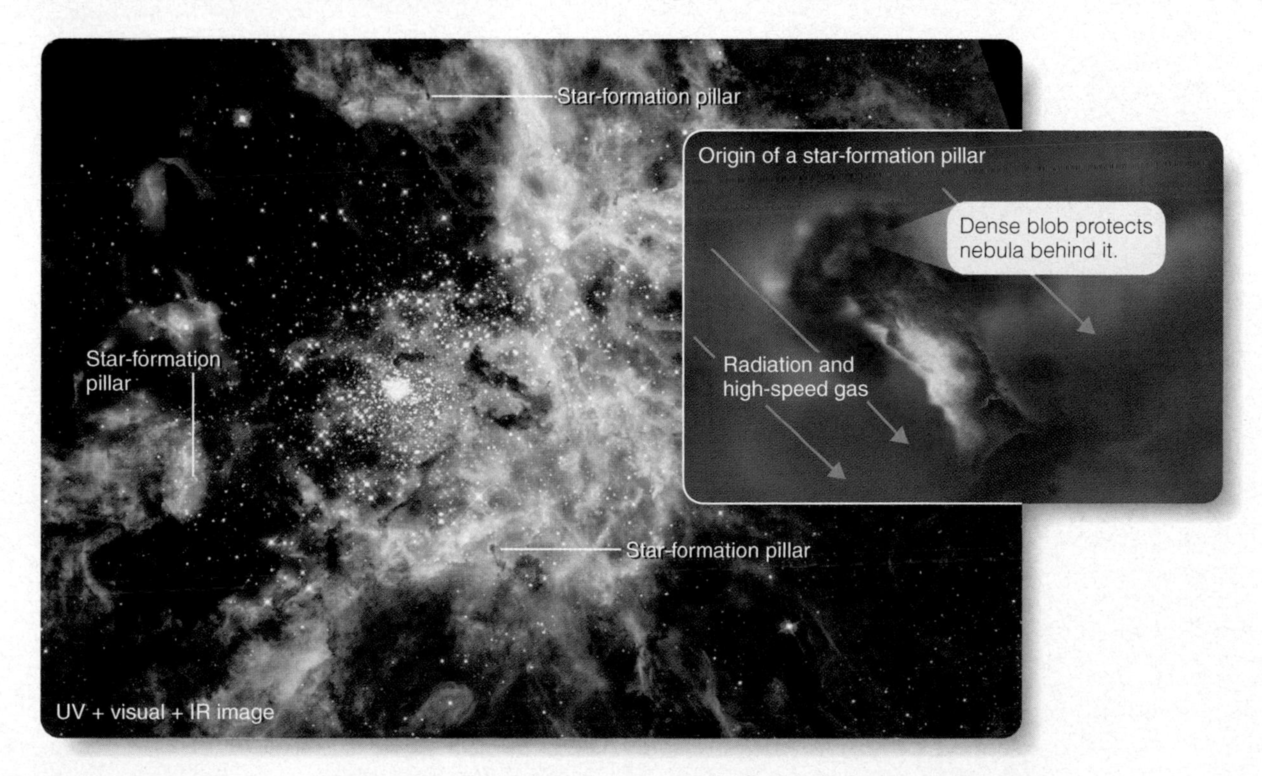

Observational Evidence of Star Formation

1 The nebula around the star S Monocerotis is bright with hot stars. Such stars live short lives of only a few million years, so they must have formed recently. Such regions of young stars are common. The entire constellation of Orion is filled with young stars and clouds of gas and dust.

Nebulae containing young stars usually contain **T Tauri** stars. These stars fluctuate irregularly in brightness, and many are bright in the infrared, suggesting they are surrounded by dust clouds and in some cases by dust disks. Doppler shifts show that gas is flowing away from many T Tauri stars. The T Tauri stars appear to be newborn stars just blowing away their dust cocoons. T Tauri stars appear to have ages ranging from 100,000 years to 100,000,000 years. Spectra of T Tauri stars show signs of an active chromosphere as we might expect from young, rapidly rotating stars with powerful dynamos and strong magnetic fields.

Visual-wavelength image

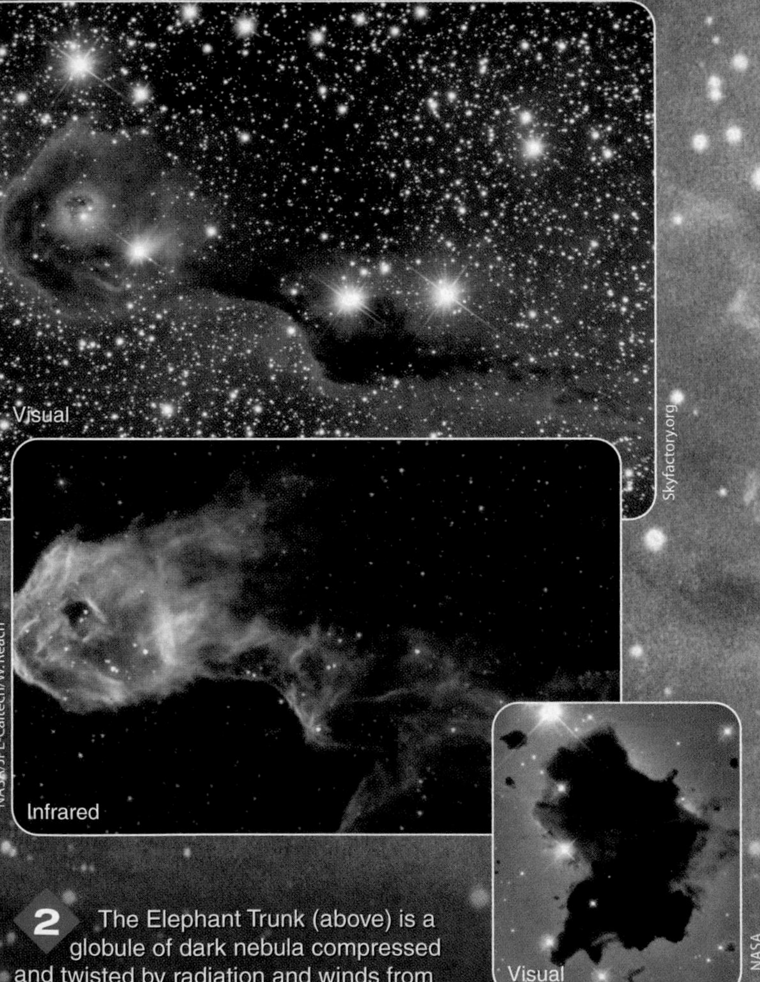

2 The Elephant Trunk (above) is a globule of dark nebula compressed and twisted by radiation and winds from a luminous star to the left of this image. Infrared observations reveal that it contains six protostars (pink images at lower edge) not detectable in visual images. The smallest dark globules are called **Bok globules** (right), named after astronomer Bart Bok. Only a light-year or so in diameter, they contain from 10 to 1000 solar masses.

3 The star cluster NGC2264, imbedded in the nebula on this page, is only a few million years old. Lower-mass stars have not yet reached the main sequence, and the cluster contains many T Tauri stars (open circles), which are found above and to the right of the main sequence, near the birth line. The faintest stars in the cluster were too faint to be observed in this study.

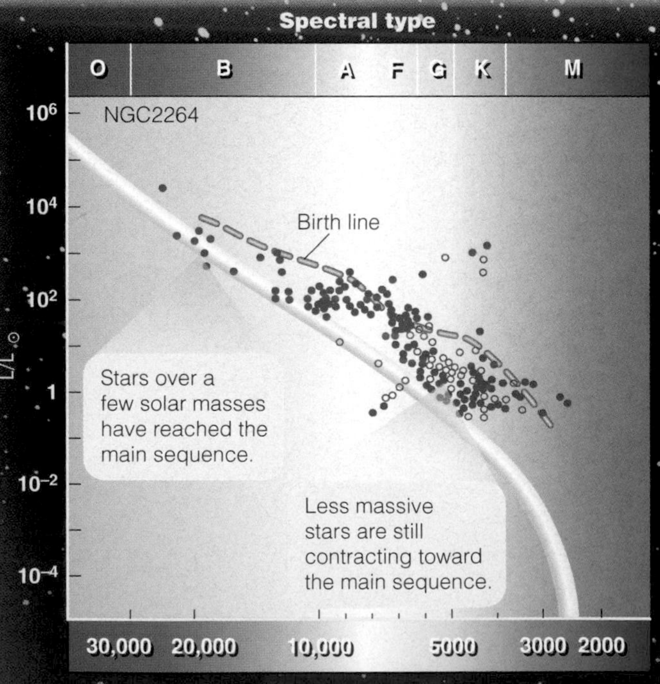

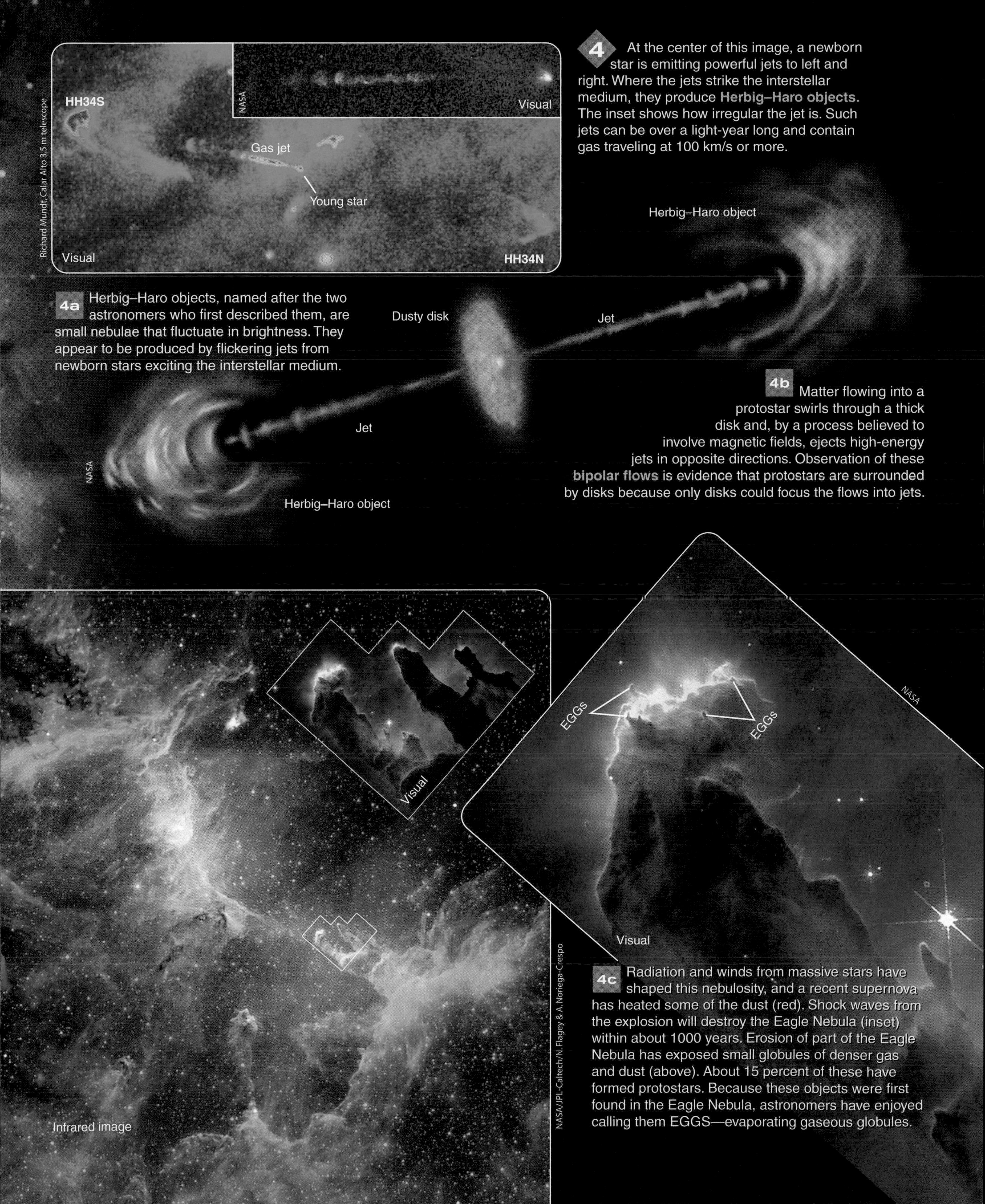

4 At the center of this image, a newborn star is emitting powerful jets to left and right. Where the jets strike the interstellar medium, they produce **Herbig–Haro objects.** The inset shows how irregular the jet is. Such jets can be over a light-year long and contain gas traveling at 100 km/s or more.

4a Herbig–Haro objects, named after the two astronomers who first described them, are small nebulae that fluctuate in brightness. They appear to be produced by flickering jets from newborn stars exciting the interstellar medium.

4b Matter flowing into a protostar swirls through a thick disk and, by a process believed to involve magnetic fields, ejects high-energy jets in opposite directions. Observation of these **bipolar flows** is evidence that protostars are surrounded by disks because only disks could focus the flows into jets.

4c Radiation and winds from massive stars have shaped this nebulosity, and a recent supernova has heated some of the dust (red). Shock waves from the explosion will destroy the Eagle Nebula (inset) within about 1000 years. Erosion of part of the Eagle Nebula has exposed small globules of denser gas and dust (above). About 15 percent of these have formed protostars. Because these objects were first found in the Eagle Nebula, astronomers have enjoyed calling them EGGS—evaporating gaseous globules.

These massive stars were triggered into formation by compression from the formation of earlier stars out of the image to the left.
New stars are forming in these dense clouds because of compression from the stars to the left.
a Visual-wavelength image
If we lived inside this monster cluster, Earth's sky would contain hundreds of stars brighter than the full moon.
b Visual + infrared-wavelength image

■ Figure 11-9

(a) An earlier generation of massive stars out of the frame to the left ejected high-speed gas that compressed nearby gas clouds and triggered the formation of young, extremely massive stars in the bright region to the left. Those stars are now triggering the birth of a third generation of stars. (NASA, ESA, The Hubble Heritage Team, AURA/STScI) (b) Compression can trigger the formation of very massive star clusters. The super star cluster, Westerlund 1, contains roughly half a million stars, some extremely massive. It must have formed no more than 5 million years ago. (ESO)

nova. You read earlier in this chapter how such an explosion can drive a shock wave through surrounding gas and can trigger more star formation (Figure 11-3).

Like a grass fire spreading through the interstellar medium, star formation can reignite itself as it creates massive stars, and astronomers can locate the remains of such episodes (■ Figure 11-9). Of course, lower-mass stars also form, but they cannot trigger further star formation because they are not hot enough to drive away vast amounts of ionized gas, nor do they explode as supernovae.

Observations provide plenty of evidence that star formation is a continuous process; you can be sure that stars are being born right now.

◄ SCIENTIFIC ARGUMENT ►

What evidence can you cite that stars are forming right now?

This is a very good example of building a scientific argument because you must cite different forms of evidence to confirm a theory. First, you should note that some extremely luminous stars can't live very long, so when you see such stars, such as the hot, blue stars in Orion, you know they must have formed in the last few million years. But you have more direct evidence when you look at T Tauri stars, which lie just above the main sequence and are often associated with gas and dust. In fact, you see entire associations of T Tauri stars as well as other associations of short-lived O stars. Many regions of gas and dust contain bright, hot stars that are blowing their nebulae apart; those stars must have formed recently.

There seems to be no doubt that star formation is an ongoing process, and you can revise your argument to discuss how it happens. **What evidence can you cite that protostars are often surrounded by disks of gas and dust?**

◄ ►

11-2 The Source of Stellar Energy

STARS ARE BORN when gravity pulls matter together, and, when the density and temperature at the centers of the stars are high enough, nuclear fusion begins making energy. In this section, you will see how stars make energy. This story will lead your imagination into a region where your body can never go—the heart of a star.

A Review of the Proton–Proton Chain

In Chapter 8 you visited the center of the sun and discovered that it manufactures energy through hydrogen fusion using a series of nuclear reactions called the proton–proton chain. The reaction must begin with the fusion of two protons. Protons are the nuclei of hydrogen atoms, have positive charges, and repel each other with a force called the Coulomb barrier. Consequently, the proton–proton chain cannot occur efficiently if the gas temperature is lower than about 10 million Kelvin. High-velocity collisions are required to penetrate the Coulomb barrier, and high velocity means high temperature.

You also discovered that the gas must be dense if the proton–proton chain is to produce significant energy. The fusion of two protons is unlikely, so a huge number of collisions are necessary to produce a few fusion reactions. Furthermore, a single cycle through the proton–proton chain produces only a tiny amount of energy, so a vast number of fusion reactions are needed to supply the energy to power a star. There will be a large number of fusion reactions only if the gas density is high.

That is why the proton–proton chain produces energy only near the sun's center, where the temperature and density are high. In fact, only about 30 percent of the sun's mass is actually hot enough to support fusion. The rest of the mass is too far from the center and isn't hot enough.

You might expect other stars to fuse hydrogen the same way the sun does, and you would be right for most stars. Some stars, however, can fuse hydrogen using a different recipe, and that makes a big difference.

ThomsonNOW Sign in at www.thomsonedu.com and go to ThomsonNOW to see Astronomy Exercise "Nuclear Fusion."

The CNO Cycle

Main-sequence stars more massive than 1.1 solar masses are hot enough to fuse hydrogen into helium using the **CNO cycle,** a hydrogen fusion process that uses carbon, nitrogen, and oxygen as stepping-stones.

Look carefully at ■ Figure 11-10 and notice the steps in the CNO cycle. The cycle begins with a carbon-12 nucleus absorbing a proton and becoming nitrogen-13, which decays to become carbon-13. The carbon-13 nucleus absorbs a second proton and becomes nitrogen-14, which absorbs a third proton and becomes oxygen-15. The oxygen-15 decays to become nitrogen-15, which absorbs a fourth proton, ejects a helium nucleus, and becomes carbon-12. Notice that carbon-12 begins the cycle and ends the cycle, so the carbon-12 nucleus can be recycled over and over. Notice also that, along the way, four protons combine to make a helium nucleus. This CNO cycle has the same outcome as the proton–proton chain, but it is different in an important way.

The CNO cycle begins with a carbon nucleus combining with a proton, a bare hydrogen nucleus. Because a carbon nucleus has a positive charge six times higher than a proton, the

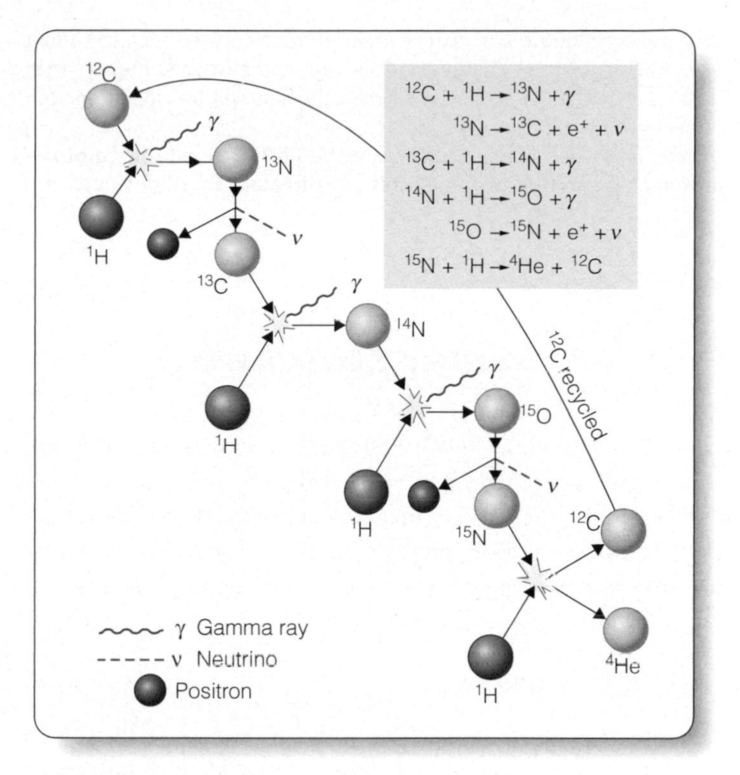

■ **Figure 11-10**

The CNO cycle uses ^{12}C as a catalyst to combine four hydrogen nuclei (^{1}H) to make one helium nucleus (^{4}He) plus energy. The carbon nucleus reappears at the end of the process, ready to start the cycle over.

Coulomb barrier is high, and temperatures higher than 16 million Kelvin are required to make the proton penetrate the Coulomb barrier. The CNO cycle needs much hotter gas than does the proton–proton chain.

The CNO cycle is so critically sensitive to temperature that virtually no reactions occur at all at temperatures below 16 million Kelvin. The proton–proton chain can make energy at these lower temperatures, which is why the sun makes nearly all of its energy using the proton–proton chain and only a little using the CNO cycle. The temperature sensitivity of the CNO cycle means that stars with higher temperature cores make nearly all of their energy using the CNO cycle, while stars with cooler cores make their energy using the proton–proton chain.

The CNO cycle and the proton–proton chain both fuse hydrogen to make helium, but the CNO cycle's sensitivity to temperature is dramatic. You will see later in this chapter how this affects the internal structure of the stars.

◄ SCIENTIFIC ARGUMENT ►

Why does the CNO cycle require higher temperatures than the proton–proton cycle?

This scientific argument requires that you discuss a bit of nuclear physics. Nuclear fusion happens when atomic nuclei collide and fuse to form a new nucleus. Inside a star, the gas is ionized, so the nuclei have no orbiting electrons, and thus the nuclei have positive charges. Two objects with positive charges repel each other in what is called the Coulomb barrier. To start fusion, the nuclei must collide at high enough velocity to penetrate that barrier. Heavier atomic nuclei such as carbon have higher positive charges, so the Coulomb barrier is higher. That requires that the gas be hotter so the collisions will be violent enough to penetrate the barrier.

Now redo your argument. **How are the CNO cycle and the proton–proton chain similar in some ways but different in other ways?**

◀ ▶

11-3 Stellar Structure

GRAVITY MAKES THE STARS CONTRACT, but why do they stop contracting? The energy generated at the center of the star flows outward to the surface, and that flow of energy stops the contraction. To understand the stable structure of a star, you must first understand how energy flows through the star and how that energy flow affects the entire star.

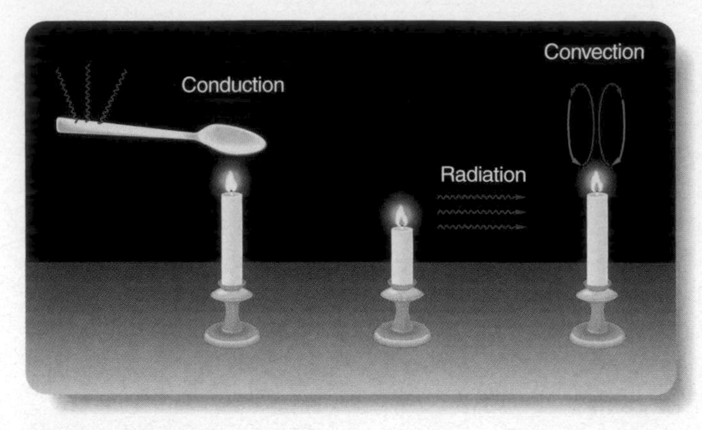

■ **Active Figure 11-11**

The three modes by which energy may be transported from the flame of a candle, as shown here, are the three modes of energy transport within a star.

Energy Transport

The sun and other stars generate nuclear energy in their deep interiors, and that energy must flow outward to their surfaces to replace the energy radiated into space as light and heat. In Chapter 8, you studied the sun and discovered that energy moves through its deep interior as radiation and through its outer layers as convection. Other stars are similar to the sun, but there can be differences. Here you can review how energy flows through the inside of a star and apply the results to stars in general.

Energy always flows from hot regions to cool regions, and the centers of stars are much hotter than their surfaces, so energy must flow outward. In the material of which stars are made, energy can move in three ways: by conduction, by radiation, or by convection.

Conduction is the most familiar form of heat flow. If you hold the bowl of a spoon in a candle flame, the handle of the spoon grows warmer. Thermal energy in the form of the motion of the particles in the metal is conducted from particle to particle up the handle, until the particles under your fingers begin to move faster and you sense heat (■ Figure 11-11). Conduction requires close contact between the particles. Matter in stars is gaseous, and the particles are not in close contact, so conduction is significant only in peculiar stars that have tremendous internal densities.

The transport of energy by radiation is another familiar process. Put your hand beside a candle flame, and you can feel the heat. What you actually feel are infrared photons radiated by the flame (Figure 11-11). Because photons are packets of energy, your hand grows warm as it absorbs them. As you saw in Chapter 8, radiation is the principal means of energy transport in the sun's interior. Photons begin near the sun's center as high-energy gamma rays and are scattered over and over as they work their way outward. The high-energy photons common near the sun's center emerge from the sun's surface as large numbers of low-energy photons of visible light.

The flow of energy by radiation depends on how difficult it is for the photons to move through the gas. When the gas is cool and dense, the photons are more likely to be absorbed or scattered, and the radiation does not penetrate the gas very well. Physicists say such a gas is opaque. In a hotter, lower-density gas, the photons can get through more easily; such a gas is less opaque. The **opacity** of a gas—its resistance to the flow of radiation—depends strongly on its temperature.

Where the opacity of the gas in a star is low, as in the sun's deep interior, energy flows outward as radiation, and the region is called a radiative zone. Where the opacity of the gas is high, radiation cannot flow through it easily. Like water behind a dam, energy builds up, raising the temperature until the gas begins to churn. Hot gas, being less dense, rises; and cool gas, being denser, sinks in convection, the third way energy can move in a star (Figure 11-11). As you learned when you studied the sun, a region inside a star where energy moves as convection is called a convective zone.

Convection is important in stars because it both carries energy and mixes the gas. Convection currents flowing through the layers of a star tend to homogenize the gas, giving it a uniform composition throughout the convective zone. As you might expect, this mixing affects the fuel supply of the nuclear reactions, just as the stirring of a campfire makes it burn more efficiently.

In the sun, the proton–proton chain converts hydrogen into helium, but the sun's interior is a radiative zone. There is no convection to mix the fuel and carry away the ashes. The hydrogen is gradually being used up, and the helium gradually accumulates. The sun is like a big pot of soup that is burning at the bottom and is being stirred only slightly by convection near its surface.

Now you are ready to understand the stability of the stars. They are elegantly simple. You can begin with our sun.

What Supports the Sun?

From its surface to its interior, the sun is gaseous. On Earth, a puff of gas, vapor from a smokestack perhaps, dissipates rapidly, driven by the motion of the air and by the random motions of the atoms in the gas. However, the sun differs from a puff of gas in a very important characteristic—its mass. The sun is over 300,000 times more massive than Earth, and that mass produces a tremendous gravitational field that draws the sun's gases into a sphere. With such a strong gravitational field, it might seem as if the gases of the sun should compress into a tiny ball, but a second force balances gravity and prevents the sun from shrinking. The gas of which the sun is made is quite hot; and, because it is hot, it has a high pressure. The pressure pushes outward; indeed, if it were not held by the sun's gravity, the pressure of the gas would blow the sun apart. The sun is balanced between two forces—gravity trying to squeeze it tighter and gas pressure trying to make it expand.

To discuss the forces inside the sun, you can imagine marking off the sun's interior into concentric shells like those in an onion (■ Figure 11-12). You can then discuss the temperature, density, pressure, and so on in each shell. Keep in mind, however, that these helpful shells are just convenient markers for discussion, much like the yard lines marked out on a football field. The concentric layers are only a convenience for discussion.

The gravity–pressure balance that supports the sun is a fundamental part of stellar structure known as the law of **hydrostatic equilibrium.** *Hydro* implies that you are discussing a fluid—the gases of the star. *Static* implies that the fluid is stable—neither expanding nor contracting. The law says that, in a stable star like the sun, the weight of the material pressing downward on a layer must be balanced by the pressure of the gas in that layer.

The law of hydrostatic equilibrium tells you that the interior of the sun must be very hot. Near the sun's surface, there is little weight pressing down on the gas, so the pressure must be low, implying a low temperature. But as you go deeper into the sun,

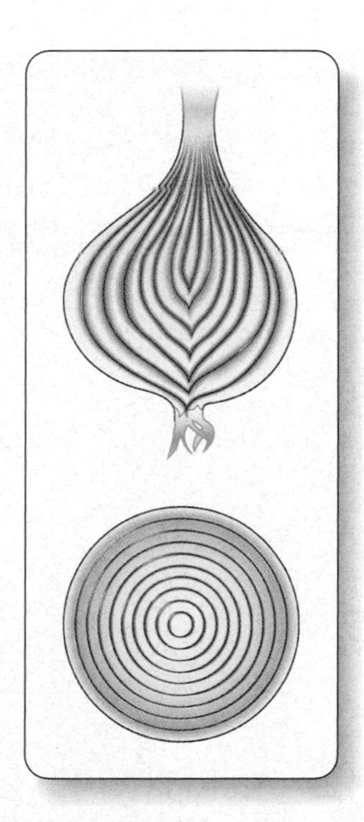

■ Figure 11-12

To discuss the structure of a star, it is helpful to divide its interior into concentric shells much like the layers in an onion. This model is, of course, only an aid to your imagination. Stars are not really divided into separable layers.

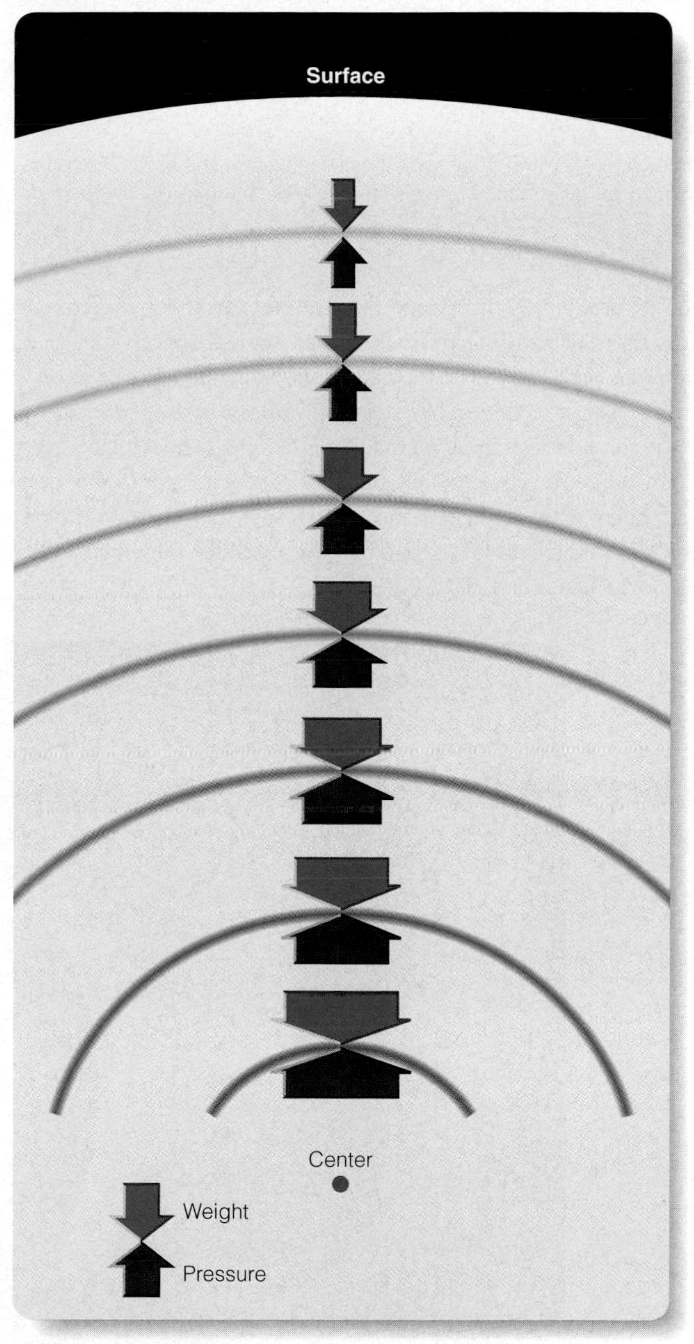

■ Figure 11-13

The law of hydrostatic equilibrium says the pressure in each layer must balance the weight on that layer. Consequently, as the weight increases from the surface of a star to its center, the pressure must also increase.

the weight becomes larger, so the pressure, and therefore the temperature, must also increase (■ Figure 11-13). Near the sun's surface, the pressure is only a few times Earth's atmospheric pressure, and the temperature is only about 5800 K. At the center of the sun, the weight pressing down equals a pressure of about 3×10^{14} times Earth's atmospheric pressure. To support that huge weight, the gas temperature has to be almost 15 million K.

The interior of the sun is kept hot by the nuclear reactions occurring at the core, and the outward flow of energy keeps each layer in the sun hot enough to support the weight pressing down from above. In a sense, the sun is supported by the flow of energy from its center to its surface. Turn off that energy, and gravity would gradually force the sun to collapse into its center.

ThomsonNOW Sign in at www.thomsonedu.com and go to ThomsonNOW to see Astronomy Exercise "Hydrostatic Equilibrium."

Inside Stars

In Chapter 9, you discovered that the stars on the main sequence are ordered according to mass, upper-main-sequence stars being most massive and lower-main-sequence stars being least massive. Combining this with what you know about hydrogen fusion reveals that there are two kinds of main-sequence stars: upper-main-sequence stars, which fuse hydrogen on the CNO cycle; and lower-main-sequence stars, which fuse hydrogen on the proton–proton chain. Viewed from the outside, these stars differ only in size, temperature, and luminosity, but inside they are quite different.

The upper-main-sequence stars are more massive and thus must have higher central temperatures to withstand their own gravity. These high central temperatures allow the star to fuse hydrogen on the CNO cycle, and that affects the internal structure of the star.

You learned earlier in this chapter that the CNO cycle is extremely temperature-sensitive. To illustrate, if the central temperature of the sun rose by 10 percent, energy production by the proton–proton chain would rise by about 46 percent, but energy production by the CNO cycle would shoot up 350 percent. This means that the more massive stars generate almost all of their energy in a tiny region at their very centers where the temperature is highest. A 10-solar-mass star, for instance, generates 50 percent of its energy in its central 2 percent of mass.

This concentration of energy production at the very center of the star causes a "traffic jam" as the energy tries to flow away from the center. Transport of energy by radiation can't drain away the energy fast enough, and the central core of the star churns in convection as hot gas rises upward and cooler gas sinks downward. Farther from the center, the traffic jam is less severe, and the energy can flow outward as radiation. This means that massive stars have convective cores at their centers and radiative envelopes extending from their cores to their surfaces (■ Figure 11-14).

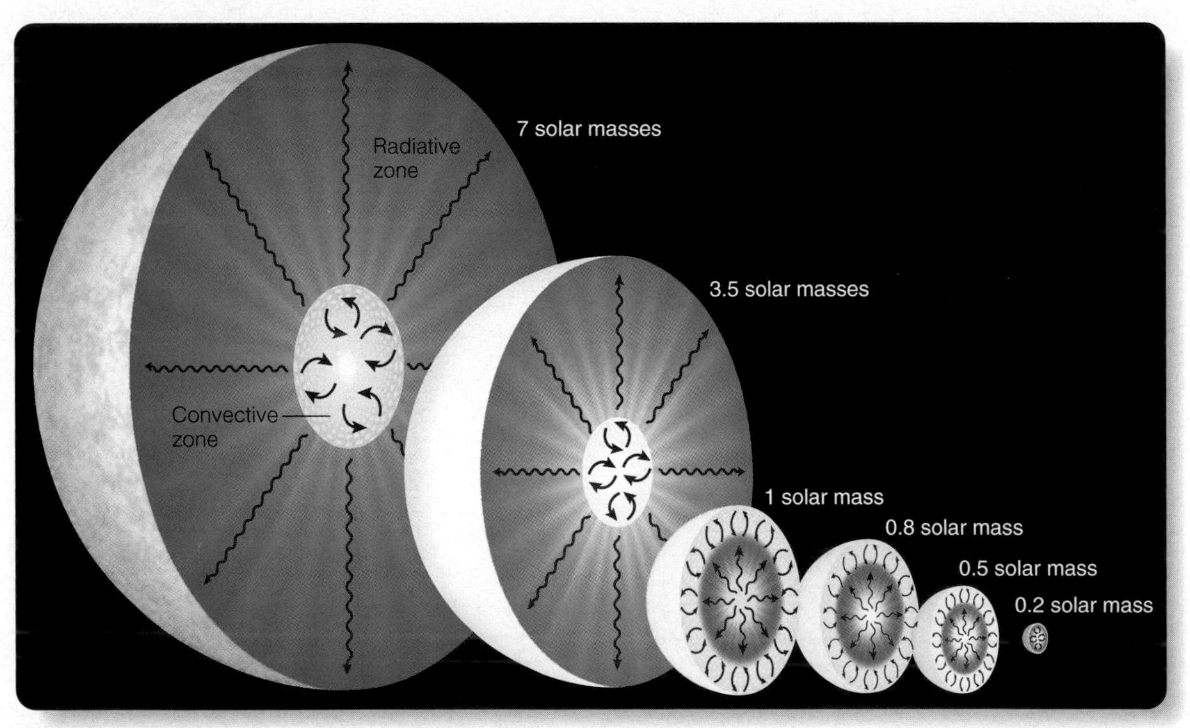

■ **Active Figure 11-14**

Inside stars. The more massive stars have small convective cores and radiative envelopes. Lower mass stars, including the sun, have radiative cores and convective envelopes. The lowest-mass stars are convective throughout. The "cores" of the stars where nuclear fusion occurs (not shown) are smaller than the interiors. (Illustration design by author)

Main-sequence stars less massive than about 1.1 solar masses cannot get hot enough to fuse much hydrogen on the CNO cycle. They generate nearly all of their energy by the proton–proton chain, which is not as sensitive to temperature, and thus the energy generation occurs in a larger region in the star's core. The sun, for example, generates 50 percent of its energy in a region that contains 11 percent of its mass. Because the energy generation is not concentrated at the very center of the star, no traffic jam develops, and the energy flows outward as radiation. Only near the surface, where the gas is cooler and therefore more opaque, does convection stir the gas. Consequently, the less massive stars on the main sequence, including the sun, have radiative cores and convective envelopes.

The lowest-mass stars have a slightly different kind of structure. For stars less than about 0.4 solar mass, the gas is relatively cool compared to the inside of more massive stars, and the radiation cannot flow outward easily. As a result, the entire bulk of these low-mass stars is stirred by convection.

The stars in the evening sky look much the same, but you have discovered that they are a diverse group. Different kinds of stars make their energy in different ways and have different internal structures. Now you are ready to answer the essential question: How do stars maintain their stability?

The Pressure–Temperature Thermostat

Newborn stars contract and heat up until nuclear fusion begins. The energy flowing outward from the core heats the layers of gas, raises the pressure, and stops the contraction. This leads to an interesting question: How does the star manage to make just the right amount of energy? The key is the relationship between pressure and temperature; it acts like a thermostat to keep the star burning steady.

Consider what would happen if the reactions began to produce too much energy. Normally, the nuclear reactions generate just enough energy to balance the inward pull of gravity. If the star made too much energy, the extra energy flowing out of the star would force its layers to expand, lowering the central temperature and density and slowing the nuclear reactions until the star regained stability. A star has a built-in regulator that keeps its nuclear reactions from occurring too rapidly.

The same thermostat keeps the reactions from dying down. Suppose the nuclear reactions began making too little energy. Then the star would contract slightly, increasing the central temperature and density, which would in turn increase the nuclear energy generation until the star regained stability.

The stability of a star depends on this relation between pressure and temperature. If an increase or decrease in temperature produces a corresponding change in pressure, the thermostat functions correctly, and the star is stable. You will see in the next chapter how the thermostat accounts for the mass–luminosity relation. In Chapter 13, you will see what happens to a star when the thermostat breaks down completely and the nuclear fires rage unregulated.

◄ SCIENTIFIC ARGUMENT ►

What would happen if the sun stopped generating energy?

Sometimes one of the best ways to test your understanding is to build an argument based on an altered situation. Stars are supported by the outward flow of energy generated by nuclear fusion in their interiors. That energy keeps each layer of the star just hot enough for the gas pressure to support the weight of the layers above. Each layer in the star must be in hydrostatic equilibrium; that is, the inward weight must be balanced by outward pressure. If the sun stopped making energy in its interior, nothing would happen at first, but over many thousands of years the loss of energy from its surface would reduce the sun's ability to withstand its own gravity, and it would begin to contract. You wouldn't notice much for 100,000 years or so, but eventually the sun would lose its battle with gravity.

Stars are elegant in their simplicity—nothing more than a cloud of gas held together by gravity and warmed by nuclear fusion. Now build a different argument. **How does the star manage to make exactly the right amount of energy to support its weight?**

◄ ►

11-4 The Orion Nebula

ON A CLEAR WINTER NIGHT, you can see with your naked eye the Great Nebula of Orion as a fuzzy wisp in Orion's sword. With binoculars or a small telescope it is striking, and through a large telescope it is breathtaking. At the center of the nebula lie four brilliant blue-white stars known as the Trapezium, the brightest in a cluster of a few hundred stars. Surrounding the stars are the glowing filaments of a nebula more than 8 pc across. Like a great thundercloud illuminated from within, the churning currents of gas and dust suggest immense power. The significance of the Orion Nebula lies hidden, figuratively and literally, beyond the visible nebula. The region is ripe with star formation.

Evidence of Young Stars

Read **Star Formation in the Orion Nebula** on pages 136–137 and note four important points:

1. The nebula you see is only a small part of a vast, dusty molecular cloud. You see the nebula because the stars born within it have ionized the gas and driven it outward, breaking out of the molecular cloud.
2. A single very hot star is almost entirely responsible for producing the ultraviolet photons that ionize the gas and make the nebula glow.
3. Infrared observations reveal clear evidence of active star formation deeper in the molecular cloud behind the visible nebula.
4. Finally, notice that many stars visible in the Orion Nebula are surrounded by disks of gas and dust. Such disks do not last long and are clear evidence that the stars are very young.

Star Formation in the Orion Nebula

Side view of Orion Nebula

Hot Trapezium stars

Protostars

To Earth

Expanding ionized hydrogen

Molecular cloud

1 The visible Orion Nebula shown below is a pocket of ionized gas on the near side of a vast, dusty molecular cloud that fills much of the southern part of the constellation Orion. The molecular cloud can be mapped by radio telescopes. To scale, the cloud would be many times larger than this page. As the stars of the Trapezium were born in the cloud, their radiation has ionized the gas and pushed it away. Where the expanding nebula pushes into the larger molecular cloud, it is compressing the gas (see diagram at right) and may be triggering the formation of the protostars that can be detected at infrared wavelengths within the molecular cloud.

Hundreds of stars lie within the nebula, but only the four brightest, those in the Trapezium, are easy to see with a small telescope. A fifth star, at the narrow end of the Trapezium, may be visible on nights of good seeing.

The cluster of stars in the nebula is less than 2 million years old. This must mean the nebula is similarly young.

Trapezium

Visual-wavelength image

Credit: NASA, ESA, M. Robberto, STScI and the Hubble Space Telescope Orion Treasury Project Team

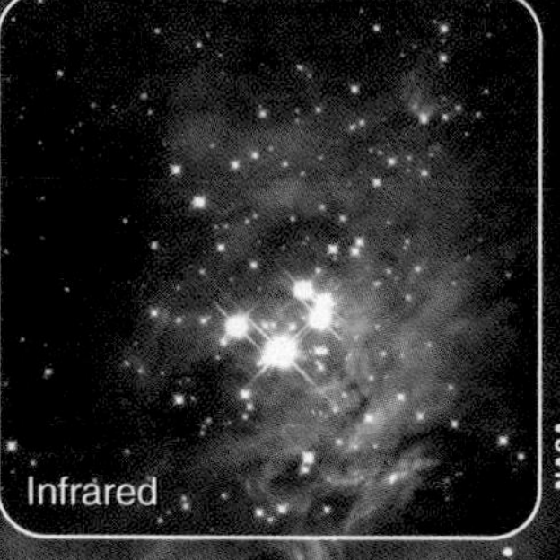

NASA

The near-infrared image above reveals about 50 low-mass, very cool stars that must have formed recently.

NASA/CXC/SAO

Roughly 1000 young stars with hot chromospheres appear in this X-ray image of the Orion Nebula.

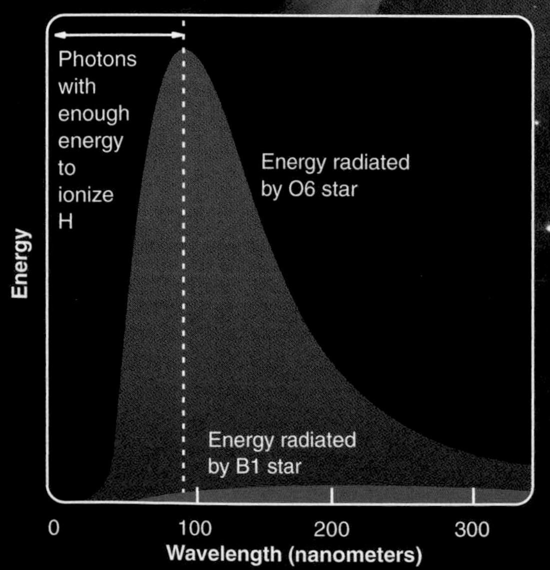

2 Of all the stars in the Orion Nebula, only one is hot enough to ionize the gas. Only photons with wavelengths shorter than 91.2 nm can ionize hydrogen. The second-hottest stars in the nebula are B1 stars, and they emit little of this ionizing radiation. The hottest star, however, is an O6 star 30 times the mass of the sun. At a temperature of 40,000 K, it emits plenty of photons with wavelengths short enough to ionize hydrogen. Remove that one star, and the nebula would turn off its emission.

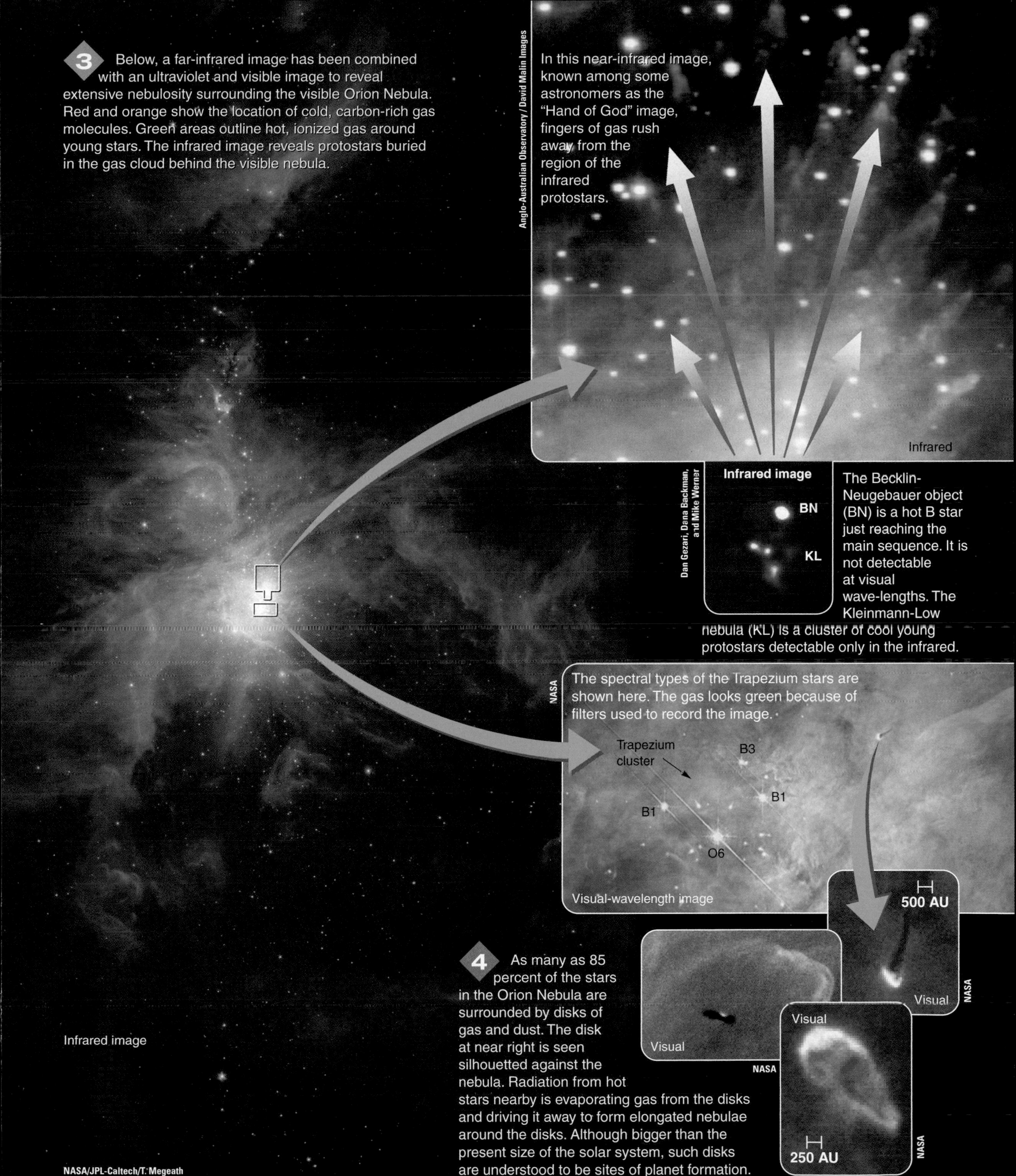
3
Below, a far-infrared image has been combined with an ultraviolet and visible image to reveal extensive nebulosity surrounding the visible Orion Nebula. Red and orange show the location of cold, carbon-rich gas molecules. Green areas outline hot, ionized gas around young stars. The infrared image reveals protostars buried in the gas cloud behind the visible nebula.
Anglo-Australian Observatory / David Malin Images
In this near-infrared image, known among some astronomers as the "Hand of God" image, fingers of gas rush away from the region of the infrared protostars.
Infrared
Dan Gezari, Dena Backman, and Mike Werner
Infrared image
BN
KL
The Becklin-Neugebauer object (BN) is a hot B star just reaching the main sequence. It is not detectable at visual wave-lengths. The Kleinmann-Low nebula (KL) is a cluster of cool young protostars detectable only in the infrared.
NASA
The spectral types of the Trapezium stars are shown here. The gas looks green because of filters used to record the image.
Trapezium cluster
B3
B1
B1
O6
Visual-wavelength image
500 AU
Visual
NASA
4
As many as 85 percent of the stars in the Orion Nebula are surrounded by disks of gas and dust. The disk at near right is seen silhouetted against the nebula. Radiation from hot stars nearby is evaporating gas from the disks and driving it away to form elongated nebulae around the disks. Although bigger than the present size of the solar system, such disks are understood to be sites of planet formation.
Visual
NASA
Visual
250 AU
NASA
Infrared image
NASA/JPL-Caltech/T. Megeath

Mythmakers and Explainers

On cold winter nights when the sky is clear and the stars are bright, Jack Frost paints icy lacework across your windowpane. That's a fairy tale, of course, but it is a graceful evocation of the origin of frost. We humans are explainers, and one way to explain the world around us is to create myths.

An ancient Aztec myth tells the story of the origin of the moon and stars. The stars, known as the Four Hundred Southerners, and the moon, the goddess Coyolxauhqui, plotted to murder their unborn brother, the great war god Huitzilopochtli. Hearing their plotting, he leaped from the womb fully armed, hacked Coyolxauhqui into pieces, and chased the stars away. You can see the Four Hundred Southerners scattered across the sky, and each month you can see the moon chopped into pieces as it passes through its phases.

Stories like these explain the origins of things and make our universe more understandable and more comfortable. Science is a natural extension of our need to explain the world. The stories have become sophisticated scientific theories and are tested over and over against reality, but we humans build those theories for the same reason people used to tell myths.

You should not be surprised to find star formation in Orion. The constellation is a brilliant landmark in the winter sky because it is marked by hot, blue stars. These stars are bright in the sky, not because they are nearby but because they are tremendously luminous. These O and B stars cannot live more than a few million years, so they must have been born recently. Furthermore, the constellation contains large numbers of T Tauri stars, which are known to be young. Orion is rich with young stars.

The history of star formation in the constellation of Orion is written in its stars. The stars at Orion's west shoulder are about 12 million years old, while the stars of Orion's belt are about 8 million years old. The stars of the Trapezium at the center of the Great Nebula are no older than 2 million years. Apparently, star formation began near Orion's west shoulder, and the massive stars that formed there triggered the formation of the stars in Orion's belt. That star formation may have triggered the formation of the stars you see in the Great Nebula. Like a grass fire, star formation has swept across Orion from northwest to southeast.

Are there other nebulae like the Orion Nebula? Such emission nebulae are quite common, and many of the emission nebulae shown in this and the previous chapter are produced by newborn stars just as is the Orion Nebula. The shape of a nebula is determined, in part, by the distribution of gas around the forming stars. The nebula around the star cluster NGC 602 is strikingly similar in shape to the Orion Nebula (■ Figure 11-15).

In the next million years, the familiar outline of the Great Nebula will change, and a new nebula may begin to form as the protostars in the molecular cloud ionize the gas, drive it away, and become visible. Throughout the cloud, centers of star formation may develop and then dissipate as massive stars are born and force the gas to expand. If enough massive stars are born, they could blow the entire molecular cloud apart and bring the successive generations of star formation to a conclusion. The Great Nebula in Orion and its invisible molecular cloud are a beautiful and dramatic example of the continuing cycles of star formation.

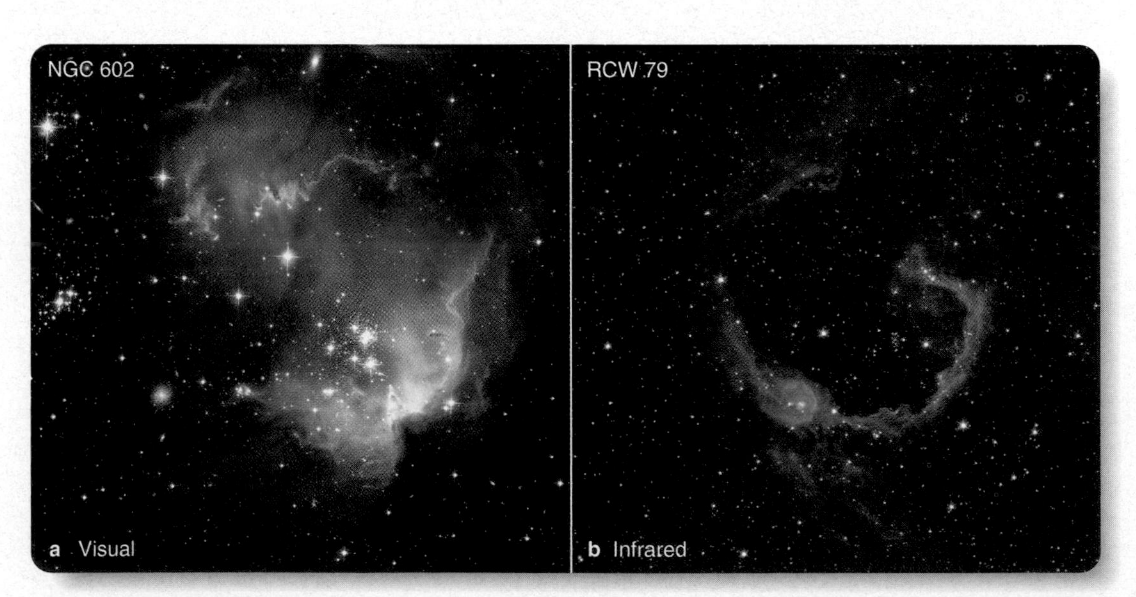

■ Figure 11-15

(a) The nebula around star cluster NGC 602 resembles the Orion Nebula. It has been inflated by radiation and winds streaming away from the cluster of hot young stars at lower right. More stars are forming where gas is compressed at upper left. Note the star-formation pillars pointing back at the cluster. (NASA/ESA, and the Hubble Heritage Team) (b) Young hot stars have inflated RCW 79, a bubble 70 ly in diameter. Its expansion is compressing gas and triggering more star formation at lower left. (NASA/JPL-Caltech/Ed Churchwell)

◄ SCIENTIFIC ARGUMENT ►

What did Orion look like to the ancient Egyptians, to the first humans, and to the dinosaurs?

Scientific arguments can do more that support a theory; they can change the way you think of the world around you. The Egyptian civilization had its beginning only a few thousand years ago, and that is not very long in terms of the history of Orion. The stars you see in the constellation are hot and young, but they are a few million years old, so the Egyptians saw the same constellation you see. (They called it Osiris.) Even the Orion Nebula hasn't changed very much in a few thousand years, and Egyptians may have admired it in the dark skies along the Nile.

Our oldest human ancestors lived about 3 million years ago, and that was about the time that the youngest stars in Orion were forming. Your earliest ancestors may have looked up and seen some of the stars you see, but some stars have formed since that time. Also, the Great Nebula is excited by the Trapezium stars, and they are not more than a few million years old, so your early ancestors probably didn't see the Great Nebula.

The dinosaurs saw something quite different. The last of the dinosaurs died about 65 million years ago, long before the birth of the brightest stars in Orion. The dinosaurs, had they the brains to appreciate the view, might have seen bright stars along the Milky Way, but they didn't see Orion. All of the stars in the sky are moving through space, and the sun is orbiting the center of our galaxy. Over many millions of years, the stars move appreciable distances across the sky. The night sky above the dinosaurs contained totally different star patterns.

The Orion Nebula is the product of a giant molecular cloud, but such a cloud can't continue spawning new stars forever. Focus your argument to answer the following: **What processes limit star formation in a molecular cloud?**

◄ ►

Study and Review

Summary

11-1 | Making Stars from the Interstellar Medium

How are stars born?

- Stars are born from the gas and dust of the interstellar medium.
- The existence of massive, hot stars such as Spica that cannot live very long is strong evidence that stars have formed recently.
- The gravity of giant molecular clouds makes them contract, but that is resisted by thermal energy in the gas, magnetic fields, rotation, and turbulence. In at least some cases, clouds are compressed by passing **shock waves,** and star formation is triggered. The birth of massive stars can produce shock waves that trigger further star formation.
- The cold gas of interstellar space heats up as it contracts because the atoms fall inward in **free-fall contraction** and pick up speed. Astronomers say the gas is converting gravitational energy into thermal energy.
- **Protostars** form deep inside dusty **cocoons** and are not directly visible at visual wavelengths until their **stellar winds** and **radiation pressure** blow their cocoons away. They become visible as they cross the **birth line** in the H–R diagram.
- Many, perhaps most, protostars form at the center of dusty **protostellar disks,** and jets of gas can be emitted as **bipolar flows** along the axis of the spinning disk. Where the jets push into the surrounding gas, they can form nebulae called **Herbig–Haro objects.**
- **T Tauri** stars have just emerged from their cocoons and are located between the birth line and the main sequence in the H–R diagram.
- **Bok globules** are small dark nebulae, some of which may be contracting to form stars.
- **Associations,** including **T associations** and **O associations,** are groups of stars born together but not bound by their mutual gravity. The presence of these associations in an area is evidence of recent star formation.
- **Star-formation pillars** are formed when hot gas rushes away from newborn stars and encounters denser blobs of gas and dust.

11-2 | The Source of Stellar Energy

How do stars make energy?

- Many stars make their energy the same way the sun does using the proton–proton chain.
- The **CNO cycle** is more efficient, but it requires a higher temperature than the proton–proton chain. Both processes combine four hydrogen nuclei to make one helium nucleus plus energy.
- Because the CNO cycle is highly temperature sensitive, stars more massive than 1.1 solar masses make energy with the CNO cycle. Less massive stars, including the sun, can only use the proton–proton chain.

11-3 | Stellar Structure

How do stars maintain their stability?

- Energy flows from the hot core to the cooler surface as radiation or as convection. The **opacity** of a gas is its resistance to the flow of radiation. In regions where the opacity of the gas does not permit radiation to carry away enough energy, the gas can churn in convection. Convection is important in stars because it stirs the material.
- Conduction is not important in most stars because the density is too low.
- The law of **hydrostatic equilibrium** says that the weight pressing down on a layer of gas in a star must be balanced by the pressure in the gas. That shows that the inner layers of stars must be hotter because they must support more weight.
- Upper-main-sequence stars, being more massive, must be hotter inside, and that allows them to use the CNO cycle. Because that cycle is so sensitive to temperature, the energy production occurs in a very small region near the center and the cores of the stars are convective zones. In their outer layers, these stars are radiative.

- Lower-main-sequence stars, being less massive, are not as hot inside and fuse hydrogen using the proton–proton chain. Because that chain is not very sensitive to temperature, the energy generation is more widely spread through the star's core, and the deep interior is radiative. The outer layers of these stars are convective.
- The lowest-mass stars are so cool that their gases are rather opaque, and they are convective throughout.

11-4 | The Orion Nebula

What evidence do astronomers have that theories of star formation are correct?

- The visible Orion Nebula is only a small part of a much larger dusty molecular cloud. Ionization by ultraviolet photons from the hottest star is lighting up the nebula and making it glow brightly.
- Only one star in the Orion Nebula is hot enough to emit the ultraviolet photons that ionize the gas and make the nebula glow.
- Infrared observations reveal clear evidence of active star formation deeper in the molecular cloud just to the northwest of the Trapezium.
- Many stars visible in the Orion Nebula are surrounded by disks of gas and dust. Such disks do not last long and are clear evidence that the stars are very young.

Review Questions

ThomsonNOW Assess your understanding of this chapter's topics with additional quizzing and animations at www.thomsonedu.com

1. What factors resist the contraction of a cloud of interstellar matter?
2. Explain the different ways a giant molecular cloud can be triggered to contract.
3. What evidence is there that (a) star formation has occurred recently? (b) Protostars really exist? (c) The Orion region is actively forming stars?
4. How does a contracting protostar convert gravitational energy into thermal energy?
5. How does the geometry of bipolar flows and Herbig–Haro objects support the hypothesis that protostars are surrounded by rotating disks?
6. How does the CNO cycle differ from the proton–proton chain? How is it similar?
7. How does the extreme temperature sensitivity of the CNO cycle affect the structure of stars?
8. How does energy get from the core of a star, where it is generated, to the surface, where it is radiated into space?
9. Describe the principle of hydrostatic equilibrium as it relates to the internal structure of a star.
10. How does the pressure–temperature thermostat control the nuclear reactions inside stars?
11. **How Do We Know?** How would you respond to someone's comment, "That's only a theory"?
12. **How Do We Know?** Why can't scientists prove a scientific theory is totally correct?

Discussion Questions

1. Ancient astronomers, philosophers, and poets assumed that the stars were eternal and unchanging. Is there any observation they could have made or any line of reasoning that could have led them to conclude that stars don't live forever?
2. How does hydrostatic equilibrium relate to hot-air ballooning?

Problems

1. The ring ejected by the supergiant star in Figure 11-1 has a radius of about 7 seconds of arc. If the cluster is 20,000 ly from Earth, what is the radius of the ring in light-years?
2. If a giant molecular cloud is 50 pc in diameter and a shock wave can sweep through it in 2 million years, how fast is the shock wave going in kilometers per second?
3. If a giant molecular cloud has a mass of 10^{35} kg, and it converts 1 percent of its mass into stars during a single encounter with a shock wave, how many stars can it make? Assume the stars each contain 1 solar mass.
4. If a protostellar disk is 200 AU in radius, and the disk plus the forming star contain 2 solar masses, what is the orbital velocity at the outer edge of the disk in kilometers per second?
5. If a contracting protostar is five times the radius of the sun and has a temperature of only 2000 K, how luminous will it be? (*Hint:* See Chapter 9.)
6. The gas in a bipolar flow can travel as fast as 100 km/s. If the length of the jet is 1 ly, how long does it take for a blob of gas to travel from the protostar to the end of the jet?
7. If a T Tauri star is the same temperature as the sun but is ten times more luminous, what is its radius? (*Hint:* See Chapter 9.)
8. Circle all of the ^{1}H and ^{4}He nuclei in Figure 11-10 and explain how the CNO cycle can be summarized by $4\,^1\text{H} \rightarrow \,^4\text{He} + \text{energy}$.
9. How much energy is produced when the CNO cycle converts 1 kg of mass into energy? Is your answer different if the mass is fused by the proton–proton chain?
10. If the Orion Nebula is 8 pc in diameter and has a density of about 600 hydrogen atoms/cm^3, what is its total mass? (*Hint:* The volume of a sphere is $\frac{4}{3}\pi R^3$.)
11. The hottest star in the Orion Nebula has a surface temperature of 40,000 K. At what wavelength does it radiate the most energy? (*Hint:* See Chapter 7.)

Learning to Look

1. In Figure 11-9, a dark globule of dusty gas is located at top right. What do you think that globule would look like if you could see it from the other side?
2. Compare the nebulae in Figure 11-1 with the image of the Orion Nebula on page 236. How are these two nebulae related?
3. Locate the star-formation pillars in Figure 11-8. What are they pointing at?
4. The star at right appears to be ejecting a jet of gas. What is happening to this star?

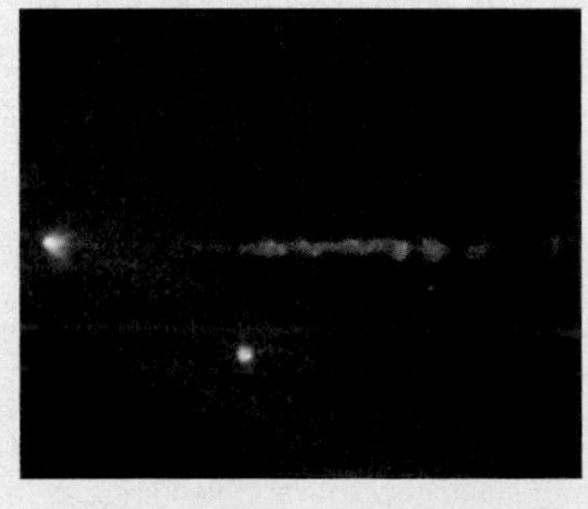
STScI and NASA

12 Stellar Evolution

Massive red supergiant star V838 Monocerotis is evolving rapidly. In early 2002, it flared up, and the expanding light is illuminating gas that was probably ejected in an earlier outburst. Spikes are produced by diffraction in the telescope. (NASA and The Hubble Heritage Team, AURA/STScI)

Guidepost

Stars form from the interstellar medium and reach stability fusing hydrogen in their cores. This chapter is about the long, stable middle age of stars on the main sequence and their old age as they swell to become giant stars. Here you will answer three essential questions:

What happens as a star uses up its hydrogen?

What happens when a star exhausts its hydrogen?

What evidence do astronomers have that stars really do evolve?

Stars evolve over billions of years because of changes deep inside. That raises an interesting question about how scientists can understand such processes:

How Do We Know? **How can astronomers study the insides of stars?**

This chapter is about how stars live. The next two chapters are about how stars die and the strange objects they leave behind.

We should be unwise to trust scientific inference very far when it becomes divorced from opportunity for observational test.

SIR ARTHUR EDDINGTON, *THE INTERNAL CONSTITUTION OF THE STARS*

EVERY STAR HAD A BEGINNING, and every star must have an ending, but in between they produce the light and energy that make our universe so beautiful. The stars above you seem eternal, but the light and warmth they emit is produced by the fusion of nuclear fuels in their cores; and, even as you look at them, they are using up their fuels and drawing closer to their ends.

In this chapter, you will see the full interplay of theory and evidence used to describe how stars change as they exhaust their nuclear fuels. As you are warned in the quotation that opens this chapter, theory alone is never enough. At each step, astronomers must compare their theories with the evidence.

12-1 Main-Sequence Stars

ONE OF THE GREATEST TRIUMPHS OF MODERN ASTRONOMY was the discovery that the stars are not eternal and that mere humans can understand them. Stars are fundamentally very simple objects, and astronomers have found ways to describe the lives of the stars using basic laws of physics.

■ **Figure 12-1**

Like acrobats in a circus stunt, the layers in a star must support the weight of everything above. Energy from the core of the star flows outward, heating each layer hot enough to produce the outward pressure to balance the inward force of gravity. Compare with Figure 11-13.

Stellar Models

Every star is balanced between gravity that tries to make it contract and internal pressure that tries to make it expand. As you learned in Chapter 11, its internal layers are balanced in hydrostatic equilibrium (■ Figure 12-1). By defining the layers mathematically, astronomers can discuss the conditions at different levels inside the star—what they call the "structure" of the star.

The internal structure of a star can be described by four simple laws of physics, two of which you have already met. In addition to the law of hydrostatic equilibrium, the law of energy transport describes how energy flows from hot to cool regions by radiation, convection, or conduction.

To these two laws you can add two basic laws of nature. The **conservation of mass law** says that the total mass of the star must equal the sum of the masses in its shells. Of course, no shell can be empty, and there is no such thing as negative mass. The **conservation of energy law** says that the amount of energy flowing out the top of a shell in the star must be equal to the amount of energy coming in at the bottom of the shell plus whatever energy is generated within the shell. This simply means that the energy leaving the surface of the star, its luminosity, must equal the sum of the energy generated in all of the shells inside the star. This is like saying that the total number of new cars driving out of a factory must equal the sum of cars manufactured on each assembly line. No car can vanish into nothing or appear from nothing. Energy in a star may not vanish without a trace or appear out of nowhere.

The four laws of stellar structure, described in general terms in ■ Table 12-1, can be written as mathematical equations. By solving those equations numerically in computers, astronomers can build a mathematical model of the inside of a star.

If you wanted to build a model of a star, you would have to divide the star into at least 100 concentric shells and then write down the four equations of stellar structure for each shell. You would then have 400 equations that would have 400 unknowns, namely, the temperature, density, mass, and energy flow in each shell. Solving 400 equations simultaneously is not easy, and the first such solutions, done by hand before the invention of electronic computers, took months of work. Now a properly programmed computer can solve the equations for a simple model in a few seconds and print a table of numbers that represent the conditions in each shell of the star. Such a table is a **stellar model.**

■ Table 12-1 | The Four Laws of Stellar Structure

1. Hydrostatic equilibrium	The weight on each layer is balanced by the pressure in that layer.
2. Energy transport	Energy moves from hot to cool by radiation, convection, or conduction.
3. Conservation of mass	Total mass equals the sum of the shell masses. No gaps are allowed.
4. Conservation of energy	Total luminosity equals the sum of the energies generated in each shell.

The table shown in ■ Figure 12-2 shows a model of the sun containing only 10 layers, but it is based on a model with many more layers. The bottom line, for radius equal to 0.00, represents the center of the sun, and the top line, for radius equal to 1.00, represents the surface. The other lines in the table show the temperature and density in each shell, the mass inside each shell, and the fraction of the sun's luminosity that is flowing outward through the shell. You can use such a model to understand many things about the sun. For example, the bottom line tells you the temperature at the center of the sun is about 15 million Kelvin. At such a high temperature, the gas is highly transparent, and energy flows as radiation. Nearer the surface, the temperature is lower, the gas is more opaque, and the energy is carried by convection.

Notice that stellar models are quantitative; that is, properties have specific numerical values. Earlier in this book, you used models that were qualitative—the Babcock model of the sun's magnetic cycle, for instance. Both kinds of models are useful, but a quantitative model can reveal deeper insights into how nature works because it incorporates the power of mathematics as a precise way of thinking **(How Do We Know? 12-1)**.

Stellar models let astronomers look into a star's past and future. In fact, models can be used as time machines to follow the evolution of stars over billions of years. To look into a star's future, for instance, you could use a stellar model to determine how fast the star uses the fuel in each shell. As the fuel is consumed, the chemical composition of the gas changes, and the amount of energy generated declines. By calculating the rate of these changes, you could calculate a new model of the star a million years in the future. Then you could repeat the process to step forward another million years. Step by step, you could follow the evolution of the star over billions of years.

Although this sounds simple, it is actually a highly challenging problem involving nuclear and atomic physics, thermodynamics, and sophisticated computational methods. Only since the 1950s have electronic computers made the rapid calculation of stellar models possible, and the advance of astronomy since then has been heavily influenced by the use of such models to study the structure and evolution of stars. The summary of star formation in this chapter is based on thousands of stellar models. You will continue to rely on theoretical models as you study

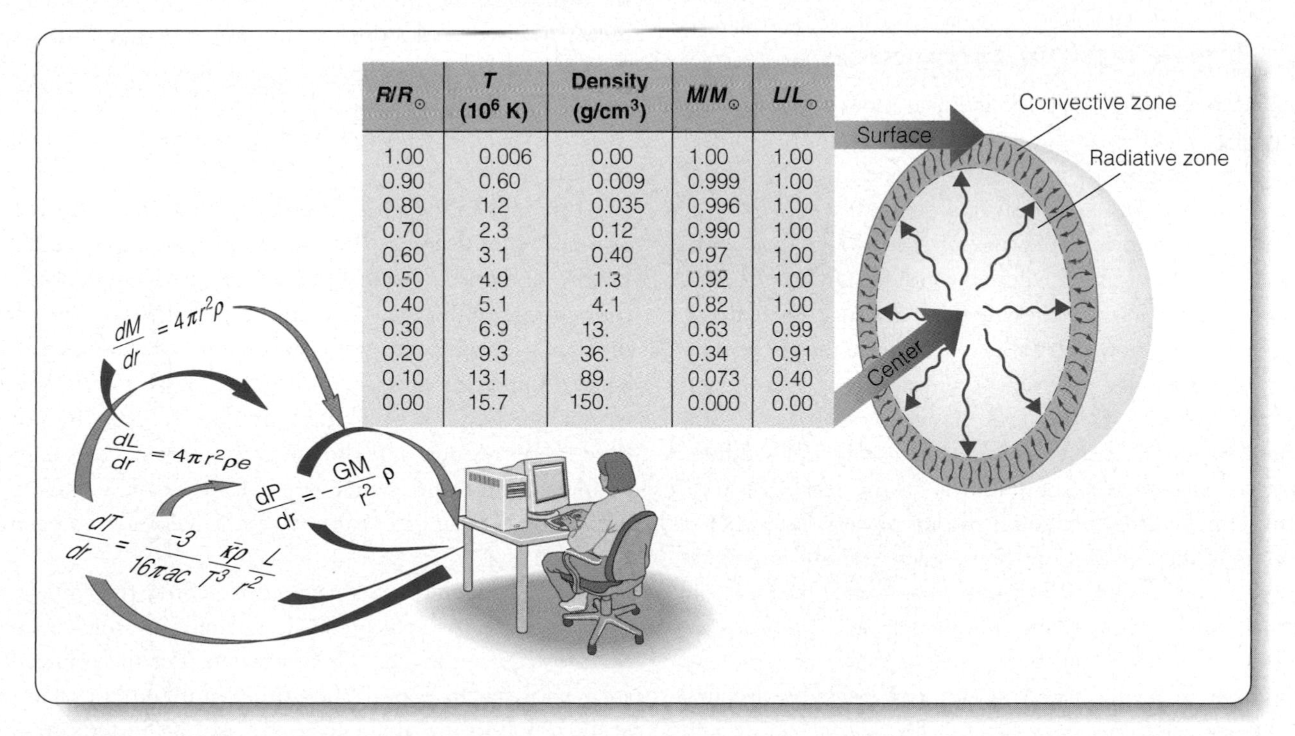

$R/R_\odot$	T (10^6 K)	Density (g/cm^3)	$M/M_\odot$	$L/L_\odot$
1.00	0.006	0.00	1.00	1.00
0.90	0.60	0.009	0.999	1.00
0.80	1.2	0.035	0.996	1.00
0.70	2.3	0.12	0.990	1.00
0.60	3.1	0.40	0.97	1.00
0.50	4.9	1.3	0.92	1.00
0.40	5.1	4.1	0.82	1.00
0.30	6.9	13.	0.63	0.99
0.20	9.3	36.	0.34	0.91
0.10	13.1	89.	0.073	0.40
0.00	15.7	150.	0.000	0.00

■ Figure 12-2

A stellar model is a table of numbers that represent conditions inside a star. Such tables can be computed using the four laws of stellar structure, shown here in mathematical form. The table in this figure describes the sun. (Illustration design by author)

How Do We Know? | 12-1

Mathematical Models

How can a test pilot climbing into an airplane for the first time know that it will fly? One of the most powerful tools in science is the mathematical model, a group of equations carefully designed to mimic the behavior of the objects and processes that scientists want to study. Astronomers build mathematical models of stars and can study the structure hidden deep inside stars. They can speed up the slow evolution of stars and slow down the rapid processes that generate energy. Stellar models are based on only four equations, but other models are much more complicated and may require many more equations.

For example, scientists and engineers designing a new airplane don't just cross their fingers, build it, and ask a test pilot to try it out. Long before any metal parts are made, mathematical models are created to test whether the wing design will generate enough lift, whether the fuselage can support the strain, and whether the rudder and ailerons can safely control the plane during takeoff, flight, and landing. Those mathematical models are put through all kinds of tests; can a pilot fly with one engine shut down, can the pilot recover from sudden turbulence, can the pilot land in a crosswind? By the time the test pilot rolls the plane down the runway for the first time, the mathematical models have flown many thousands of miles.

Scientific models, even those given mathematical form, are only as good as the assumptions that go into them and must be compared with the real world at every opportunity. If you are an engineer designing a new airplane, you can test your mathematical models by making measurements in a wind tunnel. Models of stars are much harder to test against reality. Models of stars predict the existence of a main sequence, the mass–luminosity relation, the observed numbers of giant and supergiant stars, the shapes of cluster H–R diagrams. Without mathematical models, astronomers would know little about the lives of the stars, and designing new airplanes would be a very dangerous business.

Before any new airplane flies, engineers build mathematical models to test its stability. (The Boeing Company)

main-sequence stars in the next section and the deaths of stars in the next chapter.

Why Is There a Main Sequence?

Astronomers have confidence in their stellar models because they have confidence that they understand gravity, nuclear fusion, and the behavior of hot gases. The physics that goes into the models is well understood. Another reason for confidence is that the models match the known properties of stars. With that confidence, astronomers can use the models to understand stars better. For example, the models explain why there is a main sequence.

Models of stars show that there is a main sequence because the centers of contracting protostars eventually grow hot enough to begin nuclear fusion. Deuterium, the heavy isotope of hydrogen, is the first nuclear fuel to fuse, but it is rare and produces little energy; stellar models show that it has no real effect on a contracting star. Hydrogen fusion is the big powerhouse; when it begins, it stops the contraction. Stars reach equilibrium somewhere along a line in the H–R diagram that astronomers call the main sequence. Hot stars are more luminous, and cool stars are less luminous, as you would expect. There is, however, a mystery about the main sequence that you can now solve by thinking about stellar models. Why does the luminosity of a star depend on its mass?

In Chapter 9, you used binary stars to find the masses of stars, and you discovered that the masses of main-sequence stars are ordered along the main sequence. The least massive stars are at the bottom, and the most massive stars are at the top. Further, you discovered a direct relationship between the mass of a star and its luminosity—the mass–luminosity relation. This is one of the most fundamental observations in astronomy, and stellar models can tell you why there must be a mass–luminosity relation.

The keys to the mass–luminosity relation are the law of hydrostatic equilibrium, which says that pressure must balance weight, and the pressure–temperature thermostat, which regulates energy production. You have seen that a star's internal pressure stays high because the generation of thermonuclear energy keeps its interior hot. Because more massive stars have more weight pressing down on the inner layers, their interiors must have high pressures and thus must be hotter. For example, the temperature at the center of a 15-solar-mass star is about 34,000,000 K, more than twice the central temperature of the sun.

Because massive stars have hotter cores, their nuclear reactions burn more fiercely. That is, their pressure–temperature thermostat is set higher. The nuclear fuel at the center of a 15-solar-mass star fuses over 3000 times more rapidly than the fuel at the center of the sun. The rapid reactions in the cores of massive stars produce more energy, but that energy cannot remain in the core. As the energy flows outward toward the cooler surface, it heats each level in the star and enables it to support the weight pressing inward. When all that energy reaches the surface, it radiates into space and makes the star highly luminous. So there must be a mass–luminosity relation because each star must support its weight by generating nuclear energy, and more massive stars have more weight to support.

The main sequence is elegant in its simplicity. It exists because stars balance their weight by fusing hydrogen in their cores. To understand the main sequence even better, you should next look at its top and bottom ends.

The Upper End of the Main Sequence

Models of stellar structure give astronomers a way to think about the extreme ends of the main sequence, the most massive and least massive stars. The first are rare, but the latter are common. Nevertheless, both are very difficult to study.

There are two reasons why there is an upper limit to the mass of stars. First, observations show that as gas clouds contract, they can fragment to form two or more stars. The more mass a gas cloud contains, the more likely it is to fragment, so there aren't many extremely massive stars because most of those gas clouds broke into smaller fragments and formed multiple-star systems.

Second, stellar models reveal that stars of roughly 100 solar masses are unstable at formation. To support the tremendous weight in such stars, the internal gas must be very hot, and that means it must emit floods of radiation that flow outward through the star, and the resulting radiation pressure blows gas away from the star's surface in powerful stellar winds. This mass loss from very massive stars could reduce a 60-solar-mass star to less than 30 solar masses in only a million years.

This process sets an upper limit on the masses of stars, but it is difficult to test the theory because it is hard to find truly massive stars. Most of the O and B stars in the sky have masses of 10 to 25 solar masses. The survey of stars at the end of Chapter 9 revealed that the stars at the upper end of the main sequence are very rare, so astronomers must search to great distances to find just a few. Nevertheless, a few stars are known that are thought to be very massive, and their spectra contain blueshifted emission lines. Kirchhoff's laws tell you that emission lines come from excited low-density gas, and the blueshift must be caused by a Doppler shift in gas coming toward Earth. These stars are losing mass.

■ Figure 12-3 shows a famous star, Eta Carinae. The evidence suggests it is a massive binary containing stars of 60 and 70 solar masses. The stars may have formed with about 100 solar masses each, but they are losing mass rapidly. An eruption 150 years ago made Eta Carinae the second brightest star in the sky and ejected the two expanding lobes of dusty gas. Although the star has faded, it is still very active, and more recent eruptions have ejected jets and an equatorial disk of gas and dust. These massive stars are clearly unstable.

The Lower End of the Main Sequence

The lower end of the main sequence is difficult to study, not because the stars are rare but because they are dim. If a red dwarf from the lower end of the main sequence replaced the sun, it would shine only a few times brighter than the full moon. Such stars are very common, but they are difficult to find even when they are only a few light-years away.

Stellar models predict that there are starlike objects even fainter than red dwarfs. Objects less massive than 0.08 solar mass cannot get hot enough to ignite hydrogen fusion. These **brown dwarfs** slowly contract, convert their gravitational energy into thermal energy, and radiate it away. A low-mass red dwarf containing 0.08 solar mass and fusing hydrogen has a surface temperature of about 2500 K, but brown dwarfs should have temperatures of 1000 K or so, giving them a dull muddy-red color—thus the term "brown dwarf." Brown dwarfs were difficult to find because they are so faint, but large surveys and infrared studies have turned up lots of them. Some are located in binary systems with normal stars (■ Figure 12-4a), but large numbers are free-floating objects without stellar companions (Figure 12-4b).

Brown dwarfs are clearly different from normal stars. Some brown dwarfs have methane bands in their spectra, and that means they must be quite cool. Methane molecules would be broken up at the temperatures of true stars. Color variations suggest that some brown dwarfs may be cool enough to have weather patterns.

The discovery of brown dwarfs has created a controversy. Are they failed stars, or are they planets? To discuss this, astronomers use a unit of mass related to Jupiter, the giant planet in our solar system. Jupiter is one thousand times less massive than the sun, and astronomers refer to the mass of brown dwarfs in Jupiter masses. A brown dwarf must be less than about 80 Jupiter masses. Generally, astronomers think of stars as bodies that generate energy by nuclear fusion, and a star less than 80 Jupiter masses can't heat its center to the 2.7 million Kelvin temperature needed to start minimal hydrogen fusion. Then brown dwarfs aren't stars. But astronomers tend to think of a planet as a nonluminous body that orbits a star, so can a brown dwarf floating free in space be called a planet?

Brown dwarfs more massive than 13 Jupiter masses can fuse deuterium, a heavy isotope of hydrogen, and some astronomers draw the line between stars and planets at 13 Jupiter masses. However, deuterium fusion does not last very long, does not generate much energy, and does not stop the brown dwarf from cooling. Not everyone agrees that this is a good dividing line.

Another approach to the controversy is to note how these objects form. A star forms from the contraction of a gas cloud, but astronomers know that planets form from the accumulation of solid bits of matter in disk-shaped nebulae around stars (a subject discussed in Chapter 19). In that case, free-floating brown dwarfs can't be planets.

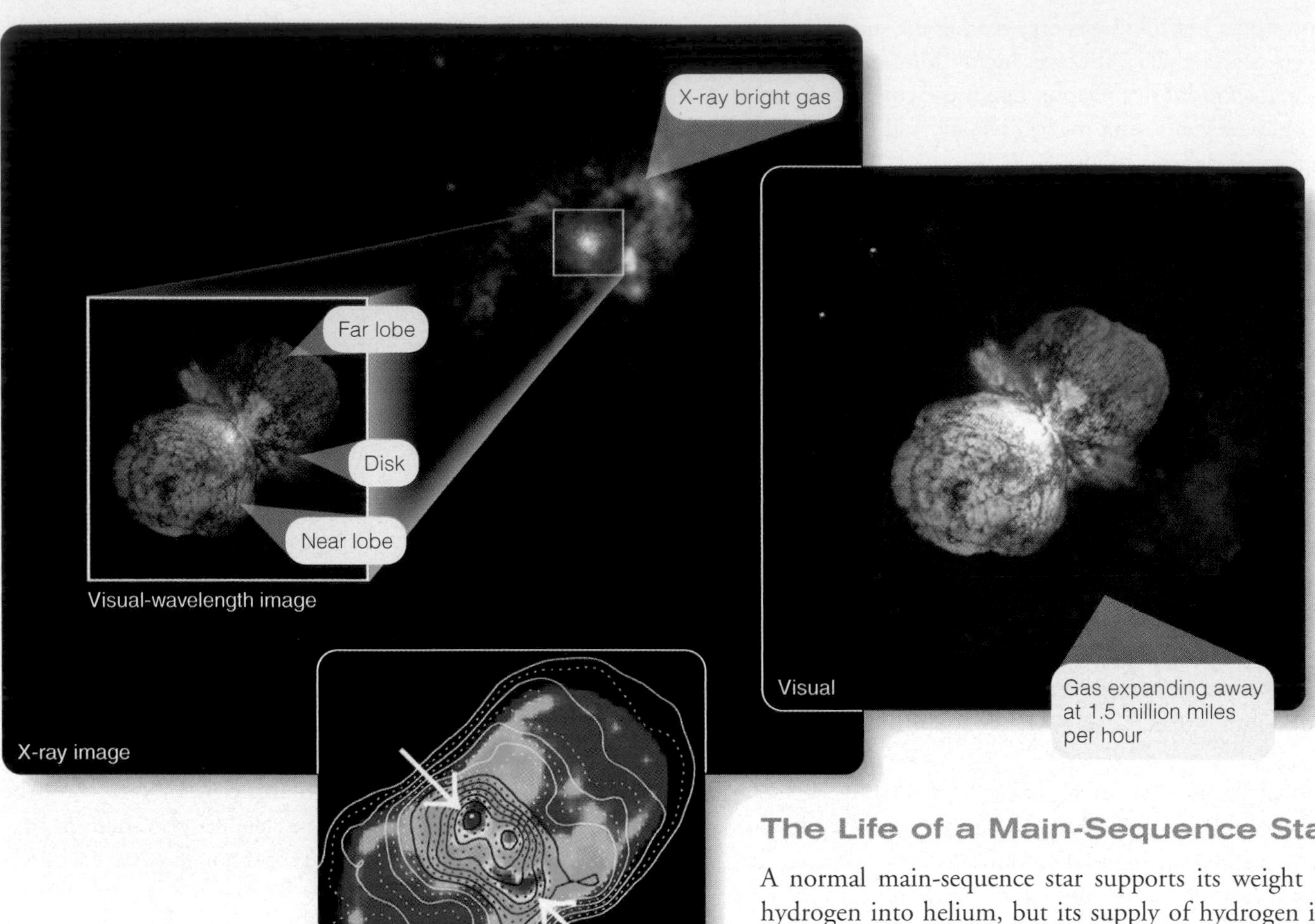

■ Figure 12-3

The star Eta Carinae is a binary containing two stars that are so massive they are rapidly losing mass. At visual wavelengths, two inflating lobes are visible with a disk of ejected material between like a plate pressed between two basketballs. Each lobe is about half a light-year in diameter. At X-ray wavelengths, very hot gas is excited by collision with high-speed gas ejected from the stars. An infrared image reveals a 15-solar-mass torus (doughnut shape) of gas and dust squeezing the outflowing gas into the two lobes. (NASA/CXC/SAO/HST; Jon Morse and J. Hester; IR image: ISO, Courtesy ESA)

The controversy over brown dwarfs is really an argument over the meaning of the words *star* and *planet*. Whatever they are called, you can understand the objects at the lower end of the main sequence. Objects less massive than 80 Jupiter masses cannot ignite hydrogen fusion; and, even if they are more massive than 13 Jupiter masses and deuterium fusion generates a little energy, they must continue contracting until the internal gas becomes so dense it cannot contract further. At that point, the object radiates its thermal energy away and slowly cools over billions of years.

The Life of a Main-Sequence Star

A normal main-sequence star supports its weight by fusing hydrogen into helium, but its supply of hydrogen is limited. As it consumes hydrogen, the chemical composition in its core changes, and the star evolves. Mathematical models of stars allow astronomers to follow that evolution.

Hydrogen fusion combines four nuclei into one. Consequently, as a main-sequence star consumes its hydrogen, the total number of nuclei in its interior decreases. Each newly made helium nucleus exerts the same pressure as one hydrogen nucleus; because the gas has fewer nuclei, its total pressure is less. This unbalances the gravity–pressure stability, and gravity squeezes the core of the star more tightly. As the core contracts, its temperature and density increase, and the nuclear reactions burn faster, releasing more energy and making the star more luminous. This additional energy flowing outward through the envelope forces the outer layers to expand and cool, so the star becomes slightly larger, brighter, and cooler.

As a result of these gradual changes in main-sequence stars, the main sequence is not a sharp line across the H–R diagram but rather a band (■ Figure 12-5). When a star begins its stable life fusing hydrogen, it settles on the lower edge of this band, the **zero-age main sequence (ZAMS).** As it combines hydrogen nuclei to make helium nuclei, the point that represents the star's luminosity and surface temperature moves upward and to the right, eventually reaching the upper edge of the main sequence just as the star exhausts nearly all of the hydrogen in its center. Astronomers find that main-sequence stars are plotted through-

■ Figure 12-4

(a) By using adaptive optics, astronomers were able to detect the brown dwarf 15 Sge B only 14 AU from its star. It is about as far from its companion star as the giant planets Saturn and Uranus are from the sun. (M. Liu, UH-IfA/W. M. Keck Observatory) (b) Too dim to see at visual wavelengths, about 50 free-floating brown dwarfs appear in an infrared image of the center of the Orion nebula. The bright stars of the Trapezium are located at the center. (STScI and NASA)

out this band showing that they are at various stages of their main-sequence lives.

These gradual changes in the sun will spell trouble for Earth. When the sun began its main-sequence life about 5 billion years ago, it was only 60 to 75 percent of its present luminosity. This, by the way, makes it difficult to explain how Earth has remained at roughly its present temperature for at least 3 billion years. Some experts suggest that Earth's atmosphere has gradually changed and compensated for the increasing luminosity of the sun.

By the time the sun leaves the main sequence in a few billion years, it will have twice its present luminosity. By this time, the average temperature on Earth will have climbed by at least 19°C (34°F). As this happens over the next few billion years, the polar caps will melt, the oceans will evaporate, and much of the atmosphere will vanish into space. Clearly, the future of Earth as the home of life is limited by the evolution of the sun.

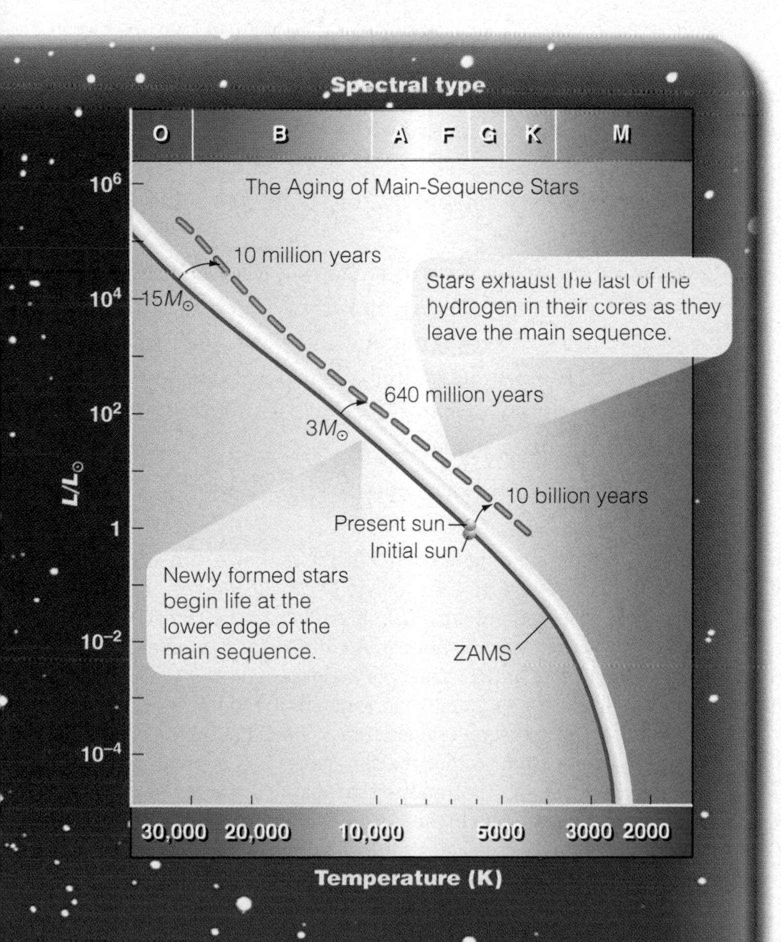

■ Active Figure 12-5

Contracting protostars reach stability at the lower edge of the main sequence, the zero-age main sequence (ZAMS). As a star converts hydrogen in its core into helium, it moves slowly across the main sequence, becoming slightly more luminous and slightly cooler. Once a star consumes all of the hydrogen in its core, it can no longer remain a stable main-sequence star. More massive stars age rapidly, but less-massive stars use up the hydrogen in their cores more slowly and live longer main-sequence lives.

Once a star leaves the main sequence, it evolves rapidly and soon dies. The average star spends 90 percent of its life fusing hydrogen on the main sequence. This explains why 90 percent of all normal stars are main-sequence stars. You are most likely to see a star during that long, stable period when it is on the main sequence.

The number of years a star spends on the main sequence depends on its mass (■ Table 12-2). Massive stars use fuel rapidly and live short lives, but low-mass stars conserve their fuel and shine for billions of years. For example, a 25-solar-mass star will exhaust its hydrogen and die in only about 7 million years. This means, for one thing, that life is very unlikely to develop on planets orbiting massive stars. These stars do not live long enough for life to get started, let alone evolve into complex creatures. You will learn about this problem in detail in Chapter 26.

Red dwarfs, very-low-mass stars, use their fuel so slowly they should survive for 200 to 300 billion years. Because the universe seems to be only about 14 billion years old, red dwarfs must still be in their infancy. None of them should have exhausted their hydrogen fuel yet.

Vast numbers of faint, low-mass stars fill the sky. Look at page 197 and notice how much more common the lower-main-sequence stars are than the massive O and B stars. Main-sequence K and M stars are so faint they are difficult to locate, but they are very common. Nature makes more low-mass stars than massive stars, but that is not sufficient to explain the excess. An additional factor is the stellar lifetimes. Because low-mass stars live long lives, there are more of them in the sky than massive stars. The O and B stars are luminous and easy to locate; but, because of their fleeting lives, there are never more than a few on the main sequence at any one time.

The Life Expectancies of Stars

To understand how nature makes stars and how stars evolve, you must be able to estimate how long they can survive, and that turns out to be easy. In general, massive stars live short lives and lower-mass stars live long lives, but you can make more accurate estimates by using simple stellar models. In fact, you can calculate the approximate life expectancy of a star from its mass.

Because main-sequence stars consume their fuel at an approximately constant rate, you can estimate the amount of time a star spends on the main sequence—its life expectancy, T—by dividing the amount of fuel by the rate of fuel consumption. This is a common calculation. If you drive a truck that carries 20 gallons of fuel and uses 5 gallons of fuel per hour, you know the truck can run for 4 hours.

The amount of fuel a star has is proportional to its mass, and the rate at which it burns its fuel is proportional to its luminosity; that means you could make a first estimate of the star's life expectancy by dividing mass by luminosity. A 2-solar-mass star is about 11 times more luminous than the sun and should live about 2/11, or 18 percent, as long as the sun. You can, however, make the calculation even easier if you remember that the mass–luminosity relation says that the luminosity of a star is approximately equal to $M^{3.5}$. The life expectancy then is:

$$T = \frac{M}{L} = \frac{M}{M^{3.5}} \text{ or } T = \frac{1}{M^{2.5}}$$

This means that you can estimate the life expectancy of a star by dividing 1 by the star's mass raised to the 2.5 power. If you express the mass in solar masses, the life expectancy will be in solar lifetimes.

For example, how long can a 4-solar-mass star live?

$$T = \frac{1}{4^{2.5}} = \frac{1}{(4 \cdot 4 \cdot \sqrt{4})} = \frac{1}{32} \text{ solar lifetimes}$$

Detailed studies of models of the sun show that the sun, presently 5 billion years old, can last another 5 billion years. So, a solar lifetime is approximately 10 billion years, and a 4-solar-mass star will last about (10 billion)/32, or about 310 million, years.

This estimation of stellar life expectancies is very approximate. For example, the model ignores mass loss, which may affect the life expectancies of very luminous and very faint stars. Nevertheless, it serves to illustrate an important point. Stars that are only slightly more massive than the sun have dramatically shorter lifetimes on the main sequence.

■ Table 12-2 | Main-Sequence Stars

Spectral Type	Mass (sun = 1)	Luminosity (sun = 1)	Approximate Years on Main Sequence
O5	40	405,000	1×10^6
B0	15	13,000	11×10^6
A0	3.5	80	440×10^6
F0	1.7	6.4	3×10^9
G0	1.1	1.4	8×10^9
K0	0.8	0.46	17×10^9
M0	0.5	0.08	56×10^9

◄ SCIENTIFIC ARGUMENT ►

Why is there a main sequence?

Most scientific arguments are simple chains of ideas that begin with a well-understood idea and lead to a less obvious conclusion. You can begin with the simple observation that weight must be balanced by pressure, the principle of hydrostatic equilibrium. A contracting protostar is not quite in equilibrium, and gravity squeezes it tighter and tighter. As it contracts, its interior heats up; and, when it gets hot enough to fuse hydrogen into helium, the pressure–temperature thermostat takes over and regulates energy production so that the star makes just enough energy to support its own weight. That means that massive stars, having more weight to support, must have higher internal pressures and therefore must make more energy. Energy must flow from hot to cool, so that energy flows from the hot core of the star out to the cooler surface and is radiated into space. Massive stars, having more weight to support, must have their pressure temperature thermostats set higher, and that makes them more luminous. They reach stability along the upper main sequence. Less massive stars support less weight and reach stability along the lower main sequence. There is a main sequence because stars fuse hydrogen to support their own weight.

Now revise your argument. **Why do massive stars have such short life expectancies?**

◄ ►

12-2 Post-Main-Sequence Evolution

IN EARLIER CHAPTERS, you probably had questions about giant stars: Why are giant stars so large? Why are they so uncommon? And why do they have such low densities? Now you are ready to answer those questions by discussing the evolution of stars after they leave the main sequence.

Expansion into a Giant

To understand how stars evolve, you must remember that they are not well mixed; that is, their interiors are not stirred. The centers of lower-mass stars like the sun are radiative, meaning the energy moves as radiation and not as circulating currents of heated gas. The gas does not move deep inside such stars, and that means they are not mixed at all. More massive stars have convective cores that mix the central regions (Figure 11-14), but these regions are not very large, and so, for the most part, these stars, too, are not mixed. (The lowest-mass stars are an exception that you will examine in the next chapter.)

In this respect, stars are like a campfire that is not stirred; the ashes accumulate at the center, and the fuel in the outer parts never gets used. Nuclear fusion consumes hydrogen nuclei and produces helium nuclei, the "ashes" at the star's center. Nothing mixes the interior of the star, so the helium nuclei remain where they are in the center of the star, and the hydrogen in the outer parts of the star is not mixed down to the center where it can be fused.

The helium ashes that accumulate in the star's core cannot fuse into heavier elements because the temperature is too low. As a result, the core eventually becomes an inert ball of helium. As this happens, the energy production in the core falls, and the weight of the outer layers forces the core to contract.

Although the contracting helium core cannot generate nuclear energy, it does grow hotter because it converts gravitational energy into thermal energy (see Chapter 11). The rising temperature heats the unprocessed hydrogen just outside the core, hydrogen that was never before hot enough to fuse. When the surrounding hydrogen becomes hot enough, it ignites in a thin, spherical shell. Like a grass fire burning outward from an exhausted campfire, the hydrogen-fusing shell fuses outward, leaving helium ash behind and increasing the mass of the helium core.

During this stage in its evolution, the star overproduces energy; that is, it produces more energy than it needs to balance its own gravity. The contracting helium core converts gravitational energy into thermal energy. Some of this energy heats the helium core, and some of that heat leaks outward through the star. At the same time, the hydrogen-fusing shell also produces energy as the contracting core brings fresh hydrogen closer to the center of the star and heats it to high temperature. The result is a flood of energy flowing outward, forcing the outer layers of the star to puff up and swelling the star into a giant (■ Figure 12-6).

The expansion of the envelope dramatically changes the star's location in the H–R diagram. As the outer layers of gas expand, energy is absorbed in lifting and expanding the gas. The loss of that energy lowers the temperature of the gas. Consequently, the point that represents the star in the H–R diagram moves quickly to the right (in less than a million years for a star of 5 solar masses but in about 150 million years for a 1-solar-mass star). A medium-mass star like the sun expands and cools to become a red giant (■ Figure 12-7). As the radius of a giant star continues to increase, the enlarging surface area makes the star more luminous, moving its point upward in the H–R diagram. One of your Favorite Stars, Aldebaran, the glowing red eye of Taurus the bull, is a red giant, with a diameter 25 times that of the sun but with only half its surface temperature.

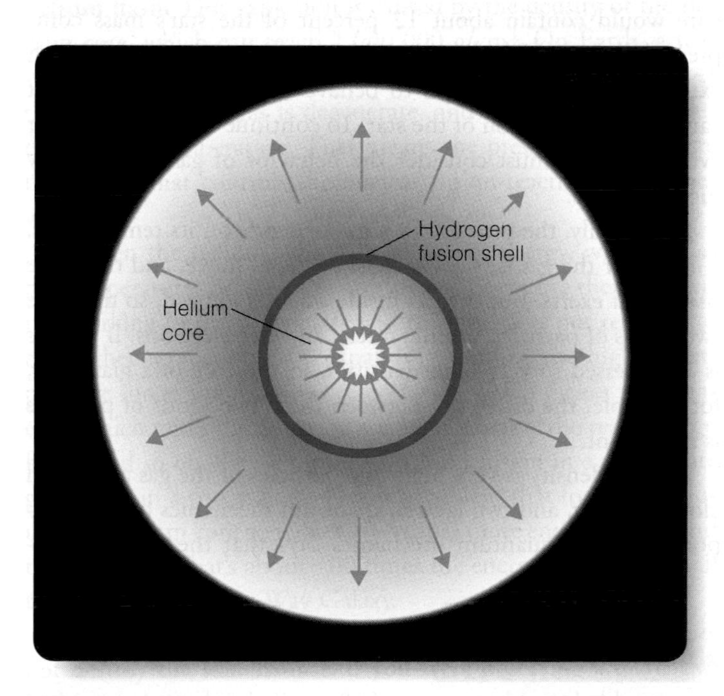

■ Figure 12-6

When a star runs out of hydrogen at its center, it ignites a hydrogen-fusing shell. The helium core contracts and heats while the envelope expands and cools. (For a scale drawing, see Figure 12-9.)

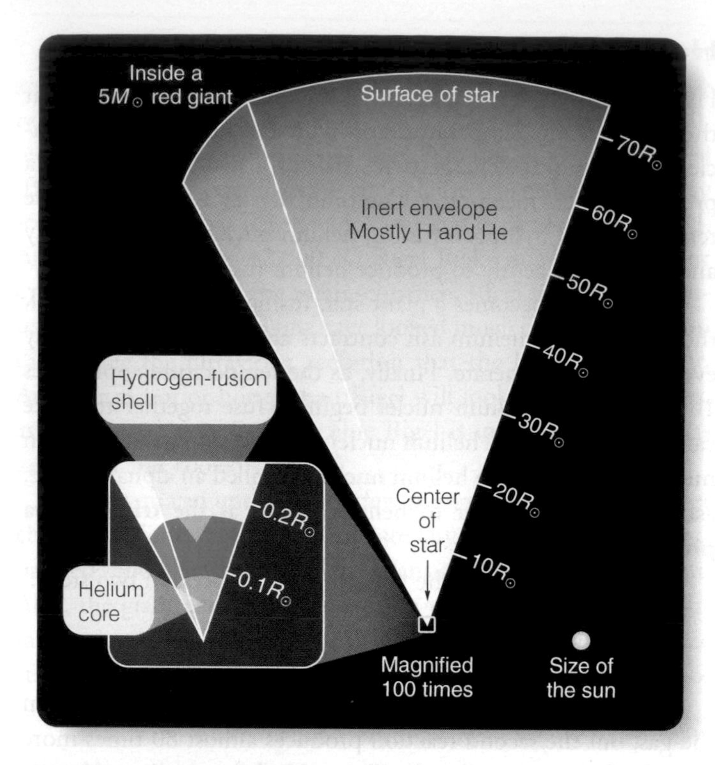

■ Active Figure 12-9

When a star runs out of hydrogen at its center, the core of helium contracts to a small size, becomes very hot, and begins nuclear fusion in a shell (blue). The outer layers of the star expand and cool. The red giant star shown here has an average density much lower than the air at Earth's surface. Here $M_\odot$ stands for the mass of the sun, and $R_\odot$ stands for the radius of the sun. (Illustration design by the author)

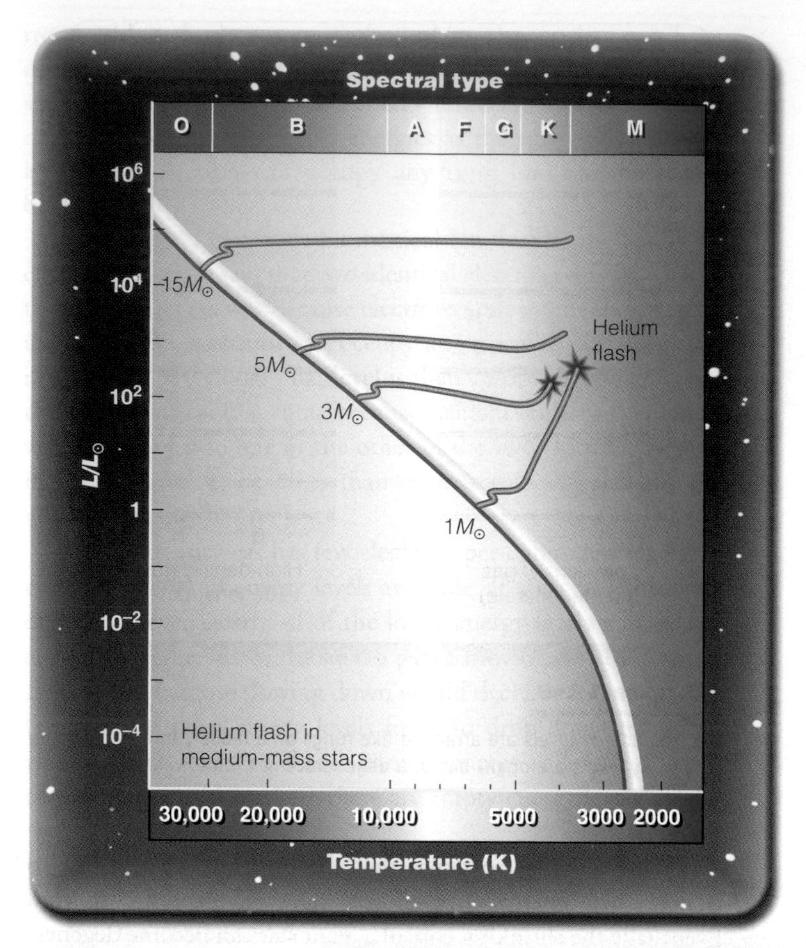

■ Figure 12-10

Stars like the sun suffer a helium flash, but more massive stars begin helium fusion without a helium flash. Stars less than 0.4 times the mass of the sun cannot get hot enough to ignite helium. The zero-age main sequence is shown here in red at the lower edge of the main sequence.

lived event. In a matter of minutes, the core of the star becomes so hot it is no longer degenerate, the pressure–temperature thermostat brings the helium fusion under control, and the star proceeds to fuse helium steadily in its core. You will learn about this post-helium-flash evolution later.

Not all stars experience a helium flash. Stars less massive than about 0.4 solar mass never get hot enough to ignite helium, and stars more massive than about 3 solar masses ignite helium before their cores become degenerate (■ Figure 12-10). In such stars, pressure depends on temperature, so the pressure–temperature thermostat keeps the helium fusion under control.

If the helium flash occurs only in some stars and is a very short-lived event that is hardly visible from outside the star, why should you worry about it? The answer is that it limits the reliability of the mathematical models astronomers use to study stellar evolution. Massive stars are not very common, and low-mass stars evolve so slowly that it is hard to find evidence of low-mass stellar evolution. For those reasons, studies of stellar evolution must concentrate on medium-mass stars, which do experience the helium flash. But the helium flash occurs so rapidly and so violently that computer programs cannot follow the changes in the star's internal structure in detail. To follow the evolution of medium-mass stars like the sun past the helium flash, astronomers must make assumptions about the way the helium flash affects stars' internal structures.

There is another reason why you should think about the helium flash. By seeing what happens when the pressure–temperature thermostat is turned off, you can appreciate its importance in maintaining the stability of stars.

Post-helium-flash evolution is generally understood. After the gas ceases to be degenerate, helium fusion proceeds under the control of the pressure–temperature thermostat. Throughout these events, the hydrogen-fusion shell continues to produce energy, but the new helium-fusion energy produced in the core makes the core expand. That expansion absorbs energy previously used to support the outer layers of the star. As a result, the outer layers contract, and the surface of the star grows slightly hotter. In the H–R diagram, the point that represents the star initially moves down toward lower luminosity and then to the left toward higher temperature. Later, as the star stabilizes fusing helium in its core, its luminosity recovers, but its surface grows hotter.

Helium fusion produces carbon and some of the carbon nuclei absorb alpha particles (helium nuclei) to form oxygen. A few of the oxygen nuclei can absorb alpha particles and form neon and then magnesium. Some of these reactions release neutrons, which, having no charge, are more easily absorbed by nuclei to gradually build even heavier nuclei. These reactions are not important as energy producers. They are slow-cooker processes that form small traces of heavier elements right up to beryllium nearly four times heavier than iron.

The nuclear "ashes," mostly carbon and oxygen, accumulate at the center of the star where they can not fuse rapidly to make energy because the core is not hot enough. As this happens, the core contracts, grows hotter, and ignites a helium-fusion shell. At this stage, the star has two shells producing energy. A hydrogen-fusion shell continues to eat its way outward, leaving behind helium ash; a helium-fusion shell eats its way outward into the helium, leaving behind carbon–oxygen ash. Because the star cannot generate energy in its carbon–oxygen core, its core contracts, its outer layers expand, and its point in the H–R diagram moves back toward the right, completing a loop. You can see that small loop in the giant region of the H–R diagram shown in ■ Figure 12-11. Expanded drawings show how the interior of the star changes as it uses up its fuels.

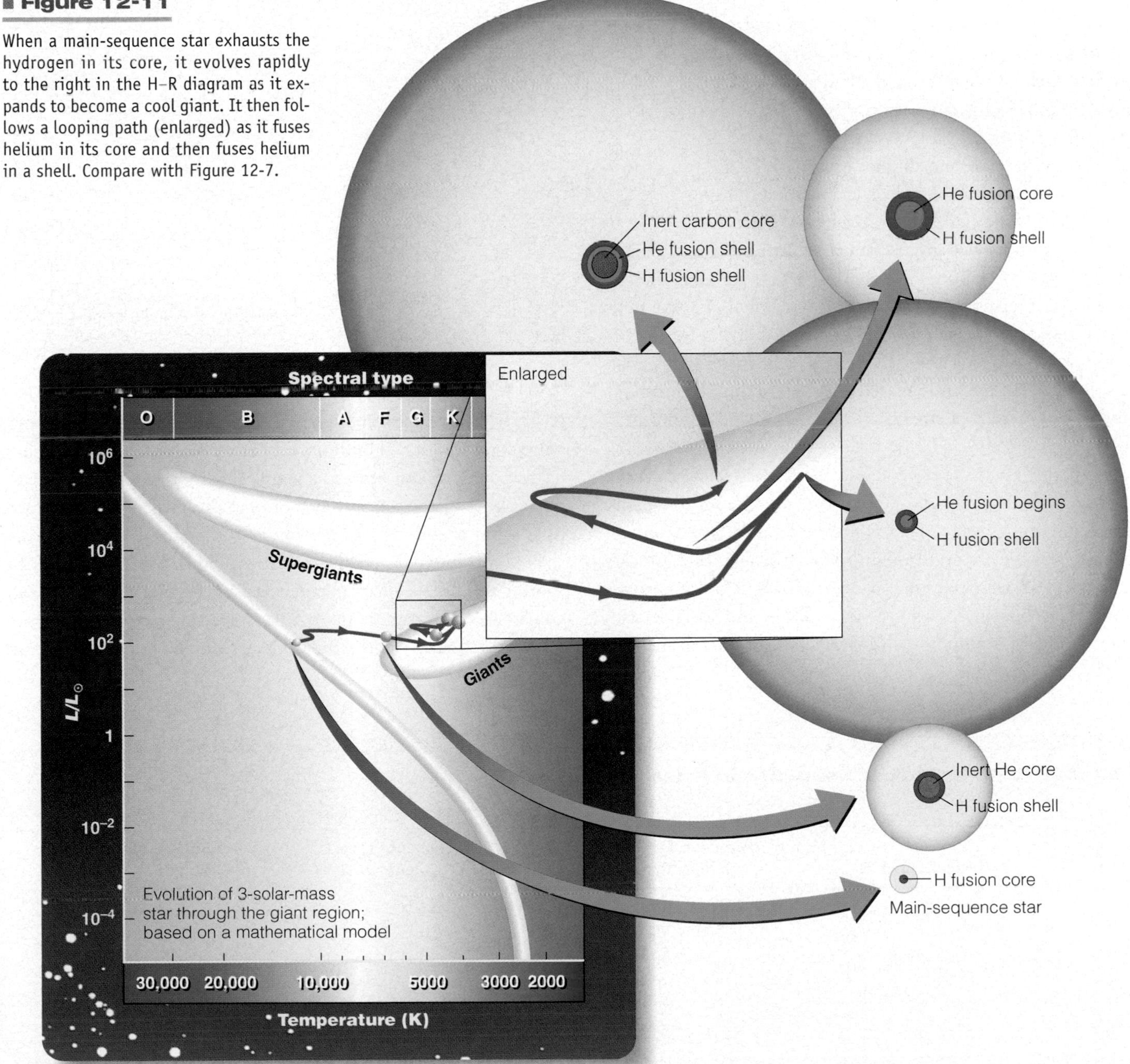

■ **Figure 12-11**

When a main-sequence star exhausts the hydrogen in its core, it evolves rapidly to the right in the H–R diagram as it expands to become a cool giant. It then follows a looping path (enlarged) as it fuses helium in its core and then fuses helium in a shell. Compare with Figure 12-7.

Fusing Elements Heavier Than Helium

Modeling the evolution of stars after helium exhaustion requires great sophistication. The inert core of carbon–oxygen ash contracts and becomes hotter. The cores of stars more massive than about 4 solar masses can reach temperatures as high as 600,000,000 K, hot enough to fuse carbon. Subsequently, these stars can fuse neon, oxygen, silicon, and other heavy elements.

Carbon fusion in a star begins a complex network of reactions illustrated in ■ Figure 12-12, where each circle represents a possible nucleus and each arrow represents a different nuclear reaction. Nuclei can react by capturing a proton, capturing a neutron, capturing a helium nucleus, or by combining directly with other nuclei. Unstable nuclei can decay by ejecting an electron, ejecting a positron, ejecting a helium nucleus, or splitting into fragments.

As you can tell from all the arrows in Figure 12-12, the fusion of elements heavier than helium is not simple. Astronomers who model the structure and evolution of massive stars must use sophisticated nuclear physics to compute the energy production and changes in chemical composition as the stars fuse carbon and heavier elements. These stars develop complex internal structure as they fuse one heavy element after another in concentric shells. Like great spherical cakes with many layers, the stars may be fusing a number of different elements simultaneously in concentric shells.

Eventually, all stars face collapse. Less massive stars cannot get hot enough to ignite the heavier nuclei. The more massive stars that can ignite heavy nuclei consume their fuels at a tremendous rate, so they run through their available fuel quickly (■ Table 12-3). No matter what the star's mass, it will eventually run out of usable fuels, and gravity will win the struggle with pressure. You will explore the ultimate deaths of stars in the next chapter.

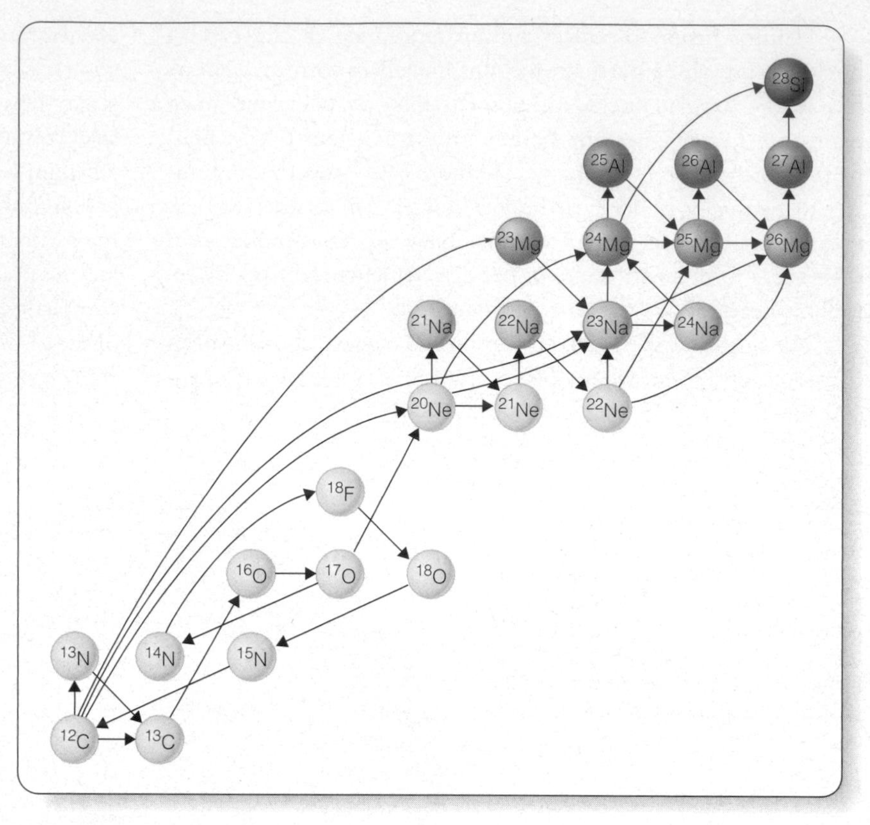

■ **Figure 12-12**

Carbon fusion involves many possible reactions that build numerous heavy nuclei.

The time a star spends as a giant or supergiant is small compared with its life on the main sequence. The sun, for example, will spend about 10 billion years as a main-sequence star but only about a billion years as a giant. The more massive stars pass through the giant stage even more rapidly. Because of the short time a star spends as a giant, you are unlikely to see many such stars. This illustrates an important principle in astronomy: The shorter the time a given evolutionary stage takes, the less likely you are to see stars in that particular stage. That explains why you see a great many main-sequence stars but few giants (see page 197).

■ Table 12-3 | Nuclear Reactions in Massive Stars

Nuclear Fuel	Nuclear Products	Minimum Ignition Temperature	Main-Sequence Mass Needed to Ignite Fusion	Duration of Fusion in a 25-$M_\odot$ Star
H	He	4×10^6K	0.1 $M_\odot$	7×10^6 yr
He	C, O	120×10^6K	0.4 $M_\odot$	0.5×10^6 yr
C	Ne, Na, Mg, O	0.6×10^9K	4 $M_\odot$	600 yr
Ne	O, Mg	1.2×10^9 K	~8 $M_\odot$	1 yr
O	Si, S, P	1.5×10^9 K	~8 $M_\odot$	~ 0.5 yr
Si	Ni to Fe	2.7 × 109 K	~8 $M_\odot$	~ 1 day

◄ SCIENTIFIC ARGUMENT ►

What uncertainties limit the accuracy of this story of stellar evolution?

A scientific argument is not meant to convince others but to test your own understanding, and that means it should include a discussion of any uncertainties in the analysis. In the case of stellar evolution, there are two things that limit the accuracy of the models. First, some events in the history of a star happen so fast that computers can't predict the results in a reasonable amount of time. In fact, some astronomers estimate that a computer following the evolution of a model star undergoing the helium flash would have to recompute the model in such short time steps that the computer program would take millions of years to predict events that occur inside the star in just seconds. Truly rapid changes such as explosions in stars make the models uncertain.

Second, you have seen how the general properties of a giant star can depend on the subatomic properties of matter. Not all of those subatomic properties are well understood. For example, neutrinos oscillate, and scientists don't entirely understand how that happens. If some subatomic particle behaves differently than expected, then stellar models may be slightly different from the real stars they are supposed to represent.

Can you add to your argument? **What other physical processes have been omitted from stellar models? What features of the sun's surface suggest motions and forces that are ignored in most stellar models?**

◄ ►

12-3 Evidence of Evolution: Star Clusters

THE THEORY OF STELLAR EVOLUTION is so complex and involves so many assumptions that astronomers would have little confidence in it were it not for H–R diagrams of clusters of stars. By observing the properties of star clusters of different ages, you can see clear evidence of the evolution of stars.

To grasp the difficulty of understanding stellar evolution, consider an analogy. Suppose a visitor to Earth who had never seen a tree wandered through a forest for an hour looking at mature trees, fallen seeds, young saplings, rotting logs, and rising sprouts. Could such an observer understand the life cycle of trees? Astronomers face the same problem when they try to understand the life story of the stars.

Humans do not live long enough to see stars evolve. You see only a momentary glimpse of the universe as it appears during your lifetime, a snapshot in which all the stages in the life cycle of the stars are represented. Unscrambling these stages and putting them in order is a difficult task. Only by looking at selected groups of stars—star clusters—can you see the pattern.

Observing Star Clusters

The H–R diagram of the cluster freezes a moment in its history and makes the evolution of the stars visible. The stars in a cluster all formed at about the same time and from the same cloud of gas, so they must be about the same age and have the same chemical composition. The differences you see among stars in a cluster must arise from differences in mass, and that makes stellar evolution visible. Study **Star Cluster H–R Diagrams** on pages 256–257 and notice three important points and four new terms:

1. There are two kinds of star clusters, *open clusters* and *globular clusters*. They look different, but they are similar in the way their stars evolve. You will learn more about these clusters in a later chapter.

2. You can estimate the age of a star cluster by observing the *turnoff point* in the distribution of the points that represent its stars in the H–R diagram.

3. Finally, notice that the shape of a star cluster's H–R diagram is governed by the evolutionary path the stars take. The H–R diagrams of older clusters are especially clear in outlining how stars evolve away from the main sequence to the giant region, then move left along the *horizontal branch* before evolving back into the giant region. By comparing clusters of different ages, you can visualize how stars evolve almost as if you were watching a film of a star cluster evolving over billions of years.

Were it not for star clusters, astronomers would have little confidence in the theories of stellar evolution. Star clusters make that evolution visible and assure astronomers that they really do understand how stars are born, live, and die.

The Evolution of Star Clusters

What you know about star formation and stellar evolution can help you understand star clusters and how they evolve.

A star cluster is formed when a cloud of gas contracts, fragments, and forms a group of stars. As you recall from Chapter 11, some groups of stars, called associations, are so widely scattered that the group is not held together by its own gravity. The stars in an association wander away from each other rather quickly. Some groups of stars are more compact; even after the stars become luminous and the gas and dust are blown away, gravity holds the group together as a star cluster.

Open clusters are not as old or as crowded as the globular clusters, and that helps explain their appearance. Close encounters between stars are rare in an open cluster where the stars are further apart, and such clusters have an irregular appearance. The globular clusters appear to be nearly perfect globes because the stars are much closer together and encounters between stars are more common. The globular clusters have had time to evenly distribute the energy of motion among all of the stars so they have settled into a more uniform, spherical shape.

As a star cluster ages, some stars traveling a bit faster than others can escape. Globular clusters are compact and massive and have survived for 11 billion years or more. A star cluster with

Star Cluster H–R Diagrams

1 An **open cluster** is a collection of 10 to 1000 stars in a region about 25 pc in diameter. Some open clusters are quite small and some are large, but they all have an open, transparent appearance because the stars are not crowded together.

In a star cluster each star follows its orbit around the center of mass of the cluster.

AURA/NOAO/NSF

Visual-wavelength image

Open Cluster
The Jewel Box

1a A **globular cluster** can contain 10^5 to 10^6 stars in a region only 10 to 30 pc in diameter. The term "globular cluster" comes from the word "globe," although globular cluster is pronounced like "glob of butter." These clusters are nearly spherical, and the stars are much closer together than the stars in an open cluster.

Globular Cluster
47 Tucanae

Visual-wavelength image

Anglo-Australian Observatory/ David Malin Images

Astronomers can construct an H–R diagram for a star cluster by plotting a point to represent the luminosity and temperature of each star.

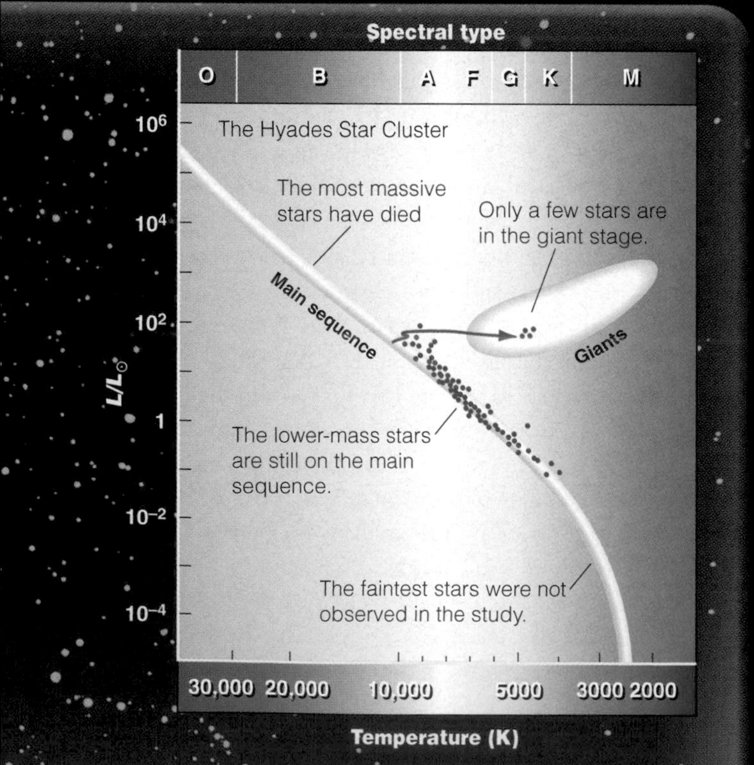

Spectral type
O B A F G K M
The Hyades Star Cluster
10^6
10^4
10^2
1
10^{-2}
10^{-4}
$L/L_\odot$
The most massive stars have died
Only a few stars are in the giant stage.
Main sequence
Giants
The lower-mass stars are still on the main sequence.
The faintest stars were not observed in the study.
30,000 20,000 10,000 5000 3000 2000
Temperature (K)

2 The H–R diagram of a star cluster can make the evolution of stars visible. The key is to remember that all of the stars in the star cluster have the same age but differ in mass. The H–R diagram of a star cluster provides a snapshot of the evolutionary state of the stars at the time you happen to be alive. The diagram here shows the 650-million-year-old star cluster called the Hyades. The upper main sequence is missing because the more massive stars have died, and our snapshot catches a few medium-mass stars leaving the main sequence to become giants.

As a star cluster ages, its main sequence grows shorter like a candle burning down. You can judge the age of a star cluster by looking at the **turnoff point**, the point on the main sequence where stars evolve to the right to become giants. Stars at the turnoff point have lived out their lives and are about to die. Consequently, the life expectancy of the stars at the turnoff point equals the age of the cluster.

ThomsonNOW™ Sign in at www.thomsonedu.com and go to ThomsonNOW to see Active Figure "Cluster Turnoff" and notice how the shape of a cluster's H–R diagram changes with time.

3 From theoretical models of stars, you could construct a film to show how the H–R diagram of a star cluster changes as it ages. You can then compare theory (left) with observation (right) to understand how stars evolve. Note that the time step for each frame in this film increases by a factor of 10.

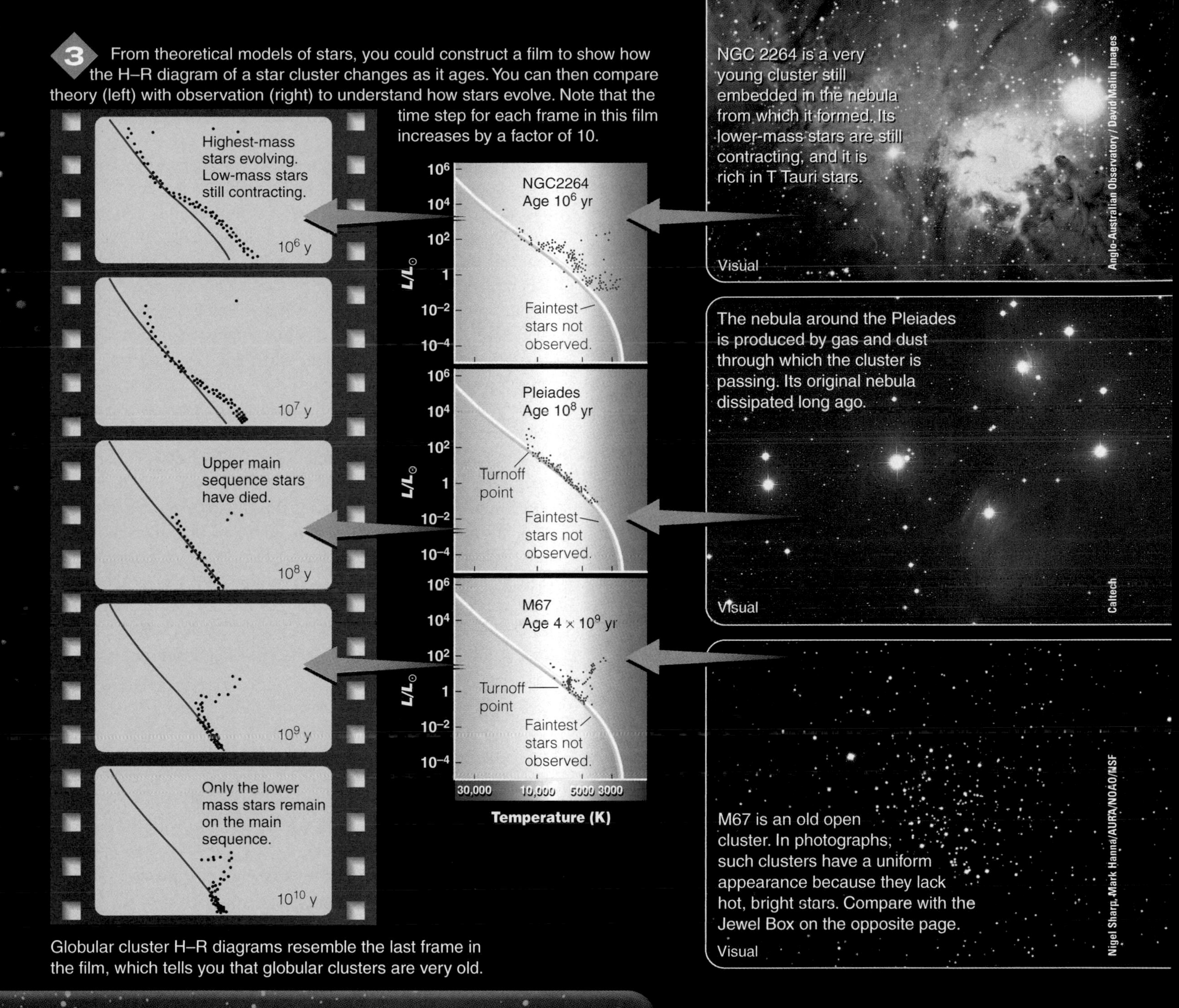

NGC 2264 is a very young cluster still embedded in the nebula from which it formed. Its lower-mass stars are still contracting, and it is rich in T Tauri stars.

The nebula around the Pleiades is produced by gas and dust through which the cluster is passing. Its original nebula dissipated long ago.

M67 is an old open cluster. In photographs, such clusters have a uniform appearance because they lack hot, bright stars. Compare with the Jewel Box on the opposite page.

Globular cluster H–R diagrams resemble the last frame in the film, which tells you that globular clusters are very old.

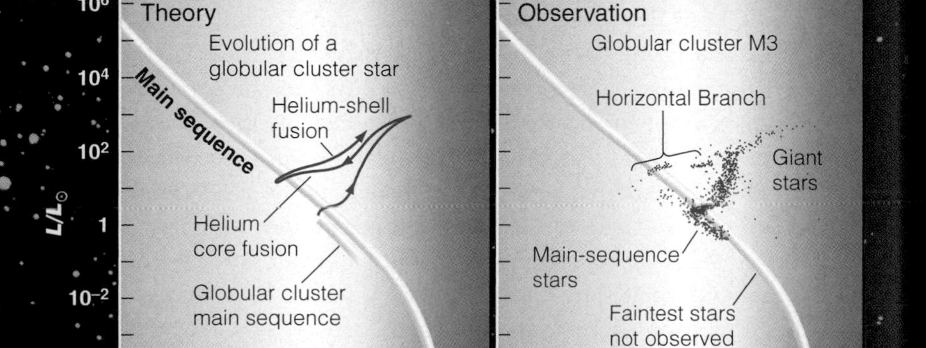

3a The H–R diagrams of globular clusters have very faint turnoff points showing that they are very old clusters. The best analysis suggests these clusters are about 11 billion years old.

The **horizontal branch** stars are giants fusing helium in their cores and then in shells. The shape of the horizontal branch outlines the evolution of these stars.

The main-sequence stars in globular clusters are fainter and bluer than the zero-age main sequence. Spectra reveal that globular cluster stars are poor in elements heavier than helium, and that means their gases are less opaque. That means energy can flow outward more easily, which makes the stars slightly smaller and hotter. Again the shape of star cluster H–R diagrams illustrates principles of stellar evolution.

only a few stars widely distributed may evaporate completely as its stars escape one by one. This is the fate of associations. Our sun probably formed in a star cluster 5 billion years ago, but there is no way to trace it back to its family home even if that star cluster still exists.

◄ SCIENTIFIC ARGUMENT ►

Why do open clusters contain only a small number of giant stars but many main-sequence stars?

Sometimes the critical factor in a scientific argument is timing. When you look at a star cluster, you see a snapshot of the cluster that freezes it in time. In the H–R diagram of an open cluster, you see a main sequence containing many stars, but typically you see significantly fewer giant stars. Those giant stars were once main-sequence stars in the cluster, but they exhausted the hydrogen fuel in their cores and expanded to become giants. A star like the sun can fuse hydrogen on the main sequence for many billions of years but can remain a giant for only a billion years or less. More massive stars evolve even faster. Because of this, at any given moment, you see most of the stars in a star cluster on the main sequence and only a few, caught in the snapshot, as giant stars.

Now create a new argument. **How do H–R diagrams reveal the age of a star cluster?**

◄ ►

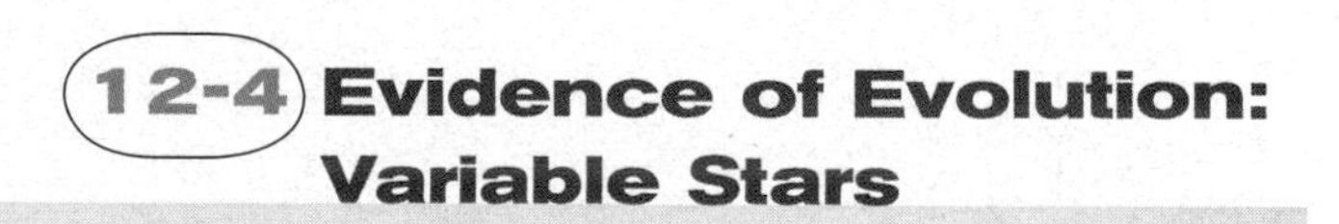

12-4 Evidence of Evolution: Variable Stars

IT IS A **Common Misconception** that the stars are constant and unchanging. Stellar models show that the stars are slowly evolving as they consume their fuels, and evidence from star clusters confirms those slow changes in structure. But can you really be sure the stars are changing and evolving? Certain stars that pulsate in brightness provide evidence of stellar evolution and illustrate once more the importance of energy flowing outward through the layers of the stars.

A **variable star** is any star that changes its brightness in a periodic way. Variable stars include many different kinds of stars. Some variable stars are eclipsing binaries (Chapter 9), but many others are stars that expand and contract, heat and cool, and grow more luminous and less luminous because of internal processes. These stars are often called **intrinsic variables** to distinguish them from eclipsing binaries. Of these intrinsic variables, one particular kind is centrally important to modern astronomy even though the first star of that class was discovered centuries ago.

Cepheid and RR Lyrae Variable Stars

In 1784, the deaf and mute English astronomer John Goodricke, then 19 years old, discovered that the star δ Cephei is variable, changing its brightness by almost a magnitude over a period of 5.37 days (■ Figure 12-13a). Goodricke died at the age of 21, but his discovery ensures his place in the history of astronomy. Since 1784, hundreds of stars like δ Cephei have been found, and they are now known as **Cepheid variable stars.**

Cepheid variables are supergiant or bright giant stars of spectral type F or G. The fastest complete a cycle—bright to faint to bright again—in about two days, whereas the slowest take as long as 60 days. A plot of a variable's magnitude versus time, known as a light curve, shows a typical rapid rise to maximum brightness and a slower decline (Figure 12-13b). Some Cepheids change their brightness by only 0.5 magnitude (about 10 percent), while others change by more than a magnitude. One of your Favorite Stars, Polaris, the North Star, is a Cepheid with a period of 3.9696 days and a very low amplitude. A related type of star, the **RR Lyrae stars,** pulsate with periods of less than a day and are fainter than the Cepheids.

Studies of Cepheids and RR Lyrae stars reveal two related facts of great importance. First, there is a **period–luminosity relation** that connects the period of pulsation of a Cepheid to its luminosity. The longer-period Cepheids are about 40,000 times more luminous than the sun, while the shortest-period Cepheids are only a few hundred times more luminous than the sun. Second, there are two types of Cepheids. Type I Cepheids, including δ Cephei itself, have chemical compositions like that of the sun, but type II Cepheids and the RR Lyrae stars are poor in elements heavier than helium. These two facts are clues that will help you understand variable stars as evidence of stellar evolution.

Pulsating Stars

Why should a star pulsate? Why should there be a period–luminosity relation? Why are there two types of Cepheids? The answers to these questions will reveal some of the deepest secrets of the stars. To find the answers, you should start with the evolution of stars in the H–R diagram.

Remember that after a star leaves the main sequence, it fuses hydrogen in a shell, and the point that represents it in the H–R diagram moves to the right as the star becomes a cool giant. When it ignites helium in its core, the star contracts and grows hotter, and the point in the diagram moves to the left. Soon, however, the helium core is exhausted; helium fusion continues only in a shell, forcing the star to expand again; and the point in the H–R diagram moves back to the right, completing a loop. Because giant stars evolve through these stages, their sizes and temperatures change in complicated ways. Certain combinations of size and temperature make a star unstable, causing it to pulsate as an intrinsic variable star. In the H–R diagram, the region where size and temperature lead to pulsation is called the **instability strip,** as shown in the H–R diagram in ■ Figure 12-14.

What makes some stars pulsate? You can start by thinking about a stable star. You could make a stable star pulsate temporarily if you squeezed it and then released it. The star would rebound outward, and it might oscillate in and out for a few cycles, but the oscillation would eventually run down because of fric-

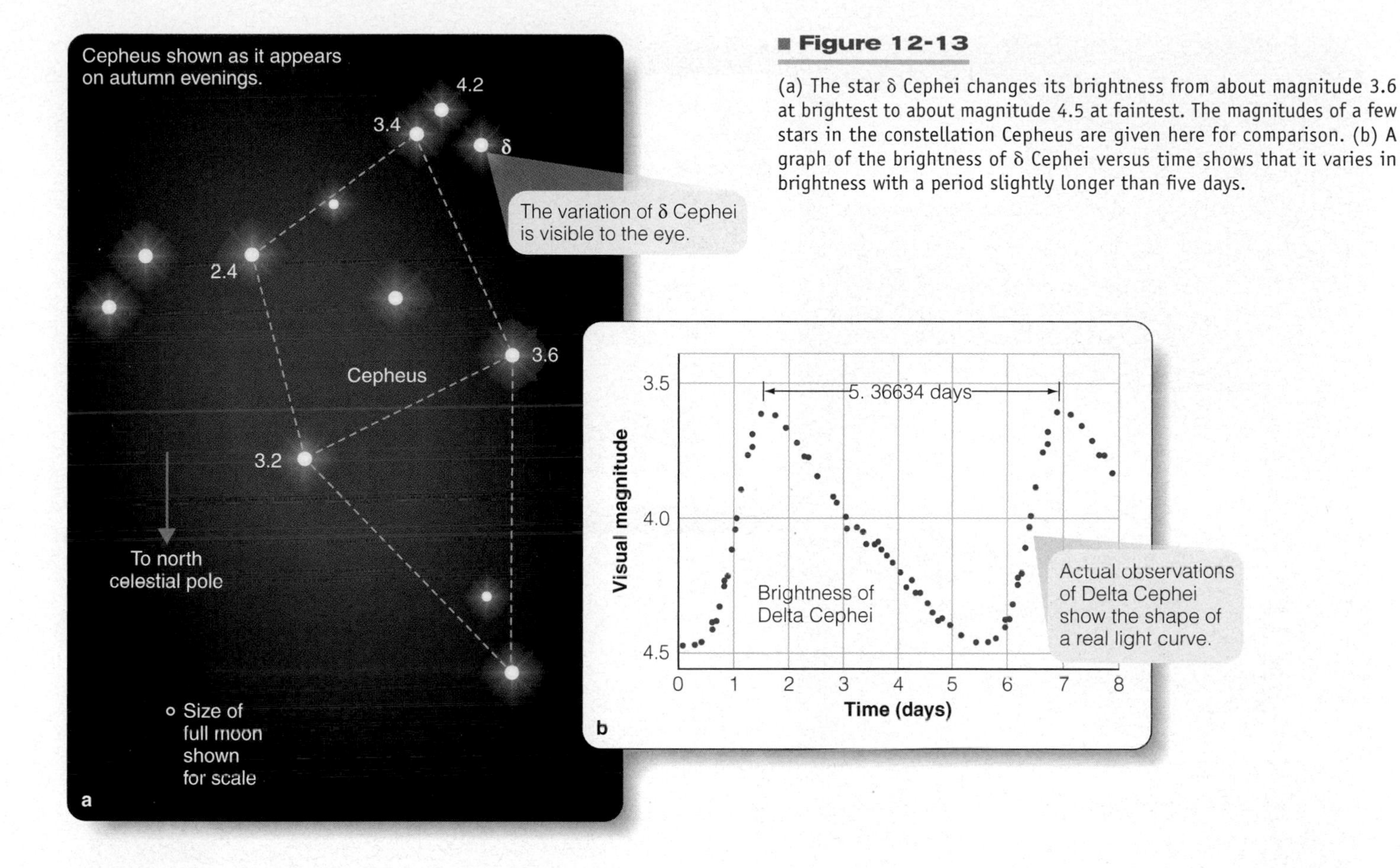

■ Figure 12-13

(a) The star δ Cephei changes its brightness from about magnitude 3.6 at brightest to about magnitude 4.5 at faintest. The magnitudes of a few stars in the constellation Cepheus are given here for comparison. (b) A graph of the brightness of δ Cephei versus time shows that it varies in brightness with a period slightly longer than five days.

tion. The energy of the moving gas would be converted to heat and radiated away, and the star would return to stability with each layer of gas in hydrostatic equilibrium. To make a star pulsate continuously, some process must drive the oscillation, just as a spring in a windup clock is needed to keep it ticking.

Studies using stellar models show that variable stars in the instability strip pulsate like beating hearts because of an energy-absorbing layer in their outer envelopes. This layer is the region where helium is partially ionized. Above this layer, the temperature is too low to ionize helium, and below the layer it is hot enough to ionize all of the helium. Like a spring, the helium ionization zone can absorb energy when it is compressed and release it when the zone expands. That is enough to keep the star pulsating.

You can follow this pulsation in your imagination and see in more detail how the helium ionization zone makes a star pulsate. As the outer layers of a pulsating star expand, the ionization zone expands, and the ionized helium becomes less ionized and releases stored energy, which forces the expansion to go even faster. The surface of the star overshoots its equilibrium position—it expands too far—and eventually falls back. As the surface layers contract, the ionization zone is compressed and becomes more ionized, which absorbs energy. Robbed of some of their energy, the layers of the star can't support the weight, and they compress further. This allows the contraction to go even faster, and the infalling layers overshoot the equilibrium point until the pressure inside the star slows them to a stop and makes them expand again.

Cepheids change their radius by 5 to 10 percent as they pulsate, and this motion can be observed as Doppler shifts in their spectra. When they expand, the surface layers approach, and astronomers see a blueshift. When the stars contract, the surface layers recede, and astronomers see a redshift. Although a 10 percent change in radius seems like a large change, it affects only the outer layers of the star. The center of the star is much too dense to be affected.

Only stars in the instability strip have their helium ionization zone at the right depth to act as a spring and drive the pulsation. A star will pulsate if the zone lies at the right depth below the surface. In cool stars, the zone is too deep and cannot overcome the weight of the layers above it. That is, the spring is overloaded and can't expand. In hot stars, the helium ionization zone lies near the surface, and there is little weight above it to compress it. That is, the spring never gets squeezed. The stars in the instability strip have the right temperature and radius for the helium ionization zone to fall exactly where it is most effective, and that is why those stars pulsate.

You can also understand why there is a period–luminosity relation. The more massive stars are larger and more luminous. When they cross the instability strip, they cross higher in the

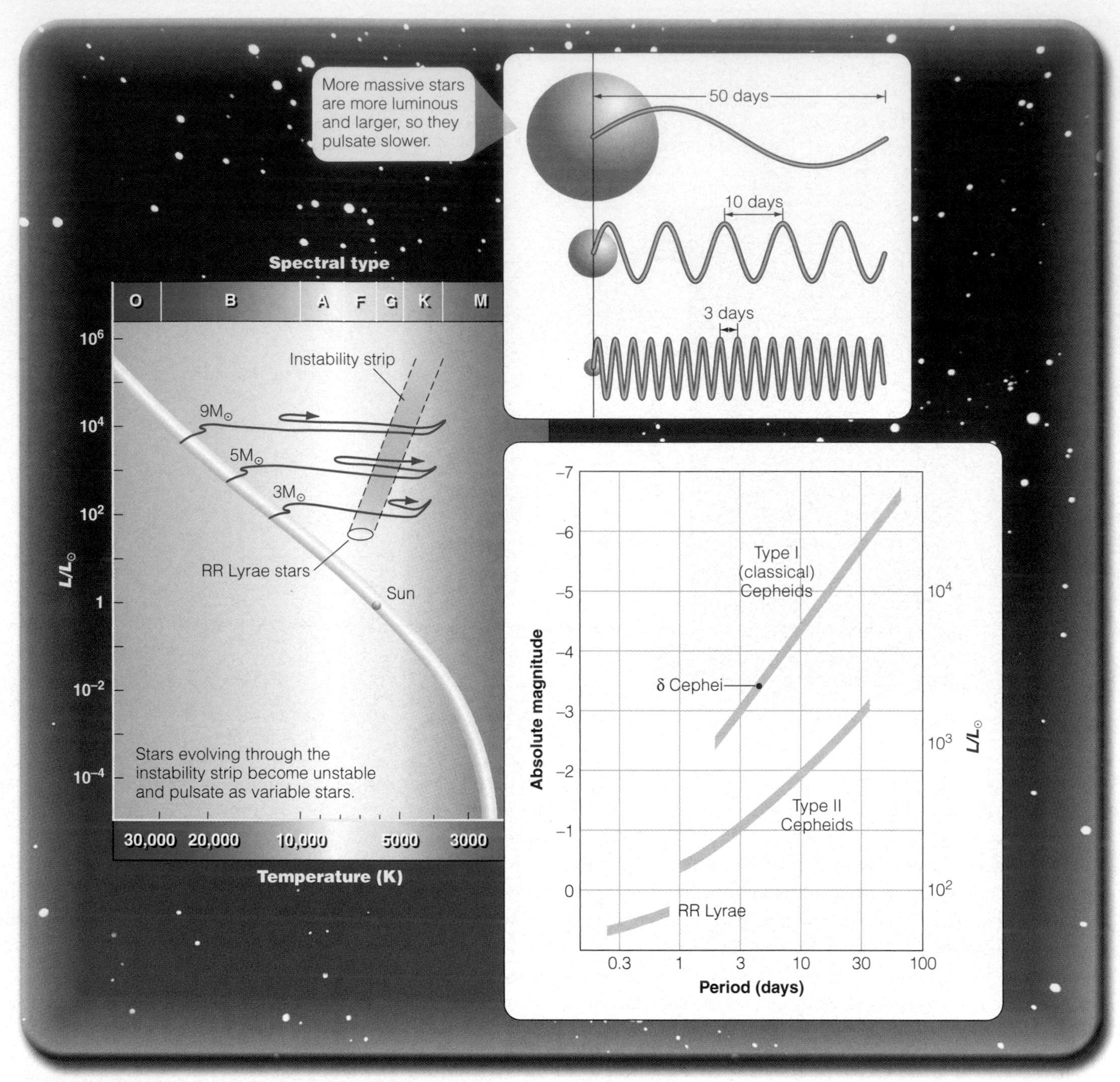

■ Figure 12-14

More massive stars are larger and more luminous and pulsate with longer periods than less massive stars. Consequently there is a period–luminosity relation: Type I and type II Cepheids have slightly different chemical compositions. The RR Lyrae stars have lower luminosities and shorter periods.

H–R diagram. But the larger a star is, the more slowly it pulsates—just as large bells go "bong" and little bells go "ding." Pulsation period depends on size, which depends on mass, and mass also determines luminosity. So the most luminous Cepheids at the top of the instability strip are also the largest and pulsate most slowly. Less luminous Cepheids are smaller and pulsate faster. If you plot the average brightness of the stars against their period of pulsation, you can clearly see the relationship between luminosity and period. Look at the right half of Figure 12-14 to see this relationship.

You know enough about stars to understand why there are two types of Cepheid variable stars. Type I Cepheids have chemical abundances roughly like those of the sun, but type II Cepheids are poor in elements heavier than helium, and that means the gases in type II Cepheids are not as opaque as the gases in type I Cepheids. The all-important energy flowing outward can escape

Snapshooters

A human lifetime is less than a century, and that is a mere flicker of a moment in the history of the universe. What we see in the sky during our lives is just a snapshot that freezes the action. The stars form gradually and evolve over billions of years. We humans see none of that action. Our snapshot shows us the many stages of star birth and evolution, but we see none of the changes through which stars pass.

Look again at Section 1-3, When Is Now? The entire 10,000-year history of human civilization is only a tiny fraction of the history of the universe. In that time, a few of the most massive stars may have evolved slightly, but the vast majority of stars have not changed since humans on Earth began building the first cities.

Only by the application of human ingenuity have astronomers figured out how stars work, how they are born, how they evolve, and, as you will see in the next chapter, how they die. Our human lifespans are only snapshots, but they reveal that we are part of an evolving universe.

more easily in type II Cepheids, and they reach a slightly different equilibrium. For stars of the same period of pulsation, type II Cepheids are less luminous, and you can see that shown in the graph in Figure 12-14.

In the H–R diagram, the RR Lyrae stars lie at the lower end of the instability strip. They are a bit smaller and hotter than Cepheids, but they pulsate for the same reason.

Period Changes in Variable Stars

Some Cepheids have been observed to change their periods of pulsation, and those changes are evidence that stars really do evolve.

The evolution of a star may carry it through the instability strip a number of times, and each time it can become a variable star. If you could watch a star as it entered the instability strip, you could see it begin to pulsate. Similarly, if you could watch a star leave the instability strip, you could watch it stop pulsating. Such an observation would be dramatic evidence that the stars are evolving. Unfortunately, stars evolve so slowly that these events are surely rare.

One famous Cepheid, RU Camelopardalis, was observed to stop pulsating temporarily. Its pulsations died away in 1966, and many astronomers assumed that it had stopped pulsating because it was leaving the instability strip. But its pulsations resumed in 1967. Remember your Favorite Star, Polaris? It may be going through a similar change now, but it does not lie near the edge of the instability strip in the H–R diagram. Perhaps it is not an example of a star evolving out of the instability strip, or perhaps there are other factors that astronomers have not discovered yet.

Even more dramatic evidence of stellar evolution can be found in the slowly changing periods of Cepheid variables. Even a tiny change in period can become easily observable because it accumulates, just as a clock that gains a second each day becomes

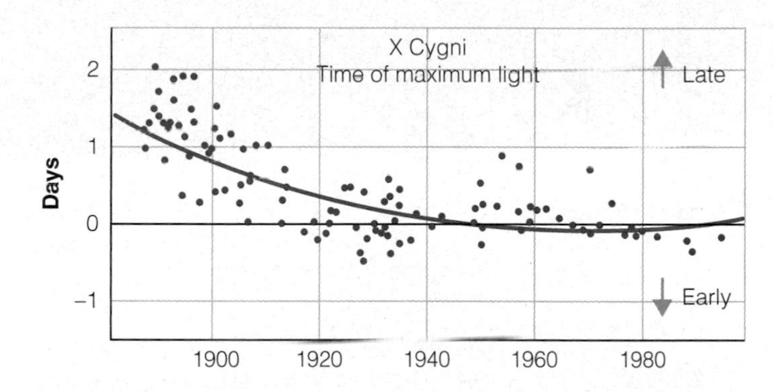

■ Figure 12-15

Like a clock running just a bit slow, the Cepheid variable star X Cygni has been reaching maximum brightness later and later for most of this century. This is shown by the upward curve in this graph of its observed minus predicted times of maximum brightness. As it evolves to the right across the instability strip, it is slowly expanding, and its period is growing longer by about 1.46 seconds per year.

obviously fast over a time as long as a year. Some Cepheids have periods that are growing shorter, and others have periods that are growing longer. These changes occur because the stars are evolving and changing their radii; contraction shortens the period, and expansion lengthens the period. The star X Cygni, for example, has a period that is gradually growing longer, and it has been "losing" since observations began in the late 19th century (■ Figure 12-15). These changes in the periods of pulsating stars provide dramatic evidence that the stars are evolving.

Some other kinds of variable stars lie outside the instability strip because their pulsation is driven by different mechanisms. It is sufficient here to discuss the Cepheids. They are common, they illustrate important ideas about stellar structure, and their changing periods make stellar evolution clearly evident. Also, you will use Cepheids in later chapters to study the distances to galaxies and the fate of the universe.

◄ SCIENTIFIC ARGUMENT ►

How do Cepheid variable stars provide evidence that stars are evolving?

The key to scientific arguments is the relationship between theory and evidence; and, in astronomy, evidence means observations. If you observed Cepheids for a few years, you might notice that some pulsations are slowing down, and others are speeding up. This is a very small change in the period of pulsation, but if you observe the stars for years, the change in their pulsation causes them to gain or lose just as a clock can gain or lose time over weeks. These changes in period are caused by the expansion or contraction of the stars as they evolve. The period of pulsation depends on the size of the star, so pulsations of a star that is expanding should slow down slightly, and those of a star that is contracting should speed up slightly. When you observed these small changes in the pulsation periods of Cepheid variable stars, you would have direct evidence that the stars are evolving.

Now build a scientific argument that is based on theory. **What makes Cepheid and RR Lyrae stars pulsate?**

Study and Review

Summary

12-1 | Main-Sequence Stars

What happens as a star uses up its hydrogen?

- Astronomers compute **stellar models** of the interiors of stars based on four simple laws of stellar structure. Two of the laws are the **conservation of mass law** and the **conservation of energy law.** The third, the law of hydrostatic equilibrium, says that the star must balance the weight of its layers by its internal pressure. The fourth law says that energy can flow outward only by conduction, convection, or radiation.
- Mathematical stellar models show how rapidly a star uses its fuel in each layer, and that allows astronomers to step forward in time and follow the evolution of the star as it ages.
- There is a main sequence because stars support their weight by hydrogen fusion. As energy flows outward, it heats the gas of the star, and pressure in the gas balances the inward pull of gravity.
- The mass–luminosity relation is explained by the requirement that a star support the weight of its layers by its internal pressure. The more massive a star is, the more weight it must support, and the higher its internal pressure must be. To keep its pressure high, it must be hot and generate large amounts of energy. Thus, the mass of a star determines its luminosity.
- Extremely massive stars are quite rare. Contracting gas clouds tend to fragment and produce pairs, groups, or clusters of stars and not extremely massive stars. Also, massive stars tend to blow matter away in strong stellar winds, and that rapidly reduces their mass.
- Stars less massive than 0.08 solar mass can't get hot enough to fuse hydrogen. Objects less massive than this become **brown dwarfs** and slowly cool as they radiate away their thermal energy. Many brown dwarfs have been observed.
- The **zero-age main sequence** is the line in the H–R diagram where contracting stars begin fusing hydrogen and reach stability. As a star ages, it moves slightly upward and to the right, making the main sequence a band.
- Massive stars, having more weight to support, reach stability higher on the main sequence than do lower-mass stars.
- The life expectancy of a main-sequence star depends on its mass. The more massive stars use their fuels rapidly and remain on the main sequence only a few million years. The sun will last a total of about 10 billion years, and the least massive stars could survive for over 100 billion years.

12-2 | Post-Main-Sequence Evolution

What happens when a star exhausts its hydrogen?

- When a main-sequence star exhausts its hydrogen, its core contracts, and it begins to fuse hydrogen in a shell around its core. The outer parts of the star—its envelope—swell, and the star becomes a giant. Because of this expansion, the surface of the star cools, and it moves toward the right in the H–R diagram. The most massive stars move across the top of the diagram as supergiants.
- Because the core of a giant star contracts and the envelope expands, the nuclear fusion is confined to a very small volume at the center of the low-density star.
- When the core of the star becomes hot enough, helium fusion, known as the **triple alpha process,** begins first in the core and then in a shell. This causes the star to describe a loop in the giant region of the H–R diagram.
- If the matter in the core becomes **degenerate** before helium ignites, the pressure of the gas does not depend on its temperature, and when helium ignites, the core explodes in the **helium flash.** Although the helium flash is violent, the star absorbs the extra energy and quickly brings the helium-fusing reactions under control.
- Helium fusion produces carbon and oxygen. Other nuclear reactions, which are not important sources of energy, slowly cook the matter during helium fusion and produce small amounts of heavy elements.
- If a star is massive enough, it can ignite carbon and other fuels after helium fusion. Each fuel fuses more rapidly, producing heavier and heavier elements.

12-3 | Evidence of Evolution: Star Clusters

What evidence do astronomers have that stars really do evolve?

- Because all the stars in a cluster have about the same chemical composition and age, you can see the effects of stellar evolution in the H–R diagram of a cluster. Massive stars evolve faster than low-mass stars, so in a given cluster the most massive stars leave the main sequence first.
- There are two types of star clusters. **Open clusters** contain 10 to 1000 stars and have an open, transparent appearance. **Globular clusters** contain 10^5 to 10^6 stars densely packed into a spherical shape. The open clusters tend to be young to middle-aged, but globular clusters tend to be very old—11 billion years or more. Also, globular clusters tend to be poor in elements heavier than helium.
- You can judge the age of a cluster by looking at the **turnoff point**, the location on the main sequence where the stars turn off to the right and become giants. The life expectancy of a star at the turnoff point equals the age of the cluster.
- H–R diagrams of old star clusters such as globular clusters show how giant stars ignite helium fusion in their cores and evolve to the left in the diagram along the **horizontal branch.**

12-4 | Evidence of Evolution: Variable Stars

- **Variable stars** are those that change in brightness. **Intrinsic variable stars** change in brightness because of internal processes and not because they are members of eclipsing binaries.
- The **Cepheid** and **RR Lyrae variable stars** provide evidence that stars evolve. These intrinsic variable stars lie in an **instability strip** in the H–R diagram; they contain a layer in their envelopes that stores and releases energy as they expand and contract. Stars outside the instability strip do not pulsate, because the layer is too deep or too shallow to make the stars unstable.
- The Cepheids obey a **period–luminosity relation** because more massive stars, which are more luminous, are larger and pulsate more slowly.
- Some Cepheids have periods that are slowly changing, showing that the evolution of the star is changing the star's radius and thus its period of pulsation.

Review Questions

ThomsonNOW Assess your understanding of this chapter's topics with additional quizzing and animations at www.thomsonedu.com

WebAssign The problems from this chapter may be assigned online in WebAssign.

1. Why is there a main sequence?
2. Why is there a lower end to the main sequence? Why is there an upper end?
3. What is a brown dwarf?
4. Why is there a mass–luminosity relation?
5. Why does a star's life expectancy depend on mass?
6. Why do expanding stars become cooler and more luminous?
7. What causes the helium flash? Why does it make it difficult for astronomers to understand the later stages of stellar evolution?
8. How do some stars avoid the helium flash?
9. Why are giant stars so low in density?
10. Why are lower-mass stars unable to ignite more massive nuclear fuels such as carbon?
11. How can you estimate the age of a star cluster?
12. How do star clusters confirm that stars evolve?
13. How do some variable stars show that stars are evolving?
14. **How Do We Know?** How can mathematical models allow scientists to study processes that are hidden from human eyes or happen too fast or too slowly for humans to experience?

Discussion Questions

1. How do you know that the helium flash occurs if it cannot be observed? Can you accept an event as real if you can never observe it?
2. Can you think of ways that chemical differences could arise in stars in a single star cluster? Consider the mechanism that triggered their formation.

Problems

1. In the model shown in Figure 12-2, how much of the sun's mass is hotter than 13,000,000 K?
2. What is the life expectancy of a 16-solar-mass star? Of a 50-solar-mass star?
3. How massive could a star be and still survive for 5 billion years?
4. If the sun expanded to a radius 100 times its present radius, what would its density be? (*Hint:* The volume of a sphere is $\frac{4}{3}\pi R^3$.)
5. If a giant star 100 times the diameter of the sun were 1 pc from Earth, what would its angular diameter be? (*Hint:* Use the small-angle formula, in Chapter 3.)
6. What fraction of the volume of a 5-solar-mass giant star is occupied by its helium core? (*Hints:* See Figure 12-9. The volume of a sphere is $\frac{4}{3}\pi R^3$.)
7. If the stars at the turnoff point in a star cluster have masses of about 4 solar masses, how old is the cluster?
8. If an open cluster contains 500 stars and is 25 pc in diameter, what is the average distance between the stars? (*Hints:* What share of the volume of the cluster surrounds the average star? The volume of a sphere is $\frac{4}{3}\pi R^3$.)
9. Repeat Problem 8 for a typical globular cluster containing a million stars in a sphere 25 pc in diameter.
10. If a Cepheid variable star has a period of pulsation of 2 days and its period increases by 1 second, how late will it be in reaching maximum light after 1 year? After 10 years? (*Hint:* How many cycles will it complete in a year?)

Learning to Look

1. In the photograph of the Pleiades on page 257 there are no bright-red stars. Use the H–R diagram to explain why the brightest stars are blue. Have there ever been bright-red stars in this cluster?
2. Look at the photograph of the star cluster M67 on page 257. Why are there not bright-blue stars in this cluster?
3. The star cluster in the photo at the right contains many hot, blue, luminous stars. Sketch its H–R diagram and discuss its probable age.

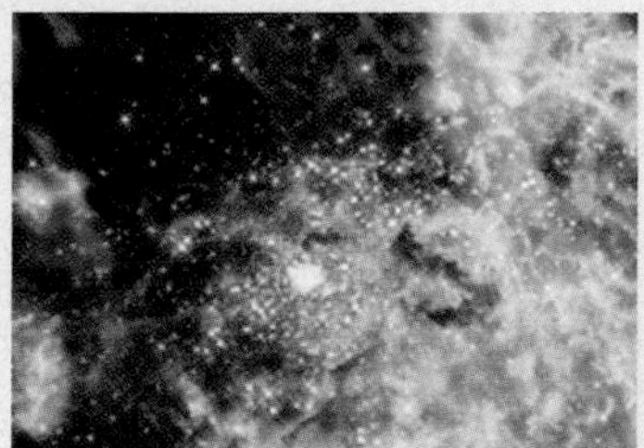

NASA/Walborn, Maíz-Apellániz, and Barba

13 The Deaths of Stars

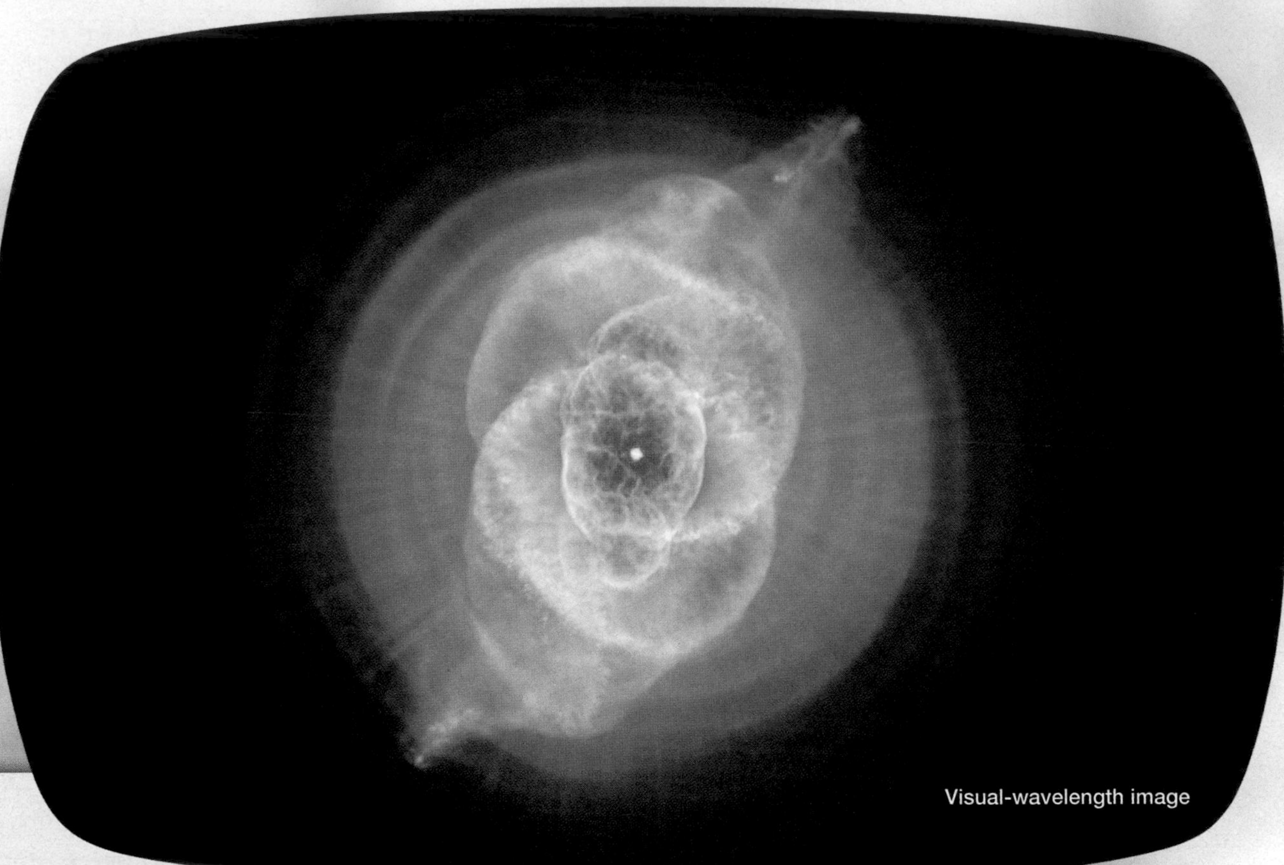

Matter ejected repeatedly from the dying star at the center has formed the nebula known as the Cat's Eye. (NASA, ESA, HEIC, and the Hubble Heritage Team)

Guidepost

Perhaps you were surprised in earlier chapters to learn that stars are born and grow old. Modern astronomers can tell the story of the stars right to the end. Here you will learn how stars die, but to follow the story you will have to proceed with care, testing theories against evidence to answer four essential questions:

How will the sun die?

Why are there so many white dwarfs?

What happens if an evolving star is in a binary system?

How do massive stars die?

This is a chapter of theory supported by evidence, and it raises an important question about how science works.

How Do We Know? How are the properties of big things explained by the properties of the smallest things?

Astronomy is exciting because it is about us. As you think about the deaths of stars, you are also thinking about the safety of Earth as a home for life and about the ultimate fate of our sun, our Earth, and the atoms of which you are made.

Natural laws have no pity.

ROBERT HEINLEIN, *THE NOTEBOOKS OF LAZARUS LONG*

GRAVITY IS PATIENT—so patient it can kill stars. In the previous chapter you saw how stars resist their own gravity by generating energy through nuclear fusion. The energy keeps their interiors hot, and the resulting high pressure balances gravity and prevents the star from collapsing. However, stars have only so much fuel, and gravity is patient. When the stars exhaust their fuel, gravity wins, and the stars die (■ Figure 13-1).

The mass of a star is critical in determining its fate. Lower-mass stars like the sun die gentle deaths, but massive stars explode violently. In addition, stars orbiting close to another star can have their evolution modified in peculiar ways. To follow the evolution of stars to their graves, you can sort the stars into three categories according to their masses: low-mass red dwarfs, medium-mass sunlike stars, and massive upper-main-sequence stars.

13-1 Lower-Main-Sequence Stars

THE STARS OF THE LOWER MAIN SEQUENCE share a common characteristic: They have relatively low masses. That means that they face similar fates as they exhaust their nuclear fuels.

As you learned in the previous chapter, when a star exhausts one nuclear fuel, its interior contracts and grows hotter until the next nuclear fuel ignites. The contracting star heats up by converting gravitational energy into thermal energy, and that's a problem for lower-main-sequence stars. Low-mass stars don't have a lot of gravitational energy, so they can't get very hot, and that limits the fuels they can ignite. The lowest-mass stars, for example, cannot get hot enough to ignite helium fusion.

Structural differences divide the lower-main-sequence stars into two subgroups—very-low-mass stars and medium-mass stars such as the sun. The critical difference between the two groups is the extent of interior convection (see Figure 11-14). If the star is convective, fuel is constantly mixed, and the resulting evolution of the star is drastically altered.

Red Dwarfs

Stars less massive than about 0.4 solar mass—the red dwarfs—can survive a long time. They have very small masses, and consequently they have very little weight to support. Their pressure–temperature thermostats are set low, and they consume their hydrogen fuel very slowly. If you calculate the life expectancy of one of these stars (Chapter 12), you will discover a red dwarf could live over ten times longer than the sun. Furthermore, they are totally convective, and that mixes the stars like pots of soup that are constantly stirred. Hydrogen is consumed, and helium accumulates uniformly throughout the stars; and that has two important consequences for red dwarfs.

First, because red dwarf stars are completely mixed and can fuse all their hydrogen, their lives are prolonged even further. Models of a 0.1-solar-mass red dwarf predict it would take 2 billion years to contract to the main sequence and 6 trillion years (6000 billion years) to use up its hydrogen fuel.

The second consequence is that they use up their hydrogen uniformly. That means they never develop a hydrogen fusion shell, and they never become giant stars. They can slowly convert their hydrogen into helium, a fuel they are not massive enough to ignite, and finally die slow, unremarkable deaths.

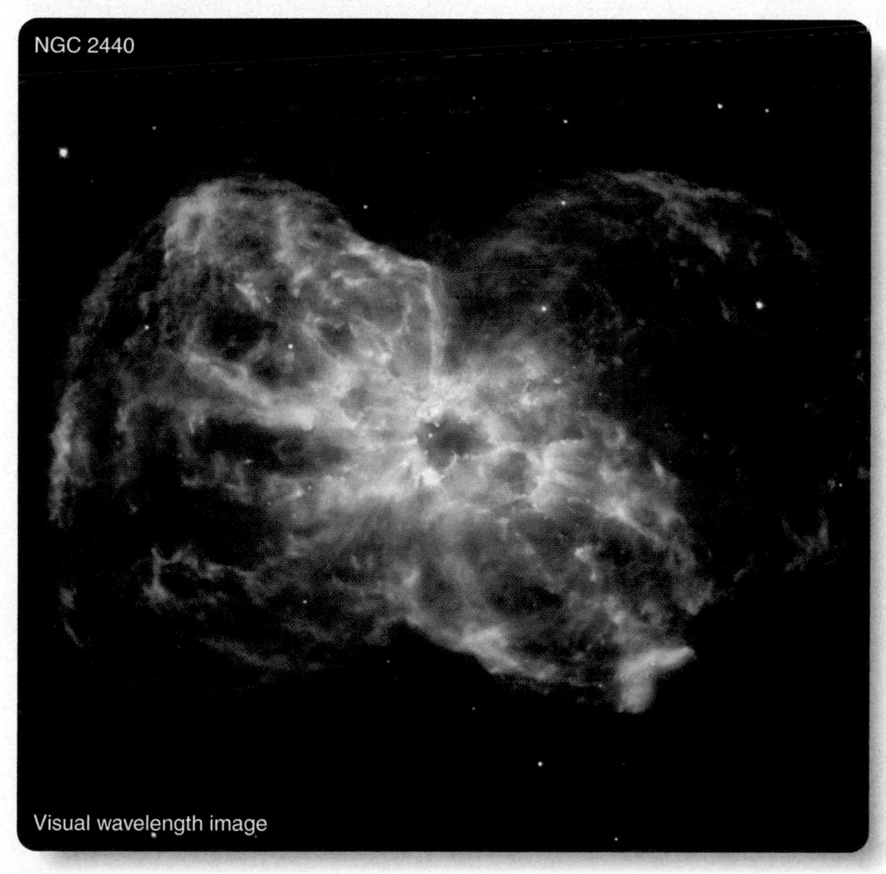

■ Figure 13-1

Astronomers find evidence of dying stars all around the sky. This small nebula formed within the last few thousand years when an aging giant star expelled its outer layers. The remains of the star collapsed to form a hot white dwarf at the center of the nebula. The sun will suffer this fate in about 5 billion years. (NASA, ESA, and K. Noll, STScI)

You will see in a later chapter that the universe is only about 14 billion years old, so none of these red dwarfs should have exhausted its fuel yet. Every red dwarf that has ever been born is still shining.

Sunlike Stars

Stars with masses between roughly 0.4 and 4 solar masses,* including the sun, ignite hydrogen and then helium and become giants, but they cannot get hot enough to ignite carbon, the next fuel in the sequence (see Table 12-3). When they reach that impasse, they can no longer maintain their stability, and their interiors contract while their envelopes expand.

To understand the fate of these stars, you need to consider two ideas, mixing and expansion. The interiors of these sunlike stars are not well mixed (see Figure 11-14). As you learned in Chapter 11, stars of 1.1 solar masses or less, including the sun, have no convection near their centers, so they are not mixed at all. Stars more massive than 1.1 solar masses have small zones of convection at their centers, but this mixes no more than about 12 percent of the star's mass. Consequently, medium-mass stars, whether they have convective cores or not, are not mixed, and the helium accumulates in an inert helium core surrounded by unprocessed hydrogen. When this core contracts, the unprocessed hydrogen just outside the core ignites in a shell, and the outer layers of the star expand and cool, transforming the star into a giant.

As a giant, the star fuses helium in its core and then in a shell surrounding a core of carbon and oxygen. This core contracts and grows hotter, but it cannot become hot enough to ignite the carbon. The carbon–oxygen core is a dead end for these medium-mass stars.

The carbon-oxygen core increases in mass as the helium-fusion shell burns outward, leaving its ashes behind. But because no nuclear reactions can begin in the carbon–oxygen core, it cannot resist the weight pressing down on it, so it contracts. The energy released by the contracting core, plus the energy generated in the helium- and hydrogen-fusing shells, flows outward and makes the envelope of the star expand and cool further.

This forces the star to become a very large giant. Its radius may become as large as the radius of Earth's orbit, and its surface can become as cool as 2000 K. Such a star can lose large amounts of mass from its surface.

Mass Loss from Sunlike Stars

It's not hard to find evidence that stars like the sun lose mass. The sun itself is losing mass. The solar wind is a gentle breeze of gas that blows outward from the solar corona and carries mass into space. The sun loses only about 0.001 solar mass per billion years. Even over the entire lifetime of the sun, that would not significantly alter the sun's mass.

*This mass limit is uncertain, as are many of the masses stated here. The evolution of stars is highly complex, and such parameters are not precisely known.

Other stars like the sun are also losing mass, and the evidence lies in their spectra. Ultraviolet and X-ray spectra of many stars reveal strong emission lines, and you can use Kirchhoff's laws to conclude that the star's outer layers must be highly ionized. That means the stars must have hot chromospheres and coronas like the sun's. If these stars have atmospheres like the sun's, they presumably have similar winds of hot gas. You can find further evidence in the spectra of some giant stars, which contain blueshifted absorption lines. The blueshifts must be Doppler shifts produced as the gas flowing out of the star comes toward Earth.

Giant stars may lose mass much more rapidly than the sun. Because the spectra of some of these stars do not show the characteristic emission, you must assume that the stars do not have hot coronas; but other processes could drive mass loss. Giant stars are so large that gravity is weak at their surfaces, and convection in the cool gas can drive shock waves outward and power mass loss. In addition, some giants are so cool that specks of carbon dust condense in their atmospheres, just as soot can condense in a fireplace. The pressure of the star's radiation can push this dust and any gas atoms that collide with the dust completely out of the star.

Another process that can cause mass loss in giant stars is periodic eruptions in the helium-fusion shell. The triple-alpha process that fuses helium into carbon is extremely sensitive to temperature, and that can make the helium-fusion shell so unstable that it experiences eruptions called **thermal pulses.** Every 200,000 years or so, the shell can erupt and suddenly produce energy equivalent to a million times the luminosity of the sun. Almost all of that energy goes into lifting the outer layers of the star and driving gas away from the surface.

Rapid mass loss can affect stars dramatically. A star expanding as its carbon–oxygen core contracts could lose an entire solar mass in only 10^5 years, which is not a long time in the evolution of a star. That means that a star that began its existence on the main sequence with a mass of 8 solar masses might reduce its mass to only 3 solar masses in half a million years once it becomes a giant.

Stellar mass loss confuses the story of stellar evolution. Astronomers would like to be able to say that stars more massive than a certain limit will evolve one way, and stars less massive will evolve another way. But giant stars may lose enough mass to alter their own evolution. As a result, you need to consider both the initial mass a star has on the main sequence and the mass it retains after mass loss. Because astronomers don't know exactly how effective mass loss is, it is difficult to be exact about the mass limits in any discussion of stellar evolution.

Planetary Nebulae

When a medium-mass star like the sun becomes a distended giant, it can expel its outer atmosphere to form one of the most interesting objects in astronomy, a **planetary nebula,** so called because through a small telescope it looks like the greenish-blue disk of a planet such as Uranus or Neptune. In fact, a planetary nebula has nothing to do with a planet. It is composed of ionized gases expelled by a dying star.

Read **The Formation of Planetary Nebulae** on pages 268–269 and notice four things:

1. You can understand what planetary nebulae are like by using simple observational methods such as Kirchhoff's laws and the Doppler effect.
2. Notice the model that astronomers have developed to explain planetary nebulae. The real nebulae are more complex than the simple model of a slow wind and a fast wind, but the model provides a way to organize the observed phenomena.
3. Oppositely directed jets (much like bipolar flows from protostars) produce many of the asymmetries seen in planetary nebulae.
4. The star itself must contract into a white dwarf.

As is always the case in science, the theories that describe the formation of planetary nebulae are open to revision. In 1996, astronomers saw the hot nucleus of a planetary nebula, a star that should have been contracting to form a white dwarf, flare back to life. Apparently, some stars flare one or more times as they push their surface layers away and collapse. Such contracting objects may be able to temporarily reignite helium fusion a number of times before they finally become white dwarfs. The resulting flares may explain some of the complex structure visible in planetary nebulae.

Most astronomy books say that the sun will form a planetary nebula, but that may not happen. The sun will certainly eject gas into space, but to light up a planetary nebula, the sun must become a white dwarf with a temperature of at least 25,000 K. Some theorists think that slightly less massive stars, such as the sun, may not get hot enough when they collapse and consequently may not be able to ionize the gases they have ejected. Time will tell, of course, but we can't wait around. Astronomers continue to refine their theories and make new observations, so they may eventually be able to say conclusively whether the sun will light up its own planetary nebula.

In any case, medium-mass stars like the sun die by losing mass and contracting into white dwarfs. This suggests another way to search for evidence of the deaths of medium-mass stars: Study white dwarfs.

White Dwarfs

White dwarfs are among the most common stars. The survey in Chapter 9 revealed that white dwarfs and red dwarfs, although very faint, are very common (page 197). Now you can recognize white dwarfs as the remains of medium-mass stars that fused hydrogen and helium, failed to ignite carbon, drove away their outer layers to form planetary nebulae, and then collapsed to form white dwarfs. The billions of white dwarfs in our galaxy are the remains of medium-mass stars.

The first white dwarf discovered was the faint companion to one of your Favorite Stars, Sirius. In that visual binary system, the bright star is Sirius A. The white dwarf, Sirius B, is 10,000 times fainter than Sirius A. The orbital motions of the stars (shown in Figure 9-16) reveal that the white dwarf's mass is 0.98 solar mass, and its blue-white color tells you that its surface is hot, about 25,000 K. Although it is very hot, it has a very low luminosity, so it must have a small surface area—in fact, its diameter is about twice Earth's. Dividing its mass by its volume reveals that it is very dense—over 3×10^6 g/cm^3. On Earth, a teaspoonful of Sirius B material would weigh more than 15 tons (■ Figure 13-2). Basic observations and simple physics lead to the conclusion that white dwarfs are astonishingly dense.

A normal star is supported by energy flowing outward from its core, but a white dwarf cannot generate energy by nuclear fusion. Although a tremendous amount of energy flows out of its hot interior, it is not the energy flow that supports the star. A white dwarf is supported against its own gravity by the inability of its degenerate electrons to pack into a smaller volume. In this case, the properties of white dwarfs, objects a big as Earth, are determined by the properties of subatomic particles. Astronomers often find that the causes of natural phenomena lead step by step into the subatomic world **(How Do We Know? 13-1)**.

The interior of a white dwarf is mostly carbon and oxygen ions floating among a whirling storm of degenerate electrons. It is

■ Figure 13-2

The degenerate matter from inside a white dwarf is so dense that a lump the size of a beach ball would, transported to Earth, weigh as much as an ocean liner.

The Formation of Planetary Nebulae

1 Simple observations tell astronomers what planetary nebulae are like. Their angular size and their distances indicate that their radii range from 0.2 to 3 ly. The presence of emission lines in their spectra assures that they are excited, low-density gas. Doppler shifts show they are expanding at 10 to 20 km/s. If you divide radius by velocity, you find that planetary nebulae are no more than about 10,000 years old. Older nebulae evidently become mixed into the interstellar medium.

Astronomers find about 1500 planetary nebulae in the sky. Because planetary nebulae are short-lived formations, you can conclude that they must be a common part of stellar evolution. Medium-mass stars up to a mass of about 8 solar masses are destined to die by forming planetary nebulae.

The Helix Nebula is 2.5 ly in diameter, and the radial texture shows how light and winds from the central star are pushing outward.

NASA/JPL-Caltech/ESA

Visual + Infrared

2 The process that produces planetary nebulae involves two stellar winds. First, as an aging giant, the star gradually blows away its outer layers in a slow breeze of low-excitation gas that is not easily visible. Once the hot interior of the star is exposed, it ejects a high-speed wind that overtakes and compresses the gas of the slow wind like a snowplow, while ultraviolet radiation from the hot remains of the central star excites the gases to glow like a giant neon sign.

Slow stellar wind from a red giant

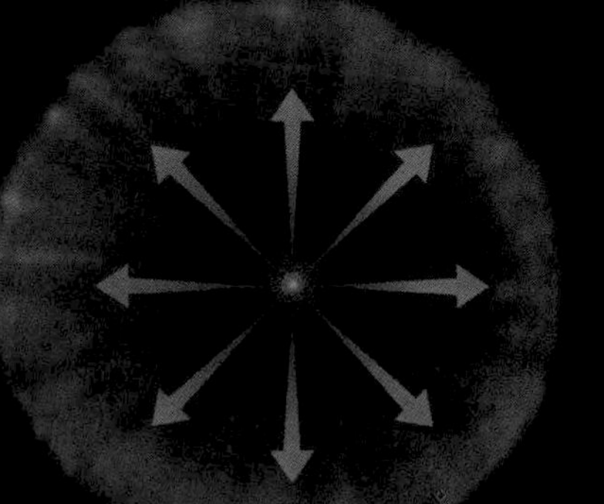

The gases of the slow wind are not easily detectable.

Fast wind from exposed interior

You see a planetary nebula where the fast wind compresses the slow wind.

2a The Cat's Eye, below, lies at the center of an extended nebula that must have been exhaled from the star long before the fast wind began forming the visible planetary nebula. See other images of the nebula on opposite page.

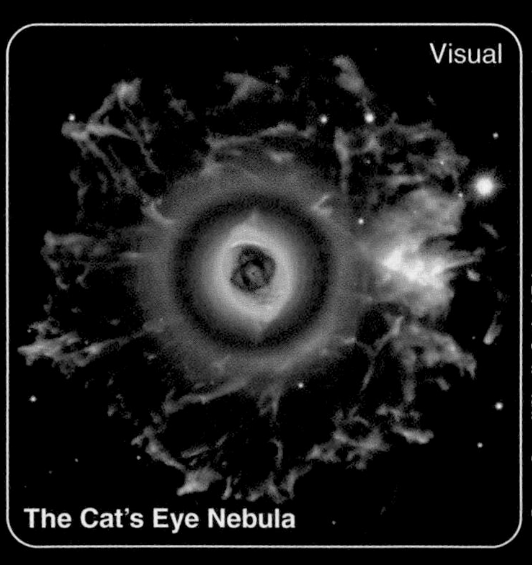

Roman Corradi/Nordic Optical Telescope

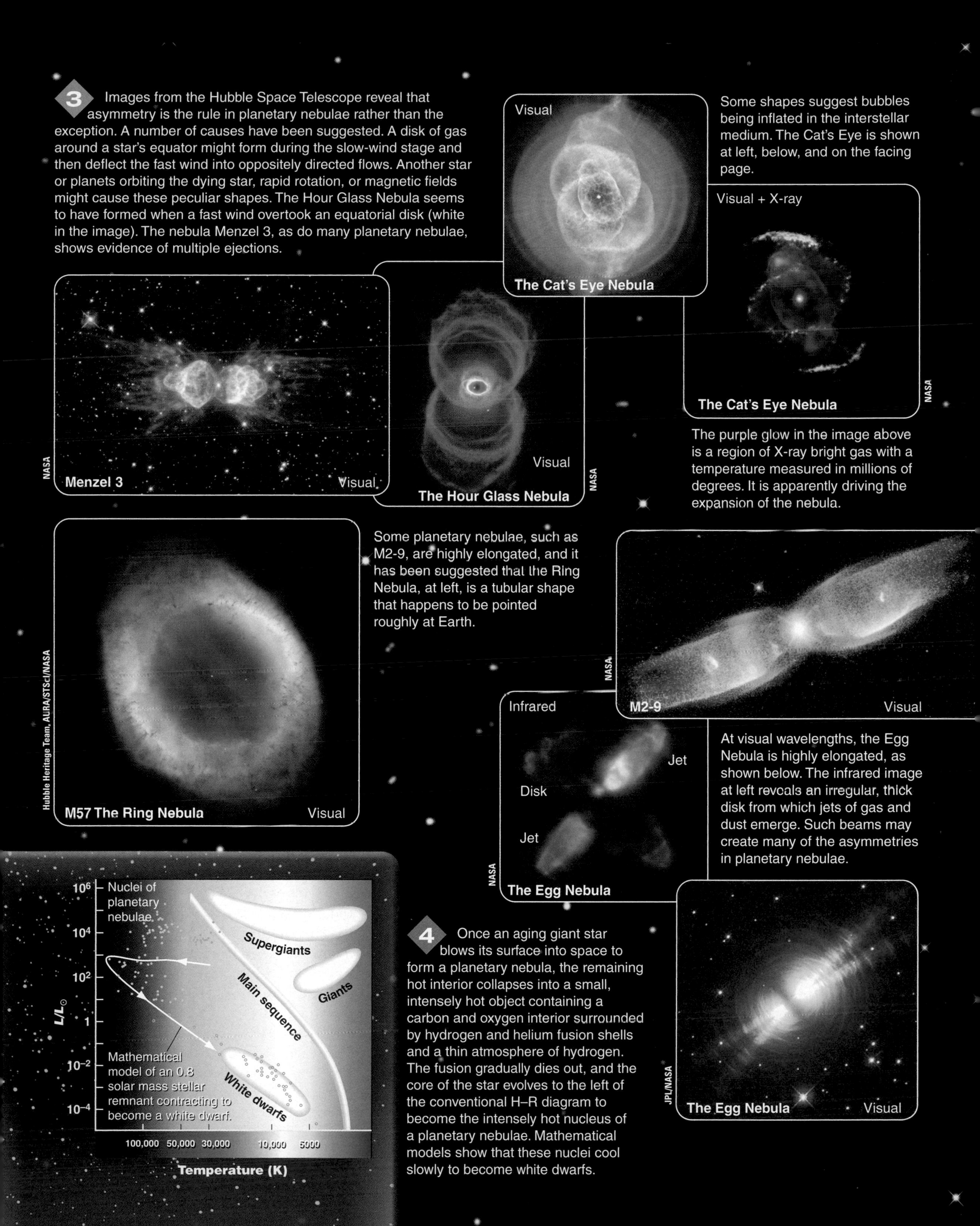

3 Images from the Hubble Space Telescope reveal that asymmetry is the rule in planetary nebulae rather than the exception. A number of causes have been suggested. A disk of gas around a star's equator might form during the slow-wind stage and then deflect the fast wind into oppositely directed flows. Another star or planets orbiting the dying star, rapid rotation, or magnetic fields might cause these peculiar shapes. The Hour Glass Nebula seems to have formed when a fast wind overtook an equatorial disk (white in the image). The nebula Menzel 3, as do many planetary nebulae, shows evidence of multiple ejections.

Some shapes suggest bubbles being inflated in the interstellar medium. The Cat's Eye is shown at left, below, and on the facing page.

The purple glow in the image above is a region of X-ray bright gas with a temperature measured in millions of degrees. It is apparently driving the expansion of the nebula.

Some planetary nebulae, such as M2-9, are highly elongated, and it has been suggested that the Ring Nebula, at left, is a tubular shape that happens to be pointed roughly at Earth.

At visual wavelengths, the Egg Nebula is highly elongated, as shown below. The infrared image at left reveals an irregular, thick disk from which jets of gas and dust emerge. Such beams may create many of the asymmetries in planetary nebulae.

4 Once an aging giant star blows its surface into space to form a planetary nebula, the remaining hot interior collapses into a small, intensely hot object containing a carbon and oxygen interior surrounded by hydrogen and helium fusion shells and a thin atmosphere of hydrogen. The fusion gradually dies out, and the core of the star evolves to the left of the conventional H–R diagram to become the intensely hot nucleus of a planetary nebulae. Mathematical models show that these nuclei cool slowly to become white dwarfs.

How Do We Know? 13-1

Toward Ultimate Causes

Why does ice float? Scientists search for causes. They are not satisfied to know that a certain kind of star dies by exploding. They want to know why it explodes. They want to find the causes for the natural events they see, and that search for ultimate causes often leads into the atomic world.

For example, why do icebergs float? When water freezes it becomes less dense than liquid water, so it floats. That answers the question, but you can search for a deeper cause.

Why is ice less dense than water? Water molecules are made up of two hydrogen atoms bonded to an oxygen atom, and the oxygen is so good at attracting electrons, the hydrogen atoms are left needing a bit more negative charge. They are attracted to atoms in nearby molecules. That means the hydrogen atoms in water are constantly trying to stick to other molecules. When water is warm, the thermal motion prevents these hydrogen bonds from forming, but when water freezes, the hydrogen atoms link the water molecules together. Because of the angles at which the bonds form, the molecules leave open spaces between molecules, and that makes ice less dense than water.

But scientists can continue their search for causes. Why do electrons have negative charge? What is charge? Nuclear particle physicists are tying to understand those properties of matter. Sometimes the properties of very large things such as supernovae are determined by the properties of the tiniest particles. Science is exciting because the simple observation that ice floats in your lemonade can lead you toward ultimate causes and some of the deepest questions about how nature works.

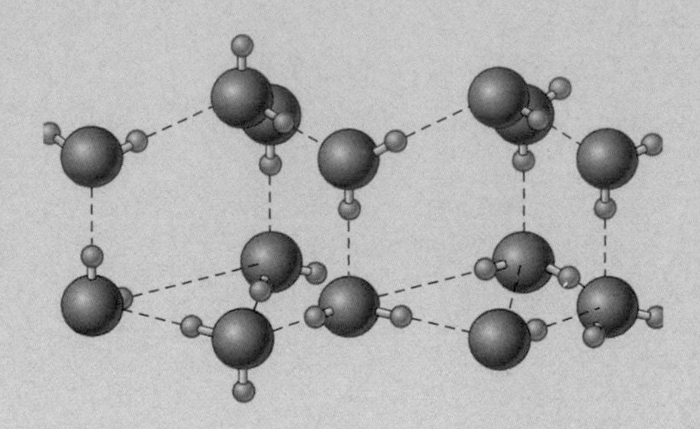

Ice has a low density and floats because of the way electrons (blue) link to oxygen (red) when water freezes.

these degenerate electrons that exert the pressure needed to support the star's weight, but most of the star's mass is represented by the carbon and oxygen ions. Theory predicts that, as the star cools, these ions lock together to form a crystal lattice, so there may be some truth in thinking of aging white dwarfs as great crystals of carbon and oxygen. Near the surface, where the pressure is lower, a layer of ionized gases makes up a hot atmosphere.

The tremendous surface gravity of white dwarfs—100,000 times that of Earth—affects its atmosphere in strange ways. The heavier atoms in the atmosphere tend to sink, leaving the lightest gases at the surface. Astronomers see some white dwarfs with atmospheres of almost pure hydrogen, whereas others have atmospheres of nearly pure helium. Still others, for reasons not well understood, have atmospheres that contain traces of heavier atoms. In addition, the powerful surface gravity pulls the white dwarf's atmosphere down into a very shallow layer. If Earth's atmosphere were equally shallow, people on the top floors of skyscrapers would have to wear oxygen masks.

Clearly, a white dwarf is not a true star. It generates no nuclear energy, is almost totally degenerate, and, except for a thin layer at its surface, contains no gas. Instead of calling a white dwarf a "star," you can call it a **compact object.** In the next chapter, you will meet two other compact objects—neutron stars and black holes.

A white dwarf's future is bleak. Degenerate matter is a very good thermal conductor, so heat flows to the surface and escapes into space, and the white dwarf gets fainter and cooler, moving downward and to the right in the H–R diagram. As it radiates energy into space, its temperature gradually falls, but it cannot shrink any smaller because its degenerate electrons cannot get closer together. Because the white dwarf contains a tremendous amount of heat, it needs billions of years to radiate that heat through its small surface area. Eventually, such objects may become cold and dark, so-called **black dwarfs.** Our galaxy is not old enough to contain black dwarfs. The coolest white dwarfs in our galaxy are just a bit cooler than the sun.

Perhaps the most interesting thing about white dwarfs has come from mathematical models. The equations predict that if you added mass to a white dwarf, its radius would *shrink,* because added mass would increase its gravity and squeeze it tighter. If you added enough to raise its total mass to about 1.4 solar masses, its radius would shrink to zero (■ Figure 13-3). This is called the **Chandrasekhar limit** after Subrahmanyan Chandrasekhar, the astronomer who discovered it. It seems to imply that a star more massive than 1.4 solar masses could not become a white dwarf unless it shed mass in some way.

As you saw earlier in this chapter, aging giant stars do lose mass (■ Figure 13-4). This suggests that stars more massive than the Chandrasekhar limit can eventually die as white dwarfs if they reduce their mass. Theoretical models show that an 8-solar-mass star should be able to reduce its mass to 1.4 solar masses before it collapses. Consequently, a wide range of medium-mass stars eventually die as white dwarfs. No wonder white dwarfs are so common.

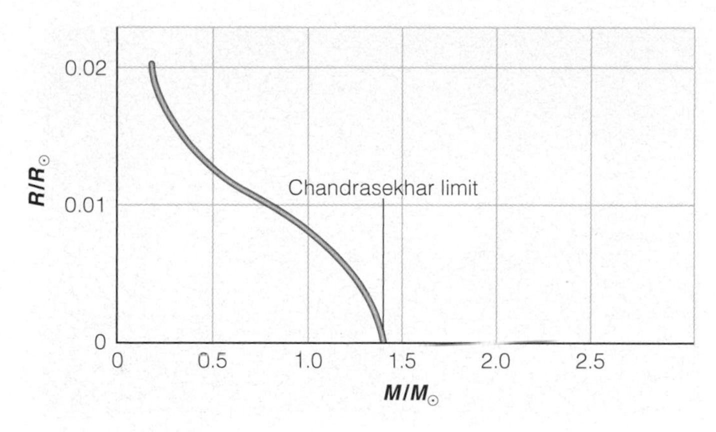

■ **Figure 13-3**

The more massive a white dwarf, the smaller its radius. Stars more massive than the Chandrasekhar limit of 1.4 solar masses cannot be white dwarfs.

◄ SCIENTIFIC ARGUMENT ►

How will the sun die?

This question seems to call for an argument based mostly on theory, but remember that all theory is tested against observations. You have seen that lower-main-sequence stars like the sun die by producing planetary nebulae and then becoming white dwarfs. For the sun, this process will begin in a few billion years, when it exhausts the hydrogen in its core. It will then expand to become a giant star. Although the surface of the sun will become cooler, it will expand until it has such a large surface that it will be about 100 times more luminous than it is now. As a giant star, the sun will have a strong solar wind carrying gas into space and pushing back the interstellar medium. Eventually, the loss of mass will expose deeper layers in the sun. These extremely hot layers will heat the gas around the sun and propel it outward to scoop up previously expelled gases. If the sun becomes hot enough, it will ionize the gas and produce a planetary nebula. In any case, the last remains of the sun will contract and slowly cool to form a small, dense white dwarf. The surface will be very hot but so small that the sun will be about 100 times fainter than it is now.

The story of stellar evolution predicts that our sun will someday become a giant star and then a white dwarf. Focus your scientific argument on a slightly different part of the story. **As the sun becomes a white dwarf, what will have become of all the hydrogen the sun once contained?**

◄ ►

13-2 The Evolution of Binary Stars

SO FAR YOU HAVE BEEN THINKING about the deaths of stars as if they were all single objects that never interact. But more than half of all stars are members of binary star systems. Most such

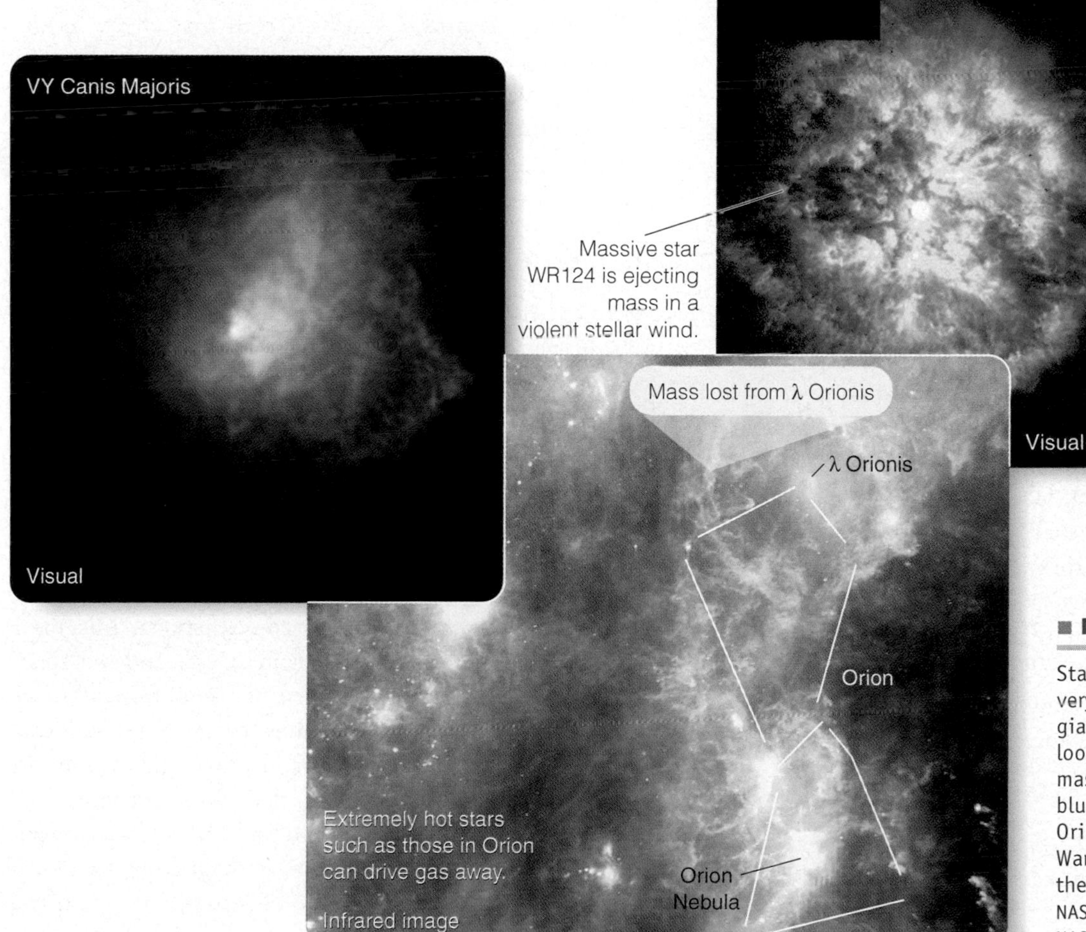

■ **Figure 13-4**

Stars can lose mass if they are very hot, very large, or both. The massive red supergiant VY Canis Majoris is ejecting gas in loops, arcs, and knots as it ages. A young massive star such as WR124 and the hot, blue stars that make up the constellation Orion constantly lose mass into space. Warmed dust in these gas clouds can make them glow in the infrared. (VY Canis Majoris: NASA, ESA, R. Humphreys; WR124: NASA; Orion: NASA/IPAC courtesy Deborah Levine)

binaries are far apart, and one of the stars can swell into a giant and eventually collapse without affecting the companion star. In some binary systems, however, the two stars orbit close together. When the more massive star begins to expand, it interacts with its companion star in peculiar ways. These interacting binary stars are fascinating objects, and they are important because they help explain observed phenomena such as nova explosions. In the next chapter, you will see how they can help astronomers find black holes.

Mass Transfer

Binary stars can sometimes interact by transferring mass from one star to the other. To understand this process, you need to think about how the gravity of the two stars controls the matter in the binary system.

Each star in a binary system is held together by its own gravity; but, because the system is rotating, there are unexpected forces acting on loose matter between the stars. If you could gently release a pebble into such a system, it might fall into one star, fall into the other star, or be ejected from the system. What happened to your pebble would depend on where you dropped it. Each star controls a region of space near it, and if you could make these regions visible they would look like two teardrop-shaped lobes that enclose the stars and meet tip to tip between the stars. These two volumes are called the **Roche lobes,** and the dumbbell-shaped surface of the two lobes is called the **Roche surface.** If you released your pebble inside one of the Roche lobes, it would belong gravitationally to the star that controls that lobe.

Because of the interaction of the gravity of the two stars and their rotation around their center of mass, there are five points of stability in the orbital plane called **Lagrangian points.*** ■ Figure 13-5 illustrates the Roche surface around the stars and the arrangement of the Lagrangian points.

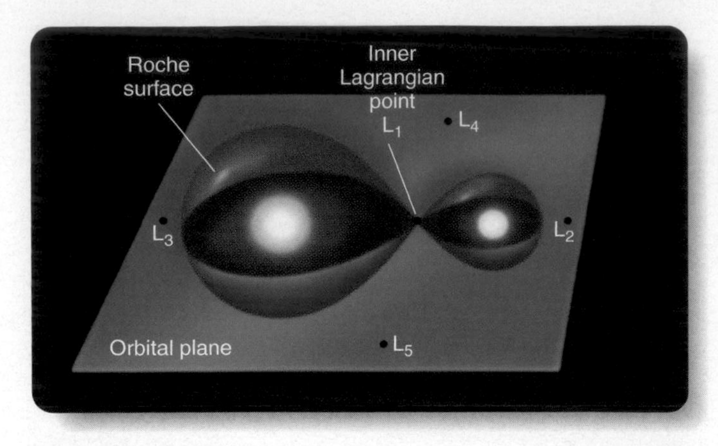

■ **Figure 13-5**

A pair of binary stars control the region of space located inside the Roche surface. The Lagrangian points are locations of stability, with the inner Lagrangian point making a connection through which the two stars can transfer matter.

The Lagrangian points are important because they are locations where matter is stable in the revolving binary system. If you could carefully release your pebble at a Lagrangian point, it would remain at that point and circle the center of mass along with the stars. Rather than a pebble, think of a bit of naturally occurring gas in the binary system. Gas at the L_2 or L_3 point is critically stable, meaning that any small disturbance will make it drift away. In contrast, gas at the L_4 or L_5 point can be trapped. (You will meet the L_4 and L_5 points again when you study planetary satellites and the orbits of asteroids in later chapters.)

In the case of a binary system, the inner Lagrangian point L_1 located between the two stars is critically important. If matter from one star can reach the inner Lagrangian point, it can flow onto the other star. That is, the stars can transfer mass through the inner Lagrangian point.

In general, there are only two ways matter can escape from a star and reach the inner Lagrangian point. First, if a star has a strong stellar wind, some of the gas blowing away from the star can pass through the inner Lagrangian point and be captured by the other star. Second, if an evolving star expands so far that it fills its Roche surface, it will be forced to take on the teardrop shape of the Roche surface. You have seen that effect in your study of binary stars (see Figure 9-22). As the star continues to expand, matter will flow through the inner Lagrangian point like water in a pond flowing over a dam and will fall into the other star. Mass transfer driven by a stellar wind tends to be slow, but mass transfer driven by an expanding star can occur rapidly.

Evolution with Mass Transfer

Mass transfer between stars can affect their evolution in surprising ways. In fact, it provides the explanation for a problem that puzzled astronomers for many years.

In some binary systems, the less massive star has become a giant, while the more massive star is still on the main sequence. That seems backward. If more massive stars evolve faster than lower-mass stars, how does the low-mass star in such binaries manage to leave the main sequence first? This is called the Algol paradox, after the binary system Algol (see Figure 9-23).

Mass transfer explains how this could happen. Imagine a binary system that contains a 5-solar-mass star and a 1-solar-mass companion (■ Figure 13-6). The two stars formed at the same time, so the more massive star must evolve faster and leave the main sequence first. When it expands into a giant, it fills its Roche surface and transfers matter to the low-mass companion. The massive star shrinks into a lower-mass star, and the companion gains mass to become a more massive main-sequence star. If you observed such a system after the mass transfer ended, you

*The Lagrangian points are named after French mathematician Joseph Louis Lagrange, who solved this famous mathematical problem around the time of the French Revolution.

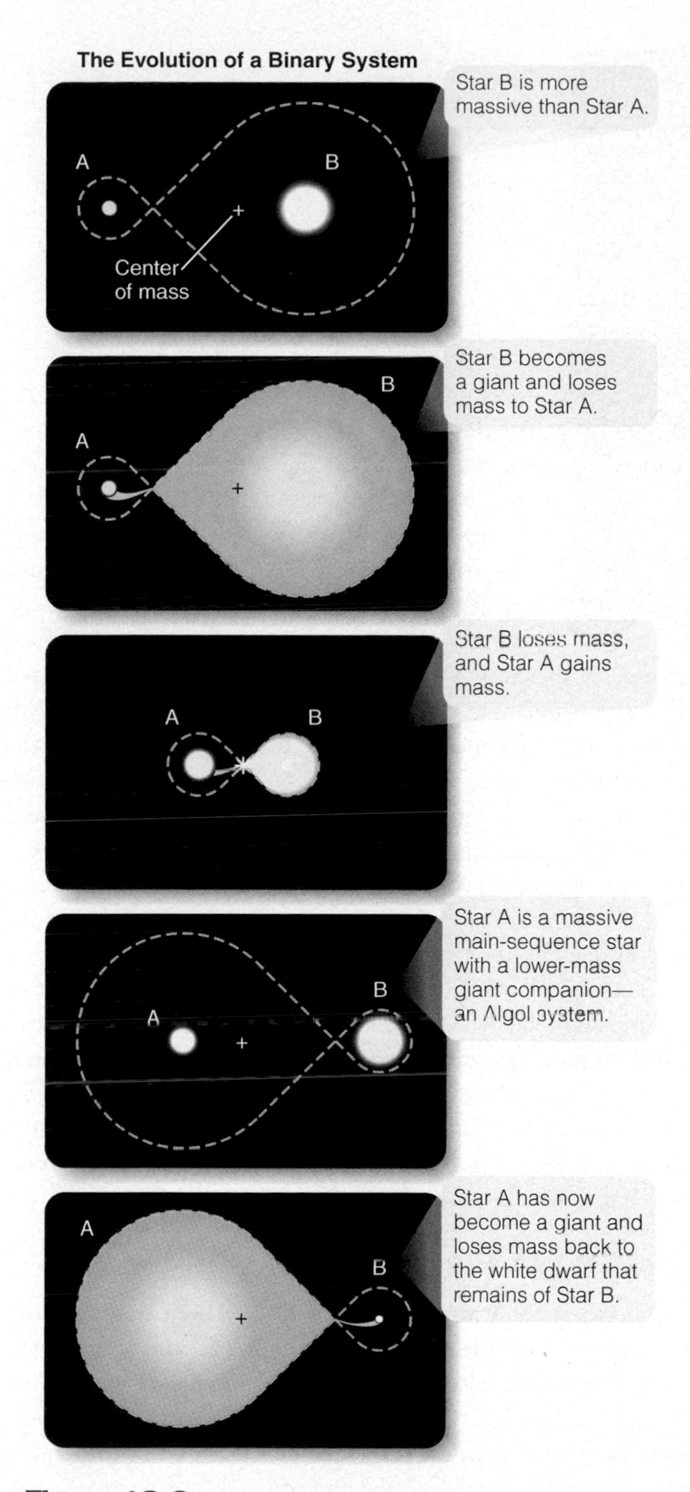

The Evolution of a Binary System
Star B is more massive than Star A.
A
B
Center of mass
Star B becomes a giant and loses mass to Star A.
B
A
Star B loses mass, and Star A gains mass.
A
B
Star A is a massive main-sequence star with a lower-mass giant companion—an Algol system.
B
A
Star A has now become a giant and loses mass back to the white dwarf that remains of Star B.
A
B

■ **Figure 13-6**

A pair of stars orbiting close to each other can exchange mass and modify their evolution.

might see a binary system such as Algol containing a 5-solar-mass main-sequence star and a 1-solar-mass giant.

Another exotic result of the evolution of close binary systems is the possible merging of the stars. Astronomers see many binaries in which both stars have expanded to fill their Roche surfaces and spill mass out into space. If the stars are close enough together and expand rapidly enough, theorists conclude, the two stars could merge into a single, rapidly rotating giant star. Most giants rotate slowly because as they expanded conservation of angular momentum slowed their rotation. Examples of rapidly rotating giant stars are known, however, and they seem to be the results of merged binary stars. Inside the distended envelope of such a star, the cores of the two stars may even continue to orbit each other until friction slows them down and they sink to the center.

Mass transfer can lead to dramatic violence. The first four frames of Figure 13-6 show how mass transfer could have produced a system like Algol. The last frame shows an additional stage in which the giant star has expelled its outer layers and collapsed to form a white dwarf. The more massive companion has expanded, transferring matter back onto the white dwarf. Such systems can become the site of tremendous explosions. To see how this can happen, you need to consider in detail how mass falls into a star.

Accretion Disks

Conservation of angular momentum will not allow matter passing through the inner Lagrangian point to fall directly into the white dwarf. Instead, it falls into a whirling disk. For a common example, consider a bathtub full of water. Gentle currents in the water give it some angular momentum, but you can't see its slow circulation until you pull the stopper. Then, as the water rushes toward the drain, conservation of angular momentum forces it to form a whirlpool. This same effect forces gas falling into a white dwarf to form a whirling disk of gas called an **accretion disk** (■ Figure 13-7).

Two important things happen in an accretion disk. First, the gas in the disk grows very hot due to friction and tidal forces. The temperature of the gas in the inner parts of an accretion disk can become very high. In some cases the gas can exceed 1,000,000 K and emit X rays. The disk also acts as a brake, ridding the gas of its angular momentum and allowing it to fall into the white dwarf.

Now you are ready to put the pieces together and explain one of the ways a star can explode.

Nova Explosions

Astronomers occasionally see a new star appear in the sky, grow brighter, then fade away after a few weeks (■ Figure 13-8). In fact, what seems to be a new star in the sky is a **nova** produced by the eruption of a very old dying star. After a nova fades, astronomers can photograph the spectrum of the object, and they invariably find a closely spaced binary star containing a normal star and a white dwarf. A nova is evidently an explosion involving a white dwarf in a binary system.

Observational evidence reveals how nova explosions occur. As the explosion begins, spectra show blueshifted absorption

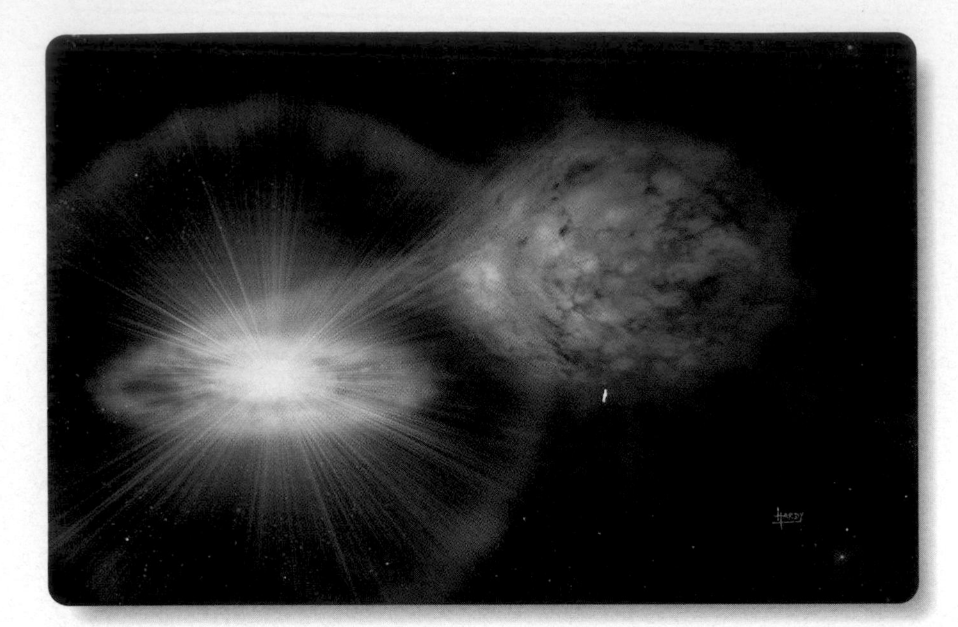

■ **Figure 13-7**

Matter from an evolving red giant falls into a white dwarf and forms a whirling accretion disk. Friction and tidal forces can make the disk very hot. Such systems can lead to nova explosions on the surface of the white dwarf as shown in this artist's impression. (David A. Hardy, www.astroart.org, and PPARC)

lines, which tell you the gas is dense and coming toward Earth at a few thousand kilometers per second. After a few days, the spectral lines change to emission lines, which tells you the gas has thinned, but the blueshifts remain, showing that a cloud of debris has been ejected into space.

Nova explosions are the result of mass transfer from a normal star through the inner Lagrangian point into an accretion disk around the white dwarf. As the matter loses its angular momentum in the accretion disk, it spirals inward and eventually settles onto the surface of the white dwarf. Because the matter came from the surface of a normal star, it is rich in unused fuel, mostly hydrogen, and when it accumulates on the surface of the white dwarf, it forms a layer of unprocessed fuel.

As the layer of fuel grows deeper, it becomes hotter and denser. Compressed by the white dwarf's gravity, the gas eventually becomes degenerate, much like the core of a sunlike star approaching the helium flash. In such a gas, the pressure–temperature thermostat does not work, so the layer of unfused hydrogen is a thermonuclear bomb waiting to explode. By the time the white dwarf has accumulated about 100 Earth masses of hydrogen, the temperature at the base of the hydrogen layer reaches millions of degrees, and the density is 10,000 times the density of water. Suddenly, the hydrogen begins to fuse on the proton–proton chain, and the energy released drives the temperature so high that the CNO cycle begins. With no pressure–temperature thermostat to control the fusion, the temperature shoots to 100 million degrees in seconds. The temperature soon rises high enough to force the gas to expand, and it stops being degenerate. But it is too late. By the time the gas begins to expand, the amount of energy released is enough to blow the surface layers of the star into space as a violently expanding shell of hot gas, which is visible from Earth as a nova.

The explosion of its surface hardly disturbs the white dwarf or its companion star. Mass transfer quickly resumes, and a new layer of fuel begins to accumulate. How fast the fuel builds up depends on the rate of mass transfer. Accordingly, you can expect novae (the plural of nova) to repeat each time an explosive layer accumulates. Many novae probably take thousands of years to build an explosive layer, but some take only a few years (■ Figure 13-9).

■ **Figure 13-8**

Nova Cygni 1975 near maximum at about second magnitude and later when it had declined to about eleventh magnitude. (Photo © UC Regents)

The End of Earth

Astronomy is about us. Although you have been reading about the deaths of medium-mass stars, white dwarfs, and novae explosions, what you have learned is related to the future of our planet. The sun is a medium-mass star and must eventually become a giant, possibly produce a planetary nebula, and collapse into a white dwarf. That will spell the end of Earth.

Mathematical models of the sun suggest that it may survive for 5 billion years or so, but it is already growing more luminous as it fuses hydrogen into helium. In a few billion years, it will exhaust the hydrogen in its core and swell into a giant star about 100 times its present radius. That giant sun will be about as large as the orbit of Earth, so the sun's expansion will mark the end of our world. Whether or not the sun becomes large enough to totally engulf Earth, its growing luminosity will certainly evaporate our oceans, drive away our atmosphere, and even vaporize most of Earth's crust.

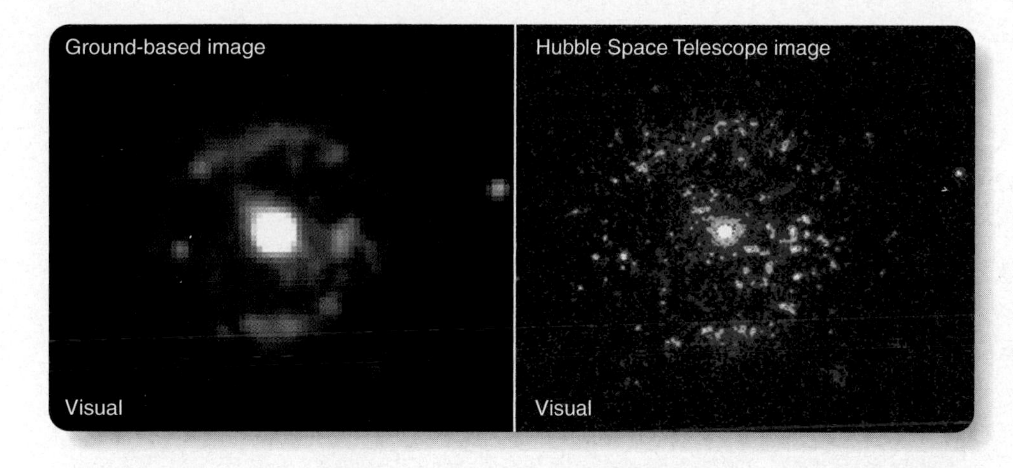

■ Figure 13-9

Nova T Pyxidis erupts about every two decades, expelling shells of gas into space. The shells of gas are visible from ground-based telescopes, but the Hubble Space Telescope reveals much more detail. The shell consists of knots of excited gas that presumably form when a new shell collides with a previous shell. (M. Shara and R. Williams, STScI; R. Gilmozzi, ESO; and NASA)

While it is a giant star, the sun will lose mass into space. This mass loss is a relatively gentle process, so any cinder that might remain of Earth will not be disturbed. Of course, when the sun finally collapses into a white dwarf, the solar system will become a much colder place, but not much of Earth will remain by then. If the collapsing sun becomes hot enough, it will ionize the expelled gas to form a planetary nebula. Some models suggest that the sun is not quite massive enough to light up a planetary nebula. An astronomer recently expressed disappointment that the dying sun might not leave behind a beautiful planetary nebula, but that embarrassment lies a few billion years in the future.

There is no danger that the sun will explode as a nova; it has no binary companion. Also, as you will see, the sun is not massive enough to die the violent death of the massive stars.

The most important lesson of astronomy is that we are part of the universe and not just observers. The atoms you are made of are destined to return to the interstellar medium in just a few billion years. That's a long time, and it is possible that the human race will migrate to other planetary systems. That might save the human race, but our planet is stardust.

◄ SCIENTIFIC ARGUMENT ►

How does modern astronomy explain the Algol paradox?

Scientific arguments combine evidence with theory to explain nature, and some of the most interesting arguments explain what seem to be impossible situations. When you encounter a paradox in nature, it is usually a warning that you don't understand things as well as you thought you did. The Algol paradox is a good example. The binary star Algol contains a lower-mass giant star and a more massive main-sequence star. Because the two stars must have formed together, they must be the same age. Then the more massive star should have evolved first and left the main sequence. But in binary systems such as Algol, the lower-mass star has left the main sequence first, or so it seems.

You can understand this paradox if you add mass transfer to your argument. The first star to leave the main sequence must be the more massive star; but if the stars are close together, the star that is initially more massive can fill its Roche surface and transfer mass back to its companion. The companion can grow more massive, and the giant can grow less massive. In this way, the lower-mass giant star may have originally been more massive and may have evolved away from the main sequence, leaving its companion behind to increase in mass as the giant decreased.

Mass transfer can explain the Algol paradox, and it can also explain the violent explosions called novae. Develop a new argument. **How does mass transfer explain why novae can explode over and over in the same binary system?**

◄ ►

13-3 The Deaths of Massive Stars

YOU HAVE SEEN that low- and medium-mass stars die relatively quietly as they exhaust their hydrogen and helium and then eject their surface layers to form planetary nebulae. In contrast, massive stars live spectacular lives and destroy themselves in violent explosions, leaving behind an expanding cloud of gas and, perhaps, a neutron star or a black hole—peculiar objects you will study in the next chapter.

Nuclear Fusion in Massive Stars

The evolution of massive stars begins like that of the sun, but, because of the higher mass, they can ignite carbon and heavier nuclear fuels.

You have seen that main-sequence stars fuse hydrogen in their cores and then swell to becomes giants when they ignite a hydrogen-fusion shell. After fusing helium in their cores and then in a shell, the stars develop carbon-rich cores. A massive star can lose significant mass as it ages; but, if it still has a mass over 4 solar masses when its carbon–oxygen core contracts, it can reach a temperature of 600 million Kelvin and ignite carbon fusion. The fusion of carbon produces heavier nuclei such as oxygen and neon. As soon as the carbon is exhausted in the core, the core contracts, and carbon ignites in a shell. This pattern of core ignition followed by shell ignition continues with fuel after fuel, and the star develops a layered structure in its core (■ Figure 13-10), with a hydrogen-fusion shell above a helium-fusion shell above a carbon-fusion shell and so on. After carbon fusion, oxygen, neon, magnesium, and heavier elements fuse right up to iron.

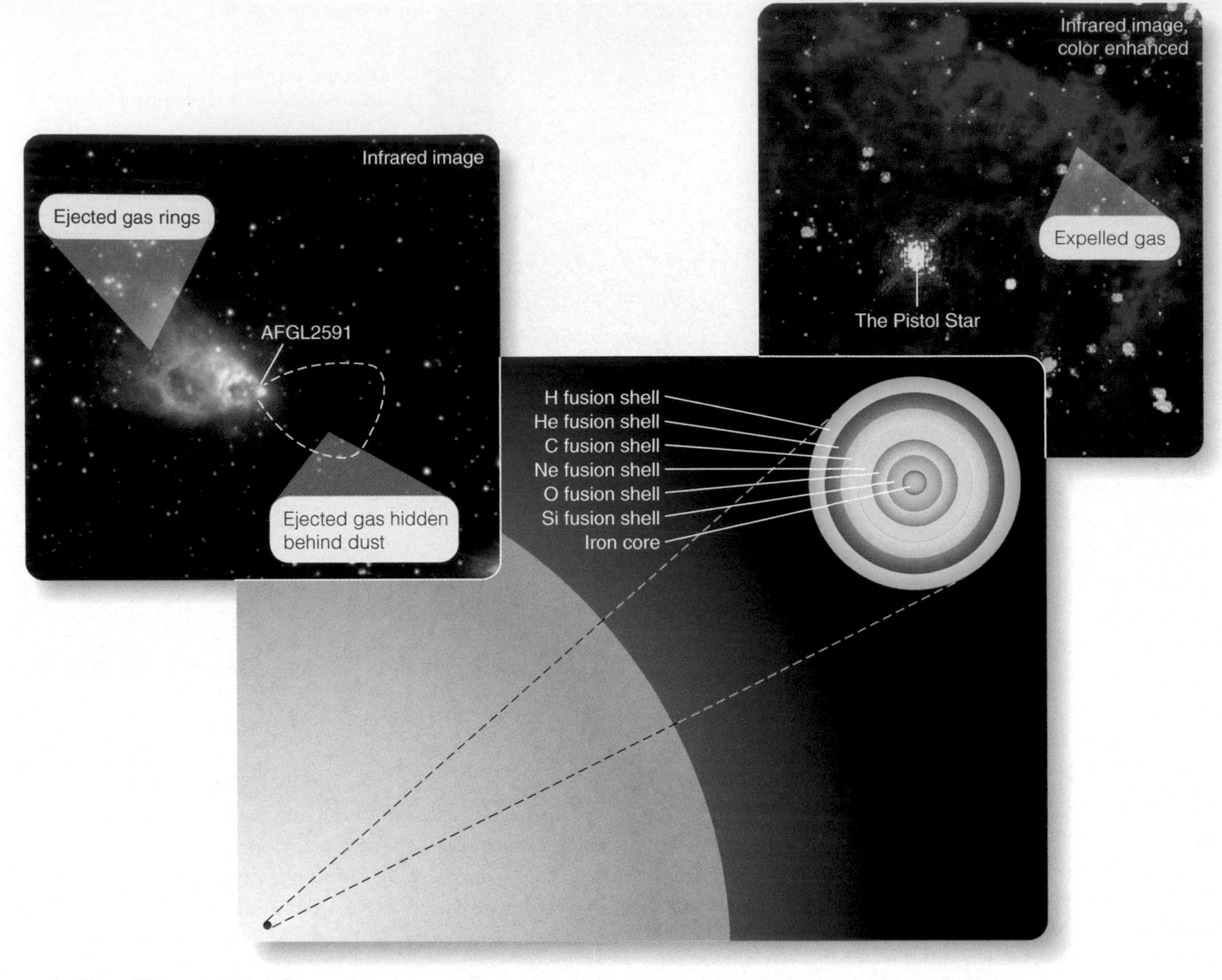

■ Active Figure 13-10

Massive stars live fast and die young. The two shown here are among the most massive stars known, containing 100 solar masses or more. They are rapidly ejecting gas into space. (See also Figure 12-3.) The centers of these massive stars develop Earth-size cores (magnified 100,000 times in this figure) composed of concentric layers of gases undergoing nuclear fusion. The iron core at the center leads eventually to a star-destroying explosion. (AFGL2591: Gemini Observatory/NSF/C. Aspin; The Pistol star: NASA)

The fusion of these nuclear fuels goes faster and faster as the massive star evolves rapidly. Recall that massive stars must consume their fuels rapidly to support their great weight, but other factors also cause the heavier fuels like carbon, oxygen, and silicon to fuse at increasing speed. For one thing, the amount of energy released per fusion reaction decreases as the mass of the fusing atom increases. To support its weight, a star must fuse oxygen much faster than it fused hydrogen. Also, there are fewer atoms in the core of the star by the time heavy atoms begin to fuse. Four hydrogens made a helium atom, and three heliums made a carbon, so there are 12 times fewer atoms of carbon available for fusion than there were of hydrogen. This means that heavy-element fusion goes very quickly in massive stars (■ Table 13-1). Hydrogen fusion can last 7 million years in a 25-solar-mass star, but that same star will fuse its oxygen in six months and its silicon in a day.

The Iron Core

Heavy-element fusion ends with iron, because nuclear reactions that use iron as a fuel cannot produce energy. Nuclear reactions can produce energy if they proceed from less tightly bound nuclei to more tightly bound nuclei. As shown in Figure 8-8, both nuclear fission and nuclear fusion produce nuclei that are more tightly bound than the starting fuel. Look again at Figure 8-8 and notice that iron is the most tightly bound nucleus of all. No nuclear reaction, fission or fusion, that starts with iron can produce a more tightly bound nucleus, and that means that iron is a dead end.

Table 13-1 | Heavy-Element Fusion in a 25-$M_{\odot}$ Star

Fuel	Time	Percentage of Lifetime
H	7,000,000 years	93.3
He	500,000 years	6.7
C	600 years	0.008
O	0.5 years	0.000007
Si	1 day	0.00000004

When a massive star develops an iron core, nuclear fusion cannot produce energy, and the core contracts and grows hotter. The shells around the core burn outward, fusing lighter elements into heavier elements and leaving behind more iron, which further increases the mass of the core. When the mass of the iron core exceeds 1.3 to 2 solar masses, it must collapse.

As the core begins to collapse, two processes can make it contract even faster. Heavy nuclei in the core can capture high-energy electrons removing thermal energy from the gas. This robs the gas of some of the pressure it needs to support the crushing weight of the outer layers. Also, temperatures are so high that many photons have gamma-ray wavelengths and can break more massive nuclei into less massive nuclei. In the process, the gamma rays are absorbed, which robs the gas of energy and allows the core to collapse even faster.

Although a massive star may live for millions of years, its iron core—about 500 km in diameter—collapses in only a few thousandths of a second triggering a star-destroying explosion.

The Supernova Deaths of Massive Stars

A **supernova,** a particularly luminous and long-lasting new star in the sky, is caused by the violent explosive death of a star. Modern astronomers find a few novae each year, but supernovae (plural) are so rare that there are only one or two supernovae each century in our galaxy. Astronomers know that supernovae occur because they occasionally flare in other galaxies and because telescopes reveal the shattered remains of these titanic explosions (■ Figure 13-11). Modern theory predicts that the collapse of a massive star can eject the outer layers of the star to produce one of the most common kinds of supernova explosions. You will learn about another kind of supernova explosion later in this section.

A supernova explosion is rare, remote, rapid, and violent. It is an event in nature that is extremely difficult to study in person, which is why astronomers have used powerful mathematical techniques and high-speed supercomputers to model the inside of a star exploding as a supernova. Such models allow astronomers to experiment on an exploding star as if the star were in a laboratory beaker.

■ Figure 13-11

Supernova explosions are rare in any one galaxy, but each year astronomers see a few erupt in other galaxies. In our own galaxy, astronomers find expanding shells filled with hot, low-density gas produced by past supernova explosions. (Pinwheel: NOAO/AURA/NSF/G. Jacoby, B. Bohannan & M. Hanna; Tycho's Supernova: NASA/CXC/Rutgers/J. Warren & J. Hughes et al.; N63A: NASA/CXC/Rutgers/J. Warren et al., STScI/U.Ill/Y. Chu; ATCA/U.Ill/J. Dickel et al.)

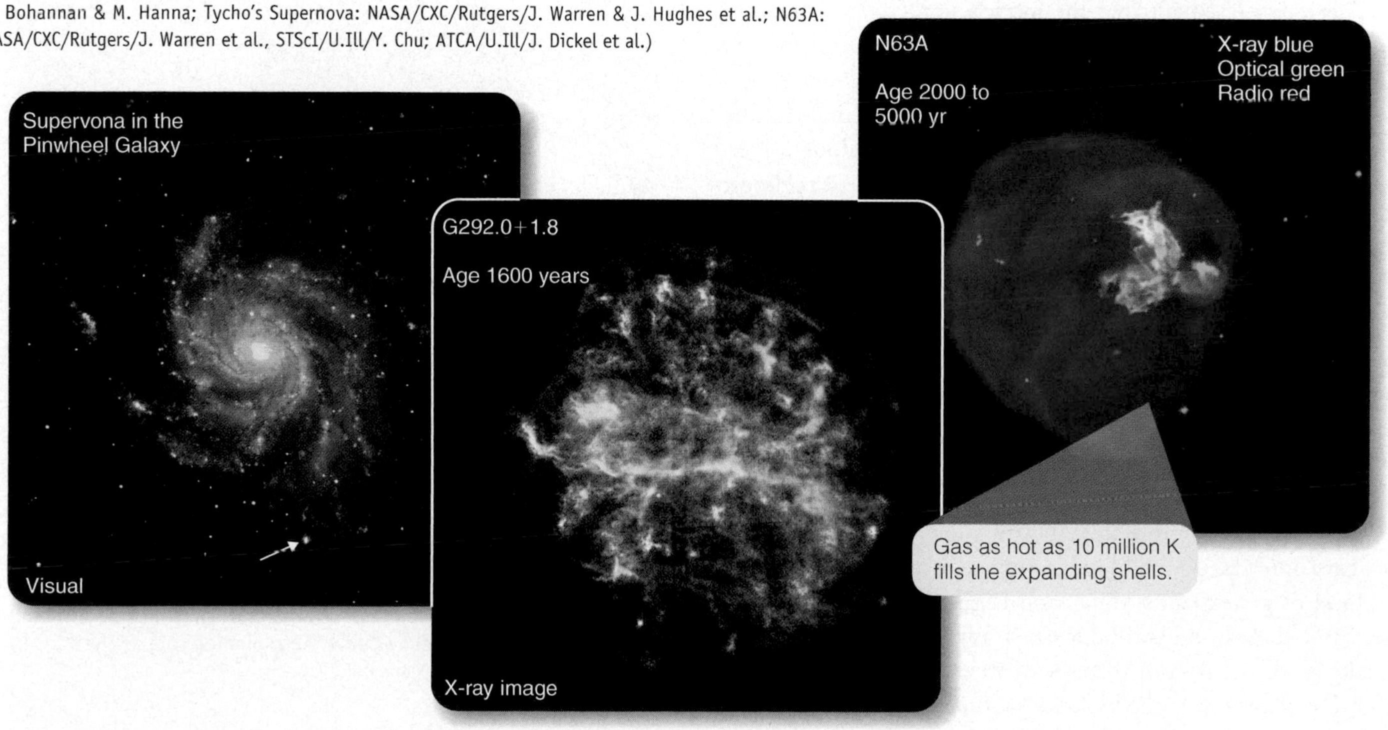

Mathematical models reveal that the key to the supernova explosion is the collapse of the iron core, which allows the rest of the interior of the star to fall inward, creating a tremendous "traffic jam" as all the nuclei fall toward the center. It is as if all the residents of Indiana suddenly tried to drive their cars as fast as possible into the center of Indianapolis. There would be a tremendous traffic jam downtown, and as more cars rushed in, the traffic jam would spread outward into the suburbs. Similarly, as the inner core of the star falls inward, a shock wave (a traffic jam) develops and begins to move outward. Containing about 100 times more energy than that necessary to destroy the star, such a shock wave was first thought to be the cause of the supernova explosion.

Computer models revealed, however, that the shock wave spreading outward through the collapsing star stalls within a few hundredths of a second. Matter flowing inward smothers the shock wave and pushes it back into the star. Those computer models wouldn't explode.

However, theory predicts that 99 percent of the energy released in the core collapse appears as neutrinos. In the sun, neutrinos zip outward unimpeded by the gas of the solar layers, but in a collapsing star the gas is billions of times denser than the gas in the sun, nearly as dense as an atomic nucleus. This gas is opaque to neutrinos, and they are partially absorbed by the gas. Not only does the tremendous burst of neutrinos remove energy from the core and allow the core to collapse even faster, but the neutrinos are absorbed outside the core and heat those layers. For some years, astronomers thought that these neutrinos could spread energy across the shock wave and, within a quarter of a second or so, reaccelerate the stalled shock wave. The computer models, however, refused to explode.

The final ingredient that the models needed was turbulent convection. When the collapse begins, the very center of the star forms a highly dense core, the beginnings of a neutron star, and the infalling material bounces off that core. As the temperature shoots up, the bouncing material produces highly turbulent convection currents that give the stalled shock wave an outward boost (■ Figure 13-12). Within a second or so, the shock wave begins to push outward, and after just a few hours it bursts out through the surface, blasting the star apart in a supernova explosion.

Of course, supernova explosions occur in silence. Science fiction movies and television have led to the **Common Misconception** that explosions in space are accompanied by deeply satisfying sounds. But space is nearly a vacuum, and there is no sound. Some of the most violent events in the universe make no sound at all.

The supernova visible from Earth is the brightening of the star as its outer layers are blasted outward. As the months pass, the cloud of gas expands, thins, and begins to fade. The way it fades can tell astronomers about the death throes of the star. Essentially all of the iron in the core of the star is destroyed when the core collapses, but the violence in the outer layers can pro-

The Exploding Core of a Supernova

The core of a massive supergiant has begun to collapse at the lower left corner of this model.

Matter continues to fall inward (blue and green) as the core expands outward (yellow) creating a shock wave.

To show the entire star at this scale, this page would have to be 30 kilometers in diameter.

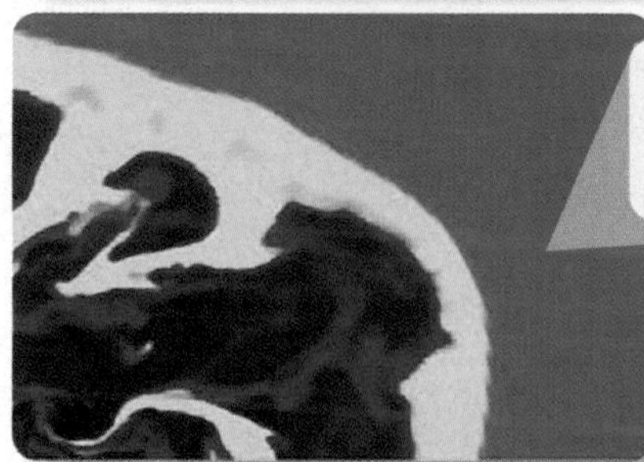

Only 0.4 s after beginning, violent convection in the expanding core (red) pushes outward.

The shock wave will blow the star apart as a neutron star forms at the extreme lower left corner.

■ Figure 13-12

As the iron core of a massive star begins to collapse, intensely hot gas triggers violent convection. Even as the outer parts of the core continue to fall inward, the turbulence blasts outward and reaches the surface of the star within hours, creating a supernova eruption. This diagram is based on mathematical models and shows only the exploding core of the star. (Courtesy Adam Burrows, John Hayes and Bruce Fryxell)

duce densities and temperatures high enough to trigger nuclear fusion reactions that produce as much as half a solar mass of radioactive nickel-56. The nickel gradually decays to form radioactive cobalt, which decays to form normal iron. The rate at which the supernova fades matches the rate at which these radioactive elements decay. Thus the destruction of iron in the core of the star is matched by the production of iron through nuclear fusion in the expanding outer layers.

The presence of nuclear fusion in the outer layers of supernovae testifies to the violence of the explosion. A typical supernova is equivalent to the explosion of 10^{28} megatons of TNT—about 3 million solar masses of high explosive.

■ Figure 13-13

Robotic telescopes search every night for supernovae flaring in other galaxies. When one is seen, astronomers can obtain spectra and record the supernova's rise in brightness and its decline to study the physics of exploding stars. If a supernova is seen in a nearby galaxy, it is sometimes possible to identify the star in earlier photos. (NASA, ESA. W. Li and S. Filippenko, Berkeley, S. Beckwith, STScI, and The Hubble Heritage Team, STScI/AURA)

Supernovae may seem powerful and remote, but you have a personal connection with the deaths of stars. All of the atoms in your body heavier than helium but lighter than iron were cooked up in stars. Some were made in medium-mass stars and expelled in planetary nebula. Even more were made inside massive stars and blasted into space in supernova explosions. Short-lived nuclear fusion reactions taking place during supernova explosions created iron and rare elements heavier than iron such as the iodine in your thyroid gland. You are made of atoms that were created by the stars.

Collapsing massive stars can trigger one kind of cosmic violence, but astronomers have observed more than one kind of supernova.

Types of Supernovae

Supernovae are rare, and only a few have been seen in our galaxy, but astronomers have been able to observe supernovae occurring in other galaxies (■ Figure 13-13). From data accumulated over decades, astronomers have noticed that there are two main types.

The Whirlpool Galaxy
Supernova 2005cs was seen exploding in a nearby galaxy in July 2005.
Ultraviolet + near infrared
The star that exploded is visible in a photo of the galaxy made months earlier.
Visual
The star that became the supernova was a 10-solar-mass red supergiant.
Near infrared

Type I supernovae have spectra that contain no hydrogen lines. They reach a maximum brightness about 4 billion times more luminous than the sun and then decline rapidly at first and then more slowly. **Type II supernovae** have spectra containing hydrogen lines. They reach a maximum brightness up to about 0.6 billion times more luminous than the sun, decline to a standstill, and then fade rapidly. The light curves in ■ Figure 13-14 summarize the behavior of the two types of supernovae.

The evidence is clear that type II supernovae occur when a massive star develops an iron core and collapses. Such supernovae occur in regions of active star formation where there is plenty of gas and dust. These are the regions where you would expect to find massive stars. Also, the spectra of type II supernovae contain hydrogen lines, as you would expect from the explosion of a massive star that contains large amounts of hydrogen in its outer layers.

Type I supernovae show no hydrogen in their spectra, which means they can't be caused by the deaths of typical massive stars. In fact there are two kinds of Type I supernovae that have dramatically different causes. Both involve binary stars. Type Ia supernovae are often found in regions where star formation ended long ago. That is further evidence that they can't be caused by massive stars; such stars don't live very long and must have formed recently.

The evidence shows that a type Ia supernova occurs when a white dwarf gaining mass in a binary star system exceeds the Chandrasekhar limit and collapses. Of course, a white dwarf gaining mass from a companion should experience nova explosions, but observations show that each nova eruption ejects only part of the matter gained from the companion. Slowly the mass of the white dwarf grows until it collapses and explodes.

The collapse of a white dwarf is different from the collapse of a massive star because the core of the white dwarf contains usable fuel. As the collapse begins, the temperature shoots up, but the gas cannot halt the collapse because it is degenerate, and the pressure–temperature thermostat is turned off in the core. Even as carbon fusion begins, the increased temperature cannot increase the pressure and make the gas expand and slow the reactions. The core of the white dwarf is a bomb. The carbon–oxygen core fuses suddenly in violent nuclear reactions called **carbon deflagration.** The word *deflagration* means to be totally destroyed by fire, in this case by nuclear fusion. In a flicker of a stellar lifetime, the entire star is consumed, with the outermost layers blasted away in a violent explosion that at its brightest is three to six times more luminous than a type II supernova. The white dwarf is entirely destroyed with nothing left behind but an expanding cloud of hot gas. Of course, you would see no hydrogen lines in the spectrum of a type Ia supernova because white dwarfs contain very little hydrogen.

Type Ib supernovae are less common. They do occur in regions where you would expect to find young stars, but their spectra also contain no hydrogen lines. They are thought to occur when a massive star in a binary system loses its hydrogen-rich outer layers to its companion star. The remains of the massive star could continue to evolve, develop an iron core, and collapse, producing a supernova explosion that lacks hydrogen lines in the spectrum. Some astronomers have referred to these as "peeled" supernovae, meaning that the massive star has had its hydrogen-rich outer layers peeled away by its binary companion.

To summarize, a type II supernova is caused by the collapse of a massive star. A type Ia is caused by the collapse of a white dwarf. A type Ib is caused by the collapse of a massive star that has lost its outer envelope of hydrogen.

Much of what you have learned so far about supernovae has been based on theory, so it is time to compare theory with observations of real supernova explosions. These frequent reality checks are a distinguishing characteristic of science.

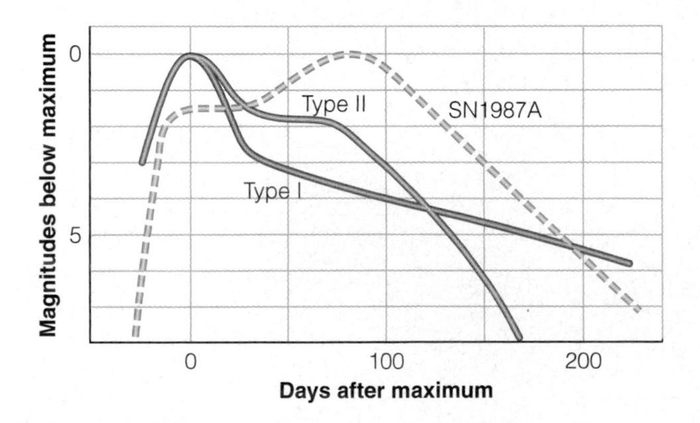

■ Figure 13-14

Type I supernovae decline rapidly at first and then more slowly, but type II supernovae pause for about 100 days before beginning a steep decline. Supernova 1987A was odd in that it did not rise directly to maximum brightness. These light curves have been adjusted to the same maximum brightness. Generally, type II supernovae are about two magnitudes fainter than type I.

Observations of Supernovae

In AD 1054, Chinese astronomers saw a "guest star" appear in the constellation now known as Taurus, the bull. The star quickly became so bright it was visible in the daytime. After a month's time, it slowly faded, taking almost two years to vanish from sight. When modern astronomers turned their telescopes to the location of the guest star, they found a cloud of gas about 1.35 pc in radius, expanding at 1400 km/s. Projecting the expansion back in time, they concluded that it must have begun about nine centuries ago, just when the guest star made its visit. From this and other evidence, astronomers conclude that the nebula, now called the Crab Nebula because of its shape (■ Figure 13-15), marks the site of the 1054 supernova.

The glowing filaments of the Crab Nebula appear to be excited gas flung outward by the explosion, but the hazy glow in the inner nebula is something else. Radio observations show that

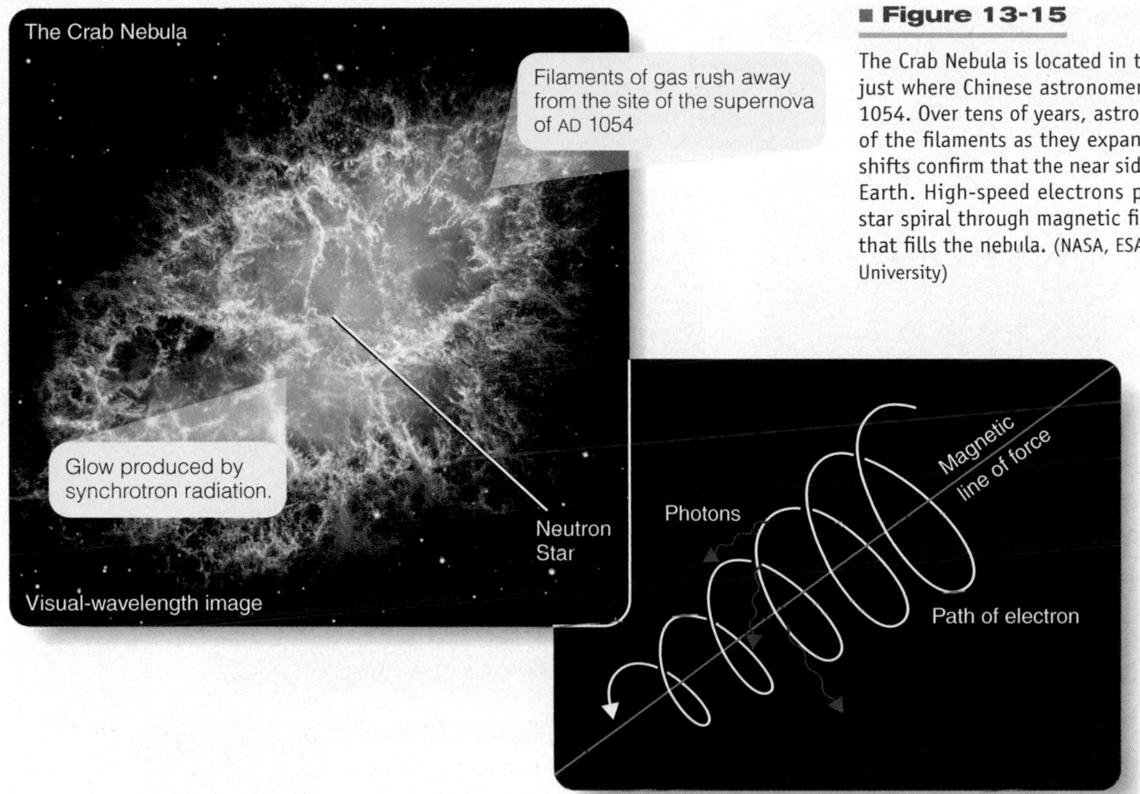

■ **Figure 13-15**

The Crab Nebula is located in the constellation Taurus the Bull, just where Chinese astronomers saw a brilliant guest star in AD 1054. Over tens of years, astronomers can measure the motions of the filaments as they expand away from the center. Doppler shifts confirm that the near side of the nebula is moving toward Earth. High-speed electrons produced by the central neutron star spiral through magnetic fields and produce the foggy glow that fills the nebula. (NASA, ESA, J. Hester and A. Loll, Arizona State University)

the gas in the nebula is emitting **synchrotron radiation**—electromagnetic energy radiated by high-speed electrons spiraling through a magnetic field. Electrons moving at slightly different velocities radiate at different wavelengths, so synchrotron radiation is spread over a wide range of wavelengths. The foggy glow of light in the Crab Nebula is synchrotron radiation at the very short wavelengths of visible light, which means the electrons must be traveling at tremendous speeds. In the nine centuries since the Crab supernova explosion, the electrons should have radiated their energy away and slowed down. Evidently they have not, so there must be an energy source in the Crab Nebula that is producing very-high-speed electrons. You will follow this clue in the next chapter and discover a neutron star at the center of the Crab Nebula.

In contrast, Kepler's supernova of 1604 left nothing behind but a cloud of gas and dust. Astronomers analyzing the chemical content of the cloud conclude the explosion was a type Ia supernova, so it would not have formed a neutron star or black hole.

A supernova fades to obscurity in a year or two, but it leaves behind an expanding shell of gas. Originally expelled at 10,000 to 20,000 km/s, the gas may carry away one-fifth of the mass of the star. As it cools, some of the gas condenses to form dust, and that makes supernovae one of the biggest sources of the dust in the interstellar medium. The collision of the expanding shell of gas and dust with the surrounding interstellar medium can sweep up even more gas and excite it to produce a **supernova remnant,** the nebulous remains of a supernova explosion.

Supernova remnants look quite delicate and do not survive very long—a few tens of thousands of years—before they gradually mix with the interstellar medium and vanish. The Crab Nebula is a young remnant, only about 950 years old, and it isn't very large, only a few parsecs in diameter. Older remnants can be larger. Some supernova remnants are detectable only at radio and X-ray wavelengths. They have become too tenuous to emit much visible light, but the collision of the expanding hot gas with the interstellar medium can generate radio and X-ray radiation and allows astronomers to create images of them at these nonvisible wavelengths. In general, supernova remnants are tenuous spheres of gas expanding into the interstellar medium (■ Figure 13-16). You saw in Chapter 11 that the compression of the interstellar medium by expanding supernova remnants can trigger star formation.

The Crab Nebula is a young supernova remnant linked to a supernova explosion that was actually seen to occur. Supernovae are rare, and only a few have been visible to the naked eye during all of recorded history. Arab astronomers recorded one in 1006, and the Chinese saw the Crab supernova in 1054. The guest stars of 185, 386, 393, and 1181 may have been supernovae. European astronomers observed two—one in 1572 (Tycho's supernova) and one in 1604 (Kepler's supernova). Most supernovae are discovered in distant galaxies, but they are faint and difficult to study.

The Great Supernova of 1987

For 383 years following Kepler's supernova in 1604, no naked-eye supernova was seen. Then, in late February 1987, the news raced around the world: Astronomers in Chile had discovered a naked-eye supernova in the Large Magellanic Cloud, a small galaxy very near our Milky Way Galaxy (■ Figure 13-17). Because the supernova was only 20 degrees from the south celestial pole, it could be studied only from southern latitudes. It was

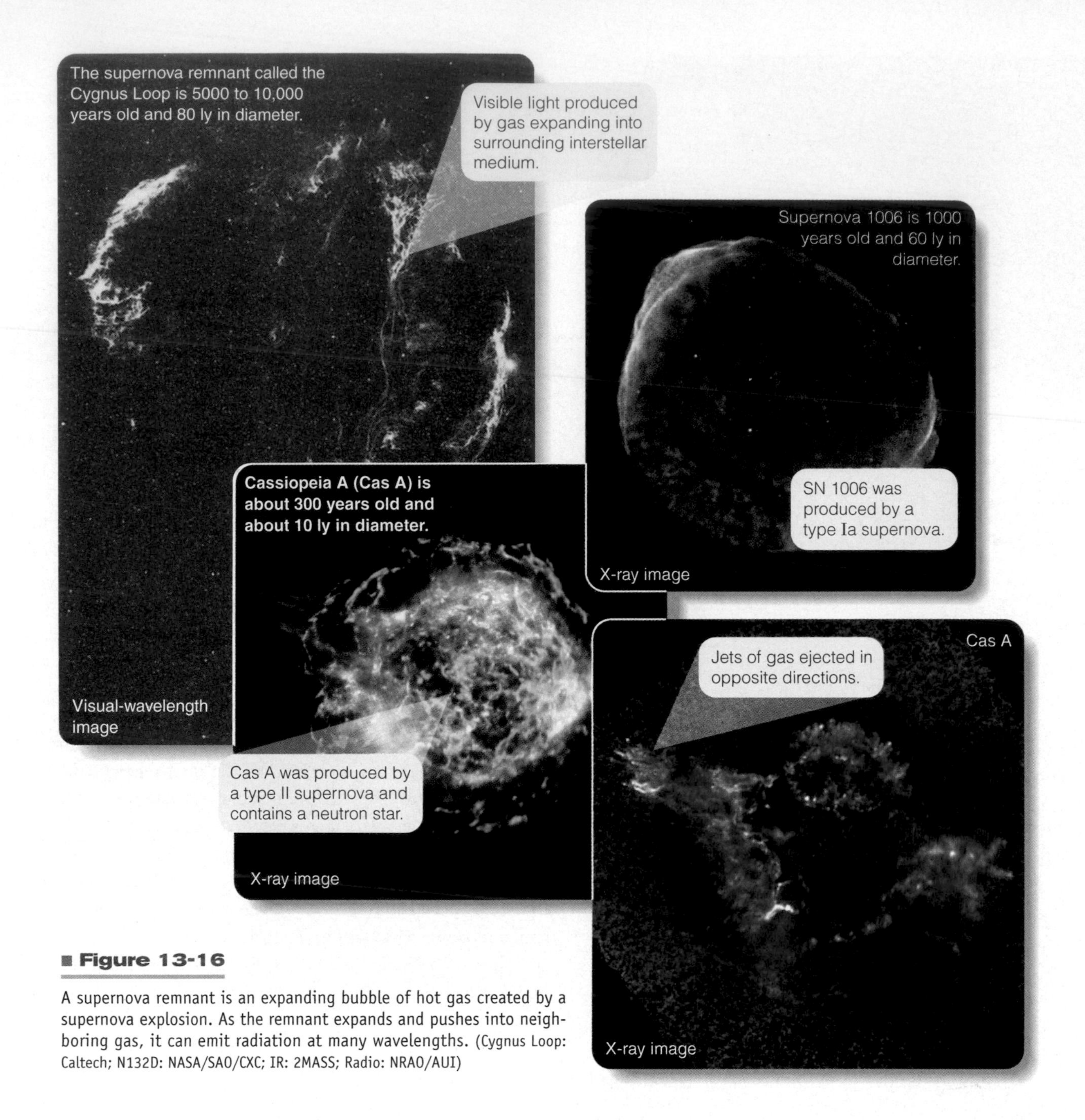

Figure 13-16

A supernova remnant is an expanding bubble of hot gas created by a supernova explosion. As the remnant expands and pushes into neighboring gas, it can emit radiation at many wavelengths. (Cygnus Loop: Caltech; N132D: NASA/SAO/CXC; IR: 2MASS; Radio: NRAO/AUI)

named SN1987A to denote the first supernova discovered in 1987.

The hydrogen-rich spectrum suggested that the supernova was a type II, caused by the collapse of the core of a massive star. As the months passed, however, the light curve proved to be odd in that it paused for a few weeks before rising to its final maximum (see Figure 13-14). From photographs of the area made some years before, astronomers were able to determine that the star that exploded, cataloged as Sanduleak −69°202, was not the expected red supergiant but rather a hot, blue supergiant of about 20 solar masses and 50 solar radii, not extreme for a supergiant. Theorists have concluded that the star was chemically poor in elements heavier than helium and had consequently contracted and heated up after a phase as a cool, red supergiant, during which it lost mass into space. The relatively small size of the supergiant may explain the pause in the light curve. Much of the energy of the explosion went into blowing apart the smaller, denser-than-usual star and making it expand.

The brightening of the expanding gases after the first few weeks seems to have been caused by the decay of radioactive nickel into cobalt, which emitted gamma rays that heated the expanding shell of gas and made it brighter. About 0.07 solar mass of nickel was produced, about 20,000 times the mass of Earth.

The cobalt atoms are also unstable, but they decay more slowly, so it was not until some time later, after much of the

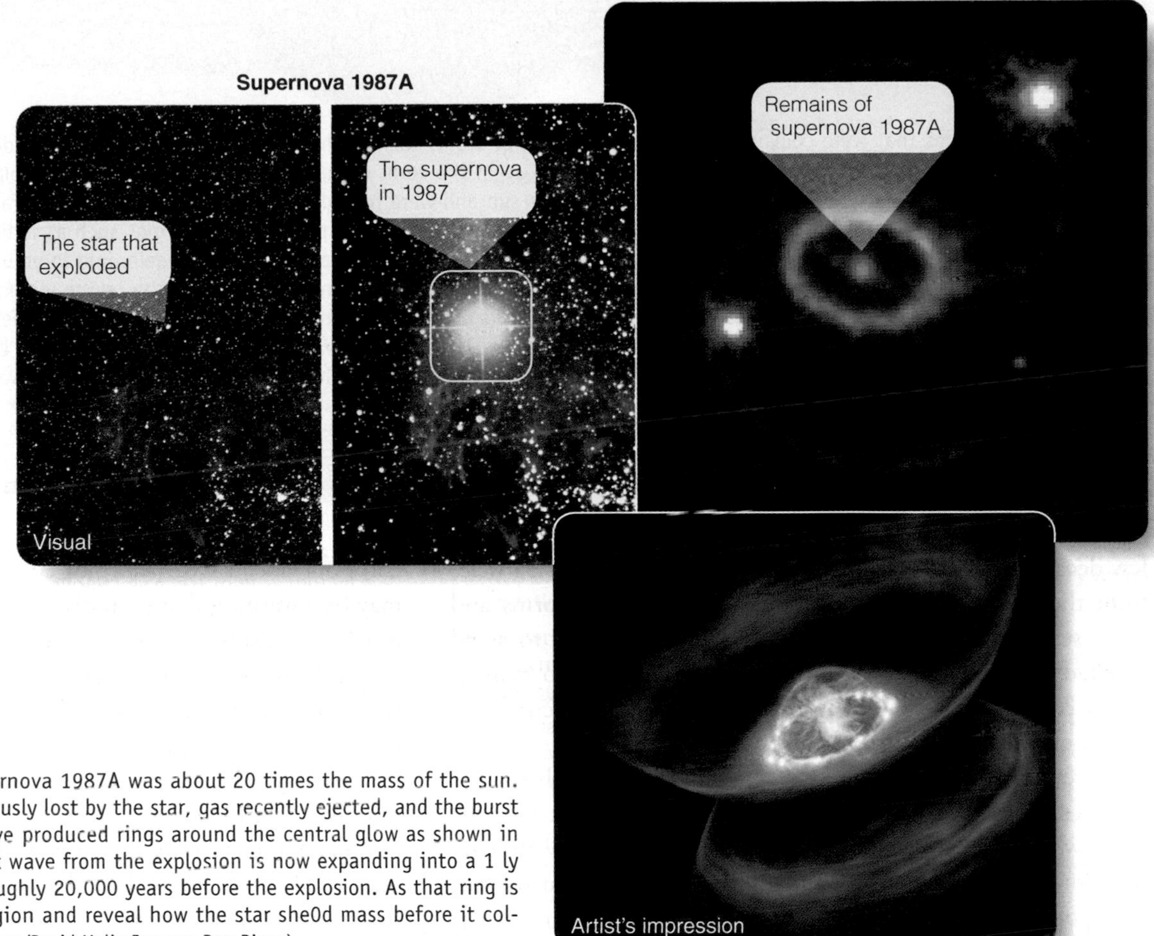

■ Figure 13-17

The star that exploded as supernova 1987A was about 20 times the mass of the sun. The interaction of matter previously lost by the star, gas recently ejected, and the burst of light from the explosion have produced rings around the central glow as shown in the artist's impression. A shock wave from the explosion is now expanding into a 1 ly diameter ring of gas ejected roughly 20,000 years before the explosion. As that ring is excited, it will light up the region and reveal how the star she0d mass before it collapsed. (Anglo-Australian Observatory/David Malin Images; Don Dixon)

nickel had decayed, that the decay of cobalt into iron began providing energy to keep the expanding gas hot and luminous. Although these processes had been predicted, they were clearly observed in SN1987A. Gamma rays from the decay of cobalt to iron were detected above Earth's atmosphere, and cobalt and iron are clearly visible in the infrared spectra of the supernova.

As the expanding gases dimmed, photographs revealed bright rings of gas (look again at Figure 13-17). Comparison with mathematical models shows that the rings were produced by two stellar winds. When the star was a red supergiant, it expelled a slow stellar wind. Later, when it became a blue supergiant, it expelled a fast stellar wind. The interaction of these two winds shaped by a magnetic field and illuminated by the supernova explosion produces the rings. Compare this with the rings seen in some of the planetary nebulae shown on page 269.

Two independent observations confirm that SN1987A probably gave birth to a neutron star. Theory predicts that the collapse of a massive star's core should liberate a tremendous blast of neutrinos that leave the star hours before the shock wave from the interior blows the star apart. Two independent neutrino detectors, one in Ohio and one in Japan, recorded a burst of neutrinos passing through Earth at 2:35:41 AM EST on February 23, 1987, about 18 hours before the supernova was seen. The data show that the neutrinos rushed toward Earth from the direction of the supernovae. The detectors caught only 19 neutrinos during a 12-second interval, but recall that neutrinos hardly ever react with normal matter. Even though only 19 were detected, the full flood of neutrinos must have been immense. Within a few seconds of that time, roughly 20 trillion neutrinos passed harmlessly through each human body on Earth. The detection of the neutrino blast confirms that the collapsing core gave birth to a neutron star.

Over time the expanding gas shell of the supernova will continue to thin and cool, and eventually astronomers on Earth will be able to peer inside and see what is left of Sanduleak −69°202. Will they see the expected neutron star, or will theories need further revision? For the first time in almost 400 years, astronomers can observe a bright supernova to test their theories.

Local Supernovae and Life on Earth

Although supernovae are rare events, they are very powerful and could affect life on planets orbiting nearby stars. In fact, supernovae explosions long ago may have affected Earth's climate and the evolution of life.

If a supernova occurred near Earth, the human race would have to abandon the surface and live below ground for at least a

Almost anything is easier to get into than out of.

AGNES ALLEN

GRAVITY ALWAYS WINS. In a star's struggle to withstand its own gravity, the star must eventually exhaust its fuel, and gravity must win. Gravity ensures that the star's last remains must eventually reach one of three final states—white dwarf, neutron star, or black hole. You studied the first of these compact objects in the previous chapter, and now you are ready to complete the story of the stars.

Theory predicts that neutron stars and black holes exist, but science depends on evidence. Can astronomers find objects in the sky that are real neutron stars and black holes? Scientists always fall back on evidence—the final reality check on their understanding of how nature works.

14-1 Neutron Stars

A **NEUTRON STAR** IS A STAR of a little over 1 solar mass compressed into a radius of about 10 km. Its density is so high that the matter is stable only as a fluid of neutrons. Theory predicts that such an object, left behind by a supernova explosion, should spin a number of times a second, be nearly as hot at its surface as the inside of the sun, and have a magnetic field a trillion times stronger than Earth's (■ Figure 14-1). Two questions should occur to you immediately. First, how could any theory predict such a wondrously unbelievable star? And second, do such neutron stars really exist?

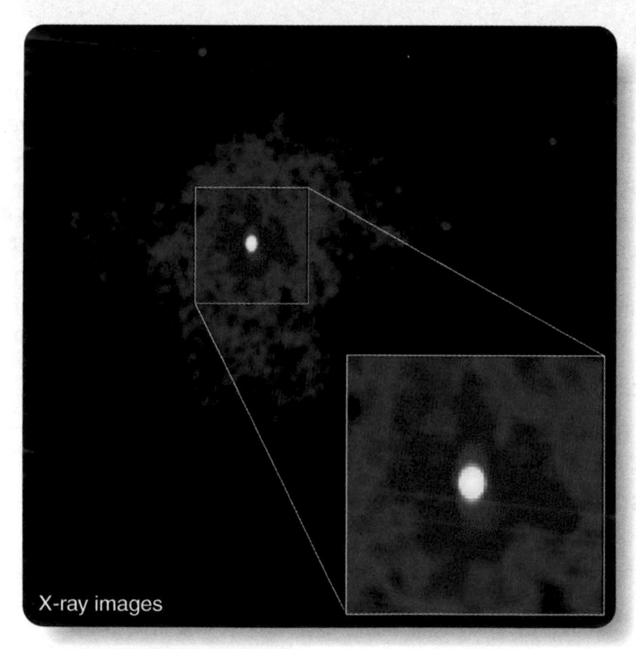

■ **Figure 14-1**

A supernova explosion seen in AD 1181 left behind an expanding supernova remnant. The Chandra X-Ray Observatory has imaged the nebula in X rays and finds a tiny hot object within—a neutron star. (NASA/SAO/CXC/P. Slane et al.)

Theoretical Prediction of Neutron Stars

The neutron was discovered in the laboratory in February 1932, and physicists were fascinated by the new particle. Only two years later, in January 1934, two Caltech astronomers published a seminal paper. Walter Baade and Fritz Zwicky suggested that some of the most luminous novae in the historical record were not true novae but were caused by the explosive collapse of a massive star in an explosion they called a supernova. The core of the star, they proposed, would form a small tremendously dense sphere of neutrons, and Zwicky coined the term *neutron star.*

Over the following years, scientists applied the principles of quantum mechanics and were able to understand how Zwicky's neutron star could support itself. Neutrons spin in much the way that electrons do, which means that neutrons must obey the Pauli exclusion principle. In that case, if neutrons are packed together tightly enough, they can become degenerate just as electrons do. White dwarfs are supported by degenerate electrons, and quantum mechanics predicts that a dense enough mass of neutrons might support itself by the pressure of degenerate neutrons. Of course, the inside of a neutron star would have to be much denser than the inside of a white dwarf.

Why would the core of a collapsing star produce a mass of neutrons? Atomic physics provides an explanation. If the collapsing core is more massive than the Chandrasekhar limit of 1.4 solar masses, then it cannot reach stability as a white dwarf. The weight is too great to be supported by degenerate electrons. The collapse of the core continues, and the atomic nuclei are broken apart by gamma rays. Almost instantly, the increasing density forces the freed protons to combine with electrons and become neutrons and neutrinos:

$$e + p \rightarrow n + \nu$$

The burst of neutrinos (ν) helps blast the envelope of the star away in a supernova explosion as the core of neutrinos collapses to become a neutron star.

As you saw in the previous chapter, a star of 8 solar masses or less could lose enough mass to die by forming a planetary nebula and leaving behind a white dwarf. More massive stars will lose mass rapidly, but they cannot shed mass fast enough to reduce their mass below the Chandrasekhar limit, so it seems likely that they must die in supernova explosions. Theoretical calculations suggest that stars that begin life on the main sequence with 8 to roughly 20 solar masses will leave behind neutron stars. Stars more massive are thought to form black holes (■ Figure 14-2).

Theoretical calculations predict that a neutron star should be only 10 or so kilometers in radius (■ Figure 14-3) and have a density of about 10^{14} g/cm^3. On Earth, a sugar-cube-sized lump of this material would weigh 100 million tons. This is roughly the

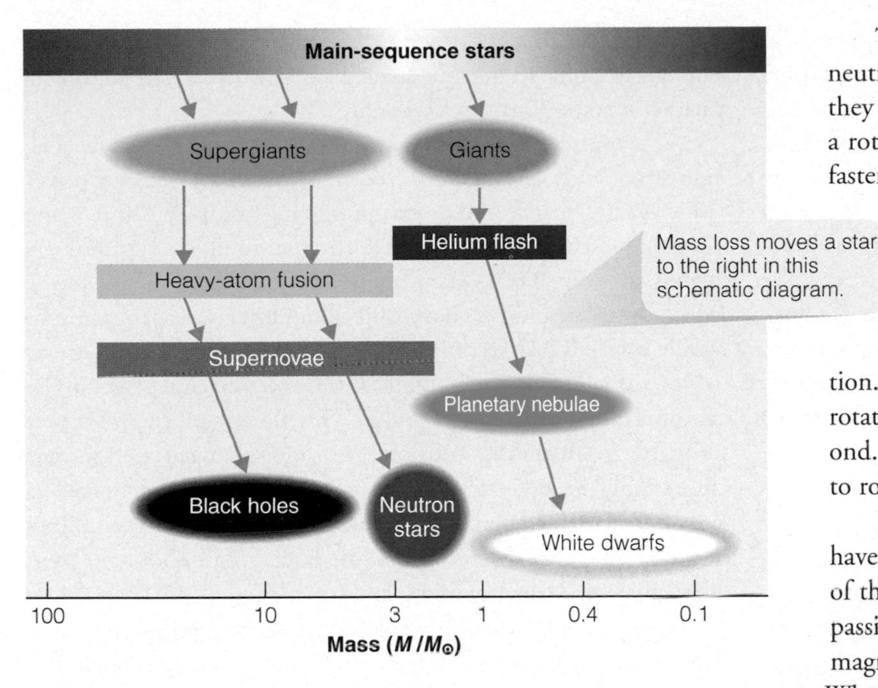

■ Figure 14-2

How a star evolves depends on its mass. Mass loss can change a star's fate by reducing its mass as it evolves. The mass limits of the categories shown here are not well known and given only for purposes of illustration. (Figure design by author)

density of the atomic nucleus, and you can think of a neutron star as matter with all of the empty space squeezed out of it.

How massive can a neutron star be? That is a critical question, and it's a difficult one to answer because physicists don't know the strength of pure neutron material. Such matter can't be made in the laboratory, so its properties must be predicted theoretically. The most widely accepted calculations suggest that a neutron star cannot be more massive than 2 to 3 solar masses. If a neutron star were more massive than that, the degenerate neutrons would not be able to support the weight, and the object would collapse (presumably into a black hole).

The sudden collapse of the core of a massive star to a radius of 10 km heats it to millions of degrees. That means that neutron stars are born very hot, but they cool rapidly at first. Neutrons are unstable and decay to produce a proton, an electron, and a neutrino, but at the densities inside neutron stars, any proton that is produced almost immediately combines with an electron to produce another neutron and a neutrino. The neutrinos carry energy out of the neutron star and cool it for the first million years or so. After that, a neutron star cools slowly because the energy can escape only from the surface, and neutron stars are so small they have little surface from which to radiate. In this way, basic theory predicts that neutron stars should be very hot for millions of years after they form.

The conservation of angular momentum predicts that neutron stars should spin rapidly. All stars rotate because they form from swirling clouds of interstellar matter. But as a rotating star collapses into a neutron star, it must rotate faster because it conserves angular momentum. Recall that you see this happen when ice skaters spin slowly with their arms extended and then speed up as they pull their arms closer to their bodies (see Figure 5-7). In the same way, a collapsing star must spin faster as it pulls its matter closer to its axis of rotation. If the sun collapsed to a radius of 10 km, its period of rotation would decrease from 25 days to about 0.001 second. You might expect the collapsed core of a massive star to rotate 10 to 100 times a second.

Basic theory also predicts that a neutron star should have a powerful magnetic field. Remember from your study of the sun's atmosphere in Chapter 8 that a magnetic field passing through an ionized gas is "frozen in." Whatever magnetic field a star has is frozen into the gas of the star. When the star collapses, the magnetic field is carried along and squeezed into a smaller area, which could make the field a billion times stronger than it was. Although the matter in a neutron star is over 90 percent neutrons, there are loose protons and electrons there so it can retain a magnetic field. Some stars have magnetic fields over 1000 times stronger than the sun's, so that means a neutron star could have a magnetic field as much as a trillion times stronger than the sun's. For comparison, that is about 10 million times stronger than any magnetic field ever produced in the laboratory.

■ Figure 14-3

A tennis ball and a road map illustrate the relative size of a neutron star. Such an object, containing slightly more than the mass of the sun, would fit with room to spare inside the beltway around Washington, DC. (Photo by author)

Theory allowed astronomers to predict the properties of neutron stars, but it also predicted that such objects should be difficult to observe. Neutron stars are very hot, so Wien's law of black body radiation (Chapter 7) told astronomers that neutron stars would radiate most of their energy in the X-ray part of the spectrum, radiation that could not be observed in the 1940s and 1950s because astronomers could not put their telescopes above Earth's atmosphere. Also, the small surface areas of neutron stars mean that they must be faint objects. Even though they are hot, they don't have much surface area from which to radiate. Consequently, astronomers of the mid-20th century were not surprised that none of the newly predicted neutron stars were found. Neutron stars were, at that point, entirely theoretical objects.

The Discovery of Pulsars

In November 1967, Jocelyn Bell, a graduate student at Cambridge University in England, found a peculiar pattern on the paper chart from a radio telescope. Unlike other radio signals from celestial bodies, this was a series of regular pulses (■ Figure 14-4). At first she and the leader of the project, Anthony Hewish, thought the signal was interference, but they found it day after day in the same place in the sky. Clearly, it was celestial in origin.

The possibility that it was a radio signal from a distant civilization led them to consider naming it LGM for Little Green Men. But within a few weeks, the team found three more objects in other parts of the sky, pulsing with different periods. The objects were clearly natural, so the team dropped the name LGM in favor of **pulsar**—a contraction of *pulsing star.* The pulsing radio source Bell had observed with her radio telescope was the first known pulsar.

As more pulsars were found, astronomers argued over their nature. Periods ranged from 0.033 to 3.75 seconds and were nearly as exact as an atomic clock. Months of observation showed that many of the periods were slowly growing longer by a few billionths of a second per day. Whatever produced the regular pulses had to be highly precise, nearly as exact as an atomic clock, but it also had to gradually slow down.

It was easy to eliminate possibilities. Pulsars could not be stars. A normal star, even a small white dwarf, is too big to pulse that fast. Nor could a star with a hot spot on its surface spin fast enough to blink so quickly. Even a small white dwarf would fly apart if it spun 30 times a second.

The pulses themselves gave the astronomers a clue. The pulses last only about 0.001 second, and this places an upper limit on the size of the object producing the pulse. If a white dwarf blinked on and then off in that interval, astronomers would not see a 0.001-second pulse. Because the near side of the white dwarf would be about 6000 km closer to Earth, its light would arrive 0.022 second before the light from the bulk of the white dwarf. The short blink 0.001 second long would be smeared out into a longer pulse. This illustrates an important principle in astronomy—an object cannot change its brightness appreciably in an interval shorter than the time light takes to cross its diameter. If pulses from pulsars are no longer than 0.001 second, then the objects cannot be larger than about 300 km (190 miles) in diameter and could be smaller.

Only a neutron star is small enough to be a pulsar. In fact, a neutron star is so small that it couldn't pulsate slowly enough, but it can spin as fast as 1000 times a second without flying apart. The missing link between pulsars and neutron stars was found in late 1968, when astronomers discovered a pulsar at the heart of the supernova remnant, the Crab Nebula (see Figure 13-15).

The short pulses and the discovery of the pulsar in the Crab Nebula strongly suggest that pulsars are neutron stars. As you learn more about pulsars, you will see that not all neutron stars are observable from Earth as pulsars, but all pulsars are neutron stars. By combining theory and observation, astronomers can devise a model of a pulsar.

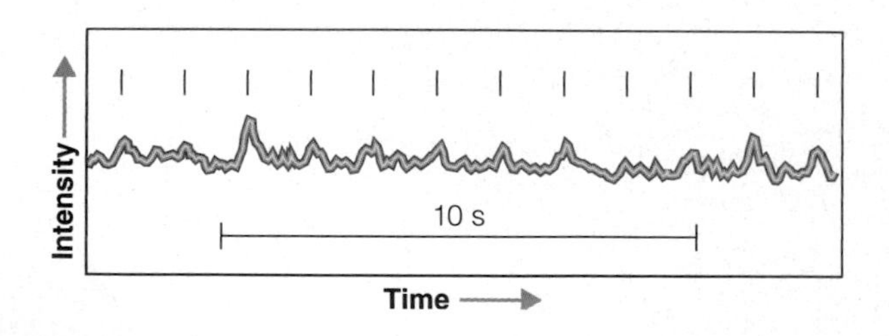

■ **Figure 14-4**

The 1967 detection of regularly spaced pulses in the output of a radio telescope led to the discovery of pulsars. This record of the radio signal from the first pulsar, CP1919, contains regularly spaced pulses (marked by ticks). The period is 1.33730119 seconds.

A Model Pulsar

Scientists often work by building a model of a natural phenomenon—not a physical model made of plastic and glue but an intellectual conception of how nature works in a specific instance. The model may be limited and incomplete, but it helps scientists organize their theories and observations. A model of a pulsar will help you draw together a lot of different ideas.

The modern model of a pulsar has been called the **lighthouse model** and is shown in **The Lighthouse Model of a Pulsar** on pages 292–293. Notice three important points:

1. A pulsar does not pulse but rather emits beams of radiation that sweep around the sky as the neutron star rotates. If the beams do not sweep over Earth, the pulses will not be detectable by Earth's radio telescopes.

2. Notice that the mechanism that produces the beams involves extremely high energies and is not fully understood.

3. Also notice how modern space telescopes observing at nonvisual wavelengths can help confirm and refine the model.

Neutron stars are not simple objects, and modern astronomers need both general relativity and quantum mechanics to try

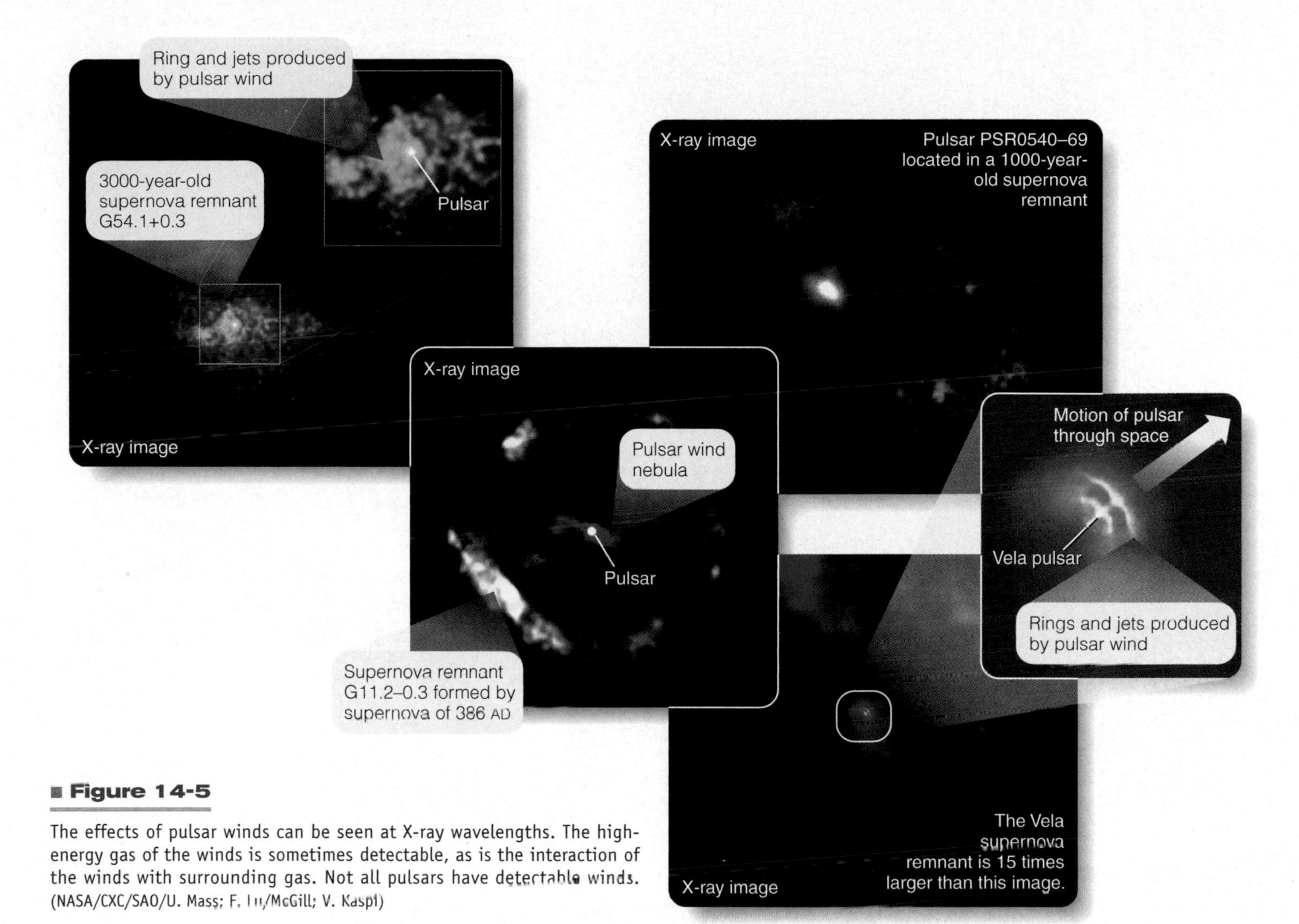

■ **Figure 14-5**

The effects of pulsar winds can be seen at X-ray wavelengths. The high-energy gas of the winds is sometimes detectable, as is the interaction of the winds with surrounding gas. Not all pulsars have detectable winds. (NASA/CXC/SAO/U. Mass; F. Lu/McGill; V. Kaspi)

to understand them. Nevertheless, the life story of pulsars can be understood in terms of the lighthouse model.

The model of a pulsar as a spinning neutron star won the support of astronomers because it explains two properties of pulsars. First, many pulsars are slowing down. Their periods are increasing by a few billionths of a second each day—a change radio astronomers can measure using atomic clocks. Evidently, the spinning neutron star is converting some of its energy of rotation into various kinds of electromagnetic energy and a powerful outflow of high-speed particles called a **pulsar wind** (■ Figure 14-5). About 99.9 percent of the energy released by the slowing of the neutron star is carried away by the pulsar wind, and only 0.1 percent goes into producing the radio beams. The energy that keeps the Crab Nebula glowing nearly 1000 years after the explosion is coming from the rotational energy of the neutron star and the pulsar wind that the neutron star produces.

The second property of pulsars that supports the neutron-star model of a pulsar is the **glitch**—a sudden increase in the pulse rate seen in some pulsars (■ Figure 14-6). Two theories have been proposed to explain these changes, and both depend on the internal structure of spinning neutron stars.

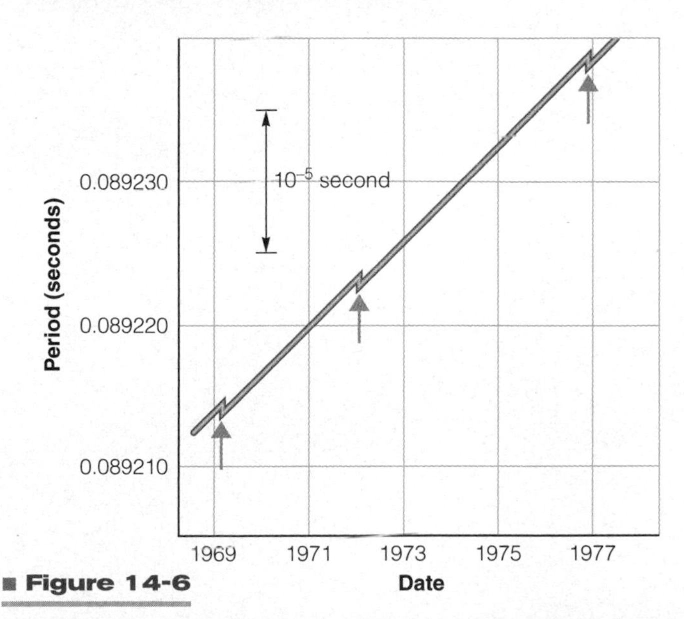

■ **Figure 14-6**

Soon after pulsars were discovered, radio astronomers accumulated enough data to show that pulsars were very gradually slowing down. That is, their pulses were growing longer. Some pulsars, such as the Vela pulsar whose data are shown here, experience glitches in which the pulsar suddenly speeds up only to resume its more leisurely decline.

One theory suggests that "starquakes" occur on the surface of a neutron star and produce glitches. To understand this theory you need to note that theoretical models of neutron stars suggest that their interiors are fluids composed mostly of neutrons. Near the surface, where the pressure is lower, matter can exist as a rigid crust composed mostly of iron nuclei. This crust might be a few hundred meters thick and is about 10^{16} times stronger than steel. The rapidly spinning neutron star would be slightly flattened, but as it slowed, its gravity would squeeze it and try to make it more spherical until the crust broke in a "starquake"—a neutron-star earthquake. When the crust breaks, the neutron star can become slightly less flattened, and the conservation of angular momentum causes it to spin slightly faster. From Earth, astronomers would see a sudden increase in the pulse rate.

The starquake theory has a drawback; it predicts that starquakes should occur every few thousand years on any given neutron star, but glitches are actually more common. The Vela pulsar has glitched over a dozen times since it was discovered. A different theory may explain most glitches. Because the liquid in the core of a neutron star can circulate with almost no friction, vortexes should develop in the fluid. Like swarms of frictionless tornadoes, they could store large amounts of angular momentum. Calculations that apply quantum mechanics to these vortices show that, as the neutron star slows, swarms of vortices should suddenly transfer their angular momentum to the crust. That would make the crust spin faster and increase the pulse rate. That is, it would produce a glitch. This theory has been confirmed in laboratory experiments with rotating fluids.

A rare glitch may be caused by a starquake, but most are probably caused by changes in internal vortices. Notice, in any case, that both explanations depend on the pulsar being a spinning neutron star. Consequently, glitches give astronomers further confidence that pulsars are spinning neutron stars.

Since the discovery of the first pulsar, over 1000 have been found. Of course, Earthbound astronomers see only those pulsars whose beams sweep over Earth. There are probably about 100 million neutron stars in our galaxy, but most are invisible from Earth.

Recognizing Neutron Stars

Neutron stars look like faint stars, so astronomers have had to develop ways of recognizing them. That has revealed that there are different kinds of neutron stars.

The link between neutron stars and type II supernova explosions is an important clue. Astronomers conclude that the explosion of Supernova 1987A left behind a neutron star because instruments on Earth detected the burst of neutrinos produced when the core collapsed. The neutron star has not been detected yet because it is hidden at the center of the expanding shells of gas ejected into space; but, as the gas expands and thins, astronomers expect to be able to see it. If its beams don't sweep over Earth, astronomers might be able to detect it from its X-ray and gamma emission.

Of course, some neutron stars can be recognized by the pulsing radiation they emit. A pulsar is powered by its rapid rotation. As it blows away its pulsar wind and blasts beams of radiation outward, its rotation slows. The average pulsar is apparently only a few million years old, and the oldest is about 10 million years old. Presumably, older neutron stars rotate too slowly to generate detectable radio beams.

The Crab Nebula is an example of a young pulsar. The supernova was seen in AD 1054, so the neutron star is only about 950 years old. The Crab pulsar is so powerful it emits photons of radio, infrared, visible, X-ray, and gamma-ray wavelengths (■ Figure 14-7).

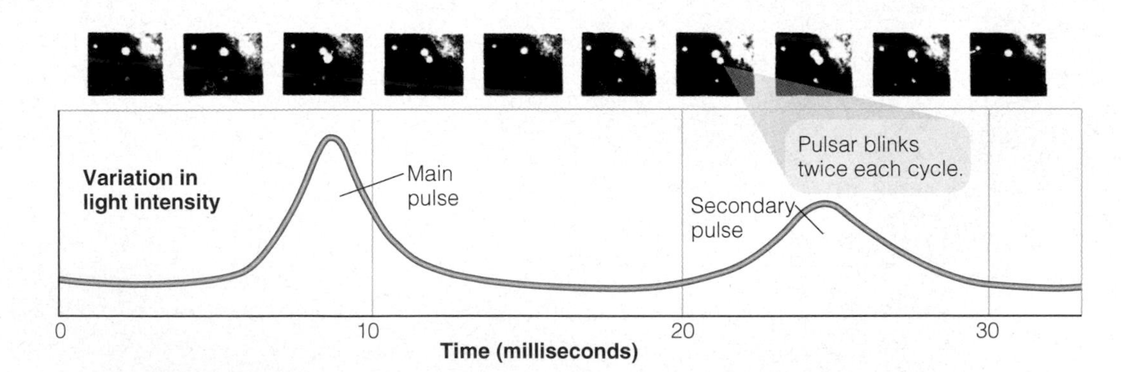

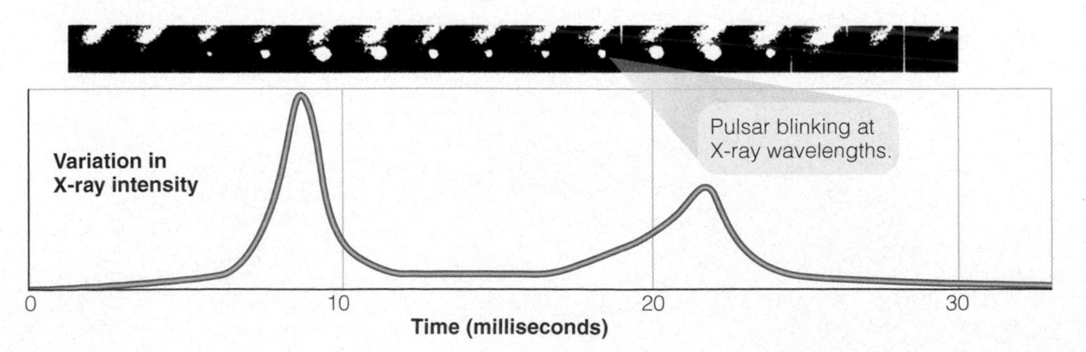

■ Figure 14-7

High-speed images of the Crab Nebula pulsar show it pulsing at visual wavelengths and at X-ray wavelengths. The period of pulsation is 33 milliseconds, and each cycle includes two pulses as its two beams of unequal intensity sweep over Earth. (Visual: © AURA, Inc., NOAO, KPNO; X-ray: F. R. Harnden, Jr., from The Astrophysical Journal, published by the University of Chicago Press; © 1984 The American Astronomical Society)

Careful measurements of its brightness with high-speed instruments show that it blinks twice for every rotation. One beam sweeps almost directly over Earth and produces a strong pulse. Half a rotation later, the edge of the other beam sweeps over Earth, producing a weaker pulse.

You would expect only the most energetic pulsars to produce short-wavelength photons and pulse at visible wavelengths. The Crab Nebula pulsar is young, fast, and powerful, and it produces visible pulses; so does a pulsar called the Vela pulsar (located in the Southern Hemisphere constellation Vela). The Vela pulsar is fast, pulsing about 11 times a second, and, like the Crab Nebula pulsar, is located inside a supernova remnant. Its age is estimated at about 20,000 to 30,000 years, young in terms of the average pulsar. This suggests that pulsars are capable of producing optical pulses only when they are young.

It is easy to recognize neutron stars such as the Crab and Vela pulsars when you see them inside a supernova remnant, but not every supernova remnant contains a pulsar, and not every pulsar is located inside a supernova remnant. In some cases the supernova remnant may contain a pulsar that we can't see; pulsar beams can be very narrow, and many supernova remnants probably contain pulsars whose beams never sweep over Earth. Additionally, pulsars are known to have such high velocities (■ Figure 14-8) that many leave their supernova remnants and even escape from the disk of our galaxy. These high velocities suggest that supernova explosions can occur asymmetrically, perhaps because of the violent turbulence in the exploding core; some supernovae that occur in binary systems probably fling the two stars apart at high velocity. Finally, although a pulsar can remain detectable for 10 million years or so, a supernova remnant cannot survive more than about 50,000 years before it is mixed into the interstellar medium. All of these factors explain why although pulsars are born in a supernova, they are not always found in a supernova remnant.

Not all neutron stars are pulsars. To some extent, this depends on whether beams of radio energy sweep over Earth, but it also depends on how the neutron star stores its energy. Pulsars are powered by their rapid rotation, and they slow down as they age. But some neutron stars emit more energy than they could generate by slowing down, so they must be powered by something other than rotation. Neutron stars called **magnetars** are powered by tremendous energy stored in magnetic fields that can be 1000 times stronger than those in a normal neutron star. **Anomalous X-ray pulsars** are magnetars that emit X rays but spin slowly with periods of 5 to 10 seconds. Evidence suggests their energy is stored in their powerful magnetic field. **Soft gamma-ray repeaters (SGRs)** are neutron stars that emit irregular bursts of lower energy (soft) gamma-rays. SGRs appear to be magnetars in which occasional shifts in the magnetic field break the rigid crust and trigger bursts of gamma rays (■ Figure 14-9). Although today only a handful of these objects are known, magnetars may make up over 10 percent of neutron stars.

Some neutron stars are difficult to recognize because they produce no radio pulses or bursts of energy. These stars seem to be powered by their internal heat. They emit X rays because they are hot. Some of these neutron stars are found isolated in space, but some are found inside supernova remnants including that inside Cassiopia A (see Figure 13-16).

One reason neutron stars are so fascinating is the peculiar physical processes found at their surfaces. When a neutron star is part of a binary system, the physics becomes even more extreme.

Binary Pulsars

Over 1500 neutron stars have been found, most of them pulsars. Those located in binary systems are of special interest because astronomers can learn more about the neutron star by studying

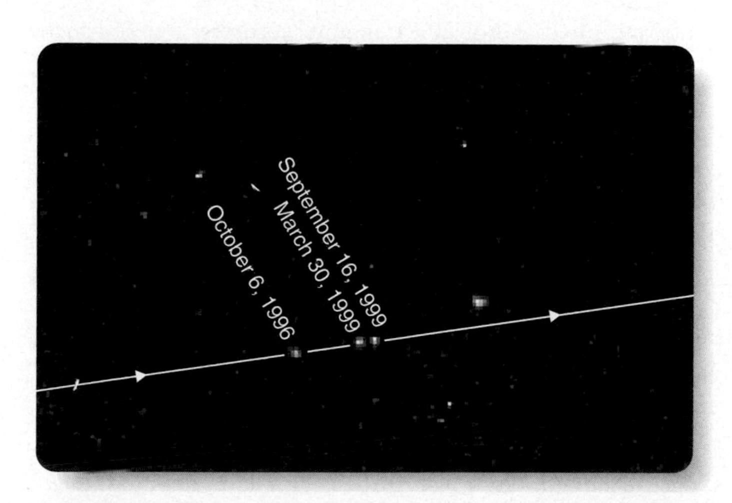

■ Figure 14-8

Many neutron stars have high velocities through space. Here the neutron star known as RX J185635-3754 was photographed on three different dates as it rushed past background stars. (NASA and F. M. Walter)

■ Figure 14-9

Some neutron stars appear to have magnetic fields up to 1000 times stronger than those in a normal neutron star. These magnetars can produce bursts of gamma rays when shifts in the magnetic field rupture the rigid crust of the neutron star. (NASA/CXC/M. Weiss)

the orbital motions of the binary. Also, in some cases, mass can flow from the companion star onto the neutron star, producing high temperatures and X rays.

The first binary pulsar was discovered in 1974 when astronomers Joseph Taylor and Russell Hulse noticed that the pulse period of the pulsar PSR 1913+16 was changing. The period first grew longer and then grew shorter in a cycle that took 7.75 hours. Thinking of the Doppler shifts seen in spectroscopic binaries, the radio astronomers realized that the pulsar had to be in a binary system with an orbital period of 7.75 hours. When the orbital motion of the pulsar carries it away from Earth, the pulses are slightly redshifted, and the period is slightly lengthened, just as the wavelength of light emitted by a receding source is lengthened. Then, when the pulsar rounds its orbit and approaches Earth, the pulse period is slightly shortened—a blueshift. From these changing Doppler shifts, the astronomers could calculate the radial velocity of the pulsar around its orbit just as if it were a spectroscopic binary star (Chapter 9). The resulting graph of radial velocity versus time could be analyzed to find the shape of the pulsar's orbit (■ Figure 14-10). When Taylor and Hulse analyzed PSR 1913+16, they discovered that the binary system consisted of two neutron stars separated by a distance roughly equal to the radius of our sun.

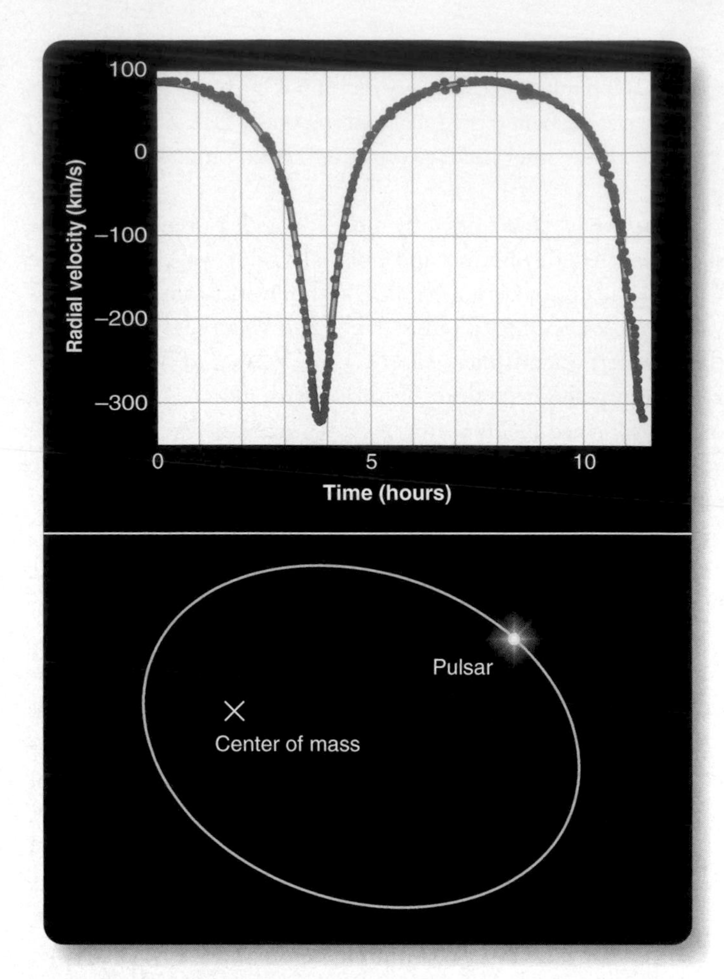

■ **Figure 14-10**

The radial velocity of pulsar PSR 1913+16 can be found from the Doppler shifts in its pulsation. Analysis of the radial velocity curve allows astronomers to determine the pulsar's orbit. Here, the center of mass does not appear to be at a focus of the elliptical orbit because the orbit is inclined. (Adapted from data by Joseph Taylor and Russell Hulse)

Yet another surprise was hidden in the motion of PSR 1913+16. Einstein's general theory of relativity, published in 1916, describes gravity as a curvature of space-time. Einstein realized that any rapid change in a gravitational field should spread outward at the speed of light as **gravitational radiation.** Gravity waves have not been detected yet, but Taylor and Hulse were able to derive indirect evidence for their existence from the binary pulsar. The orbital period of the binary pulsar is slowly growing shorter as the stars gradually spiral toward each other. They are radiating orbital energy away as gravitational radiation. Taylor and Hulse won the Nobel Prize in 1993 for their work with binary pulsars.

Dozens of binary pulsars have been found, and by analyzing the Doppler shifts in their pulse periods, astronomers can estimate the masses of the neutron stars. Typical masses are about 1.35 solar masses, in good agreement with models of neutron stars.

In 2004, radio astronomers announced the discovery of a double pulsar. The two pulsars orbit each other in only 2.4 hours, and their spinning beams sweep over Earth (■ Figure 14-11). One spins with a period of 0.023 second, and the other spins in 2.8 seconds. This system is a pulsar jackpot because the orbits are nearly edge on to Earth and the powerful magnetic fields eclipse each other, giving astronomers a chance to study their size and structure. Not only that, but the theory of general relativity predicts that they are emitting gravitational radiation and that their separation is decreasing by 7 mm per year. The two neutron stars will merge in 85 million years, presumably to trigger a violent explosion. In the meantime, the steady decrease in orbital period can be measured and gives astronomers a further test of general relativity and gravitational radiation.

Binary pulsars can emit strong gravitational waves because they contain large amounts of mass in a small volume. This also means that binary pulsars can be sites of tremendous violence because the strength of gravity at the surface of a neutron star is so extreme. An astronaut stepping onto the surface of a neutron star would be instantly smooshed into a layer of matter only 1 atom thick. Matter falling onto a neutron star can release titanic amounts of energy. If you dropped a single marshmallow onto the surface of a neutron star from a distance of 1 AU, it would hit with an impact equivalent to a 3-megaton nuclear warhead. In general, a particle falling from a large distance to the surface of a neutron star will release energy equivalent to 0.2 mc^2, where m is the particle's mass at rest. Even a small amount of matter flowing from a companion star to a neutron star can generate high temperatures and release X rays and gamma rays.

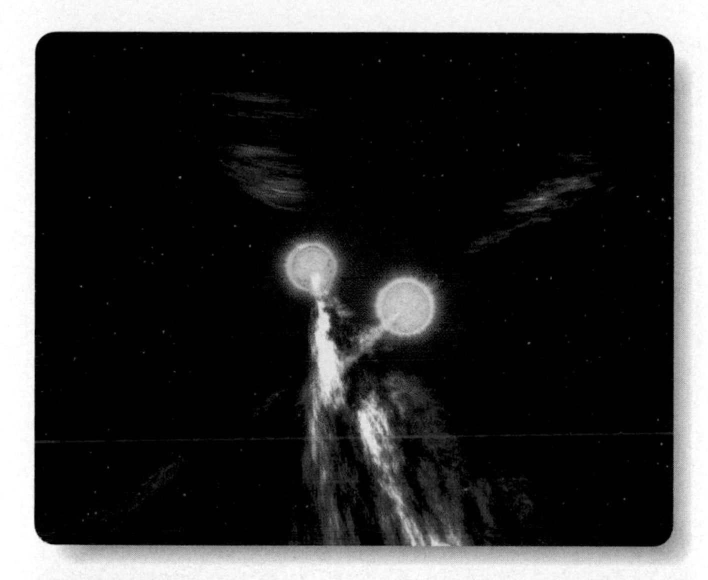

■ **Figure 14-11**

Artist's impression of the double pulsar. One star must have exploded to form a pulsar, and later the other star did the same. Gravitational radiation causes the neutron stars to drift toward each other, and they will merge in 85 million years, presumably to trigger another supernova explosion. (John Rowe Animations)

As an example of such an active system, consider Hercules X-1. It emits pulses of X rays with a period of about 1.2 seconds, but every 1.7 days the pulses vanish for a few hours (■ Figure 14-12). This behavior should remind you of an eclipsing binary star. Hercules X-1 seems to contain a 2-solar-mass star with a temperature of 7000 K and a neutron star. These two objects orbit each other with a period of 1.7 days. Matter flows from the normal star into an accretion disk around the neutron star, where it can reach temperatures of millions of degrees and emit a powerful X-ray glow. Interactions with the neutron star's magnetic field can produce beams of X rays that sweep around with the rotating neutron star. Earth receives a pulse of X rays every time a beam points our way, and the X rays shut off every 1.7 days when the neutron star is eclipsed behind the normal star. The X rays from the neutron star and its accretion disk heat the near side of the normal star to about 20,000 K. As the system rotates, astronomers on Earth alternately see the hot side of the star and then the cool side, and its brightness at visible wavelengths varies. Hercules X-1 is a complex system and is still not well understood, but this quick analysis shows you how complex and powerful such binary systems are during mass transfer.

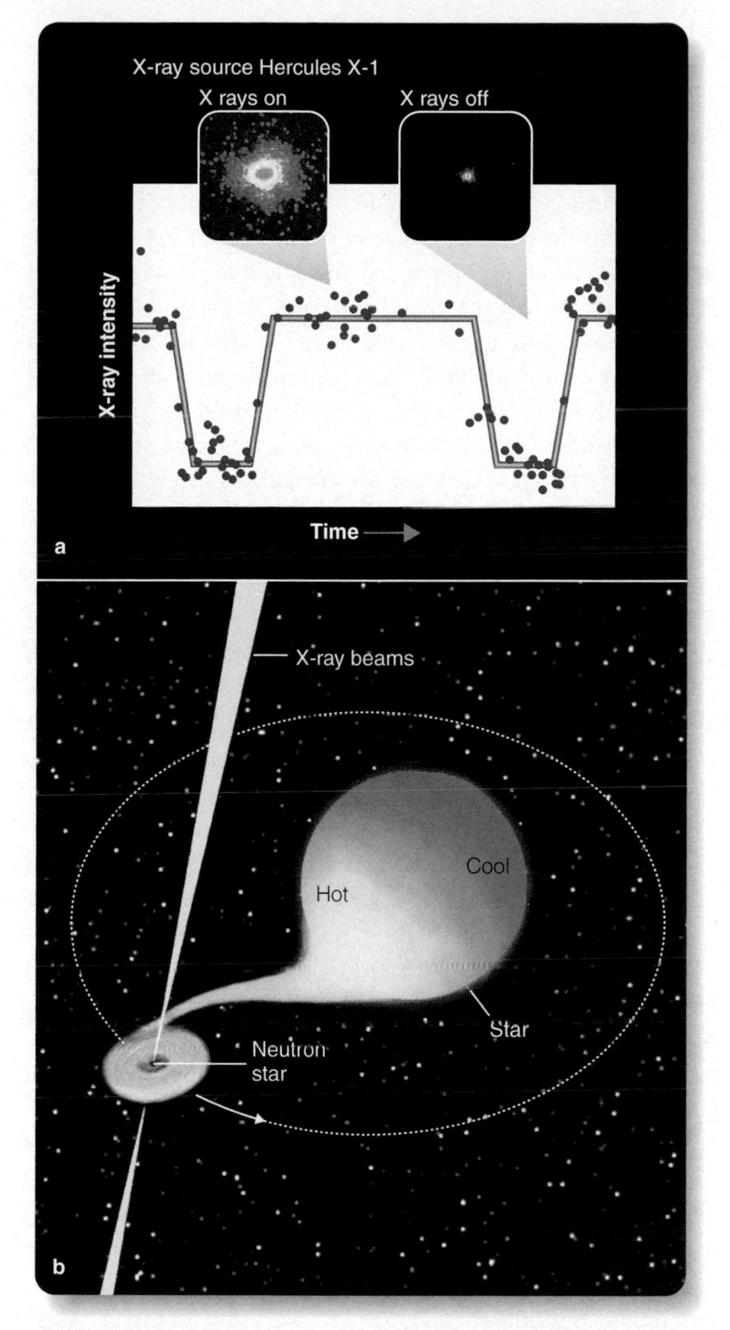

■ **Figure 14-12**

Sometimes the X-ray pulses from Hercules X-1 are on, and sometimes they are off. A graph of X-ray intensity versus time looks like the light curve of an eclipsing binary. (Insets: J. Trümper, Max-Planck Institute) (b) In Hercules X-1, matter flows from a star into an accretion disk around a neutron star producing X rays, which heat the near side of the star to 20,000 K compared with only 7000 K on the far side. X rays turn off when the neutron star is eclipsed behind the star.

The Fastest Pulsars

What you have learned about pulsars suggests that newborn pulsars should blink rapidly and old pulsars should blink slowly, but the handful that blink the fastest may be quite old. One of the fastest known pulsars is cataloged as PSR J1748-2446ad. It pulses 716 times a second and is slowing down only slightly. The energy stored in the rotation of a neutron star spinning this fast is equal to the total energy of a supernova explosion, so it seemed difficult at first to explain such fast pulsars. It now appears that such fast pulsars are old neutron stars that have gained mass and rotational energy from their companions in binary systems. Like water hitting a mill wheel, the matter falling on the neutron stars has spun them up to high speed. With their old, weak magnetic field, they slow down very gradually and will continue to spin for a very long time.

A number of these very fast pulsars have been found, and they are known generally as **millisecond pulsars** because their pulse periods are almost as short as a millisecond (0.001 s). This produces some fascinating physics because the pulse period of a pulsar equals the rotation period of the neutron star. If a neutron star 10 km in radius spins 716 times a second, as does PSR J1748-2446ad, then the period is 0.0014 second, and the equator of the neutron star must be traveling about 45,000 km/s. That is fast enough to flatten the neutron star into an ellipsoidal shape and is nearly fast enough to break it up.

All scientists should be made honorary citizens of Missouri, the "Show Me" state, because scientists demand evidence. The hypothesis that the millisecond pulsars are spun up by mass transfer from a companion star is quite reasonable, but astronomers demand reality checks, and supporting evidence has been found. For example, the pulsar PSR J1740−5340 has a period of 42 milliseconds and is orbited by a bloated red star from which it is gaining mass. This appears to be a pulsar being spun up to high speed. For another example, consider the X-ray source XTE J1751-305, a pulsar with a period of only 2.3 milliseconds. X-ray observations show that it is gaining mass from a companion star. The orbital period is only 42 minutes, and the mass of the companion star is only 0.014 solar mass, suggesting that this neutron star has devoured all but the last morsel of its binary partner.

Although some millisecond pulsars have binary companions, others are solitary. How did they get spun up if they don't have a companion star? A pulsar known as the Black Widow may provide an explanation. The Black Widow has a period of 1.6 milliseconds, meaning it is spinning 622 times per second, and it orbits with a low-mass companion. Presumably the neutron star was spun up by mass flowing from the companion star, but spectra show that the blast of radiation and high-energy particles from the neutron star are now boiling away the surface of the companion. The Black Widow has eaten its fill and is now evaporating the remains of its companion. It will soon be a solitary millisecond pulsar (■ Figure 14-13).

"Show me," say scientists; and, in the case of neutron stars, the evidence seems so strong that astronomers have great confidence that such objects really do exist. Other theories that describe how they emit beams of radiation and how they form and evolve are less certain, but continuing observations at many wavelengths are revealing more about these last embers of massive stars. In fact, observations have turned up objects no one predicted.

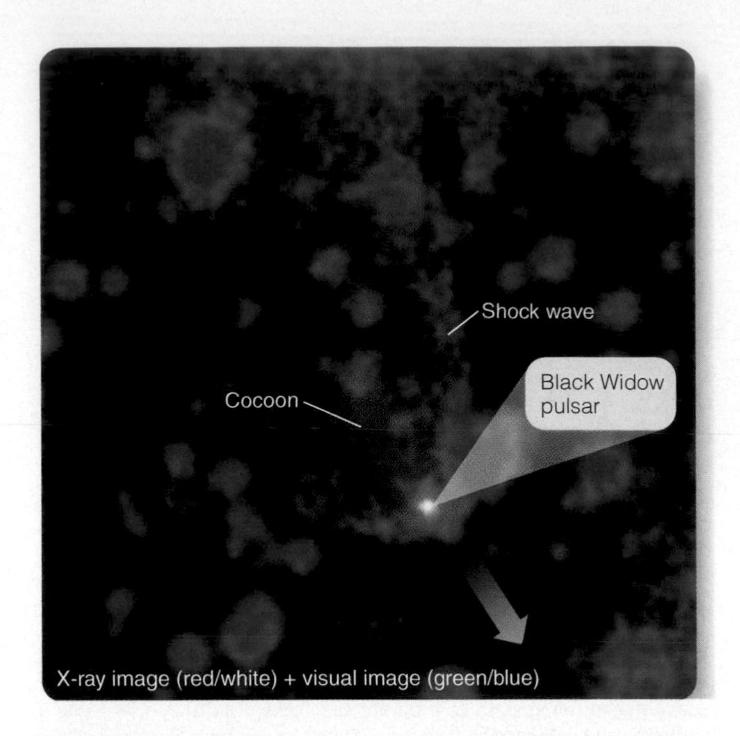

■ Figure 14-13

The Black Widow pulsar and its companion star are moving rapidly through space, creating a shock wave like the bow wave of a speedboat. The shock wave confines high-energy particles shed by the pulsar into an elongated cocoon (red). (X-ray: NASA/CXC/ASTRON/B. Stappers et al.; Optical: AAO/J. Bland-Hawthorn & H. Jones)

Pulsar Planets

Finding planets that orbit stars other than the sun is very difficult, and only about two hundred are known. Oddly, the first such planets were found orbiting a neutron star.

Because a pulsar's period is so precise, astronomers can detect tiny variations by comparison with atomic clocks. When astronomers checked pulsar PSR 1257+12, they found variations in the period of pulsation much like those observed in binary pulsars (■ Figure 14-14a). However, in the case of PSR 1257+12, the variations were much smaller; and, when they were interpreted as Doppler shifts, it became evident that the pulsar was being orbited by at least two planetlike objects of 4.3 and 3.9 Earth masses. The gravitational tugs of the planets make the pulsar wobble about the center of mass of the system by no more than 800 km producing the tiny changes in period. Compare the size of the velocity variations shown in Figure 14-14b with those shown in Figure 14-10.

Astronomers greeted this discovery with both enthusiasm and skepticism. As usual, they looked for ways to test the hypothesis. Simple gravitational theory predicts that the planets should interact and slightly modify each other's orbits. When the data were analyzed, that interaction was found, further confirming the hypothesis. In fact, further data revealed the presence of a third planet of about the mass of Earth's moon, and a fourth planet with a mass of about 100 Earth masses is now thought to follow a much larger orbit. This illustrates the astonishing precision of studies based on pulsar timing.

Astronomers wonder how a neutron star can have planets. The planets that orbit PSR 1257+12 are very close to the pulsar;

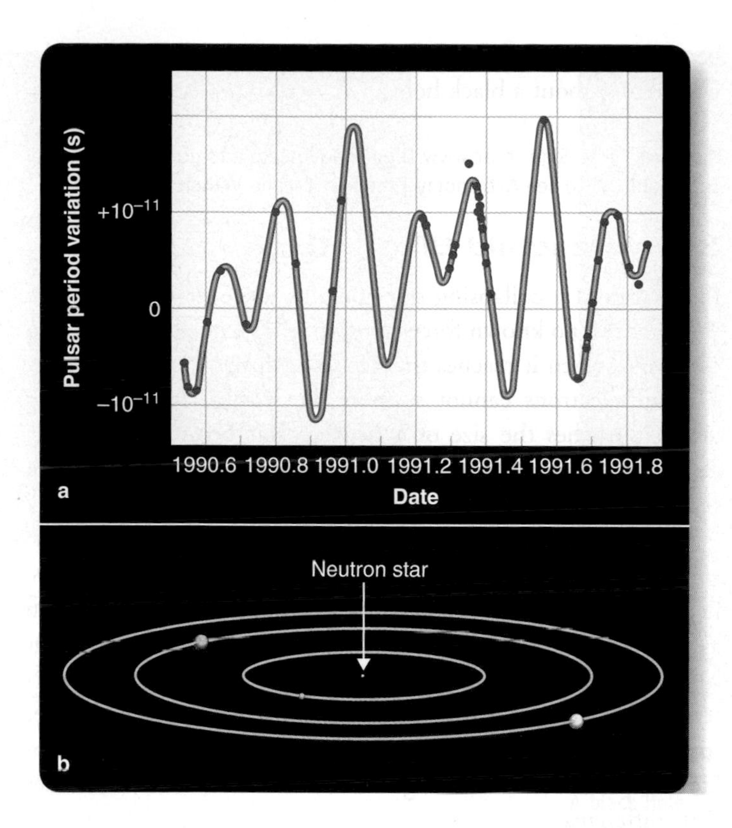

■ **Figure 14-14**

(a) The dots in this graph are observations showing that the period of pulsar PSR 1257+12 varies from its average value by a fraction of a billionth of a second. The blue line shows the variation that would be produced by planets orbiting the pulsar. (b) As the planets orbit the pulsar, they cause it to wobble by less than 800 km, a distance that is invisibly small in this diagram. (Adapted from data by Alexander Wolszczan)

the inner three orbit at 0.19 AU, 0.36 AU, and 0.47 AU—closer to the pulsar than Venus is to the sun. Planets that close should have been lost or vaporized when the star exploded. Furthermore, when the star was about to explode as a supernova it would have been a large giant or a supergiant, and these planets would have been inside the star and could not have survived. It seems more likely that the planets are the remains of a stellar companion that was devoured by the neutron star. In fact, the pulsar is very fast (162 pulses per second), suggesting that it was spun up in a binary system.

Another pulsar planet has been found in a binary system containing a neutron star and a white dwarf. Because this system is located in a very old star cluster and contains a white dwarf, astronomers suspect that the planet may be very old. Planets probably orbit other neutron stars, and small shifts in the timing of the pulses may eventually reveal their presence.

The planets formed from the remains of dying stars might be rich in heavy elements, but truly ancient planets might be poor in such elements. You can imagine visiting these worlds, landing on their surfaces, and hiking across their valleys and mountains. Above you, the neutron star would glitter in the sky, a tiny point of light.

◄ SCIENTIFIC ARGUMENT ►

Why are neutron stars detectable at X-ray wavelengths?

This argument draws together a number of ideas you know from previous chapters. First, you should remember that a neutron star is very hot because of the heat released when it contracts to a radius of 10 km. It could easily have a surface temperature of 1,000,000 K, and Wien's law (Chapter 7) tells you that such an object will radiate most intensely at a very short wavelength, typical of X rays. However, you know that the total luminosity of a star depends on its surface temperature and its surface area, and a neutron star is so small it can't radiate much energy. X-ray telescopes have found such neutron stars, but they are not easy to locate.

There is, however, a second way a neutron star can radiate X rays. If a normal star in a binary system loses mass to a neutron star companion, the inflowing matter will form a very hot accretion disk that can radiate intense X rays easily detectable by X-ray telescopes orbiting above Earth's atmosphere. Further, any matter that hits the surface of the neutron star will impact with so much energy that it will be heated to very high temperatures and will radiate X rays.

Now build a new argument as if you were seeking funds for a research project. **What observations would you make to determine whether a newly discovered pulsar was young or old, single or a member of a binary system, alone or accompanied by planets?**

◄ ►

14-2 Black Holes

YOU HAVE STUDIED WHITE DWARFS and neutron stars, two of the three end states of dying stars. Now you are ready to turn to the third—black holes.

Although the physics of black holes is difficult to describe without sophisticated mathematics, simple logic is sufficient to predict that they should exist. The problem is to consider their predicted properties and try to find objects in the heavens that could be real black holes. Even though it was more difficult than the search for neutron stars, the quest for black holes has also met with success.

To begin your study of black holes, consider a simple question. How fast must an object travel to escape from the surface of a celestial body?

Escape Velocity

Suppose you threw a baseball straight up. How fast would you have to throw it for it not to come down? Of course, gravity would pull back on the ball, slowing it, but if the ball were traveling fast enough to start with, it would never come to a stop and fall back but would instead escape from Earth. The escape velocity is the initial velocity an object needs to escape from a celestial body (■ Figure 14-15). (See Chapter 5.)

Whether you are discussing a baseball leaving Earth or a photon leaving a collapsing star, the escape velocity depends on two things, the mass of the celestial body and the distance from its center of mass to the escaping object. If the celestial body has

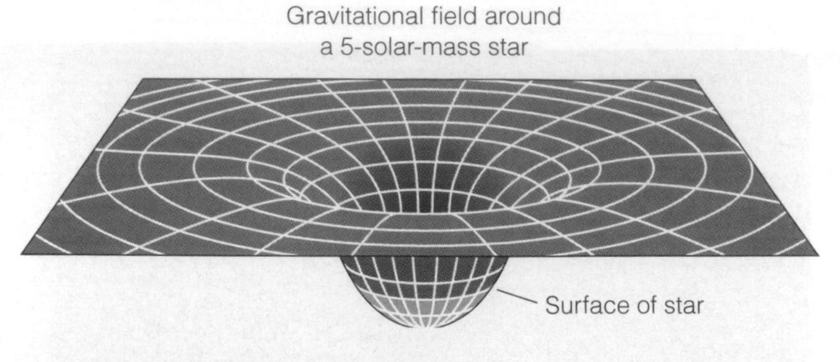

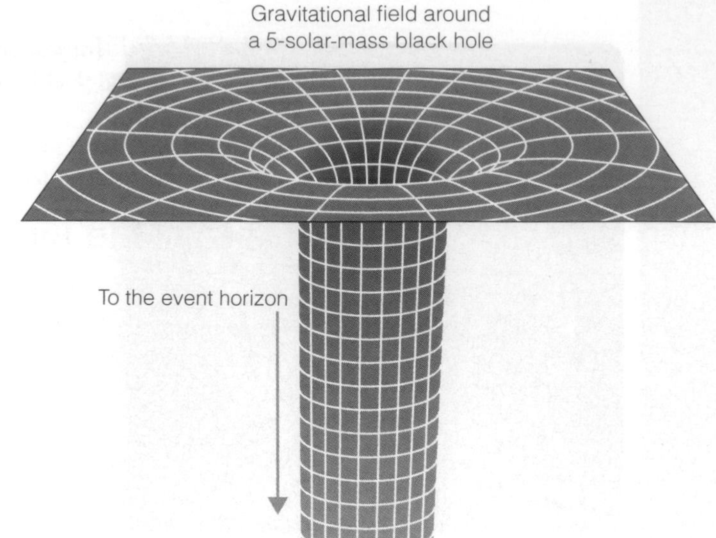

■ **Figure 14-18**

If you fell into the gravitational field of a star, you would hit the star's surface before you fell very far. Because a black hole is so small, you could fall much deeper into its gravitational field and eventually cross the event horizon. At a distance, the two gravitational fields are the same.

it. ■ Figure 14-18 illustrates this by representing gravitational fields as curvature of the fabric of space-time.

You can check off another **Common Misconception** when you look at Figure 14-18; black holes are not funnels. Physicists like to graph the strength of gravity around a black hole as a curvature in a flat sheet. The graphs look like funnels in which the depth of the funnel shows the strength of the gravitational field, but black holes themselves are not shaped like funnels. In Figure 14-18 you should note that the strength of the gravitational field around the black hole becomes extreme only if you venture too close.

ThomsonNOW Sign in at www.thomsonedu.com and go to ThomsonNOW to see Astronomy Exercise "Black Holes."

Black Holes Have No Hair

Theorists who study black holes are fond of saying, "Black holes have no hair." By that they mean that once matter falls into a black hole, it loses almost all of its normal properties. A black hole formed by a collapsed star would be indistinguishable from a black hole of the same mass made from peanut butter and fake-fur mittens. Once the matter is inside the event horizon, it retains only three properties—mass, angular momentum, and electrical charge.

The Schwarzschild black hole is represented by a solution to Einstein's equations for the special case where the object has only mass. Schwarzschild black holes do not rotate or have charge. The solutions for rotating or charged black holes (or for rotating, charged black holes) are more difficult and have been found in only the last few decades. Generally, rotating, charged black holes are similar to Schwarzschild black holes.

It seems that astronomers need not worry about charged black holes because stars, whose collapse presumably forms black holes, cannot have large electrostatic charges. If you could somehow give the sun a large positive charge, it would begin to repel protons in its corona and attract electrons and would soon return to neutral charge. For this reason, you can expect stars and black holes to be electrically neutral.

Charge is not important, but everything in the universe rotates, and collapsing stars spin rapidly as they conserve angular momentum. Consequently, you should expect black holes to have angular momentum. In 1963, New Zealand mathematician Roy P. Kerr found a solution to Einstein's equations that describes a rotating black hole. This is now known as the **Kerr black hole.**

The mass of a black hole curves neighboring space-time, and the Kerr solution predicts that the rotation of a black hole drags space-time around with it. The **ergosphere** is a region outside the event horizon in which space-time rotates with the rotating black hole so powerfully that nothing could avoid being dragged along. No one has ever approached a rotating black hole, so no one has any idea what it might feel like to enter a region where space-time was whirling past.

The word *ergosphere* comes from the Greek word *ergo,* meaning "work," because the rotating space-time in the ergosphere can do work on a particle; that is, the particle can gain energy. In particular, the Kerr solution shows that a particle that enters the ergosphere can break into two pieces, one that falls into the black hole and another that escapes with more energy than it had when it entered. In this way, energy can be extracted from a rotating black hole, and, as a result, the black hole slows its rotation very slightly.

The Kerr solution has an important application in astronomy. Almost certainly, black holes rotate, so matter falling into a black hole must pass through the ergosphere. This suggests that you should expect to find situations where energy is extracted from rotating black holes.

A Leap into a Black Hole

Before you can search for real black holes, you need to understand what theory predicts about their appearance. To begin, imagine that you leap, feet first, into a Schwarzschild black hole.

If you were to leap into a black hole of a few solar masses from a distance of an astronomical unit, the gravitational pull would not be very large, and you would fall slowly at first. Of course, the longer you fell and the closer you came to the center, the faster you would travel. By the time you approached the event horizon, your wristwatch would tell you that you had been falling for about 65 days.

Your friends who stayed behind would see something different. They would see you falling more slowly as you came closer to the event horizon because, as explained by general relativity, clocks slow down in curved space-time. This is known as **time dilation.** In fact, your friends would never actually see you cross the event horizon. To them, you would fall more and more slowly until you seemed hardly to move. Generations later, your descendants could focus their telescopes on you and see you still inching closer to the event horizon. You, however, would sense no slowdown and would conclude that you had crossed the event horizon after only about 65 days.

Other relativistic effects would make it difficult for your descendents to see you. As light travels out of a gravitational field, it loses energy, and its wavelength grows longer. This is known as a **gravitational redshift** (■ Table 14-2). Light leaving you and traveling away from the black hole would suffer a larger and larger gravitational redshift. In addition, as your inward velocity grew higher and higher, a relativistic effect would cause more and more of the light leaving you to be emitted in the forward direction into the black hole. This would reduce the amount of light leaving you and traveling outward, making you even more difficult to see. Although you would notice none of these effects as you fell toward the black hole, your friends would need to observe at longer wavelengths and with larger telescopes to detect you.

While these relativistic effects seem merely peculiar, other effects would be quite unpleasant. Your feet, which would be closer to the black hole, would be pulled in more strongly than your head. This is a tidal force, and at first it would be minor. But as you fell closer, the tidal force would become very large. Another tidal force would compress you as your left side and your right side both fell toward the center of the black hole. For any black hole with a mass like that of a star, the tidal forces would crush you laterally and stretch you longitudinally long before you reached the event horizon (■ Figure 14-19). The friction from such severe distortions of your body would heat you to millions of degrees, and you would emit X rays and gamma rays.

■ Table 14-2 | The Gravitational Redshift

Object	Redshift (percent)
Sun	0.0002
White dwarf	0.01
Neutron star	20
Black hole event horizon	Infinite

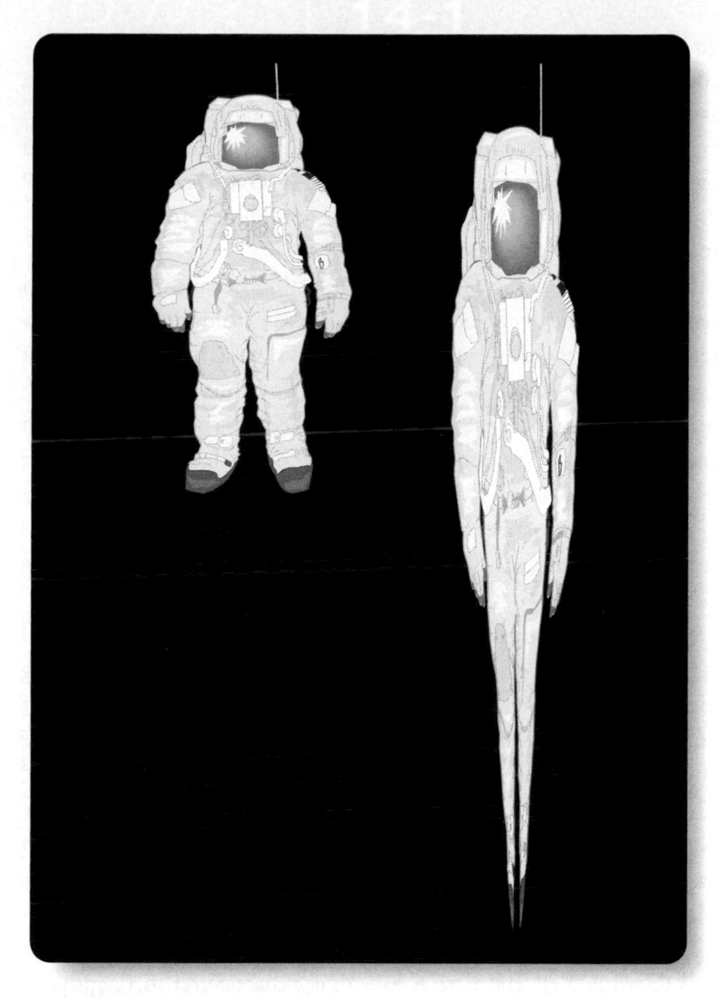

■ Figure 14-19

Leaping feet first into a black hole, a person of normal proportions (left) would be distorted by tidal forces (right) long before reaching the event horizon around a typical black hole of stellar mass. Tidal forces would stretch the body lengthwise while compressing it laterally. Friction from this distortion would heat the body to high temperatures.

(Needless to say, this would render you inoperative as a thoughtful observer.)

Some years ago a popular book suggested that you could travel through the universe by jumping into a black hole in one place and popping out of another somewhere far across space. That might make for good science fiction, but tidal forces would make it an unpopular form of transportation even if it worked. You would certainly lose your luggage.

Your imaginary leap into a black hole was not frivolous. You now know how to find a black hole: Look for a strong source of X rays. It may be a black hole into which matter is falling.

The Search for Black Holes

Do black holes really exist? Beginning in the 1970s, astronomers searched for observational evidence that their theories were correct. They tried to find one or more objects that were obviously

hole will probably not have much matter falling in, but black holes in binary systems may have large amounts of matter flowing in from the companion star. Consequently, the best place to search for black holes is in X-ray binaries.

A good scientific argument includes both theory and evidence. Expand your argument to include observations. **What observations would you make of an X-ray binary system to distinguish between a black hole and a neutron star?**

◄ ►

14-3 Compact Objects with Disks and Jets

NEUTRON STARS AND BLACK HOLES seem to be exotic objects, and they generate equally exotic phenomena. By studying those phenomena, you can learn more about the strange objects.

X-Ray Bursters

Beginning in the 1970s, X-ray telescopes revealed that some objects emit irregular bursts of X rays. Typically, bursts that follow a long quiet period are especially large (■ Figure 14-21), and this suggests that some mechanism accumulates energy that is then released by the bursts. The longer the quiet phase, the more energy accumulates.

Dozens of these **X-ray bursters** are known, and they are thought to be binaries in which hydrogen-rich matter flows from a normal star into an accretion disk and then onto the surface of a neutron star. When the gas reaches the neutron star, the hydrogen fuses into helium, which accumulates in a degenerate layer. When the helium reaches a depth of about a meter, it fuses explosively into carbon and produces an X-ray burst. The rapid increase in brightness (in a few seconds) and the total amount of energy produced (up to 100,000 times the luminosity of the sun) fit well with theoretical models of an explosion on the surface of an object as small as a neutron star. Of course, mass continues to accumulate, producing burst after burst. Notice the similarity with the mechanism that produces nova explosions on the surfaces of white dwarfs.

Observations of rapid oscillations during bursts suggest that the fusion begins at a small spot on the spinning neutron star and is confined by the same forces that confine storms on Earth. As this nuclear fusion hurricane spreads over the surface, the rapid rotation causes the high frequency oscillations.

X-ray telescopes are revealing more about these strange systems. Rare superbursts 1000 times brighter and longer than normal bursts appear to be caused by the fusion of accumulated carbon left over from helium fusion. The neutron star in burster 4U 1820-30 (■ Figure 14-22) is pulling mass away from a white dwarf, so it is accumulating helium and not hydrogen.

Accretion Disk Observations

When material falls into a neutron star or black hole it forms an accretion disk, and high-speed observations have been able to reveal some of the processes that go on in such disks.

When a blob of matter gets caught in an accretion disk, it can orbit so fast that its orbital period is measured in thousandths of a second, and because it is hot, it emits X rays. X-ray telescopes can detect these blobs as a rapid X-ray flicker. Because the blob of matter loses energy through friction, it spirals inward, its orbital period grows shorter, and the frequency of the flicker increases. In seconds, the blob falls into the central object, and the burst of flickers ends. Some accretion disks produce flickers with a period as short as 0.00075 second, as blobs of material orbit only tens of kilometers from the center of the compact object. Because the flickers don't last long and because the period grows rapidly shorter, they are called **quasi-periodic oscillations (QPOs).** Do not confuse these rapid flickering pulses with the regular click of pulses emitted by a pulsar.

QPOs can be observed in J1550-564, a star shedding mass into an accretion disk around a black hole. The flow of mass is irregular and causes X-ray flares. Short flickering strings of pulses are seen with periods as short as 0.003 second, and the period decreases rapidly. This suggests the energy is being emitted by material in the accretion disk moving rapidly inward toward orbits of shorter and shorter period.

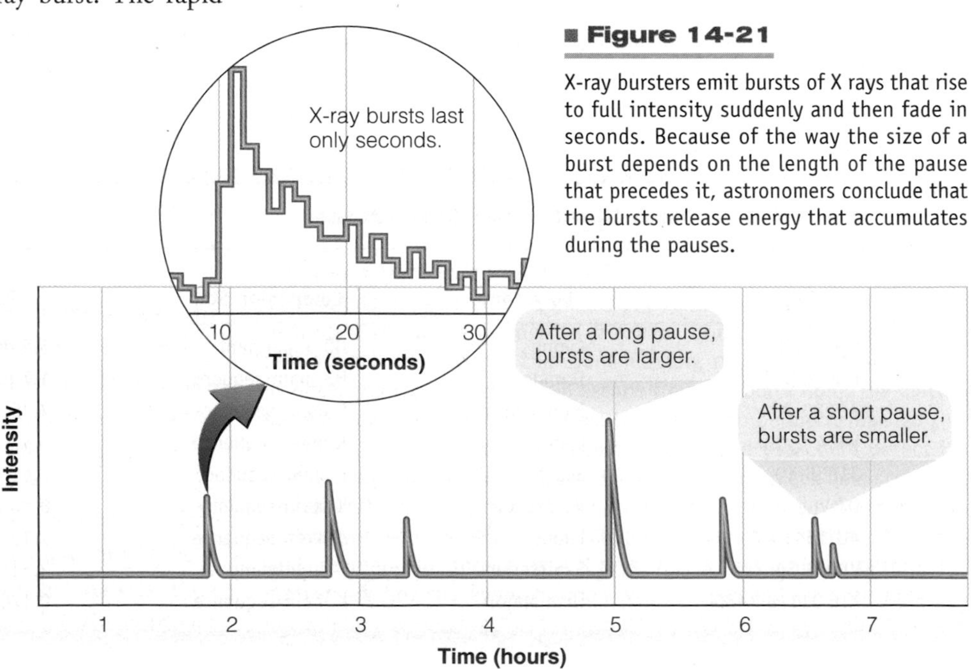

■ **Figure 14-21**

X-ray bursters emit bursts of X rays that rise to full intensity suddenly and then fade in seconds. Because of the way the size of a burst depends on the length of the pause that precedes it, astronomers conclude that the bursts release energy that accumulates during the pauses.

■ **Figure 14-22**

(a) At visible wavelengths, the center of star cluster NGC6624 is crowded with stars. (b) In the ultraviolet, one object stands out, the X-ray burster 4U 1820–30, consisting of a neutron star orbiting a white dwarf. (c) An artist's conception shows matter flowing from the white dwarf into an accretion disk around the neutron star. (a and b, Ivan King and NASA/ESA; c, Dana Berry, STScI)

Cyg X-1 appears to be a star shedding mass into a black hole of between 3.8 and 10 solar masses. Flickering pulse trains have been detected with periods of 0.018 second. The pulses grow dimmer and faster, which seems to mean the material is spiraling into the black hole. But no burst of energy is seen from an impact, as you would expect of material falling into a neutron star. Instead, the pulses fade away as the material approaches the event horizon, and a powerful gravitational redshift stretches the photons to such long wavelengths they cannot be detected.

Further evidence of an event horizon was found in a detailed study of 12 X-ray binaries. Six consisted of binary systems containing neutron stars. In these systems, bursts of energy were produced by infalling material striking the surfaces of the neutron stars. In contrast, six of the novae were systems containing black holes, and no bursts were seen from impacts on a surface. Apparently, in these systems, the material spiraled inward and disappeared as it approached the event horizons (■ Figure 14-23).

Because space-time is so strongly curved near a compact object, very small orbits are unstable. Once material spirals inward to that smallest possible orbit, it must quickly fall into the object. Yet the Rossi X-ray Timing Explorer satellite found an X-ray binary containing a black hole in which a QPO had a period of 0.002 second. The smallest possible orbit around the black hole in that system has a period of a little over 0.003 second. This appears to be evidence that the spinning black hole is dragging space-time with it as theory predicts and making smaller, shorter-period orbits stable. Astronomers are searching for more of these very-short-period QPOs, because they confirm the Kerr solution's prediction that spinning black holes drag space-time.

High-speed observations at many wavelengths are allowing astronomers to follow the rapid processes that occur inside accretion disks. Those same accretion disks can eject high-energy jets out into space.

Jets of Energy from Compact Objects

It is a **Common Misconception** that it is impossible to get any energy out of a black hole. As you have seen both black holes and neutron stars have intense gravitational fields, and matter flowing into those fields forms accretion disks heated to such high temperatures by friction that the inner regions can emit X rays and gamma rays. Those same accretion disks can emit high-energy jets.

Although the process isn't well understood yet, it seems to involve magnetic fields that get caught in the accretion disk and are twisted into tightly wound tubes that squirt gas and high energy radiation out of the disk and confine it in narrow beams (■ Figure 14-24).

You have seen in the X-ray image on page 293 that the Crab Nebula pulsar is ejecting jets of highly excited gas. The Vela pulsar does the same (look again at Figure 14-5). Systems containing black holes can also eject jets. The black hole candidate J1655-40 is observed at radio wavelengths to be sporadically ejecting oppositely directed jets at 92 percent the speed of light.

High-energy jets produced by accretion disks are a common phenomenon, and many such systems are known. The process that produces them is almost certainly related to the lower-energy process that produces bipolar flows from the disks around

A Hypernova Explosion

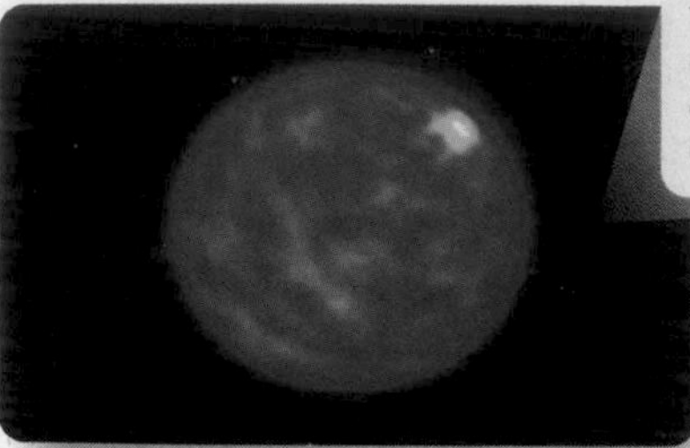

The collapsing core of a massive star drives its energy along the axis of rotation because. . .

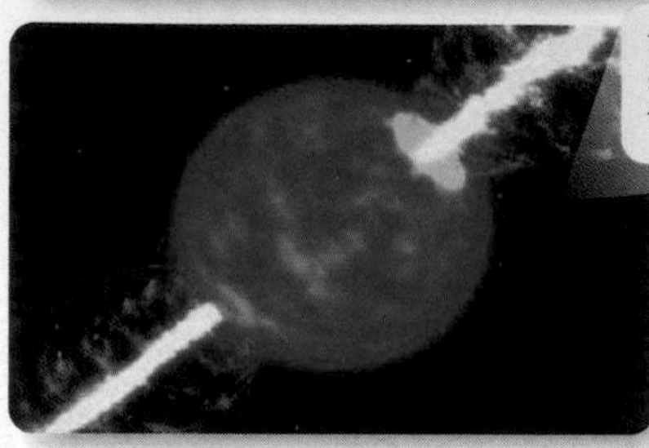

the rotation of the star slows the collapse of the equatorial regions.

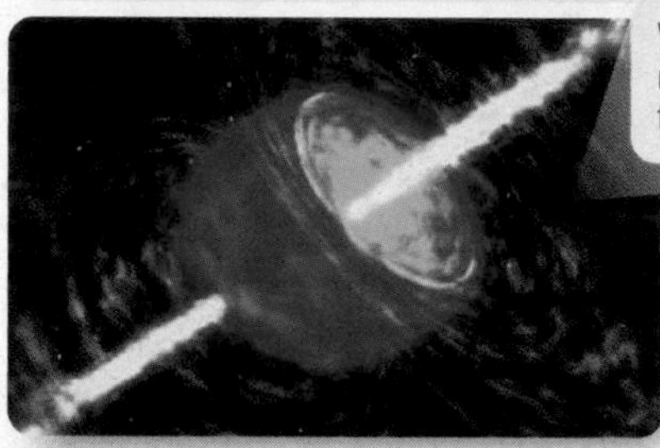

Within seconds, the remaining parts of the star fall in.

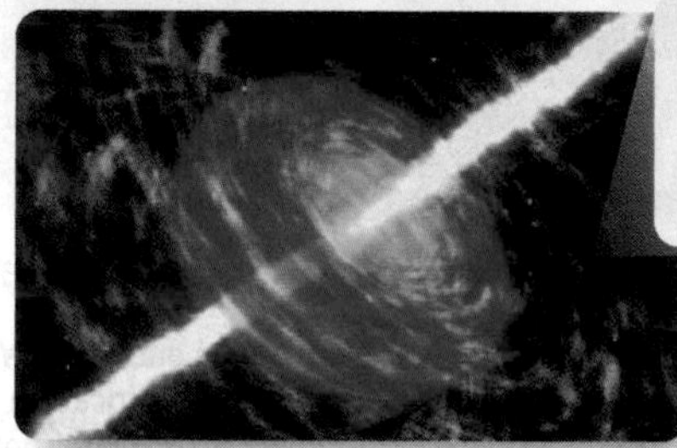

Beams of gas and radiation strike surrounding gas and generate beams of gamma rays.

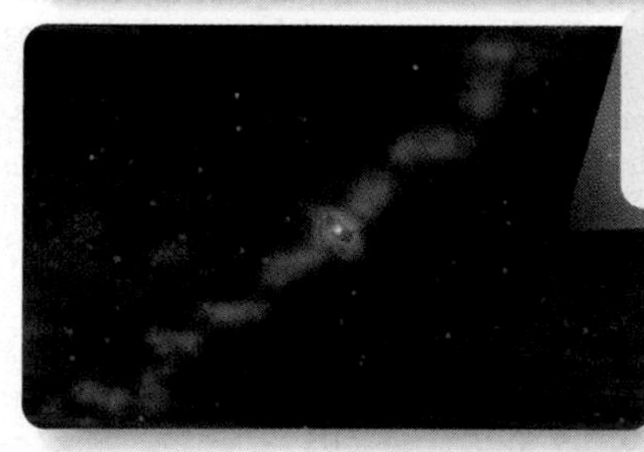

The gamma-ray burst fades in seconds, and a hot accretion disk is left around the black hole.

■ Active Figure 14-26

The collapse of the cores of extremely massive stars can produce hypernova explosions, which are thought to be the source of at least some gamma-ray bursts. (NASA/Skyworks Digital)

megaton nuclear blast. (The largest bombs ever made were a few megatons.) The gamma rays could create enough nitric oxide in the atmosphere to produce intense acid rain and could destroy the ozone layer, exposing life on Earth to deadly levels of solar ultraviolet radiation. Gamma-ray bursts may occur relatively near the Earth as often as every few hundred million years and could be one of the causes of the mass extinctions that show up in the fossil record.

It may seem odd that hypernovae and merging neutron stars are so common that gamma-ray telescopes observe one or more every day. There may be 30,000 neutron star binaries in each galaxy, so mergers must occur now and then in each galaxy. Massive stars explode as hypernovae about once in 10 million years in any one galaxy. These are rare events, but remember that gamma-ray bursts are so powerful they can be detected over huge distances, and that includes billions of galaxies. The evidence shows that the 1000 or so gamma-ray bursts that the Compton Observatory detected each year appear to be coming from the violent deaths of stars far across space.

◄ SCIENTIFIC ARGUMENT ►

Why do fluctuations from accretion disks have such short periods?

"Show me the evidence," say scientists, so scientific arguments always fall back on observations. When astronomers see an X-ray flicker coming from an accretion disk, they are seeing the effects of a blob of hot gas orbiting the neutron star or black hole at a very small radius. Earth takes a year to orbit the sun once, but it is very far from the sun. A blob of gas in an accretion disk around a neutron star is orbiting an object with a mass of about a solar mass, but it may have an orbit only a few tens of kilometers in radius. The orbital period must be very short, as you would expect from Kepler's third law.

The flickers coming from accretion disks tell of fantastic processes going on there. Extend your argument. **Why do these trains of pulses become more rapid as they fade away?**

◄ ►

What Are We?

Abnormal

Look around. What do you see? A table, a chair, a tree? It's all normal stuff. The world we live in is familiar and comfortable, but astronomy reveals that "normal" isn't normal at all. The universe is, for the most part, utterly unlike anything you have ever experienced.

Throughout the universe, gravity makes clouds of gas form stars, and in turn the stars generate energy through nuclear fusion in their cores, which delays gravity's final victory. Gravity always wins. You have learned that stars of different masses die in different ways, but you have also discovered that they always reach one of three end states: white dwarfs, neutron stars, or black holes. However strange these compact objects seem to you, they are very common.

The physics of compact objects is extreme and violent. You are not accustomed to objects as hot as the surface of a neutron star, and you have never experienced a black hole, a place where gravity is so strong it would pull you to pieces.

The universe is filled with things that are so violent and so peculiar they are almost unimaginable, but they are so common they deserve the label "normal." Next time you are out for a walk, look around and notice how beautiful Earth is and recall how unusual it is compared to the rest of the universe.

Study and Review

Summary

14-1 | Neutron Stars

How did scientists predict the existence of neutron stars?

- When a supernova explodes, the core collapses to very small size. Theory predicts that protons and electrons will combine to form a degenerate neutron gas.
- The collapsing core cannot support itself as a white dwarf if its mass is greater than 1.4 solar masses, the Chandrasekhar limit. If its mass lies between 1.4 solar masses and about 3 solar masses, it can halt its contraction and form a **neutron star.**
- A neutron star is supported by the pressure of the degenerate gas of neutrons. Theory predicts that a neutron star should be about 10 km in radius, spin very fast because it conserves angular momentum as it contracts, have a high temperature, and have a powerful magnetic field.

What is the evidence that neutron stars really exist?

- **Pulsars,** rapidly pulsing radio sources, were discovered in 1967 and were eventually understood to be spinning neutron stars. The discovery of a pulsar in the supernova remnant called the Crab Nebula was a key link in the story.
- Pulsars do not really blink. As described by the **lighthouse model,** pulsars are spinning neutron stars that emit beams of radiation that sweep around the sky; if the beams sweep over Earth, pulses can be detected.
- A spinning neutron star slows as it radiates its energy into space. Most of the energy emitted by a pulsar is carried away as a **pulsar wind.**
- Sudden decreases in period called **glitches** appear to be caused by breaks in the rigid neutron star crust or by changes in internal circulation.
- Some neutron stars called **magnetars** are powered by intense magnetic fields. This may explain the **anomalous X-ray pulsars,** which are energetic but rotate slowly. Shifts in these magnetic fields can break the rigid crust and may explain the **soft gamma-ray repeaters (SGR).**
- Many pulsars have been found in binary systems. In some, mass flows into a hot accretion disk around the neutron star and causes the emission of X rays.
- Observations of the first binary containing two neutron stars revealed that the system is losing energy by radiating **gravitational radiation.**
- The fastest pulsars, the **millisecond pulsars,** appear to be old pulsars that have been spun up to high speed by mass flowing from binary companions.
- Planets have been found orbiting at least one neutron star. They may be the remains of a companion star that was mostly devoured by the neutron star.

14-2 | Black Holes

How did scientists predict the existence of black holes?

- If the collapsing core of a supernova has a mass greater than 3 solar masses, then degenerate neutrons cannot stop the contraction, and it must contract to a very small size—perhaps to a **singularity,** an object of zero radius. Near such an object, gravity is so strong that not even light can escape, and the region is called a **black hole.**
- The outer boundary of a black hole is the **event horizon;** no event inside is detectable. The radius of the event horizon is the **Schwarzschild radius,** amounting to only a few kilometers for a black hole of a few solar masses.
- Once matter falls into a black hole, it loses all of its properties except for mass, electrical charge, and angular momentum.
- Rotating black holes are called **Kerr black holes** after the mathematician who solved the equations that describe them. The solution predicts a region called the **ergosphere.** A particle breaking up inside the ergosphere can extract energy from the black hole if half falls in and half escapes.

What is the evidence that black holes really exist?

- If you were to leap into a black hole, your friends who stayed behind would see two relativistic effects. They would see your clock slow relative to their own clock because of **time dilation.** Also, they would see your light redshifted to longer wavelengths because of a **gravitational redshift.**
- You would not notice these effects, but you would feel powerful tidal forces that would deform and heat your mass until you grew hot enough to emit X rays. Any X rays your mass emitted before your mass reached the event horizon could escape.
- To search for black holes, astronomers look for binary star systems in which mass flows into a compact object and emits X rays. If the mass of the compact object is greater than about 3 solar masses, then the object cannot be a neutron star and is presumably a black hole. A number of such objects have been located.

14-3 | Compact Objects with Disks and Jets

What happens when matter falls into a neutron star or black hole?

- **X-ray bursters** appear to be binary systems in which mass transfer deposits matter on the surface of a neutron star. When helium fusion ignites, the surface explodes and produces a burst of X rays.
- Astronomers can observe rapid flickering called **quasi-periodic oscillations (QPOs)** produced by blobs of gas orbiting very rapidly and spiraling inward in accretion disks. Bursts of energy are seen when such matter hits the surface of neutron stars, but no bursts are seen if the matter approaches the event horizon around a black hole.
- Rapidly spinning accretion disks around neutron stars or black holes can twist magnetic fields into tubes and eject narrow, powerful jets of radiation and matter in a process that is not yet well understood.
- **Gamma-ray bursters** appear to be related to violent events involving neutron stars and black holes. Many bursts appear to arise during **hypernovae** (also called **collapsars**), the collapse of the most massive stars to form black holes and eject powerful beams of gamma rays. Some bursts may be caused by the merger of two neutron stars, but such events have not been confirmed by direct observations.

Review Questions

ThomsonNOW™ Assess your understanding of this chapter's topics with additional quizzing and animations at www.thomsonedu.com.

WebAssign The problems from this chapter may be assigned online in WebAssign.

1. Why is there an upper limit to the mass of neutron stars? Why is that upper limit not well known?
2. Explain in detail why you would expect neutron stars to be hot, spin fast, and have strong magnetic fields.
3. Why can't astronomers use visual-wavelength telescopes to locate neutron stars?
4. Why does the short length of pulsar pulses eliminate normal stars as possible pulsars?
5. What do you mean when you say, "Every pulsar is a neutron star, but not every neutron star is a pulsar"?
6. According to the modern model of a pulsar, if a neutron star formed with no magnetic field at all, could it be a pulsar? Why or why not?
7. Why did astronomers first assume that the millisecond pulsar was very young?
8. Why would you suspect that only very fast pulsars can emit visible pulses?
9. If the sun were replaced by a 1-solar-mass black hole, how would Earth's orbit change?
10. If the sun has a Schwarzschild radius, why isn't it a black hole?
11. How can a black hole emit X rays?
12. What do theorists mean when they say, "Black holes have no hair"?
13. What evidence can you cite that black holes really exist?
14. How can mass transfer into a compact object produce jets of high-speed gas?
15. How does an X-ray burster resemble a nova?
16. What observational evidence can you cite to show that black holes do have event horizons?
17. Discuss the possible causes of gamma-ray bursts.
18. **How Do We Know?** Why do scientists conclude that good science must be repeatable?
19. **How Do We Know?** How does peer review make fraud rare in science?

Discussion Questions

1. Has the existence of neutron stars been sufficiently tested to be called a theory, or should it be called a hypothesis? What about the existence of black holes?
2. Why would you expect an accretion disk around a star the size of the sun to be cooler than an accretion disk around a compact object?
3. In this chapter, you imagined what would happen if you jumped into a Schwarzschild black hole. From what you have read, what do you think would happen to you if you jumped into a Kerr black hole?

Problems

1. If a neutron star has a radius of 10 km and rotates 716 times a second, what is the speed of the surface at the neutron star's equator in terms of the speed of light?
2. Suppose that a neutron star has a radius of 10 km and a temperature of 1,000,000 K. How luminous is it? (*Hint:* See Chapter 9.)
3. A neutron star and a white dwarf have been found orbiting each other with a period of 11 minutes. If their masses are typical, what is their average separation? Compare the separation with the radius of the sun, 7×10^5 km. (*Hint:* See Chapter 9.)
4. If the accretion disk around a neutron star has a radius of 2×10^5 km, what is the orbital velocity of a particle at its outer edge? (*Hint:* Use circular velocity, Chapter 5.)
5. What is the escape velocity from the surface of a typical neutron star? How does that compare with the speed of light? (*Hint:* See Chapter 5.)
6. If Earth's moon were replaced by a typical neutron star, what would the angular diameter of the neutron star be as seen from Earth? (*Hint:* Use the small-angle formula, Chapter 3.)
7. If the inner accretion disk around a black hole has a temperature of 1,000,000 K, at what wavelength will it radiate the most energy? What part of the spectrum is this in? (*Hint:* Use Wien's law, Chapter 7.)
8. What is the orbital period of a bit of matter in an accretion disk 2×10^5 km from a 10-solar-mass black hole? (*Hint:* Use circular velocity, Chapter 5.)
9. If an X-ray binary consists of a 20-solar-mass star and a neutron star orbiting each other every 13.1 days, what is their average separation? (*Hint:* See Chapter 9.)

Learning to Look

1. The X-ray image at the right shows the supernova remnant G11.2−0.3 and its central pulsar in X rays. The blue nebula near the pulsar is caused by the pulsar wind. How old do you think this system is? Discuss the appearance of this system a million years from now.

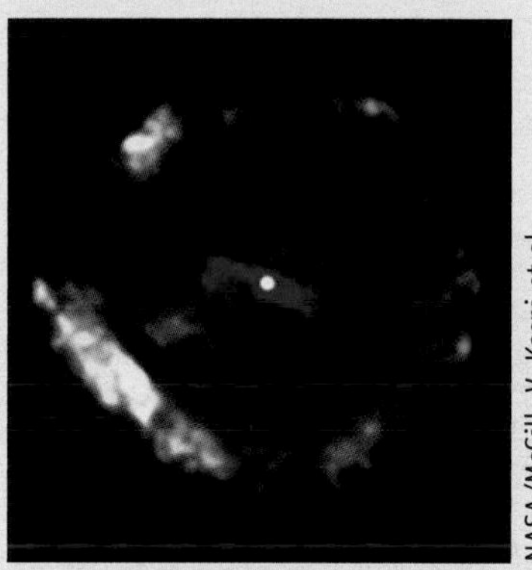

NASA/McGill, V. Kaspi et al.

2. What is happening in the artist's impression at the right? How would you distinguish between a neutron star and a black hole in such a system?

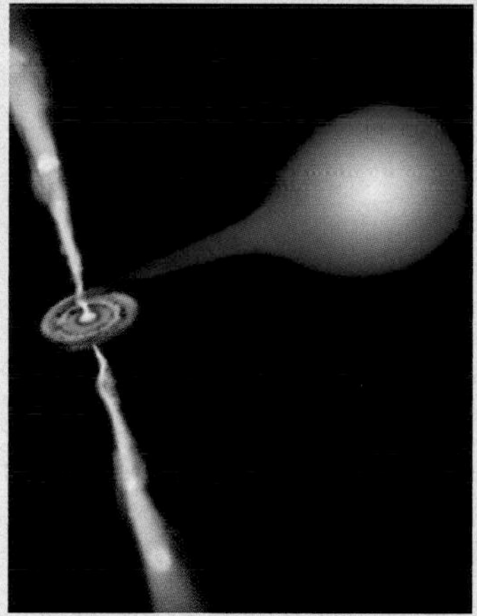

CXC/M. Weiss

Virtual Astronomy Labs

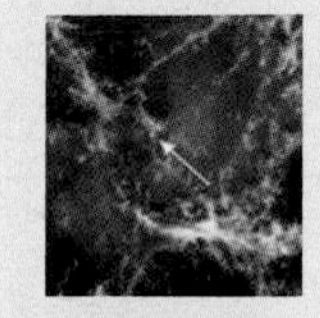

Lab 14: Neutron Stars and Pulsars
This lab looks at the extraordinary properties of neutron stars, the dense balls of neutrons that may remain after some stars have exploded. Later the lab examines pulsars, neutron stars that appear to emit radiation in rapid pulses.

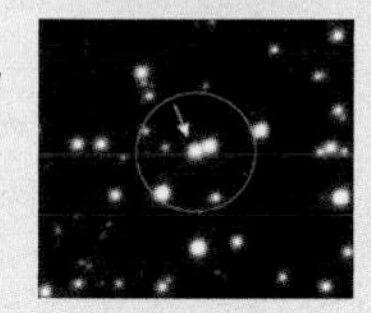

Lab 15: General Relativity and Black Holes
This lab explores the properties of black holes. It includes an exercise illustrating Einstein's general theory of relativity and an exercise on binary quasars. The lab concludes with a discussion and an exercise on black hole detection.

15 The Milky Way Galaxy

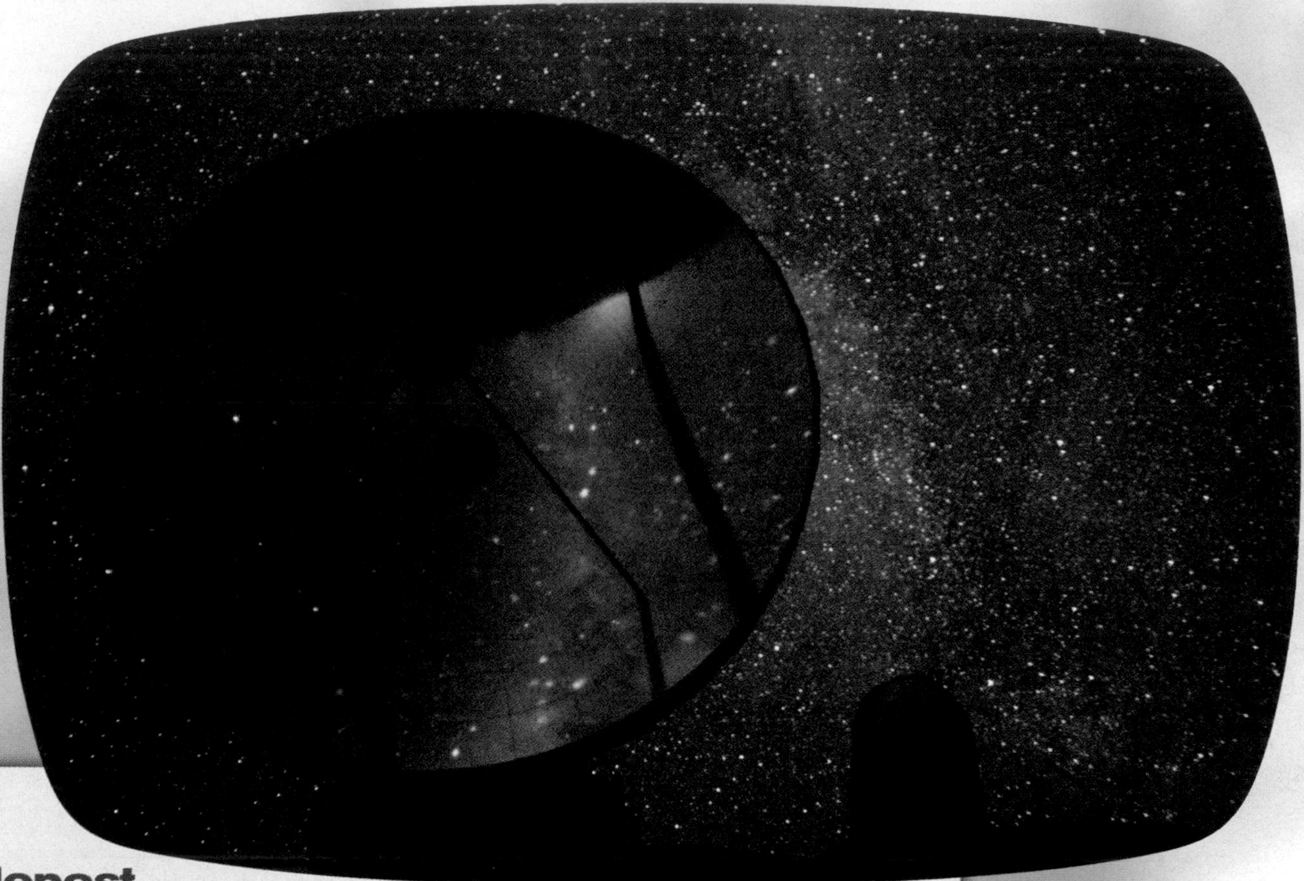

The stars of our home galaxy, the Milky Way, rise behind a telescope dome and the highly polished surface of a submillimeter telescope at the La Silla European Southern Observatory in Chile. (ESO and Nico Housen)

Guidepost

You have traced the life story of the stars from their birth in clouds of gas and dust, to their deaths as white dwarfs, neutron stars, or black holes. Now you are ready to see stars in their vast communities called galaxies. This chapter discusses our home galaxy, the Milky Way Galaxy, and attempts to answer four essential questions:

- **How do astronomers describe our galaxy?**
- **How did our galaxy form and evolve?**
- **What are the spiral arms?**
- **What lies at the very center?**

This chapter illustrates how scientists work and think. It will help you answer two questions about science as a study of nature:

- How Do We Know? **How can scientists simplify complex measurements?**
- How Do We Know? **How do scientists organize their understanding of natural events?**

Answering these questions will prepare you to leave our home galaxy behind and voyage out among the billions of galaxies that fill the depths of the universe.

The Stars Are Yours.

JAMES S. PICKERING

THE STARS ARE YOURS is the title of a popular astronomy book written by James S. Pickering in 1948. The point of the title is that the stars belong to everyone equally, and you can enjoy the wonder of the night sky as if you owned it.

Next time you admire the night sky, recall that every star you see is part of the star system in which you live. You will see in this chapter how the evidence reveals that we live inside a great wheel of stars, a galaxy. The Milky Way Galaxy is over 75,000 ly in diameter and contains over 100 billion stars. It is your galaxy because you live in it, but you are also in part a product of it, because the stars in the Milky Way Galaxy made the atoms in your body. As you begin this chapter it may seem that the stars belong to you; but, by the end of this chapter, you may decide that you belong to the stars.

15-1 The Nature of the Milky Way Galaxy

IT ISN'T OBVIOUS that you live in a galaxy, so you might well ask, "How do we know what our galaxy is like?" Because all scientific knowledge is based on evidence, you need to know a few facts about our galaxy to get started. But the more interesting question is how astronomers know those facts, and that story makes up one of the great adventures in astronomy.

First Studies of the Milky Way Galaxy

Since ancient times, humanity has been aware of a hazy band of light around the sky (■ Figure 15-1). The ancient Greeks named that band *galaxies kuklos,* the "milky circle." The Romans changed the name to *via lactia,* "milky road" or "milky way." It was not until Galileo looked at it with his telescope in 1610 that anyone knew that the Milky Way was made of stars. Little more was learned during the two centuries after Galileo lived.

By the mid-18th century, astronomers generally understood that the stars were other suns, but they had little understanding of how the stars were distributed in space. One of the first people to study this problem was the English astronomer Sir William Herschel (1738–1822). He and his sister Caroline (1750–1848) tried to map the distribution of stars in three dimensions; they assumed that the sun was located inside a great cloud of stars and that they could see to the edges of the cloud in any direction. They hypothesized that by counting the number of stars that were visible in different directions, they could gauge the relative distance to the edge of the cloud. If their telescope revealed few stars in one direction, they assumed that the edge was not far away; and, if they saw many stars in another direction, they assumed that the edge was very distant. Calling their method "star gauges," they counted stars in 683 directions in the sky and outlined a model of the star cloud. Their data showed that the cloud was a disk with the sun near the center; and, using the technology of their day as an analogy, they called it the "grindstone model" (■ Figure 15-2).

■ Figure 15-1

Nearby stars look bright, and peoples around the world group them into constellations. Nevertheless, the vast majority of the stars in our galaxy merge into a faintly luminous path that circles the sky, the Milky Way. This artwork shows the location of a portion of the Milky Way near a few bright winter constellations. (See Figure 15-3 and the star charts at the end of this book to further locate the Milky Way in your sky.)

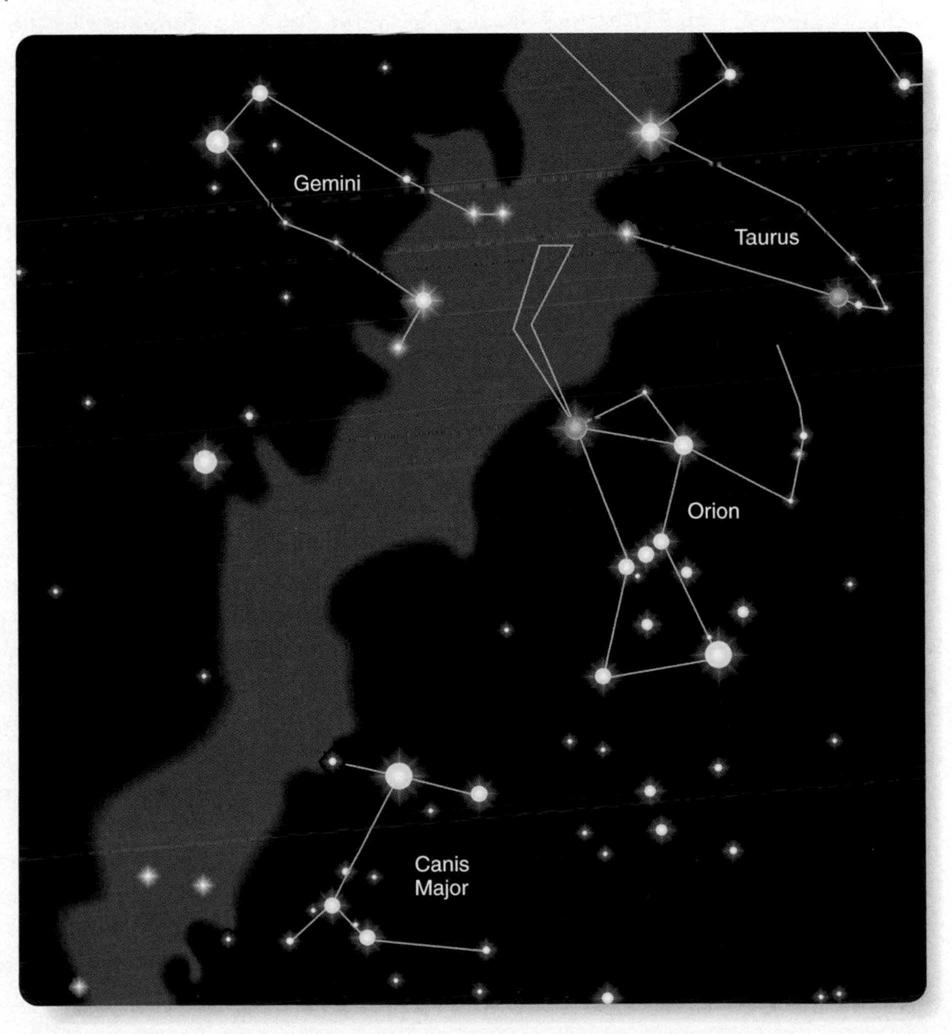

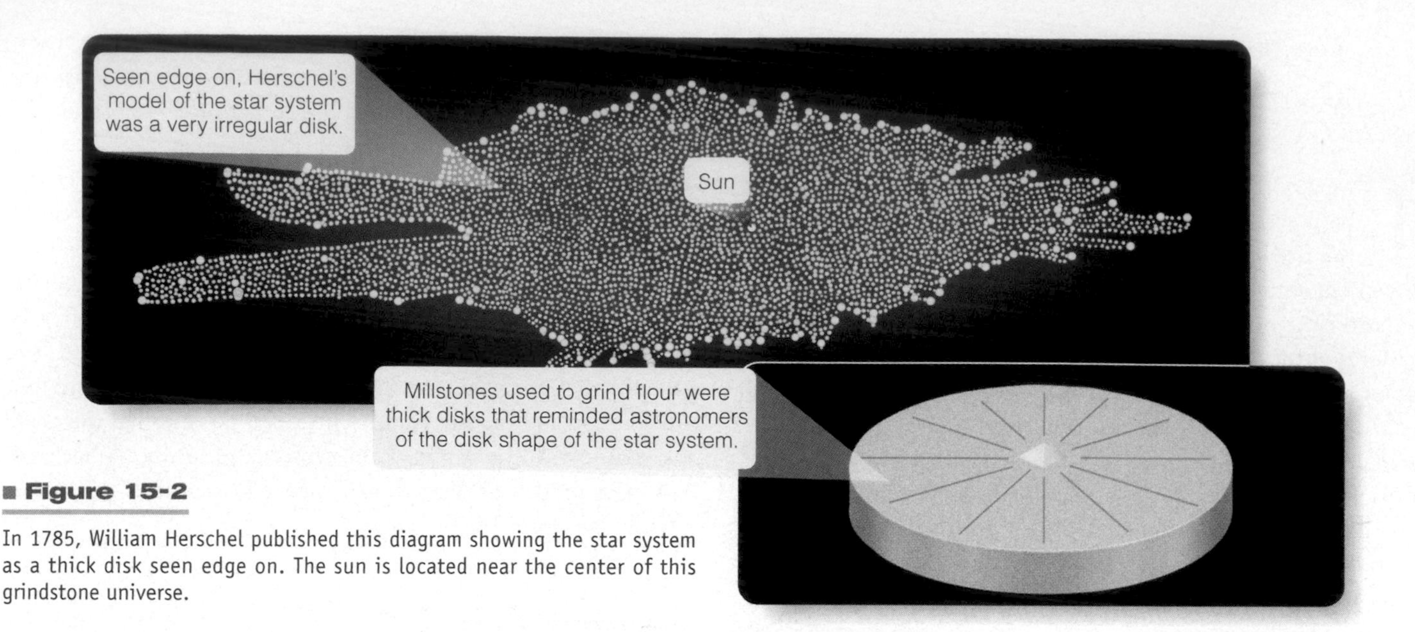
Seen edge on, Herschel's model of the star system was a very irregular disk.
Sun
Millstones used to grind flour were thick disks that reminded astronomers of the disk shape of the star system.

■ **Figure 15-2**

In 1785, William Herschel published this diagram showing the star system as a thick disk seen edge on. The sun is located near the center of this grindstone universe.

This model of the star system explains the most obvious feature of the Milky Way—it is a glowing path that circles the sky. The grindstone model explains this as a disk of stars seen from the location of the sun near the center. Unfortunately, there was no way for the Herschels to find the diameter of the disk.

Generations of astronomers studied the size of our star system, culminating with Jacobus C. Kapteyn (1851–1922). In the early 20th century, he analyzed the magnitudes, number, and motions of stars and concluded that our star system is a disk about 10 kiloparsecs in diameter. A **kiloparsec (kpc)** is 1000 pc. So far as Kapteyn could tell, the disk is about 2 kpc thick with the sun near the center.

Modern astronomers know that the Milky Way Galaxy is much larger than this and that the sun is not at its center. In the next section you will see how astronomers discovered the true size of the Milky Way and the true location of the sun. Then you will understand why earlier astronomers mistakenly thought the star system was so small.

Discovering the Galaxy

It seems odd to say that astronomers discovered something that is all around us, but until early in the 20th century no one knew that we live in a galaxy. That began to change in the second decade of the 20th century, when a young astronomer named Harlow Shapley (1885–1972) discovered how big our star system really is. Besides being one of the turning points of modern astronomy, Shapley's study illustrates one of the most common techniques in astronomy. If you want to know how astronomers know things about the universe, then Shapley's story is well worth tracing in detail.

Shapley began by noticing that although open star clusters are scattered all along the Milky Way, more than half of all globular clusters lie in or near the constellation Sagittarius. The globular clusters seem to be distributed in a great cloud whose center is located in the direction of Sagittarius (■ Figure 15-3). Shapley assumed that the orbital motion of these clusters was controlled by the gravitation of the entire star system, and, for that reason, the center of the star system could not be near the sun but must lie somewhere toward Sagittarius.

To find the distance to the center of the star system and thus the size of the star system as a whole, Shapley needed to find the distance to the star clusters, but that was difficult to do. The clusters are much too far away to have measurable parallaxes. They do, however, contain variable stars (Chapter 12), and those were the lampposts Shapley needed to find the distances to the clusters.

Shapley knew of the work of Henrietta S. Leavitt (1868–1921), who in 1912 had shown that a Cepheid variable star's period of pulsation was related to its brightness. Leavitt worked on stars whose distances were unknown, so she was unable to find their absolute magnitudes or luminosities, but astronomers realized that the Cepheids could be a powerful tool in astronomy if their true luminosities could be discovered.

Cepheids are giant and supergiant stars and so are relatively rare. None lies close enough to Earth to have a measurable parallax, but their proper motions—slow movements across the sky—can be measured. The more distant a star is, the smaller its proper motion tends to be, so proper motions contain clues to distance. Ejnar Hertzsprung (codiscoverer of the H–R diagram) made an early attempt to determine the absolute magnitudes of the Cepheids, but Shapley went further. He found 11 Cepheids with measured proper motions and used a statistical process to find their average distance and thus their average absolute magnitude. Then he could erase Leavitt's *apparent* magnitudes from the period–luminosity diagram (see Figure 12-14) and write in

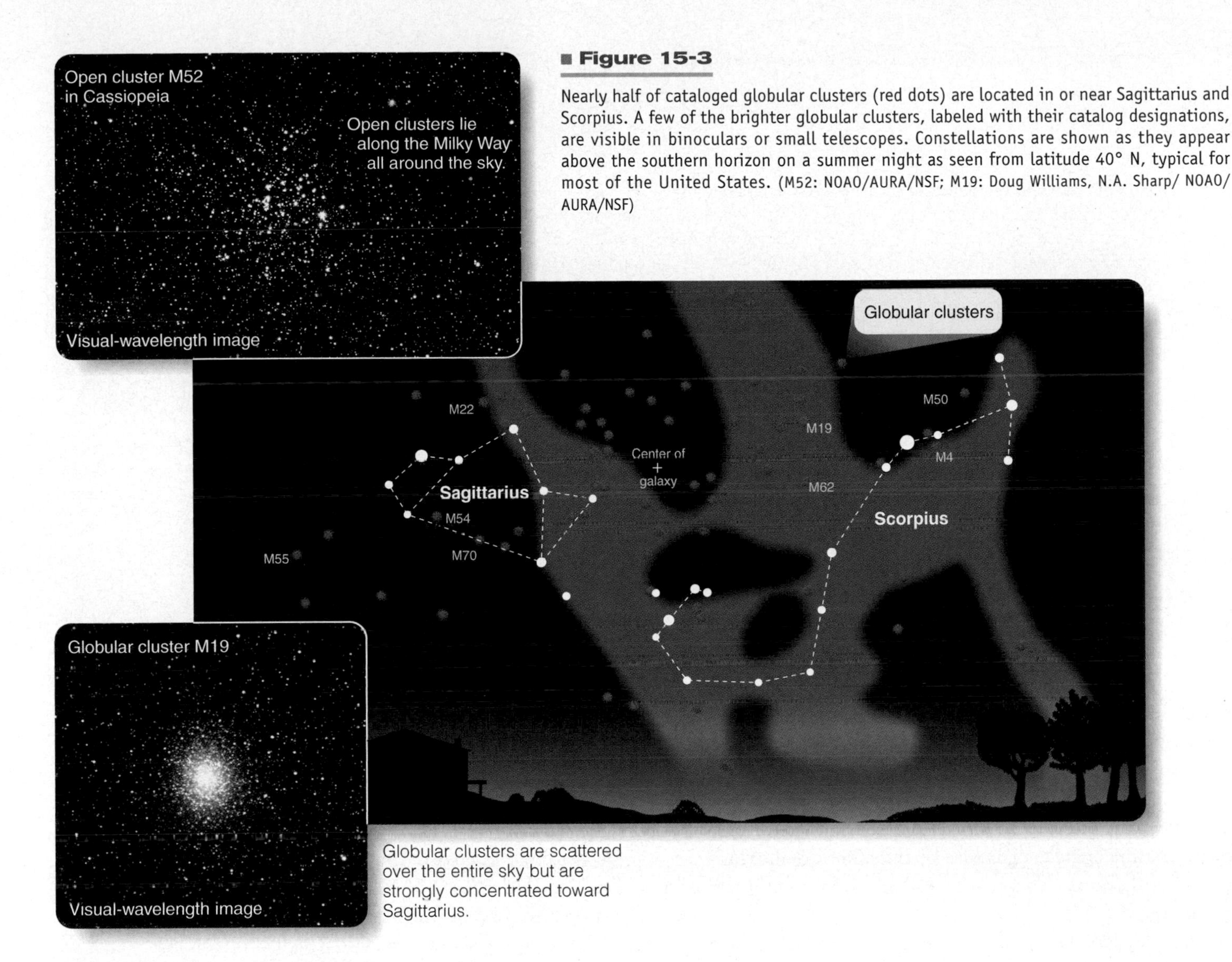

■ **Figure 15-3**

Nearly half of cataloged globular clusters (red dots) are located in or near Sagittarius and Scorpius. A few of the brighter globular clusters, labeled with their catalog designations, are visible in binoculars or small telescopes. Constellations are shown as they appear above the southern horizon on a summer night as seen from latitude 40° N, typical for most of the United States. (M52: NOAO/AURA/NSF; M19: Doug Williams, N.A. Sharp/ NOAO/ AURA/NSF)

absolute magnitudes. All of the stars in the diagram were thus calibrated as lampposts that astronomers could use to find distances. Calibrations like this one are important tools in science **(How Do We Know? 15-1)**.

Finding distances using Cepheid variable stars is so important in astronomy that you should pause to examine the process. Suppose you studied an open star cluster and discovered that it contained a type I Cepheid with a period of 10 days and an average apparent magnitude of 10.5. How far away is the cluster? From the period–luminosity diagram (see Figure 12-14), you can see that the absolute magnitude of the star must be about -3. Then the distance modulus (Chapter 9) is:

$$m - M_V = 10.5 - (-3) = 13.5$$

You could use this distance modulus and a table such as Table 9-1 to estimate that the distance to the cluster is between 1000 and 16,000 pc. You can be more accurate if you solve the magnitude–distance formula (Chapter 9) for distance:

$$d = 10^{(m_V - M_V + 5)/5}$$

Now you can substitute the apparent magnitude and absolute magnitude and determine that the distance is equal to $10^{3.7}$ pc, which equals about 5000 pc. This is one of the most common calculations in astronomy.

Once Shapley had calibrated the period–luminosity relation, he could use it to find the distance to any cluster in which he could identify variable stars. By taking a series of photographic plates over a number of nights, Shapley was able to pick out the variable stars in a cluster, measure their average apparent magnitude, and find their periods of pulsation. Because the periods were very short, he knew the variable stars he saw in the clusters were the lowest-luminosity variable stars (known today as RR Lyrae stars). Knowing that allowed him to read their absolute magnitude off the period–luminosity diagram. Once he knew the apparent magnitude and the absolute magnitude, he could calculate the distance to the star cluster.

This worked well for the nearer globular clusters, but variable stars in the more distant clusters are too faint to detect. Shapley estimated the distances to these more distant clusters by

How Do We Know? | 15-1

Calibration

How do you take the temperature of a vat of molten steel? Astronomers often say that Shapley "calibrated" the Cepheids for the determination of distance, meaning that he did all the detailed background work so that the Cepheids could be used to find distances. Other astronomers could then use Shapley's calibrated diagram to find the distance to other Cepheids without repeating the detailed calibration.

Calibration is actually very common in science because it saves a lot of time and effort. For example, engineers in steel mills must monitor the temperature of molten steel, but they can't dip in a thermometer. Instead, they can use handheld devices that measure the color of molten steel. You recall from Chapter 7 that the color of black body radiation is determined by its temperature. Molten steel emits visible and infrared radiation that is nearly perfect black body radiation, so the manufacturer can calibrate the engineer's devices to convert the measured color to a temperature displayed on digital readouts. The engineers don't have to repeat the calibration every time; they just point their instrument at the molten steel and read off the temperature. (Astronomers have made the same kind of color–temperature calibration for stars.)

As you read about any science, notice how calibrations are used to simplify common measurements. But notice, too, how important it is to get the calibration right. An error in calibration can throw off every measurement made with that calibration.

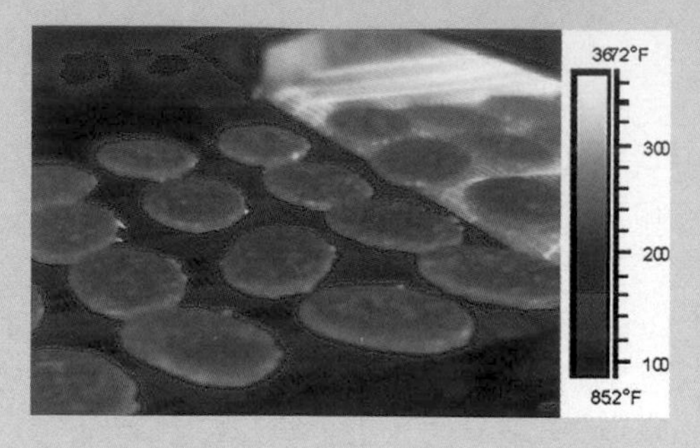

An infrared video camera calibrated to measure temperature allows bakers to monitor the operation of their ovens. (Courtesy of FLIR Systems, Inc.)

calibrating the diameters of the clusters. For the clusters whose distance he knew, he could use their angular diameters and the small-angle formula to calculate linear diameters in parsecs. He found that the nearby clusters are about 25 pc in diameter, which he assumed is the average diameter of all globular clusters. He then used the angular diameters of the more distant clusters to find their distances. This is another illustration of calibration in astronomy.

Shapley later wrote that it was late at night when he finally plotted the directions and distances to the globular clusters on graph paper and found that, just as he had supposed, they formed a great swarm whose center lay many thousands of light-years in the direction of Sagittarius, confirming his suspicion that the center of the star system was not near the sun but was far away in Sagittarius (■ Figure 15-4). He found the only other person in the building, a cleaning lady, and the two stood looking at his graph as he explained that they were the only two people on Earth who understood that humanity lives, not at the center of a small star system, but in the suburbs of a vast wheel of stars.

Why did astronomers before Shapley think we lived near the center of a small star system? Space is filled with gas and dust that dim distant stars. When astronomers look into the band of the Milky Way, they can see only the neighborhood near the sun. Most of the star system is invisible, so, like travelers in a fog, we seem to be at the center of a small region. Shapley was able to see the globular clusters at greater distances because they lie outside the plane of the star system and are not dimmed very much by the gas and dust.

Building on Shapley's work, other astronomers began to suspect that some of the faint patches of light visible through telescopes were other star systems. Within a few years they found evidence that the faint patches of light were indeed other galaxies much like our own Milky Way Galaxy. Today the largest telescopes can detect an estimated 100 billion galaxies similar to our own. Our home galaxy is special only in that it is our home. The next chapter will continue this story and discuss galaxies in general.

Shapley's study of star clusters led to the discovery that we live in a galaxy and that the universe is filled with similar galaxies. Like Copernicus, Shapley moved us from the center to the suburbs. To get a true impression of our galaxy, you must conduct a careful analysis.

The Structure of Our Galaxy

Astronomers commonly give the diameter of our galaxy as 25,000 pc or 75,000 ly, and place the sun about two-thirds of the way from the center to the edge (■ Figure 15-5). The diameter of our galaxy is not known very accurately, as you can see from the two numbers given above. Convert 25,000 pc into light-years, and you get significantly more than 75,000 ly. The fact that these two numbers don't quite match is a warning that the size of our galaxy is not known to better than about 10 percent.

The disk of the galaxy is often referred to as the **disk component.** It contains most of the galaxy's stars and nearly all of its gas and dust. Because the disk is home to the giant molecular clouds within which stars form (Chapter 11), nearly all star formation in our galaxy takes place in the disk. Most of the stars in

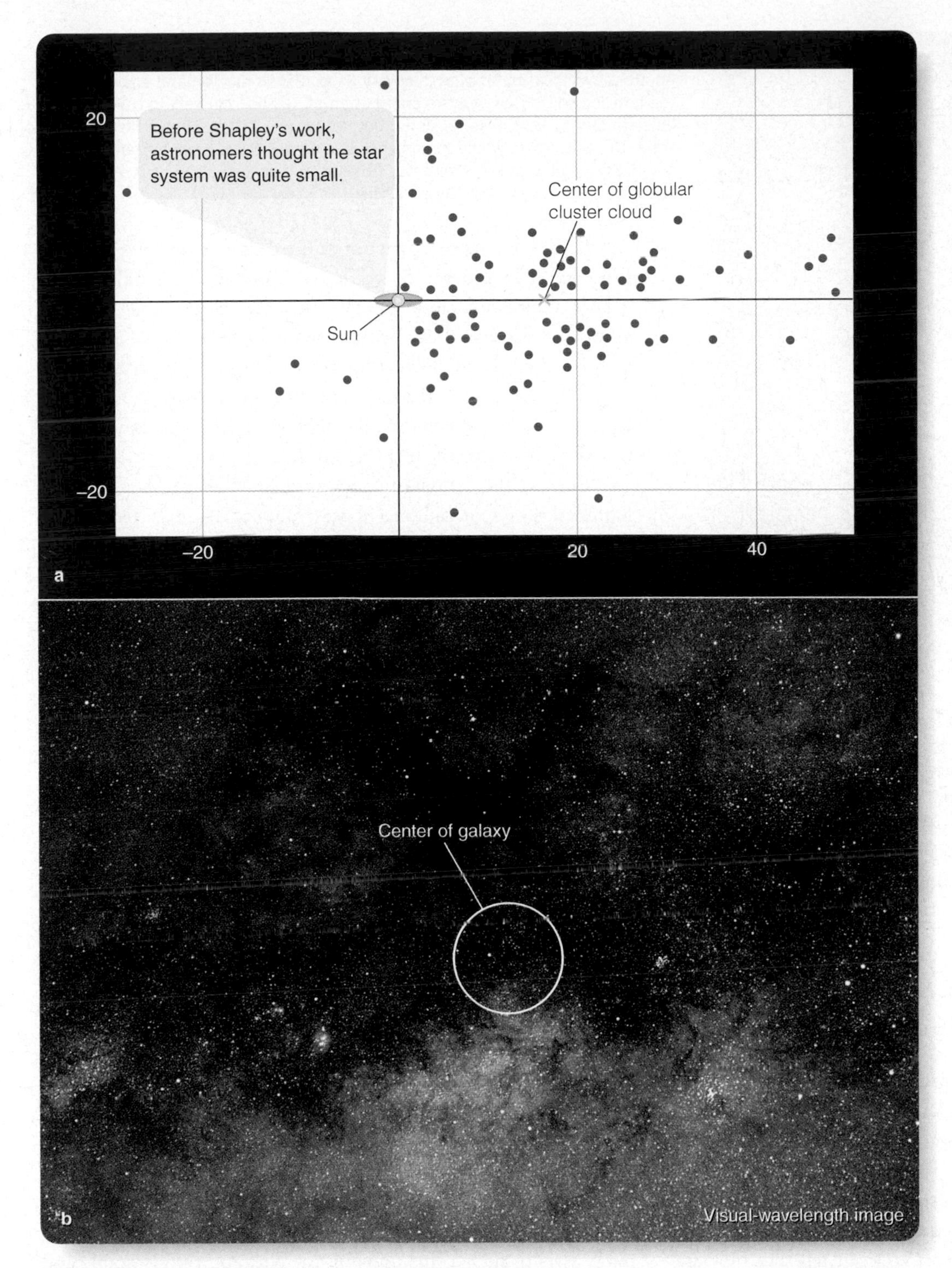

Figure 15-4

(a) Shapley's study of globular clusters showed that they were not centered on the sun, at the origin of this graph, but rather formed a great cloud centered far away in the direction of Sagittarius. Distances on this graph are given in thousands of parsecs. (b) Looking toward Sagittarius, you see nothing to suggest that this is the center of the galaxy. Gas and dust block your view. Only the distribution of globular clusters told Shapely the sun lay far from the center of the star system. (Daniel Good)

the disk are middle- to lower-main-sequence stars like the sun, a few are giants, and fewer still are brilliant O and B stars. Although these hot, luminous stars are rare, they produce so much light that they are the main source of illumination; they light up the disk and make it bright. It is difficult to judge just how big the disk is, because it is filled with gas and dust, which dim starlight and make it difficult to see distant stars in the disk. Also, floating through the thinner gas and dust are large, dense, dusty clouds that block the view in many directions within the disk. In most directions in the plane of the disk, astronomers can't see farther than a few kiloparsecs.

Although astronomers can survey the locations and distances of the stars, they cannot cite a single number for the thickness of the disk because it lacks sharp boundaries. Stars become less crowded farther from the central plane of the galaxy. Also, the thickness of the disk depends on the kind of object studied. Stars like the sun with ages of a few billion years lie within about 500 pc above and below the central plane. But the youngest stars,

■ Figure 15-5

An artist's conception of our Milky Way Galaxy, seen face on and edge on. Note the position of the sun and the distribution of globular clusters in the halo. Hot blue stars light up the spiral arms. Only the inner halo is shown here. At this scale, the entire halo would be larger than a dinner plate. From far out in space, our galaxy would probably look much like the Andromeda Galaxy. (left: ©2003, 2004 Three Rivers Foundation; right: Robert Gendler)

If you look up or down out of the disk, you are looking away from the dust and gas, so you can see out into the **halo** of our galaxy, a spherical cloud of stars and star clusters that contains almost no gas and dust. Because the halo contains no dense gas clouds, it cannot make new stars. Halo stars are old, cool, lower-main-sequence stars, red giants, and white dwarfs. It is difficult to judge the extent of the halo, but it could be as much as 10 times the diameter of the visible disk.

Around the center of our galaxy lies the **nuclear bulge,** a flattened cloud of billions of stars about 3 kpc in radius (see Figure 15-5). Like the halo, it contains little gas and dust. Astronomers often refer to the halo and the nuclear bulge as the **spherical component** of the galaxy. The nuclear bulge is the most crowded part of the spherical component. Visible above and below the plane of the galaxy and through gaps in the obscuring dust, the nuclear bulge contains stars that are old and cool like the stars in the halo.

including the O and B stars, and the gas and dust from which these young stars are forming are confined to a thin disk extending only about 50 pc above and below the plane (■ Figure 15-6). With a diameter of 25 kpc, the disk is, in proportion to its diameter, thinner than a thin pizza crust.

The disk of the galaxy contains two kinds of star clusters. Associations (Chapter 11) are groups of 10 to a few hundred stars so widely scattered in space that their mutual gravity cannot hold the association together. From the turnoff points in their H–R diagrams (Chapter 12), you can tell that they are very young groups of stars. The stars move together through space (■ Figure 15-7) because they formed from a single gas cloud and haven't had time to wander apart.

The second kind of cluster in the disk is the open cluster (see Figure 15-3), a group of 100 to a few thousand stars in a region about 25 pc in diameter. Because they have more stars in less space than associations, open clusters are more firmly bound by gravity. Although they lose stars occasionally, they can survive for a long time, and the turnoff points in their H–R diagrams give ages from a few million to a few billion years.

The halo contains roughly 200 globular clusters, each of which contains 50,000 to a million stars in a sphere about 25 pc in diameter (refer again to Figure 15-3). Because they contain so many stars in such a small region, the clusters are very stable and have survived for billions of years. From the turnoff points in their H–R diagrams, you can tell they average about 11 billion years old.

Orbital motions are dramatically different in the two components of the galaxy. Disk stars follow nearly circular orbits that lie in the plane of the galaxy (■ Figure 15-8a). Halo stars and globular clusters, however, follow highly elongated orbits tipped steeply to the plane of the disk (Figure 15-8b). Although the diagram shows these orbits as elliptical, the gravitational influence of the thick bulge forces them into rosettes that do not quite

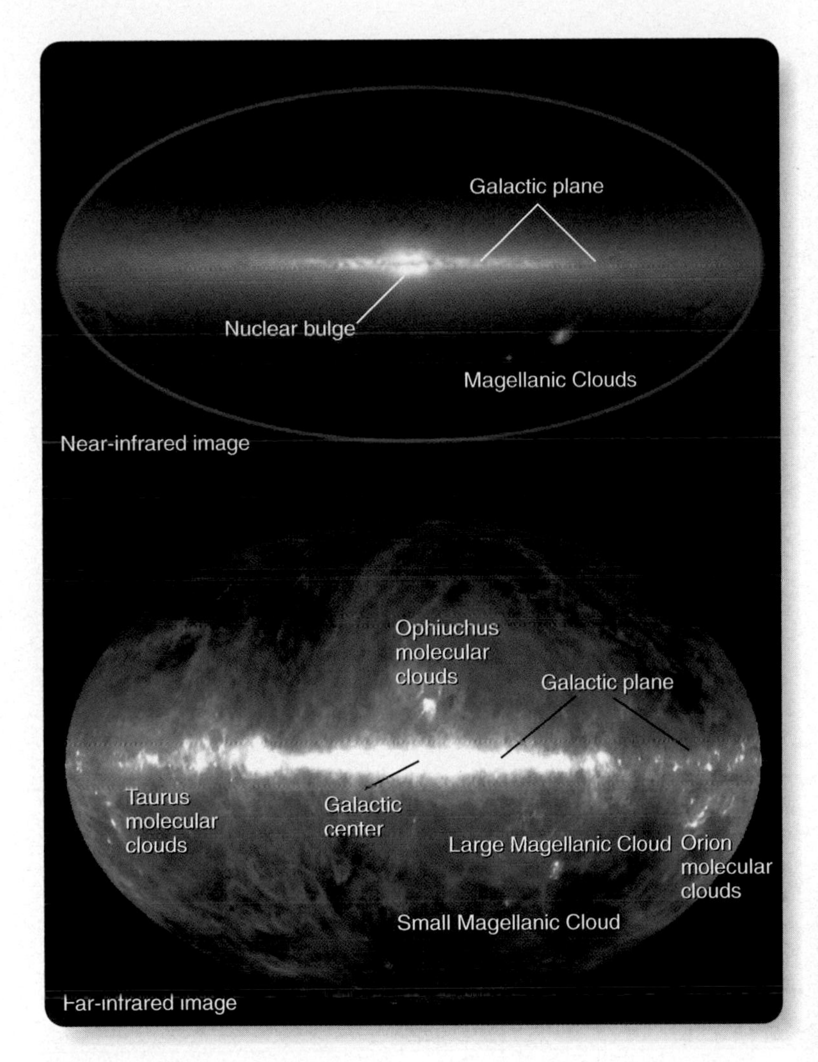

■ Active Figure 15-6

In these infrared images, the entire sky has been projected onto ovals with the center of the galaxy at the center. The Milky Way extends from left to right. In the near-infrared, the nuclear bulge is prominent, and dust clouds block your view. At longer wavelengths, the dust emits black body radiation and glows brightly. (Near-IR: 2Mass; Far-IR: DIRBE Image courtesy Henry Freudenreich)

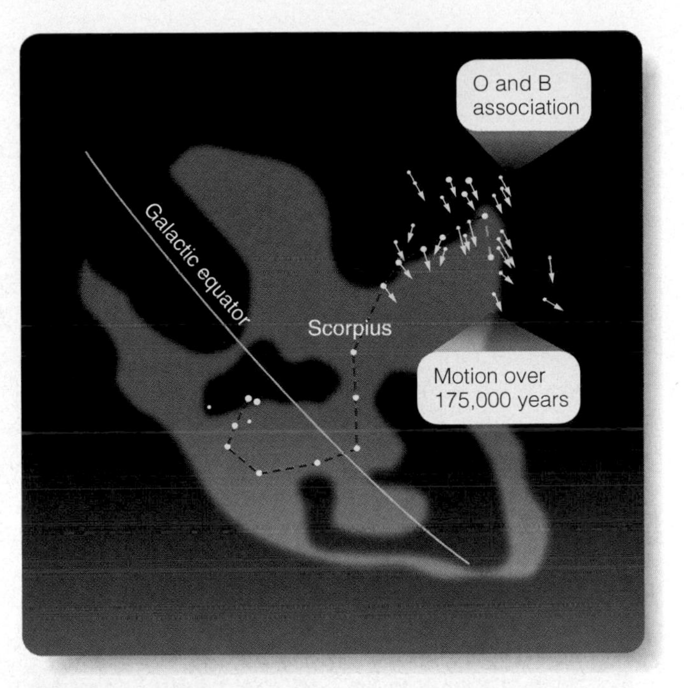

■ Figure 15-7

Many of the stars in the constellation Scorpius are members of an O and B association that has formed recently from a single cloud of gas. As the stars orbit the center of our galaxy, they are moving together southwest along the Milky Way.

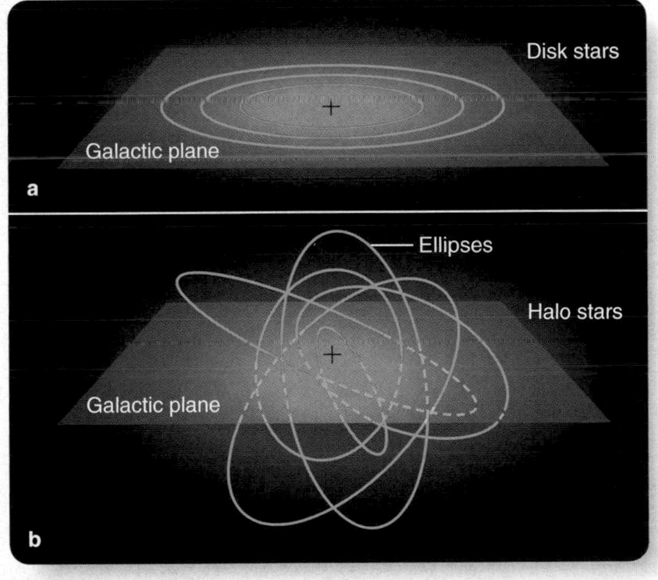

■ Figure 15-8

(a) Stars in the galactic disk have nearly circular orbits that lie in the plane of the galaxy. (b) Stars in the halo have randomly oriented, highly elongated orbits.

return to the same starting point. The dramatic difference between the motions of halo stars and disk stars will be important evidence when you consider the formation of the galaxy later in this chapter.

From the shape and size of our galaxy, astronomers conclude that we live in a spiral galaxy (■ Figure 15-9). The most dramatic features of these disk galaxies are the spiral arms—long spiral patterns of bright stars, HII regions, star clusters, and clouds of gas and dust. The sun is located on the inside edge of one of these spiral arms. You will see later in this chapter how astronomers can observe the spiral arms in our own galaxy.

This analysis of the components of our galaxy may leave you wondering about one critical question: How much matter does our galaxy contain? To answer that question, you must watch our galaxy rotate.

The Mass of the Galaxy

To find the mass of an object, astronomers must observe its orbital motion as in a binary star system. Humans don't live long enough to see stars move significantly along their orbits around the galaxy, but astronomers can observe the radial velocities, proper motions, and distances of stars and then calculate the sizes and periods of their orbits. The results can reveal the mass of the galaxy.

It is a **Common Misconception** to imagine that the sun is drifting slowly through space. Stars in the disk of the galaxy follow nearly circular orbits that lie in the plane of the disk (look again at Figure 15-8a). By observing the radial velocities of other galaxies in various directions, astronomers can conclude that as the sun orbits the center of our galaxy it is moving at about 220 km/s in the direction of Cygnus. The evidence suggests the sun's orbit is nearly circular, so given the distance to the center of our galaxy, 8.5 kpc, you can find the circumference of the sun's orbit by multiplying by 2π. If you divide the circumference of its orbit by its orbital velocity, you will discover that the sun has an orbital period of about 240 million years.

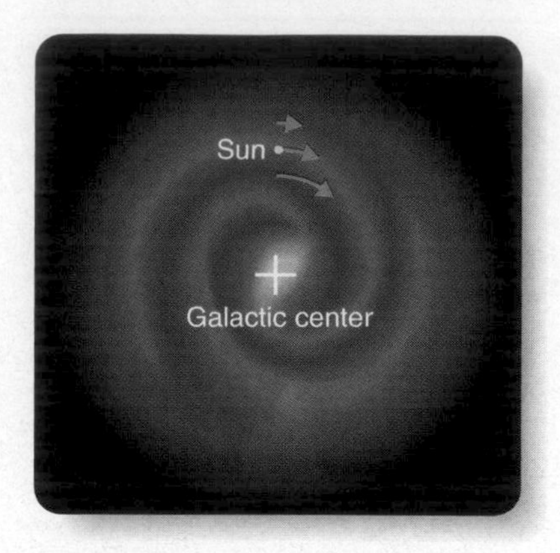

■ Figure 15-10

The differential rotation of the galaxy means that stars at different distances from the center have different orbital periods. In this example, the star just inside the sun's orbit has a shorter period and pulls ahead of the sun, while the star outside falls behind.

■ Figure 15-9

Our disk galaxy resembles the spiral galaxies visible through large telescopes. (a) Seen edge on, spiral galaxies have nuclear bulges and dust-filled disks. (Todd Boroson/NOAO/AURA/NSF) (b) Seen face on, the spiral arms are dramatically outlined by hot O and B stars, emission nebulae, and dust. (ESO)

When you studied binary stars in Chapter 9, you saw that the total mass (in solar masses) of a binary star system equals the cube of the separation of the stars a (in AU) divided by the square of the period P (in years):

$$M = \frac{a^3}{P^2}$$

You can use the same equation here to find the mass of the galaxy. The radius of the sun's orbit around the galaxy is about 8500 pc, and each parsec contains 206,265 AU. Multiplying, you find that the radius of the sun's orbit is 1.75×10^9 AU. The orbital period is 240 million years, so the mass is:

$$M = \frac{(1.75 \times 10^9)^3}{(240 \times 10^6)^2} = 0.93 \times 10^{11}\ M_\odot$$

This is only a rough estimate because it does not include the mass that lies outside the orbit of the sun. Correcting for this overlooked mass yields a total mass for our galaxy of about 4×10^{11} solar masses.

The rotation of our galaxy is actually the orbital motion of each of its stars around the center of mass. Stars at different distances from the center revolve around the center of the galaxy with different periods, so three stars near each other will draw apart as time passes (■ Figure 15-10). This is called differential rotation. (Recall that differential rotation was defined in Chapter 8.)

To fully describe the rotation of our galaxy, astronomers graph orbital velocity versus radius producing a **rotation curve** (■ Figure 15-11). If all of the mass of the galaxy were concentrated near its center, then you would expect to see orbital velocities fall as you moved away from the center. This is what you see in our solar system, where nearly all of the mass is concentrated in the sun, and it is called **Keplerian motion** (a reference to Kepler's laws). In contrast, the best observations of the rotation curve of the Milky Way show that orbital velocities are constant or rising in the outer disk. These higher orbital velocities indicate that the larger orbits enclose more mass and suggest that our galaxy is more massive than it appears to be.

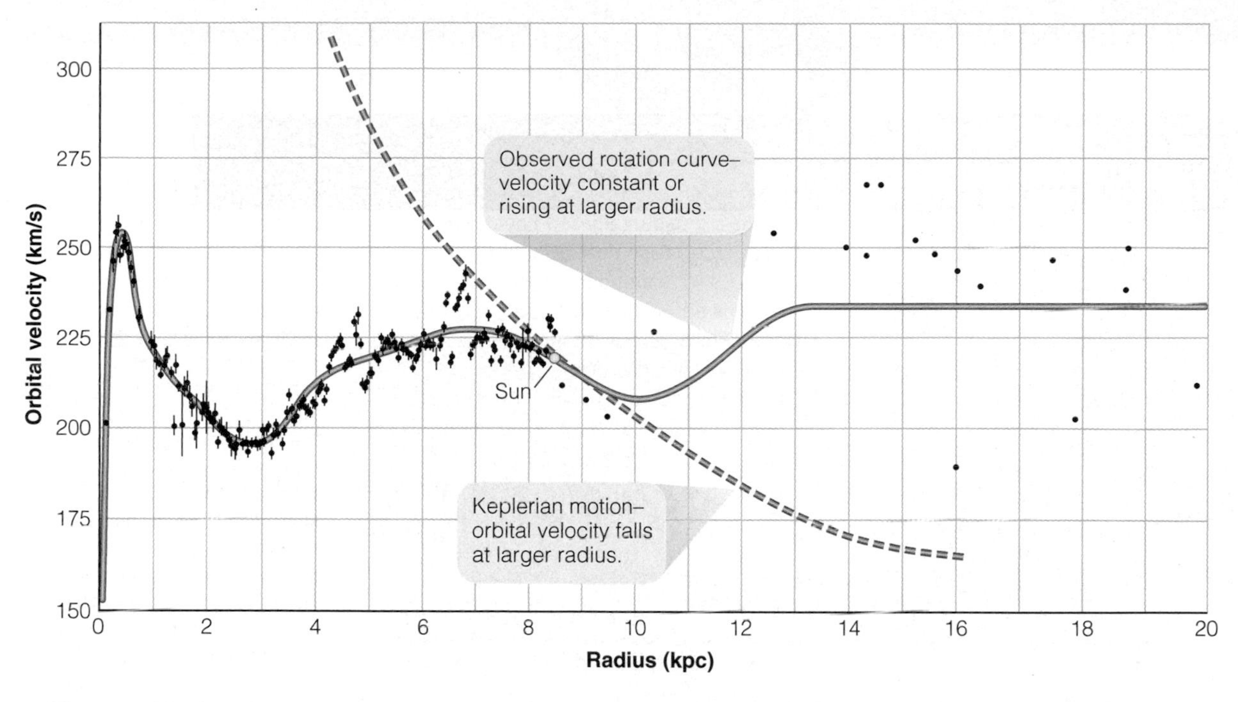

■ **Figure 15-11**

The rotation curve of our galaxy is plotted here as orbital velocity versus radius. Data points show measurements made by radio telescopes. Observations outside the orbit of the sun are much more uncertain, and the data points scatter widely. Orbital velocities do not decline outside the orbit of the sun, as you would expect if most of the mass of the galaxy were concentrated toward the center (Keplerian motion). Rather, the curve is approximately flat at great distances, suggesting that the galaxy contains significant mass outside the orbit of the sun. (Adapted from a diagram by Françoise Combes)

Both observational evidence and mathematical models show that the extra mass lies in an extended halo sometimes called a **galactic corona.** It may extend up to 10 times farther than the edge of the visible disk and could contain a trillion solar masses. Much of this mass is invisible, so astronomers conclude that it is not emitting or absorbing light and refer to it as **dark matter.** At least some of the mass in the galactic corona is made up of low-luminosity stars and white dwarfs, but much of the mass must be some other form of matter. You will learn more about the problem of the dark matter in the following two chapters. It is one of the fundamental problems of modern astronomy.

◄ **SCIENTIFIC ARGUMENT** ►

If gas and dust block the view, how do astronomers know how big our galaxy is?

Because scientific arguments depend ultimately on evidence, they must explain how scientists know what they know. The gas and dust in our galaxy block the view only in the plane of the galaxy. When telescopes look away from the plane of the galaxy, they look out of the gas and dust and can see to great distances. The globular clusters are scattered through the halo with a strong concentration in the direction of Sagittarius. When astronomers look above or below the plane of the galaxy, they can see those clusters, and careful observations reveal variable stars in the clusters. By using a modern calibration of the period–luminosity diagram, they can find the distance to those clusters. If they assume that the distribution of the clusters is controlled by the gravitation of the galaxy as a whole, then they can find the distance to the center of the galaxy by finding the center of the distribution of globular clusters.

In any logical argument, it is important to ask "How do we know?" and that is especially true in science. Create a new argument: **What measurements and assumptions reveal the total mass of our galaxy?**

◄ ►

15-2 The Origin of the Milky Way

JUST AS DINOSAURS LEFT BEHIND fossilized footprints, our galaxy has left behind a fossil footprint of its youth. The stars of the spherical component are old and must have formed long ago when the galaxy was very young.

Stellar Populations

In the 1940s, astronomers realized that there are two families of stars in the galaxy. They form and evolve in similar ways, but they differ in the abundances of atoms that are heavier than helium—atoms that astronomers call **metals.** (Note that this is not the way the word *metal* is defined by nonastronomers.) **Population I stars** are metal rich, containing 2 to 3 percent metals, whereas **population II stars** are metal poor, containing only about 0.1 percent metals or less. The difference may seem small, but it is dramatically evident in spectra (■ Figure 15-12).

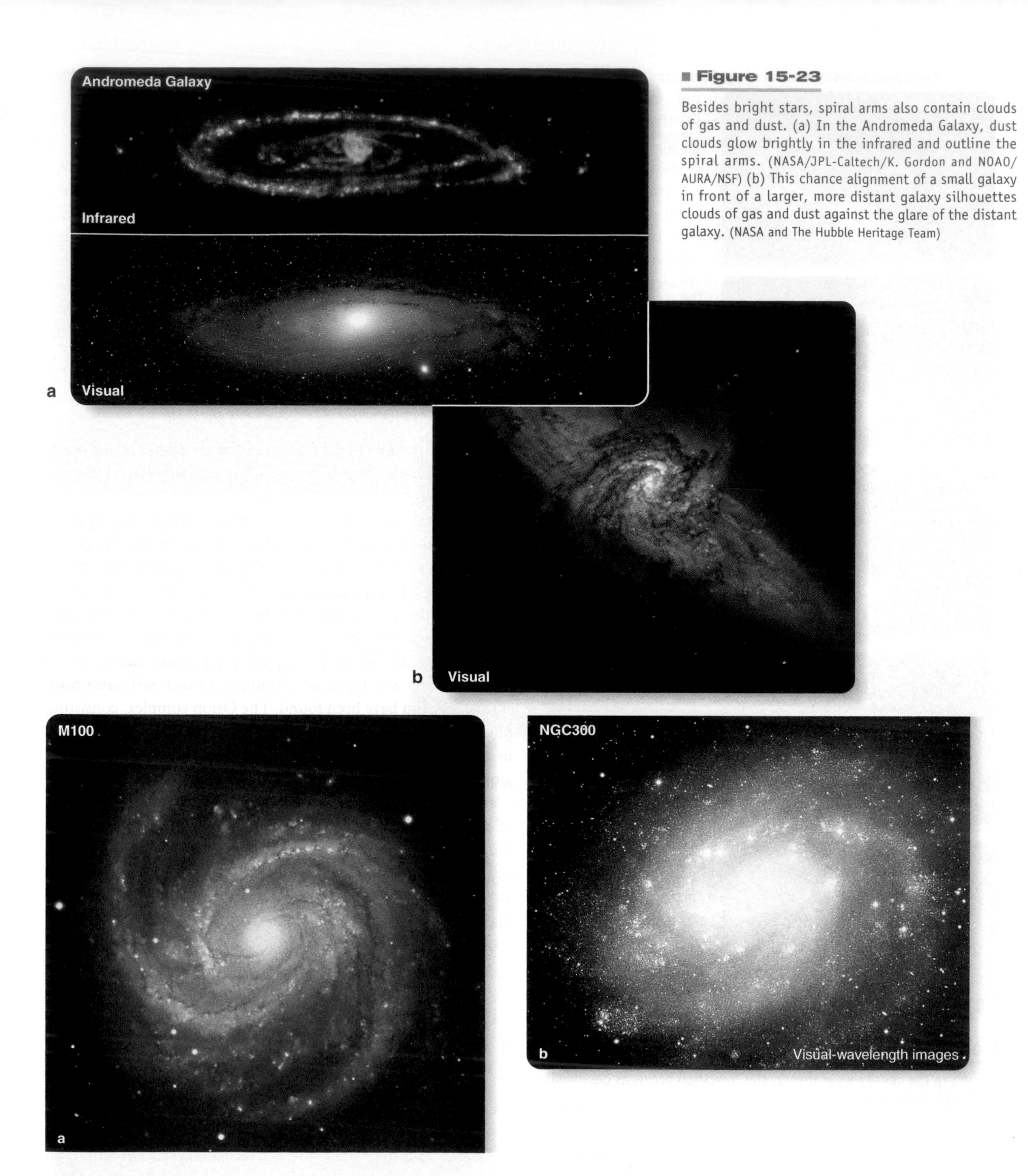

■ **Figure 15-23**

Besides bright stars, spiral arms also contain clouds of gas and dust. (a) In the Andromeda Galaxy, dust clouds glow brightly in the infrared and outline the spiral arms. (NASA/JPL-Caltech/K. Gordon and NOAO/AURA/NSF) (b) This chance alignment of a small galaxy in front of a larger, more distant galaxy silhouettes clouds of gas and dust against the glare of the distant galaxy. (NASA and The Hubble Heritage Team)

■ **Figure 15-24**

(a) Some galaxies are dominated by two spiral arms; but, even in these galaxies, minor spurs and branches are common. The spiral density wave can generate the two-armed, grand-design pattern, but self-sustained star formation may be responsible for the irregularities. (b) Many spiral galaxies do not appear to have two dominant spiral arms. Spurs and branches suggest that star formation is proceeding rapidly in such galaxies. (Anglo-Australian Observatory/David Malin Images)

Gas cloud

Intense radiation from a hot star compresses a nearby gas cloud.

Protostars

Forming star cluster

Newborn massive star

■ **Figure 15-25**

Most stars in a star cluster are too small and too cool to affect nearby gas clouds; but, once a massive star forms, it becomes so hot and so luminous its radiation can push gas away and compress a nearby gas cloud. In the densest regions of the compressed cloud, new stars can begin forming.

◄ SCIENTIFIC ARGUMENT ►

Why can't astronomers use solar-type stars as spiral tracers?

Sometimes the timing of events is the critical factor in a scientific argument. In this case, you need to think about the evolution of stars and their orbital periods around the galaxy. Stars like the sun live about 10 billion years, but the sun's orbital period around the galaxy is 240 million years. The sun almost certainly formed when a gas cloud passed through a spiral arm, but since then the sun has circled the galaxy many times and has passed through spiral arms often. That means the sun's present location has nothing to do with the location of any spiral arms. An O star, however, lives only a few million years. It is born in a spiral arm and lives out its entire lifetime before it can leave the spiral arm. Short-lived stars such as O stars are found only in spiral arms, but G stars are found all around the galaxy.

The spiral arms of our galaxy would make it beautiful if it could be photographed from a distance, but we are trapped inside it. Create an argument based on evidence: **How do astronomers know that the spiral arms mapped out near us by spiral tracers actually extend across the disk of our galaxy?**

◄ ►

Self-Sustaining Star Formation

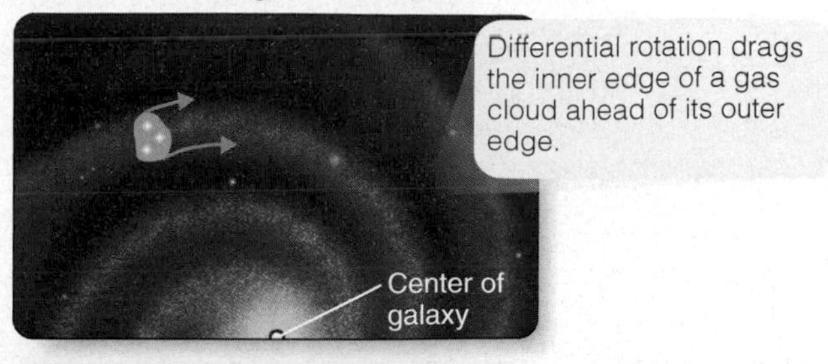

Differential rotation drags the inner edge of a gas cloud ahead of its outer edge.
Center of galaxy

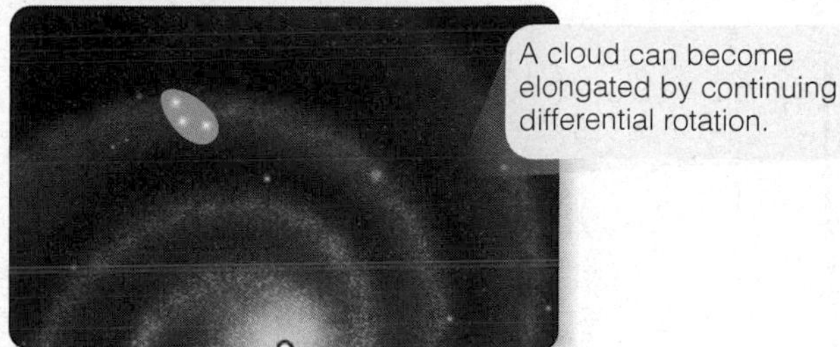

A cloud can become elongated by continuing differential rotation.

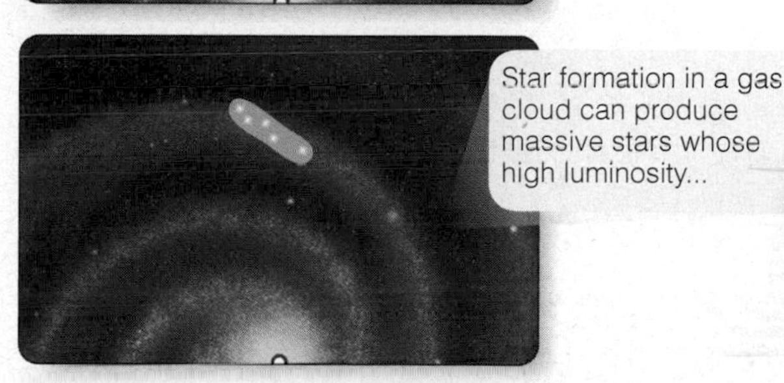

Star formation in a gas cloud can produce massive stars whose high luminosity...

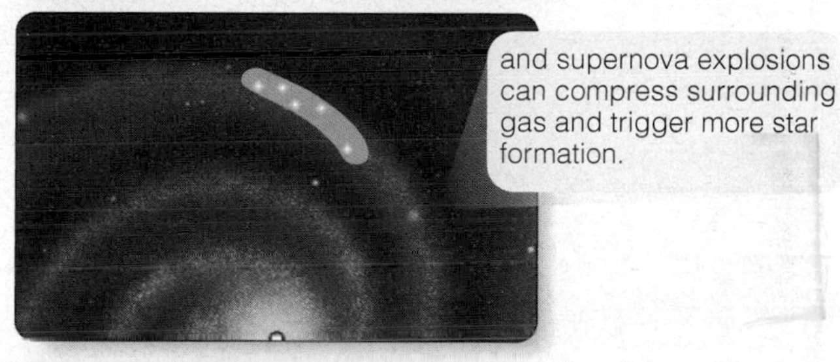

and supernova explosions can compress surrounding gas and trigger more star formation.

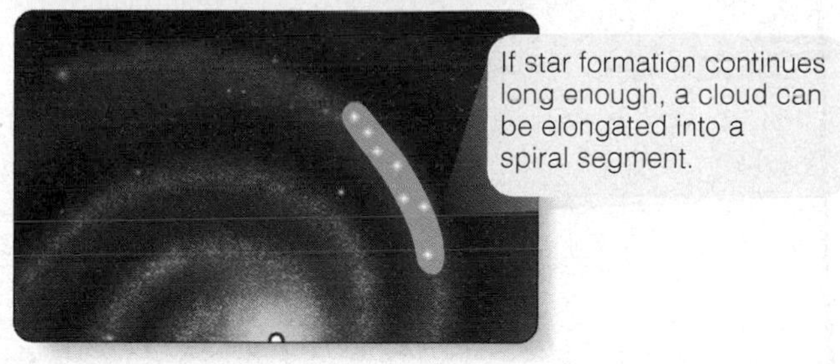

If star formation continues long enough, a cloud can be elongated into a spiral segment.

■ **Figure 15-26**

Continuing star formation may be able to produce long clouds of young stars that look like segments of spiral arms.

Sagittarius A*

1 The constellation of Sagittarius is so filled with stars and with gas and dust you can see nothing at visual wavelengths of the center of our galaxy.

The image below is a wide-field radio image of the center of our galaxy. Many of the features are supernova remnants (SNR), and a few are clouds of star formation. Peculiar features such as threads, the Arc, and the Snake may be gas trapped in magnetic fields. At the center lies Sagittarius A, the center of our galaxy.

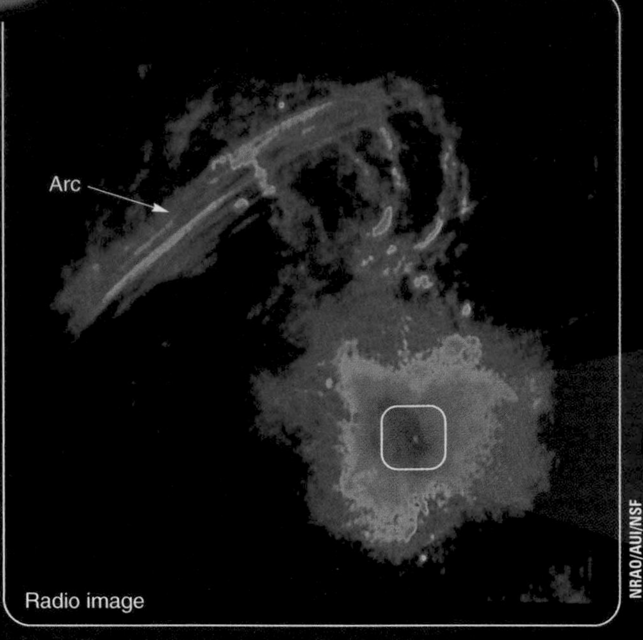

The radio map above shows Sgr A and the Arc filaments, 50 parsecs long. The image was made with the VLA radio telescope. The contents of the white box are shown on the opposite page.

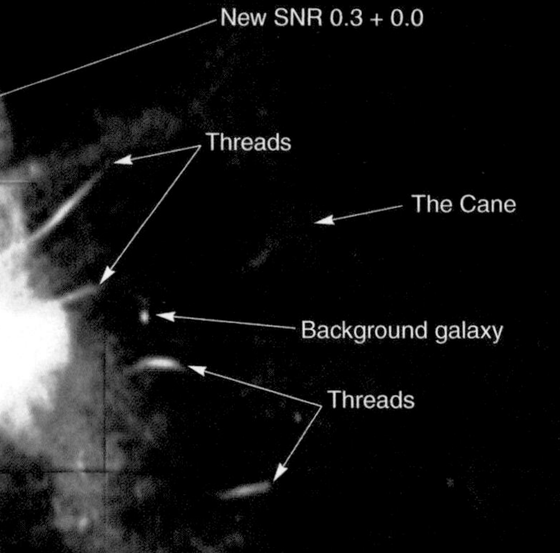

2 Infrared photons with wavelengths longer than 4 microns (4000 nm) come almost entirely from warm interstellar dust. The radiation at these wavelengths coming from Sagittarius is intense, and that indicates that the region contains lots of dust and is crowded with stars that warm the dust.

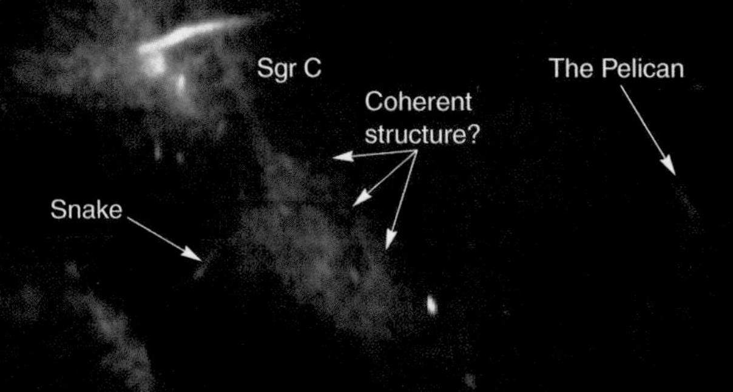

Infrared image

2MASS

1a This high-resolution radio image of Sgr A (the white boxed area on the opposite page) reveals a spiral swirl of gas around an intense radio source known as Sgr A*, the presumed central object in our galaxy. About 3 pc across, this spiral lies in a low-density cavity inside a larger disk of neutral gas. The arms of the spiral are thought to be streams of matter flowing into Sgr A* from the inner edge of the larger disk (drawing at right).

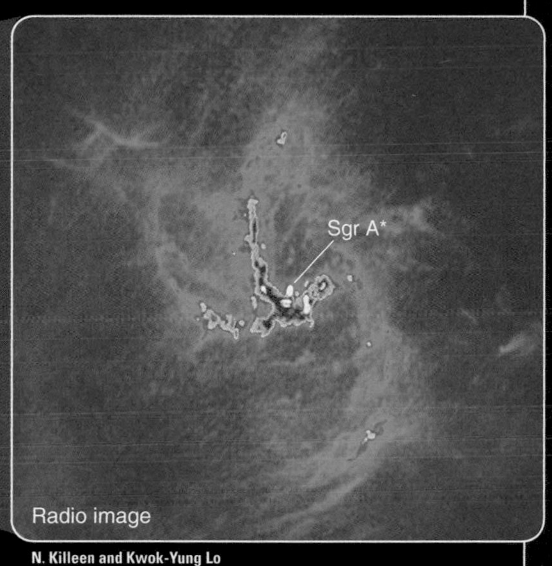

N. Killeen and Kwok-Yung Lo

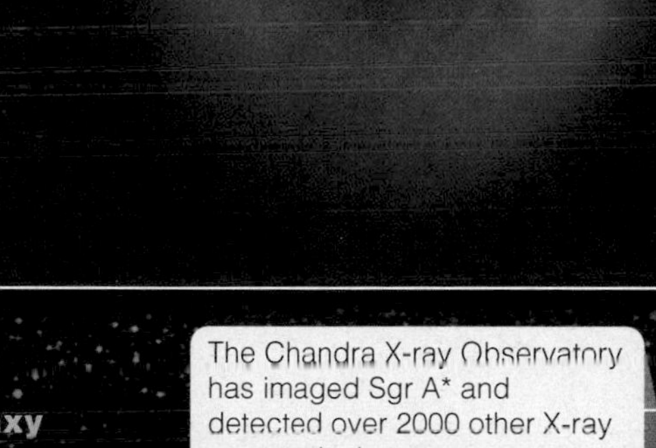

Evidence of a Black Hole at the Center of Our Galaxy

3 Since the middle 1990s, astronomers have been able to use large infrared telescopes and active optics to follow the motions of stars orbiting around Sgr A*. A few of those orbits are shown here. The size and period of the orbit allows astronomers to calculate the mass of Sgr A* using Kepler's third law. The orbital period of the star SO-2, for example, is 15.2 years and the semimajor axis of its orbit is 950 AU. The combined motions of the observed stars suggest that Sgr A* has a mass of 2.6 million solar masses.

The Chandra X-ray Observatory has imaged Sgr A* and detected over 2000 other X-ray sources in the area.

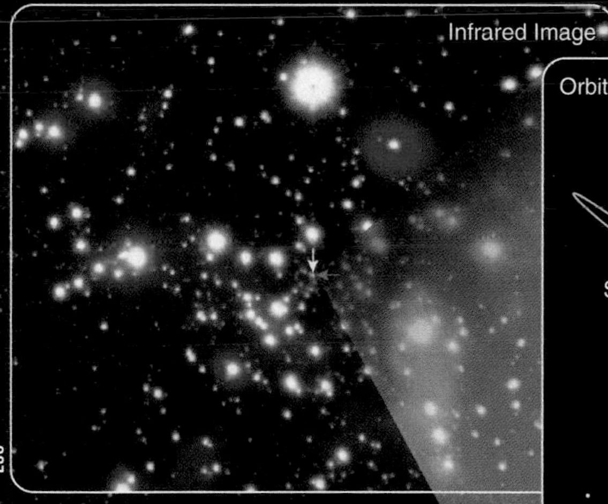

ESO

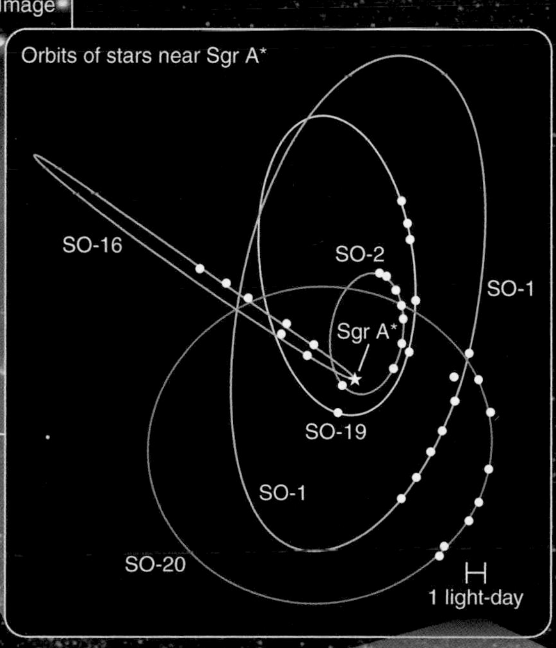

At its closest, SO-2 comes within 17 light-hours of Sgr A*. Alternative theories that Sgr A* is a cluster of stars, of neutron stars, or of stellar black holes are eliminated. Only a single black hole could contain so much mass in so small a region.

Our solar system is half a light-day in diameter.

3a A black hole with a mass of 2.6 million solar masses would have an event horizon smaller than the smallest dot in this diagram. A slow dribble of only 0.0002 solar masses of gas per year flowing into the black hole could produce the observed energy. A sudden increase as when a star falls in could produce a violent eruption.

The evidence of a massive black hole at the center of our galaxy seems conclusive. It is much too massive to be the remains of a dead star, however, and astronomers conclude that it probably formed as the galaxy first took shape.

Figure 15-27

The beautiful symmetry of the grand-design spiral pattern is clear in this image of the barred spiral galaxy NGC1300. Young star clusters containing hot, bright stars and ionized hydrogen are located along the arms, while dark lanes of dust mark the inner edges of the arms. (NASA and The Hubble Heritage Team. STScI/AURA)

15-4 The Nucleus

THE MOST MYSTERIOUS REGION OF OUR GALAXY is its very center, the nucleus. At visual wavelengths, this region is totally hidden by dust that dims the light it emits by 30 magnitudes. If a trillion (10^{12}) photons of light left the center of the galaxy on a journey to Earth, only one would make it through the dust. Consequently, visual-wavelength photos reveal nothing about the nucleus. Observations at radio and infrared wavelengths can see through the dust, and they paint a picture of tremendously crowded stars orbiting the center at high velocity. To understand what is happening at the center of our galaxy, you need to carefully compare observations and theories.

Observations

If you look up at the Milky Way on a dark night, you might notice a slight thickening in the direction of the constellation Sagittarius, but nothing specifically identifies this as the direction of the heart of the galaxy. Even Shapley's study of globular clusters could identify the center only approximately. When radio astronomers turned their telescopes toward Sagittarius, they found a complex collection of radio sources, with one, **Sagittarius A*** (abbreviated Sgr A* and usually pronounced "sadge A-star"), lying at the expected location of the galactic core.

Observations show that Sgr A* is less than an astronomical unit in diameter but is a strong source of radio energy. The tremendous amount of infrared radiation coming from the central area appears to be produced by crowded stars and by dust warmed by those stars. But what could be as small as Sgr A* and produce so much energy?

Read **Sagittarius A*** on pages 336–337 and notice three important points:

1. Observations at radio wavelengths reveal complex structures near Sgr A* caused by magnetic fields and by rapid star formation. Supernova remnants show that massive stars have formed there recently and died supernova deaths.

2. The center is crowded. Tremendous numbers of stars heat the dust, which emits strong infrared radiation.

3. Finally, there is evidence that Sgr A* is a supermassive black hole into which gas is flowing.

A supermassive black hole is an exciting idea, but scientists must always be aware of the difference between adequacy and necessity. A supermassive black hole is adequate to explain the observations, but is it necessary? Could there be other explanations? For example, astronomers have suggested that gas flowing inward could trigger tremendous bursts of star formation. Such theories have been considered and tested against the evidence, but none appears to be adequate to explain the observations. So far, the only theory that seems adequate is that our galaxy is home to a supermassive black hole.

Meanwhile, observations are allowing astronomers to refine their models. For instance, Sgr A* is not as bright in X rays as it should be if it has a hot accretion disk with matter constantly flowing into the black hole. Observations of X-ray and infrared flares lasting only a few hours suggest that mountain-size blobs

What Are We?

Children of the Milky Way

Hang on tight. The sun, with Earth in its clutch, is ripping along at about 220 km/sec (that's 490,000 mph) as it orbits the center of the Milky Way Galaxy. We live on a wildly moving ball of rock in a large galaxy that some people call our home galaxy, but the Milky Way is more than just our home. Perhaps "parent galaxy" would be a better name.

Except for hydrogen atoms, which have survived unchanged since the universe began, you and Earth are made of metals—atoms heavier than helium. There is no helium in your body, but there is plenty of carbon, nitrogen, and oxygen. There is calcium in your bones and iron in your blood. All of those atoms and more were cooked up inside stars or in their supernova deaths.

Stars are born when clouds of gas orbiting the center of the galaxy slam into the gas in spiral arms and are compressed. That process has given birth to generations of stars, and each generation has produced elements heavier than helium and spread them back into the interstellar medium. The abundance of metals has grown slowly in the galaxy. About 4.6 billion years ago a cloud of gas enriched in those heavy atoms slammed into a spiral arm and produced the sun, the Earth, and you. You have been cooked up by the Milky Way Galaxy—your parent galaxy.

of matter may occasionally fall into the black hole and be ripped apart and heated by tides. But the black hole may be mostly dormant and lack a fully developed hot accretion disk because little matter is flowing into it at the present time.

Such a supermasssive black hole could not be the remains of a single dead star. It contains too much mass. It probably formed when the galaxy first formed over 13 billion years ago. In later chapters you will see that such supermassive black holes are found at the centers of most galaxies.

◄ SCIENTIFIC ARGUMENT ►

Why do astronomers think the center of our galaxy contains a large mass?

Because scientific arguments hinge so often on evidence, they can involve discussions of measurement—one of the keys to science. The best way to measure the mass of an astronomical object is to watch something orbit around it. Then you can use Kepler's third law to find the mass inside the orbit. Because of gas and dust, astronomers can't see to the center of our galaxy at visual wavelengths, but infrared observations can detect individual stars orbiting Sgr A*. The star S2 has been particularly well observed, but a number of stars can be followed as they orbit the center. The sizes and periods of these orbits, interpreted using Kepler's laws, reveal that Sgr A* contains roughly 2.6 million solar masses.

Now build a new argument to analyze a different observation. **The strong infrared radiation at wavelengths longer than 4 microns (4000 nm) implies vast numbers of stars crowded into the center. Why?**

◄ ►

Study and Review

Summary

15-1 | The Nature of the Milky Way Galaxy

How do astronomers describe our galaxy?

- The hazy band of the Milky Way is our wheel-shaped galaxy seen from within, but its size and shape are not obvious. William and Caroline Herschel counted stars at many locations over the sky to show that our star system seemed to be shaped like a grindstone with the sun near the center.
- Later astronomers studied the distributions of stars, but, because gas and dust in space blocked their view of distant stars, they concluded the star system was only about 10 **kiloparsecs** in diameter with the sun at the center.
- In the early 20th century, Harlow Shapley calibrated Cepheid variable stars to find the distance to globular clusters and demonstrated that our galaxy is much larger than what we can see and that the sun is not at the center.
- Modern observations suggest that our galaxy contains a **disk component** about 75,000 ly in diameter and that the sun is two-thirds of the way from the center to the visible edge. The **nuclear bulge** around the center and an extensive **halo** containing old stars and little gas and dust make up the **spherical component.**
- The mass of the galaxy can be found from its **rotation curve.** Kepler's third law reveals that the galaxy contains over 100 billion solar masses. If stars orbited in **Keplerian motion,** more distant stars would orbit more slowly. They do not, and that shows that the halo may contain much more mass than is visible. Because the mass in this **galactic corona** is not emitting detectable electromagnetic radiation, astronomers call it **dark matter.**

15-2 | The Origin of the Milky Way

How did our galaxy form and evolve?

- The oldest star clusters reveal that the disk of our galaxy is younger than the halo, and the oldest globular clusters appear to be about 13 billion years old. So our galaxy must have formed about 13 billion years ago.
- Stellar populations are an important clue to the formation of our galaxy. The first stars to form, termed **population II stars,** were poor in elements heavier than helium—elements that astronomers call **metals.** As generations of stars manufactured metals in a process called **nucleosynthesis** and spread them back into the interstellar medium, the metal abundance of more recent generations increased. **Population I stars,** including the sun, are richer in metals.
- **Galactic fountains** produced by expanding supernova remnants may help spread metals throughout the disk.
- Because the halo is made up of population II stars and the disk is made up of population I stars, astronomers conclude that the halo formed first and the disk later. A theory that the galaxy formed from a single, roughly spherical cloud of gas and gradually flattened into a disk has been amended to include mergers with other galaxies and infalling gas contributing to the disk.

15-3 | Spiral Arms

What are the spiral arms?

- You can trace the spiral arms through the sun's neighborhood by using **spiral tracers** such as O and B stars; but, to extend the map over the entire galaxy, astronomers must use radio telescopes to see through the gas and dust.
- The most massive stars live such short lives they don't have time to move from their place of birth. Because they are found scattered along the spiral arms, astronomers conclude that the spiral arms are sites of star formation.
- The spiral **density wave theory** suggests that the spiral arms are regions of compression that move around the disk. When an orbiting gas cloud overtakes the compression wave, the gas cloud is compressed and forms stars. A density wave produces a two-armed spiral galaxy.
- Another process, **self-sustaining star formation,** may act to modify the arms with branches and spurs as the birth of massive stars triggers the formation of more stars by compressing neighboring gas clouds. This may account for the wooly appearance of **flocculent** galaxies.

15-4 | The Nucleus

What lies at the very center?

- The nucleus of the galaxy is invisible at visual wavelengths, but radio, infrared, and X-ray radiation can penetrate the gas and dust. These wavelengths reveal crowded central stars and warmed dust.
- The very center of the Milky Way Galaxy is marked by a radio source, **Sagittarius A*.** The core must be less than an astronomical unit in diameter, but the motions of stars around the center show that it must contain roughly 2.6 million solar masses. A supermassive black hole is the only object that could contain so much mass in such a small space.

Review Questions

ThomsonNOW Assess your understanding of this chapter's topics with additional quizzing and animations at www.thomsonedu.com.

WebAssign The problems from this chapter may be assigned online in WebAssign.

1. Why is it difficult to specify the dimensions of the disk and halo?
2. Why didn't astronomers before Shapley realize how large the galaxy is?
3. What evidence can you cite that our galaxy has a galactic corona?
4. Explain why some star clusters lose stars more slowly than others.
5. Contrast the motion of the disk stars and that of the halo stars. Why do their orbits differ?
6. Why are metals less abundant in older stars than in younger stars?
7. Why do metal-poor stars have a wider range of orbital shapes than metal-rich stars like the sun?
8. What evidence contradicts the traditional theory for the origin of our galaxy?
9. Why are all spiral tracers young?
10. Why couldn't spiral arms be physically connected structures? What would happen to them?
11. What kind of galaxy would the spiral density wave produce if it acted alone?
12. Why does self-sustaining star formation produce clouds of stars that look like segments of spiral arms?
13. Describe the kinds of observations you would make to study the galactic nucleus.
14. Why must astronomers use infrared telescopes to observe the motions of stars around Sgr A*?
15. What evidence can you cite that the nucleus of the galaxy contains a supermassive energy source that is very small in size?

16. **How Do We Know?** Calibration simplifies complex measurements, but how does that make the work of later astronomers dependent on the expertise of the astronomer who did the calibration?
17. **How Do We Know?** The story of a process makes the facts easier to remember, but that is not the true goal of the scientist. What is the real value of understanding a scientific process?

Discussion Questions

1. How would this chapter be different if interstellar dust did not scatter light?
2. Why doesn't the Milky Way circle the sky along the celestial equator or the ecliptic?

Problems

1. Make a scale sketch of our galaxy in cross section. Include the disk, sun, nucleus, halo, and some globular clusters. Try to draw the globular clusters to scale.
2. Because of dust, astronomers can see only about 5 kpc into the disk of the galaxy. What percentage of the galactic disk does that include? (*Hint:* Consider the area of the entire disk and the area visible from Earth.)
3. If the fastest passenger aircraft can fly 0.45 km/s (1000 mph), how long would it take to reach the sun? The galactic center? (*Hint:* 1 pc = 3×10^{13} km.)
4. If a typical halo star has an orbital velocity of 250 km/s, how long does it take to pass through the disk of the galaxy? Assume that the disk is 1000 pc thick.
5. If the RR Lyrae stars in a globular cluster have apparent magnitudes of 14, how far away is the cluster? (*Hint:* See Figure 12-14.)
6. If interstellar dust makes an RR Lyrae variable star look 1 magnitude fainter than it should, by how much will you overestimate its distance? (*Hint:* Use the magnitude–distance formula or Table 9-1.)
7. If a globular cluster is 10 minutes of arc in diameter and 8.5 kpc away, what is its diameter? (*Hint:* Use the small-angle formula.)
8. If you assume that a globular cluster 4 minutes of arc in diameter is actually 25 pc in diameter, how far away is it? (*Hint:* Use the small-angle formula.)
9. If the sun is 5 billion years old, how many times has it orbited the galaxy?
10. If the true distance to the center of the galaxy is found to be 7 kpc and the orbital velocity of the sun is 220 km/s, what is the minimum mass of the galaxy? (*Hint:* Use Kepler's third law.)
11. What temperature would interstellar dust have to have to radiate most strongly at 100 μm? (*Hints:* 1μm = 1000 nm. Use Wien's law, Chapter 7.)
12. Infrared radiation from the center of our galaxy with a wavelength of about 2 μm (2×10^{-6} m) comes mainly from cool stars. Use this wavelength as λ_{max} and find the temperature of the stars.
13. If an object at the center of our galaxy has a linear diameter of 10 AU, what will its angular diameter be as seen from Earth? (*Hint:* Use the small-angle formula, Chapter 3.)

Learning to Look

1. Why does the galaxy shown at the right have so much dust in its disk? How big do you suppose the halo of this galaxy really is?

NASA/Hubble Heritage Team/STScI/AURA

2. Why are the spiral arms in the galaxy at the right blue? What color would the halo be if it were bright enough to see in this photo?

NASA/Hubble Heritage Team and A. Riess, STScI

Virtual Astronomy Labs

Lab 16: Astronomical Distance Scales
In this lab you explore some methods for determining distances in astronomy.

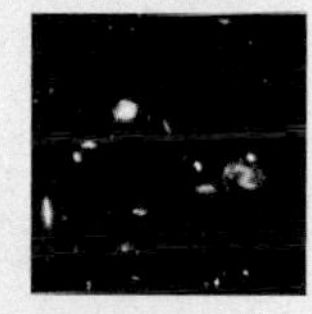

16 Galaxies

The galaxy M101 is 25 million light-years from Earth. Even if your spaceship could travel at the speed of light, it would take you 25 million years to reach this galaxy, which is one of the closer galaxies to our Milky Way Galaxy. (NASA and ESA/ Canada-France-Hawaii Telescope/NOAO/AURA/NSF)

Guidepost

Our Milky Way Galaxy is only one of the many billions of galaxies visible in the sky. This chapter will expand your horizon to discuss the different kinds of galaxies and their complex histories. Here you can expect answers to five essential questions:

What do galaxies look like?

How do astronomers measure the distances to galaxies?

How do galaxies differ in size, luminosity, and mass?

Do other galaxies contain supermassive black holes and dark matter, as does our own galaxy?

Why are there different kinds of galaxies?

As you begin studying galaxies, you will discover they are classified into different types, and that will lead you to insights into how galaxies form and evolve. It will also answer an important question about scientific methods:

How Do We Know? How does classification help scientists understand nature?

In the next chapter, you will discover that some galaxies are violently active, and that will give you more clues to the evolution of galaxies.

A hypothesis or theory is clear, decisive, and positive, but it is believed by no one but the man who created it. Experimental findings, on the other hand, are messy, inexact things which are believed by everyone except the man who did that work.

HARLOW SHAPLEY, *THROUGH RUGGED WAYS TO THE STARS*

SCIENCE FICTION HEROES FLIT EFFORTLESSLY between the stars, but almost none voyages between the galaxies. As you leave your home galaxy, the Milky Way, behind, you will voyage out into the depths of the universe, out among the galaxies, into space so deep it is unexplored even in fiction.

Before you can begin to understand the life stories of the galaxies, you must gather some basic data. How many kinds of galaxies are there? How big are they? How massive are they? That is, you need to characterize the family of galaxies just as you characterized the family of stars in Chapter 9.

16-1 The Family of Galaxies

LESS THAN A CENTURY AGO, astronomers did not understand that there were galaxies. Nineteenth-century telescopes revealed faint nebulae scattered among the stars, and some were spiral. Astronomers argued about the nature of these faint nebulae, but it was not understood until the 1920s that some were other galaxies much like our own; and it was not until recent decades that astronomical telescopes could reveal their tremendous beauty and intricacy (■ Figure 16-1).

The Discovery of Galaxies

Galaxies are faint objects, so they were not noticed until telescopes had grown large enough to gather significant amounts of light. In 1845, William Parsons, third Earl of Rosse in Ireland, built a telescope 72 in. in diameter. It was, for some years, the largest telescope in the world. Parsons lived before the invention of astronomical photography, so he had to view the faint nebulae directly at the eyepiece and sketch their shapes. He noticed that some have a spiral shape, and they became known as **spiral nebulae.** Parsons immediately concluded that the spiral nebulae were great spiral clouds of stars. The German philosopher Immanuel Kant in 1755 had proposed that the universe was filled with great wheels of stars that he called **island universes.** Parsons adopted that term for the spiral nebulae.

Not everyone agreed that the spiral nebulae were clouds of stars lying outside our own star system. An alternative point of view was that our star system was alone in an otherwise empty universe. Sir William Herschel had counted stars in different directions and had shown that our star system was shaped approximately like a grindstone (Chapter 15). Beyond the edge of this grindstone star system, space was supposed to be a limitless

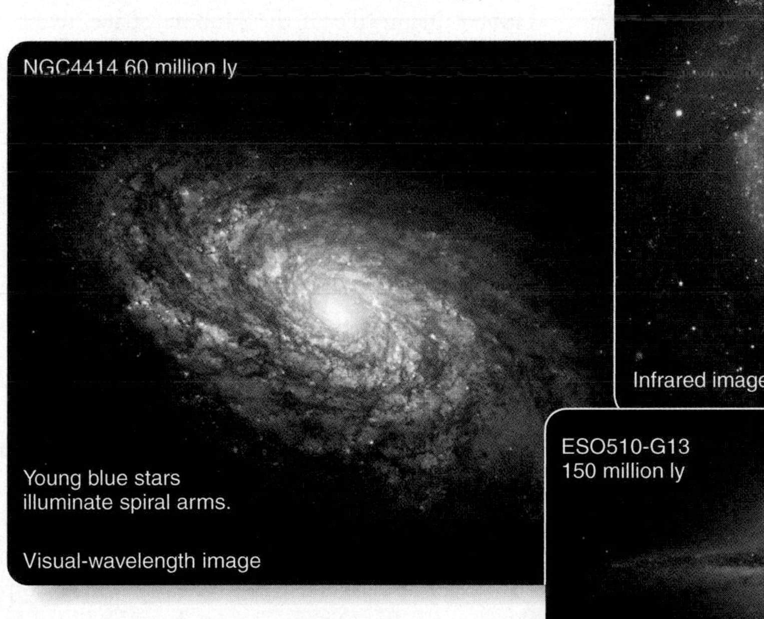

NGC4414 60 million ly
Young blue stars illuminate spiral arms.
Visual-wavelength image

M83 12 million ly
Dust clouds glow red in this infrared image.
Infrared image
ESO510-G13
150 million ly
Dusty disk of galaxy warped by interaction with another galaxy
New stars forming in dust clouds.
Visual-wavelength image

■ **Figure 16-1**

A century ago, photos of galaxies looked like spiral clouds of haze. Modern images of these relatively nearby galaxies reveal dramatically beautiful objects filled with newborn stars and clouds of gas and dust. (NGC4414 and ESO 510-G13: Hubble Heritage Team, Aura/STScI/NASA; M83: ESO)

■ Figure 16-2

An apparently empty spot on the sky only 1/30 the diameter of the full moon contains over 1500 galaxies in this extremely long time exposure known as the Northern Hubble Deep Field. Only four stars are visible in this image; they are sharp points of light with diffraction spikes produced by the telescope optics. Presumably the entire sky is similarly filled with galaxies. (R. Williams and the Hubble Deep Field Team, STScI, NASA)

void. The spiral nebulae, in this view, were nothing more than whirls of gas and faint stars within our star system.

The debate over the spiral nebulae could not be resolved in the 19th century because the telescopes were not large enough and photographic plates, when they became available in the late 1800s, were not sensitive enough. The spiral nebulae remained foggy swirls even in the best photographs, and astronomers continued to disagree over their true nature. In April 1920, two astronomers debated the issue at the National Academy of Science in Washington, D.C. Harlow Shapley of Mt. Wilson Observatory had recently shown that the Milky Way was a much larger star system than had been thought. He argued that the spiral nebulae were just nearby nebulae within the Milky Way star system. Heber D. Curtis of Lick Observatory argued that the spiral nebulae were island universes far outside the Milky Way star system. Historians of science mark this **Shapley–Curtis Debate** as a turning point in modern astronomy, but it was inconclusive. Shapley and Curtis cited the right evidence but drew incorrect conclusions. The disagreement was finally resolved, as is often the case in astronomy, by a bigger telescope.

On December 30, 1924, Mt. Wilson astronomer Edwin Hubble (namesake of the Hubble Space Telescope) announced that he had taken photographic plates of a few bright galaxies using the new 100-in. telescope. Not only could he detect individual stars in the spiral nebulae, but he could identify some of the stars as Cepheid variables with apparent magnitudes of about 18. For the brightest Cepheids, which are supergiants, to look that faint, they had to be very distant, and that meant the spiral nebulae had to be outside the Milky Way star system. They were galaxies.

How Many Galaxies Are There?

Like leaves on the forest floor, galaxies carpet the sky. Pick any spot on the sky away from the dust and gas of the Milky Way, and you are looking deep into space. Photons that have traveled for billions of years enter your eye, but they are too few to register on your retina. Only the largest telescopes can gather enough light to detect distant galaxies.

When astronomers picked a seemingly empty spot on the sky near the Big Dipper and used the Hubble Space Telescope to record an extremely long time exposure of 10 days' duration, they found thousands of galaxies crowded into the image. That image, now known as a Hubble Deep Field, contains a few relatively nearby galaxies along with many others that are over 10 billion light-years away (■ Figure 16-2).

How Do We Know? | 16-1

Classification in Science

What does a flamingo have in common with a *T. rex*? Classification is one of the most basic and most powerful of scientific tools. Establishing a system of classification is often the first step in studying a new aspect of nature, and it can produce unexpected insights.

Charles Darwin sailed around the world from 1831 to 1836 with a scientific expedition aboard the ship HMS *Beagle*. Everywhere he went, he studied the living things he saw and tried to classify them. For example, he classified different types of finches he saw on the Galapagos Islands based on the shapes of their beaks. He found that those that fed on seeds with hard shells had thick, powerful beaks, whereas those that picked insects out of deep crevices had long, thin beaks. His classifications of these and other animals led him to think about how natural selection shapes creatures to survive in their environment, which led him to understand how living things evolve.

Years after Darwin's work, paleontologists classified dinosaurs into two orders, lizard-hipped and bird-hipped dinosaurs. This classification, based on the shapes of dinosaur hip joints, helped the scientists understand patterns of evolution of dinosaurs. It also led to the conclusion that modern birds like the finches that Darwin saw on the Galapagos evolved from dinosaurs.

Astronomers use classifications of galaxies, stars, moons, and many other objects to help them see patterns, trace evolutions, and generally make sense of the astronomical world. Whenever you encounter a scientific discussion, look for the classifications on which it is based. Classifications are the orderly framework on which much of science is built.

The careful classification of living things has revealed that the birds, including this flamingo, are descended from dinosaurs. (M. Seeds)

Since the first Hubble Deep Field was recorded, other deep fields have been imaged in other parts of the sky and at other wavelengths. The GOODS program (Great Observatories Origins Deep Survey) has used the Hubble Space Telescope with other space telescopes such as the Spitzer Space Telescope and the Chandra X-ray Observatory and with some of the largest ground-based telescopes to image selected deep fields at many wavelengths. The GOODS images show that the first Hubble Deep Field was not unusual; the entire sky appears to be just as thickly covered with galaxies.

The GOODS study of distant galaxies is especially amazing when you think that, less than a century ago, humanity did not know that we live in a galaxy or that the universe contained other galaxies. Today telescopes are capable of detecting a few 100 billion galaxies, and new, larger telescopes will reveal even more. The discovery of galaxies is one of the turning points in astronomy.

The Shapes of Galaxies

Look at the galaxies in the Hubble Deep Field, and you will see galaxies of different shapes. Some are spiral, some are elliptical shapes, and some are irregular. Astronomers classify galaxies according to their shapes in photographs made at visual wavelengths using a system developed by Edwin Hubble in the 1920s. Such systems of classification are a fundamental technique in science (**How Do We Know? 16-1**).

Read **Galaxy Classification** on pages 346–347 and notice three important points and four new terms that describe the main types of galaxies:

1. Many galaxies have no disk, no spiral arms, and almost no gas and dust. These *elliptical galaxies* range from huge giants to small dwarfs.
2. Disk-shaped galaxies usually have spiral arms and contain gas and dust. Many of these *spiral galaxies* have a nucleus shaped like an elongated bar and are called *barred spiral galaxies*. A few disk galaxies contain little gas and dust.
3. Finally, notice the *irregular galaxies,* which are highly irregular in shape and tend to be rich in gas and dust.

The amount of gas and dust in a galaxy strongly influences its appearance. Galaxies rich in gas and dust usually have active star formation and emission nebulae and contain hot, bright stars. That gives the disk galaxies a blue tint. Galaxies that are poor in gas and dust contain few or none of these highly luminous stars and no emission nebulae and consequently look redder and have a much more uniform appearance.

Spiral galaxies are clearly disk shaped, but the true, three-dimensional shape of elliptical galaxies isn't obvious from images. Some are spherical, but the more elongated elliptical galaxies could be shaped like flattened spheres (bun shaped) or like elongated spheres (football shaped). Some may even have three different diameters; they are longer than they are thick and thicker than they

Galaxy Classification

1 **Elliptical galaxies** are round or elliptical, contain no visible gas and dust and have few or no bright stars. They are classified with a numerical index ranging from 1 to 7; E0s are round, and E7s are highly elliptical. The index is calculated from the largest and smallest diameter of the galaxy used in the following formula and rounded to the nearest integer.

$$\frac{10(a-b)}{a}$$

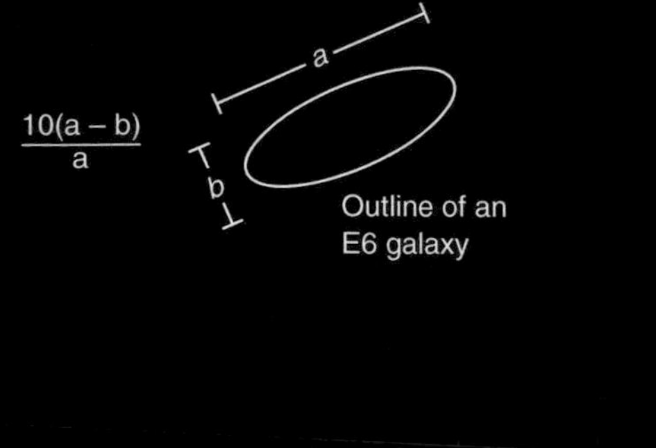

Outline of an E6 galaxy

AURA/NOAO/NSF

Visual-wavelength image

M87 is a giant elliptical galaxy classified E1. It is a number of times larger in diameter than our own galaxy and is surrounded by a swarm of over 500 globular clusters.

Anglo-Australian Telescope Board

The Leo 1 dwarf elliptical galaxy is not many times bigger than a globular cluster.

2 **Spiral galaxies** contain a disk and spiral arms. Their halo stars are not visible, but presumably all spiral galaxies have halos. Spirals contain gas and dust and hot, bright O and B stars, as shown at right. The presence of short-lived O and B stars alerts us that star formation is occurring in these galaxies. Sa galaxies have larger nuclei, less gas and dust, and fewer hot, bright stars. Sc galaxies have small nuclei, lots of gas and dust, and many hot, bright stars. Sb galaxies are intermediate.

Anglo-Australian Telescope Board

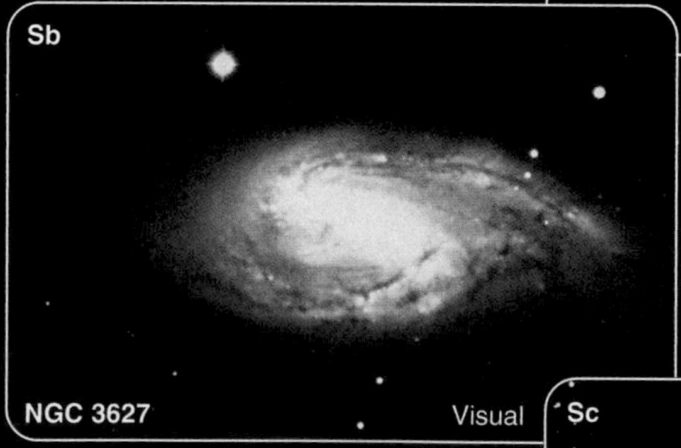

Sb

NGC 3627 Visual

Anglo-Australian Telescope Board

2a Roughly 2/3 of all spiral galaxies are **barred spiral galaxies** classified SBa, SBb, and SBc. They have an elongated nucleus with spiral arms springing from the ends of the bar, as shown at left. Our own galaxy is a barred spiral.

ThomsonNOW

Sign in at www.thomsonedu.com and go to ThomsonNOW to see Active Figure "Galaxy Types" and review the classification of galaxies.

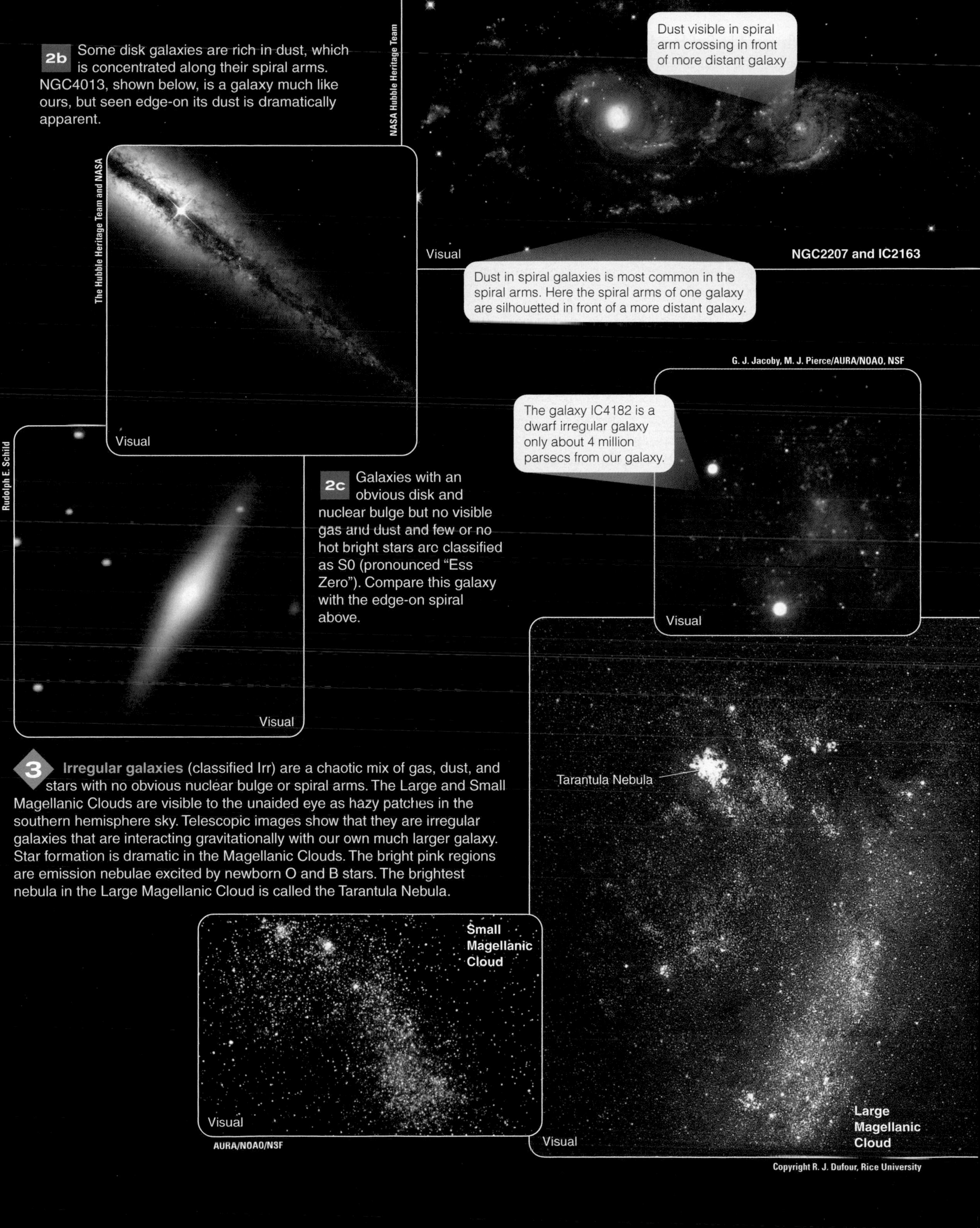
NASA Hubble Heritage Team
Dust visible in spiral arm crossing in front of more distant galaxy
Visual
NGC2207 and IC2163
Dust in spiral galaxies is most common in the spiral arms. Here the spiral arms of one galaxy are silhouetted in front of a more distant galaxy.
The Hubble Heritage Team and NASA
Visual
Rudolph E. Schild
Visual
G. J. Jacoby, M. J. Pierce/AURA/NOAO, NSF
The galaxy IC4182 is a dwarf irregular galaxy only about 4 million parsecs from our galaxy.
Visual
Tarantula Nebula
Small Magellanic Cloud
Visual
AURA/NOAO/NSF
Large Magellanic Cloud
Visual
Copyright R. J. Dufour, Rice University

2b Some disk galaxies are rich in dust, which is concentrated along their spiral arms. NGC4013, shown below, is a galaxy much like ours, but seen edge-on its dust is dramatically apparent.

2c Galaxies with an obvious disk and nuclear bulge but no visible gas and dust and few or no hot bright stars are classified as S0 (pronounced "Ess Zero"). Compare this galaxy with the edge-on spiral above.

3 **Irregular galaxies** (classified Irr) are a chaotic mix of gas, dust, and stars with no obvious nuclear bulge or spiral arms. The Large and Small Magellanic Clouds are visible to the unaided eye as hazy patches in the southern hemisphere sky. Telescopic images show that they are irregular galaxies that are interacting gravitationally with our own much larger galaxy. Star formation is dramatic in the Magellanic Clouds. The bright pink regions are emission nebulae excited by newborn O and B stars. The brightest nebula in the Large Magellanic Cloud is called the Tarantula Nebula.

Visual-wavelength image

■ **Figure 16-8**

The galaxy M33 is near our Milky Way Galaxy in space, and it can be studied in detail. The velocities at the center of the galaxy are low, showing that it does not contain a supermassive black hole at its center. It also lacks a nuclear bulge, and that confirms the observation that the mass of a supermassive black hole is related to the size of a galaxy's nuclear bulge. (Bill Schoening/AURA/NOAO/NSF)

Dark Matter in Galaxies

Given the size and luminosity of a galaxy, astronomers can make a rough guess as to the amount of matter it should contain. They know how much light stars produce, and they know about how much matter there is between the stars, so it is quite possible to estimate very roughly the mass of a galaxy from its luminosity. But when astronomers compare their estimates with measured masses of galaxies, they find that the measured masses are much too large. Measured masses of galaxies amount to 10 times more mass than can be seen. This must mean that nearly all galaxies contain dark matter. Theorists conclude that dark matter must be made up of some as yet undiscovered subatomic particles that do not interact with normal matter, with each other, or with light. Dark matter is detectable only through its gravitational field.

X-ray observations reveal more evidence of dark matter. X-ray images of galaxy clusters show that many of them are filled with very hot, low-density gas. The amount of gas present is much too small to account for the dark matter. Rather, the gas is important because it is very hot and its rapidly moving atoms have not leaked away. Evidently the gas is held in the cluster by a strong gravitational field. To have a high enough escape velocity to hold the hot gas, the cluster must contain much more matter than what astronomers see. The detectable galaxies in the Coma cluster, for instance, amount to only a small fraction of the total mass of the cluster (■ Figure 16-9).

■ **Figure 16-9**

(a) The Coma cluster of galaxies contains at least 1000 galaxies and is especially rich in E and S0 galaxies. Two giant galaxies lie near its center. Only the central area of the cluster is shown in this image. If the cluster were visible in the sky, it would span eight times the diameter of the full moon. (Gregory Bothun, University of Oregon) (b) In false colors, this X-ray image of the Coma cluster shows it filled and surrounded by hot gas. Note that the two brightest galaxies are visible in the X-ray image. (NASA/CXC/SAO/A.Vikhlinin et al.)

Gas in the Coma Cluster
has a temperature over
100 million K.
b
X-ray image
a Visual-wavelength image

Dark matter is not an insignificant issue. Observations of galaxies and clusters of galaxies show that approximately 90 percent of the matter in the universe is dark matter. The universe you see—the kind of matter that you and the stars are made of—has been compared to the foam on an invisible ocean. You will find further evidence of dark matter when you study cosmology in Chapter 18.

Gravitational Lensing and Dark Matter

When you solve a math problem and get an answer that doesn't seem right, you check your work for a mistake. Astronomers using Newton's laws (orbital velocity and escape velocity) to measure the mass of galaxies find evidence of large amounts of dark matter, so they look for ways to check their work. Fortunately, there is another way to detect dark matter that does not depend on Newton's laws. Astronomers can follow a light beam.

Albert Einstein described gravity as a curvature of space-time (Chapter 5). The presence of mass actually distorts space-time, and that is what you feel as gravity. He predicted that a light beam traveling through a gravitational field would be deflected by the curvature of space-time much as a golf ball is deflected as it rolls over a curved putting green. That effect has been observed and is a strong confirmation that Einstein's theories are correct.

Gravitational lensing occurs when light from a distant object passes a nearby massive object and is deflected by the gravitational field. The gravitational field of the nearby object is actually a region of curved space-time that acts as a lens to deflect the passing light. Astronomers can use gravitational lensing to detect dark matter when light from very distant galaxies passes through a cluster of galaxies on its way to Earth and is deflected by the strong curvature. The distortion can produce multiple images of the distant galaxies and distort them into arcs. The amount of the distortion depends on the mass of the cluster of galaxies (■ Figure 16-10a). Observations of gravitational lensing made with very large telescopes reveal that clusters of galaxies contain far more matter than what can be seen. That is, they contain large amounts of dark matter. This confirmation of the existence of dark matter is independent of Newton's laws and gives astronomers much greater confidence that dark matter is real.

Dark matter is difficult to detect, and it is even harder to explain. The halos of galaxies contain faint objects such as white

■ Figure 16-10

(a) The gravitational lens effect is visible in galaxy cluster 0024+1654 as its mass bends the light of a much more distant galaxy to produce arcs that are actually distorted images of the distant galaxy. This reveals that the galaxy cluster must contain large amounts of dark matter. (W. N. Colley and E. Turner, Princeton, and J. A. Tyson, Bell Labs, and NASA) (b) When two galaxy clusters passed through each other, normal matter (pink) collided and was swept out of the clusters, but the dark matter (purple), detected by gravitational lensing, was not affected. (NASA/CXC/CfA/STScI/Magellan/ESO WFI)

reradiates the energy in the infrared. Some of these galaxies are a hundred times more luminous than our Milky Way Galaxy but so deeply shrouded in dust that they are very dim at visible wavelengths. Such **ultraluminous infrared galaxies** show evidence of tidal tails and are probably the result of the merger of three or more galaxies that triggered firestorms of star formation and generated tremendous clouds of dust. The Antennae (page 361) contain over 15 billion solar masses of hydrogen gas and will become a starburst galaxy as the merger triggers rapid star formation. As these galaxies use up the last of their gas and dust making stars, they will probably become normal elliptical galaxies or merge to form a single elliptical.

Another process can help clear the gas and dust out of elliptical galaxies. Supernovae in a starburst galaxy can blow away the remaining gas and dust, and a burst of rapid star formation produces lots of supernova explosions.

In this way, a few collisions and mergers could leave a galaxy with no gas and dust from which to make new stars and could scramble the orbits of the remaining stars to produce an elliptical shape. Astronomers now suspect that most elliptical galaxies are formed by the merger of at least two or three galaxies of comparable size.

In contrast, spirals have never suffered major collisions. Their thin disks are delicate and would be destroyed by the tidal forces generated during a collision with a massive galaxy. Also, they retain plenty of gas and dust and continue making stars, so they have never experienced a major starburst triggered by a merger.

Of course, spiral galaxies can safely cannibalize smaller galaxies with no ill effects. The small galaxies do not generate strong tides to distort a full-sized galaxy. You have seen plenty of evidence of cannibalism in our own galaxy, and some astronomers suspect that much of the halo consists of the remains of cannibalized galaxies. But our Milky Way Galaxy has never collided with a similarly large galaxy. Such a collision would trigger star formation, use up gas and dust, scramble the orbits of stars, and destroy the thin disk. In a few billion years, our galaxy may merge with the approaching Andromeda Galaxy, and the final result will be an elliptical galaxy.

Just as people do, spiral galaxies may need some interaction to fully develop. Only one in 10,000 galaxies is isolated from its neighbors, and, although most of these are disk galaxies, many are flocculent without a strong two-armed spiral pattern. Some isolated galaxies are rich in gas and dust but have little star formation. All of this suggests that a gentle interaction with a neighboring galaxy is necessary to stimulate spiral structure and star formation.

Earlier, the Virgo cluster was cited as a rich cluster, so you may find it surprising to learn that it contains lots of spiral galaxies. Rich clusters are supposed to contain few spirals. Astronomers understand that the cluster is so extremely massive that the galaxies orbit the center of mass of the cluster at very high velocities; and, when they encounter each other, they rush past too quickly to interact strongly. The galaxies are whittled down a bit, but many are still disk galaxies with spiral patterns.

Barred spiral galaxies may also be the products of tidal interactions. Mathematical models show that bars are not stable and eventually dissipate. Tidal interactions with other galaxies may regenerate the bars. Well over half of all spiral galaxies have bars, suggesting that these tidal interactions are common.

Other processes can alter galaxies. The S0 galaxies retain a disk shape but lack gas and dust. Although they could have lost gas and dust during starburst episodes, they may also have lost it as they orbited through the gas in their clusters of galaxies. A galaxy moving rapidly through the thin gas filling a cluster would encounter a tremendous wind that could blow the galaxy's gas and dust away. In this way, a disk-shaped spiral galaxy could be reduced to an S0 galaxy. For example, X-ray observations show that the Coma cluster contains thin, hot gas between the galaxies (Figure 16-9), and astronomers have located, in a similar cluster, a galaxy in the act of plunging through such gas and being stripped of its gas and dust (■ Figure 16-15).

Small galaxies may be produced in a number of ways. The dwarf ellipticals are too small to be produced by mergers, but they could be fragments of galaxies ripped free during interactions. Another possibility is that the dwarf ellipticals are small galaxies that have plunged through the gas in a cluster of galaxies and lost their gas and dust. In contrast, the irregular galaxies may

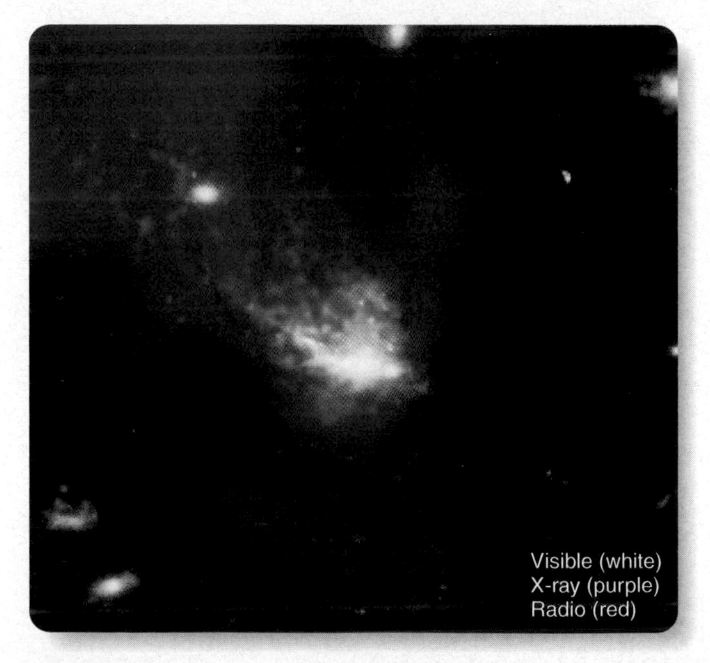

■ Figure 16-15

The distorted galaxy C153 is orbiting through the thin gas in its home cluster of galaxies at 4.5 million miles per hour. At that speed, it feels a tremendous wind stripping gas out of the galaxy in a trail 200,000 ly long. Such galaxies could quickly lose almost all of their gas and dust. (NASA, W. Keel, F. Owen, M. Ledlow, and D. Wang)

What Are We?

Small and Proud

Do you feel insignificant yet? You are riding a small planet orbiting a humdrum star that is just one of at least 100 billion in the Milky Way Galaxy, and you have just learned that there are at least a few 100 billion galaxies visible with existing telescopes.

When you look at galaxies you are looking across voids deeper than human imagination. You can express such distances with numbers and say a certain galaxy is 5 billion light-years from Earth, but the distance is truly beyond human comprehension. Furthermore, looking at distant galaxies also leads you back in time. You see distant galaxies as they were billions of years ago. The realm of the galaxies is deep space and deep time.

Some people say astronomy makes them feel humble, but before you agree, consider that you can feel small without feeling humble. We humans live out our little lives on our little planet, but we are figuring out some of the most profound mysteries of the universe. We are exploring deep space and deep time and coming to understand what galaxies are and how they evolve. Most of all, we humans are beginning to understand what we are. That's something to be proud of.

be small fragments splashed from larger galaxies during collisions but retaining enough gas and dust to continue forming stars.

Other factors must influence the evolution of galaxies. The hot gas in galaxy clusters warns that galaxies rarely form in isolation. Cooler gas clouds may fall into galaxies and add material for star formation. These processes are just beginning to be understood.

A good theory helps you understand how nature works, and astronomers are beginning to understand the exciting and complex story of the galaxies. Nevertheless, it is already clear that galaxy evolution is much like a pie-throwing contest and just about as neat.

The Farthest Galaxies

Observations made with the largest and most sophisticated telescopes are taking astronomers back to the age of galaxy formation. At great distances the look-back time is so great that they see the universe as it was soon after the galaxies began to form. There were more spirals then and fewer ellipticals. On the whole, galaxies long ago were more compact and more irregular than they are now. The observations also show that galaxies were closer together long ago; about a third of all distant galaxies are in close pairs, but only 7 percent of nearby galaxies are in pairs. The observational evidence clearly supports the hypothesis that galaxies have evolved by merger.

At great distances, astronomers see tremendous numbers of small, blue, irregularly shaped galaxies that have been called blue dwarfs. The blue color indicates that they are rapidly forming stars, but the role of these blue dwarfs in the formation of galaxies is not clear. Blue dwarf galaxies are not found at small look-back times, so they no longer exist in the present universe. They may be clouds of gas and stars that were absorbed long ago in the formation of larger galaxies, or they may have used up their gas and dust and faded into obscurity.

At the limits of the largest telescopes, astronomers see faint red galaxies (■ Figure 16-16). They look red from Earth because

■ **Figure 16-16**

Arrows point to three very distant red galaxies. The look-back time to these galaxies is so large they appear as they were when the universe was only a billion years old. Such highly redshifted galaxies are thought to be among the first to form stars and begin shining after the beginning of the universe about 14 billion years ago. (NASA, H.-J. Yan, R. Windhorst and S. Cohen, Arizona State University)

of their great redshifts. The floods of ultraviolet light emitted by these star-forming galaxies have been shifted into the far-red part of the spectrum. Their look-back times are so great that they appear as they were when the universe was only a billion years old. These galaxies seem to be among the first to begin shining after the beginning of the universe, a story told in Chapter 18.

◄ SCIENTIFIC ARGUMENT ►

How did elliptical galaxies get that way?

Telling the story of a natural object is a goal that lies at the heart of science, and many scientific arguments tell such tales. A growing body of evidence suggests that elliptical galaxies have been subject to collisions in their past and that spiral galaxies have not. During collisions, a galaxy can be driven to use up its gas and dust in a burst of star formation, and the resulting supernova explosions can help drive gas and dust out of the galaxy. This explains why elliptical galaxies now contain little star-making material. The beautiful disk typical of spiral galaxies is very orderly, with all the stars following similar orbits. When galaxies collide, the stellar orbits get scrambled, and an orderly disk galaxy could be converted into a chaotic swarm of stars typical of elliptical galaxies. It seems likely that elliptical galaxies have had much more complex histories than spiral galaxies have had.

When scientists create a scientific argument to tell the story of a natural object, they use evidence to make the story as true as it can be. Expand your argument. **What evidence can you cite to support the story in the preceding paragraph?**

◄ ►

Study and Review

Summary

16-1 | The Family of Galaxies

What do galaxies look like?

- Through 19th-century telescopes, galaxies looked like hazy **spiral nebulae.** Some astronomers said they were other star systems sometimes called **island universes,** but others said they were clouds of gas inside the Milky Way system. The controversy culminated in the **Shapley-Curtis Debate** in 1920.
- A few years later, with the construction of larger telescopes, astronomers could identify stars, including Cepheid variable stars, in the spiral nebulae. That showed that the spiral nebulae were galaxies.
- Astronomers divide galaxies into three classes—**elliptical, spiral,** and **irregular**—with subclasses specifying the galaxy's shape.
- Elliptical galaxies contain little gas and dust and cannot make new stars. Consequently, they lack hot, blue stars and have a reddish tint.
- Spiral galaxies contain more gas and dust in their disks and support active star formation, especially along the spiral arms. Some of the newborn stars are massive, hot, and blue, and that gives the spiral arms a blue tint. About two-thirds of spirals are **barred spiral galaxies.**
- The halo and nuclear bulge of a spiral galaxy usually lack gas and dust and contain little star formation. The halos and nuclear bulges have a reddish tint because they lack hot, blue stars.
- Irregular galaxies have no obvious shape but contain gas and dust and support star formation.

16-2 | Measuring the Properties of Galaxies

How do astronomers measure the distances to galaxies?

- Galaxies are so distant astronomers measure their distances in **megaparsecs**—millions of parsecs.
- Astronomers find the distance to galaxies using **distance indicators,** sometimes called **standard candles,** objects of known luminosity. The most accurate distance indicators are the Cepheid variable stars. Globular clusters and type Ia supernovae explosions have also been calibrated as distance indicators.
- By calibrating additional distance indicators using galaxies of known distance, astronomers have built a **distance scale.** The Cepheid variable stars are the most dependable.
- When astronomers look at a distant galaxy, they see it as it was when it emitted the light now reaching Earth. The **look-back time** to distant galaxies can be a significant fraction of the age of the universe.
- According to the **Hubble law,** the apparent velocity of recession of a galaxy equals its distance times the **Hubble constant.** Astronomers can estimate the distance to a galaxy by observing its redshift, calculating its apparent velocity of recession, and then dividing by the Hubble constant.

How do galaxies differ in size, luminosity, and mass?

- Once the distance to a galaxy is known, its diameter can be found from the small-angle formula and its luminosity from the magnitude-distance relation.
- Astronomers measure the masses of galaxies in two basic ways. The **rotation curve** of a galaxy shows the orbital motion of its stars, and astronomers can use the **rotation curve method** to find the galaxy's mass.
- The **cluster method** uses the velocities of the galaxies in a cluster to find the total mass of the cluster. The **velocity dispersion method** uses the velocities of the stars in a galaxy to find the total mass of the galaxy.
- Galaxies come in a wide range of sizes and masses. Some dwarf ellipticals and dwarf irregular galaxies are only a few percent the size and luminosity of our galaxy, but some giant elliptical galaxies are five times larger than the Milky Way Galaxy.

Do other galaxies contain supermassive black holes and dark matter, as does our own galaxy?

- Stars near the centers of galaxies are following small orbits at high velocities, which suggests the presence of supermassive black holes in the centers of most galaxies.
- The mass of a galaxy's supermassive black hole is proportional to the mass of its nuclear bulge. That shows that the supermassive black holes must have formed when the galaxy formed.
- Observations of individual galaxies show that galaxies contain 10 to 100 times more dark matter than visible matter.
- The hot gas held inside some clusters of galaxies and the **gravitational lensing** caused by the mass of galaxy clusters reveal that the clusters must be much more massive than can be accounted for by the visible matter—further evidence of dark matter.

16-3 | The Evolution of Galaxies

Why are there different kinds of galaxies?

- **Rich clusters** of galaxies contain thousands of galaxies with fewer spirals and more ellipticals. **Poor clusters** of galaxies contain few galaxies with a larger proportion of spirals. This is evidence that galaxies evolve by collisions and mergers.
- When galaxies collide, tides twist and distort their shapes and can produce **tidal tails.**
- Large galaxies can absorb smaller galaxies in what is called **galactic cannibalism.** You can see clear evidence that our own Milky Way Galaxy is devouring some of the small galaxies that orbit nearby and that our galaxy has consumed other small galaxies in the past.
- Shells of stars, counterrotating parts of galaxies, streams of stars in the halos of galaxies, and multiple nuclei are evidence that galaxies can merge.
- **Ring galaxies** are produced by high-speed collisions in which a small galaxy plunges through a larger galaxy perpendicular to its disk.
- The compression of gas clouds can trigger bursts of star formation, producing **starburst galaxies.** The rapid star formation can produce lots of dust, which is warmed by the stars to emit infrared radiation, making the galaxy an **ultraluminous infrared galaxy.**
- The merger of two larger galaxies can scramble star orbits and drive bursts of star formation to use up gas and dust. Most larger ellipticals have evidently been produced by past mergers.
- Spiral galaxies have thin, delicate disks and appear not to have suffered mergers with large galaxies.
- A galaxy moving through the gas in a cluster of galaxies can be stripped of its own gas and dust and may become an S0 galaxy.
- Rare isolated galaxies tend to be spirals and lack a bar or a strong two-armed spiral pattern, which suggests that gentle interactions with neighbors are needed to stimulate the formation of bars and spiral arms.
- At great distance and great look-back times, the largest telescopes reveal that galaxies were smaller, more irregular, and closer together. There were more spirals and fewer ellipticals long ago.
- At the largest distances, astronomers find small irregular clouds of stars that may be the objects that fell together to begin forming galaxies when the universe was very young.

Review Questions

ThomsonNOW™ Assess your understanding of this chapter's topics with additional quizzing and animations at www.thomsonedu.com.

WebAssign The problems from this chapter may be assigned online in WebAssign.

1. If a civilization lived on a planet in an E0 galaxy, do you think they would have a Milky Way in their sky? Why or why not?
2. Why can't the evolution of galaxies go from elliptical to spiral? From spiral to elliptical?
3. If all elliptical galaxies had three different diameters, you would never see an elliptical galaxy with a circular outline on a photograph. True or false? Explain your answer. (*Hint:* Can a football ever cast a circular shadow?)
4. What is the difference between an Sa and an Sb galaxy? Between an S0 and an Sa galaxy? Between an Sb and an SBb galaxy? Between an E7 and an S0 galaxy?
5. Why wouldn't white dwarfs make good distance indicators?
6. Why isn't the look-back time important among nearby galaxies?
7. Explain how the rotation curve method of finding a galaxy's mass is similar to the method used to find the masses of binary stars.
8. Explain how the Hubble law allows you to estimate the distances to galaxies.
9. How can collisions affect the shape of galaxies?
10. What evidence can you cite that galactic cannibalism really happens?
11. Describe the future evolution of a galaxy that astronomers now see as a starburst galaxy. What will happen to its interstellar medium?
12. Why does the gas held in a cluster of galaxies help determine the nature of the galaxies in a cluster?
13. **How Do We Know?** Classification helped Darwin understand how creatures evolve. How has classification helped you understand how galaxies evolve?

Discussion Questions

1. From what you know about star formation and the evolution of galaxies, do you think the Infrared Astronomy Satellite should have found irregular galaxies to be bright or faint in the infrared? Why or why not? What about starburst galaxies? What about elliptical galaxies?
2. Imagine that you could observe a few gas clouds at such a high look-back time that they are just beginning to form one of the first galaxies. Further, suppose you discovered that the gas was metal rich. Would that support or contradict your understanding of galaxy formation?

Problems

1. If a galaxy contains a type I (classical) Cepheid with a period of 30 days and an apparent magnitude of 20, what is the distance to the galaxy?
2. If you find a galaxy that contains globular clusters that are 2 seconds of arc in diameter, how far away is the galaxy? (*Hints:* Assume that a globular cluster is 25 pc in diameter and use the small-angle formula.)
3. If a galaxy contains a supernova that at its brightest has an apparent magnitude of 17, how far away is the galaxy? (*Hint:* Assume that the absolute magnitude of the supernova is -19.)
4. If you find a galaxy that is the same size and mass as our Milky Way Galaxy, what orbital velocity would a small satellite galaxy have if it orbited 50 kpc from the center of the larger galaxy?
5. Find the orbital period of the satellite galaxy described in Problem 4.
6. If a galaxy has an apparent radial velocity of 2000 km/s and the Hubble constant is 70 km/s/Mpc, how far away is the galaxy? (*Hint:* Use the Hubble law.)
7. If you find a galaxy that is 20 minutes of arc in diameter and you measure its distance to be 1 Mpc, what is its diameter?
8. Suppose you found a galaxy in which the outer stars have orbital velocities of 150 km/s. If the radius of the galaxy is 4 kpc, what is the orbital period of the outer stars? (*Hints:* 1 pc $= 3.08 \times 10^{13}$ km, and 1 yr $= 3.15 \times 10^{7}$ seconds.)
9. A galaxy has been found that is 5 kpc in radius and whose outer stars orbit the center with a period of 200 million years. What is the mass of the galaxy? On what assumptions does this result depend?
10. Among the globular clusters orbiting a distant galaxy, the fastest is traveling 420 km/s and is located 11 kpc from the center of the galaxy. Assuming the globular cluster is just barely gravitationally bound to the galaxy, what is the mass of the galaxy? (*Hint:* The galaxy had a slightly faster globular cluster, but it escaped some time ago. What is the escape velocity?)

How Do We Know? | 17-1

Statistical Evidence

How can statistics help you decide where to buy a house? Some scientific evidence is statistical. Observations suggest, for example, that Seyfert galaxies are three times more likely to have a nearby companion than a normal galaxy is. This is statistical evidence because you can't be certain that any specific Seyfert galaxy will have a companion. How can scientists use statistical evidence to learn about nature when statistics contains built-in uncertainty?

Meteorologists use statistics to determine how frequently storms of a certain size are likely to occur. Small storms happen every year, but medium-sized storms may happen on average only every ten years. Hundred-year storms are much more powerful, but occur much less frequently—on average only once in a hundred years.

Those meteorological statistics can help you make informed decisions—as long as you understand the powers and limitations of statistics. Would you buy a house protected by a river levee that was not designed to withstand a hundred-year storm? In any one year, the chance of your house being destroyed would be only 1 in 100. You know the storm will hit eventually, but you don't know when. If you buy the house, a storm might destroy the levee the next year, but you might own the house for your whole life and never see a hundred-year storm. The statistics can't tell you anything about a specific year.

Before you buy that house, there is an important question you should ask the meteorologists. "How much data do you have on storms?" If they only have 10 years of data, then they don't really know much about hundred-year storms. If they have three centuries of data, then their statistical data are significant.

Sometimes people dismiss important warnings by saying, "Oh, that's only statistics." Scientists can use statistical evidence if it passes two tests. It cannot be used to draw conclusions about specific cases, and it must be based on large enough data samples so the statistics is significant. With these restrictions, statistical evidence can be a powerful scientific tool.

Statistics can tell you that a bad storm will eventually hit, but it can't tell you when. (Marko Georgiev/Getty Images)

evidence (**How Do We Know? 17-1**) hints that the activity in Seyfert galaxies may have been triggered by collisions or interactions with companions. Some Seyferts are expelling matter in oppositely directed flows (Figure 17-2b), a geometry you saw on smaller scales when you studied matter falling into accretion disks around neutron stars and black holes.

Encounters with other galaxies could throw matter into a black hole; and, as you have seen in Chapter 14, large amounts of energy can be liberated by matter flowing into a black hole. Later in this chapter, you will follow this idea further, but for now you need to search for more evidence of supermassive black holes by turning your attention to a different kind of active galaxy that emits powerful radio signals.

Double-Lobed Radio Sources

Beginning in the 1950s, radio astronomers found that some sources of radio energy in the sky consisted of pairs of radio-bright regions. When optical telescopes studied the locations of these **double-lobed radio sources,** they revealed galaxies located between the two radio lobes. (See Fornax A on page 367.) Apparently, the central galaxies were producing the radio lobes.

Read **Cosmic Jets and Radio Lobes** on pages 372–373 and notice four important points and two new terms:

1. The shapes of radio lobes suggest that they are inflated by jets of excited gas emerging from the central galaxy. This has been called the *double-exhaust model,* and the presence of *hot spots* and synchrotron radiation shows that the jets are very powerful.

2. Active galaxies with jets and radio lobes are often deformed or interacting with other galaxies.

3. The complex shapes of some jets and radio lobes can be explained by the motions of the active galactic nuclei. A good example of this is 3C31 (the 31st source in the *Third Cambridge Catalog of Radio Sources*) with its twisting radio lobes.

4. These jets seem to be produced when matter flows into an accretion disk around a central black hole. You have seen similar jets produced by accretion disks around protostars, neutron stars, and stellar mass black holes. Although the details are not understood, the same process seems to be producing all of these jets.

All of the evidence suggests that supermassive black holes at the center of galaxies can eject powerful jets and inflate radio lobes. Now it is time to try to understand what these supermassive energy machines are like.

Exploring Supermassive Black Holes

Supermassive black holes with masses of millions or billions of solar masses sound fantastic, but they are real, and you can learn more about them by studying a few specific galaxies.

The motion of stars and gas near the centers of galaxies is important because it can tell you the amount of mass located there. Consider the giant elliptical galaxy M87 (page 346). It has a very small, bright nucleus and a visible jet of matter 1800 pc long racing out of its core (■ Figure 17-3). Radio observations show that the nucleus must be no more than a light-week in diameter. This evidence is suggestive, but observations of motions around the center are decisive. A high-resolution image shows that the core lies at the center of a spinning disk and the jet lies along the axis of the disk. Only 60 ly from the center, the gas in the disk is orbiting at 750 km/s. If you substitute radius and velocity into the equation for circular velocity (Chapter 5), it will tell you that the central mass must be roughly 2.4 billion solar masses.

For two other examples, look at Figure 17-4. NGC4261 is a distorted galaxy ejecting jets into a pair of radio lobes. High-resolution images made by the Hubble Space Telescope reveal a dusty disk surrounding the bright central core. The axis of the disk points along the jets into the radio lobes. In the case of NGC4261, you can see the energy source in the act of producing jets and radio lobes. NGC7052, the second galaxy in ■ Figure 17-4, is a strong source of radio energy and is ejecting jets in opposite directions. It seems normal at visual wavelengths, but images made by the Hubble Space Telescope show that a dust disk rotates at high speed around the core. The rotation of the disk reveals that the central object contains 300 million solar masses, evidently a supermassive black hole.

The appearance of active galaxies varies dramatically from galaxy to galaxy. The spiral galaxy NGC1068, the brightest and nearest Seyfert galaxy, has an elongated glow of X rays emerging from its central region (■ Figure 17-5). High velocities around the center reveal that the galaxy must contain a 5-million-solar-mass black hole at its core. Radio wavelength observations show that a dusty doughnut of gas orbits from a few light-years out to about 300 ly from the central black hole.

The evidence seems conclusive: Active galactic nuclei are powered by supermassive black holes. That probably leaves you with a few questions, and astronomers are struggling to answer those same questions. How do supermassive black holes and their accretion disks produce these eruptions? Furthermore, where did these supermassive black holes come from?

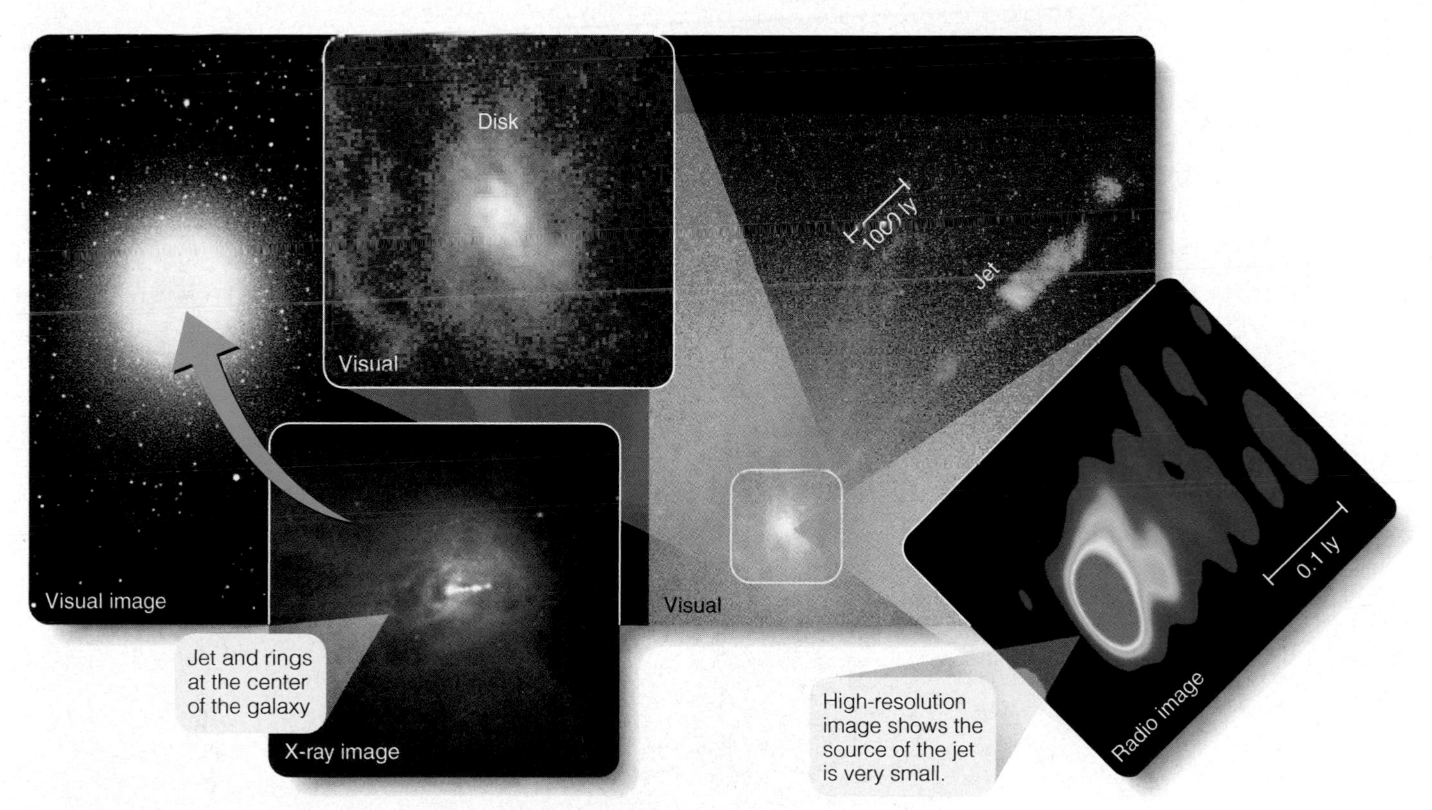

■ Figure 17-3

The jet at the center of giant elliptical galaxy M87 is only a few percent the diameter of the entire galaxy. Images reveal a small, rapidly spinning disk at the center of the galaxy with the jet emerging at nearly half the speed of light along the axis of the disk. X-ray images show rings and twisted plumes of hot gas produced in previous eruptions. (M87: AURA/NOAO/NSF; jet: NASA/STScI; radio image: NRAO and J. Biretta; X-ray: NASA/CXC/W. Worman et al.)

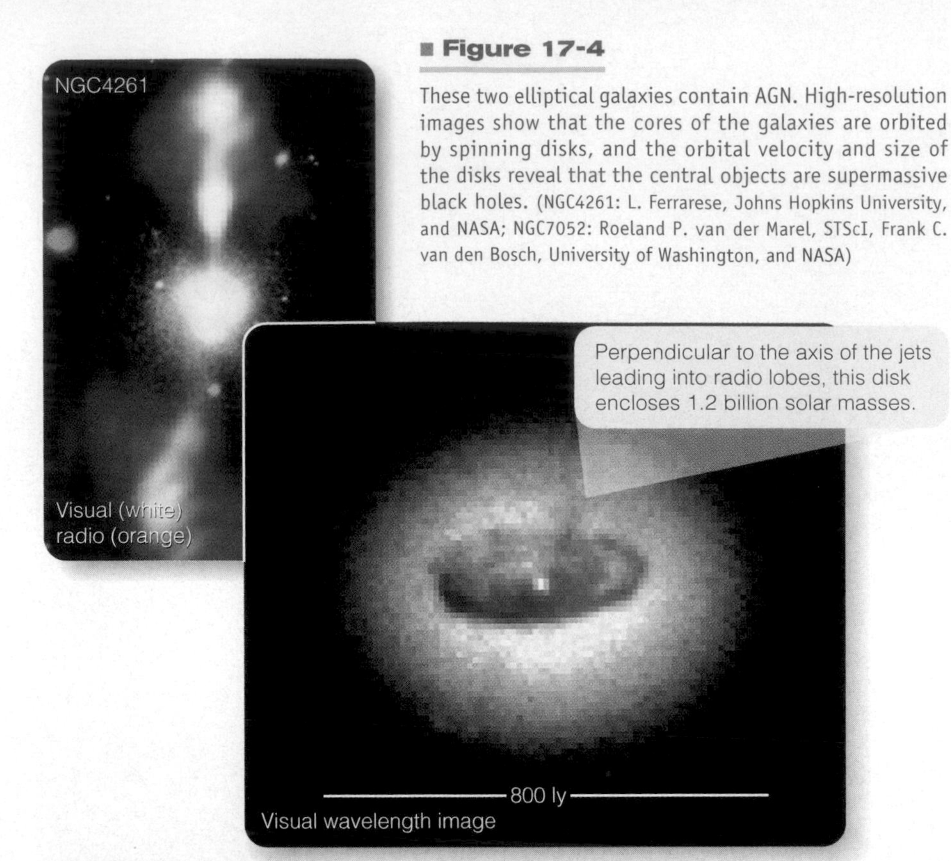

■ **Figure 17-4**

These two elliptical galaxies contain AGN. High-resolution images show that the cores of the galaxies are orbited by spinning disks, and the orbital velocity and size of the disks reveal that the central objects are supermassive black holes. (NGC4261: L. Ferrarese, Johns Hopkins University, and NASA; NGC7052: Roeland P. van der Marel, STScI, Frank C. van den Bosch, University of Washington, and NASA)

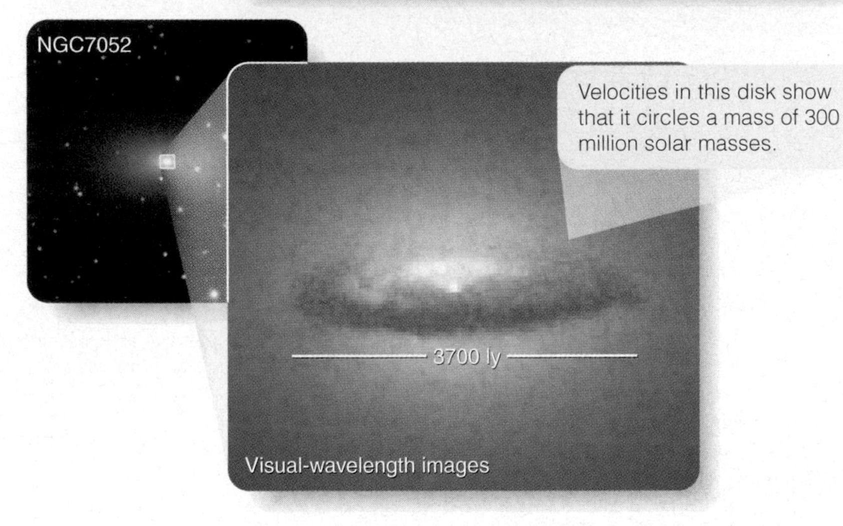

NGC7052
Velocities in this disk show that it circles a mass of 300 million solar masses.
3700 ly
Visual-wavelength images

The Search for a Unified Model

When a field of research is young, scientists often find many seemingly unrelated phenomena, such as double-lobed radio galaxies, Seyfert galaxies, cosmic jets, and so on. As the research matures, scientists begin seeing connections and eventually are able to unify the different phenomena as different aspects of a single process. This is the real goal of science, to organize evidence and theory in logical arguments that explain how nature works (look again at How Do We Know? 1-2). Astronomers studying active galaxies are now developing a **unified model** of active galaxy cores. A monster black hole is the centerpiece.

Even a supermassive black hole is quite small. A ten-million-solar-mass black hole would be only one-fifth the diameter of Earth's orbit. That means matter in an accretion disk can get very close to the black hole, orbit very fast, and grow very hot. Theoretical calculations predict that the central cavity in the disk around the black hole is very small but that the disk there is "puffed up" and thick. This means the black hole may be hidden deep inside this central well. The hot inner disk seems to be the source of the jets often seen coming out of active galaxy cores, but the process by which jets are generated is not understood. The outer part of the disk, according to calculations, is a fat, cold torus (doughnut shape) of dusty gas.

According to the unified model, what you see when you view the core of an active galaxy must depend on how this accretion disk is tipped with respect to your line of sight. You should note, at this point, that the accretion disk may be tipped at a steep angle to the plane of its galaxy, so just because you see a galaxy face-on doesn't mean you are looking at the accretion disk face-on. The important factor is not the inclination of the galaxy but the inclination of the accretion disk.

If you view the accretion disk from the edge, you cannot see the central area at all because the thick dusty torus blocks your view. Instead you see radiation emitted by gas lying above and below the central disk. Because this gas is farther from the center, it is cooler, orbits more slowly, and has smaller Doppler shifts. Thus you see narrower spectral lines coming from this narrow line region (■ Figure 17-6). This might account for the Seyfert 2 galaxies.

If the accretion disk is tipped slightly, you may be able to see some of the intensely hot gas in the central cavity. This broad line region emits broad spectral lines because the gas is so hot and is orbiting at high velocities, and the high Doppler shifts smear out the lines. Seyfert 1 galaxies may be explained by this phenomenon.

What happens if you look directly into the central cavity? According to the unified model, you see something that can explain a long-standing problem. In 1968 astronomers realized that an object they had thought was an irregular variable star in the constellation Lacerta was actually the core of an active galaxy. The visible spectrum is featureless, but the spectrum of faint, surrounding nebulosity is that of a giant elliptical galaxy. The object, and others like it known as **BL Lac objects** or **blazars,** are 10,000 times more luminous than the Milky Way Galaxy and

Figure 17-5

At visual wavelengths, the galaxy NGC1068 looks like a normal spiral, but X-ray observations reveal hot gas blowing outward at high speed from a supermassive black hole hidden deep inside a thick doughnut of gas and dust. (X-ray: NASA/CXC/MIT/UCSB/P. Ogle et al.; Optical: NASA/STScI/A. Capetti et al.)

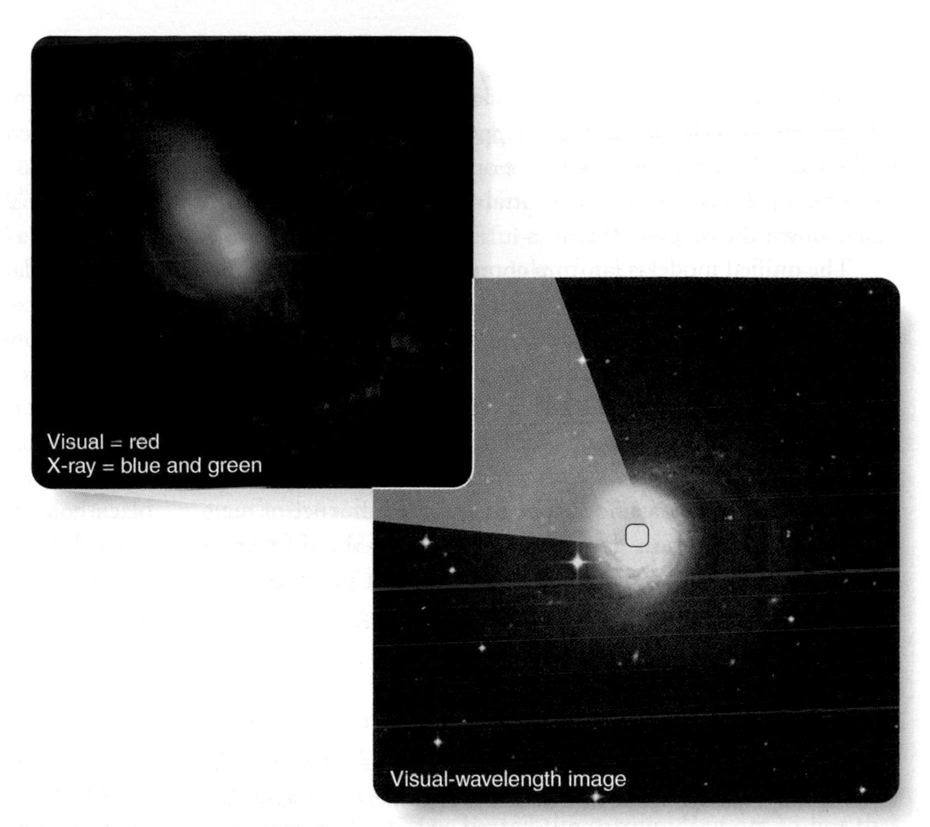
Visual = red
X-ray = blue and green
Visual-wavelength image

Artist's impression
Gas clouds
Narrow line region: Cooler gas above and below the disk is excited by radiation from the hot, inner disk.
Thin accretion disk
Black hole
Jet
Broad line region: Gas is hottest in the deep, hot cavity around the black hole
Dense torus of dusty gas

Figure 17-6

The features visible in the spectrum of an AGN depend on the angle at which it is viewed. The unified model, shown in cross section, suggests that matter flowing inward passes first through a large, opaque torus; then into a thinner, hotter disk; and finally into a small, hot cavity around the black hole. Telescopes viewing such a disk edge-on would see only narrow spectral lines from cooler gas, but a telescope looking into the central cavity would see broad spectral lines formed by the hot gas there. This diagram is not to scale. The central cavity may be only 0.01 pc in radius, while the outer torus may be 1000 pc in radius. (Top: A. Hobart, CXC; bottom: adapted from NASA)

Figure 17-12

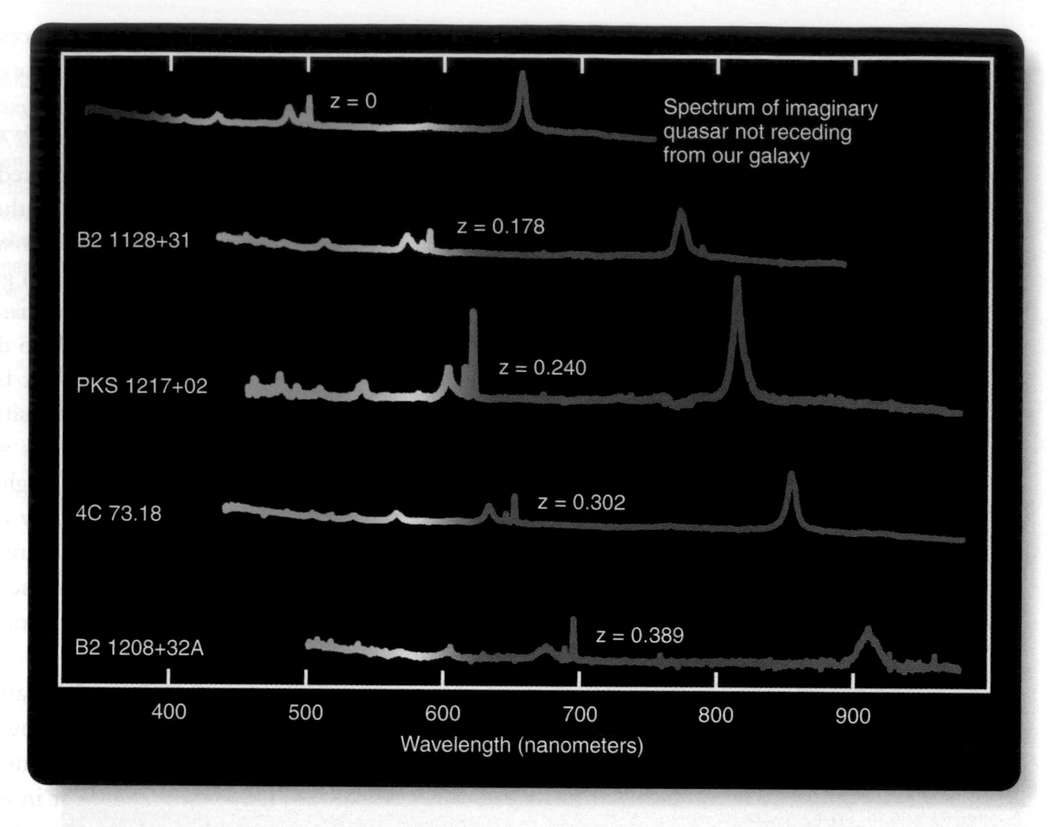

Spectra of four quasars are compared here with an idealized spectrum for an imaginary quasar that has no redshift. The first three Balmer lines of hydrogen are visible, plus lines of other atoms, but the redshifts of the quasars move these lines to longer wavelengths. Nevertheless, the relative spacing of the spectral lines is unchanged, and astronomers can recognize the lines even with a large redshift. (C. Pilachowski, M. Corbin, AURA/NOAO/NSF)

Doppler formula you studied in Chapter 7, you get velocities greater than the velocity of light, and that is supposed to be impossible.

First, you can eliminate a **Common Misconception.** For over a generation, astronomy textbooks have listed a version of the Doppler formula called the relativistic Doppler formula. Einstein derived it to describe the Doppler effect for objects traveling at velocities approaching the speed of light. That is the wrong formula because the redshifts of the galaxies and the quasars are not really Doppler shifts, even though astronomers often express the redshifts in kilometers per second as if they were true velocities. Rather you should refer to these as *apparent* velocities of recession. In the next chapter you will discover a much more sophisticated way of understanding these redshifts; but, for the moment, you can just note that modern astronomers understand that the redshifts of the galaxies are caused by the expansion of the universe and that the Hubble law allows them to find the distance to a galaxy by dividing its apparent velocity of recession by the Hubble constant.

The Hubble law works well for relatively nearby galaxies, but it is difficult to extend to very distant objects. The exact mathematical relationship between the redshift of a very distant galaxy or quasar and its apparent velocity of recession depends on parameters that are not yet well known. That means you can't calculate the precise distance to a quasar from its redshift, but astronomers can make good approximations. Certainly the very large redshifts of the quasars assure you that the quasars are very far away. Some quasars are over 10 billion light-years away, and because of their large look-back times they appear as they were when the universe was only 10 percent of its present age.

Perhaps you have another question about quasar distances: How can astronomers be sure quasars really are that far away? Astronomers faced with explaining how a small object could produce so much energy asked themselves the same question. In the 1970s, using some of the biggest telescopes on Earth, astronomers were able to photograph faint objects near some quasars. The spectra of those objects look like the spectra of normal galaxies, and they have the same redshift as the quasar. That implied that the quasar was located in a galaxy that was a member of a very distant cluster of galaxies. In the early 1980s, astronomers were able to photograph faint nebulosity surrounding some quasars, which was called quasar fuzz. The spectra of quasar fuzz looked like the spectra of normal but very distant galaxies.

A discovery made in 1979 provides a dramatic confirmation that quasars lie at tremendous distances. The object cataloged as 0957+561 lies just a few degrees west of the bowl of the Big Dipper and consists of two quasars separated by only 6 seconds of arc. The spectra of these quasars are identical and even have the same redshift of 1.4136. Quasar spectra do resemble each other, but in detail they are as different as fingerprints. When two quasars so close together were discovered to have the same redshift and identical spectra, astronomers concluded that they were two separate images of the same quasar.

The two images are formed by a gravitational lens. You met gravitational lenses in Chapter 13. In this case, the gravitational field of a galaxy located between Earth and the quasar bends the light from the quasar and forms multiple images. Today, astronomers know of dozens of quasars that are being distorted by gravitational lenses (■ Figure 17-13); and, in most cases, the lensing galaxy is so far away it is difficult to detect. This provides strong evidence that quasars really are distant objects, as their large redshifts imply.

For more evidence that quasars are very distant, look again at Figure 17-10. Just above the image of the quasar lies the small, oval image of an elliptical galaxy, which, as evidenced by its redshift, is about 7 billion light-years away. The spectrum of the quasar contains absorption lines with the same redshift as the elliptical galaxy. Evidently the lines are produced as light from the quasar in the far distance passes through the thin gas in the outer fringes of the elliptical galaxy. From this you can conclude that the quasar, though very bright, must be farther away than the distant elliptical galaxy.

About 10 percent of quasars have absorption lines as well as emission lines in their spectra. These absorption lines have a

■ Figure 17-13

Gravitational lensing can occur when a relatively nearby lensing galaxy is aligned with a distant quasar. This can produce multiple images of the quasar. In the case of the quasar at upper right, the host galaxy, which contains the quasar, is visible as a faint, distorted ring around the lensing galaxy. Gravitational lenses provide direct evidence that quasars are at the great distances suggested by their redshifts. (Q0957+561: George Rhee, NASA, STScI; color composite courtesy Bill Keel; PG1115+080: Chris Impy, Univ. of Arizona and NASA; Q2237+030: Image by W. Keel from data in the NASA/ESA Hubble Space Telescope archive, originally obtained with J. Westphal as Principal Investigator)

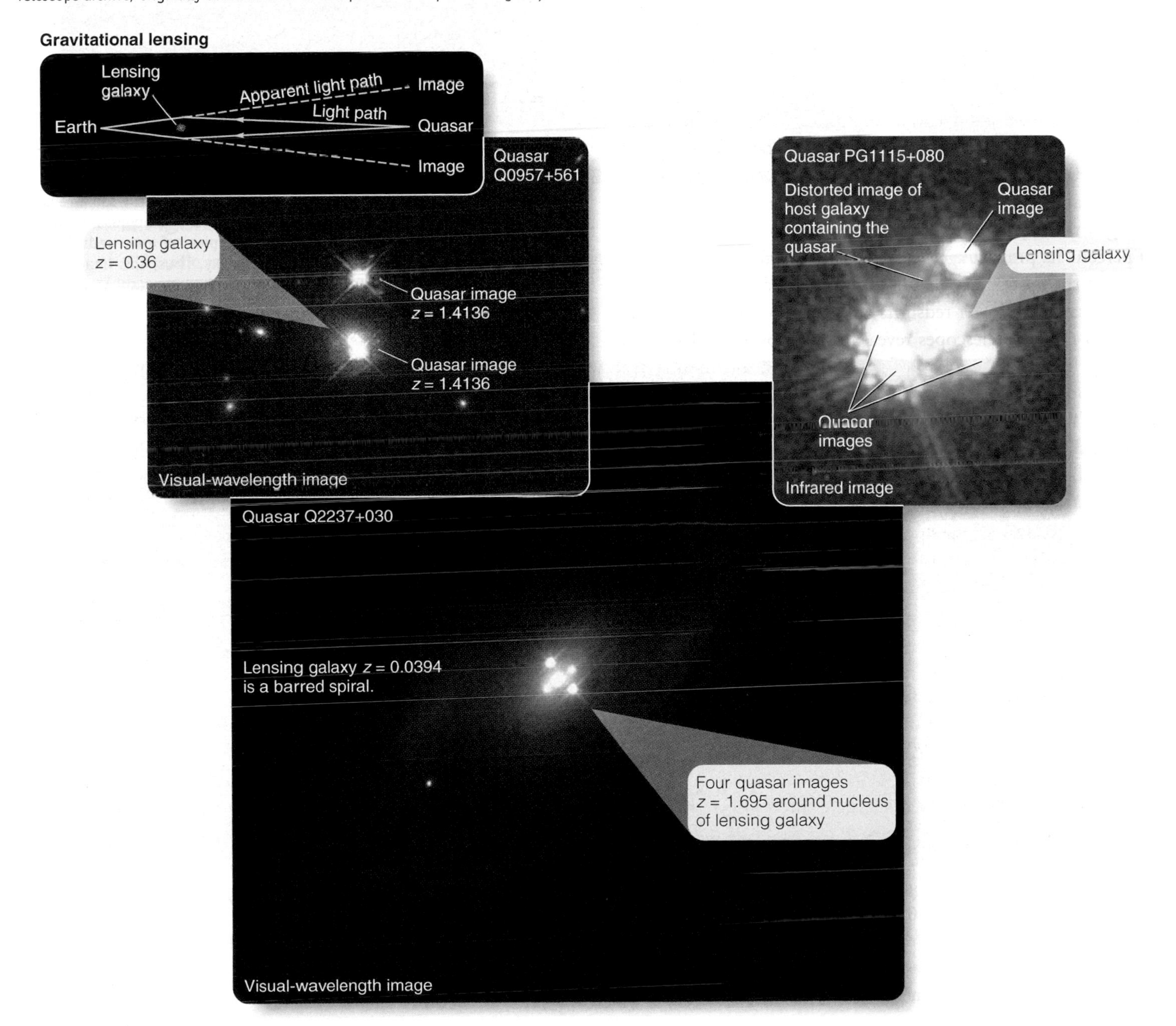

How Do We Know? | 18-2

Science: A System of Knowledge

What is the difference between believing in the big bang and understanding it? If you ask a scientist, "Do you believe in the big bang?" he or she may hesitate before responding. The question implies something about science that isn't true.

Science is a system of knowledge, not a system of belief. Science is an attempt to understand logically how nature works, and it is based on observations and experiments used to test and confirm theories. A scientist does not really *believe* in a theory. Rather the scientist understands the theory and recognizes how the different pieces of evidence support or contradict the theory.

There are other ways to know things, and systems of belief are not unusual. Religion, for example, is a system of belief and is not based on observation and experiment. In some cases, politics is a system of belief; many people believe that democracy is the best form of government and do not ask for or expect evidence supporting their belief. A system of belief can be powerful and lead to deep insights, but it is different from science.

Scientists try to be careful with words, so thoughtful scientists would not say they *believe* in the big bang. They would say that the evidence is overwhelming that the big bang really did occur, and that they are forced by the logical analysis of the theory and observations to conclude that the theory is very likely correct. In this way, scientists try to be objective and reason without distortion from personal feelings and prejudices.

A scientist once referred to "the terrible rule of evidence." Sometimes the evidence forces a scientist to an unpleasant conclusion, but science is not a system of belief, so the personal preferences of the scientist must take second place to the rule of evidence.

Scientific knowledge is based objectively on evidence such as that gathered by spacecraft. (NASA/WMAP Science Team)

Do you believe in the big bang? Or do you have confidence the theory is right because of your analysis of the evidence? There is a big difference.

◄ SCIENTIFIC ARGUMENT ►

How do you know there was a big bang?

A good scientific argument combines evidence and theory to describe how nature works, and this question calls for a detailed argument. The cosmic microwave background radiation consists of photons emitted by the hot gas of the big bang, so when astronomers detect those photons, they are "seeing" the big bang. Of course, all scientific evidence must be interpreted, so you must understand how the big bang could produce radiation all around the sky before you can accept the background radiation as evidence. First, you must remember that the big bang event filled all of the universe with hot, dense gas. The big bang didn't happen in a single place; it happened everywhere. At recombination, the expansion of the universe reached the stage where the matter became transparent and the radiation was free to travel through space. Today that radiation from the age of recombination arrives from all over the sky. It is all around you because you are part of the big bang event, and as you look out into space to great distance, you look back in time and see the hot gas in any direction you look. You can't see the radiation as light because of the large redshift that has lengthened the wavelengths by a factor of 1100 or so, but you could detect the radiation as photons with infrared and short radio wavelengths.

With this interpretation, the cosmic microwave background radiation is powerful evidence that there was a big bang. That tells you how the universe began, but your argument hinges on an important point. **Why do you conclude that the universe cannot have a center or an edge?**

◄ ►

18-2 The Shape of Space and Time

WHAT CAUSES THE COSMIC REDSHIFT? Answering that question will introduce you to some peculiar properties of space-time and show you how space-time dominates the evolution of the universe.

Looking at the Universe

The universe is **isotropic,** meaning that it looks about the same whichever way you look. Of course, there are local differences. If you look toward a galaxy cluster you see more galaxies, but that is only a local variation. On the average, you see similar numbers of galaxies in every direction. Furthermore, the background radiation is also almost perfectly uniform across the sky. Certainly the universe is highly isotropic.

The universe also seems **homogeneous,** the same everywhere. Of course there are local variations. Some regions contain more galaxies and some less. Also, if the universe evolves, then at large look-back times, you see galaxies at an earlier stage. If you account for these well-understood variations, then the universe seems to be, on average, the same everywhere.

Isotropy and homogeneity lead to the **cosmological principle,** which says that any observer in any galaxy sees the same general features of the universe. Again, you must account for local and evolutionary variations. The cosmological principle assures that there are no special places in the universe. What you see from the Milky Way Galaxy is typical of what all intelligent creatures see from their galaxies. Furthermore, the cosmological principle assures that there can be no center or edge. Such locations would be special places, and the cosmological principle says there are no special places.

The Cosmic Redshift

Einstein's theory of general relativity, published in 1916, describes space and time as the fabric of the universe called space-time. That idea will give you a new insight into the meaning of the phrase *expanding universe.*

General relativity describes space-time as if it were made of stretching rubber, and that explains one of the most important observations in cosmology—the redshifts. It is a **Common Misconception** that the expansion of the universe makes the galaxies move rapidly through space. Except for small, local motions as galaxies orbit each other or orbit within a cluster of galaxies, the galaxies are at rest. They are being *carried* away from each other as space-time expands. Like dots painted on a sheet of rubber, the galaxies move away from each other as space-time stretches, and there is no center to the expansion.

Furthermore, as space-time expands, it stretches any photon traveling through space, gradually making its wavelength longer. Photons from distant galaxies spend more time traveling through space and are stretched more than photons from nearby galaxies. That is why redshift is proportional to distance (■ Figure 18-13).

Astronomers often express redshifts as if they were radial velocities, but the redshifts of the galaxies are not Doppler shifts. That is why this book is careful to refer to a galaxy's *apparent* velocity of recession. As you will recall from the previous chapter, Einstein's relativistic Doppler formula applies to motion through space, so it does not apply to the recession of the galaxies. The Hubble law relates redshift to distance for nearby galaxies and can be used to estimate the distance to distant galaxies. Nevertheless, the exact relationship at great distances is not precisely known because it depends on exactly how the expansion of space-time has occurred over the history of the universe.

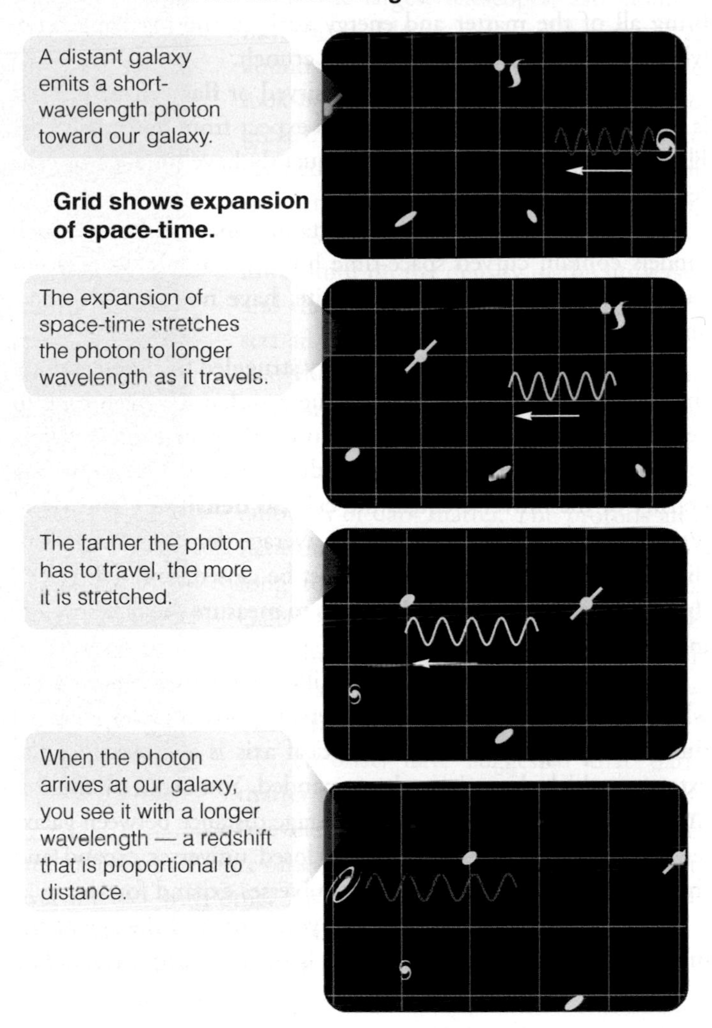

What are the cosmological redshifts?
A distant galaxy emits a short-wavelength photon toward our galaxy.
Grid shows expansion of space-time.
The expansion of space-time stretches the photon to longer wavelength as it travels.
The farther the photon has to travel, the more it is stretched.
When the photon arrives at our galaxy, you see it with a longer wavelength — a redshift that is proportional to distance.

■ **Figure 18-13**

Like a rubber sheet, space-time stretches, moving the galaxies away from each other and increasing the wavelength of photons as they travel through space-time.

Model Universes

Almost immediately after Einstein published his theory, theorists were able to solve the highly sophisticated mathematics to compute simplified descriptions of the behavior of space-time and matter. Those model universes dominated cosmology throughout the 20th century.

General relativity showed that space-time did not have to be flat; it might be curved. You are accustomed to thinking of space-time as flat. But the equations of general relativity showed that space-time might seem flat over small distances like tennis courts, star clusters, and nearby galaxies but might have a curvature that would make parallel lines diverge or converge at great distances. Cosmologists built mathematical models of the universe that incorporated this curvature of space-time. Most people find these curved model universes difficult to imagine, and modern observations have shown that only one, the simplest, is correct; so you don't have to warp your brain around these curved models. But you do need to know a few of their most important properties.

Some models predict that space-time is curved back on itself to form a **closed universe.** You would not notice this curvature in daily life; it would only be evident if you measured distances to galaxies very far away. Throughout the 20th century, observations could not eliminate closed models, so they were considered a real possibility. Closed models are finite, but because space-time is curved back on itself closed models have no edge and no center. Such closed models predicted that the expansion of the

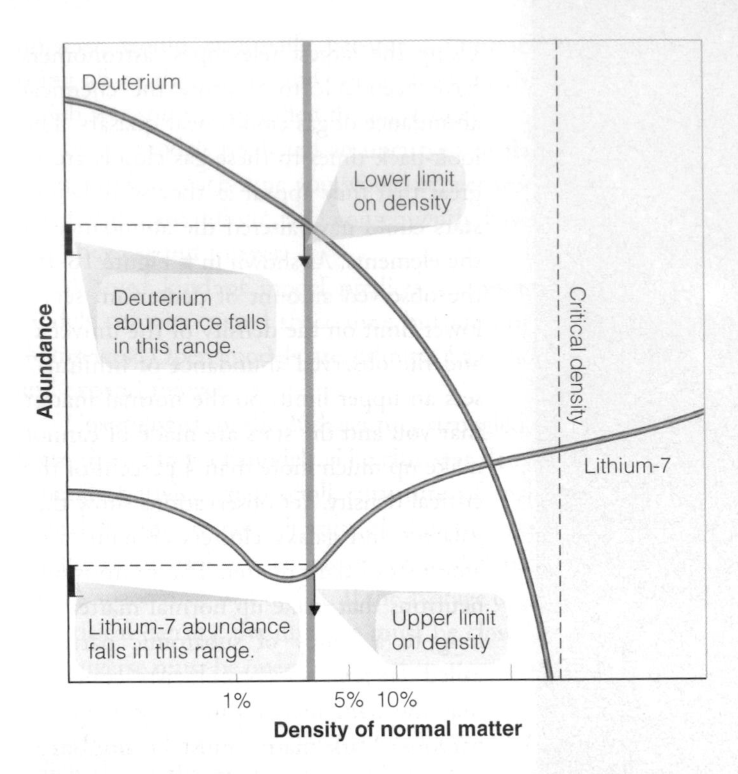

■ **Figure 18-16**

This diagram compares observation with theory. Theory predicts how much deuterium and lithium-7 you would observe for different densities of normal matter (red and blue curves). The observed density of deuterium falls in a narrow range shown at upper left and sets a lower limit on the possible density of normal matter. The observed density of lithium-7 shown at lower left sets an upper limit. This means the true density of normal matter must fall in a narrow range represented by the green column. Certainly, the density of normal matter is much less than the critical density.

moving and, if they exist, could clump together in the early universe and help pull matter into galaxy-size clouds.

Nonbaryonic dark matter does not interact significantly with normal matter or with photons, which is why you can't see it. But that means that dark matter was not affected by the intense radiation that dominated the universe when it was very young. So long as radiation dominated the universe, it prevented normal matter from contracting to begin forming galaxies. But the dark matter was immune to the radiation, and the dark matter could contract to form clouds while the universe was very young. Once the density of radiation fell low enough, normal matter could begin falling into the clouds of dark matter to form the first galaxies. Dark matter could have given galaxy formation a head start soon after the big bang. Models with cold, nonbaryonic dark matter are most successful at forming galaxies and clusters of galaxies early in the history of the universe.

Although there is good evidence that dark matter exists, it has proven difficult to identify. No form of dark matter has been found that is abundant enough to provide the critical density needed to make the universe flat. In fact, all forms of dark matter appear to add up to less than 30 percent of the critical density. As you will see later in this chapter, there is more to the universe than meets the eye, and more even than the dark matter.

◄ SCIENTIFIC ARGUMENT ►

Why do astronomers think that dark matter can't be baryonic?

Good scientific arguments always fall back on evidence. In this case, the evidence is very strong. Small amounts of isotopes like deuterium and lithium-7 were produced in the first minutes of the big bang, and the abundance of those elements depends strongly on the density of protons and neutrons. Because these particles belong to the family of particles called baryons, astronomers refer to normal matter as baryonic. Measurements of the abundance of deuterium and lithium-7 show that the universe cannot contain more baryons than about 4 percent of the critical density. Yet observations of galaxies and galaxy clusters show that dark matter must make up almost 30 percent of the critical density. Consequently, astronomers conclude that the dark matter must be made up of nonbaryonic particles.

Finding the dark matter is important because the density of matter in the universe determines its curvature. Now build a new argument. **How does the modern understanding of space-time explain cosmic redshifts?**

◄ ►

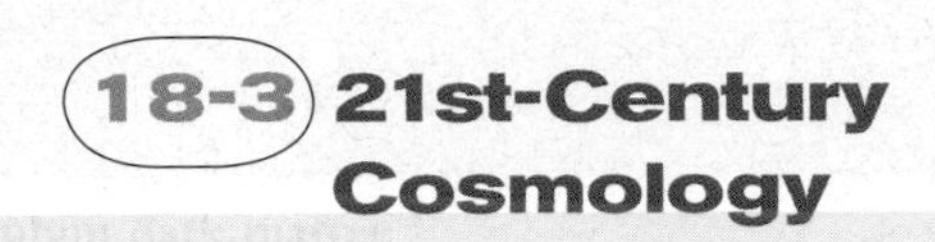

18-3 21st-Century Cosmology

IF YOU ARE A LITTLE DIZZY from the weirdness of curved space-time and dark matter, make sure you are sitting down before you read much further. As the 21st century began, cosmologists made a startling discovery, and all around the world astronomers looked at each other and said, "What? What!" The most amazing thing about these amazing discoveries is that they fit so well with some of the things you have been learning in this chapter. To get a running start on these new discoveries, you'll have to go back a couple of decades.

Inflation

In 1980, the big bang model was widely accepted, but it faced two problems that led to the development of a new version—a big bang model with an astonishing addition.

One of the problems is called the **flatness problem.** The universe seems to be balanced near the boundary between an open and a closed universe. That is, it seems nearly flat. Given the vast range of possibilities, from zero to infinity, it seems peculiar that the density of the universe is within a factor of 10 of the critical density that would make it flat. If dark matter is as common as it seems, the density may be even closer than a factor of three to being perfectly flat.

Even a small departure from critical density when the universe was young would be magnified by subsequent expansion. To be so near critical density now, the density of the universe during its first moments must have been within 1 part in 10^{49} of the critical density. So the flatness problem is: Why is the universe so nearly flat?

The second problem with the original big bang theory is the isotropy of the primordial microwave background radiation. When astronomers correct for the motion of Earth, they see the same background radiation in all directions to at least 1 part in 1000. Yet when you look at background radiation coming from two points in the sky separated by more than a degree, you look at two parts of the big bang that were not causally connected when the radiation was emitted. That is, when recombination occurred and the gas of the big bang became transparent to the radiation, the universe was not old enough for any signal to have traveled from one of these regions to the other. Thus, the two spots you look at did not have time to exchange heat and even out their temperatures. So how did every part of the entire big bang universe get to be so nearly the same temperature by the time of recombination? This is called the **horizon problem** because the two spots in this example are said to lie beyond their respective light-travel horizons.

The key to these two problems and to others involving subatomic physics may lie with the theory called the **inflationary universe.** It predicts a sudden inflation when the universe was very young, a violent expansion even more extreme than that predicted by the big bang theory.

To understand the inflationary universe, you need to recall from Chapter 8 that physicists know of only four forces—gravity, the electromagnetic force, the strong force, and the weak force (Chapter 8). The strong force holds atomic nuclei together, and the weak force is involved in certain kinds of radioactive decay.

For many years, theorists have tried to unify these forces; that is, they have tried to describe the forces with a single mathematical law. A century ago, James Clerk Maxwell showed that the electric force and the magnetic force were really the same effect, and physicists now count them as a single electromagnetic force. In the 1960s, theorists succeeded in unifying the electromagnetic force and the weak force in what they called the electroweak force, effective only for processes at very high energy. At lower energies, the electromagnetic force and the weak force behave differently. Now theorists have found ways of unifying the electroweak force and the strong force at even higher energies. These new theories are called **grand unified theories,** or **GUTs.**

According to the inflationary universe, the universe expanded and cooled until about 10^{-35} second after the big bang, when it became so cool that the electroweak force and the strong force began to disconnect from each other and behave in different ways. This released tremendous energy, which suddenly inflated the universe by a factor between 10^{20} and 10^{30} (■ Figure 18-17). At that time the part of the universe that is now visible from Earth, the entire observable universe, was no larger than the volume of an atom, but it suddenly inflated to the volume of a cherry pit and then continued its slower expansion to its present extent.

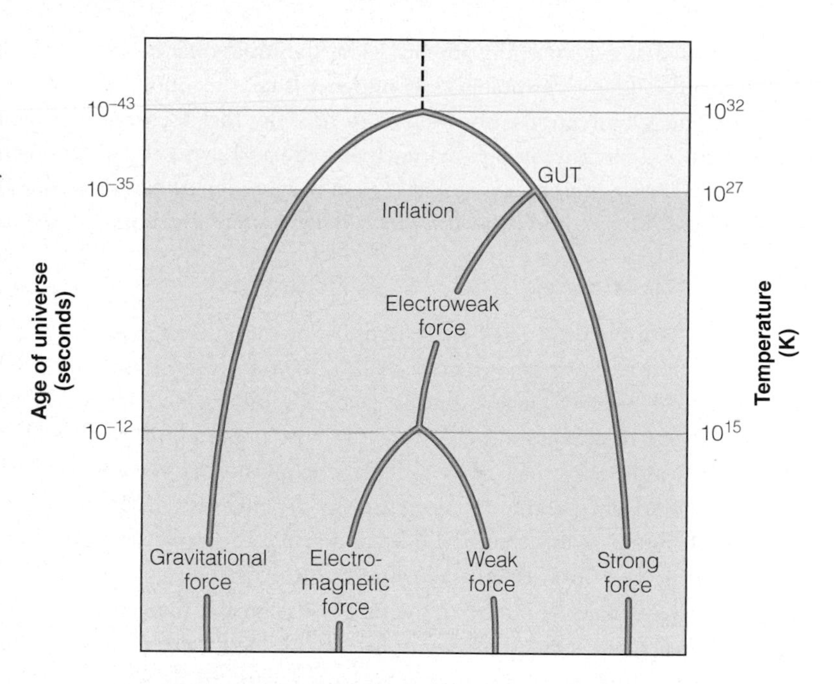

■ Figure 18-17

When the universe was very young and hot (top), the four forces of nature were indistinguishable. As the universe began to expand and cool, the forces separated and triggered a sudden inflation in the size of the universe.

That sudden inflation can solve the flatness problem and the horizon problem. The sudden inflation of the universe would have forced whatever curvature it had toward zero, just as inflating a balloon makes a small spot on its surface flatter. You now live in a universe that is almost perfectly flat because of that sudden inflation long ago. In addition, because the observable part of the universe was once no larger in volume than an atom, it had plenty of time to equalize its temperature before inflation occurred. That is how the inflationary universe theory explains the fact that you now live in a universe where the background radiation is the same temperature in all directions.

The inflationary universe is based, in part, on quantum mechanics, and a slightly different aspect of quantum mechanics may explain why there was a big bang at all. Theorists conclude that a universe totally empty of matter could be unstable and decay spontaneously by creating pairs of particles until it was filled with the hot, dense state called the big bang. This theoretical discovery has led some cosmologists to propose that the universe could have been created by a chance fluctuation in space-time. In the words of physicist Frank Wilczyk, "The reason there is something instead of nothing is that 'nothing' is unstable."

The inflationary theory predicts that the universe is flat. That is, the true density must equal the critical density. A theory can never be used as evidence, but the beauty of the inflationary theory has given many cosmologists confidence that the universe must be flat. Observations, however, seem to show that the universe does not contain enough matter (baryonic and dark) to be flat. Can there be more to the universe than baryonic matter and dark matter? What could be weirder than dark matter? Read on.

The Acceleration of the Universe

Ever since Hubble discovered the expansion of the universe, astronomers have known what to expect. Both common sense and mathematical models suggest that as galaxies recede from each other, the expansion should be slowed as gravity tries to pull the galaxies toward each other. How much the expansion is slowed should depend on the amount of matter in the universe. If the density of matter is less than the critical density, the expansion should be slowed only slightly, allowing the universe to expand forever. If the density of matter in the universe is greater than the critical density, the expansion should be slowing down dramatically, eventually causing the universe to stop expanding and begin contracting. Notice that this is the same as saying a low-density universe should be open and a high-enough-density universe should be closed.

For decades, astronomers struggled to measure the Hubble constant at great look-back times accurately enough to detect the slowing of the expansion. A direct measurement of the rate of slowing would reveal the true curvature of the universe. This was one of the key projects for the Hubble Space Telescope, and two teams of astronomers spent years making the measurements.

Detecting a change in the rate of expansion is a difficult project because it requires accurate measurements of the distances to very remote galaxies, and both teams used the same technique: They calibrated type Ia supernovae as distance indicators. A type Ia supernova occurs when a white dwarf gains matter from a companion star, exceeds the Chandrasekhar limit, and collapses in a supernova explosion. Because all such white dwarfs should collapse at the same mass limit, they should all produce explosions of the same luminosity, and that makes them good distance indicators.

In Chapter 16, you read that astronomers refer to distance indicators such as Cepheid variable stars as *standard candles,* objects of known luminosity. Type Ia supernovae have been described as *standard bombs.* They are titanic explosions, but they all reach the same peak brightness, and that makes them good distance indicators.

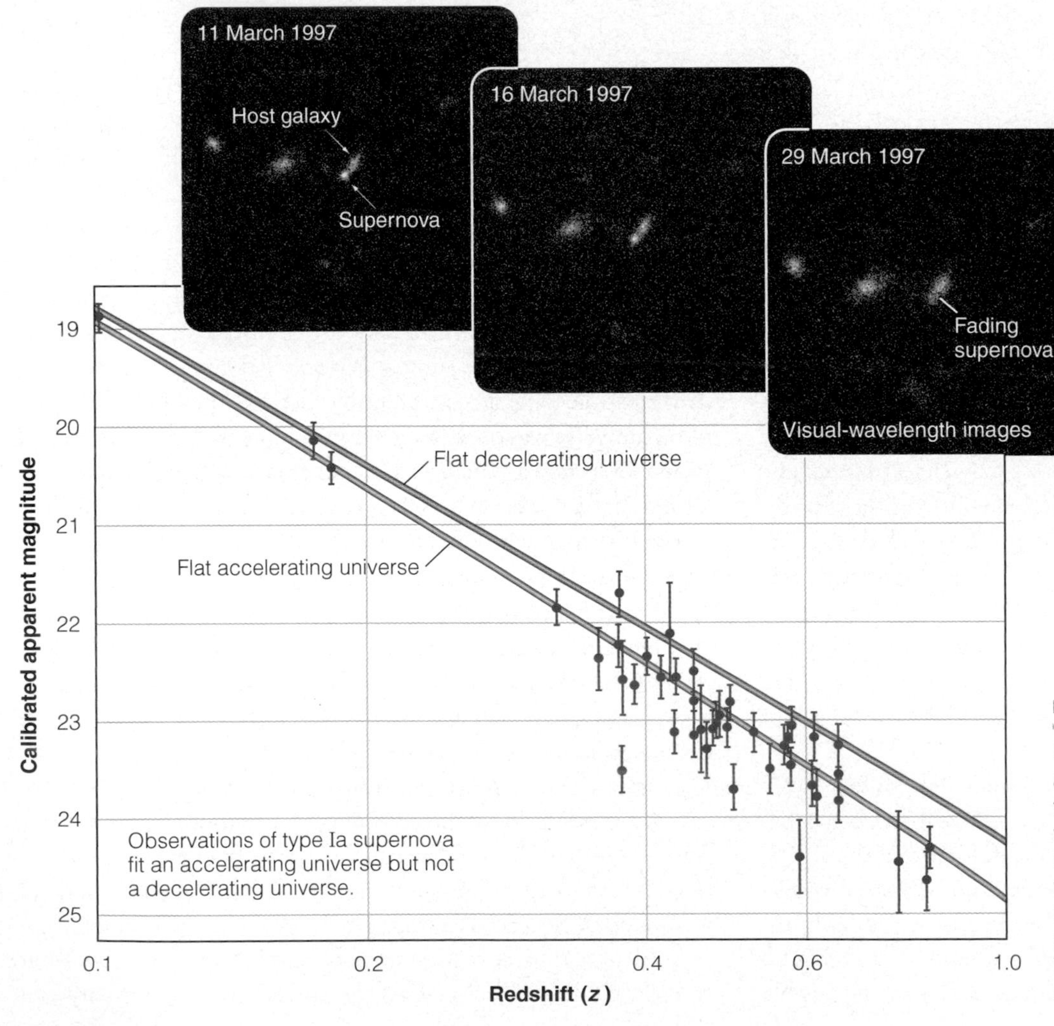

■ Figure 18-18

From the way a supernova fades over time, astronomers could determine which were type Ia. Once calibrated, those supernovae could be compared with their redshift revealing that distant type Ia supernovae were about 25 percent fainter than expected. That must mean they are farther away than expected given their redshifts. This is strong evidence that the universe is accelerating. (ESO)

The two teams calibrated type Ia supernovae by locating such supernovae occurring in nearby galaxies whose distance was known from Cepheid variables and other reliable distance indicators. Once the peak luminosity of type Ia supernovae had been determined, they could be used to find the distance to much more distant galaxies.

Both teams announced their results in 1998. The expansion of the universe is not slowing down. It is speeding up! Contrary to all expectations, the expansion of the universe is accelerating (■ Figure 18-18).

The announcement that the expansion of the universe is accelerating made astronomers stop and stare at each other. It was totally unexpected. It was simultaneously an exciting, puzzling, and revealing discovery, so astronomers immediately began testing it. It depends critically on the calibration of type Ia supernovae as distance indicators, and some astronomers suggested that the calibration might be wrong (look again at How Do We Know? 15-1). Some astronomers began testing the calibration, and others searched for more type Ia supernovae in even more distant galaxies. The results confirm the discovery (■ Figure 18-19). The universe really does seem to be expanding faster and faster.

Dark Energy and Acceleration

If the expansion of the universe is accelerating, then there must be a force of repulsion in the universe, and astronomers are struggling to understand what it could be. One possibility leads back to 1916.

When Albert Einstein published his theory of general relativity in 1916, he recognized that his equations describing space-time implied that space had to contract or expand. The galaxies could not float unmoving in space because their gravity would pull them toward each other. The only solutions seemed to be a universe that was contracting under the influence of gravity or a universe in which the galaxies were rushing away from each other so rapidly that gravity could never slow them to a stop and pull them back. In 1916, astronomers did not yet know that the universe was expanding, so Einstein made what he later said was a mistake.

To balance the attractive force of gravity, Einstein added a constant to his equations called the **cosmological constant,** represented by an uppercase lambda (Λ). The constant represents a force of repulsion that balances gravitation so the universe does not have to contract or expand. Thirteen years later, in 1929, Edwin Hubble announced that the universe was expanding, and Einstein said introducing the cosmological constant was his biggest blunder. Modern astronomers aren't so sure.

One explanation for the acceleration of the universe is that the cosmological constant really does represent a force that drives continuing acceleration in the expansion of the universe. Because the cosmological constant remains constant with time, the universe would have to have experienced this acceleration throughout its history.

Another solution is to suppose that totally empty space, the vacuum, contains energy, which drives the acceleration. This is an interesting possibility because for years theoretical physicists have discussed energy inherent in empty space. Astronomers have begun referring to this energy of the vacuum as **quintessence.** Unlike the cosmological constant, quintessence need not remain constant over time.

Whichever explanation is right, the cosmological constant or quintessence, acceleration is evidence that some form of energy is spread throughout space. Astronomers refer to this energy as **dark energy,** the energy that drives the acceleration of the universe but does not contribute to the formation of starlight or the cosmic microwave background radiation.

You will recall that acceleration and dark energy were first discovered when astronomers found that supernovae just a few billion light-years away were slightly fainter than expected. The acceleration of the expansion made those supernovae a bit farther away than expected, and so they looked fainter. Since then, astronomers have continued to find even more distant type Ia supernovae, some as distant as 12 billion light-years. The more distant of those supernovae are not too faint; they are too bright! That reveals even more about dark energy.

The very distant supernovae are a bit too bright because they are not as far away as expected, and that confirms a theoretical prediction based on dark energy. When the universe was young, galaxies were closer together, and their gravitational pull on each other could overpower dark energy and slow the expansion. That makes the very distant supernovae a bit too bright. As the universe expanded, galaxies moved farther

■ Figure 18-19

Follow-up observations of extremely distant galaxies located in the GOODs deep fields have detected type Ia supernovae. The upper images here show the supernovae, and the lower images show the galaxies before the supernovae erupted. These supernovae are so distant they confirm the original discovery that the expansion of the universe is accelerating and eliminate concerns that the calibration of type Ia supernovae was wrong. (NASA, ESA, and Adam Riess, STScI)

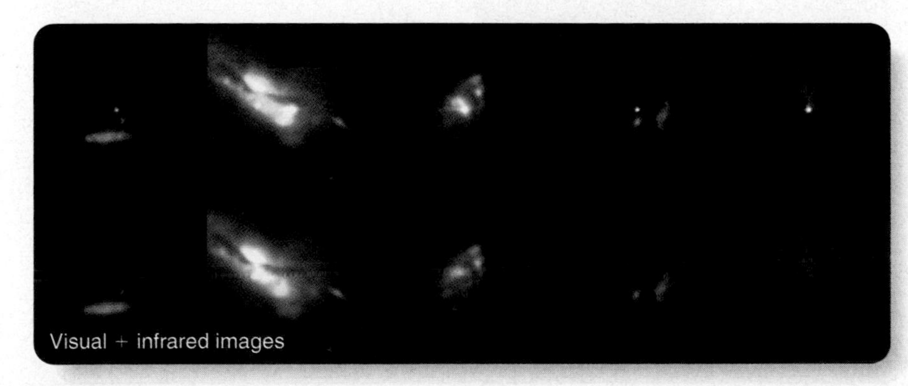

apart, their gravitational pull on each other became weaker than dark energy, and acceleration began. That makes the less distant supernovae a bit too faint. Sometime about 6 billion years ago, the universe shifted gears from deceleration to acceleration. The calibration of type Ia supernovae allows astronomers to observe this change from deceleration to acceleration.

Furthermore, dark energy can help you understand the curvature of the universe. The theory of inflation makes the specific prediction that the universe is flat. Dark energy seems to confirm that prediction. According to Einstein's most famous equation, $E=mc^2$, you know that energy and matter are equivalent. That means the dark energy is equivalent to a mass spread through space. Baryonic matter plus dark matter makes up about a third of the critical density, and dark energy appears to make up two-thirds. That is, to the accuracy of the measurements, the total density of the universe equals the critical density, which means that the geometry of the universe is flat.

Step by step you have been climbing the cosmological pyramid. Each step has been small and logical, but look where it has led you. You know some of the deepest secrets of the universe, but there are still more steps above and more secrets to explore.

The Age and Fate of the Universe

Acceleration helps astronomers solve another problem. Earlier in this chapter you calculated the Hubble time using a Hubble constant of 70 km/s/Mpc and found an approximate age of about 14 billion years. Then you calculated the age of the universe assuming it was flat and got an answer of about 9 billion years. That was a problem because the globular clusters are older than that. Now you are ready to solve that age problem.

If the expansion of the universe has been accelerating, then it must have been expanding more slowly in the past, and that means its age can be more than two-thirds of the Hubble time. The latest estimates suggest that acceleration increases the age of a flat universe from 9 billion years back up to about 14 billion years, which is clearly older than the oldest known star clusters.

For many years cosmologists have enjoyed saying, "Geometry is destiny." By that they meant that the destiny of the universe is determined entirely by its geometry. An open universe must expand forever, and a closed universe must fall back. But that is true only if the universe is ruled by gravity. If acceleration dominates gravity, then geometry is not destiny, and even a closed universe might expand forever.

The ultimate fate of the universe depends on the nature of dark energy. If dark energy is described by the cosmological constant, then the force driving acceleration does not change with time, and our flat universe will expand forever with the galaxies getting farther and farther apart and using up their gas and dust making stars and the stars dying until each galaxy is isolated, burnt out, dark and alone. If, however, dark energy is described by quintessence, then it may be increasing with time, and the universe may accelerate faster and faster as space pulls the galaxies away from each other, eventually pulls the galaxies apart, then pulls the stars apart, and finally rips individual atoms apart. This has been called the **big rip.** Don't worry. Even if a big rip is in our future, nothing will happen for at least 30 billion years.

Maybe there will be no big rip; critically important observations made by the Chandra X-Ray Observatory and the Hubble Space Telescope have confirmed acceleration is real and tend to support the cosmological constant. The observations are not accurate enough to totally rule out quintessence, but they have given astronomers more confidence that dark energy does not change with time and that there will be no big rip (■ Figure 18-20).

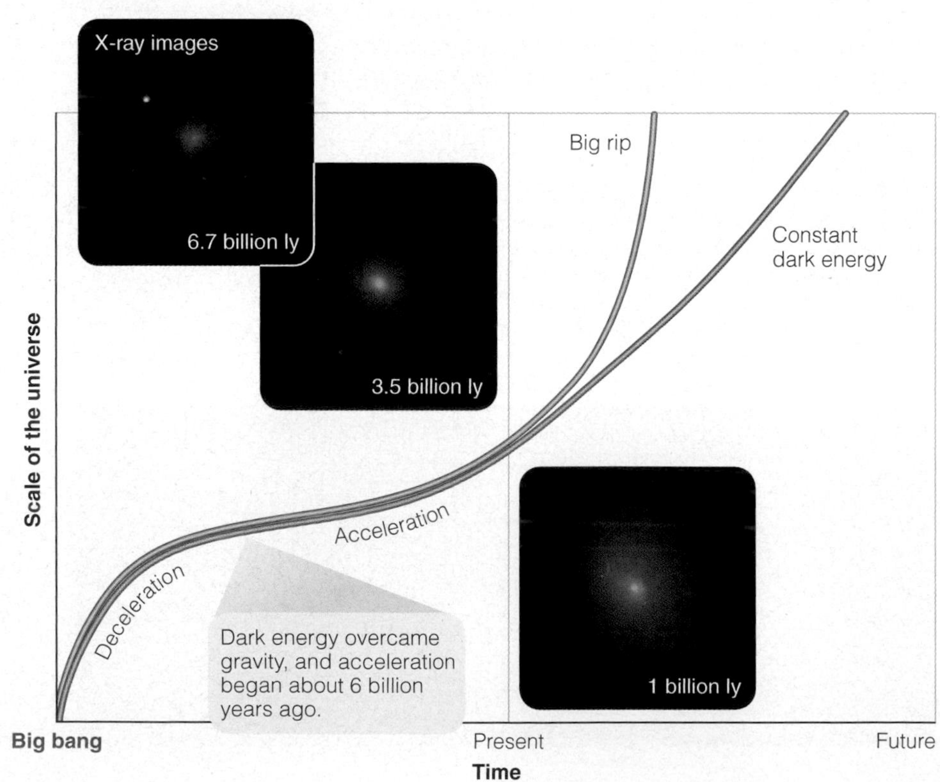

■ Figure 18-20

X-ray observations of hot gas in galaxy clusters confirm that in its early history the universe was decelerating because gravity was stronger than the dark energy. As expansion weakened the influence of gravity, dark energy began to force an acceleration. The evidence is not conclusive, but it most directly supports the cosmological constant and weighs against quintessence, which means the universe may not face a big rip. This diagram is only schematic, and the two curves are drawn separated for clarity; at the present the two curves have not diverged from each other. (NASA/CXC/IoA/S. Allen et al.)

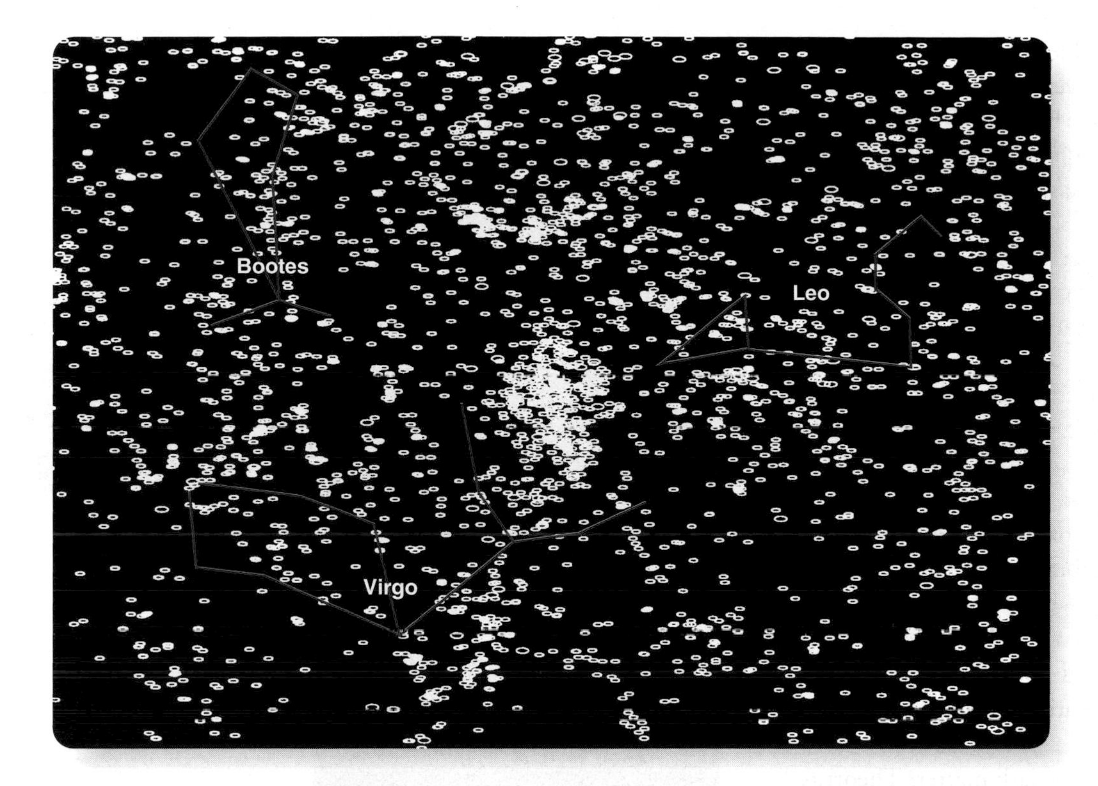

■ **Figure 18-21**

The distribution of brighter galaxies in the sky reveals the great Virgo cluster (center), containing over 1000 galaxies only about 17 Mpc away. Other clusters fill the sky, such as the more distant Coma cluster just above the Virgo cluster in this diagram. The Virgo cluster is linked with others to form the Local Supercluster.

The Origin of Structure and the Curvature of the Universe

Do you believe this stuff? Or, rather, does the evidence give you confidence that these theories correctly describe the universe? Science is not a system of belief, but rather it is a system of knowledge that is based on evidence. Scientists work by building confidence in theories, and astronomers like to say, "Extraordinary claims require extraordinary evidence." Faced with an amazing discovery like the acceleration of the universe, astronomers search for other ways to confirm that the universe is flat and accelerating. One way to measure the curvature of space-time is to measure the size of things at great look-back times, and that measurement is yielding yet more exciting results.

On the largest scales, the universe is isotropic. That is, it looks the same in all directions. But, on smaller scales, there are irregularities. The sky is filled with galaxies and clusters of galaxies that seem to be related to their neighbors in even larger aggregations that astronomers call **large-scale structure.** Studies of large-scale structure lead to astonishing insights that are changing cosmology.

When you look at galaxies in the sky, you see them in clusters ranging from a few galaxies to thousands, and those clusters appear to be grouped into **superclusters** (■ Figure 18-21). The Local Supercluster, in which we live, is a roughly disk-shaped swarm of galaxy clusters 50 to 75 Mpc in diameter. By measuring the redshifts and positions of over 100,000 galaxies in great slices across the sky, astronomers have been able to create maps revealing that the superclusters are not scattered at random. They are distributed in long, narrow filaments and thin walls that outline great voids nearly empty of galaxies (■ Figure 18-22). These filaments and walls of superclusters stretching half a billion light-years are the largest structures in the universe. Nothing larger exists.

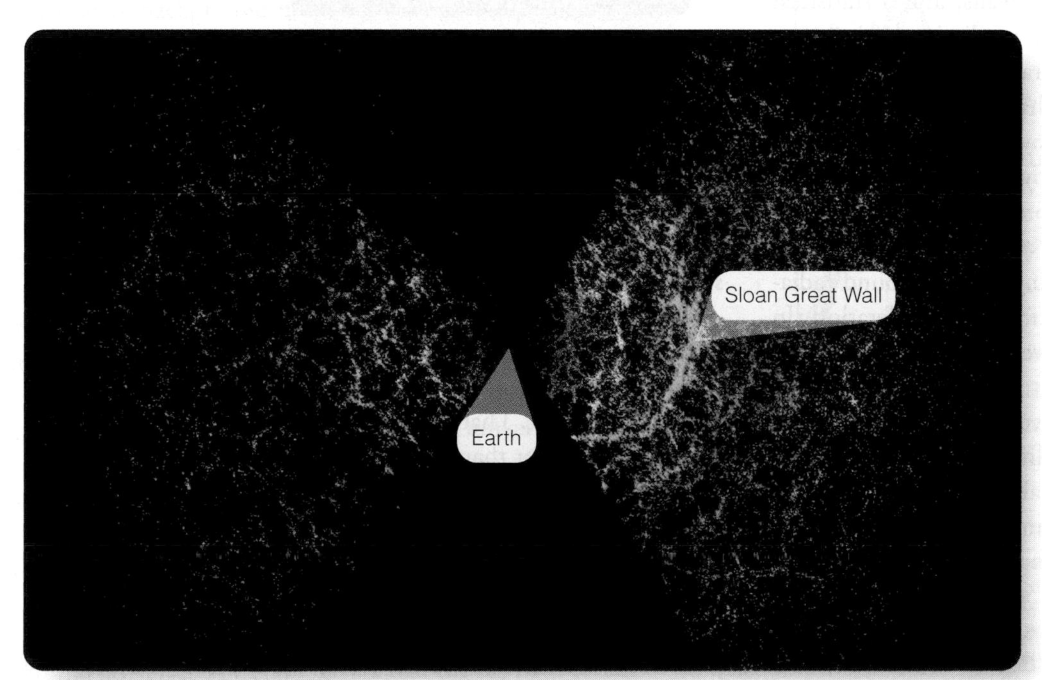

■ **Figure 18-22**

Nearly 70,000 galaxies are plotted in this double slice of the universe extending outward in the plane of Earth's equator. The nearest galaxies are shown in red and the more distant in green and blue. The galaxies form filaments and walls enclosing empty voids. The Sloan Great Wall is almost 1.4 billion light-years long and is the largest known structure in the universe. The most distant galaxies in this diagram are roughly 3 billion light-years from Earth. (Sloan Digital Sky Survey)

■ **Figure 19-1**

(a) The passing star hypothesis was catastrophic. It proposed that the sun was hit by or had a very close encounter with a passing star and that matter torn from the sun and the star formed planets orbiting the sun. The theory is no longer accepted. (b) Laplace's nebular hypothesis was evolutionary. It suggested that a contracting disk of matter conserved angular momentum, spun faster, and shed rings of matter that then formed planets.

and the sun, however, they found that the sun rotated slowly and that the planets moving in their orbits had most of the angular momentum in the solar system. In fact, the rotation of the sun contains only about 0.3 percent of the angular momentum of the solar system. Because the nebular hypothesis could not explain this **angular momentum problem,** it was never fully successful, and astronomers toyed with various versions of the passing star hypothesis for over a century. Today astronomers have a consistent theory for the origin of our solar system, and they are refining the details.

The Solar Nebula Theory

By 1940, astronomers were beginning to understand how stars form and how they generate their energy, and it became clear that the origin of the solar system was linked to that story.

The modern theory of the origin of the planets is the **solar nebula theory,** which proposes that the planets were formed in the disk of gas and dust that surrounded the sun as it formed (■ Figure 19-2). LaPlace's nebular hypothesis included a disk, but his process depended on rings of matter left behind as the disk contracted. Also, it was not based on a clear understanding of how gas and dust behave in such a disk. In the solar nebula theory, the planets grew within the disk by carefully described physical processes.

You have seen clear evidence that disks of gas and dust are common around young stars. Bipolar flows from protostars (Chapter 11) were the first evidence of such disks, but modern techniques can image the disks directly. Our own planetary system formed in such a disk-shaped cloud around the sun. When the sun became luminous enough, the remaining gas and dust were blown away into space, leaving the planets orbiting the sun.

According to the solar nebula theory, our Earth and the other planets of the solar system formed billions of years ago as the sun condensed from a cloud of gas and dust. This means that planet formation is a natural part of star formation, and most stars should have planets.

◄ SCIENTIFIC ARGUMENT ►

Why does the solar nebula theory imply planets are common?

Often, the implications of a theory are more important in building a scientific argument than the theory's own conjecture about nature. The solar nebula theory is an evolutionary theory; and, if it is correct, the planets of our solar system formed from the disk of gas and dust that surrounded the sun as it condensed from the interstellar medium. That suggests it is a common process. Most stars form with disks of gas and dust around them, and planets should form in such disks. Planets should be very common in the universe.

Now build a new scientific argument to consider the old catastrophic theory. **Why did the passing star hypothesis suggest that planets are very rare?**

◄ ►

The Solar Nebula Hypothesis

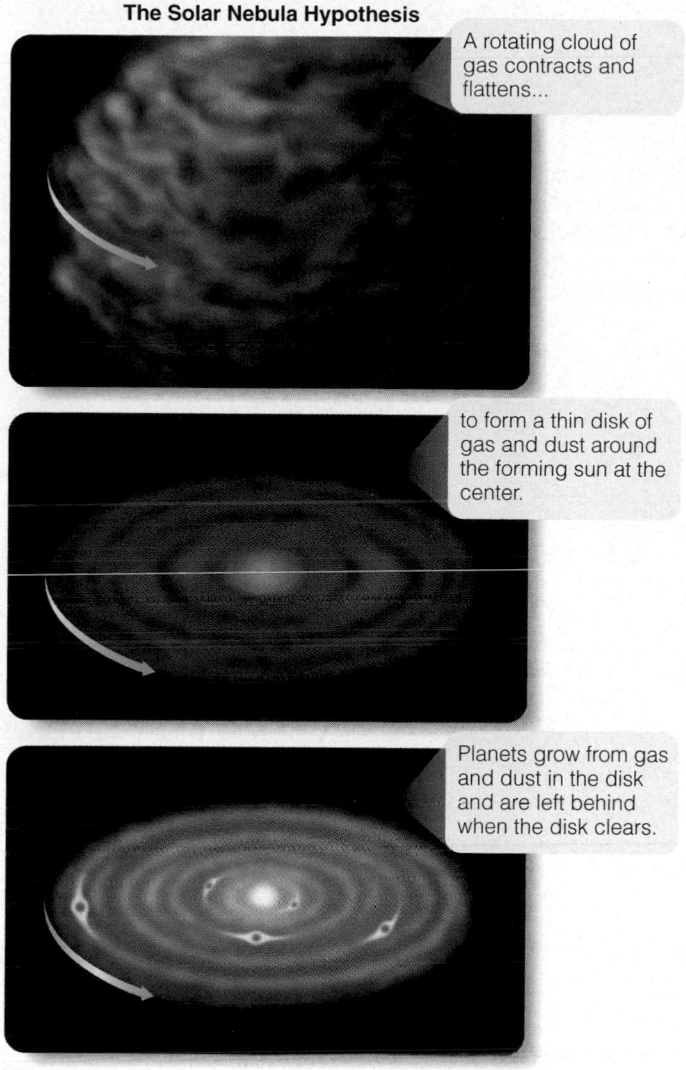

A rotating cloud of gas contracts and flattens...

to form a thin disk of gas and dust around the forming sun at the center.

Planets grow from gas and dust in the disk and are left behind when the disk clears.

Figure 19-2

The solar nebula theory proposes that the planets formed along with the sun.

19-2 A Survey of the Solar System

TO TEST THEIR THEORIES, astronomers must search the present solar system for evidence of its past. In this section, you will survey the solar system and compile a list of its most significant characteristics, which you can use as clues to how it formed.

A General View

The solar system is almost entirely empty space (look back again at Figure 1-7). The planets are small and are scattered far apart in a large disk around the sun.

To see how widely scattered the planets are, imagine that you reduce the solar system until Earth is the size of a grain of table salt, about 0.3 mm (0.01 in.) in diameter. The moon is a speck of pepper about 1 cm (0.4 in.) away, and the sun is the size of a small plum 4 m (13 ft) from Earth. Mercury, Venus, and Mars are grains of salt. Jupiter is an apple seed 20 m (66 ft) from the sun, and Saturn is a smaller seed over 36 m (120 ft) away. Uranus and Neptune are slightly larger than average salt grains, and Neptune is over 150 m (500 ft) from the central plum. Compared to the size of the solar system, the planets are small worlds scattered in a huge disk around the sun—the last remains of the solar nebula.

The motions of the planets follow the disk shape of the solar system. The planets revolve* around the sun in orbits that lie close to a common plane. The orbit of Mercury, the closest planet to the sun, is tipped 7° to the plane of Earth's orbit. The rest of the planets' orbital planes are inclined by no more than 3.4°. As you can see, the solar system is basically disk shaped.

The rotation of the sun and planets on their axes is also related to this disk shape. The sun rotates with its equator inclined only 7.25° to the plane of Earth's orbit, and most of the other planets' equators are tipped less than 30°. The rotations of Venus and Uranus are peculiar, however. Venus rotates backward compared with the other planets, and Uranus rotates on its side (with its equator almost perpendicular to its orbit). You will explore these planets in detail in Chapters 22 and 24, but later in this chapter you will be able to understand how they could have acquired their peculiar rotations.

Apparently, the preferred direction of motion in the solar system—counterclockwise as seen from the north—is also related to its disk shape. All the planets revolve in the same direction (counterclockwise) around the sun and remain in the plane of the disk. Also, with the exception of Venus and Uranus, they rotate counterclockwise on their axes. Furthermore, nearly all of the moons in the solar system, including Earth's moon, orbit around their planets counterclockwise. With only a few exceptions, most of which are understood, revolution and rotation in the solar system follow a disk theme.

Two Kinds of Planets

Perhaps the most striking clue to the origin of the solar system comes from the division of the planets into two categories, the small Earthlike worlds and the giant Jupiter-like worlds. The difference is so dramatic that it is hard to keep from saying, "Aha, this must mean something!" Study **Terrestrial and Jovian Planets** on pages 422–423 and notice three important points and three new terms:

1. Notice how the two kinds of planets are distinguished by their location. The four inner planets are the small, rocky

*Recall from Chapter 2 that the words *revolve* and *rotate* refer to different motions. A planet revolves around the sun but rotates on its axis. Cowboys in the old west didn't carry revolvers. They carried rotators.

Evolution and Catastrophe

How do mountains grow? The modern solar nebula theory proposes that the planets evolved gradually as material in the solar nebula accumulated. That evolutionary theory is much more successful at explaining the properties of the solar system than the old catastrophic theory that the planets formed when a star smashed into the sun. Such catastrophic theories were popular a few centuries ago.

In the very early 1800s, the French scientist Georges Cuvier proposed that the fossils people were finding were the remains of plants and animals destroyed by floods. Other scientists and philosophers, especially in England, picked up his ideas and proposed that the flood had been catastrophic and covered the entire world—the biblical flood of Noah. This led to other catastrophic theories, such as that mountain ranges are pushed up suddenly or can be washed away by more devastating floods. Even the mythical sinking of Atlantis seemed fit in with catastrophic theories.

In the late 1800s, scientists began to understand that changes on Earth take place gradually over long periods of time. There are floods, but they cover local areas for short periods. Mountains rise slowly millimeter by millimeter and are eroded away one bit of rock at a time. Slowly changing climate patterns can change a swamp into a desert, and species can gradually go extinct while new species evolve.

In the 20th century, scientists recognized that catastrophic events do occur but that they are very rare. Astronomers know that large planetesimals can hit planets, and that may cause extinctions. Geologists understand that massive volcanic eruptions can suddenly alter Earth's climate and caused some extinctions. In general, however, modern scientists think of nature evolving gradually and quietly over many millions of years.

Mountains evolve to great heights by rising slowly, not catastrophically. (Janet Seeds)

terrestrial planets and the four outer planets are the large *Jovian planets.*

2. Also notice how common craters are. Almost every solid surface in the solar system is covered with craters.

3. Finally, notice how the planets are accompanied by rings and moons. Jupiter's *Galilean satellites* are large, but most moons are quite small.

The division of the planets into two families is a clue to how our solar system formed, and the craters on Earth's moon hold a further clue. Most of the craters are quite old, and that suggests that the moon, and presumably all of the planets, suffered from a **heavy bombardment** of meteorite impacts when they first formed. As the meteorites were swept up, the bombardment declined only to surge again in an episode of intense cratering called the **late heavy bombardment** lasting from about 4.1 to 3.8 billion years ago. After that, cratering gradually declined to its present low level. Later in this chapter you will see what might have triggered this late heavy bombardment.

Space Debris

The sun and planets are not the only remains of the solar nebula. The solar system is littered with small bodies such as asteroids and comets. Although these objects represent a tiny fraction of the mass of the system, they are a rich source of information about the origin of the planets.

The **asteroids** are small rocky worlds, most of which orbit the sun in a belt between the orbits of Mars and Jupiter. More than 100,000 asteroids have well-charted orbits, of which at least 2000 follow orbits that bring them into the inner solar system or into the outer solar system, where they can occasionally collide with a planet. Earth has been struck many times in its history. Some asteroids are located in Jupiter's orbit, while some have been found beyond the orbit of Saturn.

About 200 asteroids are more than 100 km (60 mi) in diameter, and tens of thousands are estimated to be more than 10 km (6 mi) in diameter. There are probably a million or more that are larger than 1 km (0.6 mi) and billions that are smaller. Because even the largest asteroids are only a few hundred kilometers in diameter, Earth-based telescopes can detect no details on their surfaces, and the Hubble Space Telescope can image only the largest features.

Even so, astronomers have clear evidence that asteroids are irregularly shaped cratered worlds. A number of spacecraft have visited asteroids and sent back photos. For instance, the NEAR spacecraft rendezvoused with the asteroid Eros, went into orbit, and studied it in detail. Like most asteroids, Eros is an irregular, rocky body pocked by craters (■ Figure 19-3). These observations will be discussed in detail in Chapter 25, but in this quick

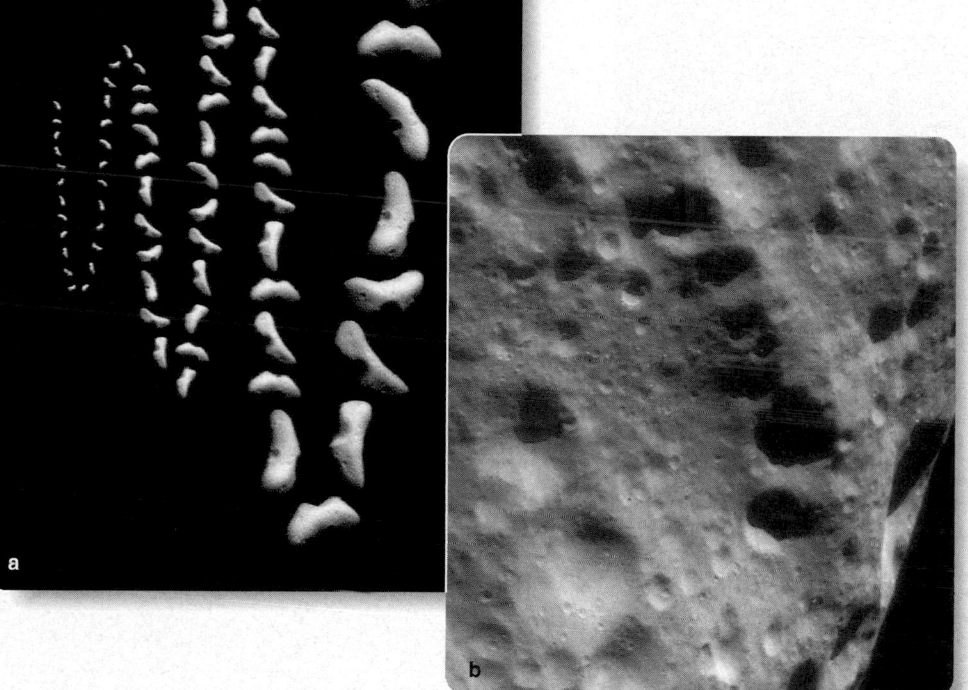

■ **Figure 19-3**

(a) Over a period of three weeks, the NEAR spacecraft approached the asteroid Eros and recorded a series of images arranged here in an entertaining pattern showing the irregular shape and 5-hour rotation of the asteroid. Eros is 34 km (21 mi) long. (b) This close-up of the surface of Eros shows an area about 11 km (7 mi) from top to bottom. (Johns Hopkins University, Applied Physics Laboratory, NASA)

survey of the solar system you can note that all the evidence suggests the asteroids have suffered many impacts from collisions with other asteroids.

It is a **Common Misconception** that the asteroids are the remains of a planet that broke up. In fact, planets are held together very tightly by their gravity and do not "break up." Modern astronomers recognize the asteroids as the debris left over by a planet that failed to form at a distance of 2.8 AU from the sun. When you study the formation of planets later in this chapter, you will see why material in the solar nebula failed to form a planet at the location of the asteroid belt between Mars and Jupiter.

Since 1992, astronomers have discovered roughly a thousand small, dark, icy bodies orbiting in the outer fringes of the solar system beyond Neptune. This collection of objects is called the **Kuiper belt** after the Dutch-American astronomer Gerard Kuiper (pronounced *KI-per*), who predicted their existence in the 1950s. Some are over 1000 km in diameter, but most are smaller. There are probably 100 million bodies larger than 1 km in the Kuiper belt, and any successful theory should explain how they came to orbit so far from the sun.

In contrast to the rocky asteroids and dark Kuiper belt objects, the brightest **comets** are impressively beautiful objects in the sky (■ Figure 19-4). A comet may take months to sweep through the inner solar system, during which time it appears as a glowing head with an extended tail of gas and dust. Most comets, however, never become bright enough to see with the unaided eye.

What produces a comet? The tail of a comet can be longer than an astronomical unit, but it is produced by an icy nucleus only a few tens of kilometers in diameter. The nucleus follows a long, elliptical orbit around the sun and remains frozen and inactive while it is far from the sun. As its orbit carries the nucleus into the inner solar system, the sun's heat begins to vaporize the ices, releasing gas and dust. The pressure of sunlight and the solar wind pushes the gas and dust away, forming a long tail. The motion of the nucleus along its orbit, the pressure of sunlight, and the outward flow of the solar wind can create comet tails that are long and straight or gently curved, but in either case the tails of comets always point approximately away from the sun (Figure 19-4b).

For decades astronomers described comet nuclei as dirty snowballs, meaning that they were thought to be icy bodies with a little bit of embedded rock and dust. Starting with the passage of Comet Halley in 1986, astronomers have found growing evidence that comet nuclei are not made of dirty ice but rather of icy dirt. That is, the nuclei are at least 50 percent rock and dust. One astronomer has suggested replacing the dirty snowball model with the icy mudball model.

The nuclei of comets are ice-rich bodies left over from the formation of the planets. From this you can conclude that at least some parts of the solar nebula were rich in ices. You will see later in this chapter that these ices were important in the formation of the Jovian planets, and you will have a chance to study comets in more detail in Chapter 25.

Unlike the stately comets, **meteors** flash across the sky in momentary streaks of light (■ Figure 19-5). They are commonly called "shooting stars" even though they have nothing to do with stars but are rather small bits of rock and metal falling into Earth's atmosphere. They travel so fast that friction with the air

Terrestrial and Jovian Planets

1 The distinction between the terrestrial planets and the Jovian planets is dramatic. The inner four planets, Mercury, Venus, Earth, and Mars, are **terrestrial planets,** meaning they are small, dense, rocky worlds with little or no atmosphere. The outer four planets, Jupiter, Saturn, Uranus, and Neptune, are **Jovian planets,** meaning they are large, low-density worlds with thick atmospheres and liquid interiors. Pluto does not fit this scheme, being small but low density; you will see in a later chapter that it is a very special world and is no longer considered a planet at all.

The planets and the sun to scale. Saturn's rings would just reach from Earth to the moon.

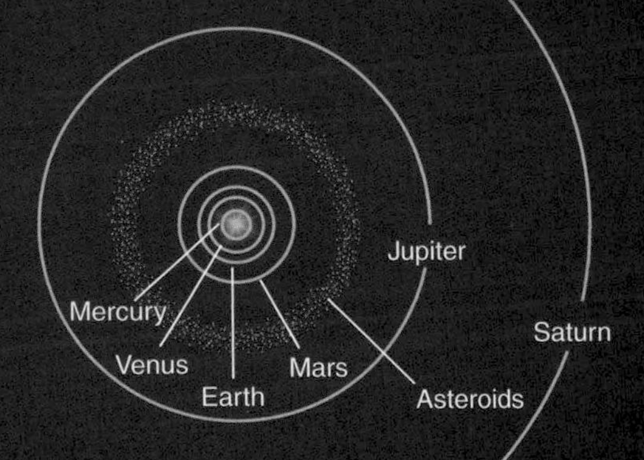

Planetary orbits to scale. The terrestrial planets lie quite close to the sun, whereas the Jovian planets are spread far from the sun outside the asteroid belt.

1a Of the terrestrial planets, Earth is most massive, but the Jovian planets are much more massive. Jupiter is over 300 Earth masses, and Saturn is nearly 100 Earth masses. Uranus and Neptune are 15 and 17 Earth masses.

Mercury is only 40 percent larger than Earth's moon, and its weak gravity cannot retain a permanent atmosphere. Like the moon, it is covered with craters from meteorite impacts.

Mercury

Earth's moon

NASA

UCO/Lick Observatory

2 Craters are common on all of the surfaces in the solar system that are strong enough to retain them. Earth has about 150 impact craters, but many more have been erased by erosion. Besides the planets, the asteroids and nearly all of the moons in the solar system are scarred by craters. Ranging from microscopic to hundreds of kilometers in diameter, most of these craters were produced soon after the planets formed. But some have accumulated over billions of years, made by a continuous, low level rain of meteorites. When astronomers see a rocky or icy surface that contains few craters, they know that the surface is young.

Mercury is so close to the sun it is difficult to study from Earth. The Mariner 10 spacecraft flew past Mercury in 1974, but it was not able to take detailed photos of all of the planet's surface.
Mercury
Moon
The surface of Venus is not visible through its cloudy atmosphere, but radar maps reveal a dry desert world of craters and volcanoes.
These five worlds are shown in proper relative size.
Earth
Moon, UCO/Lick Observatory; all planets, NASA
3
The terrestrial planets have densities like that of rock or metal. The Jovian planets all have low densities, and Saturn's density is only 70 percent the density of water. It would float in a big-enough bathtub.
The atmospheres of the Jovian planets are turbulent, and some are marked by great storms such as the Great Red Spot on Jupiter. The gaseous atmospheres are not deep. If Jupiter were shrunk to a few centimeters in diameter, its outer layers of gas would be no deeper than the fuzz on a badly worn tennis ball.
Venus (radar image)
Mars
Mars has a thin atmosphere and little water. Craters and volcanoes are common on its desert surface.
These Jovian worlds are shown in proper relative size.
3a
The interiors of the Jovian planets contain small cores of heavy elements such as metals, surrounded by a liquid. Jupiter and Saturn contain hydrogen forced into a liquid state by the high pressure. Less-massive Uranus and Neptune contain heavy-element cores surrounded by partially frozen water mixed with heavy material such as rocks and minerals.
The cloudy atmosphere of Venus at visual wavelengths
The terrestrial planets are drawn here to the same scale as the Jovian planets.
Great Red Spot
Jupiter
The Jovian planets have large systems of satellites, and Jupiter is orbited by four large moons known as the **Galilean satellites** because they were discovered by Galileo in 1610.
Saturn's rings seen through a small telescope.
Grundy Observatory
Neptune
Uranus
3b
All four Jovian planets have ring systems. Saturn's rings are made of ice particles. The rings of Jupiter, Uranus, and Neptune are made of dark rocky particles. Terrestrial planets have no detectable rings.
NASA
Saturn

ets, forming at different distances, accumulated from different kinds of materials. The inner planets formed from high-density metal oxides and metals, and the outer planets formed from low-density ices.

People who have read a little bit about the origin of the solar system may hold the **Common Misconception** that the matter in the solar nebula became sorted by density, with the heavy rock and metal sinking toward the sun and the low-density gases being blown outward. That is not the case. The chemical composition of the solar nebula was roughly similar throughout the solar nebula. The important factor was temperature. The inner nebula was hot, and only metals and rock could condense. The cold outer nebula could form lots of ices. The ice line seems to have been between Mars and Jupiter, and it separates the formation of the dense terrestrial planets from that of the low-density Jovian planets.

The Formation of Planetesimals

In the development of the planets, three groups of processes operate to collect gas and dust into larger objects, combine those objects to build the planets, and finally clear away leftover gas and dust. The study of planet building is the study of these three groups of processes.

The first group of processes converted the gas into dust and then collected the dust into large particles. A particle grows by **condensation** when it adds matter one atom at a time from a surrounding gas. Snowflakes, for example, grow by condensation in Earth's atmosphere. In the solar nebula, dust grains were continuously bombarded by atoms of gas, and some of these stuck to the grains. A microscopic grain capturing a single layer of gas atoms increases its mass by a much larger fraction than a gigantic boulder capturing a single layer of atoms. That is why condensation can increase the mass of a small grain rapidly; but, as the grain grows larger, condensation becomes less effective.

Small particles stuck together to form bigger particles in a process called **accretion.** You may have seen accretion in action if you have walked through a snowstorm with big, fluffy flakes. If you caught one of those "flakes" on your mitten and looked closely, you saw that it was actually made up of many tiny, individual flakes that had collided as they fell and accreted to form larger particles. In the solar nebula the dust grains were, on the average, no more than a few centimeters apart, so they collided frequently. Their mutual gravitation was too small to hold them to each other, but other effects may have helped. Static electricity generated by their passage through the gas could have held them together, as could compounds of carbon that might have formed a sticky surface on the grains. Ice grains might have stuck together better than some other types. Of course, some collisions might have broken up clumps of grains; on the whole, however, accretion must have increased grain size. If it had not, the planets would not have formed.

The growth of dust specks by condensation and then by accretion formed larger bodies called **planetesimals.** There is no clear distinction between a very large grain and a very small planetesimal, but you can consider an object a planetesimal when its diameter becomes a kilometer or so (■ Figure 19-7). Objects larger than a centimeter were subject to new processes that tended to concentrate them and help them grow larger. As they grew larger, the objects collapsed into the plane of the solar nebula. Dust grains could not fall into the plane because the turbulent motions of the gas kept them stirred up, but the larger objects had more mass, and the gas motions could not have prevented them from settling into the plane of the spinning nebula. This concentrated the solid particles into a thin plane about 0.01 AU thick and made further planetary growth more rapid.

The collapse of the planetesimals into the plane of the solar nebula is analogous to the flattening of a forming galaxy, and a related process may have become important once the thin disk of planetesimals formed. Computer models show that the rotating disk of particles should have been gravitationally unstable and would have been disturbed by spiral density waves much like those found in spiral galaxies. This would have further concentrated the planetesimals and helped them coalesce into larger objects.

Through this first group of processes the nebula became filled with trillions of planetisimals ranging in size from pebbles to tiny planets. As the largest began to exceed 100 km in diameter, a new group of processes began to alter them, and a new stage in planet building began—the collection of planetisimals to form planets.

■ **Figure 19-7**

What did the planetesimals look like? You can get a clue from this photo of the 5-km-wide nucleus of Comet Wild2. Whether rocky or icy, the planetesimals must have been small, irregular bodies, pocked by craters from collisions with other planetesimals. (NASA)

The Growth of Protoplanets

The coalescing of planetesimals eventually formed **protoplanets,** massive objects destined to grow into planets. As these larger bodies grew, new processes began making them grow ever faster and altering their physical structure.

If planetesimals collided at orbital velocities, it is unlikely that they could have stuck together. The average orbital velocity in the solar system is about 10 km/s (22,000 mph), and head-on collisions at this velocity would have vaporized the material. However, the planetesimals were moving in the same direction in the nebular plane and didn't collide head on. Instead, they merely rubbed shoulders at low relative velocities. Such gentle collisions were more likely to fuse them than to shatter them.

In addition, some adhesive effects probably helped. Sticky coatings and electrostatic charges on the surfaces of the smaller planetesimals probably aided their growth. On larger planetesimals, collisions would have fragmented some of the surface rock; but, if the planetesimals were large enough, their gravity would have held on to some fragments, forming a layer of soil composed entirely of crushed rock. Such a layer may have been effective in trapping smaller bodies.

The largest planetesimals grew the fastest because they had the strongest gravitational field. Not only could they hold on to a cushioning layer to trap fragments, but their stronger gravity could also attract additional material. A few of the largest of these planetesimals grew quickly to protoplanetary dimensions, sweeping up more and more material.

Protoplanets had to begin growing by accumulating solid bits of rock, metal, and ice because they did not have enough gravity to capture and hold large amounts of gas. In the gases of the solar nebula, the atoms and molecules were traveling at velocities much larger than the escape velocities of modest-size protoplanets. Only when a protoplanet approached a mass of 10 to 15 Earth masses could it begin to grow by **gravitational collapse,** the rapid accumulation of large amount of infalling gas.

In its simplest form, the theory of protoplanet growth supposes that all the planetesimals building a protoplanet had about the same chemical composition. The planetesimals accumulated gradually to form a planet-size ball of material that was of homogeneous composition throughout. But once the planet formed, heat began to accumulate in its interior from the decay of short-lived radioactive elements, and this heat eventually melted the planet and allowed it to differentiate. **Differentiation** is the separation of material according to density. When the planet melted, the heavy metals such as iron and nickel settled to the core, while the lighter silicates floated to the surface to form a low-density crust. The story of planet formation from planetesimals of similar composition is shown in the left half of ■ Figure 19-8.

Two Models of Planet Building
Planetesimals contain both rock and metal.
The first planetesimals contain mostly metals.
A planet grows slowly from the uniform particles.
Later the planetesimals contain mostly rock.
The resulting planet is of uniform composition.
A rock mantle forms around the iron core.
Heat from radioactive decay causes differentiation.
Heat from rapid formation can melt the planet.
The resulting planet has a metal core and low-density crust.
The resulting planet has a metal core and low-density crust.

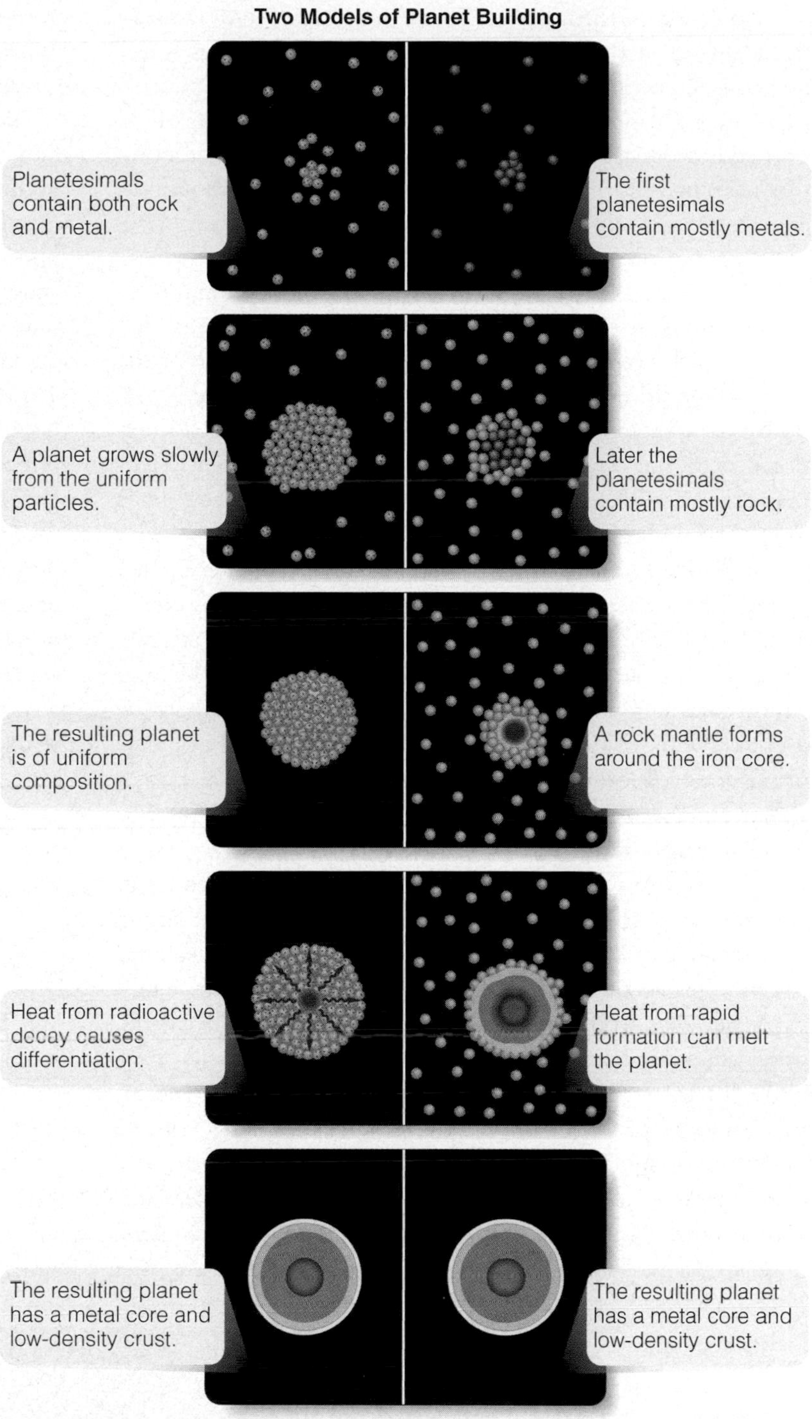

■ **Figure 19-8**

If the temperature of the solar nebula changed during planet building, the composition of the planetesimals may have changed. The simple model at left assumes no change occurred, but the model at the right incorporates a change from metallic to rocky planetesimals.

This process depends on the presence of short-lived radioactive elements in the solar nebula. Astronomers know such elements were present because very old minerals found in meteorites contain daughter isotopes such as magnesium-26. That isotope is produced by the decay of aluminum-26 in a reaction that has a half-life of only 0.74 million years. The aluminum-26 and similar short-lived radioactive isotopes are gone now. Where did they come from?

Short-lived radioactive elements, including aluminum-26, are produced in supernova explosions, and that suggests that such an explosion occurred shortly before the formation of the solar nebula. In fact, many astronomers suspect that the supernova explosion compressed nearby gas and triggered the formation of stars, one of which became the sun. Thus our solar system may exist because of a supernova explosion that occurred about 4.6 billion years ago.

If planets formed in this gradual way and were later melted by radioactive decay, then Earth's present atmosphere was not its first. That first atmosphere consisted of small amounts of gases trapped from the solar nebula—mostly hydrogen and helium. Those gases were later driven off by the heat, aided perhaps by outbursts from the infant sun, and new gases released from the planet's interior formed a secondary atmosphere. This creation of a planetary atmosphere by gases exhaled from a planet's interior is called **outgassing.**

This simple theory of planet formation can be improved in two ways. First, it seems likely that the solar nebula cooled during the formation of the planets, so any given planet did not accumulate from planetesimals of the same composition. As planet building began, the first particles to condense in the inner solar system were rich in metals and metal oxides, so the protoplanets may have begun by accreting metallic cores. Later, as the nebula cooled, more silicates could form, and the protoplanets added silicate mantles. A second improvement in the theory proposes that the planets grew so rapidly that the heat released by the violent impacts of infalling particles, the **heat of formation,** did not have time to escape. This heat rapidly accumulated and melted the protoplanets as they formed. If planets formed this rapidly, then they must have differentiated as they formed. This improved story of planet formation is shown in the right half of Figure 19-8.

If the Earth formed rapidly, then it was mostly molten from the beginning, and there was never a time when it could have captured a primitive atmosphere of hydrogen and helium accumulated from the solar nebula. Rather, the gases of the atmosphere were released by the molten rock as the protoplanet grew. Those gases would not have included much water, however, so some astronomers now think that Earth's water and much of its present atmosphere accumulated late in the formation of the planets as Earth swept up volatile-rich planetesimals forming in the cooling solar nebula. These icy planetesimals must have formed in the outer parts of the solar nebula and may have been scattered by encounters with the Jovian planets.

How large a planet could grow would depend on where it formed in the solar nebula. The terrestrial planets formed in the inner nebula where it was so hot that only metals and rock could form solid particles. Chemical compounds such as water, methane, and ammonia could not form solids (ices) and were not incorporated into the planets in large amounts. In the outer solar system, beyond the ice line, there was metal and rock, but there were also lots of ice particles. The Jovian planets grew rapidly and became massive enough to grow by gravitational collapse as they drew in large amounts of gas from the solar nebula. The terrestrial planets never became massive enough to grow by gravitational collapse.

Is There a Jovian Problem?

The Jovian planets (■ Figure 19-9) are so massive that they dominate the outer solar system. Any description of the origin of the solar system needs to account for their size and position, and this has posed a problem for astronomers in recent years.

For some time, astronomers have known that almost all of the T Tauri stars that have gas disks are younger than 10 million years; older T Tauri stars have blown their gas disks away. Presumably the sun blew its disk away in the same amount of time.

■ Figure 19-9

Saturn is a beautiful planet, but the Jovian worlds are a problem for modern astronomers. Planet-forming nebulas are blown away in only a few million years by nearby stars, so Jovian planets must form quickly. Direct gravitational collapse could have formed them quickly under certain conditions, but newer research suggests that accretion followed by gravitational collapse could build Jovian planets in only a million years. (JPL/NASA)

Jovian planets grow by accumulating gas, so they must have been complete within 10 million years or so. The terrestrial planets grew from solids and not from the gas, so they could have continued to grow by accretion from solid debris left behind when the gas was blown away. The terrestrial planets were nearly complete within 10 million years but could have continued to grow for another 20 million years or so.

New research upset this picture of leisurely planet formation. Studies of star formation show that they form when a large gas cloud collapses and forms a star cluster. Telescopes can image gas and dust disks around newborn stars, stars in the Orion Nebula for example, and those disks are being destroyed by the ultraviolet radiation and strong winds blowing away from the hottest stars in the cluster. Furthermore, the gravitational influence of nearby stars should quickly strip away the outer parts of disks. It seems likely that most disks around newborn stars don't last more than 7 million years at most and some may hardly survive for one million years. That didn't seem to be long enough to grow Jovian planets by accreting a core and then drawing in gas; astronomers began to search for alternative theories to explain the growth of the Jovian planets.

Mathematical models of the solar nebula have been computed using specially built computers running programs that take weeks to finish a calculation. The results show that the rotating gas and dust of the solar nebula may have become unstable and formed outer planets by **direct gravitational collapse.** That is, massive planets may have been able to form from the gas without first forming a dense core by accretion. Planets the size of Jupiter or Saturn form in these mathematical models within a few hundred years.

This has produced a controversy among astronomers. Do Jovian planets form by direct gravitational collapse and terrestrial planets by accretion? Newer research suggests an explanation that does not require a new theory of planet formation. At the end of this chapter, you will read that astronomers have detected many planets orbiting other stars. These observations show that Jupiter-sized planets are common; the Jovian planets in our solar system are not unusual. Furthermore, the observations of these other planets suggest that planets can migrate as they form. A giant planet sweeping up planetesimals and gas may creep closer to its star. Also, gravitational interactions among planets can shift some orbits inward closer to the star and move others outward.

The newest models of planet formation take these effects into account. If a Jovian planet does not change its orbit, it can sweep up all of the material near its orbit, and then it can grow only as fast as new material spreads toward it. But if it migrates as it forms, it can move into new regions with more material. It can migrate to a fresh "feeding zone," astronomers say. These models show that a Jovian planet could form in only 1 to 2 million years.

Also, improved models of conventional planet formation show that a Jovian planet can radiate away its heat faster than previously thought, and that means it can pull in gas faster and form more quickly. Even without orbit migration, this can reduce the formation time from 7 to 2 million years.

Is there a Jovian problem? Improved models including orbit migration and rapid cooling suggest that the Jovian planets in our solar system could have formed quickly by accreting a core of solid particles and, when they were massive enough, pulled in gas by gravitational collapse. Direct gravitational collapse, which skips the accretion of a core, does not seem to be necessary. On the other hand, direct gravitational collapse is possible in some of the denser disks of gas and dust orbiting other stars, so some of the Jovian planets orbiting other stars may have formed directly.

Orbit migration may help explain a different problem that astronomers have had explaining the formation of Uranus and Neptune. Those two planets are so far from the sun that accretion could not have built them rapidly. The gas and dust of the solar nebula must have been sparse out there, and Uranus and Neptune orbit so slowly they would not have swept up material very rapidly. Theoretical calculations show that the four Jovian planets may have formed closer to the sun in the region of Jupiter and Saturn. In the models, gravitational interactions shift Jupiter slightly inward and Saturn outward. That forces Uranus and Neptune to migrate rapidly outward to their present orbits.

The migration of the Jovian planets would have happened over many millions of years, but it would have had dramatic effects on the orbits of smaller bodies in the solar system. The outward migration of Neptune would push the Kuiper belt objects to larger orbits, and the combined influences of all four Jovian planets would fling many remaining planetesimals into highly elliptical orbits, and they could hit planets. That may explain the late heavy bombardment that battered the planets and their moons about 3.9 billion years ago.

Improved models of planet formation are revealing some of the details of our solar system's history. That is enough to allow you to explain the distinguishing characteristics of the solar system.

Explaining the Characteristics of the Solar System

Now you have learned enough to put all the pieces of the puzzle together and explain the distinguishing characteristics of the solar system in Table 19-1.

As you have seen, the disk shape of the solar system was inherited from the solar nebula. The sun and planets revolve and rotate in the same direction because they formed from the same rotating gas cloud. The orbits of the planets lie in the same plane because the rotating solar nebula collapsed into a disk, and the planets formed in that disk.

The solar nebula theory is evolutionary in that it calls on continuing processes to gradually build the planets. To explain the rotation of Venus and Uranus, however, you may need to

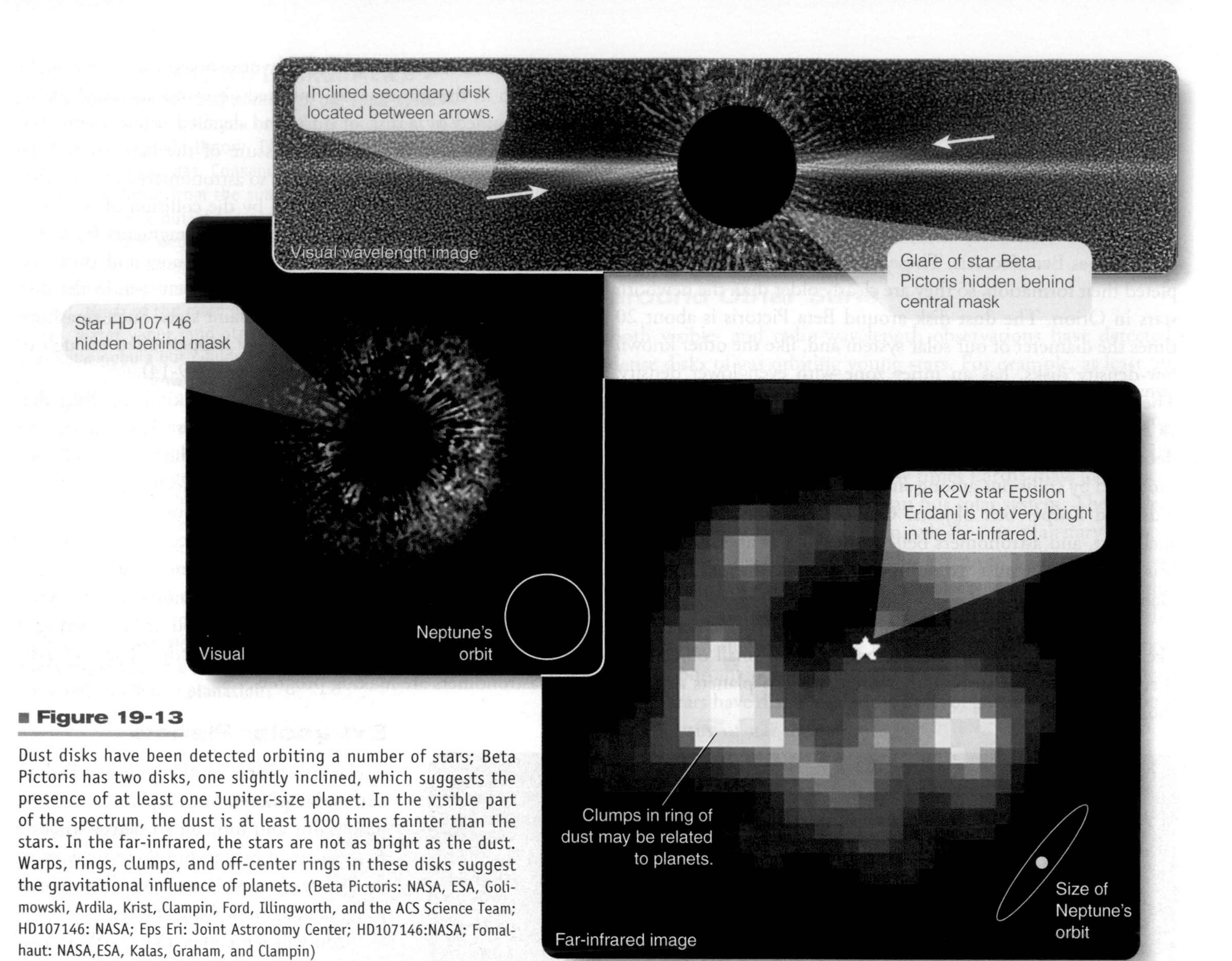
Inclined secondary disk
located between arrows.
Visual wavelength image
Glare of star Beta
Pictoris hidden behind
central mask
Star HD107146
hidden behind mask
Visual
Neptune's
orbit
The K2V star Epsilon
Eridani is not very bright
in the far-infrared.
Clumps in ring of
dust may be related
to planets.
Size of
Neptune's
orbit
Far-infrared image

■ **Figure 19-13**

Dust disks have been detected orbiting a number of stars; Beta Pictoris has two disks, one slightly inclined, which suggests the presence of at least one Jupiter-size planet. In the visible part of the spectrum, the dust is at least 1000 times fainter than the stars. In the far-infrared, the stars are not as bright as the dust. Warps, rings, clumps, and off-center rings in these disks suggest the gravitational influence of planets. (Beta Pictoris: NASA, ESA, Golimowski, Ardila, Krist, Clampin, Ford, Illingworth, and the ACS Science Team; HD107146: NASA; Eps Eri: Joint Astronomy Center; HD107146:NASA; Fomalhaut: NASA,ESA, Kalas, Graham, and Clampin)

owner was jerked back and forth. Astronomers can detect a planet orbiting another star by watching how the star moves as the planet tugs on it.

The first planet detected in this way was discovered in 1995. It orbits the star 51 Pegasi. As the planet circles the star, the star wobbles slightly, and these very small motions of the star are detectable as Doppler shifts in the star's spectrum (■ Figure 19-15a). From the motion of the star and estimates of the star's mass, astronomers can deduce that the planet has half the mass of Jupiter and orbits only 0.05 AU from the star. Half the mass of Jupiter amounts to 160 Earth masses, so this is a large planet.

Astronomers were not surprised by the announcement that a planet orbited 51 Pegasi; for years astronomers had assumed that many stars had planets. Nevertheless, they greeted the discovery with typical skepticism **(How Do We Know? 19-2)**. That skepticism led to careful tests of the data and further observations that confirmed the discovery. In fact, well over 200 planets have been discovered orbiting other stars, including true planetary systems such as the three planets orbiting the star Upsilon Andromedae (Figure 19-15b). Roughly 20 such planetary systems have been found.

Another way to search for planets is to look for changes in the brightness of the star as the orbiting planet crosses in front of or behind the star. The change in brightness is very small, but it is detectable, and astronomers have used this technique to detect planets. All of these planets are roughly the size of Jupiter. The Hubble Space Telescope has successfully detected planets as they cross in front of their stars, and the Spitzer Infrared Space Telescope has detected planets when they pass behind their stars. The planets are hot because they are close to their stars and emit significant infrared radiation. When they pass behind the stars they orbit, the total infrared brightness of the system decreases. These observations further confirm the existence of extrasolar planets.

■ Figure 19-14

Collisions between asteroids are rare events, but they generate lots of dust and huge numbers of fragments, as in this artist's conception. Further collisions between fragments can continue to produce dust. Because such dust is blown away quickly, astronomers treat the presence of dust as evidence that objects of asteroidal size are also present. (J. Lomberg/Gemini Observatory)

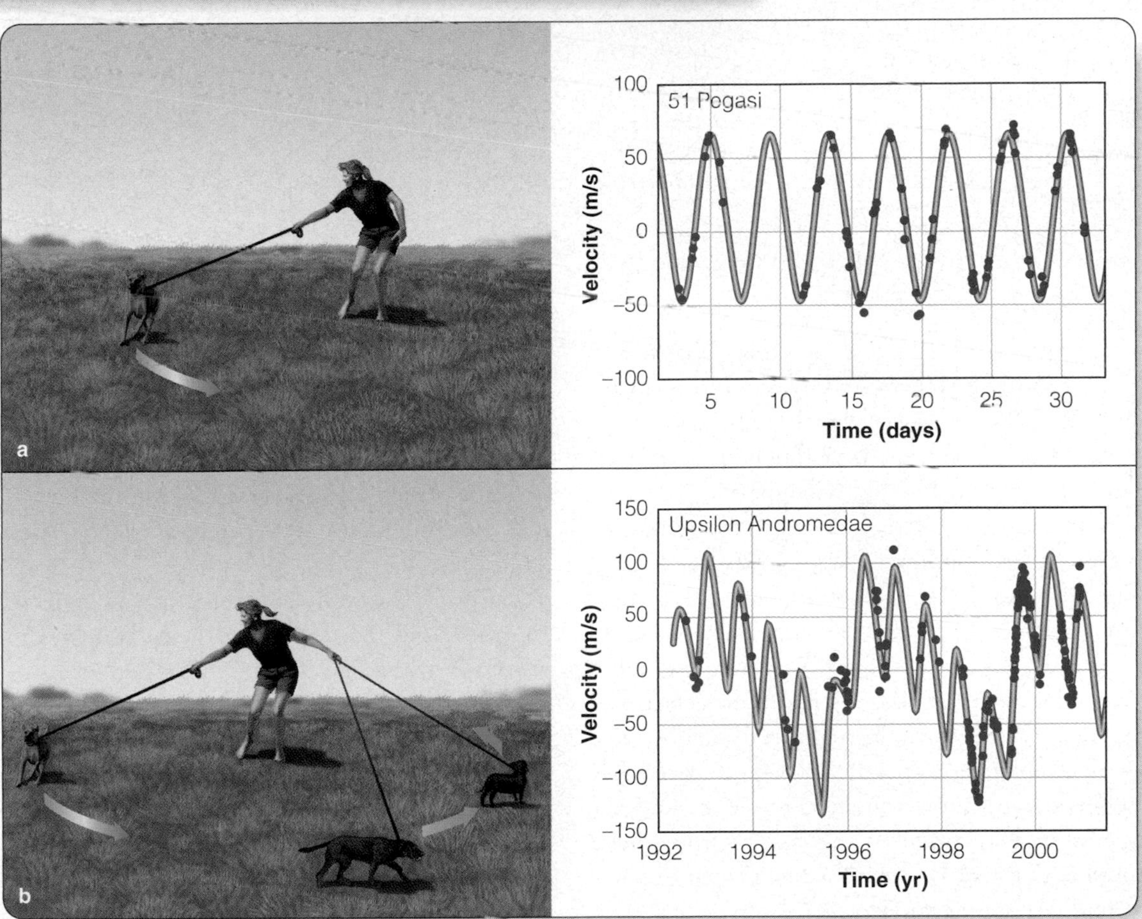

■ Figure 19-15

Just as someone walking a lively dog is tugged around, the star 51 Pegasi is tugged around by the planet that orbits it every 4.2 days. The wobble is detectable in precision observations of its Doppler shift. Someone walking three dogs is pulled about in a more complicated pattern, and you can see something similar in the Doppler shifts of Upsilon Andromedae. The influence of its shortest-period planet has been removed in this graph to reveal the orbital influences of two additional planets.

How Do We Know? | 19-2

Courteous Skeptics

Why do scientists treat every new idea with suspicion? "Scientists are just a bunch of skeptics who don't believe in anything." That is a common complaint about scientists, but it misinterprets the fundamental characteristic of the scientist. Yes, scientists are skeptical about new ideas and discoveries, but that is because they test every new idea.

For example, in 1989 two physicists announced that they had generated energy by triggering nuclear fusion in a tabletop apparatus. This was tremendously important because it promised cheap power generated without the extremely high temperatures thought necessary for nuclear fusion. The excitement over this cold fusion was intense. Politicians gave speeches about the wonders of cheap energy, legislatures appropriated money for new research centers, business leaders announced new divisions to bring the benefits of cold fusion to the market place; but not everyone jumped on the bandwagon.

Scientists were skeptical, not because they didn't want cheap power for the public good, but because scientists test every new idea. Among scientists it is not bad manners to say, "Really, how do you know that?" or "Show me the evidence." Skepticism is the way scientists weed out mistakes.

Around the world scientists built their own cold fusion cells; a few thought they could detect low-level energy generation, but many could not. No one could detect neutrons, which had to be released during nuclear fusion. From many such experiments, scientists concluded that nuclear fusion was not occurring in the cells, and that the cells were not generating energy. It wasn't a hoax; it was a mistake. All the speeches and public relations couldn't make it true, and in the end, scientific skepticism saved the day.

Nearly three decades later, a few scientists are still trying to find a way to generate energy through cold fusion, but they are following good scientific methods. They are skeptical of even their own ideas and are testing every new result. If there is ever a new announcement of cold fusion, there will probably be fewer speeches and more heat.

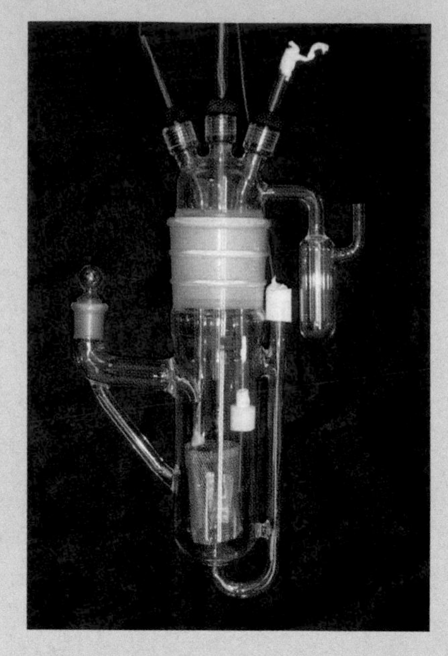

A laboratory cell for the study of cold fusion. (Photo by Steven Krivit of device at US Navy SPAWAR Systems Center in San Diego)

Notice how the techniques used to detect these planets resemble techniques used to study binary stars. Most of the planets were discovered using the same observational methods used to study spectroscopic binaries, but a few were found by observing the stars as if they were eclipsing binaries (Chapter 9).

The extrasolar planets discovered so far tend to be massive and have short periods because lower-mass planets or longer-period planets are harder to detect. Low-mass planets don't tug on their stars very much, and present-day spectrographs can't detect the very small velocity changes that these gentle tugs produce. Also, planets with longer periods are harder to detect because astronomers have not been making high-precision observations for many years. Jupiter takes nearly 12 years to circle the sun once, so it will take years for astronomers to see the longer-period wobbles produced by planets lying farther from their stars. So you should not be surprised that the first planets discovered are massive and have short periods.

The new planets may seem odd for another reason. In our own solar system, the large planets formed farther from the sun where the solar nebula was colder and ices could condense. Yet many extrasolar planets lie close to their stars. In fact, some of them are strongly heated by their star and are referred to as **hot Jupiters.** How could big planets form so near their stars? Mathematical models show that planets forming in an especially dense disk of matter can spiral inward as they sweep up planetesimals. That orbital migration means it is possible for some planets to become massive in the outer parts of a disk and then migrate inward to become the short-period, hot Jupiters.

Many of the newly discovered extrasolar planets have elliptical orbits, and that seems odd when compared with our solar system, in which the planetary orbits are nearly circular. Theorists point out, however, that planets can interact with each other in some young planetary systems and can be thrown into elliptical orbits. This is probably rare in planetary systems, but astronomers find these extreme systems more easily because they tend to produce big wobbles.

The preceding paragraphs should reassure you that massive planets in small or elliptical orbits do not contradict the solar nebula theory, and in fact, as astronomers continue to refine their instruments to detect smaller velocity shifts in stars, they find lower-mass planets. An Earthlike planet was found in 2007. It is about 5 times the mass of Earth, and it orbits its red dwarf star in only 13 Earth-days at a distance of 0.07 AU. With a diameter

What Are We?

Planetwalkers

The matter you are made of came from the big bang, and it has been cooked into a wide range of atoms inside stars. Now you can see how those atoms came to be part of Earth. Your atoms were in the solar nebula 4.6 billion years ago, and nearly all of that matter contracted to form the sun. Only a small amount was left behind to form planets. In the process, your atoms became part of Earth.

You are a planetwalker, and you have evolved to live on the surface of Earth. Are there other planetwalkers in the universe? Now you know that planets are common, and you can reasonably suppose that there are more planets in the universe than there are stars. However elegant and intricate the formation of the solar system was, it is a common process, so there may indeed be more planetwalkers living on other worlds.

What are those creatures like? They have been shaped by their home planet just as you have been, and as you explore further in the following chapters you will discover that planets are diverse, and some are highly unlikely homes for living creatures. But some are not such unwelcoming places. It is time to pack your spacesuit and voyage out among the planets of our solar system, visit them one by one, and search for the natural principles that govern planets. Your journey begins in the next chapter.

1.5 times larger than Earth and a surface temperature that would permit liquid water, it is clearly not a Jovian planet.

Among the planets found so far, most are orbiting stars that are metal rich rather than metal poor. This supports planet formation by the accretion of a core of solids and the later accumulation of gas. It is evidence against formation by direct gravitational collapse, which does not require the presence of solids such as metals and silicates to start planet formation.

Detecting extrasolar planets is an exciting part of modern astronomy, but actually photographing a planet orbiting another star is about as difficult as photographing a bug crawling on the bulb of a searchlight miles away. Planets are small and dim and get lost in the glare of the stars they orbit. Nevertheless, a few objects have been photographed that may be planets (■ Figure 19-16). Searches for more are being conducted, and space telescopes are being developed that will eventually be able to image Earth-size planets orbiting nearby stars.

The discovery of extrasolar planets gives astronomers added confidence in the solar nebula theory. The theory predicts that planets are common, and astronomers are finding them orbiting many stars.

■ Figure 19-16

Infrared observations reveal an object of about 5 Jupiter masses orbiting a brown dwarf in an orbit roughly twice as large as that of Neptune around the sun. Spectra showing water vapor and the object's infrared colors suggest it is relatively cool and is probably a planet. In an artist's impression, the planet orbits around a brown dwarf, still surrounded by its dusty disk. (ESO)

Planet 2M1207b orbits 77AU from its brown dwarf "sun."
Brown dwarf
Infrared image
Artist's conception

second question is more complex. Where did these atmospheres come from? To answer that question, you will have to study the geological history of these worlds.

You are studying Earth in this chapter to use it as a basis for comparison with other worlds. Of course, four of the worlds in our solar system, the Jovian planets, are dramatically different from Earth. But the four terrestrial planets are rocky worlds much like Earth, and many of the moons in the solar system have geology that you can compare with Earth.

◄ SCIENTIFIC ARGUMENT ►

Why do you expect the inner planets to be high-density worlds?

A good scientific argument gives you a way to see nature—a way to understand why the universe behaves as it does. The best questions in science always begin with "why?" The inner planets formed from the hotter regions of the solar nebula where no ice could form. Only metals and rocky minerals could condense to form solid solids. So you expect the inner planets to have accumulated mostly rock and metal, which are dense materials.

As you begin studying planets one by one, keep thinking in scientific arguments. They will help you organize all of the information you will meet. Now build an argument to review an important point. **What made all of the craters that are spread through the solar system?**

◄ ►

20-2 The Early History of Earth

LIKE ALL THE TERRESTRIAL PLANETS, Earth formed from the inner solar nebula about 4.6 billion years ago. Even as it took form, it began to change.

Four Stages of Planetary Development

The Earth has passed through four stages of planetary development (■ Figure 20-2). All terrestrial planets pass through these same stages to some degree, but some planets evolved further or were affected in different ways. The first stage of planetary evolution is *differentiation,* the separation of material according to density. Earth now has a dense core and a lower-density crust, and that structure must have originated very early in its history. Differentiation would have occurred easily if Earth was molten when it was young. Two sources of energy could have heated Earth. First, heat of formation was released by infalling material. A meteorite hitting Earth at high velocity converts most of its energy of motion into heat, and the impacts of a large number of meteorites would have released tremendous heat. If Earth formed rapidly, this heat would have accumulated much more rapidly than it could leak away, and Earth was probably molten when it formed. A second source of heat requires more time to develop. The decay of radioactive elements trapped in the Earth releases heat gradually; but, as soon as Earth formed, that heat began to accumulate and helped melt Earth. That would have helped the planet differentiate.

Four Stages of Planetary Development
Differentiation produces a dense core, thick mantle, and low-density crust.
The young Earth was heavily bombarded in the debris-filled early solar system.
Flooding by molten rock and later by water can fill lowlands.
Slow surface evolution continues due to geological processes, including erosion.

■ **Figure 20-2**

The four stages of planetary development are illustrated for Earth.

While Earth was still in a molten state, meteorites could leave no trace, but in the second stage in planetary evolution, *cratering,* the young Earth was battered by meteorites that pulverized the newly forming crust. The largest meteorites blasted out crater basins hundreds of kilometers in diameter. As the solar nebula cleared, the amount of debris decreased, and after the late heavy bombardment, the level of cratering fell to its present low level. Although meteorites still occasionally strike Earth and dig craters, cratering is no longer the dominant influence on Earth's geology. As you compare other worlds with Earth, you will discover traces of this intense period of cratering on every old surface in the solar system.

The third stage, *flooding,* no doubt began while cratering was still intense. The fracturing of the crust and the heat produced by radioactive decay allowed molten rock just below the crust to well up through fissures and flood the deeper basins. You will find such basins filled with solidified lava flows on other worlds, such as the moon, but all traces of the early lava flooding on Earth have been destroyed by later geological activity. On Earth, flooding continued as the atmosphere cooled and water fell as rain, filling the deepest basins to produce the first oceans. Notice that on Earth flooding involved both lava and water, a circumstance that you will not find on most worlds.

The fourth stage, *slow surface evolution,* has continued for the last 3.5 billion years or more. Earth's surface is constantly changing as sections of crust slide over each other, push up mountains, and shift continents. At the same time, moving air and water erode the surface and wear away geological features. Almost all traces of the first billion years of Earth's geology have been destroyed by the active crust and erosion.

Earth as a Planet

All terrestrial planets pass through these four stages, but some have emphasized one stage over another, and some planets have failed to progress fully through the four stages. Earth is a good standard for comparative planetology, because every major process on any rocky world in our solar system is represented in some form on Earth.

Nevertheless, Earth is peculiar in two ways. First, it has large amounts of liquid water on its surface. Fully 75 percent of its surface is covered by this liquid; no other planet in our solar system is known to have such extensive liquid water on its surface. Not only does water fill the oceans, but it evaporates into the atmosphere, forms clouds, and then falls as rain. Water falling on the continents flows downhill to form rivers that flow back to the sea, and, in so doing, the water produces intense erosion. Entire mountain ranges can literally dissolve and wash away in only a few tens of millions of years, less than 1 percent of Earth's total age. You will not see such intense erosion on most worlds. Liquid water is, in fact, a rare material on most planets.

Your home planet is special in a second way. Some of the matter on the surface of this world is alive, and a small part of that living matter is aware. No one is sure how the presence of living matter has affected the evolution of Earth, but this process seems to be totally missing from other worlds in our solar system. Furthermore, the thinking part of the life on Earth, humankind, is actively altering our planet. Your use of Earth as a standard for the study of other worlds will also give you new insight into your own planet and how modern society may be altering it.

◄ SCIENTIFIC ARGUMENT ►

Why should you think Earth went through an early stage of cratering?

When you build a scientific argument, take great care to distinguish between theory and evidence. Recall from the previous chapter that the planets formed by the accretion of planetesimals from the solar nebula. The proto-Earth may have been molten as it formed, but as soon as it grew cool enough to form a solid crust, the remaining planetesimal impacts would have formed craters. So you can reason from the solar nebula hypothesis that Earth should have been cratered. But you can't use a theory as evidence to support some other theory. To find real observational evidence, you need only look at the moon. The moon has craters, and so does every other old surface in our solar system. There must have been a time, when the solar system was young, when there were large numbers of objects striking all the planets and moons and blasting out craters. If it happened to other worlds in our solar system, it must have happened to Earth, too.

The best evidence to support your argument would be lots of craters on Earth, but, of course, there are few craters on Earth. Extend your argument. **Why don't you see lots of craters on Earth today?**

◄ ►

20-3 The Solid Earth

ALTHOUGH YOU MIGHT THINK OF EARTH as solid rock, it is in fact neither entirely solid nor entirely rock. The thin crust seems solid, but it floats and shifts on a semiliquid layer of molten rock just below the crust. Below that lies a deep, rocky mantle surrounding a core of liquid metal. Much of what you see on the surface of Earth is determined by its interior.

Earth's Interior

The theory of the origin of planets from the solar nebula predicts that Earth should have melted and differentiated into a dense metallic core and a dense mantle with a low-density silicate crust. But did it? Where's the evidence? Clearly, Earth's surface is made of lower-density silicates, but what of the interior?

High temperature and tremendous pressure in Earth's interior make any direct exploration impossible. Even the deepest oil wells extend only a few kilometers down and don't reach through the crust. It is impossible to drill far enough to sample Earth's core. Yet Earth scientists have studied the interior and found clear evidence that Earth did differentiate **(How Do We Know? 20-1)**.

This exploration of Earth's interior is possible because earthquakes produce vibrations called **seismic waves,** which travel through the crust and interior and eventually register on sensitive detectors called **seismographs** all over the world (■ Figure 20-3). Two kinds of seismic waves are important to this discussion. The **pressure** (P) **waves** are much like sound waves in that they travel as a region of compression. As a *P* wave passes, particles of matter vibrate back and forth parallel to the direction of wave travel (■ Figure 20-4a). In contrast, the **shear** (S) **waves** move as displacements of particles perpendicular to the waves' direction of travel (Figure 20- 4b). That means that *S* waves distort the material but do not compress it. Normal sound waves are pressure waves, whereas the vibrations you see in a bowl of jelly are shear waves. Because *P* waves are compression waves, they can move through a liquid, but *S* waves cannot. A glass of water can't

The Magnetic Field

Apparently, Earth's magnetic field is a direct result of its rapid rotation and its molten metallic core. Internal heat forces the liquid core to circulate with convection while Earth's rotation turns it about an axis. The core is a highly conductive iron–nickel alloy, an even better electrical conductor than copper, the material commonly used for electrical wiring. The rotation of this convecting, conducting liquid generates Earth's magnetic field in a process called the dynamo effect (■ Figure 20-7). This is the same process that generates the solar magnetic field in the convective layers of the sun, and you will see it again when you explore other planets.

Earth's magnetic field protects it from the solar wind. Blowing outward from the sun at about 400 km/s, the solar wind consists of ionized gases carrying a small part of the sun's magnetic field. When the solar wind encounters Earth's magnetic field, it is deflected like water flowing around a boulder in a stream. The surface where the solar wind is first deflected is called the **bow shock,** and the cavity dominated by Earth's magnetic field is called the **magnetosphere** (■ Figure 20-8a). High-energy particles from the solar wind leak into the magnetosphere and become trapped within Earth's magnetic field to produce the

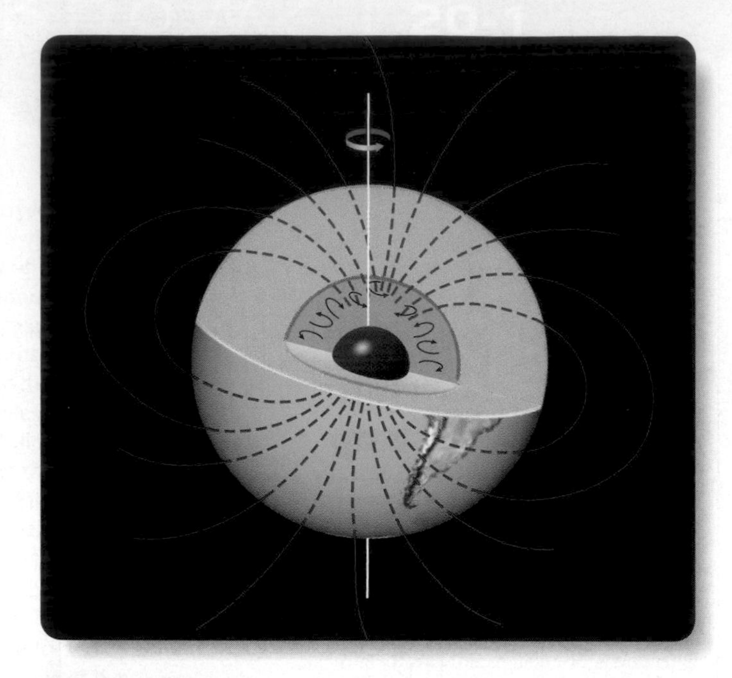

■ Figure 20-7

The dynamo effect couples convection in the liquid core with Earth's rotation to produce electric currents that are believed to be responsible for Earth's magnetic field.

■ Figure 20-8

Earth's magnetic field dominates space around Earth by deflecting the solar wind and trapping high-energy particles in radiation belts. Around the north and south magnetic poles, where the magnetic field enters Earth's atmosphere, powerful currents can flow down and excite gas atoms to emit photons, which produces auroras. Colors are produced as different atoms are excited. Note the meteor (shooting star). (Jimmy Westlake)

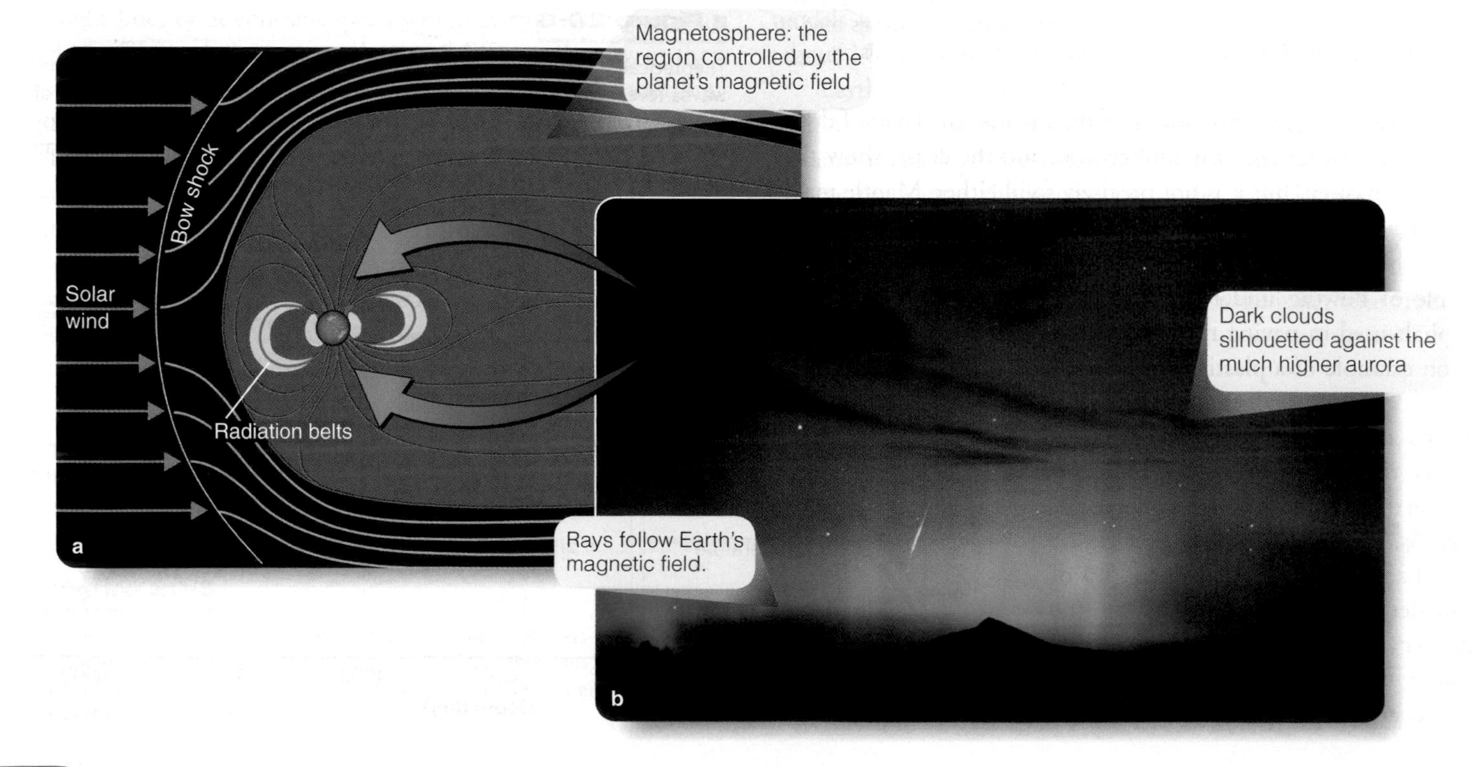

Van Allen belts of radiation. You will see in later chapters that all planets that have magnetic fields have bow shocks, magnetospheres, and radiation belts.

Earth's magnetic field produces the dramatic and beautiful auroras, glowing rays and curtains of light in the upper atmosphere (Figure 20-8b). The solar wind carries charged particles past Earth's extended magnetic field, and this generates tremendous electrical currents that flow into Earth's atmosphere near the north and south magnetic poles. The currents ionize gas atoms in Earth's atmosphere, and when the ionized atoms capture electrons and recombine, they emit light as if they were part of a vast "neon" sign. The spectrum of an aurora is an emission spectrum.

Although you can be confident that Earth's magnetic field is generated within its molten core, many mysteries remain. For example, rocks retain traces of the magnetic field in which they solidify, and some contain fields that point backward. That is, they imply that Earth's magnetic field was reversed at the time they solidified. Careful analysis of such rocks indicates that Earth's field has reversed itself every million years or so, with the north magnetic pole becoming the south magnetic pole and vice versa. These reversals are poorly understood, but they may be related to changes in the core's convection.

Convection in Earth's core is important because it generates the magnetic field. As you will see in the next section, convection in the mantle constantly remakes Earth's surface.

ThomsonNOW Sign in at www.thomsonedu.com and go to ThomsonNOW to see Astronomy Exercise "Convection and Magnetic Fields."

Earth's Active Crust

Earth's crust is composed of low-density rock that floats on the mantle. The image of a rock floating may seem odd, but recall that the rock of the mantle is very dense. Also, just below the crust, the mantle rock tends to be highly plastic, so great sections of low-density crust do indeed float on the semiliquid mantle like great lily pads floating on a pond.

The motion of the crust and the erosive action of water make Earth's crust highly active. Read **The Active Earth** on pages 452–453 and notice three important points and six new terms:

1. *Plate tectonics*, the motion of crustal plates, produces much of the geological activity on Earth. Plates spreading apart can form *rift valleys*, or, on the ocean floor, *midocean rises* where molten rock solidifies to form *basalt*. A plate sliding into a *subduction zone* can trigger volcanism, and the collision of plates can produce *folded mountain ranges*.

2. Notice how the continents on Earth's surface have moved and changed over periods of hundreds of millions of years. A hundred million years is only 0.1 billion years, so sections of Earth's crust are in rapid motion.

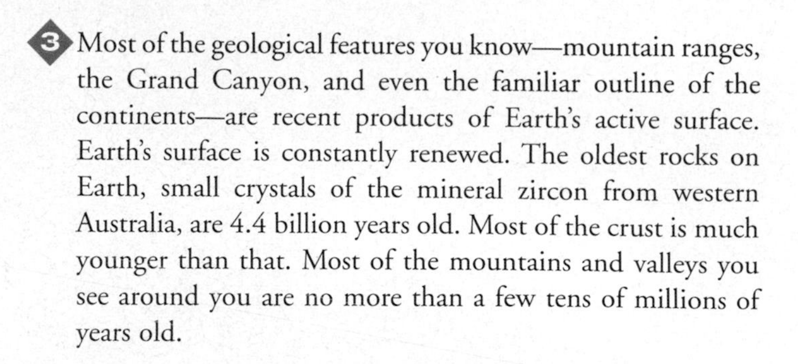

3. Most of the geological features you know—mountain ranges, the Grand Canyon, and even the familiar outline of the continents—are recent products of Earth's active surface. Earth's surface is constantly renewed. The oldest rocks on Earth, small crystals of the mineral zircon from western Australia, are 4.4 billion years old. Most of the crust is much younger than that. Most of the mountains and valleys you see around you are no more than a few tens of millions of years old.

The average speed of plate movement is slow but sudden movements do occur. Plate margins can stick, accumulate stress and then release it suddenly. That's what happened on December 26, 2004, along a major subduction zone in the Indian Ocean. The total motion was as much as 15 m, and the resulting earthquake caused devastating tidal waves. Every day, minor earthquakes occur on moving faults, and the stress that builds in those faults that are sticking will eventually be released in major earthquakes.

Earth's active crust explains why Earth contains so few impact craters. The moon is richly cratered, but Earth contains only about 150 impact craters. Plate tectonics and erosion have destroyed all but the most recent craters on Earth.

You can see that Earth's geology is dominated by two dramatic forces. Heat rising from the interior drives plate tectonics; just below the thin crust of solid rock lies a churning molten interior that rips the crust to fragments and pushes the pieces about like bits of algae on a pond. The second force modifying the crust is water. It falls as rain and snow and tears down the mountains, erodes the river valleys, and washes any raised ground into the sea. Tectonics builds mountains and continents, and then erosion rips them to nothing.

ThomsonNOW Sign in at www.thomsonedu.com and go to ThomsonNOW to see Astronomy Exercise "Convection and Plate Tectonics."

◄ SCIENTIFIC ARGUMENT ►

What evidence can you cite that Earth has a liquid core?

In a scientific argument, the critical analysis of ideas must eventually return to evidence. In this case, the evidence is indirect because you can never visit Earth's core. Seismic waves from distant earthquakes pass through Earth, but a certain kind of wave, the *S* waves, cannot pass through the core. Because the *S* waves cannot propagate through a liquid, you can conclude that Earth's core is a liquid. Earth's magnetic field gives you evidence of a molten metallic core. The theory for the generation of magnetic fields, the dynamo effect, requires a rotating liquid core composed of a conducting material and stirred by convection. If the core were not a liquid, it would not be able to generate a magnetic field. That gives you two different kinds of evidence that our planet has a liquid core.

Now build a new argument again focusing on evidence. **Why do scientists conclude that Earth's crust is broken into moving plates?**

◄ ►

The Active Earth

1 Our world is an astonishingly active planet. Not only is it rich in water and therefore subject to rapid erosion, but its crust is divided into moving sections called plates. Where plates spread apart, lava wells up to form new crust; where plates push against each other, they crumple the crust to form mountains. Where one plate slides over another, you see volcanism. This process is called **plate tectonics**, referring to the Greek word for "builder." (An architect is literally an arch builder.)

A typical view of planet Earth

Mountains are common on Earth, but they erode away rapidly because of the abundant water.

A **rift valley** forms where continental plates begin to pull apart. The Red Sea has formed where Africa has begun to pull away from the Arabian peninsula.

Midocean rise

Red Sea

Midocean rise

National Geophysical Data Center

1a Evidence of plate tectonics was first found in ocean floors, where plates spread apart and magma rises to form **midocean rises** made of rock called **basalt**, a rock typical of solidified lava. Radioactive dating shows that the basalt is younger near the midocean rise. Also, the ocean floor carries less sediment near the midocean rise. As Earth's magnetic field reverses back and forth, it is recorded in the magnetic fields frozen into the basalt. This produces a magnetic pattern in the basalt that shows that the seafloor is spreading away from the midocean rise.

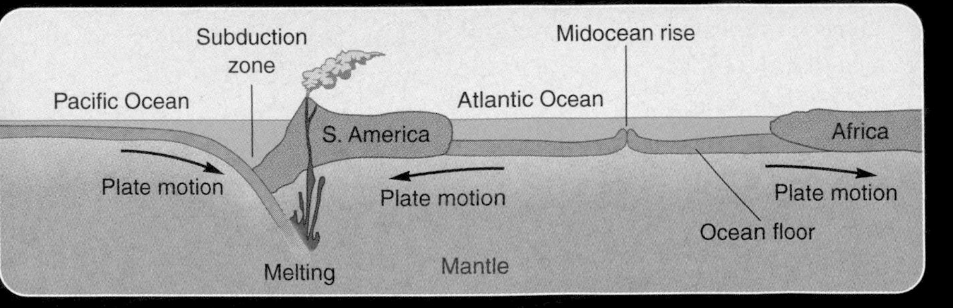

1b A **subduction zone** is a deep trench where one plate slides under another. Melting releases low-density magma that rises to form volcanoes such as those along the northwest coast of North America, including Mt. St. Helens.

1c Hot spots caused by rising magma in the mantle can poke through a plate and cause volcanism such as that in Hawaii. As the Pacific plate has moved northwestward, the hot spot has punched through to form a chain of volcanic islands, now mostly worn below sea level. **Folded mountain ranges** can form where plates push against each other. For example, the Ural Mountains lie between Europe and Asia, and the Himalaya Mountains are formed by India pushing north into Asia. The Appalachian Mountains are the remains of a mountain range pushed up when North America was pushed against Africa.

ThomsonNOW™

Sign in at www.thomsonedu.com and go to ThomsonNOW to see Active Figure "Hot Spot Volcanoes." Notice how the moving plate can produce a chain of volcanic peaks, mostly under water in the case of Earth.

1d The floor of the Pacific Ocean is sliding into subduction zones in many places around its perimeter. This pushes up mountains such as the Andes and triggers earthquakes and active volcanism all around the Pacific in what is called the Ring of Fire. In places such as southern California, the plates slide past each other, causing frequent earthquakes.

Hawaii

National Geophysical Data Center

Yellow lines on this globe mark plate boundaries. Red dots mark earthquakes since 1980. Earthquakes within the plate, such as those at Hawaii, are related to volcanism over hot spots in the mantle.

Continental Drift

Not long ago, Earth's continents came together to form one continent.

200 million years ago

Pangaea broke into a northern and a southern continent.

Laurasia

Gondwanaland

135 million years ago

Notice India moving north toward Asia.

65 million years ago

The continents are still drifting on the highly plastic upper mantle.

Today

2 The floor of the Atlantic Ocean is not being subducted. It is locked to the continents and is pushing North and South America away from Europe and Africa at about 3 cm per year, a motion called *continental drift*. Radio astronomers can measure this motion by timing and comparing radio signals from pulsars using European and American radio telescopes. Roughly 200 million years ago, North and South America were joined to Europe and Africa. Evidence of that lies in similar fossils and similar rocks and minerals found in the matching parts of the continents. Notice how North and South America fit against Europe and Africa like a puzzle.

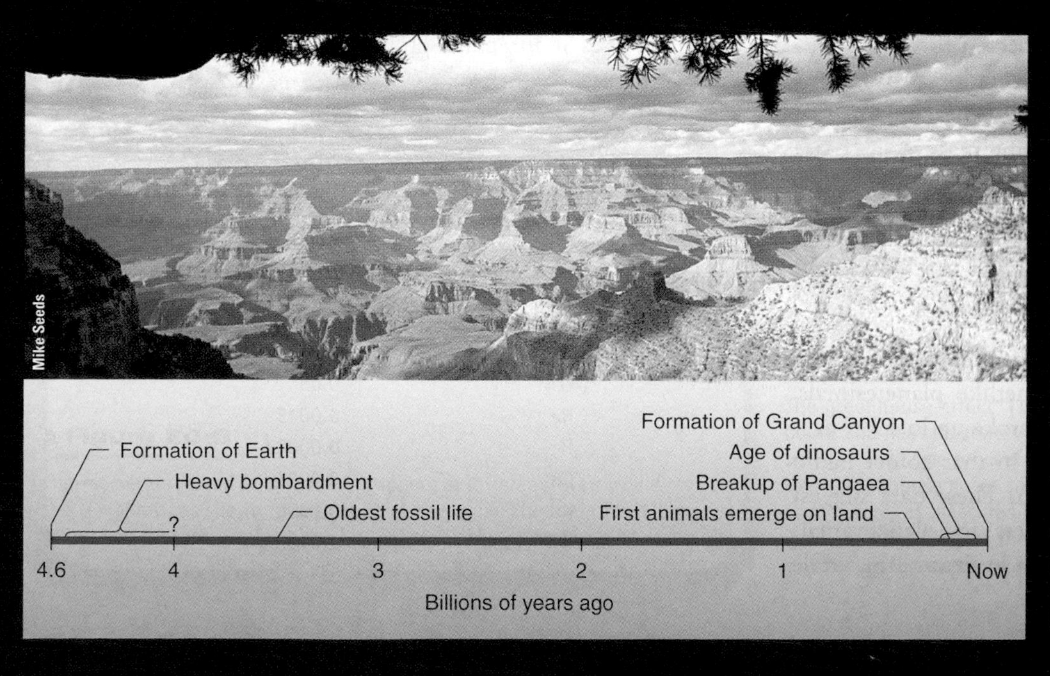

3 Plate tectonics pushes up mountain ranges and causes bulges in the crust, and water erosion wears the rock away. The Colorado River began cutting the Grand Canyon only about 10 million years ago when the Colorado plateau warped upward under the pressure of moving plates. That sounds like a long time ago, but it is only 0.01 billion years. A mile down, at the bottom of the canyon, lie rocks 0.57 billion years old, the roots of an earlier mountain range that stood as high as the Himalayas. It was pushed up, worn away to nothing, and covered with sediment long ago. Many of the geological features we know on Earth have been produced by very recent events.

the glass roof of a greenhouse, it heats the benches and plants inside. The warmed interior radiates infrared radiation, but the glass is opaque to infrared. Warm air in the green house cannot mix with cooler air outside, so heat is trapped within the greenhouse, and the temperature climbs until the glass itself grows warm enough to radiate heat away as fast as the sunlight enters. This is the same process that heats a car when it is parked in the sun with the windows rolled up.

Earth's atmosphere is transparent to sunlight, and when the ground absorbs the sunlight, it grows warmer and radiates infrared. However, CO_2 makes the atmosphere less transparent to infrared radiation, so infrared radiation from the warm surface is absorbed by the atmosphere and cannot escape back into space. That traps heat and makes Earth warmer (Figure 20-10b).

It is a **Common Misconception** that the greenhouse effect is bad. Without the greenhouse effect, Earth would be at least 30 K (54°F) colder and uninhabitable for humans. The problem is that human civilization is adding CO_2 to the atmosphere and increasing the intensity of the greenhouse effect.

CO_2 is not the only greenhouse gas. Water vapor, methane, and other gases also help warm Earth, but CO_2 is the most important. For 4 billion years, natural processes on Earth have removed CO_2 from the atmosphere and buried the carbon in the form of limestone, coal, oil, and natural gas. Since the beginning of the industrial revolution in the mid-19th century, industry has been digging up carbon-rich fuels, burning them to get energy, and releasing CO_2 back into the atmosphere. At the same time, many nations are cutting down large parts of the forests that absorb CO_2. Estimates are that the amount of CO_2 in Earth's atmosphere will double during the 21st century.

The increased concentration of CO_2 is increasing the greenhouse effect and warming Earth in what is known as **global warming.** The actual amount of warming in the future is difficult to predict. The best mathematical models predict a warming between 1.1 and 6.4 °C (2.0 and 11.5 °F) by 2100. Predictions are uncertain because Earth's climate depends on so many factors. A slight warming should increase water vapor in the atmosphere, and water vapor is another greenhouse gas that would enhance the warming. But increased water vapor could increase cloud cover, increase Earth's albedo, and partially reduce the warming. On the other hand, high icy clouds tend to enhance the greenhouse effect. Even small changes in temperature can alter circulation patterns in the atmosphere and in the oceans, and the consequences of such changes are very difficult to predict.

Even though the future is uncertain, current evidence clearly shows that Earth is growing warmer. Studies of the growth rings in very old trees show that Earth's climate had been cooling for the last 1000 years, but the 20th century reversed that trend with a rise of 0.56 to 0.92 °C (0.98 to 1.62 °F). General trends now point to warming. Mountain glaciers have melted back dramatically since the 19th century. Measurements show that polar ice in the form of permafrost, ice shelves, and ice on the open Arctic Ocean is melting.

Although changes are small now, it is a serious issue for the future. Even a small rise in temperatures will dramatically affect agriculture, not only through rising temperatures but also through changes in rainfall. It is a **Common Misconception** that all of Earth will warm. Models predict that although most of North American will grow warmer and dryer, Europe will grow cooler and wetter. Also, the melting of ice on polar landmasses such as Greenland can cause a rise in sea levels that will flood coastal regions and alter shore environments. A modest rise will cover huge low-lying areas such as much of the state of Florida.

There is no doubt that civilization is warming Earth through an enhanced greenhouse effect, but a remedy is difficult to imagine. Reducing the amount of CO_2 and other greenhouse gases released to the atmosphere is difficult because modern society depends on burning fossil fuels for energy. Conserving forests is difficult because growing populations, especially in developing countries with large forest reserves, demand the wood and the agricultural land produced when forests are cut. Political, business, and economic leaders argue that the issue is uncertain, but all around the world scientists of stature have reached agreement: Global warming is real and will change Earth. What humanity can or will do about it is uncertain.

Human influences on Earth's atmosphere go beyond the greenhouse effect. Our modern industrial civilization is also reducing ozone in Earth's atmosphere. Ozone (O_3) is an unstable molecule and is chemically active. You may have heard of ozone because it is produced in auto emissions, and it is a pollutant in city air. But ozone is produced naturally in Earth's atmosphere at an altitude of about 25 km, and there it is beneficial. The ozone layer absorbs ultraviolet photons and prevents them from reaching the ground.

Certain chemicals called chlorofluorocarbons (CFCs), used in industrial processes and in refrigeration and air conditioning, can destroy ozone. As these CFCs escape into the atmosphere, they become mixed into the ozone layer and convert the ozone into normal oxygen molecules. Oxygen does not block ultraviolet radiation, so depleting the ozone layer causes an increase in ultraviolet radiation at Earth's surface. In small doses, ultraviolet radiation can produce a suntan, but in larger doses it can cause skin cancers.

The ozone layer over the Antarctic may be especially sensitive to CFCs. Since the late 1970s, the ozone concentration has been falling over the Antarctic, and a hole in the ozone layer now develops over the continent each October at the time of the Antarctic spring (■ Figure 20-11). Satellite and ground-based measurements show that the same thing is happening at higher northern latitudes and that the amount of ultraviolet radiation reaching the ground is increasing. This is an early warning that human activity is modifying Earth's atmosphere in a potentially dangerous way.

What Are We?

Scientific Imagineers

One of the most fascinating aspects of science is its power to reveal the unseen. That is, it reveals regions you can never visit. You saw this in earlier chapters when you studied the inside of the sun and stars, the surface of neutron stars, the event horizon around black holes, the cores of active galaxies, and more. In this chapter, you have seen Earth's core.

An engineer is a person who builds things, so you can call a person who imagines things an imagineer. Most creatures on Earth cannot imagine situations that do not exist, but humans have evolved the ability to say, "What if?" Our ancient ancestors could imagine what would happen if a tiger was hiding in the grass, and we can imagine the inside of Earth.

A poet can imagine the heart of Earth, and a great writer can imagine a journey to the center of Earth. In contrast, scientists use their imagination in a carefully controlled way. Guided by evidence and theory, they can imagine the molten core of our planet. As you read this chapter you saw the yellow-orange glow and felt the heat of the liquid iron, and you were a scientific imagineer.

Human imagination makes science possible and provides one of the great thrills of science—exploring beyond the limits of normal human experience.

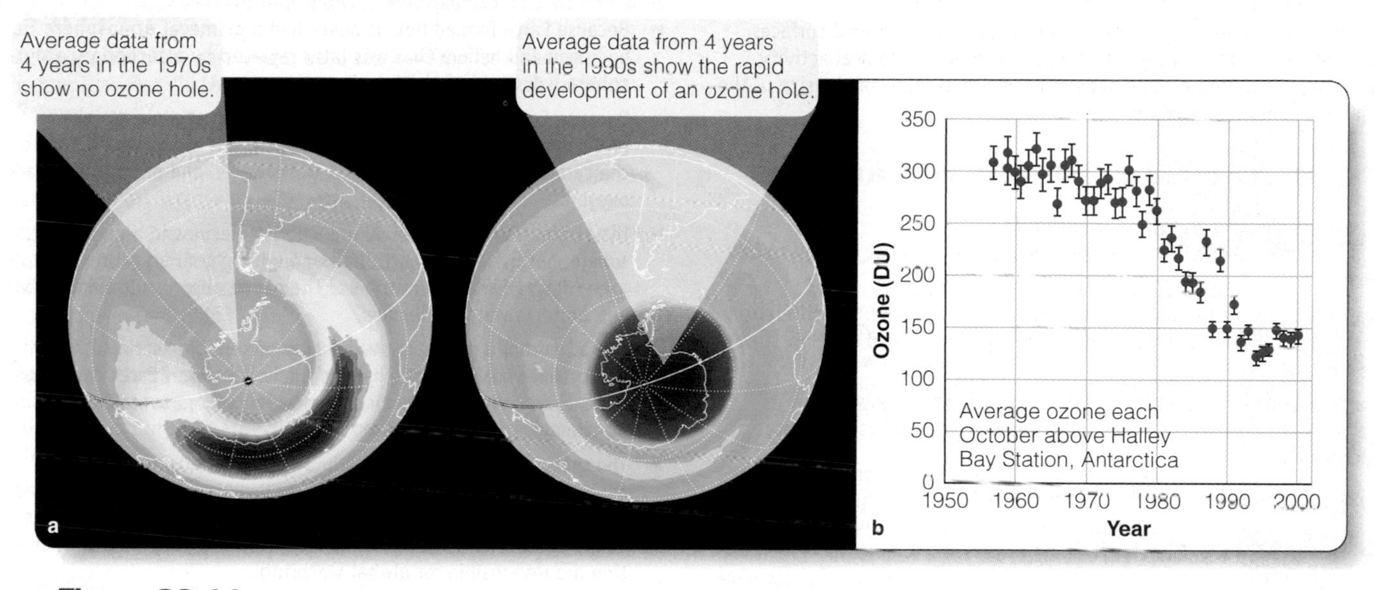

Figure 20-11

(a) Satellite observations of ozone concentrations over Antarctica are shown here as red for highest concentration and violet for lowest. Since the 1970s, a hole in the ozone layer has developed over the South Pole. (b) Although ozone depletion is most dramatic above the South Pole, ozone concentrations have declined at all latitudes. (NASA/GSFC/TOMS and Glenn Carver)

The CO_2 and ozone problems in Earth's atmosphere are paralleled on Venus and Mars. When you study Venus in Chapter 22, you will discover a runaway greenhouse effect that has made the surface of the planet hot enough to melt lead. On Mars you will discover an atmosphere without an ozone layer. A few minutes of sunbathing on Mars would kill you. Once again, you will learn more about your own planet by studying other planets.

ThomsonNOW Sign in at **www.thomsonedu.com** and go to ThomsonNOW to see Astronomy Exercise "The Greenhouse Effect."

◄ SCIENTIFIC ARGUMENT ►

Why does Earth's atmosphere contain little carbon dioxide and lots of oxygen?

Sometimes as you build a scientific argument, you must contradict what seems, at first glance, a simple truth. In this case, because outgassing releases mostly carbon dioxide, CO_2, and water vapor, you might expect Earth's atmosphere to be very rich in CO_2. Luckily for the human race, CO_2 is highly soluble in water, and Earth's surface temperature allows it to be covered with liquid water. The CO_2 dissolves in the oceans and combines with minerals in seawater to form deposits of silicon dioxide, limestone, and other mineral deposits. In this way, the CO_2 is removed from the atmosphere and buried in Earth's crust. Oxygen, in contrast, is highly reactive and forms oxides so easily you might expect it to be rare in the atmosphere. Happily it is continually replenished as green plants release oxygen into Earth's atmosphere faster than chemical reactions can remove it. Were it not for liquid-water oceans and plant life, Earth would be choking in a thick CO_2 atmosphere with no free oxygen.

Now follow up on your argument. **Why would an excess of CO_2** and a deficiency of free oxygen be harmful to all life on Earth in ways that go beyond mere respiration?

◄ ►

21 The Moon and Mercury: Comparing Airless Worlds

When astronauts stepped onto the surface of the moon, they found an unearthly world with no air, no water, weak gravity, and a dusty, cratered surface. Through comparative planetology, the moon reveals a great deal about our own beautiful Earth. (JSC/NASA)

Guidepost

Want to fly to the moon? You will need to pack more than your lunch. There is no air and no water, and the sunlight is strong enough to kill you. Mercury is the same kind of world. Take shelter in the shade, and you will freeze to death in moments. You have never left Earth, so our planet seems normal to you, and other worlds are, well, unearthly. Exploring these two airless worlds will answer four essential questions:

- **Why is the moon airless and cratered?**
- **How did the moon form and evolve?**
- **Why is Mercury different from the moon?**
- **How did Mercury form and evolve?**

As you begin exploring other worlds, you may feel buried under a landslide of details, but the nature of scientific knowledge will come to your rescue. You will see how as you answer an important question about scientific knowledge:

- **How Do We Know? How do theories consolidate details into understanding?**

Once you feel comfortable exploring airless worlds, you will be ready for bigger planets with atmospheres. They are not necessarily more interesting places, but they are just a tiny bit less unearthly.

That's one small step for a man . . .
one giant leap for mankind.

NEIL ARMSTRONG, ON THE MOON

Beautiful, beautiful. Magnificent desolation.

EDWIN E. (BUZZ) ALDRIN JR. ON THE MOON

IF YOU HAD BEEN THE FIRST PERSON to step onto the surface of the moon, what would you have said? Neil Armstrong responded to the historic significance of the first human step onto the surface of another world. Buzz Aldrin was second, and he responded to the moon itself. It *is* desolate, and it *is* magnificent. But it is not unusual. Most planets in the universe probably look like Earth's moon, and astronauts may someday walk on such worlds and compare them with our moon.

In this chapter, you will use comparative planetology to study the moon and Mercury, and you will discover three important themes of planetary astronomy: impact cratering; internal heat; and giant impacts. These three themes will help you organize the flood of details astronomers have learned about the moon and Mercury.

21-1 The Moon

ONLY 12 PEOPLE HAVE STOOD on the moon, but planetary astronomers know it well. The photographs, measurements, and samples brought back to Earth paint a picture of an airless, ancient, battered crust.

The View from Earth

A few billion years ago, the moon must have rotated faster than it does today, but Earth is over 80 times more massive than the moon (**Celestial Profile 3**), and its tidal forces on the moon are strong. Earth's gravity raised tidal bulges on the moon, and friction in the bulges has slowed the moon until it now rotates once each orbit, keeping the same side facing Earth. A moon whose rotation is locked to its planet is said to be **tidally coupled.** That is why we always see the same side of the moon; the back of the moon is never visible from Earth. The moon's familiar face has shone down on Earth since long before there were humans (■ Figure 21-1).

Based on what you already know, you can predict that the moon should have no atmosphere. It is a small world with an escape velocity too low to keep gas atoms and molecules from escaping into space. You can confirm your theory with even a small telescope. You see no clouds or other obvious traces of an atmosphere, and shadows near the **terminator,** the dividing line between daylight and darkness, are sharp and black. There is no

Earth's moon is about one-fourth the diameter of Earth. Its low density indicates that it contains little iron, but the size of its iron core and the amount of remaining heat are unknown. (NASA)

Celestial Profile 3: The Moon

Motion:

Average distance from Earth	384,400 km (center to center)
Eccentricity of orbit	0.055
Maximum distance from Earth	405,500 km
Minimum distance from Earth	363,300 km
Inclination of orbit to ecliptic	5°9′
Average orbital velocity	1.022 km/s
Orbital period (sidereal)	27.321661 days
Orbital period (synodic)	29.5305882 days
Inclination of equator to orbit	6°41′

Characteristics:

Equatorial diameter	3476 km
Mass	7.35×10^{22} kg (0.0123 $M_{\oplus}$)
Average density	3.36 g/cm^3 (3.35 g/cm^3 uncompressed)
Surface gravity	0.167 Earth gravity
Escape velocity	2.38 km/s (0.21 $V_{\oplus}$)
Surface temperature	−170° to 130°C (−274° to 266°F)
Average albedo	0.07

Personality Point:

Lunar superstitions are common. The words *lunatic* and *lunacy* come from *luna,* the moon. Someone who is *moonstruck* is supposed to be a bit nutty. Because the moon affects the ocean tides, many superstitions link the moon to water, to weather, and to women's cycle of fertility. According to legend, moonlight is supposed to be harmful to unborn children; but, on the plus side, moonlight rituals are said to remove warts.

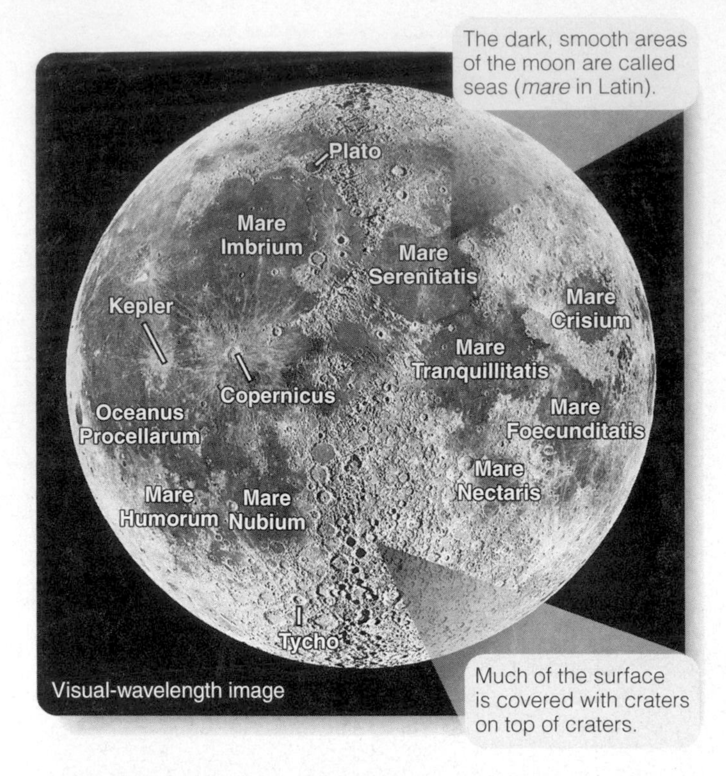

■ **Figure 21-1**

The side of the moon that faces Earth is a familiar sight. Craters have been named for famous scientists and philosophers, and the so-called seas have been given romantic names. Mare Imbrium is the Sea of Rains, and Mare Tranquillitatis is the Sea of Tranquillity. There is, in fact, no water on the moon. (Photo © UC Regents/Lick Observatory)

air on the moon to scatter light and soften shadows. Also, with even a small telescope you could watch stars disappear behind the **limb** of the moon—the edge of its disk—without dimming. Clearly, the moon is an airless (and soundless) world.

The surface of the moon is divided into two dramatically different kinds of terrain. The lunar highlands are filled with jumbled mountains, but there are no folded mountain ranges as on Earth. The mountains are pushed up by millions of impact craters one on top of the other. In fact, the highlands are saturated with craters, meaning that it would be impossible to form a new crater without destroying the equivalent of one old crater. In contrast, the lowlands, about 3 km lower than the highlands, are smooth, dark plains called **maria,** Latin for "seas." (The singular of *maria* is *mare.*) A small telescope shows that the maria are marked by ridges, faults, smudges, and scattered craters and can't be water. Rather, the maria are ancient lava flows that have apparently covered the older, cratered lowlands.

These lava flows suggest volcanism, but only small traces of past volcanic activity are visible from Earth. No major volcanic peaks are visible on the moon, and no active volcanism has ever been detected. The lava flows that created the maria happened long ago and were much too fluid to build peaks. With a good telescope and some diligent searching you could see a few small domes pushed up by lava below the surface, and you could see long, winding channels called **sinuous rilles** (■ Figure 21-2). These channels are often found near the edges of the maria and were evidently cut by flowing lava. In some cases, such a channel may once have had a roof of solid rock, forming a lava tube. After the lava drained away, meteorite impacts collapsed the roof to form a sinuous rille. The view from Earth provides only hints of ancient volcanic activity associated with the maria.

Lava flows and impact cratering have dominated the history of the moon. Study **Impact Cratering** on pages 464–465 and notice three important points and five new terms:

1. Impact craters have certain distinguishing characteristics, such as their shape and the *ejecta, rays,* and *secondary craters* around them.
2. Lunar impact craters range from tiny pits formed by *micrometeorites* to giant *multiringed basins.*
3. Most of the craters on the moon are old; they were formed long ago when the solar system was young.

Meteorites strike the moon all the time, but large impacts are rare today. Meteorites a few tens of meters in diameter probably strike the moon every 50 years or so, but no one has ever seen such an impact with certainty. Small flashes of light have been seen on the dark side of the moon during showers of meteors, but those impacts must have been made by very small objects. No significant change has been seen on the moon since the invention of the telescope. Large impacts on the moon and Earth are quite rare, and nearly all of the lunar craters seen through telescopes date from the solar system's youth.

It is difficult to estimate the age of any specific crater. In some cases, you can find **relative ages** by noting that a crater partially covers another crater. Clearly the crater on top must be younger than the crater on the bottom. From studies of the way the cratering rate fell when the moon was young, astronomers can study the size and number of craters on a section of the moon's surface and estimate the section's **absolute age** in years. The maria, containing few craters, appear to be three to four billion years old, and the highlands are older.

The lunar features visible from Earth allow you to construct a tentative theory to explain the history of the moon. Such a theory provides a framework that will help you organize all of the details and observations (**How Do We Know? 21-1**). As the moon formed, its crust was heavily cratered by the debris in the solar nebula, including the late heavy bombardment possibly caused by the migration of the outer planets to their present orbits (Chapter 20). Sometime after the cratering subsided, lava welled up from below the crust and flooded the lowlands, covering the craters there and forming the smooth maria. You can locate a few large craters on the maria such as Kepler and Copernicus in Figure 21-1, but note that the bright rays around them show that they are young. The maria are only lightly scarred by impacts and must be younger than the cratered highlands.

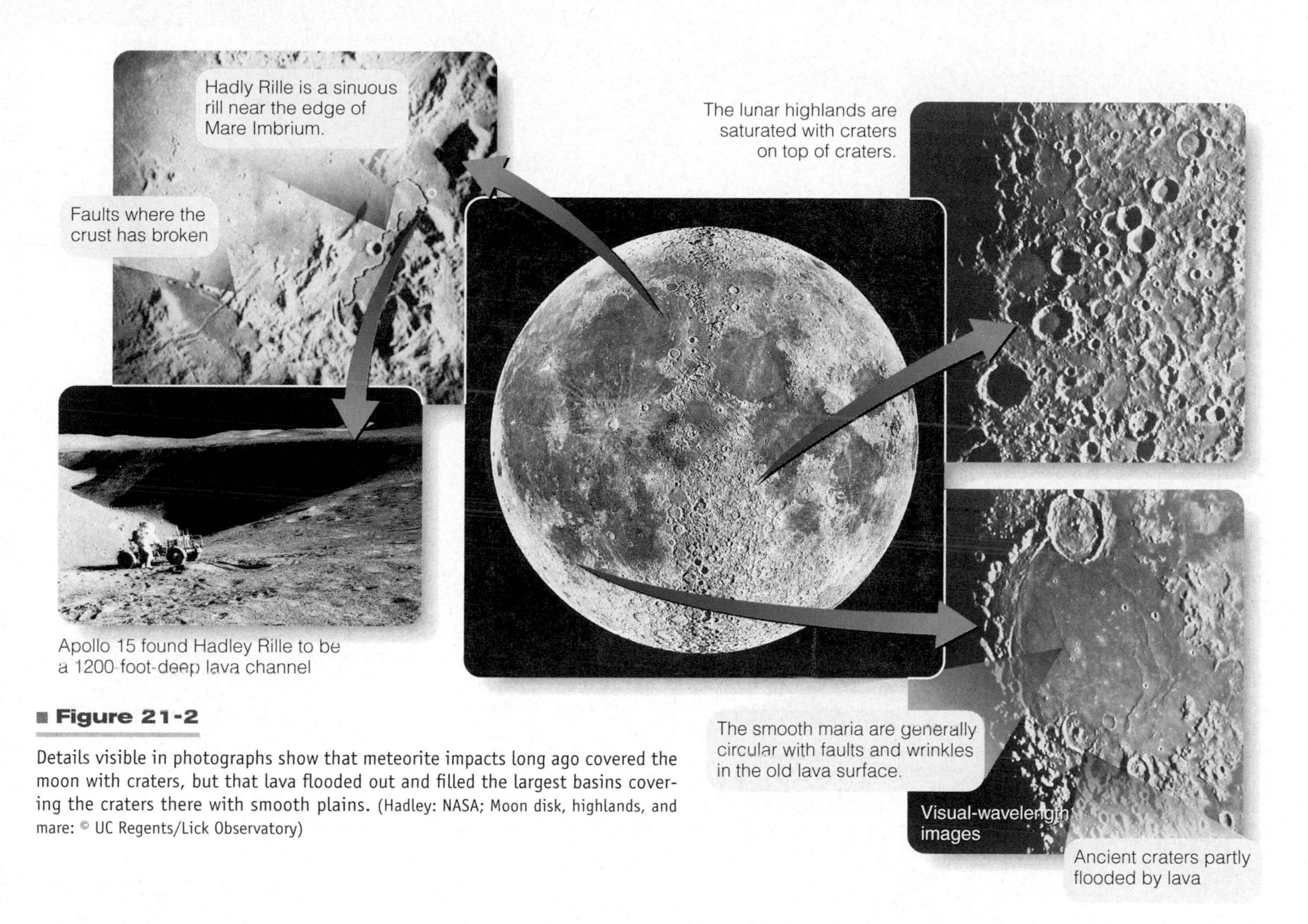

■ **Figure 21-2**

Details visible in photographs show that meteorite impacts long ago covered the moon with craters, but that lava flooded out and filled the largest basins covering the craters there with smooth plains. (Hadley: NASA; Moon disk, highlands, and mare: © UC Regents/Lick Observatory)

Earth-based observations allowed astronomers to begin telling the story of the moon's surface, but the view from Earth does not provide enough evidence. To really understand the lunar surface, humans had to go there.

The Apollo Missions

On May 25, 1961, President John Kennedy committed the United States to landing a human being on the moon by 1970. Although the reasons for that decision related more to economics, international politics, and the stimulation of technology than to science, the Apollo program became a fantastic adventure in science, a flight to the moon that changed how we all think about Earth.

Flying to the moon is not particularly difficult; with powerful enough rockets and enough food, water, and air, it is a straightforward trip. Landing on the moon is more difficult but not impossible. The moon's gravity is only one-sixth that of Earth, and there is no atmosphere to disturb the trajectory of the spaceship. Getting to the moon isn't too hard, and landing is possible; the difficulty is doing both on one trip. The spaceship must carry food, water, and air for a number of days in space plus fuel and rockets for midcourse corrections and for a return to Earth. All of this adds up to a ship that is too massive to make a safe landing on the lunar surface. The solution was to take two spaceships to the moon, one to ride in and one to land in (■ Figure 21-3).

The command module was the long-term living space and command center for the trip. Three astronauts had to live in it for a week, and it had to carry all the life-support equipment, navigation instruments, computers, power packs, and so on for a week's jaunt in space. The lunar landing module (LM for short) was tacked to the front of the command module like a bicycle strapped to the front of the family camper. It carried only enough fuel and supplies for the short trip to the lunar surface, and it was built to minimize weight and maximize maneuverability.

The weaker gravity of the moon made the design of the LM simpler. Landing on Earth requires reclining couches for the astronauts, but the trip to the lunar surface involved smaller accelerations. In an early version of the LM, the astronauts sat on what looked like bicycle seats, but these were later scrapped to save weight. The astronauts had no seats at all in the LM, and once they began their descent and acquired weight, they stood at the controls held by straps, riding the LM like daredevils riding a rocket surfboard.

Impact Cratering

1 The craters that cover the moon and many other bodies in the solar system were produced by the high-speed impact of meteorites of all sizes. Meteorites striking the moon travel 10 to 60 km/s and can hit with the energy of many nuclear bombs.

A meteorite striking the moon's surface can deliver tremendous energy and can produce an impact crater 10 or more times larger in diameter than the meteorite. The vertical scale is exaggerated at right for clarity.

Impact Cratering

A meteorite approaches the lunar surface at high velocity.

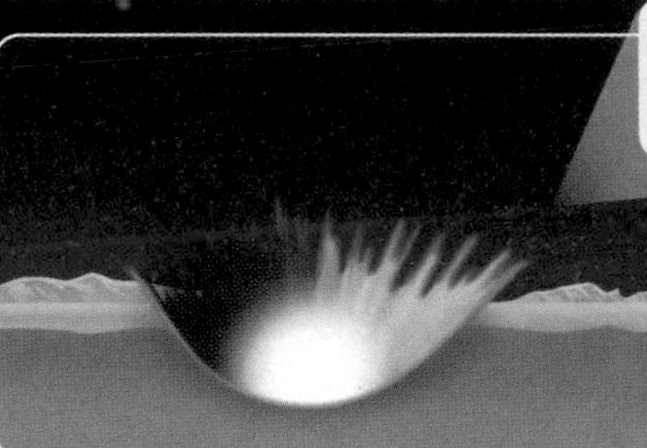

On impact, the meteorite is deformed, heated, and vaporized.

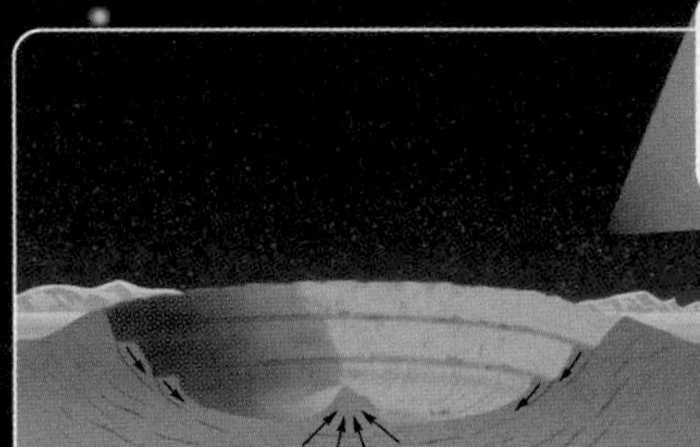

The resulting explosion blasts out a round crater.

Slumping produces terraces in crater walls, and rebound can raise a central peak.

1a Lunar craters such as Euler, 27 km (17 mi) in diameter, look deep when you see them near the terminator where shadows are long, but a typical crater is only a fifth to a tenth as deep as its diameter, and large craters are even shallower.

Because craters are formed by shock waves rushing outward, by the rebound of the rock, and by the expansion of hot vapors, craters are almost always round, even when the meteorite strikes at a steep angle.

Debris blasted out of a crater is called **ejecta**, and it falls back to blanket the surface around the crater. Ejecta shot out along specific directions can form bright **rays**.

Euler

Visual-wavelength image

NASA

ThomsonNOW™

Sign in at www.thomsonedu.com and go to ThomsonNOW to see Active Figure "The Moon's Craters." Notice that the structure of the craters depends on their size.

Visual

NASA

1b Rock ejected from distant impacts can fall back to the surface and form smaller craters called **secondary craters**. The chain of craters here is a 45-km-long chain of secondary craters produced by ejecta from the large crater Copernicus 200 km out of the frame to the lower right.

Bright ejecta blankets and rays gradually darken as sunlight darkens minerals and [illegible] meteorites stir the dusty surface. Bright rays are signs of youth. Rays from the crater Tycho, perhaps only 100 million years old, extend halfway around the moon.

Rays

Tycho

Visual

NASA

2 Plum Crater (right), 40 m (130 ft) in diameter, was visited by Apollo 16 astronauts. Note the many smaller craters visible. Lunar craters range from giant impact basins to tiny pits in rocks struck by **micrometeorites**, meteorites of microscopic size.

Lunar rover
Sun glare in camera lens
NASA

Mare Orientale

Solidified lava

Visual-wavelength images

NASA

2a In larger craters, the deformation of the rock can form one or more inner rings concentric with the outer rim. The largest of these craters are called **multiringed basins.** In Mare Orientale on the west edge of the visible moon, the outermost ring is almost 900 km (550 mi) in diameter.

2b The energy of an impact can melt rock, some of which falls back into the crater and solidifies. When the moon was young, craters could also be flooded by lava welling up from below the crust.

A few meteorites found on Earth have been identified chemically as fragments of the moon's surface blasted into space by cratering impacts. The fragmented nature of these meteorites indicates that the moon's surface has been battered by impact craters.

Meteorite from moon
NASA

3 Most of the craters on the moon were produced long ago when the solar system was filled with debris from planet building. As that debris was swept up, the cratering rate fell rapidly, as shown schematically below.

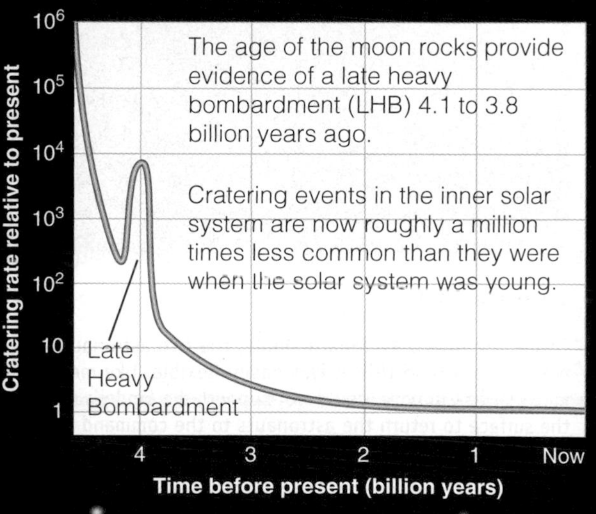
Rate of Crater Formation
The age of the moon rocks provide evidence of a late heavy bombardment (LHB) 4.1 to 3.8 billion years ago.
Cratering events in the inner solar system are now roughly a million times less common than they were when the solar system was young.
Late Heavy Bombardment
Cratering rate relative to present
10^6
10^5
10^4
10^3
10^2
10
1
4
3
2
1
Now
Time before present (billion years)

followed the cracks up to the surface and flooded the giant basins with successive lava flows of dark basalts from 3.8 to about 2 billion years ago. This formed the maria (■ Figure 21-6).

It is a **Common Misconception** that the lava that floods out on the surfaces of planets comes from the molten core. The lava comes from the lower crust and upper mantle. The pressure is lower there, and that lowers the melting point of the rock enough that radioactive decay can melt portions of the rock. If there are faults and cracks, the magma can reach the surface and form volcanoes and lava flows. Whenever you see lava flows on a planet, you can be sure heat is flowing out of the interior (one of the themes of this chapter), but the lava did not come all the way from the core.

Some maria on the moon, such as Mare Imbrium, Mare Serenitatis, Mare Humorum, and Mare Crisium, retain their round shapes, but others are irregular because the lava overflowed the edges of the basin or because the shape of the basin was modified by further cratering. The floods of lava left other characteristic features frozen into the maria. In places, the lava formed channels that are seen from Earth as sinuous rills. Also, the weight of the maria pressed the crater basins downward, and the solidified lava was compressed and formed wrinkle ridges visible even in small telescopes. The tension at the edges of the maria broke the hard lava to produce straight fractures and faults. All of these features are visible in Figure 21-2. As time passed, further cratering and overlapping lava floods modified the maria. Consequently, you should think of the maria as accumulations of features reflecting the moon's complex history.

Mare Imbrium is a dramatic example of how the great basins became the maria. Its story can be told in detail in part because of evidence gathered by Apollo 14 astronauts, who landed on ejecta from the Imbrium impact (■ Figure 21-7). Near the end of the heavy bombardment, roughly 4 billion years ago, a planetesimal the size of Rhode Island struck the moon and blasted out a giant multiringed basin. The impact was so violent the ejecta blanketed 16 percent of the moon's surface. After the cratering rate fell at the end of the heavy bombardment, lava flows welled up time after time and flooded the Imbrium Basin, burying all but the highest parts of the giant multiringed basin. The Imbrium Basin is now a large, generally round mare marked by only a few craters that have formed since the last of the lava flows (■ Figure 21-8).

This story of the moon might suggest that it was a violent place during the cratering phase, but large impacts were in fact rare; the moon was, for the most part, a peaceful place even during the heavy bombardment. Had you stood on the moon at that

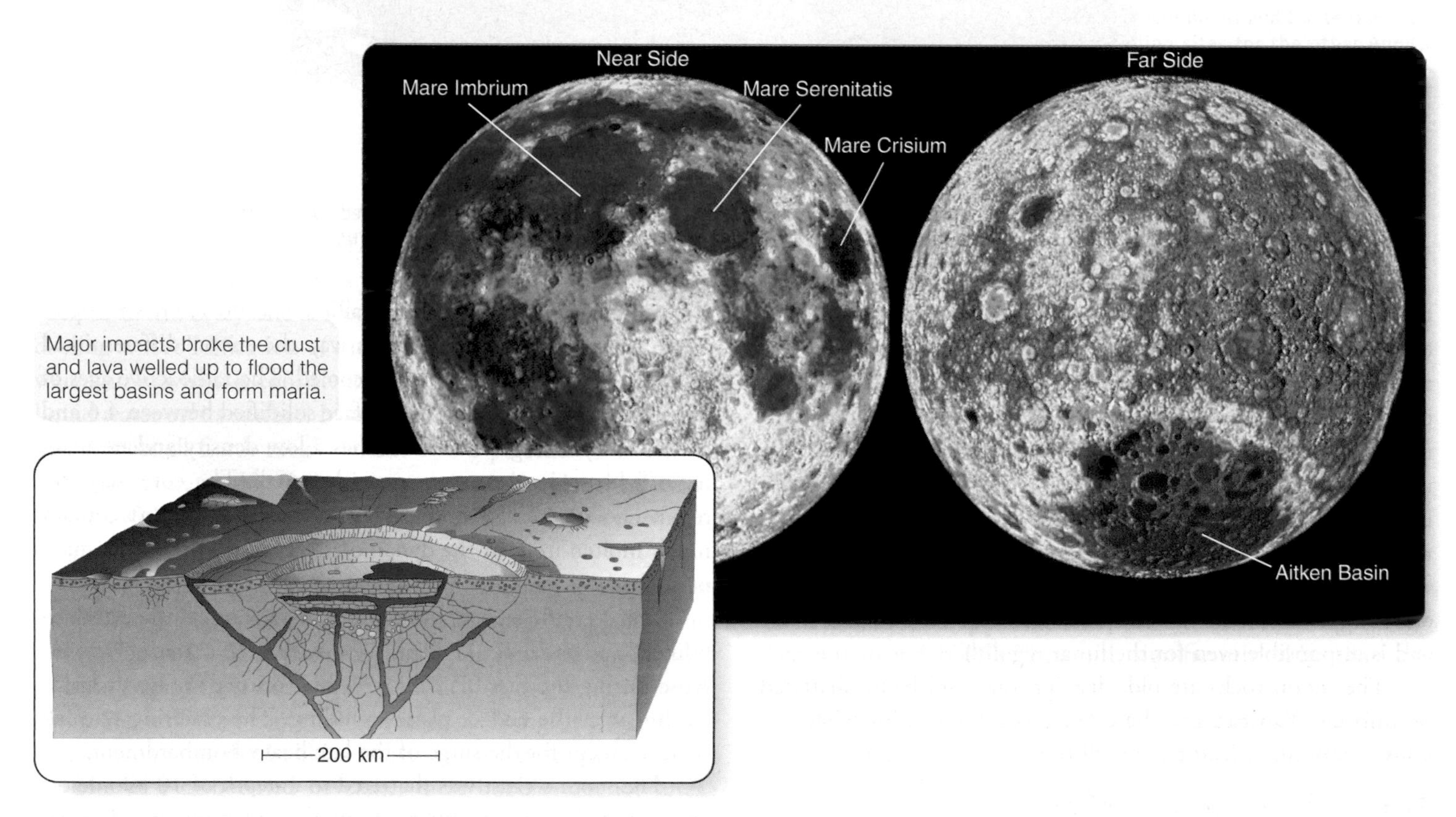

■ **Figure 21-6**

Much of the near side of the moon is marked by great, generally circular lava plains called maria. The crust on the far side is thicker, and there is much less flooding. Even the huge Aitken Basin contains little lava flooding. In these maps, color marks elevation, with red the highest regions and purple the lowest. (Diagram adapted from a diagram by William Hartmann; NASA/Clementine)

■ **Figure 21-7** ▶

Apollo 14 landed on rugged terrain suspected of being ejecta from the Imbrium impact. The large boulder here is ejecta that, at some time in the past, fell here from an impact far beyond the horizon. (NASA)

■ **Figure 21-8**

Lava flooding after the end of the heavy bombardment filled a giant multiringed basin and formed Mare Imbrium. (Don Davis)

Origin of Mare Imbrium

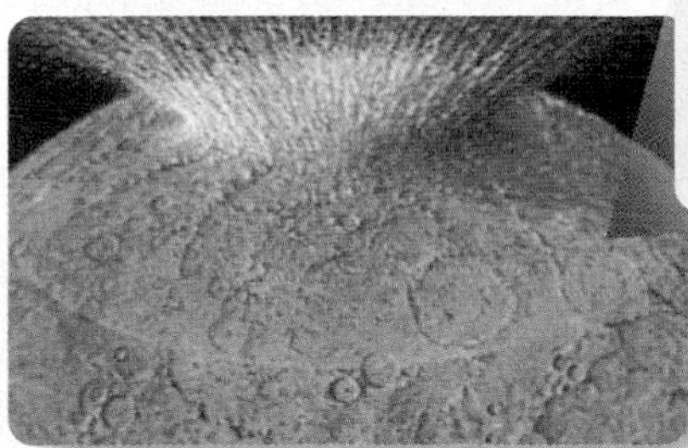

Four billion years ago, an impact forms a multiringed basin over 1000 km (700 mi) in diameter.

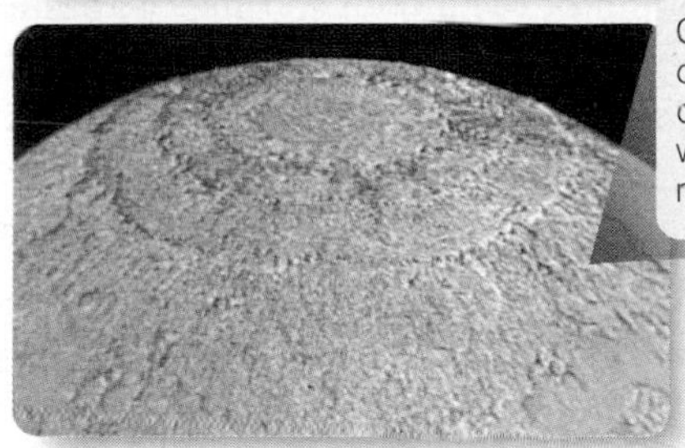

Continuing impacts crater the surface but do not erase the high walls of the multiringed basin.

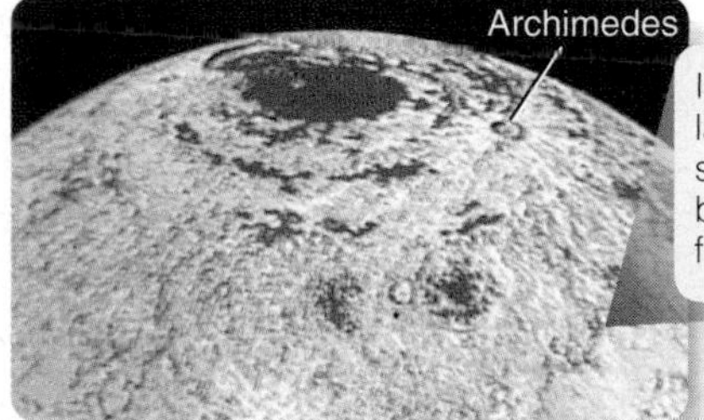

Archimedes
Impacts form a few large craters, and, starting about 3.8 billion years ago, lava floods low regions.

Repeated lava flows cover most of the inner rings and overflow the basin to merge with other flows.

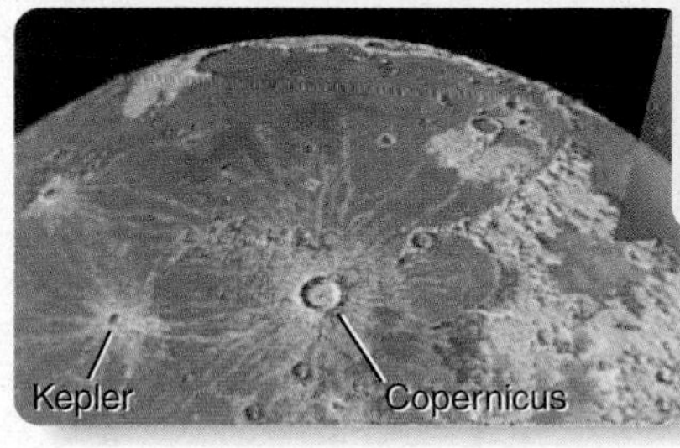

Impacts continue, including those that formed the relatively young craters Copernicus and Kepler.
Kepler
Copernicus

time you would have experienced a continuous rain of micrometeorites and much less common pebble-size impacts. Centuries might pass between major impacts. Of course, when a large impact did occur far beyond the horizon, it might have buried you under ejecta or jolted you by seismic shocks. You could have felt the Imbrium impact anywhere on the moon, but had you been standing on the side of the moon directly opposite that impact, you would have been at the focus of seismic waves traveling around the moon from different directions. When the waves met under your feet, the surface would have jerked up and down by as much as 10 m. The place on the moon opposite the Imbrium Basin is a strangely disturbed landscape called **jumbled terrain.** You will see similar effects of large impacts on other worlds.

Studies of our moon show that its crust is thinner on the side facing Earth, perhaps due to tidal effects. Consequently, while lava flooded the basins on the Earthward side, it was unable to rise through the thicker crust to flood the lowlands on the far side. The largest impact basin in the solar system is the Aitkin Basin near the moon's south pole (Figure 21-6b). It is about 2500 kilometers (1500 miles) in diameter and as deep as 13 kilometers (8 miles) in places, but flooding has never filled it with smooth lava flows.

The moon is small, and small worlds cool rapidly because they have a large ratio of surface area to volume. The rate of heat loss is proportional to the surface area, and the amount of heat in a world is proportional to the volume. The smaller a world is, the easier it is for the heat to escape. That is why a small cupcake fresh from the oven cools more rapidly than a large cake. The moon lost much of its internal heat when it was young, and it is the outward flow of heat that drives geological activity, so the moon is mostly inactive today. The crust of the moon rapidly grew thick and never divided into moving plates. There are no rift valleys or folded mountain chains on the moon. The last lava flows on the moon ended about two billion years ago when the moon's temperature fell too low to maintain subsurface lava.

The overall terrain on the moon is almost fixed. On Earth a billion years from now, plate tectonics will have totally changed the shapes of the continents, and erosion will have long ago worn away the Rocky Mountains.

On the moon, with no atmosphere and no water, there is no Earth-like erosion. Over the next billion years, impacts will have formed only a few more large craters, and nearly all of the lunar scenery will be unchanged. Micrometeorites are the biggest influence; they will have blasted the soil, erasing the footprints left by the Apollo astronauts and reducing the equipment they left behind to peculiar chemical contamination in the regolith at the six Apollo landing sites.

You have studied the story of the moon's evolution in detail for later comparison with other planets and moons in our solar system, but the story has skipped one important question: Where did Earth get such a large satellite?

The Origin of Earth's Moon

During the last two centuries, astronomers developed three different hypotheses for the origin of Earth's moon, but these traditional ideas have failed to survive comparison with the evidence. A relatively new theory proposed in the 1970s may hold the answer. You can begin by testing the three unsuccessful theories against the evidence to see why they failed.

The first of the three traditional theories, the **fission hypothesis,** supposes that the moon formed by the fission of Earth. If the young Earth spun fast enough, tides raised by the sun might break into two parts (■ Figure 21-9a). If this separation occurred after Earth differentiated, the moon would have formed from crust material, which would explain the moon's low density.

But the fission theory has problems. No one knows why the young Earth should have spun so fast, nearly ten times faster than today, nor where all that angular momentum went after the fission. In addition, the moon's orbit is not in the plane of Earth's equator, as it would be if it had formed by fission.

The second traditional theory is the **condensation** (or double-planet) **hypothesis.** It supposes that Earth and the moon condensed as a double planet from the same cloud of material (Figure 21-9b). However, if they formed from the same material, they should have the same chemical composition and density, which they don't. The moon is very poor in certain heavy elements like iron and titanium, and in volatiles such as water vapor and sodium. Yet the moon contains almost exactly the same ratios of oxygen isotopes as does Earth's mantle. The condensation theory cannot explain these compositional differences.

The third theory is the **capture hypothesis.** It supposes that the moon formed somewhere else and was later captured by Earth (Figure 21-9c). If the moon formed inside the orbit of Mercury, the heat would have prevented the condensation of solid metallic grains, and only high-melting-point metal oxides could have solidified. According to the theory, a later encounter with Mercury could have "kicked" the moon out to Earth.

The capture theory was never popular because it requires highly unlikely events involving interactions with Mercury and Earth to move the moon from place to place. Scientists are always suspicious of explanations that require a chain of unlikely coincidences. Also, on encountering Earth, the moon would have been moving so rapidly that Earth's gravity would have been unable to capture it without ripping the moon to fragments through tidal forces.

Until recently, astronomers were left with no acceptable theory to explain the origin of the moon, and they occasionally joked that the moon could not exist. But during the 1970s, planetary astronomers developed a new theory that combines the best aspects of the fission hypothesis and the capture hypothesis.

The **large-impact theory** supposes that the moon formed from debris ejected into a disk around Earth by the impact of a large body. The impacting body may have been twice as large as Mars. In fact, instead of saying that Earth was hit by a large body, it may be more nearly correct to say that Earth and the moon resulted from the collision and merger of two very large planetesimals. The resulting large body became Earth, and the ejected debris formed the moon (■ Figure 21-10). Such an impact would have melted the proto-Earth, and the material falling together to form the moon would have been heated hot enough to melt. This theory fits well with the evidence from moon rocks that shows the moon formed as a sea of magma.

This theory would explain other things. The collision must have occurred at a steep angle to eject enough matter to make the moon. The objects could not have collided head-on. A glancing collision would have spun the material rapidly enough to explain the observed angular momentum in the Earth–moon system. And if the two colliding planetesimals had already differentiated,

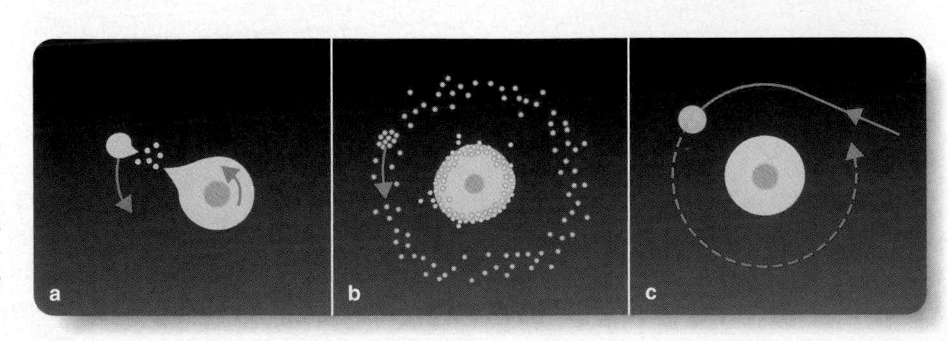

■ **Figure 21-9**

Three traditional theories for the moon's origin. (a) Fission theories suppose that Earth and the moon were once one body and broke apart. (b) Condensation theories suppose that the moon formed at the same time and from the same material as Earth. (c) Capture theories suggest that the moon formed elsewhere and was captured by Earth. None of these theories explains all the facts.

The Large-Impact Hypothesis

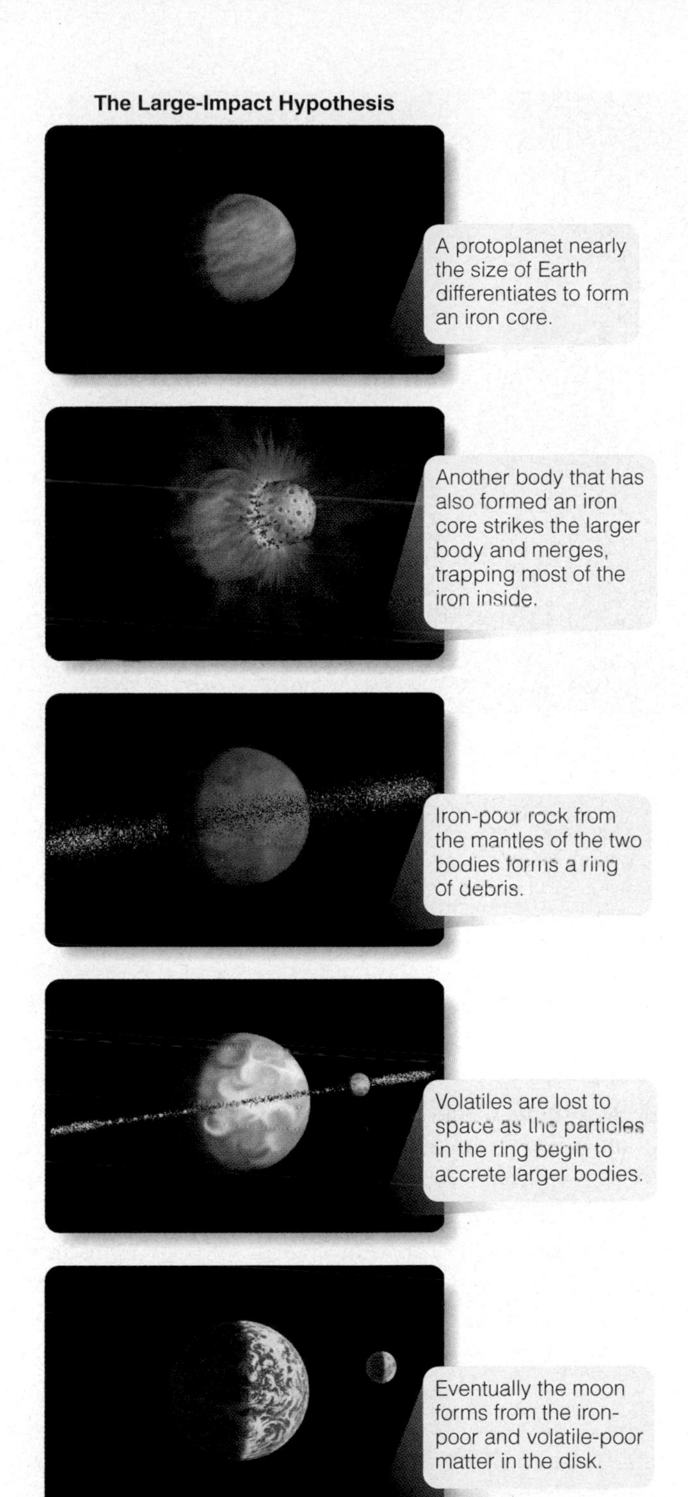

A protoplanet nearly the size of Earth differentiates to form an iron core.
Another body that has also formed an iron core strikes the larger body and merges, trapping most of the iron inside.
Iron-poor rock from the mantles of the two bodies forms a ring of debris.
Volatiles are lost to space as the particles in the ring begin to accrete larger bodies.
Eventually the moon forms from the iron-poor and volatile-poor matter in the disk.

■ **Figure 21-10**

Sometime before the solar system was 50 million years old, a collision produced Earth and the moon in its inclined orbit.

the ejected material would be mostly iron-poor mantle and crust. Calculations show that the iron core of the impacting body could have fallen into the larger body that became Earth. This would explain why the moon is so poor in iron and why the abundances of other elements are so similar to those in Earth's mantle. Finally, the material that eventually became the moon would have remained in a disk long enough for volatile elements, which the moon lacks, to be lost to space.

The moon may be the result of a giant impact. Until recently, astronomers have been reluctant to consider such catastrophic events, but a number of lines of evidence suggest that some planets may have been affected by giant impacts. Consequently, the third theme identified in the introduction to this chapter, giant impacts, has the potential to help you understand other worlds. Catastrophic events are rare, but they can occur.

◄ SCIENTIFIC ARGUMENT ►

If the moon was intensely cratered by the heavy bombardment and then formed great lava plains, why didn't the same thing happen on Earth?

Is this argument obvious? It is still worth reviewing as a way to test your understanding. In fact, the same thing did happen on Earth. Although the moon has more craters than Earth, the moon and Earth are the same age, and both were battered by meteorites during the heavy bombardment. Some of those impacts on Earth must have been large and dug giant multiringed basins. Lava flows must have welled up through Earth's crust and flooded the lowlands to form great lava plains much like the lunar maria.

Earth, however, is a larger world and has more internal heat, which escapes more slowly than the moon's heat did. The moon is now geologically dead, but Earth is very active, with heat flowing outward from the interior to drive plate tectonics. The moving plates long ago erased all evidence of the cratering and lava flows dating from Earth's youth.

Comparative planetology is a powerful tool in that it allows you to see similar processes occurring under different circumstances. For example, expand your argument to explain a different phenomenon. **Why doesn't the moon have a magnetic field?**

◄ ►

21-2 Mercury

EARTH'S MOON AND MERCURY ARE GOOD SUBJECTS for comparative planetology. They are similar in a number of ways. Most important, they are small worlds **(Celestial Profile 4)**; the moon is only a fourth of Earth's diameter, and Mercury is just over a third of Earth's diameter. Their rotation has been altered by tides, their surfaces are heavily cratered, their lowlands are flooded in places by ancient lava flows, and both are airless and have ancient, inactive surfaces. Yet the impressive differences between them will help you understand the nature of these airless worlds.

Mercury is the innermost planet in the solar system, and thus its orbit keeps it near the sun in the sky. It is sometimes visible near the horizon in the evening sky after sunset or in the dawn sky just before sunrise. An Earth-based telescope shows the tiny disk of Mercury passing through phases like the moon, but no surface detail is visible. The Mariner 10 spacecraft looped through the inner solar system in 1974 and 1975 and photographed parts of Mercury (■ Figure 21-11). A new spacecraft

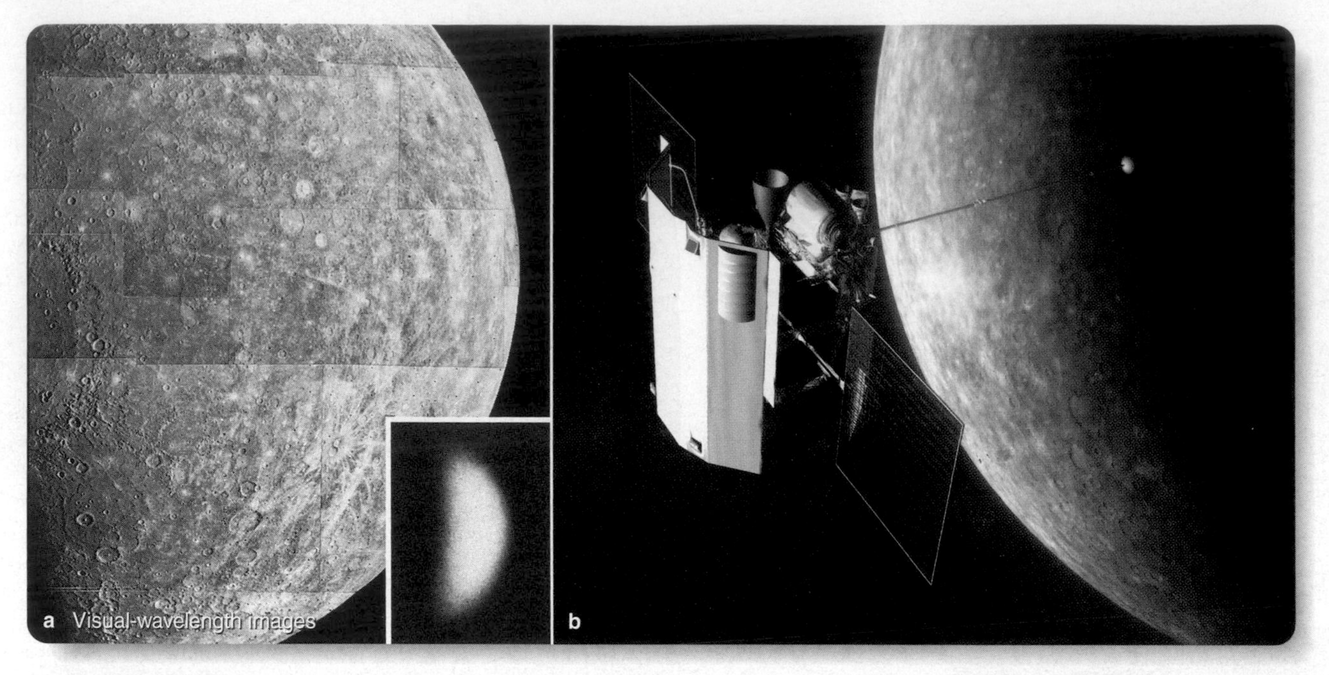

■ Figure 21-11

(a) No surface detail on Mercury is visible from Earth (inset). This photomosaic of Mercury was made by the Mariner 10 spacecraft in the 1970s. Caloris Basin is in shadow at left. (b) The MESSENGER spacecraft now on its way to Mercury will spend a year orbiting the planet and observing from behind a ceramic fabric sunscreen. (a. NASA, inset courtesy Lowell Observatory; b. NASA/Johns Hopkins Univ. Applied Physics Lab./Carnegie Institution of Washington)

called MESSENGER (MErcury Surface, Space ENvironment, GEochemistry, and Ranging mission) will begin a yearlong study of Mercury when it goes into orbit in 2011. Until then, astronomers must use the Mariner 10 data to try to understand Mercury.

Rotation and Revolution

During the 1880s, the Italian astronomer Giovanni Schiaparelli sketched the faint features he thought he saw on the disk of Mercury and concluded that the planet was tidally locked to the sun and kept the same side facing the sun throughout its orbit. This was actually a very good guess because, as you will see, tidal coupling between rotation and revolution is common in the solar system. You have already seen that the moon is tidally locked to Earth. But the rotation of Mercury is more complex than Schiaparelli thought.

In 1962, radio astronomers detected radio emissions from the planet and concluded that the dark side was not as cold as it should have been if the planet kept one side in perpetual darkness. In 1965, radio astronomers made radar contact with Mercury by using the 305-m Arecibo dish (see Figure 6-19) to transmit a pulse of radio energy at Mercury and then waiting for the reflected signal to return. Doppler shifts in the reflected radio pulse showed that the planet was rotating with a period of only about 59 days, much shorter than the orbital period of 87.969 days.

Mercury is tidally coupled to the sun but in a more complex way than the moon is coupled to Earth. Mercury rotates not once per orbit but 1.5 times per orbit. That is, its period of rotation is two-thirds its orbital period, or 58.65 days. This means that a mountain on Mercury directly below the sun at one place in its orbit will point away from the sun one orbit later and toward the sun after the next orbit (■ Figure 21-12).

If you flew to Mercury and landed your spaceship in the middle of the day side, the sun would be high overhead, and it would be noon. Your watch would show almost 44 Earth days passing before the sun set in the west, and a total of 88 Earth days would pass before the sun reached the midnight position. In those 88 Earth days, Mercury would have completed one orbit around the sun (Figure 21-12). It would require another entire orbit of Mercury for the sun to return to the noon position overhead. So a full day on Mercury is two Mercury years long!

The complex tidal coupling between the rotation and revolution of Mercury is an important illustration of the power of tides. Just as the tides in the Earth–moon system have slowed the moon's rotation and locked it to Earth, so have the sun–Mercury tides slowed the rotation of Mercury and coupled its rotation to its revolution. Astronomers refer to such a relationship as a **resonance.** You will see many such resonances as you explore the solar system.

Like its rotation, Mercury's orbital motion is complex. Recall from Chapter 5 that Mercury's orbit is modestly elliptical

The Rotation of Mercury

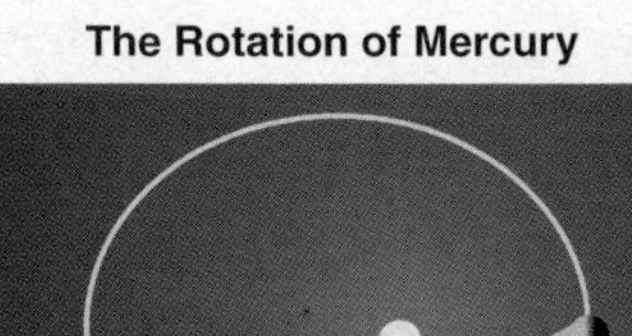

Imagine a mountain on Mercury that points at the sun. It is noon at the mountain.

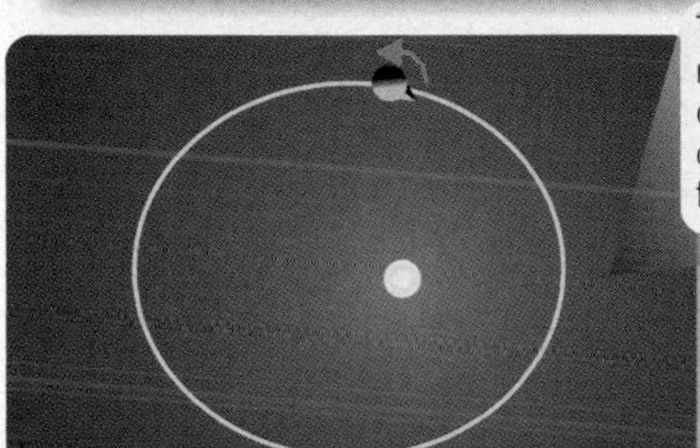

The planet orbits and rotates in the same direction, counter-clockwise as seen from the north.

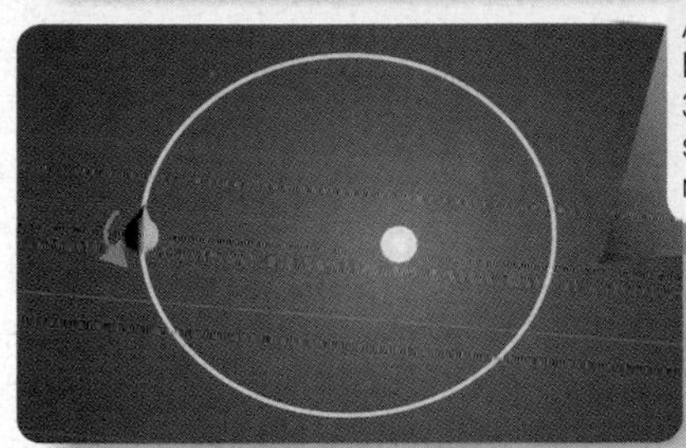

After half an orbit, Mercury has rotated 3/4 of a turn, and it is sunset at the mountain.

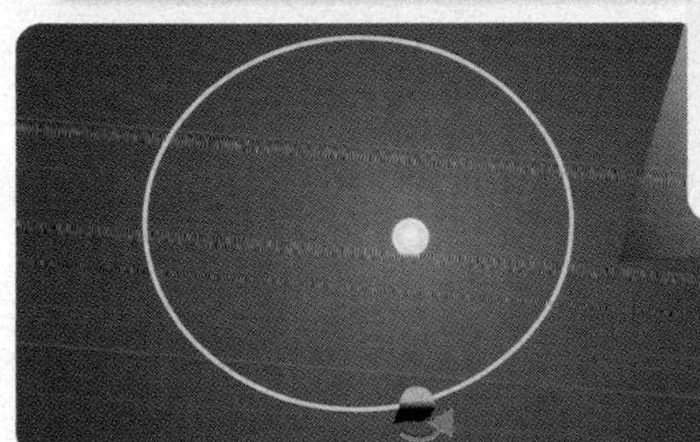

As the planet continues along its orbit, rotation carries the mountain into darkness.

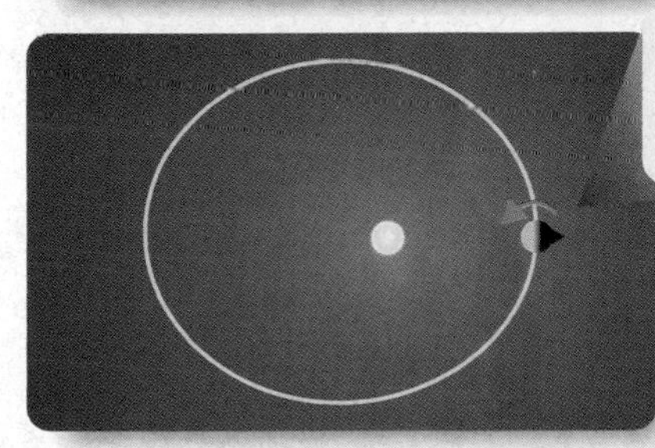

After one orbit, Mercury has rotated 1.5 times, and it is midnight at the mountain.

■ Figure 21-12

Mercury's rotation is in resonance with its orbital motion. It orbits the sun in 88 days and rotates on its axis in two-thirds of that time. One full day on Mercury from noon to noon takes two full orbits.

and precesses faster than can be explained by Newton's laws but at just the rate predicted by Einstein's theory of general relativity. The orbital motion of Mercury is taken as strong confirmation of the curvature of space-time as predicted by general relativity.

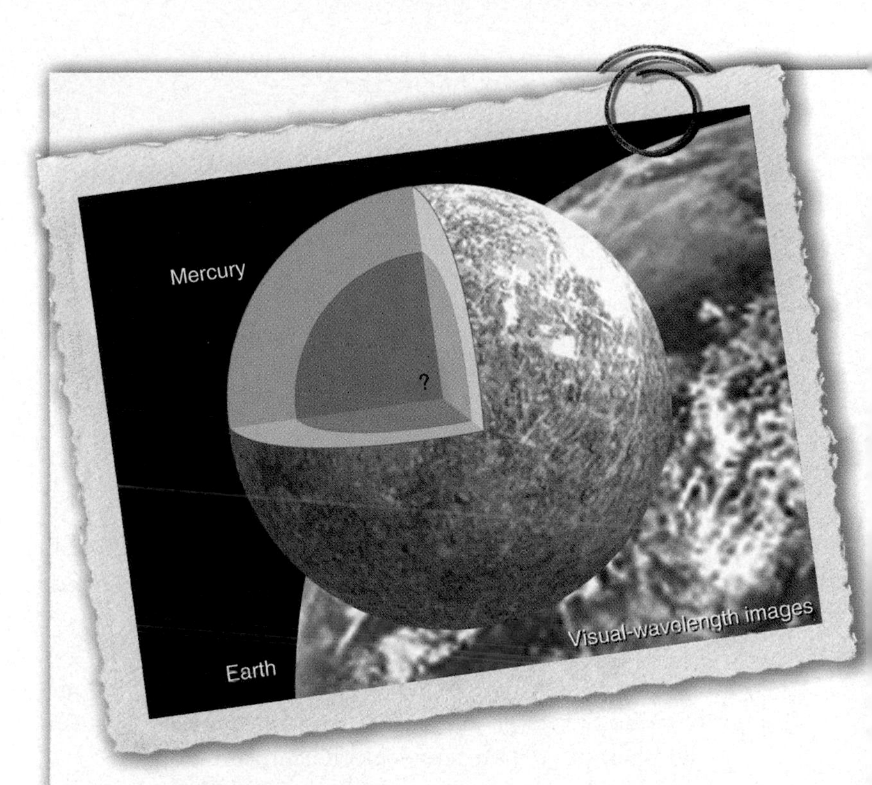

Mercury is a bit over one-third the diameter of Earth, but its high density must mean it has a large iron core. The amount of heat it retains is unknown.

Celestial Profile 4: Mercury

Motion:

Average distance from the sun	0.387 AU (5.79×10^7 km)
Eccentricity of orbit	0.2056
Maximum distance from the sun	0.467 AU (6.97×10^7 km)
Minimum distance from the sun	0.306 AU (4.59×10^7 km)
Inclination of orbit to ecliptic	7°00′16″
Average orbital velocity	47.9 km/s
Orbital period	87.969 days (0.24085 y)
Period of rotation	58.646 days (direct)
Inclination of equator to orbit	0°

Characteristics:

Equatorial diameter	4878 km (0.382 $D_\oplus$)
Mass	3.31×10^{23} kg (0.0558 $M_\oplus$)
Average density	5.44 g/cm^3 (5.4 g/cm^3 uncompressed)
Surface gravity	0.38 Earth gravity
Escape velocity	4.3 km/s (0.38 $V_\oplus$)
Surface temperature	−173° to 430°C (−280° to 800°F)
Average albedo	0.1
Oblateness	0

Personality Point:

Mercury lies very close to the sun and completes an orbit in only 88 days. For this reason, the ancients named the planet after Mercury, the fleet-footed messenger of the gods. The name is also applied to the element mercury, which is also known as quicksilver because it is a heavy, quickly flowing silvery liquid at room temperatures.

The Surface of Mercury

Because Mercury is close to the sun, the temperatures on Mercury are extreme. If you stood in direct sunlight on Mercury, you would hear your spacesuit's cooling system cranking up to top speed as it tried to keep you cool. Daytime temperatures can exceed 700 K (800°F), although about 500 K is a more usual high temperature. If you stepped into shadow on Mercury or took a walk at night, with no atmosphere to distribute heat, your spacesuit heaters would struggle to keep you warm. The surface can cool to 100 K (−280°F). Nights on Mercury are bitter cold. Don't go to Mercury in a cheap spacesuit.

Nights are cold on Mercury because it has almost no atmosphere. It borrows hydrogen and helium atoms from the solar wind, and atoms such as oxygen, sodium, potassium, and calcium have been detected in a cloud above the planet's surface that has such a low density that the atoms do not collide with each other. They just bounce from place to place on the surface, and, because of the low escape velocity, eventually disappear into space. Some of these atoms are probably baked out of the crust.

In photographs, Mercury looks much like Earth's moon. It is heavily battered, with craters of all sizes, including some large basins. Some craters are obviously old and degraded; others seem quite young and have bright rays of ejecta. However, a quick glance at photos of Mercury shows no large, dark maria like the moon's flooded basins.

When planetary scientists began looking at the Mariner photographs in detail, they discovered something not seen on the moon. Mercury is marked by great curved cliffs called **lobate scarps** (■ Figure 21-13). These seem to have formed when the planet cooled and shrank in diameter by a few kilometers, wrinkling its crust as a drying apple wrinkles its skin. Some of these scarps are as high as 3 km and reach hundreds of kilometers across the surface. Other faults in Mercury's crust are straight and may have been produced by tidal stresses generated when the sun slowed Mercury's rotation.

The largest basin on Mercury is called Caloris Basin after the Latin word for "heat," recognition of its location at one of the two "hot poles," which face the sun at alternate perihelions. At the times of the Mariner encounters, the Caloris Basin was half in shadow (■ Figure 21-14a). Although half cannot be seen, the low angle of illumination is ideal for the study of the lighted half because it produces dramatic shadows.

The Caloris Basin is a gigantic multiringed impact basin 1300 km in diameter with concentric mountain rings up to 3 km high. The impact threw ejecta 600 to 800 km across the planet, and the focusing of seismic waves on the far side produced peculiar terrain that looks much like the jumbled surface of the moon that lies opposite the Imbrium basin (Figure 21-14b and c). The Caloris Basin is partially filled with lava flows. Some of this lava may be material melted by the energy of the impact, but some may be lava from below the crust that leaked up through cracks. The weight of this lava and the sagging of the crust have produced deep cracks in the central lava plains. The geophysics of such large, multiringed crater basins is not well understood at present, but Caloris Basin seems to be the same kind of structure

■ Figure 21-13

A lobate scarp (arrow) crosses craters, indicating that Mercury cooled and shrank, wrinkling its crust, after many of its craters had formed. (NASA)

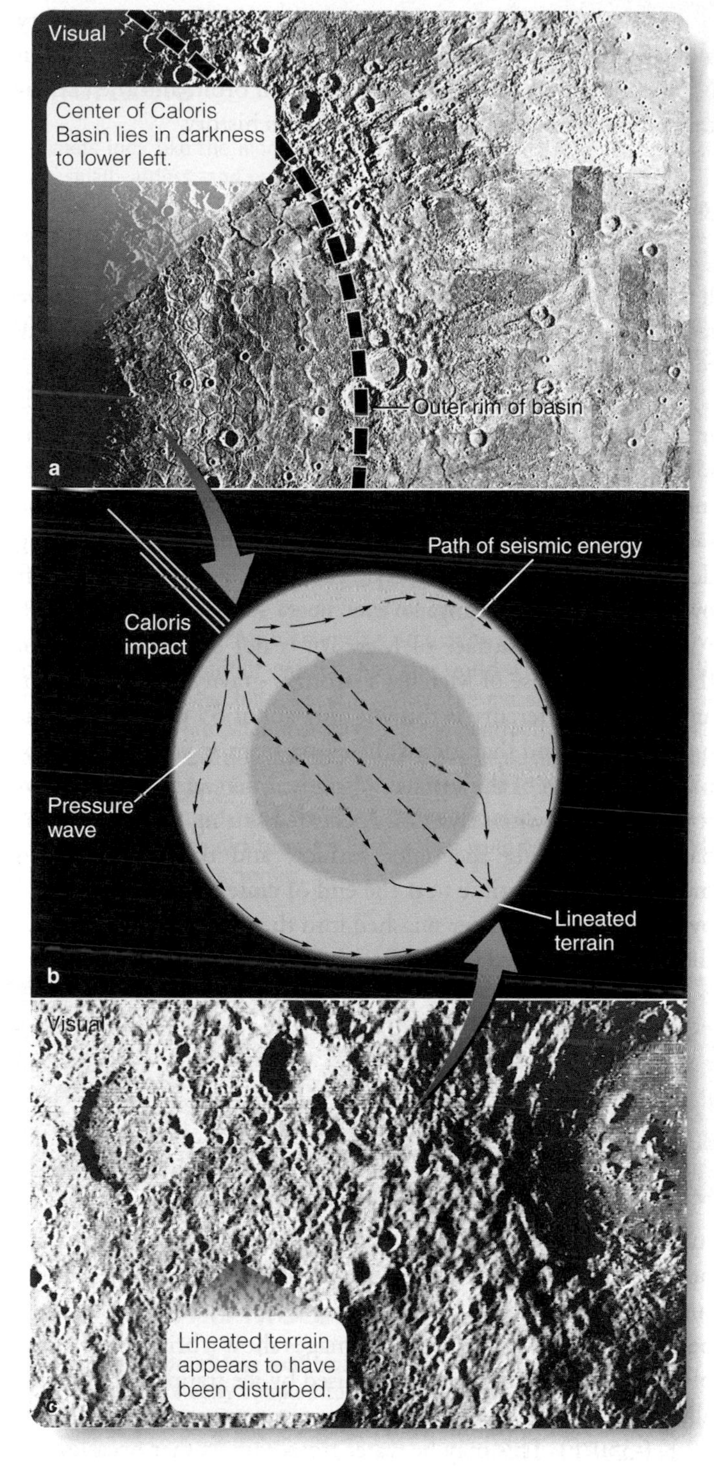

■ **Figure 21-14**

The huge impact that formed the Caloris Basin on Mercury sent seismic waves through the planet. Where the waves came together on the far side, they produced the lineated terrain, which resembles the jumbled terrain on Earth's moon opposite the Imbrium impact. (NASA)

as the Imbrium Basin on the moon, although it has not been as deeply flooded with lava.

When the MESSENGER spacecraft reaches Mercury in 2011, it will photograph nearly all the planet's surface at a much higher resolution than did Mariner 10. Those new photographs will help planetary scientists build a modern understanding of the surface.

The Plains of Mercury

The most striking difference between Mercury and the moon is that Mercury lacks the great dark lava plains so obvious on the moon. Under careful examination, the Mariner 10 photographs show that Mercury has plains, two different kinds, in fact, but they are different from the moon's. Understanding these differences is the key to understanding the history of Mercury.

Much of Mercury's surface is old, cratered terrain (■ Figure 21-15), but other areas called **intercrater plains** are less heavily cratered. These plains are marked by meteorite craters less than 15 km in diameter and secondary craters produced by chunks of ejecta from larger impacts. Unlike the heavily cratered regions, the intercrater plains are not totally saturated with craters. As an expert in comparative planetology, you can recognize that this means that the intercrater plains were produced by later lava flows that buried older terrain.

Smaller regions called **smooth plains** appear to be even younger than the intercrater plains. They have even fewer craters and appear to be ancient lava flows that occurred after most cratering had ended. Much of the region around the Caloris Basin is composed of these smooth plains (Figure 21-15), and they appear to have formed soon after the Caloris impact.

Given the available evidence, planetary astronomers conclude that the plains of Mercury are solidified lava flows much like the maria on the moon. Unlike the maria, Mercury's lava

■ **Figure 21-15**

Study this region of Mercury carefully, and you will notice the plains between the craters. This surface is not saturated as are the lunar highlands, but it contains more craters than the surface of the maria on Earth's moon. This shows that these plains formed after the heaviest of the cratering in the early solar system. (NASA)

The Rotation of Venus

Nearly all of the planets in our solar system rotate counterclockwise as seen from the north. Uranus is an exception, and so is Venus.

In 1962, radio astronomers were able to transmit a radio pulse of precise wavelength toward Venus and detect the echo returning some minutes later. That is, they detected Venus by radar. But the echo was not a precise wavelength. Part of the reflected signal had a longer wavelength, and part had a shorter wavelength. Evidently the planet was rotating; radio energy reflected from the receding edge was redshifted, and radio energy reflected from the approaching edge was blueshifted. From this Doppler effect, the radio astronomers could tell that Venus was rotating once every 243.01 days. Furthermore, because the western edge of Venus produced the blueshift, the planet had to be rotating in the backward direction.

Why does Venus rotate backward? For decades, textbooks have suggested that proto-Venus was set spinning backward when it was struck off-center by a large planetesimal. That is a reasonable possibility; you have seen that a similar collision probably gave birth to Earth's moon. But there is an alternative. Sophisticated mathematical models suggest that the rotation of a terrestrial planet with a molten core and a dense atmosphere can be gradually reversed by solar tides in its atmosphere. (Notice the contrast between the catastrophic theory of a giant impact and the evolutionary theory of atmospheric tides.)

The Atmosphere of Venus

Although Venus is Earth's twin in size, its atmosphere is truly unearthly. The composition, temperature, and density of Venus' atmosphere make it the most inhospitable of planets. About 96 percent of its atmosphere is carbon dioxide, and 3.5 percent is nitrogen. The remaining 0.5 percent is water vapor, sulfuric acid (H_2SO_4), hydrochloric acid (HCl), and hydrofluoric acid (HF). In fact, the thick clouds that hide the surface are composed of sulfuric acid droplets and microscopic sulfur crystals.

Soviet and American spacecraft have dropped probes into the atmosphere of Venus, and those probes have radioed data back to Earth as they fell toward the surface. These studies show that Venus's cloud layers are much higher and much more stable than those on Earth. The highest layer of clouds, the layer visible from Earth, extends from 68 to 58 km above the surface (■ Figure 22-1). For comparison, the highest clouds on Earth do not extend higher than about 16 km.

These cloud layers are highly stable because the atmospheric circulation on Venus is much more regular than that on Earth. The heated atmosphere at the **subsolar point,** the point on the planet where the sun is directly overhead, rises and spreads out in the upper atmosphere. Convection circulates this gas toward the dark side of the planet and the poles, where it cools and sinks. This circulation produces 300-km/h jet streams in the upper atmosphere, which move from east to west (the same direction the

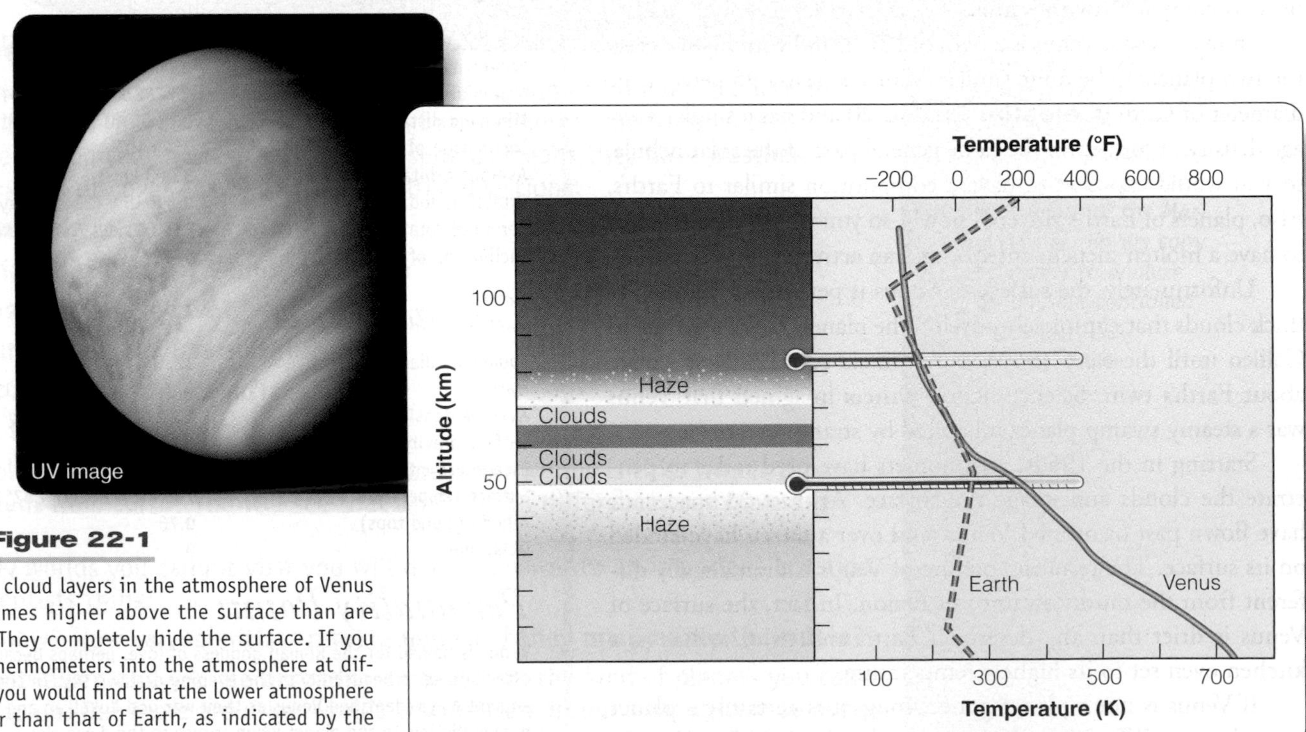

■ **Active Figure 22-1**

The four main cloud layers in the atmosphere of Venus are over 10 times higher above the surface than are Earth clouds. They completely hide the surface. If you could insert thermometers into the atmosphere at different levels, you would find that the lower atmosphere is much hotter than that of Earth, as indicated by the red line in the graph. (NASA)

planet rotates) so rapidly that the entire atmosphere seems to rotate with a period of only four days.

The details of this atmospheric circulation are not well understood, but it seems that the slow rotation of the planet is an important factor. On Earth, large-scale circulation patterns are broken into smaller cyclonic disturbances by Earth's rapid rotation. Because Venus rotates more slowly, its atmospheric circulation is not broken up into small cyclonic storms but instead is organized as a planetwide wind pattern.

Although the upper atmosphere is cool, the lower atmosphere is quite hot (Figure 22-1b). Instrumented probes that have reached the surface report that the temperature is 745K (880°F), and the atmospheric pressure is 90 times that of Earth. Earth's atmosphere is 1000 times less dense than water, but on Venus the air is only 10 times less dense than water. If you could survive the unpleasant composition, intense heat, and high pressure, you could strap wings to your arms and fly.

The present atmosphere of Venus is extremely dry, but there is evidence that it once had significant amounts of water. As a Pioneer Venus spacecraft descended through the atmosphere of Venus in 1978, it discovered that deuterium is about 150 times more abundant compared to normal hydrogen atoms than it is on Earth. This abundance of deuterium, the heavy isotope of hydrogen, could have developed because Venus has no ozone layer to absorb the ultraviolet radiation in sunlight. These UV photons broke water molecules into hydrogen and oxygen. The oxygen would have formed oxides in the soil, and the hydrogen would have leaked away into space. The heavier deuterium atoms would leak away slower than normal hydrogen atoms, which would increase the ratio of deuterium to normal hydrogen. Venus has essentially no water now, but the amount of deuterium in the atmosphere suggests that it may have once had enough water to make a planetwide ocean up to 25 meters deep. (For comparison, the water on Earth would make a uniform planetwide ocean 3000 meters deep.) Venus is now a deadly dry world with only enough water vapor in its atmosphere to make an ocean 0.3 meter deep.

Even if Venus once had lots of water, it was probably too warm for that water to fall from the atmosphere as rain and form oceans. Rather, much of the water may have remained in the atmosphere. That lack of oceans is the biggest difference between Earth and Venus.

The Venusian Greenhouse

You saw in Chapter 20 how the greenhouse effect warms Earth. Carbon dioxide (CO_2) is transparent to light but opaque to infrared (heat) radiation. That means energy can enter the atmosphere as light and warm the surface, but the surface cannot radiate an equal amount of energy back to space because the carbon dioxide is opaque to infrared radiation. Venus also has a greenhouse effect, but on Venus the effect is fearsome. Whereas Earth's atmosphere contains only about 0.03 percent carbon dioxide, the atmosphere of Venus contains 96 percent carbon dioxide, and the surface of Venus is nearly twice as hot as a hot kitchen oven.

Planetary astronomers think they know how Venus got into such a jam. When Venus was young, it may have been cooler than it is now; but, because it formed 30 percent closer to the sun than did Earth, it was warmer than Earth, and that unleashed processes that made it even hotter. As you saw in the previous section, even if Venus once had water, it probably never had large liquid-water oceans because of the heat. Carbon dioxide is highly soluble in water, but without large oceans to dissolve carbon dioxide and remove it from the atmosphere, Venus was trapped with a carbon dioxide–rich atmosphere, and the greenhouse effect made the planet warmer. As the planet warmed, any small oceans that did exist evaporated, and Venus lost any ability it had to cleanse its atmosphere of carbon dioxide. Volcanoes on Earth vent gases that are mostly carbon dioxide and water vapor, and presumably the volcanoes on Venus did the same. The water vapor disappeared as it was broken up by ultraviolet radiation, and the carbon dioxide accumulated in the atmosphere. The high temperature baked even more carbon dioxide out of the surface, and the atmosphere became even less transparent to infrared radiation, causing the temperature to rise even further. This runaway greenhouse effect has made the surface so hot that even sulfur, chlorine, and fluorine have baked out of the rock and formed sulfuric, hydrochloric, and hydrofluoric acid vapors.

Earth avoided this runaway greenhouse effect because it was farther from the sun and cooler. Consequently, it could form and preserve liquid-water oceans, which absorbed the carbon dioxide and left an atmosphere of nitrogen that was relatively transparent in some parts of the infrared. If all of the carbon in Earth's sediments was put back into the atmosphere as carbon dioxide, our air would be as dense as that of Venus, and Earth would suffer from a terrifying greenhouse effect. Recall from Chapter 20 how the use of fossil fuels and the destruction of forests is increasing the carbon dioxide concentration in our atmosphere and warming the planet. Venus warns us of what a greenhouse effect can do.

ThomsonNOW Sign in at www.thomsonedu.com and go to ThomsonNOW to see Astronomy Exercise "The Greenhouse Effect."

The Surface of Venus

Given that the surface of Venus is perpetually hidden by clouds, is hot enough to melt lead, and suffers under crushing atmospheric pressure, it is surprising how much planetary scientists know about the geology of Venus. Early radar maps made from Earth penetrated the clouds and showed that it had mountains, plains, and some craters. The Soviet Union launched a number of spacecraft that landed on Venus, and although the spacecraft failed within an hour or so of landing, they did analyze the rock and transmit a few

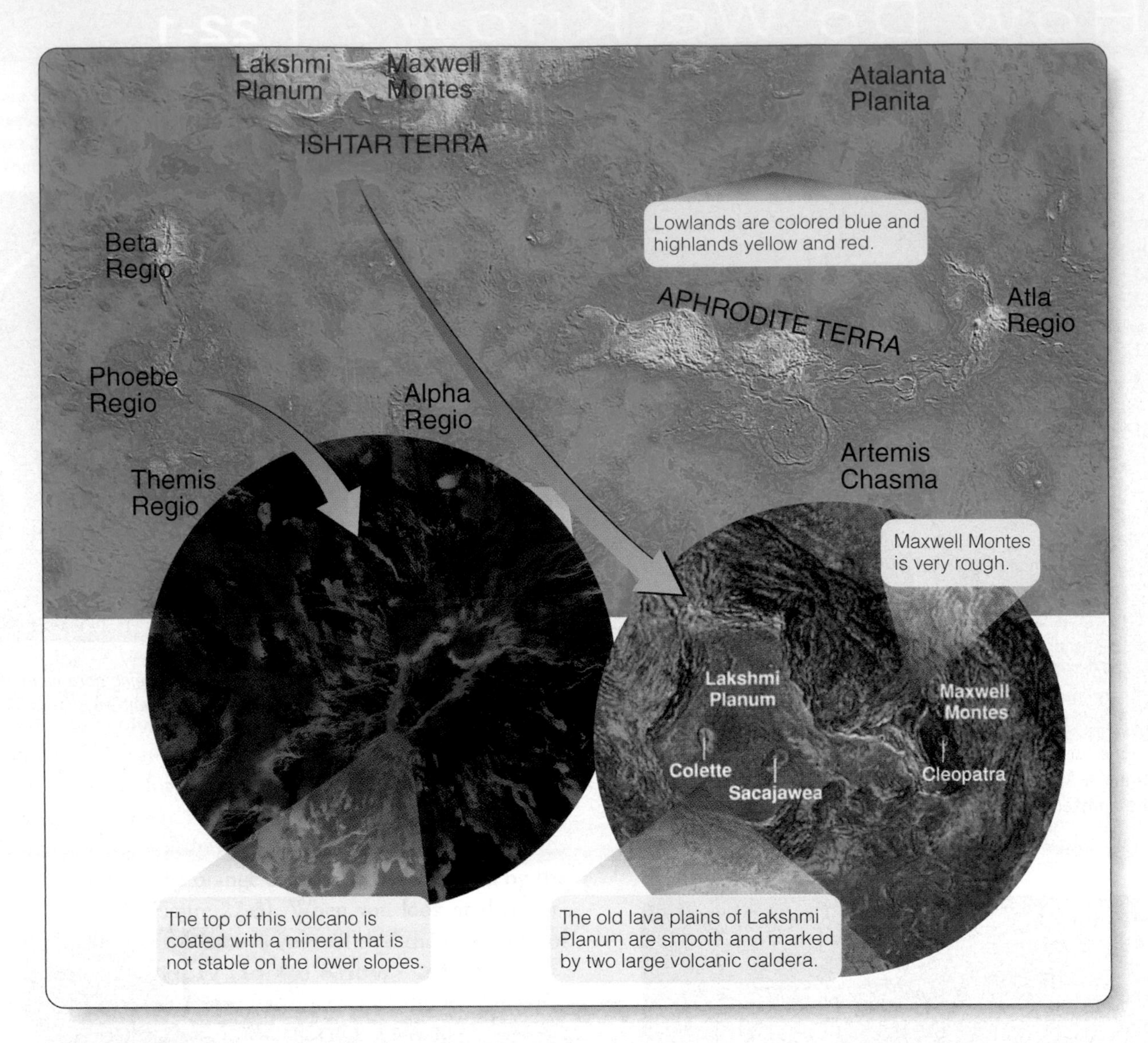
Lakshmi
Planum
Maxwell
Montes
ISHTAR TERRA
Atalanta
Planita
Lowlands are colored blue and highlands yellow and red.
Beta
Regio
APHRODITE TERRA
Atla
Regio
Phoebe
Regio
Alpha
Regio
Artemis
Chasma
Themis
Regio
Maxwell Montes is very rough.
Lakshmi
Planum
Maxwell
Montes
Colette
Sacajawea
Cleopatra
The top of this volcano is coated with a mineral that is not stable on the lower slopes.
The old lava plains of Lakshmi Planum are smooth and marked by two large volcanic caldera.

■ Figure 22-5

Notice how these three radar maps show different things. The main map here shows elevation over most of the surface of Venus. Only the polar areas are not shown. The inset map at left shows an electrical property of surface minerals related to chemical composition. The detailed map of Maxwell Montes and Lakshmi Planum at right is color coded to show degree of roughness, with purple smooth and orange rough. The rock on Venus is, in fact, dark gray. (Maxwell and Lakshmi Planum map: USGS; other maps: NASA)

Regio and Beta Regio, later found to be volcanic peaks, were also discovered by radar before the naming convention was adopted. Other names on Venus are feminine, such as the highland regions Ishtar Terra and Aphrodite Terra, named for the Babylonian and Greek goddesses of love. The insets in Figure 22-5 show roughness and composition of the surface. These maps were made by different satellites and color coded in different ways. All of these radar maps paint a picture of a hot, violent, desert world.

Radar maps show that the surface of Venus consists of low, rolling plains and highland regions. The rolling plains appear to be large-scale smooth lava flows, and the highlands are regions of deformed crust.

Just as in the case of the lunar landscape, craters are the key to figuring out the age of the surface. With nearly 1000 impact craters on its surface, Venus has more craters than Earth but not nearly as many as the moon. The craters are uniformly scattered over the surface and look sharp and fresh (■ Figure 22-6). With no water and a thick, sluggish atmosphere, there is little erosion on Venus, and the thick atmosphere protects the surface from small meteorites. Consequently, there are no small craters. Planetary scientists conclude that the surface of Venus is older than Earth's surface but not nearly as old as the moon's. Unlike the moon, there are no old, cratered highlands. Lava flows have covered the entire surface within the last half-billion years.

planet rotates) so rapidly that the entire atmosphere seems to rotate with a period of only four days.

The details of this atmospheric circulation are not well understood, but it seems that the slow rotation of the planet is an important factor. On Earth, large-scale circulation patterns are broken into smaller cyclonic disturbances by Earth's rapid rotation. Because Venus rotates more slowly, its atmospheric circulation is not broken up into small cyclonic storms but instead is organized as a planetwide wind pattern.

Although the upper atmosphere is cool, the lower atmosphere is quite hot (Figure 22-1b). Instrumented probes that have reached the surface report that the temperature is 745K (880°F), and the atmospheric pressure is 90 times that of Earth. Earth's atmosphere is 1000 times less dense than water, but on Venus the air is only 10 times less dense than water. If you could survive the unpleasant composition, intense heat, and high pressure, you could strap wings to your arms and fly.

The present atmosphere of Venus is extremely dry, but there is evidence that it once had significant amounts of water. As a Pioneer Venus spacecraft descended through the atmosphere of Venus in 1978, it discovered that deuterium is about 150 times more abundant compared to normal hydrogen atoms than it is on Earth. This abundance of deuterium, the heavy isotope of hydrogen, could have developed because Venus has no ozone layer to absorb the ultraviolet radiation in sunlight. These UV photons broke water molecules into hydrogen and oxygen. The oxygen would have formed oxides in the soil, and the hydrogen would have leaked away into space. The heavier deuterium atoms would leak away slower than normal hydrogen atoms, which would increase the ratio of deuterium to normal hydrogen. Venus has essentially no water now, but the amount of deuterium in the atmosphere suggests that it may have once had enough water to make a planetwide ocean up to 25 meters deep. (For comparison, the water on Earth would make a uniform planetwide ocean 3000 meters deep.) Venus is now a deadly dry world with only enough water vapor in its atmosphere to make an ocean 0.3 meter deep.

Even if Venus once had lots of water, it was probably too warm for that water to fall from the atmosphere as rain and form oceans. Rather, much of the water may have remained in the atmosphere. That lack of oceans is the biggest difference between Earth and Venus.

The Venusian Greenhouse

You saw in Chapter 20 how the greenhouse effect warms Earth. Carbon dioxide (CO_2) is transparent to light but opaque to infrared (heat) radiation. That means energy can enter the atmosphere as light and warm the surface, but the surface cannot radiate an equal amount of energy back to space because the carbon dioxide is opaque to infrared radiation. Venus also has a greenhouse effect, but on Venus the effect is fearsome. Whereas Earth's atmosphere contains only about 0.03 percent carbon dioxide, the atmosphere of Venus contains 96 percent carbon dioxide, and the surface of Venus is nearly twice as hot as a hot kitchen oven.

Planetary astronomers think they know how Venus got into such a jam. When Venus was young, it may have been cooler than it is now; but, because it formed 30 percent closer to the sun than did Earth, it was warmer than Earth, and that unleashed processes that made it even hotter. As you saw in the previous section, even if Venus once had water, it probably never had large liquid-water oceans because of the heat. Carbon dioxide is highly soluble in water, but without large oceans to dissolve carbon dioxide and remove it from the atmosphere, Venus was trapped with a carbon dioxide–rich atmosphere, and the greenhouse effect made the planet warmer. As the planet warmed, any small oceans that did exist evaporated, and Venus lost any ability it had to cleanse its atmosphere of carbon dioxide. Volcanoes on Earth vent gases that are mostly carbon dioxide and water vapor, and presumably the volcanoes on Venus did the same. The water vapor disappeared as it was broken up by ultraviolet radiation, and the carbon dioxide accumulated in the atmosphere. The high temperature baked even more carbon dioxide out of the surface, and the atmosphere became even less transparent to infrared radiation, causing the temperature to rise even further. This runaway greenhouse effect has made the surface so hot that even sulfur, chlorine, and fluorine have baked out of the rock and formed sulfuric, hydrochloric, and hydrofluoric acid vapors.

Earth avoided this runaway greenhouse effect because it was farther from the sun and cooler. Consequently, it could form and preserve liquid-water oceans, which absorbed the carbon dioxide and left an atmosphere of nitrogen that was relatively transparent in some parts of the infrared. If all of the carbon in Earth's sediments was put back into the atmosphere as carbon dioxide, our air would be as dense as that of Venus, and Earth would suffer from a terrifying greenhouse effect. Recall from Chapter 20 how the use of fossil fuels and the destruction of forests is increasing the carbon dioxide concentration in our atmosphere and warming the planet. Venus warns us of what a greenhouse effect can do.

ThomsonNOW Sign in at www.thomsonedu.com and go to ThomsonNOW to see Astronomy Exercise "The Greenhouse Effect."

The Surface of Venus

Given that the surface of Venus is perpetually hidden by clouds, is hot enough to melt lead, and suffers under crushing atmospheric pressure, it is surprising how much planetary scientists know about the geology of Venus. Early radar maps made from Earth penetrated the clouds and showed that it had mountains, plains, and some craters. The Soviet Union launched a number of spacecraft that landed on Venus, and although the spacecraft failed within an hour or so of landing, they did analyze the rock and transmit a few

■ Figure 22-2

The Venera lander touched down on Venus in 1982 and carried a camera that swiveled from side to side to photograph the surface. The orange glow is produced by the thick atmosphere; when that is corrected, you can see that the rocks are dark gray. Isotopic analysis suggests they are basalts.

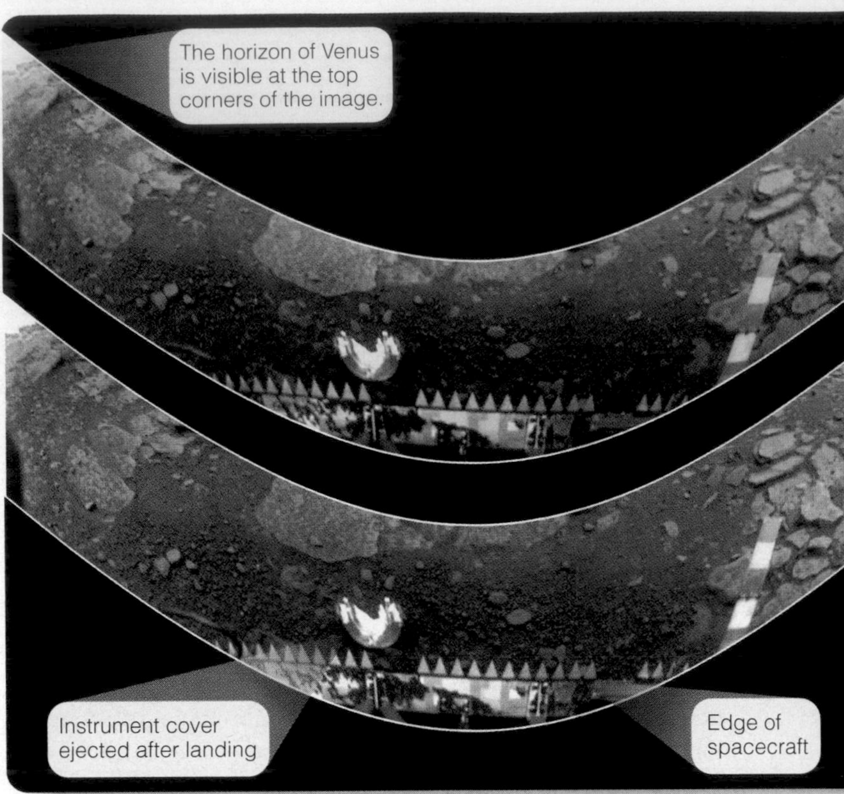

images back to Earth. The rock seems to be basalt, a typical product of volcanism, The images revealed dark-gray rocky plains bathed in a deep-orange glow caused by sunlight filtering down through the thick atmosphere (■ Figure 22-2).

Both the United States and the Soviet Union sent spacecraft to Venus that used radar to penetrate the clouds and map the surface. The NASA Pioneer Venus probe orbited Venus in 1978 and made radar maps showing features as small as 25 km in diameter. Later, two Soviet Venera spacecraft mapped the north polar regions with a resolution of 2 km. From 1992 to 1994, the NASA Magellan spacecraft orbited Venus and created radar maps showing details as small as 100 m. These radar maps provide a unique look below the clouds.

The color of Venus radar maps is mostly arbitrary. Human eyes can't see radio waves, so the radar maps must be given some false colors to distinguish height or roughness or composition. Some maps use blues and greens for lowlands and yellows and reds for highlands. Remember when you look at these maps that there is no liquid water on Venus. Magellan scientists chose to use yellows and oranges for their radar maps in an effort to mimic the orange color of daylight caused by the thick atmosphere (■ Figure 22-3). When you look at these orange images, you need to remind yourself that the true color of the rock is dark gray. **(How Do We Know? 22-1).**

Radar maps of Venus reveal a number of things about the surface. If you transmit a radio signal down through the clouds and measure the time until you hear the echo coming back up, you can measure the altitude of the surface. Part of the Magellan data is a detailed altitude map of the surface. You can also measure the amount of the signal that is reflected from each spot on the surface. Much of the surface of Venus is made up of old, smooth lava flows that do not look bright in radar maps, but faults and uneven terrain look brighter. Some young, rough lava flows are very rough and contain billions of tiny crevices (■ Figure 22-4) that bounce the radar signal around and shoot it back the way it came. These rough lava flows look very bright in radar maps, and certain mineral deposits are also bright. Radar maps do not show how the surface would look to human eyes but rather provide information about altitude, roughness, and, in some cases, chemical composition.

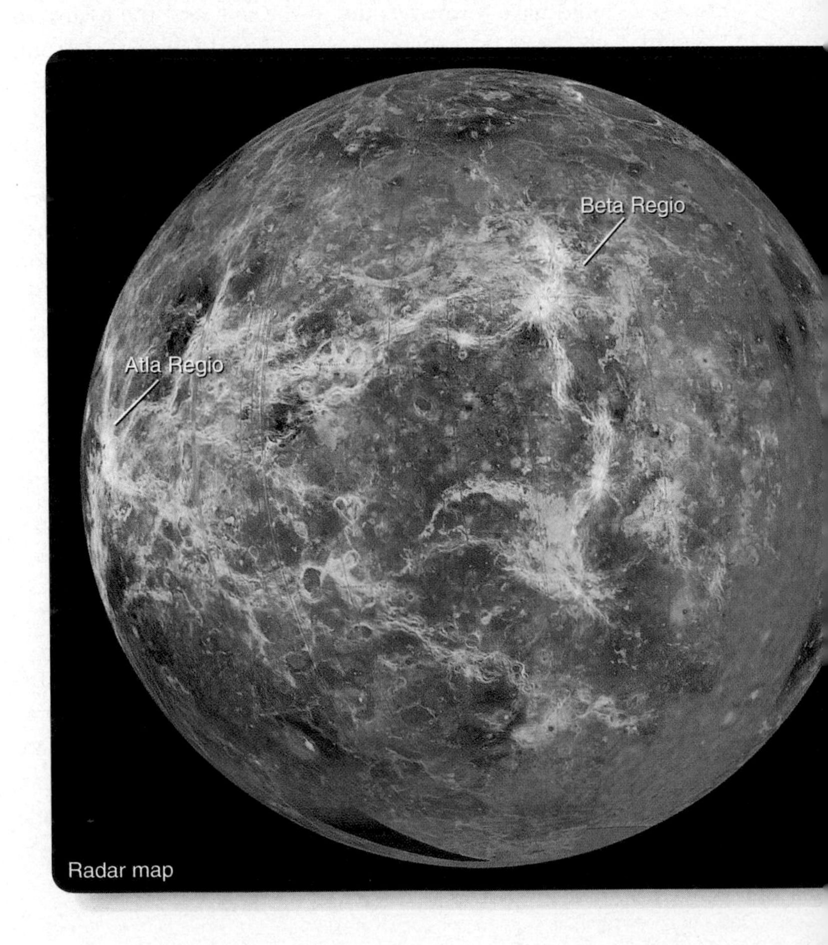

■ Figure 22-3

Venus without its clouds: This mosaic of Magellan radar maps has been given an orange color to mimic the sunset coloration of daylight at the surface of the planet. The image shows scattered impact craters and volcanic regions such as Beta Regio and Atla Regio. (NASA)

How Do We Know? 22-1

Data Manipulation

Why do scientists think it is OK to enhance their data? Planetary astronomers studying Venus change the colors of radar maps and stretch the height of mountains. If they were making political TV commercials and were caught digitally enhancing a politician's voice, they would be called dishonest, but scientists often manipulate and enhance their data. It's not dishonest because the scientists are their own audience.

Research physiologists studying knee injuries, for instance, can use magnetic resonance imaging (MRI) data to study both healthy and damaged knees. By placing a patient in a powerful magnetic field and irradiating his or her knee with precisely tuned radio frequency pulses, the MRI machine can force one in a million hydrogen atoms to emit radio frequency photons. The intensity and frequency of the emitted photons depends on how the hydrogen atoms are bonded to other atoms, so bone, muscle, and cartilage emit different signals. An antenna in the machine picks up the emitted signals and stores huge masses of data in computer memory as tables of numbers.

The tables of numbers are meaningless to the physiologists, but by manipulating the data, they can form images that reveal the anatomy of a knee. By enhancing the data, they can distinguish between bone and cartilage and see how tendons are attached. They can filter the data to see fine detail or smooth the data to eliminate distracting textures. Because the physiologists are their own audience, they know how they are manipulating the data and can use it to devise better ways to treat knee injuries.

When scientists say they are "massaging the data," they mean they are filtering, enhancing, and manipulating it to bring out the features they need to study. If they were presenting that data to a television audience to promote a cause or sell a product, it would be dishonest, but scientists are their own audience, so their manipulation of the data is just another way to better understand how nature works.

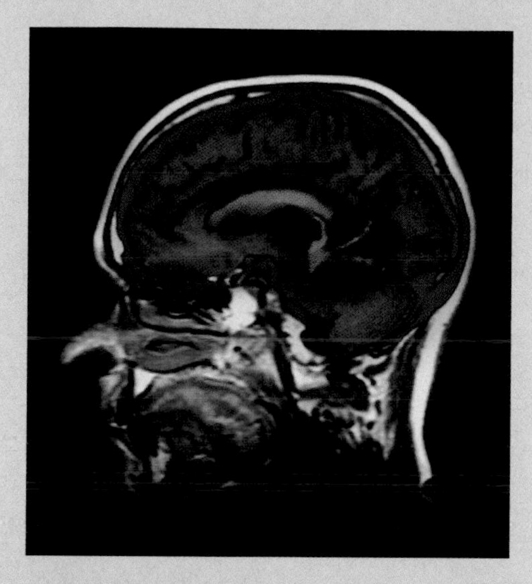

You are accustomed to seeing data manipulated and presented in convenient ways. (PhotoDisc/Getty Images)

■ **Figure 22-4**

Although it is nearly 1000 years old, this lava flow near Flagstaff, Arizona, is still such a rough jumble of sharp rock that it is dangerous to venture onto its surface. Rough surfaces are very good reflectors of radio waves and look bright in radar maps. Solidified lava flows on Venus show up as bright regions in the radar maps because they are rough. (M. A. Seeds)

The big map in ■ Figure 22-5 is a map of all of Venus except the polar regions. (Note that this map has been color coded by altitude.) By international agreement, the names of celestial bodies and features on celestial bodies are assigned by the International Astronomical Union, which has decided that all names on Venus should be feminine. There are only a few exceptions, such as the mountain Maxwell Montes, 50 percent higher than Mt. Everest. It was discovered during early Earth-based radar mapping and named for James Clark Maxwell, the 19th-century physicist who first described electromagnetic radiation. Alpha

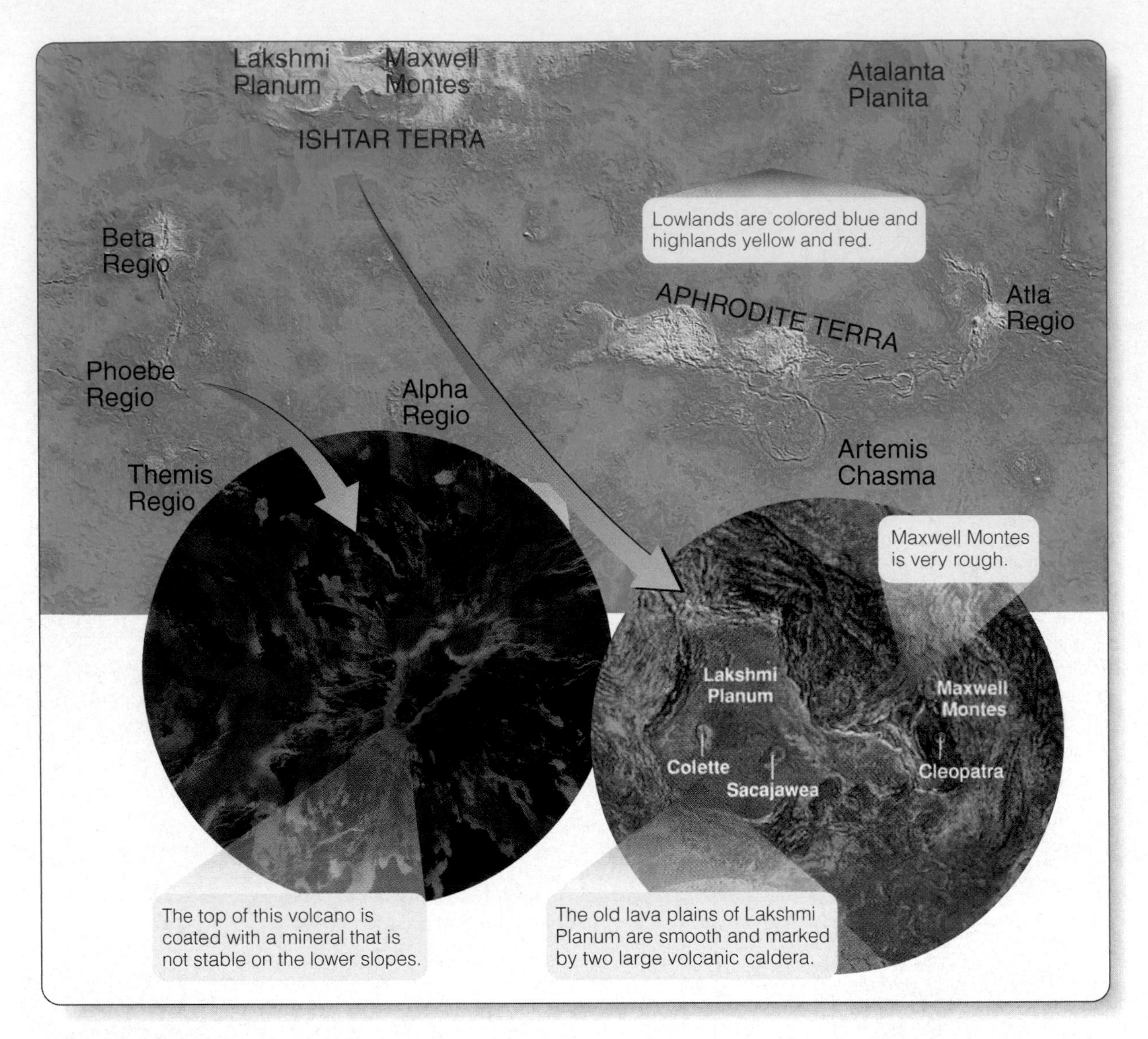
Lakshmi Planum
Maxwell Montes
Atalanta Planita
ISHTAR TERRA
Lowlands are colored blue and highlands yellow and red.
Beta Regio
APHRODITE TERRA
Atla Regio
Phoebe Regio
Alpha Regio
Artemis Chasma
Themis Regio
Maxwell Montes is very rough.
Lakshmi Planum
Maxwell Montes
Colette
Sacajawea
Cleopatra
The top of this volcano is coated with a mineral that is not stable on the lower slopes.
The old lava plains of Lakshmi Planum are smooth and marked by two large volcanic caldera.

■ **Figure 22-5**

Notice how these three radar maps show different things. The main map here shows elevation over most of the surface of Venus. Only the polar areas are not shown. The inset map at left shows an electrical property of surface minerals related to chemical composition. The detailed map of Maxwell Montes and Lakshmi Planum at right is color coded to show degree of roughness, with purple smooth and orange rough. The rock on Venus is, in fact, dark gray. (Maxwell and Lakshmi Planum map: USGS; other maps: NASA)

Regio and Beta Regio, later found to be volcanic peaks, were also discovered by radar before the naming convention was adopted. Other names on Venus are feminine, such as the highland regions Ishtar Terra and Aphrodite Terra, named for the Babylonian and Greek goddesses of love. The insets in Figure 22-5 show roughness and composition of the surface. These maps were made by different satellites and color coded in different ways. All of these radar maps paint a picture of a hot, violent, desert world.

Radar maps show that the surface of Venus consists of low, rolling plains and highland regions. The rolling plains appear to be large-scale smooth lava flows, and the highlands are regions of deformed crust.

Just as in the case of the lunar landscape, craters are the key to figuring out the age of the surface. With nearly 1000 impact craters on its surface, Venus has more craters than Earth but not nearly as many as the moon. The craters are uniformly scattered over the surface and look sharp and fresh (■ Figure 22-6). With no water and a thick, sluggish atmosphere, there is little erosion on Venus, and the thick atmosphere protects the surface from small meteorites. Consequently, there are no small craters. Planetary scientists conclude that the surface of Venus is older than Earth's surface but not nearly as old as the moon's. Unlike the moon, there are no old, cratered highlands. Lava flows have covered the entire surface within the last half-billion years.

■ **Figure 22-6**

Impact crater Howe in the foreground of this Magellan radar image is 37 km in diameter. Craters in the background are 47 km and 63 km in diameter. This radar map has been digitally modified to represent the view from a spacecraft flying over the craters. (NASA)

Volcanism on Venus

Volcanism seems to dominate the surface of Venus. Much of Venus is covered by lava flows such as those photographed by the Venera landers (Figure 22-2). Also, volcanic peaks and other volcanic features are evident in radar maps.

As usual, you can learn more about other worlds by comparing them with each other and with Earth. Read **Volcanoes** on pages 490–491 and notice three important ideas and two new terms:

1. There are two main types of volcanoes found on Earth. *Composite volcanoes* tend to be associated with plate motion, and *shield volcanoes* are associated with hot spots.
2. Notice that you can recognize the volcanoes on Venus and Mars by their shapes, even when the images are manipulated in computer mapping. The shield volcanoes found on Venus and Mars are produced by hot-spot volcanism and not by plate tectonics.
3. Also notice the large size of volcanoes on Venus and Mars. They have grown very large because of repeated eruptions at the same place in the crust. This is evidence that neither Venus nor Mars has been dominated by plate tectonics as has Earth.

The radar image of Sapas Mons in ■ Figure 22-7 shows a dramatic overhead view of this volcano, which is 400 km (250 mi) in diameter at its base and 1.5 km (0.9 mi) high. Many bright, young lava flows radiate outward, covering older, darker flows. Remember that the colors in this image are artificial; if you could walk across these lava flows, you would find them solid, dark gray

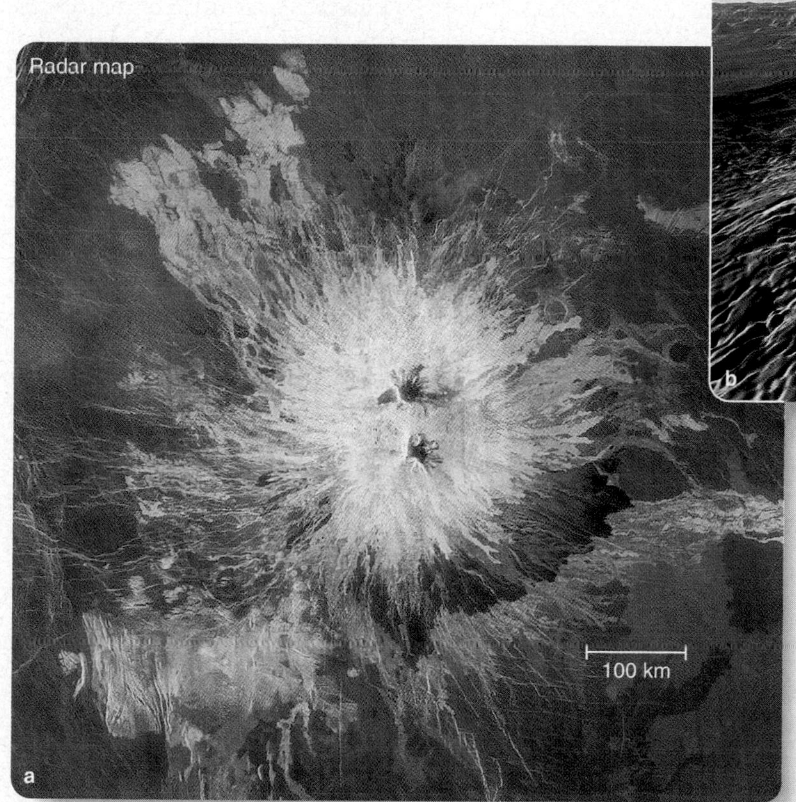

■ **Figure 22-7**

(a) Volcano Sapas Mons, lying along a major fracture zone, is topped by two lava-filled calderas and flanked by rough lava flows. The orange color of this radar map mimics the orange light that filters through the thick atmosphere. (NASA) (b) Seen by light typical of Earth's surface, Sapas Mons might look more like this computer-generated landscape. Volcano Maat Mons rises in the background. The vertical scale has been exaggerated by a factor of 15 to reveal the shape of the volcanoes and lava flows. (Copyright © 1992, David P. Anderson, Southern Methodist University)

Volcanoes

1 Molten rock (magma) is less dense than the surrounding rock and tends to rise. Where it bursts through Earth's crust, you see volcanism. The two main types of volcanoes on Earth provide good examples for comparison with those on Venus and Mars.

On Earth, **composite volcanoes** form above subduction zones where the descending crust melts and the magma rises to the surface. This forms chains of volcanoes along the subduction zone, such as the Andes along the west coast of South America.

Magma rising above subduction zones is not very fluid, and it produces explosive volcanoes with sides as steep as 30°.

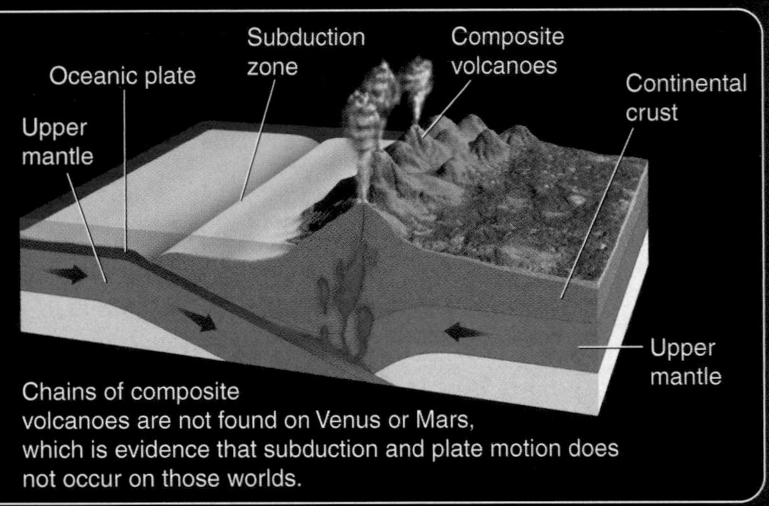

Chains of composite volcanoes are not found on Venus or Mars, which is evidence that subduction and plate motion does not occur on those worlds.

Based on *Physical Geology*, 4th edition, James S. Monroe and Reed Wicander, Wadsworth Publishing Company. Used with permission.

Mount St. Helens exploded northward on May 18, 1980, killing 63 people and destroying 600 km^2 (230 mi^2) of forest with a blast of winds and suspended rock fragments that moved as fast as 480 km/hr (300 mph) and had temperatures as hot as 350°C (660°F). Note the steep slope of this composite volcano.

USGS

Shield volcano

Lava flow

Magma chamber

Oceanic crust

1a A **shield volcano** is formed by highly fluid lava (basalt) that flows easily and creates low-profile volcanic peaks with slopes of 3° to 10°. The volcanoes of Hawaii are shield volcanoes that occur over a hot spot in the middle of the Pacific plate.

Magma collects in a chamber in the crust and finds its way to the surface through cracks.

Magma forces its way upward through cracks in the upper mantle and causes small, deep earthquakes.

A hot spot is formed by a rising convection current of magma moving upward through the hot, deformable (plastic) rock of the mantle.

The Cascade Range composite volcanoes are produced by an oceanic plate being subducted below North America and partially melting.

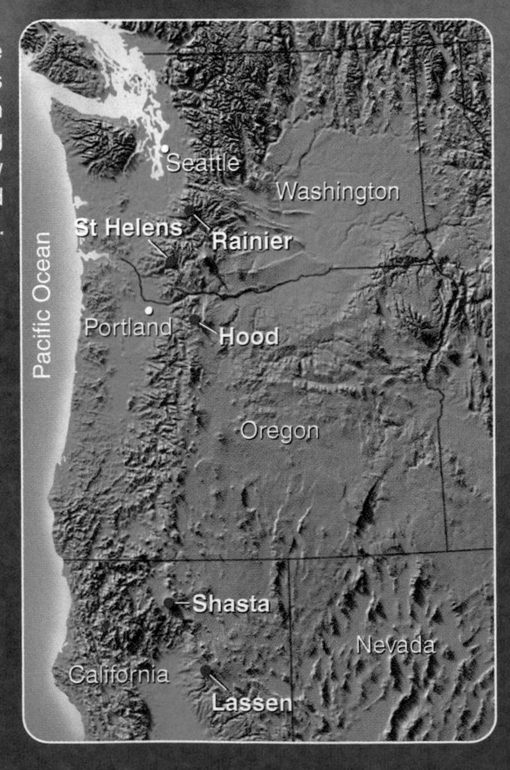

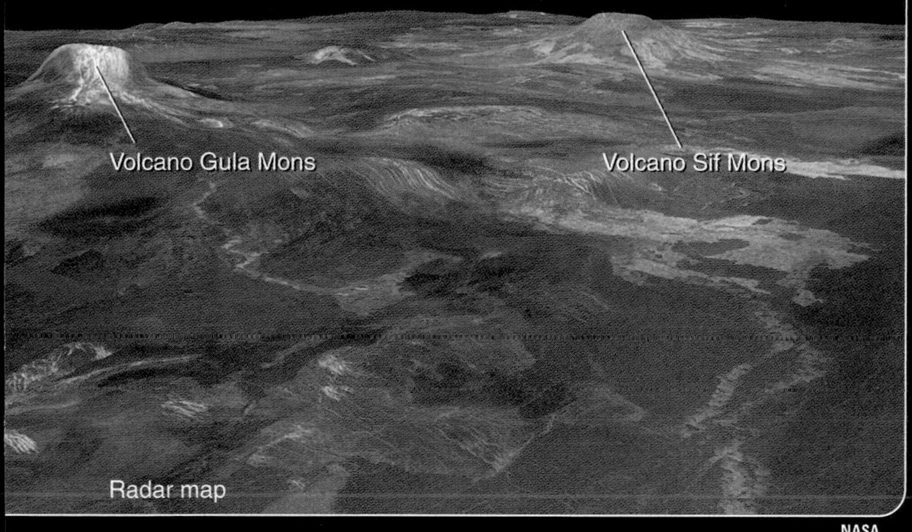

NASA

2 Volcanoes on Venus are shield volcanoes. They appear to be steep sided in some images created from Magellan radar maps, but that is because the vertical scale has been exaggerated to enhance detail. The volcanoes of Venus are actually shallow-sloped shield volcanoes.

2a This computer model of a mountain with the vertical scale magnified 10 times appears to have steep slopes such as those of a composite volcano.

A true profile of the computer model shows the mountain has very shallow slopes typical of shield volcanoes.

Mike Seeds

3 Volcanism over a hot spot results in repeated eruptions that build up a shield volcano of many layers. Such volcanoes can grow very large.

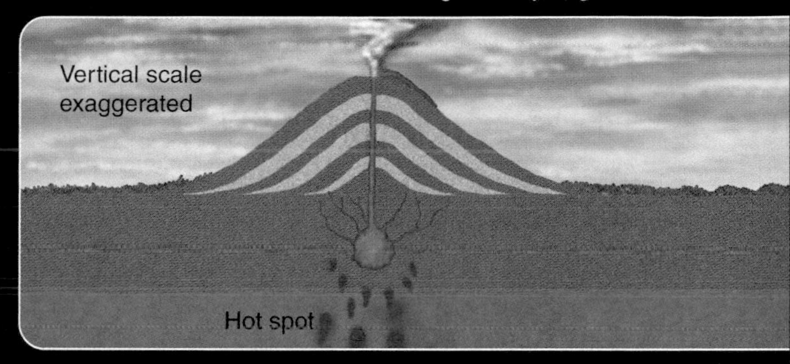

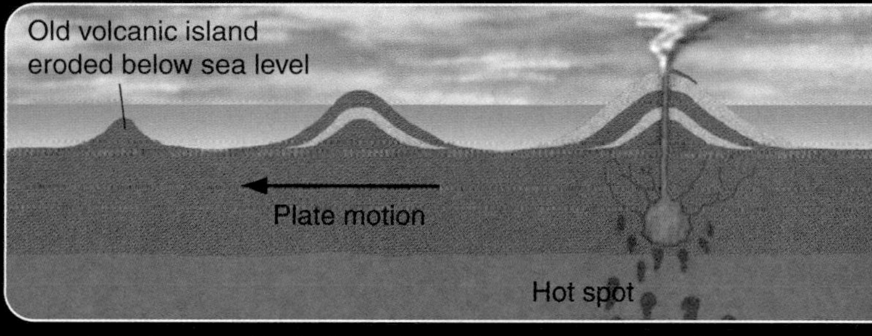

If the crustal plate is moving, magma generated by the hot spot can repeatedly penetrate the crust to build a chain of volcanoes. Only the volcanoes over the hot spot are active. Older volcanoes slowly erode away. Such volcanoes cannot grow large because the moving plate carries them away from the hot spot.

ThomsonNOW™

Sign in at www.thomsonedu.com and go to ThomsonNOW to see the Active Figure "Hot Spot Volcanoes" and compare volcanism on Earth with that on Venus.

Time since last eruption (million years)
5 3 1.5 1 0
Kauai
Oahu
Molokai
Maui
Hawaii
Active volcanoes
Plate motion
Newborn underwater volcano

3a The volcanoes that make up the Hawaiian Islands as shown at left have been produced by a hot spot poking upward through the middle of the moving Pacific plate.

NASA

3b The plate moves about 9 cm/yr and carries older volcanic islands northwest, away from the hot spot. The volcanoes cannot grow extremely large because they are carried away from the hot spot. New islands form to the southeast over the hot spot.

Olympus Mons at right is the largest volcano on Mars. It is a shield volcano 25 km (16 mi) high and 700 km (440 mi) in diameter at its base. Its vast size is evidence that the crustal plate must have remained stationary over the hot spot. This is evidence that Mars has not had plate tectonics.

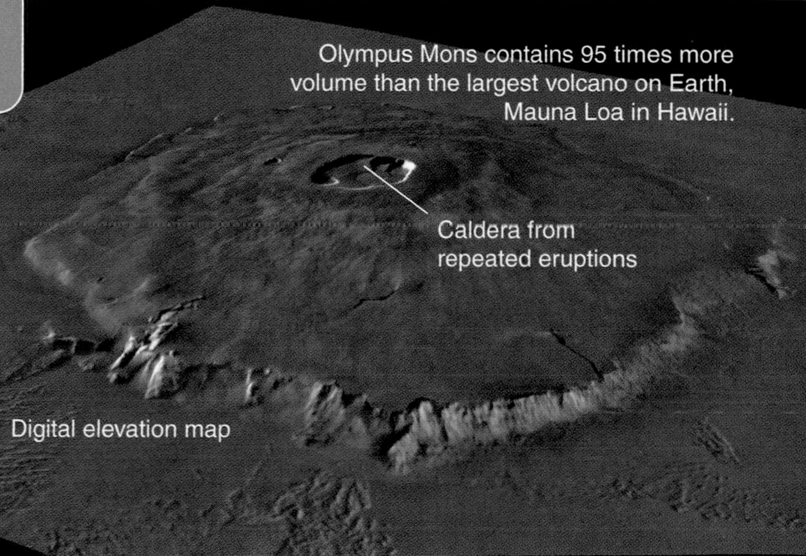

Olympus Mons contains 95 times more volume than the largest volcano on Earth, Mauna Loa in Hawaii.
Caldera from repeated eruptions
Digital elevation map

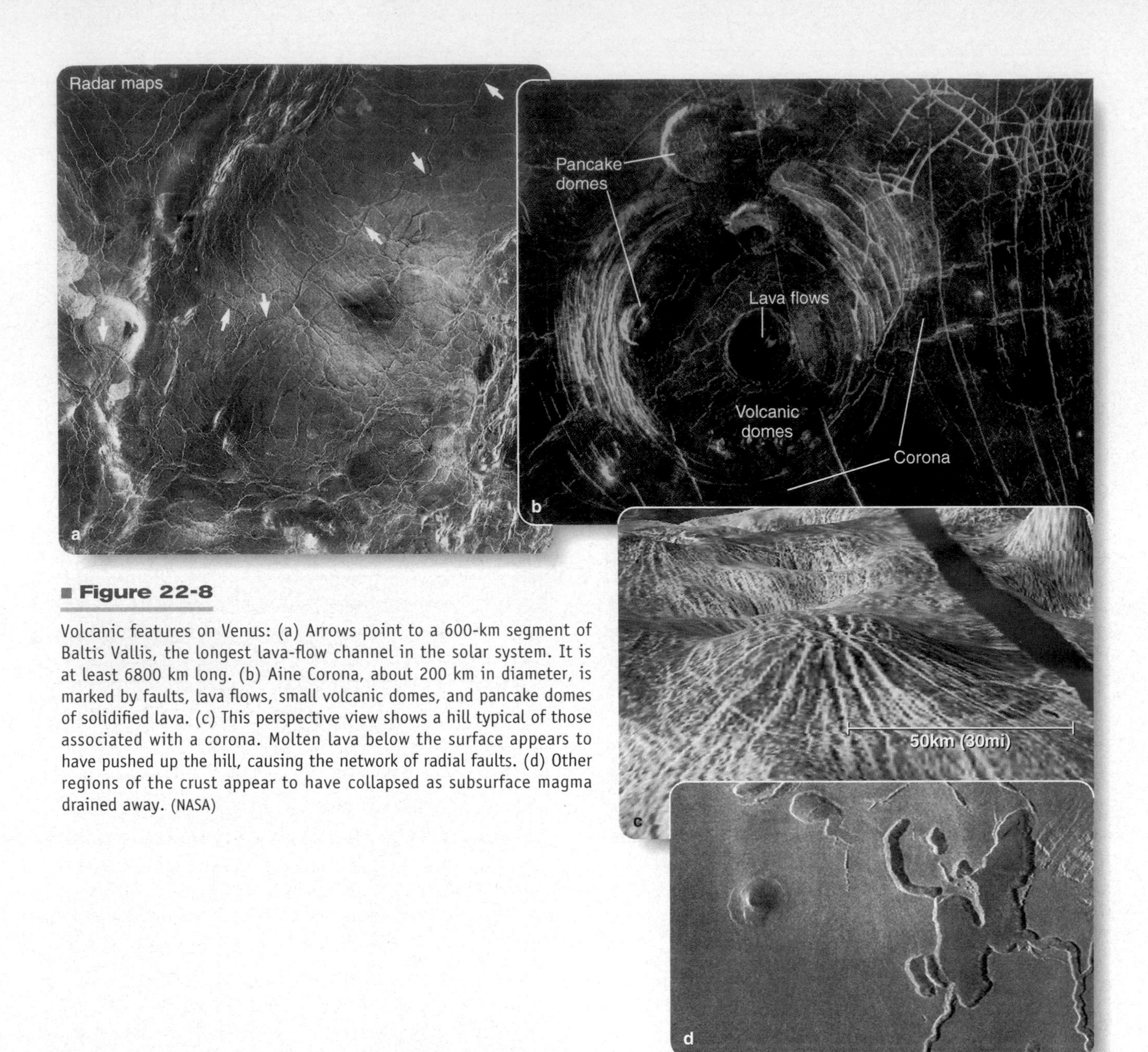

■ **Figure 22-8**

Volcanic features on Venus: (a) Arrows point to a 600-km segment of Baltis Vallis, the longest lava-flow channel in the solar system. It is at least 6800 km long. (b) Aine Corona, about 200 km in diameter, is marked by faults, lava flows, small volcanic domes, and pancake domes of solidified lava. (c) This perspective view shows a hill typical of those associated with a corona. Molten lava below the surface appears to have pushed up the hill, causing the network of radial faults. (d) Other regions of the crust appear to have collapsed as subsurface magma drained away. (NASA)

stone. Sapas Mons and other nearby volcanoes are located along a system of faults where rising magma broke through the crust.

In addition to the volcanoes, radar images reveal other volcanic features on the surface of Venus. Lava channels are common, and they appear similar to the sinuous rills visible on Earth's moon. The longest channel on Venus is also the longest known lava channel in the solar system. It stretches 6800 km (4200 miles), roughly twice the distance from Chicago to Los Angeles. These channels are 1 to 2 km wide and can sometimes be traced back to collapsed areas where lava appears to have drained from beneath the crust.

For further evidence of volcanism on Venus, you can look at circular features called **coronae,** circular bulges up to 2100 km in diameter containing volcanic peaks and lava flows. The coronae appear to be caused by rising currents of molten magma below the crust that create an uplifted dome and then withdraw to allow the surface to subside and fracture. Coronae are sometimes accompanied by features called pancake domes, circular outpourings of viscous lava, and by domes and hills pushed up by molten rock below the surface. All of these volcanic features are shown in ■ Figure 22-8.

There is no reason to suppose that all of these volcanoes are extinct, so volcanoes may be erupting on Venus right now. However, the radar maps caught no evidence of actual eruptions in progress.

If you want to learn more of the secrets of Venus, you will want to visit the huge land mass called Ishtar Terra. Western Ishtar consists of a high volcanic plateau called Lakshmi Planum. It rises 4 km above the rolling plains and appears to have formed from lava flows from volcanic vents such as Collette and Sacajawea (see Figure 22-5). Although folded mountain ranges do not occur on Venus as they do on Earth, the areas north and west of Lakshmi Planum appear to be ranges of wrinkled mountains where horizontal motion of the crust has pushed up against Ishtar Terra. Furthermore, faults and deep chasms are widespread over the surface of Venus, and that suggests stretching of the crust in those areas. So, although no direct evidence of plate motion is visible, there has certainly been compression and wrinkling of the crust in some areas and stretching and faulting in other areas, and that suggests some limited crustal motion.

While Earth's crust has broken into rigid moving plates, the surface of Venus seems more pliable and does not break easily into plates. The history of Venus has been dominated by volcanism. The story of Venus is a fiery tale.

ThomsonNOW Sign in at www.thomsonedu.com and go to ThomsonNOW to see Astronomy Exercise "Convection and Plate Tectonics."

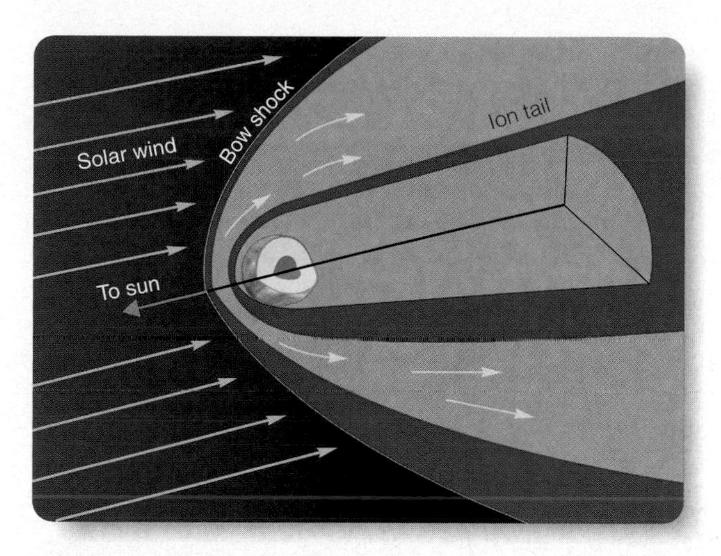

■ **Figure 22-9**

By analogy with Earth, the interior of Venus should contain a molten core (estimated here), but no spacecraft has detected a planetary magnetic field. Thus, Venus is unprotected from the solar wind, which strikes the planet's upper atmosphere and is deflected into an ion tail.

A History of Venus

Earth passed through four major stages in its history (Chapter 20), and you have seen how the moon and Mercury were affected by the same stages. Venus, however, has had a peculiar passage through the four stages of planetary development, and its history is difficult to understand. Planetary scientists are not sure how the planet formed and differentiated, how it was cratered and flooded, or how its surface has evolved.

Venus formed only slightly closer to the sun than did Earth, so you might expect it to be a similar planet that has differentiated into a silicate mantle and a molten iron core. The density and size of Venus require that it have a dense interior much like Earth's; but, if the core is indeed molten, then you would expect the dynamo effect to generate a magnetic field. No spacecraft has detected a magnetic field around Venus. The magnetic field must be at least 25,000 times weaker than Earth's. Some theorists wonder if the core of the planet is solid. If it is solid, then how did Venus get rid of its internal heat so much faster than Earth did?

Because the planet lacks a magnetic field, it is not protected from the solar wind. The solar wind slams into the uppermost layers of Venus' atmosphere, forming a bow shock where the wind is slowed and deflected (■ Figure 22-9). Planetary scientists know little about the differentiation of the planet into core and mantle, so the size of the core shown in Figure 22-9 is estimated by analogy with Earth's. The magnetic field carried by the solar wind drapes over Venus like seaweed over a fishhook, forming a long tail within which ions flow away from the planet. You will see in Chapter 25 that comets, which also lack magnetic fields, interact with the solar wind in the same way.

Studies of moon rocks show that the moon formed as a sea of magma, and presumably, Venus and Earth formed in the same way and never had primeval atmospheres rich in hydrogen. Rather they outgassed carbon dioxide atmospheres as they formed. Calculations show that Venus and Earth have outgassed about the same amount of carbon dioxide, but Earth's oceans have dissolved it and converted it to sediments such as limestone. The main difference between Earth and Venus is the lack of water on Venus. Venus may have had small oceans when it was young; but, being closer to the sun, it was warmer, and the carbon dioxide in the atmosphere created a greenhouse effect that made the planet even warmer. That process could have dried up any oceans that did exist and reduced the ability of the planet to clear its atmosphere of carbon dioxide. As more carbon dioxide was outgassed, the greenhouse effect grew even more severe. Venus became trapped in a runaway greenhouse effect.

Fully 70 percent of the heat from Earth's interior flows outward through volcanism along midocean ridges. But Venus lacks crustal rifts, and even its numerous volcanoes cannot carry much heat out of the interior. Rather, Venus seems to get rid of its interior heat through large currents of hot magma that rise beneath the crust. Coronae, lava flows, and volcanism occur above such currents. The surface rock on Venus is the same kind of dark-gray basalt found in ocean crust on Earth.

True plate tectonics is not important on Venus. For one thing, the crust is very dry and is consequently about 12 percent less dense than Earth's crust. This low-density crust is more buoyant than Earth's crust and resists being pushed into the interior. Also, the crust is so hot it is halfway to its melting point. Such hot rock is not very stiff, so it cannot form the rigid plates typical of plate tectonics on Earth.

There is no sign of plate tectonics on Venus, but there is evidence that convection currents below the crust are deforming the crust to create coronae and push up mountains such as Maxwell. Detailed measurements of the strength of gravity over

Venus's mountains show that some must be held up not by deep roots like mountains on Earth but by rising currents of magma. Other mountains, like those around Ishtar Terra, appear to be folded mountains caused by limited horizontal motions in the crust, driven perhaps by convection currents in the mantle.

The small number of craters on the surface of Venus hints that the entire crust has been replaced within the last half-billion years or so. This may have occurred in a planetwide overturning as the old crust broke up and sank and lava flows created a new crust. This could happen periodically on Venus, or the planet may have had a geology more like Earth's until a single resurfacing not too long ago. In either case, unearthly Venus may eventually reveal more about how our own world works.

ThomsonNOW Sign in at www.thomsonedu.com and go to ThomsonNOW to see Astronomy Exercise "Convection and Magnetic Fields."

◄ SCIENTIFIC ARGUMENT ►

What evidence can you point to that Venus does not have plate tectonics?

Sometimes a scientific argument can be helpful by eliminating a possibility. On Earth, plate tectonics is identifiable by the worldwide network of faults, subduction zones, volcanism, and folded mountain chains that outline the plates. Although some of these features are visible on Venus, they do not occur in a planetwide network that outlines plates. Volcanism is widespread, but folded mountain ranges occur in only a few places, such as near Lakshmi Planum and Maxwell Montes, and they do not make up long mountain chains as on Earth. Also, the large size of the shield volcanoes on Venus show that the crust is not moving over the hot spots as the Pacific seafloor is moving over the Hawaiian hot spot.

At first glance, you might think that Earth and Venus should be sister worlds, but comparative planetology reveals that they can be no more than distant cousins. You can blame the thick atmosphere of Venus for altering its geology, but that calls for a new scientific argument: **Why isn't Earth's atmosphere similar to that of Venus?**

◄ ►

22-2 Mars

MERCURY AND THE MOON ARE SMALL. Venus and Earth are, for terrestrial planets, large. But Mars occupies an intermediate position. It is twice the diameter of the moon but only 53 percent Earth's diameter **(Celestial Profile 6).** Its small size has allowed it to cool faster than Earth, and much of its atmosphere has leaked away. Its present carbon-dioxide atmosphere is only 1 percent as dense as Earth's.

The Canals on Mars

Long before the space age, the planet Mars was a mysterious landscape in the public mind. In the century following Galileo's first astronomical use of the telescope, astronomers discovered dark markings on Mars as well as bright polar caps. Timing the motions of the markings, they concluded that a Martian day was about 24 hours 40 minutes long. Its axis is tipped 23.5°, so it has seasons, and its year is only 1.88 Earth years long. The similarity with Earth encouraged the belief that Mars might be inhabited.

In 1858, the Jesuit astronomer Angelo Secchi referred to a region on Mars as *Atlantic Canale.* This is the first use of the Italian word *canale* (channel) to refer to a feature on Mars. Then, in the late summer of 1877, the Italian astronomer Giovanni

■ Figure 22-10

(a) Early in the 20th century, Percival Lowell mapped canals over the face of Mars and concluded that intelligent life resided there. (b) Modern images recorded by spacecraft reveal a globe of Mars with no canals. Instead, the planet is marked by craters and, in some places, volcanoes. Both of these images are reproduced with south at the top, as they appear in telescopes. Lowell's globe is inclined more nearly vertically and is rotated slightly to the right compared with the modern globe. (a, Lowell Observatory; b, U.S. Geological Survey)

Virginio Schiaparelli, using a telescope only 8.75 in. in diameter, thought he glimpsed fine, straight lines on Mars. He too used the Italian word *canali* (plural) for these lines, and the word was translated into English not as "channel," a narrow body of water, but as "canal," an artificially dug channel. The "canals of Mars" were born. Many astronomers could not see the canals at all, but others drew maps showing hundreds (■ Figure 22-10).

In the decades that followed Schiaparelli's discovery, many people assumed that the canals were watercourses built by an intelligent race to carry water from the polar caps to the lower latitudes. Much of this excitement was generated by Percival Lowell, a wealthy Bostonian who, in 1894, founded Lowell Observatory principally for the study of Mars. He not only mapped hundreds of canals but also popularized his results in books and lectures. Although some astronomers claimed the canals were merely illusions, by 1907 the general public was so sure that life existed on Mars that the *Wall Street Journal* suggested that the most extraordinary event of the previous year had been "the proof by astronomical observations . . . that conscious, intelligent human life exists upon the planet Mars." Further sightings of bright clouds and flashes of light on Mars strengthened this belief, and some urged that gigantic geometrical diagrams be traced in the Sahara Desert to signal to the Martians that Earth, too, is inhabited. All seemed to agree that the Martians were older and wiser than humans.

This fascination with men from Mars was not a passing fancy. Beginning in 1912, Edgar Rice Burroughs wrote a series of 11 novels about the adventures of the Earthman John Carter, lost on Mars. Burroughs made the geography of Mars, named by Schiaparelli after Mediterranean lands both real and mythical, into household words. He also made his Martians small and gave them green skin.

By Halloween night of 1938, people were so familiar with life on Mars that they were ready to believe that Earth could be invaded. When a radio announcer repeatedly interrupted dance music to report the landing of a spaceship in New Jersey, the emergence of monstrous creatures, and their destruction of whole cities, thousands of otherwise sensible people fled in panic, not knowing that Orson Welles and other actors were dramatizing H. G. Wells's book *The War of the Worlds.*

Public fascination with Mars, its canals, and its little green men lasted right up until July 15, 1965, when Mariner 4, the first spacecraft to fly past Mars, radioed back photos of a dry, cratered surface and proved that there are no canals and no Martians. The canals are optical illusions produced by the human brain's astounding ability to assemble a field of disconnected marks into a coherent image. If your brain could not do this, the photos on these pages would be nothing but swarms of dots, and the images on the screen of a television set would never make sense. The brain of an astronomer looking for something at the edge of visibility is capable of connecting faint, random markings on Mars into the straight lines of canals.

Mars is only half the diameter of Earth and probably retains some internal heat, but the size and composition of its core are not well known. (NASA)

Celestial Profile 6: Mars

Motion:

Average distance from the sun	1.5237 AU (2.279×10^8 km)
Eccentricity of orbit	0.0934
Maximum distance from the sun	1.6660 AU (2.492×10^8 km)
Minimum distance from the sun	1.3814 AU (2.066×10^8 km)
Inclination of orbit to ecliptic	1°51′09″
Average orbital velocity	24.13 km/s
Orbital period	1.8808 y (686.95 days)
Period of rotation	$24^h37^m22.6^s$
Inclination of equator to orbit	25°11′

Characteristics:

Equatorial diameter	6792 km (0.53 $D_\oplus$)
Mass	0.6424×10^{24} kg (0.1075 $M_\oplus$)
Average density	3.94 g/cm³ (3.3 g/cm³ uncompressed)
Surface gravity	0.379 Earth gravity
Escape velocity	5.0 km/s (0.45 $V_\oplus$)
Surface temperature	−140° to 20°C (−220° to 68°F)
Average albedo	0.16
Oblateness	0.009

Personality Point:

Mars is named for the god of war. Minerva was the goddess of defensive war, but Bullfinch's *Mythology* refers to Mars's "savage love of violence and bloodshed." You can see how the planet glows blood red in the evening sky because of iron oxides in its soil.

Even today, Mars holds some fascination for the general public. Grocery store tabloids regularly run stories about a giant face carved on Mars by an ancient race. Although planetary scientists recognize it as nothing more than chance shadows in a photograph and dismiss the issue as a silly hoax, the stories persist. A hundred years of speculation has raised high expectations for Mars. If there were intelligent life on Mars and its representatives came to Earth, they would probably be a big disappointment to the readers of the tabloids.

The Atmosphere of Mars

If you visited Mars, your first concern, even before you opened the door of your spaceship, would be the atmosphere. Is it breathable? Even for the astronomer observing safely from Earth, the atmosphere of Mars is a major concern. The gases that cloak Mars are critical to understanding the history of the planet.

The air on Mars is 95 percent carbon dioxide, with a few percent each of nitrogen and argon. The reddish color of the soil is caused by iron oxides (rusts), and that warns that the oxygen humans would prefer to find in the atmosphere is locked in chemical compounds in the soil. The Martian atmosphere contains almost no water vapor or oxygen, and its density at the surface of the planet is only 1 percent that of Earth's atmosphere. This does not provide enough pressure to prevent liquid water from boiling into vapor. If you stepped outdoors on Mars without a spacesuit, your own body heat would make your blood boil.

Although the air is thin, it is dense enough to be visible in photographs (■ Figure 22-11). Haze and clouds come and go, and occasional weather patterns are visible. Winds on Mars can be strong enough to produce dust storms that envelop the entire planet. The polar caps visible in photos are also related to the Martian atmosphere. The ices in the polar caps are frozen carbon dioxide ("dry ice") with frozen water underneath.

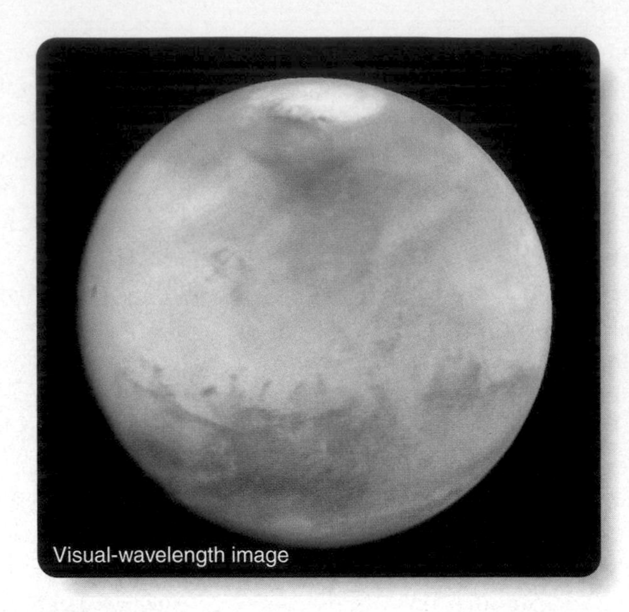

■ Figure 22-11

The atmosphere of Mars is evident in this image made by the Hubble Space Telescope. The haze is made up of high, water-ice crystals in the thin CO_2 atmosphere. The spot at extreme left is the volcano Ascraeus Mons, 25 km (16 mi) high, poking up through the morning clouds. Note the north polar cap at the top. (Philip James, University of Toledo; Steven Lee, University of Colorado, Boulder; and NASA)

If you could visit Mars you would find it a reddish, airless, bone-dry desert (■ Figure 22-12). To understand Mars, you should ask why its atmosphere is so thin and dry and why the surface is rich in oxides. To find those answers you must consider the origin and evolution of the Martian atmosphere.

Presumably, the gases in the Martian atmosphere were mostly outgassed from its interior. Volcanism on terrestrial planets typically releases carbon dioxide and water vapor, plus other gases. Because Mars formed farther from the sun, you might expect that it would have incorporated more volatiles when it formed. But

■ Figure 22-12

Mars is a red desert planet, as shown in this true-color photo made by the Rover Opportunity. The rock outcrop is a meter-high crater wall. After taking this photo, Opportunity descended into the crater to study the wall. Dust suspended in the atmosphere colors the sky red-orange. (NASA © Calvin J. Hamilton)

Mars is smaller than Earth, so it has had less internal heat to drive geological activity, and that would lead you to suspect that it has not outgassed as much as Earth. In any case, whatever outgassing took place occurred early in the planet's history, and Mars, being small, cooled rapidly and now releases little gas.

How much atmosphere a planet has depends on how rapidly it loses gas to space, and that depends on the planet's mass and temperature. The more massive the planet, the higher its escape velocity (Chapter 5), and the more difficult it is for gas atoms to leak into space. Mars has a mass less than 11 percent that of Earth, and its escape velocity is only 5 km/s, less than half Earth's. Consequently, gas atoms can escape from it much more easily than they can escape from Earth.

The temperature of a planet's atmosphere is also important. If a gas is hot, its molecules have a higher average velocity and are more likely to exceed escape velocity. That means a planet near the sun is less likely to retain an atmosphere than a more distant, cooler planet. The velocity of a gas molecule, however, also depends on the mass of the molecule. On average, a low-mass molecule travels faster than a massive molecule. For that reason, a planet loses its lowest-mass gases more easily because those molecules travel fastest.

You can see this principle of comparative planetology if you plot a diagram such as that in ■ Figure 22-13. The points show the escape velocity versus temperature for the larger objects in our solar system. The temperature used in the diagram is the temperature of the gas that is in a position to escape. For the moon, which has essentially no atmosphere, this is the temperature of the sunlit surface. For Mars, the temperature that is important is that at the top of the atmosphere.

The lines in Figure 22-13 show the typical velocities of the fastest-traveling examples of various molecules. At any given temperature, some water molecules, for example, travel faster than others, and it is the highest-velocity molecules that escape from a planet. The diagram shows that the Jovian planets have escape velocities so high that very few molecules can escape. Earth and Venus can't hold hydrogen, and Mars can hold only the more massive molecules. Earth's moon is too small to keep any gases from leaking away. Refer to this diagram again whenever you study the atmospheres of other worlds.

Over the 4.6 billion years since Mars formed, it has lost some of its lower-mass gases. Water molecules are massive enough for Mars to keep, but ultraviolet radiation can break them up. The hydrogen escapes, and the oxygen, a very reactive element, forms more oxides in the soil—the oxides that make Mars the red planet. Recall that on Earth, the ozone layer protects water vapor from ultraviolet radiation, but Mars never had an oxygen-rich atmosphere, so it never had an ozone layer. Ultraviolet photons from the sun can penetrate deep into the atmosphere and

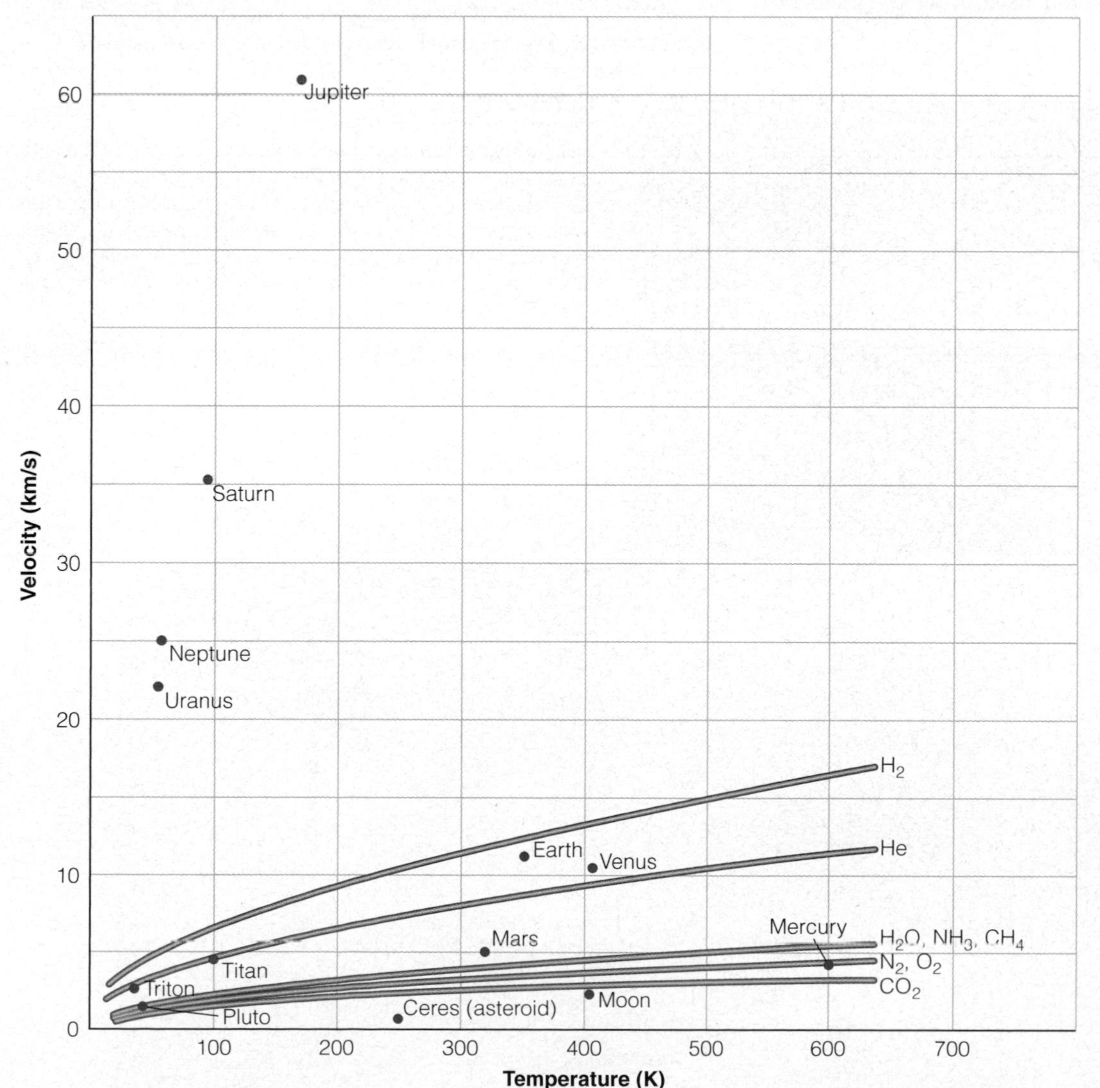

■ Figure 22-13

The loss of planetary gases. Dots represent the escape velocity and temperature of various solar-system bodies. The lines represent the typical highest velocities of molecules of various masses. The Jovian planets have high escape velocities and can hold on to even the lowest-mass molecules. Mars can hold only the more massive molecules, and the moon has such a low escape velocity that even the most massive molecules can escape.

break up molecules. In this way, molecules too massive to leak into space can be lost if they break into lower-mass fragments.

The argon in the Martian atmosphere is evidence that there once was a denser blanket of air. Argon atoms are massive, almost as massive as a carbon dioxide molecule, and would not be lost easily. In addition, argon is inert and cannot form compounds in the soil. The 1.6 percent argon in the atmosphere of Mars is evidently left over from an ancient atmosphere that was 10 to 100 times denser than the present Martian air.

Finally, you should consider the interaction of the solar wind with the atmosphere of Mars. This is not an important process for Earth because Earth has a magnetic field that deflects the solar wind. But Mars has no magnetic field, and the solar wind interacts directly with the Martian atmosphere. Detailed calculations show that significant amounts of carbon dioxide could have been carried away by the solar wind over the history of the planet. This process would have been most efficient long ago when the sun was more active and the solar wind was stronger. However, you should also keep in mind that Mars probably had a magnetic field when it was younger and still retained significant internal heat. A magnetic field would have protected its atmosphere from the solar wind.

The polar caps contain large amounts of carbon dioxide ice, and as spring comes to a hemisphere, that ice begins to vaporize and returns to the atmosphere. Meanwhile, at the other pole, carbon dioxide is freezing out and adding to the polar cap there. Dramatic evidence of this cycle appeared when the camera aboard the Mars Odyssey probe sent back images of dark markings on the south polar cap. Evidently as spring comes to the polar cap and the sun begins to peek above the horizon, sunlight penetrates the meter-thick ice and vaporizes carbon dioxide, which bursts out in geysers a few tens of meters high carrying dust and sand. Local winds push the debris downwind to form the fan-shaped dark markings (■ Figure 22-14). These dark markings appear each spring but last only a few months as frozen carbon dioxide returns to the atmosphere.

Although planetary scientists remain uncertain as to how much of an atmosphere Mars has had in its past and how much it has lost, it is a good example for your study of comparative planetology. When you look at Mars, you see what can happen to the atmosphere of a medium-size world. Like its atmosphere, the geology of Mars is typical of medium-size worlds.

ThomsonNOW Sign in at www.thomsonedu.com and go to ThomsonNOW to see Astronomy Exercise "Primary Atmospheres."

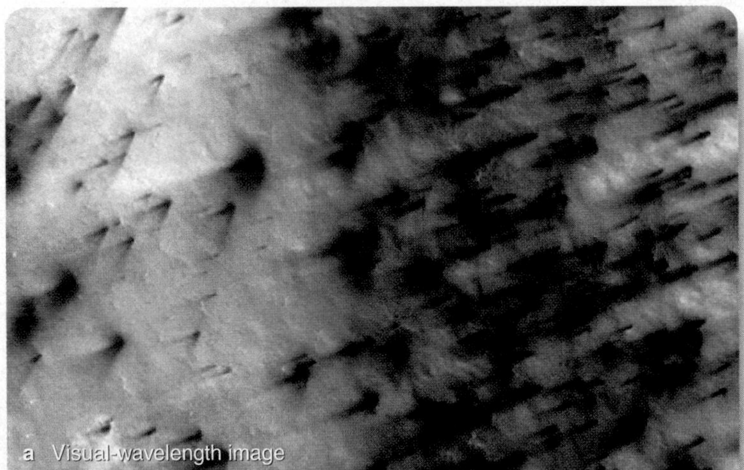

■ Figure 22-14

(a) Each spring spots and fans appear on the ice of the south polar cap on Mars. (b) Studies show the ice is frozen carbon dioxide in a nearly clear layer about a meter thick with high-pressure carbon dioxide gas vaporized by spring sunlight bursting out of the ice in geysers. At speeds of 100 miles an hour, the gas carries sand and dust hundreds of feet into the air. (NASA; NASA/Arizona State University/Ron Miller)

The Geology of Mars

If you ever decide to visit another world, Mars may be your best choice. You will need a heated, pressurized spacesuit with air, water, and food, but even so Mars is not as inhospitable as the moon. The nights on Mars are deadly cold, but a hot summer day would be comfortable (Celestial Profile 6). Mars is also more interesting than the moon because it has weather, complex geology, and signs that water once flowed over its surface. You might even hope to find traces of ancient life hidden in the rocks.

Spacecraft have been visiting Mars for almost 40 years, but the pace has picked up recently. A small armada of spacecraft has gone into orbit around Mars to photograph and analyze its surface, and five spacecraft have landed. Two Viking landers touched down in 1976, and three rovers have landed in recent years. Rovers have an advantage because they are wheeled robots that can be controlled from Earth and directed to travel from feature to feature and make detailed measurements. Pathfinder and its rover, Sojourner, landed in 1997. Rovers Spirit and Opportunity landed in January 2004 carrying sophisticated instruments to explore the rocky surface.

Photographs made by rovers on the surface of Mars, such as Figure 22-12, show reddish deserts of broken rock. These appear to be rocky plains fractured by meteorite impacts, but they don't look much like the surface of Earth's moon. The atmosphere of Mars, thin though it is, protects the surface from the blast of micrometeorites that grinds moon rocks to dust. Also, Martian dust storms may sweep fine dust away from some areas leaving larger rocks exposed.

Spacecraft orbiting Mars have imaged the surface and measured elevations to reveal that all of Mars is divided into two parts. The southern highlands are heavily cratered, and the number of craters there shows that they must be old. In contrast, the northern lowlands are smooth (■ Figure 22-15) and so remarkably free of craters that they must have been resurfaced about a billion years ago. Some astronomers have suggested that volcanic floods filled the northern lowlands and buried the craters there. Growing evidence, however, suggests that the northern lowlands were once filled by an ocean of liquid water. This is an exciting theory and will be discussed later when you consider the history of water on Mars.

The cratering and volcanism on Mars fit with what you already know of comparative planetology. Mars is larger than Earth's moon, so it cooled more slowly, and its volcanism has continued longer. But Mars is smaller than Earth and less geologically active, so some of its ancient cratered terrain has survived undamaged by volcanism and plate tectonics.

The Martian volcanoes are shield volcanoes with shallow slopes, showing that the lava flowed easily. Such volcanoes occur over hot spots of rising magma below the crust and are not related to plate tectonics. The largest volcano in the solar system is Olympus Mons on Mars (see page 491). The volcano Mauna Loa in Hawaii is so heavy it has sunk into Earth's crust to form an undersea depression like a moat around a castle. Olympus Mons is much larger than Mauna Loa but has not sunk into the crust of Mars, which is evidence that the crust of Mars is much thicker than the crust of Earth. You can see the evidence in ■ Figure 22-16.

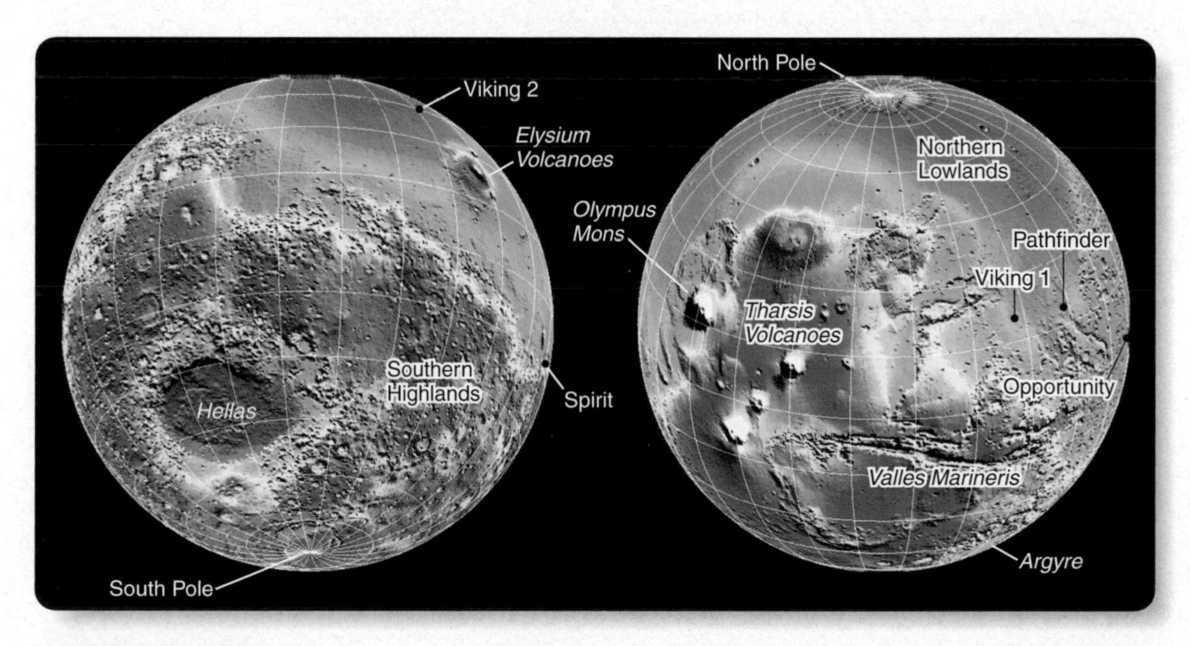

■ Figure 22-15

These globes of Mars are color coded to show elevation. The northern lowlands lie about 4 km below the southern highlands. Volcanoes are very high (white), and the giant impact basins, Hellas and Argyre, are low. Note the depth of the canyon Valles Marineris. The two Viking spacecraft landed on Mars in 1976. Pathfinder landed in 1997. Rovers Spirit and Opportunity landed in 2004. (NASA)

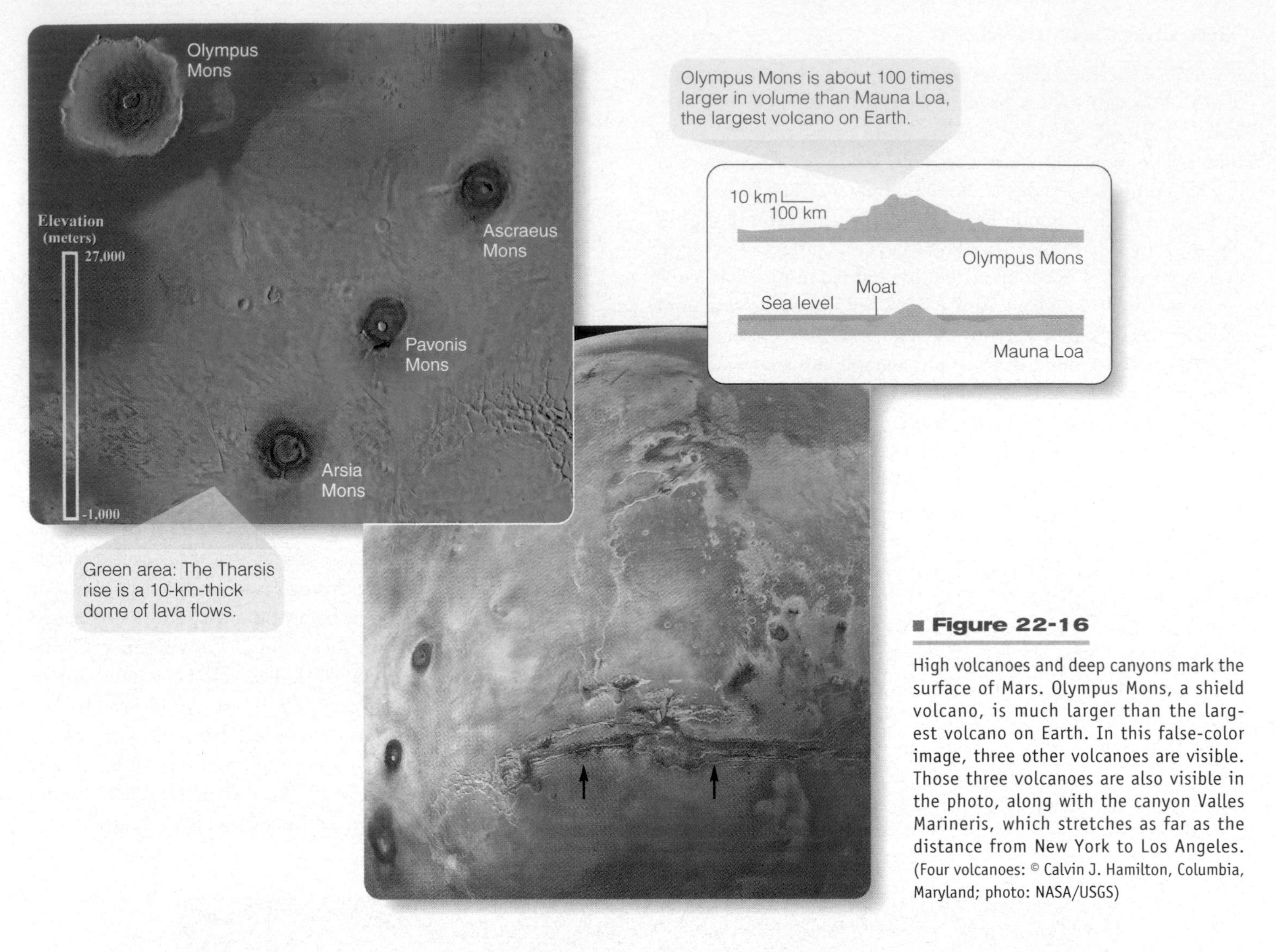
Olympus Mons
Ascraeus Mons
Pavonis Mons
Arsia Mons
Elevation (meters)
27,000
-1,000
Green area: The Tharsis rise is a 10-km-thick dome of lava flows.
Olympus Mons is about 100 times larger in volume than Mauna Loa, the largest volcano on Earth.
10 km
100 km
Olympus Mons
Moat
Sea level
Mauna Loa

Figure 22-16

High volcanoes and deep canyons mark the surface of Mars. Olympus Mons, a shield volcano, is much larger than the largest volcano on Earth. In this false-color image, three other volcanoes are visible. Those three volcanoes are also visible in the photo, along with the canyon Valles Marineris, which stretches as far as the distance from New York to Los Angeles. (Four volcanoes: © Calvin J. Hamilton, Columbia, Maryland; photo: NASA/USGS)

Other evidence shows that the Martian crust has been thinner and more active than the moon's. Valles Marineris is a network of canyons 4000 km (2500 mi) long and up to 600 km (400 mi) wide (Figure 22-16). At its deepest, it is four times deeper than the Grand Canyon on Earth, and it is long enough to stretch from New York to Los Angeles. The canyon has been produced by faults in the crust that allowed great blocks to sink. Later landslides and erosion modified the canyon further. Although Valles Marineris is an old feature, it does show that the crust of Mars has been more active than the crusts of the moon or Mercury, worlds that lack such dramatic canyons.

The faults that created Valles Marineris seem to be linked at the western end to a great volcanic bulge in the crust of Mars called the Tharsis rise. Nearly as large as the United States, the Tharsis rise extends 10 km (6 mi) above the mean radius of Mars. Tharsis is home to many smaller volcanoes, but on its summit lie three giants, and just off of its northwest edge lies Olympus Mons (Figure 22-16). The origin of the Tharsis rise is not well understood, but it appears that magma rising from below the crust has pushed the crust up and broken through repeatedly to build a giant bulge of volcanic deposits. This bulge is large enough to have modified the climate and seasons on Mars and may be critical in understanding the history of the planet.

A similar uplifted volcanic bulge, the Elysium region, visible in Figure 22-15, lies halfway around the planet. It appears to be similar to the Tharsis rise, but it is more heavily cratered and so must be older.

The vast sizes of features like the Tharsis rise and Olympus Mons show that the crust of Mars has not been broken into moving plates. If a plate were moving over a hot spot, the rising magma would produce a long chain of shield volcanoes and not a single large peak (page 491). On Earth, the hot spot that creates the volcanic Hawaiian Islands has punched through the moving Pacific plate repeatedly to produce the Hawaiian-Emperor island chain extending 7500 km (4700 mi) northwest across the Pacific seafloor (page 452). No such chains of volcanoes are evident on Mars, so you can conclude that the crust is not divided into moving plates.

No spacecraft has ever photographed an erupting volcano on Mars, but it is possible that some of the volcanoes are still active.

Impact craters in the youngest lava flows in the Elysium region show that the volcanoes may have been active as recently as a few million years ago. Mars may still retain enough heat to trigger an eruption, but the interval between eruptions could be very long.

Finding the Water on Mars

The quest for water on Mars is exciting because water has been deeply involved in the evolution of the planet, but it is also exciting because life depends on water. If life managed to begin on Mars, then the planet must have had water, and if life survives there, water must be hidden somewhere on the desert planet.

The two Viking spacecraft reached orbit around Mars in 1976 and photographed exciting hints that water once flowed over the surface. You would not expect liquid water on the surface now because it would boil away under the extremely low atmospheric pressure, so the Viking photos were evidence that water once flowed on Mars long ago. More recent missions to Mars such as Mars Global Surveyor, which reached Mars in 1997, Mars Odyssey (2001), Mars Express (2003), and Mars Reconnaissance Orbiter (2005) have identified additional features related to water.

Two kinds of formations hint at water flowing over the surface. **Outflow channels** appear to have been cut by massive floods carrying as much as 10,000 times the water flowing down the Mississippi River (■ Figure 22-17a). In a matter of hours or days, such floods swept away geological figures and left outflow channels. The number of craters formed on top of the outflow channels show that they are billions of years old. In contrast, the **valley networks** look like meandering riverbeds that may have formed over long periods (Figure 22-17b). The valley networks are located in the old, cratered, southern hemisphere, and they are very old.

Many flow features lead into the northern lowlands, and the smooth terrain there has been interpreted as ancient ocean floor. Features along the edges of the lowlands have been compared to shorelines, and many planetary scientists conclude that the lowlands were filled by an ocean when Mars was younger. Large, generally circular depressions such as Hellas and Arqyre appear to be impact basins once flooded by water.

Mars Orbiter photographed the eroded remains of a river delta in an unnamed crater in the old highlands (■ Figure 22-18). Details show that the river flowed for long periods of time, shifting its channel to form meanders and braided channels. The shape of the delta suggests it formed when the river flowed into deeper water and dropped its sediment, much as the Mississippi drops its sediment and builds its delta in the Gulf of Mexico.

Did Mars once have that much water? Deuterium is 5.5 times more abundant on Mars than on Earth relative to normal hydrogen, and that suggests that Mars once had about 20 times more water than it has now. Presumably, much of the water was broken up and the hydrogen lost to space.

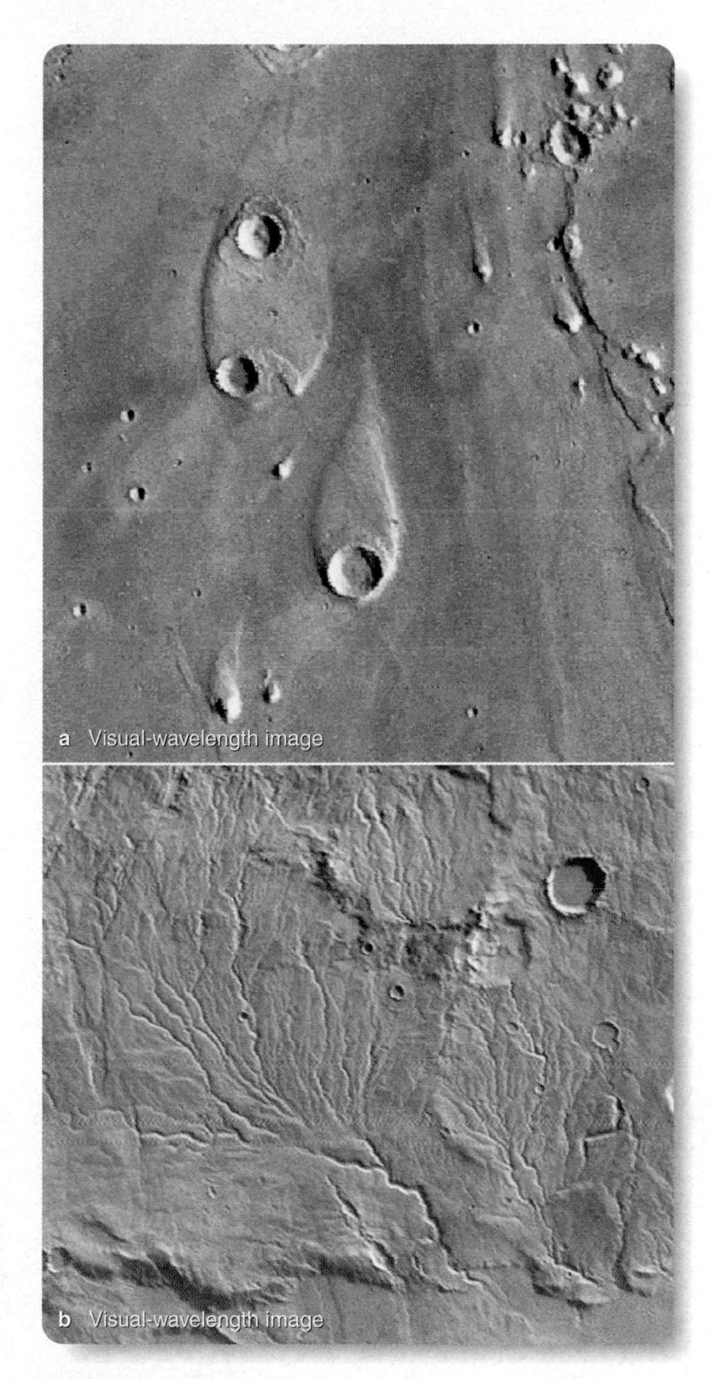

■ **Figure 22-17**

These visual-wavelength images made by the Viking orbiters show some of the features that suggest liquid water on Mars. (a) Outflow channels are broad and shallow and deflect around obstructions such as craters. They appear to have been produced by sudden floods. (a) Valley networks resemble drainage patterns and suggest water flowing over long periods. Crater counts show that both formations are old, but valley networks are older than outflow channels. (NASA © Calvin J. Hamilton)

The remaining water on Mars could survive if it were frozen in the crust. High resolution images and measurements made from orbit reveal features that suggest subsurface ice. Gullies leading downhill appear to have been eroded recently, judging from their lack of craters; these may have been formed by water seeping from below the surface. Some regions of collapsed terrain appear to be places where subsurface water has drained away. Photos taken over a period of years reveal the appearance of recent landslides that may have been caused by water gushing from crater walls and carrying debris downhill before vaporizing completely. Instruments aboard the Mars Odyssey spacecraft detected water frozen in the soil over large areas of the planet. At latitudes farther than 60 degrees from the equator, water ice may make up more than 50 percent of the surface soil.

If you added a polar bear, changed the colors, and hid the craters in ■ Figure 22-19a, it would look like the broken pack ice on Earth's Arctic Ocean. Mars Express photographed these dust-covered formations near the Martian equator, and the shallow depth of the craters suggests that the ice is still there just below the surface.

Much of the ice on Mars may be hidden below the polar caps. Radar aboard the Mars Express orbiter was able to penetrate 3.7 kilometers below the surface and map ice deposits hidden below the south polar region (Figure 22-19b). At least 90 percent pure, the water could cover the entire planet to a depth of 11 meters.

■ Figure 22-18

This distributary fan was formed where an ancient stream flowed into a flooded crater. Sediment in the moving water was deposited in the still water to form a lobed delta that later became sedimentary rock. Detailed analysis reveals that the stream changed course time after time, and that shows that the stream flowed for an extended period and was not a short-term flood. Also, the shape of the fan is evidence that a lake persisted in the crater while the fan developed. (NASA/JPL/Malin/Space Science Systems)

■ Figure 22-19

(a) Like broken pack ice, these formations near the equator of Mars suggest floating ice that broke up and drifted apart. Scientists propose that the ice was covered by a protective layer of dust and volcanic ash and may still be present. (b) Radar aboard the Mars Express satellite probed beneath the surface to image water ice below the south polar cap. The black circle is the area that could not be studied from the satellite's orbit. (a. ESA/DLR/F. U. Berlin/G. Neukum; b. NASA/JPL/ESA/Univ. of Rome/MOLA Science Team/USGS)

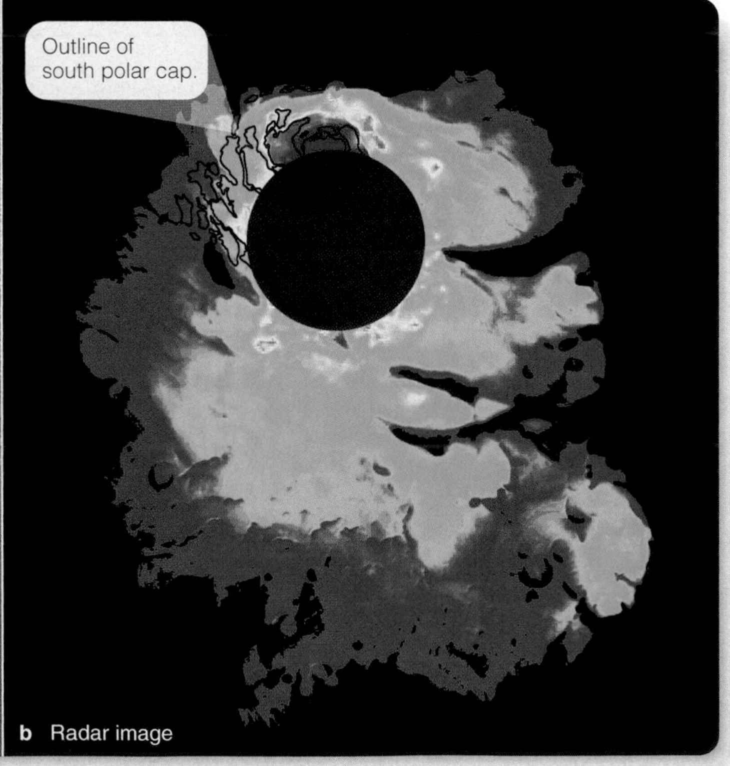

Rovers Spirit and Opportunity were targeted to land in areas suspected of having had water on their surfaces. Images made from orbit showed flow features at the Spirit landing site, and hematite, a mineral that forms in water, was detected from orbit at the Opportunity landing site. Both rovers reported exciting discoveries, including evidence of past water. Using its analytic instruments, Opportunity found small spherical concretions (dubbed blueberries, although they are, in fact, hematite) that must have formed in water. Later, Spirit found similar concretions in its area. In other rocks, Opportunity found layers of sediments with ripple marks and crossed layers showing they were deposited in moving water (■ Figure 22-20). Chemical analysis of the rocks at the Opportunity site showed the presence of sulfates much like Epsom salts plus bromides and chlorides. On Earth, these compounds are left behind when bodies of water dry up. Halfway around Mars, Spirit found the mineral goethite, clear evidence that water was once present.

One astonishing bit of evidence of water on Mars is the analysis of rock samples from the planet. Of course, no astronaut has ever visited Mars and brought back a rock, but over the history of the solar system occasional impacts by asteroids have blasted giant craters on Mars and ejected bits of rock into space. A few of those bits of rock have fallen to Earth as meteorites, and over 30 have been found and identified (■ Figure 22-21). These meteorites include basalts, as you might expect from a planet so heavily covered by lava flows. They also contain small traces of water and minerals that are deposited by water. Chemical analysis shows that the magma from which the rocks solidified must have contained up to 1.8 percent water. If all of the lava flows on Mars contained that much water and it was all outgassed, it could create a planetwide ocean 20 m deep. That isn't quite enough to explain all the flood features on Mars, but it does assure that the planet once had more water on its surface than it does now.

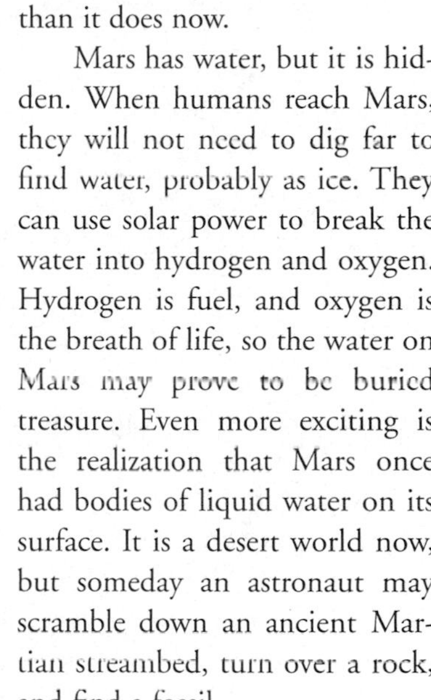

Mars has water, but it is hidden. When humans reach Mars, they will not need to dig far to find water, probably as ice. They can use solar power to break the water into hydrogen and oxygen. Hydrogen is fuel, and oxygen is the breath of life, so the water on Mars may prove to be buried treasure. Even more exciting is the realization that Mars once had bodies of liquid water on its surface. It is a desert world now, but someday an astronaut may scramble down an ancient Martian streambed, turn over a rock, and find a fossil.

■ Figure 22-20

(a) Rover Opportunity photographed these hematite concretions in a rock near its landing site. The spheres appear to have grown as minerals collected around small crystals in the presence of water. Similar concretions are found on Earth. (b) The layers in this rock were deposited as sand and silt in rapidly flowing water. From the way the layers curve and cross each other, geologists can estimate that the water was at least ten centimeters deep. A few "blueberries" and one small pebble are also visible in this image. (NASA. JPL/Cornell/USGS)

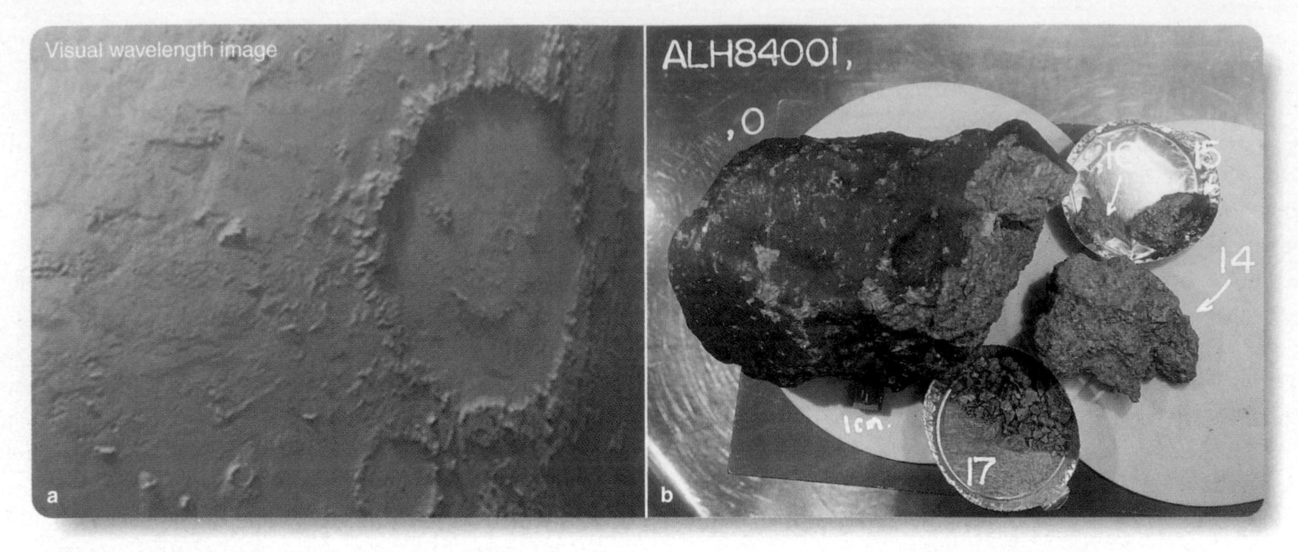

■ **Figure 22-21**

(a) The impact crater Galle, known for obvious reasons as the Happy Face Crater, is 210 km in diameter. Such large impacts may occasionally eject fragments of the crust into space. A few fragments from Mars have fallen to Earth as meteorites. (Malin Space Science Systems/NASA) (b) Meteorite ALH84001, found in Antarctica, has been identified as an ancient rock from Mars. Minerals found in the meteorite were deposited in water and thus suggest that the Martian crust was once richer in water than it is now. (NASA)

A History of Mars

The history of Mars is like a play where all the exciting stuff happens in the first act. After the first two billion years on Mars, most of the activity was over, and it has gone downhill ever since.

Planetary scientists have good evidence that Mars differentiated when it formed and had a hot, molten core. Some of the evidence comes from exquisitely sensitive Doppler-shift measurements of radio signals coming from spacecraft orbiting Mars. Measuring the shifts allowed scientists to map the gravitational field and study the shape of Mars in such detail that they could detect tides on Mars caused by the sun's gravity. Those tides are less than a centimeter high; but, by comparing them with models of the interior of Mars, the scientists can show that Mars has a very dense core, a dense mantle, and a low-density crust.

Observations made from orbit show that Mars has no overall magnetic field, but it does have traces of magnetism frozen into some sections of old crust. This shows that soon after Mars formed, it had a hot metallic core in which the dynamo effect generated a magnetic field. Because Mars is small, it lost its heat rapidly, and the central part of its core gradually froze solid. That may be why the dynamo finally shut down. Today Mars probably has a large solid core surrounded by a thin shell of liquid core material in which the dynamo effect is unable to generate a magnetic field.

No one is sure what produced the dramatic difference between the southern highlands and northern lowlands. Powerful convection in the planet's mantle may have pushed crust together to form the southern highlands. The suggestion that a catastrophic impact produced the northern lowlands does not seem to fit with the structure of the highlands, but it is not impossible.

Planetary scientists divide the history of Mars into three periods. The first, the **Noachian period,** extends from the formation of the crust about 4.3 billion years ago until roughly 3.7 billion years ago. During this time the crust was battered by the heavy bombardment as the last of the debris in the young solar system was swept up. The old southern hemisphere survives from this period. The largest impacts blasted out the great basins like Hellas and Argyre very late in the cratering, and there is no trace of magnetic field in those basins. Evidently the dynamo had shut down by then.

The Noachian period included flooding by great lava flows that smoothed some regions. Volcanism in the Tharsis and Elysium regions was very active, and the Tharsis rise grew into a huge bulge on the side of Mars (■ Figure 22-22). Some of the oldest lava flows on Mars are in the Tharsis rise, but it also contains some of the most recent lava flows. It has evidently been a major volcanic area for most of the planet's history.

The valley networks found in the southern highlands were formed during the Noachian period as water fell as rain or snow and drained down slopes. This requires a higher temperature and higher atmospheric pressure to keep the water liquid. Violent volcanism could have vented gases, including more water vapor that kept the air pressure high. This may have produced episodes in which water flowed over the surface and collected in the northern lowlands and in the deep basins to form oceans and lakes, but it's not known how long those bodies of water survived. A planetwide magnetic field may have protected the atmosphere from the solar wind.

■ **Figure 22-22**

The Tharsis rise is a bulge on the side of Mars consisting of stacked lava flows extending up to 10 km high and over a third the diameter of Mars. It dominates this image of Mars, with a few clouds clinging to the four largest volcanoes. This image was recorded in late winter for the southern hemisphere. Notice the size of the southern polar cap. (NASA/JPL/Malin Space Science Systems)

Because Mars is small, it lost its internal heat quickly, and atmospheric gases escaped into space. The **Hesperian period** extends from roughly 3.7 billion years ago to about 3 billion years ago. During this time massive lava flows covered some sections of the surface. Most of the outflow channels date from this period, which suggests that the loss of atmosphere drove Mars to become a deadly cold desert world with its water frozen in the crust. When volcanic heat or large impacts melted subsurface ice, the water could have produced violent floods and shaped the outflow channels.

The history of Mars may hinge on climate variations. Recently calculated models suggest that Mars may have once rotated at a much steeper angle to its orbit, as much as 45°. This could have supported a warmer climate and kept more of the carbon dioxide from freezing out at the poles. The rise of the Tharsis bulge could have tipped the axis to its present 25° and cooled the climate. Some calculations suggest that Mars goes through cycles as its rotational inclination fluctuates, and this may cause short-term variations in climate, much like the ice ages on Earth.

The third period in the history of Mars, the **Amazonian period,** extends from about 3 billion years ago to the present and is mostly uneventful. The planet has lost much of its internal heat, and the core no longer generates a planetwide magnetic field. The crust of Mars is too thick to be active with plate tectonics, and consequently there are no folded mountain ranges on Mars as there are on Earth. The huge size of the Martian volcanoes clearly shows that crustal plates have not moved on Mars. Repeated eruptions have built the volcanoes to vast size. Volcanism may still occur occasionally on Mars, but the crust has grown too thick for much geological activity beyond slow erosion by wind-borne dust and the occasional meteorite impact.

Planetary scientists cannot tell the story of Mars in great detail, but it is clear that the size of Mars has influenced both its atmosphere and its geology. Neither small nor large, Mars is a medium world.

◄ SCIENTIFIC ARGUMENT ►

Why doesn't Mars have coronae like those on Venus?

This argument is a good opportunity to apply the principles of comparative planetology. The coronae on Venus are caused by rising currents of molten magma in the mantle pushing upward under the crust and then withdrawing to leave the circular scars called coronae. Earth, Venus, and Mars have had significant amounts of internal heat, and there is plenty of evidence that they have had rising convection currents of magma under their crusts. Of course, you wouldn't expect to see coronae on Earth; its surface is rapidly modified by erosion and plate tectonics. Furthermore, the mantle convection on Earth seems to produce plate tectonics rather than coronae. Mars, however, is a smaller world and must have cooled faster. There is no evidence of plate tectonics on Mars, and giant volcanoes suggest rising plumes of magma erupting up through the crust at the same point over and over. Perhaps there are no coronae on Mars because the crust of Mars rapidly grew too thick to deform easily over a rising plume. On the other hand, perhaps you could think of the entire Tharsis bulge as a single giant corona.

Planetary scientists haven't explored enough planets yet to see all the fascinating combinations nature has in store. But it does seem likely that the geology of Mars is typical of medium-size worlds. Of course, Mars is not medium in terms of its location. Of the terrestrial planets, Mars is the farthest from the sun. Build a scientific argument to analyze that factor. **How has the location of Mars affected the evolution of its atmosphere?**

◄ ►

22-3 The Moons of Mars

IF YOU COULD CAMP OVERNIGHT ON MARS, you might notice its two small moons, Phobos and Deimos. Phobos, shaped like a flattened loaf of bread measuring 20 km × 23 km × 28 km, would appear less than half as large as Earth's full moon. Deimos, only 12 km in diameter and three times farther from Mars, would look only $\frac{1}{15}$ the diameter of Earth's moon.

Both moons are tidally locked to Mars, keeping the same side facing the planet as they orbit. Also, both moons revolve around Mars in the same direction that Mars rotates, but Phobos follows such a small orbit that it revolves faster than Mars rotates. If you camped overnight on Mars, you would see Phobos rise in the west, drift eastward across the sky and set in the east 6 hours later.

Origin and Evolution

Deimos and Phobos are typical of the small, rocky moons in our solar system (■ Figure 22-23). Their albedos are only about 0.06, making them look as dark as coal. They have low densities, about 2 g/cm^3.

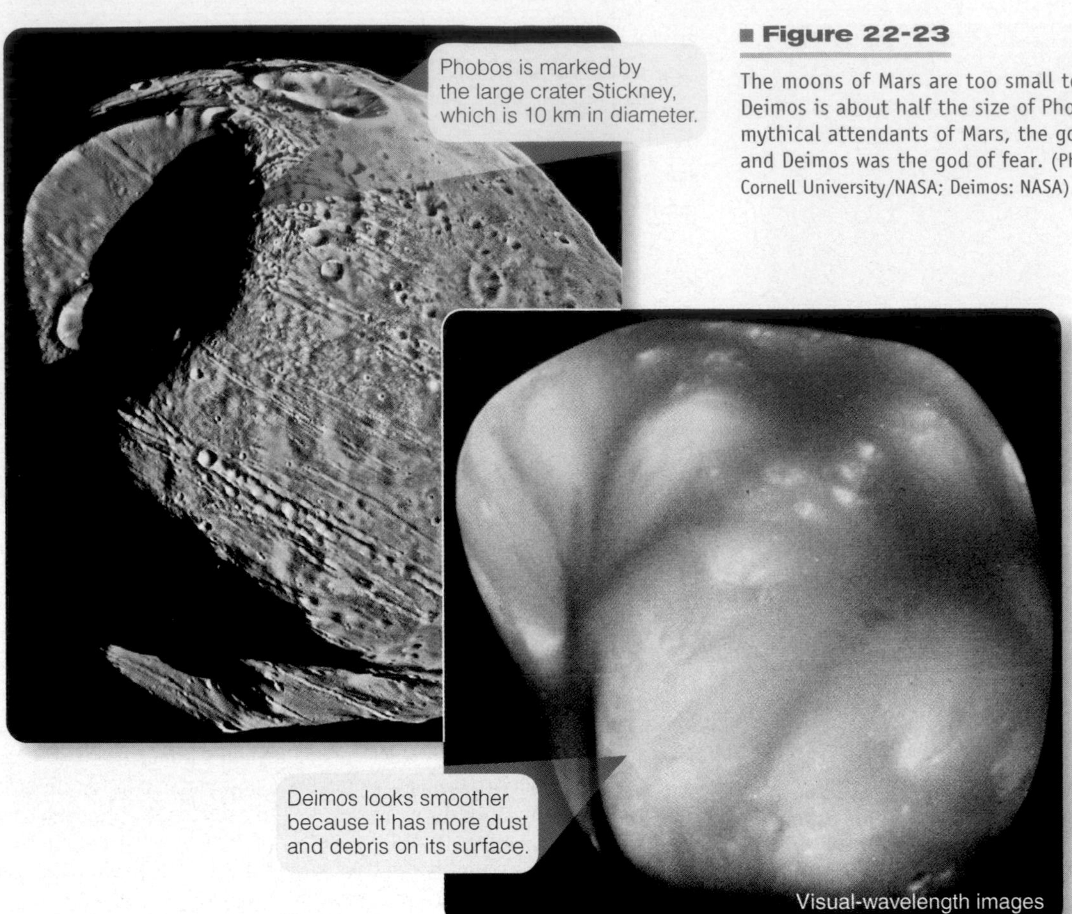

Phobos is marked by the large crater Stickney, which is 10 km in diameter.
Deimos looks smoother because it has more dust and debris on its surface.
Visual-wavelength images

■ Figure 22-23

The moons of Mars are too small to pull themselves into spherical shape. Deimos is about half the size of Phobos. The two moons were named for the mythical attendants of Mars, the god of war. Phobos was the god of panic, and Deimos was the god of fear. (Phobos: Damon Simonelli and Joseph Ververka, Cornell University/NASA; Deimos: NASA)

Many of the properties of these moons suggest that they are captured asteroids. In the outer parts of the asteroid belt, almost all asteroids are dark, low-density objects. Massive Jupiter, orbiting just outside the asteroid belt, can scatter such bodies throughout the solar system, so you should not be surprised if Mars, the closest terrestrial planet to the asteroid belt, has captured a few of these as satellites.

However, capturing a passing asteroid into a closed orbit is not so easy that it happens often. An asteroid approaches a planet along a hyperbolic (open) orbit and, if it is unimpeded, swings around the planet and disappears back into space. To convert the hyperbolic orbit into a closed orbit, the planet must slow the asteroid as it passes. Tidal forces might do this, but in the case of Mars they would be rather weak. Interactions with other moons or grazing collisions with a thick atmosphere might also slow the asteroid enough so it could be captured.

Both satellites have been photographed by spacecraft, and those photos show that the satellites are heavily cratered. Such cratering could have occurred either while the moons were still in the asteroid belt or while they were in orbit around Mars. In any case, the heavy battering has broken the satellites into irregular chunks of rock, and they cannot pull themselves into smooth spheres because their gravity is too weak to overcome the structural strength of the rock. You will discover that low-mass moons are typically irregular in shape, whereas more massive moons are more spherical.

Images of Phobos reveal a unique set of narrow, parallel grooves (see Figure 22-23). Averaging 150 m wide and 25 m deep, the grooves run from Stickney, the largest crater, to an oddly featureless region on the opposite side of the satellite. One theory suggests that the grooves are deep fractures produced by the impact that formed the crater. The featureless region opposite Stickney may be similar to the jumbled terrain found on Earth's moon and on Mercury. All these regions were produced by the focusing of seismic waves from a major impact on the far side of the body. High-resolution photographs show that the grooves are lines of pits, suggesting that the pulverized rock material on the surface has drained into the fractures or that gas, liberated by the heat of impact, escaped through the fractures and blew away the dusty soil.

Observations made with the Mars Global Surveyor's infrared spectrometer show that the moon's surface cools quickly from −4°C to −112°C (from 25°F to −170°F) as it passes from sunlight into the shadow of Mars. Solid rock would retain heat and cool more slowly. To cool as quickly as it does, the dust must be at least a meter deep and very fine. In most photos made by the spacecraft camera, the dust blankets the terrain, but some photos show boulders a few meters in diameter that are thought to be ejecta from impacts.

Deimos looks even smoother than Phobos because of a thicker layer of dust on its surface (Figure 22-23). This material partially fills craters and covers minor surface irregularities. It seems certain that Deimos experienced collisions in its past, so fractures may be hidden below the debris.

The debris on the surfaces of the moons raises an interesting question. How can the weak gravity of small bodies hold on to fragments from meteorite impacts? Escape velocity on Phobos is only 12 m/s. An athletic astronaut could almost jump into space.

What Are We?

Earthfolk

Space travel isn't easy. We humans made it to the moon, but it took everything we had in the late 1960s. Going back to the moon will be easier next time because the technology will be better, but it will still be expensive and will require people with heroic talent to design, build, and fly the spaceships. Going beyond the moon will be even more difficult.

Going to Mercury or Venus doesn't seem worth the effort. Mercury is barren and dangerous, and the heat and air pressure on Venus may prevent any astronaut from ever visiting its surface. In the next two chapters, you will discover that the Jovanian planets and their moons are not places humans are likely to visit soon. The stars are so far away they may be forever beyond the reach of human spaceships. But Earth has a neighbor.

Astronomically Mars is just up the street, and it isn't such a bad place. You would need a good spacesuit and a pressurized colony to live there, but it isn't impossible. Solar energy and water are abundant. It seems inevitable not only that humans will walk on Mars but that they will someday live there. We Earthfolk have an exciting future. We are the Martians.

Certainly, most fragments from impacts should escape, but some do fall back and accumulate on the surface.

Deimos, smaller than Phobos, has a smaller escape velocity. But it has more debris on its surface because it is farther from Mars. Phobos is close enough to Mars that most ejecta from impacts on Phobos will be drawn into Mars. Deimos, being farther from Mars, is able to keep a larger fraction of its ejecta. Phobos is so close to Mars that tides are making its orbit shrink, and it will fall into Mars or be ripped apart by tidal forces within about 100 million years.

Deimos and Phobos illustrate three principles of comparative planetology that you will find helpful as you explore farther from the sun. First, some satellites are probably captured asteroids. Second, small satellites tend to be irregular in shape and heavily cratered. And third, tidal forces can affect small moons and gradually change their orbits. You will find even stronger tidal effects in Jupiter's satellite system (Chapter 23).

◄ SCIENTIFIC ARGUMENT ►

Why would you be surprised if you found volcanism on Phobos or Deimos?

This is another obvious argument, isn't it? But remember, the purpose of a scientific argument is to test your own understanding, so it is a good way to review. In discussing Earth's moon, Mercury, Venus, Earth, and Mars, you have seen illustrations of the principle that the larger a world is, the more slowly it loses its internal heat. It is the flow of that heat from the interior through the surface into space that drives geological activity such as volcanism and plate motion. A small world, like Earth's moon, cools quickly and remains geologically active for a shorter time than a larger world like Earth. Phobos and Deimos are not just small, they are tiny. However they formed, any interior heat would have leaked away very quickly; with no energy flowing outward, there can be no volcanism.

Some futurists suggest that the first human missions to Mars will not land on the surface of the planet but will build a colony on Phobos or Deimos. These plans speculate that there may be water deep inside the moons that colonists could use. Build an argument based on what you know about water on Mars. **What would happen to water released in the sunlight on the surface of such small worlds?**

◄ ►

23 Comparative Planetology of Jupiter and Saturn

Saturn, nearly 10 times the diameter of Earth, is a planet of liquid hydrogen hidden below its cloudy atmosphere. Its rings are composed of billions of ice particles. (NASA, ESA and Erich Karkoschka [University of Arizona])

Guidepost

As you begin this chapter, you leave behind the psychological security of planetary surfaces. You can imagine standing on the moon, on Mars, or even on Venus, but Jupiter and Saturn have no surfaces. Here you face a new challenge—to use comparative planetology to study worlds so unearthly you cannot imagine being there. As you study these worlds you will find answers to five essential questions:

- **How do the outer planets compare with the inner planets?**
- **How is Jupiter different from Earth?**
- **How did Jupiter and its system of moons and rings form and evolve?**
- **How is Saturn different from Jupiter?**
- **How did Saturn and its system of moons and rings form and evolve?**

Your study of Jupiter and Saturn will give you a chance to answer two important questions about science.

- **How Do We Know? How is science different from technology?**
- **How Do We Know? Who pays to gather scientific knowledge?**

There's no place to stand on Jupiter or Saturn, but be sure to bring your spacesuit. Both planets have big systems of moons, and when you visit them you will be able to watch erupting volcanoes, stroll through a methane rain storm, and swim in Saturn's rings. It will be interesting, but it is no place like home.

There is something fascinating about science. One gets such wholesale returns of conjecture out of such a trifling investment of fact.

MARK TWAIN, *LIFE ON THE MISSISSIPPI*

When Mark Twain wrote the sentences that open this chapter, he was poking gentle fun at science, but he was right. The exciting thing about science isn't the so-called facts, the observations in which scientists have greatest confidence. Rather, the excitement lies in the understanding that scientists get by rubbing a few facts together. Science can take you to strange new worlds such as Jupiter and Saturn, and you can get to know them by combining the available observations with known principles of comparative planetology.

23-1 A Travel Guide to the Outer Planets

If you travel much, you know that some cities make you feel at home, and some do not. In this and the next chapter, you will visit worlds that are truly unearthly. You will not feel welcome, so this travel guide will warn you what to expect.

The Outer Planets

The outermost planets in our solar system are Jupiter, Saturn, Uranus, and Neptune—the Jovian planets, meaning they are like Jupiter. ■ Figure 23-1 compares the four outer worlds, and one striking feature is diameter. Jupiter is the largest of the Jovian worlds, over 11 times the diameter of Earth. Saturn is a bit smaller, but Uranus and Neptune are quite a bit smaller than Jupiter.

The other feature you will notice immediately when you look at Figure 23-1 is Saturn's rings. They are bright and beautiful and composed of billions of ice particles, each particle following its own orbit around the planet. Jupiter, Uranus, and Neptune also have rings, but they are not easily detected from Earth and are not visible in this figure. As you visit these worlds you will be able to compare four different sets of planetary rings.

Atmospheres and Interiors

All four Jovian worlds have hydrogen-rich atmospheres filled with clouds. On Jupiter and Saturn, you can see that the clouds form belts and zones that circle the planets like the stripes on a child's ball. This form of atmospheric circulation is called **belt–zone circulation.** You will find traces of belts and zones on Uranus and Neptune, but they are not very distinct.

The gaseous atmospheres of the Jovian planets are not very deep. Jupiter's atmosphere makes up only about one percent of its radius. Below that, Jupiter and Saturn are composed of liquid hydrogen, so the older term for these planets, the *gas giants*, reflects a **Common Misconception.** In fact they are made mostly of liquid rather than gas and could more correctly be called the *liquid giants*. Only near their centers could these worlds contain dense material with the composition of rock and metal, but the sizes of these cores are poorly known.

Uranus and Neptune are sometimes called the ice giants because they contain a great deal of water, both as a liquid and as a solid. Uranus and Neptune contain denser material in their cores.

On your visits to the Jovian planets, notice that they are low-density worlds that are rich in hydrogen. Jupiter and Saturn are mostly liquid hydrogen, and even Uranus and Neptune contain a much larger proportion of hydrogen than does Earth. Recall from Chapter 19 that these are hydrogen-rich, low-density worlds because they formed in the outer solar nebula where water vapor could freeze to form tremendous amounts of tiny ice particles. These hydrogen-rich ice particles accreted to begin forming the planets, and once the growing planets became massive enough, they could draw in more hydrogen gas directly by gravitational collapse.

Satellite Systems

All of the Jovian worlds have large satellite systems. As you visit the moons of the Jovian worlds, look for two processes. The orbits of some moons may have been modified by interactions with other moons, so that they now revolve in synchronism around their planet. The same process may allow moons to affect the orbital motions of particles in planetary rings.

The second process allows tides and the decay of radioactive elements in some moons to heat their interiors and produce geological activity on their surfaces including volcanoes and lava flows. You have learned that cratered surfaces are old, so when you see a section of a moon's surface that has few craters, you know that the moon must have been geologically active since the end of the heavy bombardment. Of course, geological activity depends on heat, so be alert for possible sources of internal heat as you visit these moons.

◄ SCIENTIFIC ARGUMENT ►

Why do you expect the outer planets to be low-density worlds?

To build this scientific argument, you need to think about how the planets formed from the solar nebula. In Chapter 19, you discovered that the inner planets could not incorporate ice when they formed because it was too hot near the sun; but, in the outer solar nebula, the growing planets could accumulate lots of ice. Eventually they grew massive enough to grow by gravitational collapse, and that pulled in hydrogen and helium gas. That makes the outer planets low-density worlds.

The outer planets may be unearthly, but they are understandable. For example, extend your argument. **Why do you expect the outer planets to have rings and moons?**

◄ ►

Jupiter, 11.2 times Earth's diameter, is the largest planet in our solar system.
The cloud belts and zones on Saturn are less distinct than those on Jupiter.
Uranus is four times Earth's diameter.
Shadow of one of Jupiter's many moons
Earth is the largest of the terrestrial worlds, but it is small compared with the Jovian planets.
Neptune, like Uranus, is generally blue because of small amounts of methane in its hydrogen-rich atmosphere.

■ **Figure 23-1**

The worlds of the outer solar system consist of the four Jovian planets, which are large, low-density worlds rich in hydrogen. (NASA/JPL/Space Science Institute/University of Arizona)

23-2 Jupiter

JUPITER IS THE MOST MASSIVE of the Jovian planets. Over 11 times the diameter of Earth, it contains nearly three-fourths of all the planetary matter in our solar system. This high mass accentuates some processes that are less obvious or nearly absent on the other Jovian worlds. Just as you used Earth as the basis of comparison for your study of the terrestrial planets, you should examine Jupiter in detail so you can use it as a standard in your comparative study of the other Jovian planets.

Surveying Jupiter

Jupiter is interesting because it is big, massive, mostly liquid hydrogen, and very hot inside. The preceding facts are common knowledge among astronomers, but you should demand an explanation of how they know these facts. Often the most interesting thing about a fact isn't the fact itself but how it is known.

You can be sure that Jupiter is a big planet because it looks big. At its closest point to Earth, Jupiter is about eight times farther away than Mars, but even a small telescope will reveal that the disk of Jupiter appears more than twice as big as the disk of Mars. If you use the small-angle formula, you can compute the diameter of Jupiter—a bit over 11 times Earth's diameter **(Celestial Profile 7)**. That's a big planet!

You can see that Jupiter is massive by watching its moons race around it at high speed. Io is the innermost of the four Galilean moons, and its orbit is just a bit larger than the orbit of our moon around Earth. Io streaks around its orbit in less than two days, whereas Earth's moon takes a month. Jupiter has to be a very massive world to hold on to such a rapidly moving moon (■ Figure 23-2). In fact, you can use the radius of Io's orbit and its orbital period in Newton's version of Kepler's third law (Chapter 5) to calculate the mass of Jupiter—almost 318 times Earth's mass.

Learning the size and mass of Jupiter is relatively easy, but you might wonder how astronomers know that it is made mostly of hydrogen. The first step is to divide mass by volume to find Jupiter's average density, about 1.34 g/cm^3. Of course, it is denser at the center and less dense near the surface, but this average density reveals that it can't be made totally of rock. Rock has a density of 2.5 to 4 g/cm^3, so Jupiter must contain material of lower density, such as hydrogen.

Spectra recorded from Earth and from spacecraft visiting Jupiter show that the composition of Jupiter is much like that of the sun—it is mostly hydrogen and helium. This fact was confirmed in 1995 when a probe from the Galileo spacecraft fell into the atmosphere and radioed its results back to Earth. Jupiter is mostly hydrogen and helium, with traces of heavier atoms that form molecules such as methane (CH_4), ammonia (NH_3), and water (■ Table 23-1).

Just as astronomers can build mathematical models of the interiors of stars, they can use the equations that describe gravity, energy, and the compressibility of matter to build mathematical models of the interior of Jupiter. These models reveal that the interior of the planet is mostly liquid hydrogen containing small amounts of suspended heavier elements. The pressure and temperature are higher than the **critical point** for hydrogen, and that means there is no difference between gaseous hydrogen and liquid hydrogen. If you parachuted into Jupiter, you would fall through the gaseous atmosphere and notice the density of the surrounding fluid increasing until you were in a liquid. You would never splash into a liquid surface.

Roughly a quarter of the way to the center, you would notice a change in the liquid hydrogen. At that point, the pressure is high enough to force the hydrogen to change into **liquid metallic hydrogen,** which is a very good electrical conductor. Because liquid metallic hydrogen is difficult to study in the laboratory, its properties are poorly understood, and the models are uncertain about the depth of the transition from normal to metallic liquid hydrogen.

The models are also uncertain about the presence of a heavy element core in Jupiter. The planet contains about 30 Earth masses of elements heavier than helium, but much of that is probably suspended in the convectively stirred liquid hydrogen. No more than 10 Earth masses could form a heavy element core. Some astronomy books refer to this as a rocky core, but, if it exists, it cannot be anything like the rock you know on Earth. The center of Jupiter is five or six times hotter than the surface of the sun and is prevented from exploding into vapor only by the tremendous pressure there. If there is a core, it is "rocky" only in that it contains heavy elements.

How do astronomers know Jupiter is hot inside? Infrared observations show that Jupiter is glowing in the infrared and radiating 1.7 times more energy than it receives from the sun. That

■ Table 23-1 | Composition of Jupiter and Saturn (by Number of Molecules)

Molecule	Jupiter (%)	Saturn (%)
H_2	86	93
He	13	5
H_2O	0.1	0.1
CH_4	0.1	0.2
NH_3	0.02	0.01

Jupiter is mostly a liquid planet. It may have a small core of heavy elements not much bigger than Earth. (NASA/JPL/University of Arizona)

Celestial Profile 7: Jupiter

Motion:

Average distance from the sun	5.2028 AU (7.783×10^8 km)
Eccentricity of orbit	0.0484
Maximum distance from the sun	5.455 AU (8.160×10^8 km)
Minimum distance from the sun	4.951 AU (7.406×10^8 km)
Inclination of orbit to ecliptic	1°18′29″
Average orbital velocity	13.06 km/s
Orbital period	11.867 y (4334.3 days)
Period of rotation	$9^h55^m30^s$
Inclination of equator to orbit	3°5′

Characteristics:

Equatorial diameter	142,900 km (11.20 $D_\oplus$)
Mass	1.899×10^{27} kg (317.83 $M_\oplus$)
Average density	1.34 g/cm^3
Gravity at base of clouds	2.54 Earth gravities
Escape velocity	61 km/s (5.4 $V_\oplus$)
Temperature at cloud tops	−130°C (−200°F)
Albedo	0.51
Oblateness	0.0637

Personality Point:

Jupiter is named for the Roman king of the gods, and it is the largest planet in our solar system. It can be very bright in the night sky, and its cloud belts and four largest moons can be seen through even a small telescope. Its moons are visible even with a good pair of binoculars mounted on a tripod or braced against a wall.

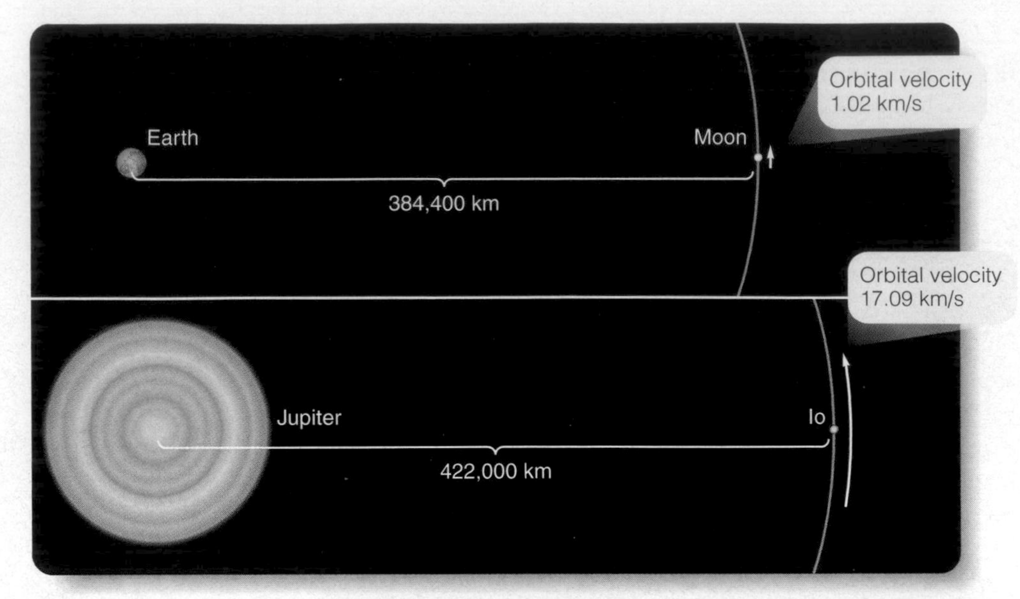

■ **Figure 23-2**

It is obvious that Jupiter is a very massive planet when you compare Jupiter's moon Io with Earth's moon. Although Io is 10 percent farther from Jupiter, it travels 17 times faster in its orbit than does Earth's moon around Earth. Clearly, Jupiter's gravitational field is much stronger than Earth's, and that means Jupiter must be very massive.

observation, combined with models of its interior, provides an estimate of its internal temperature.

You can tell that Jupiter is mostly a liquid just by looking at it. If you measure a photograph of Jupiter, you will discover that it is slightly flattened; it is a bit over 6 percent larger in diameter through its equator than through its poles. This is referred to as Jupiter's **oblateness.** The amount of flattening depends on the speed of rotation and on the rigidity of the planet. Its flattened shape shows that Jupiter cannot be as rigid as a terrestrial planet and must be mostly liquid.

Basic observations and the known laws of physics can tell you a great deal about Jupiter. Its vast magnetic field can tell you even more.

Jupiter's Magnetic Field

As early as the 1950s, astronomers detected radio noise coming from Jupiter and recognized it as synchrotron radiation. That form of radio energy is produced by fast electrons spiraling in a magnetic field, so it was obvious that Jupiter had a magnetic field.

In 1973 and 1974, two Pioneer spacecraft flew past Jupiter, followed in 1979 by two Voyager spacecraft. Those probes found that Jupiter has a magnetic field about 14 times stronger than the Earth's field. Evidently the field is produced by the dynamo effect operating in the highly conductive liquid metallic hydrogen as it is circulated by convection and spun by the rapid rotation of the planet. This powerful magnetic field dominates a huge magnetosphere around the planet. Compare the size of Jupiter's field with that of Earth in ■ Figure 23-3.

Jupiter's magnetic field deflects the solar wind and traps high-energy particles in radiation belts much more intense than Earth's. The radiation is more intense because Jupiter's magnetic field is stronger and can trap and hold higher-energy particles, mostly electrons and protons. The spacecraft passing through the radiation belts received radiation doses equivalent to a billion chest X rays—at least 100 times the lethal dose for a human. Some of the electronics on the spacecraft were damaged by the radiation.

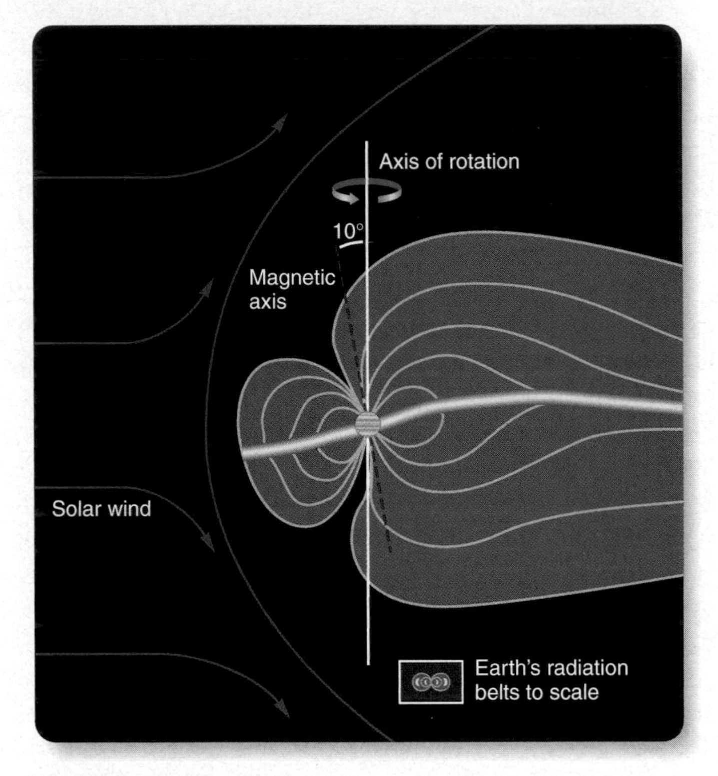

■ **Figure 23-3**

Jupiter's magnetic field is large and powerful. It traps particles from the solar wind to form powerful radiation belts. The rapid rotation of the planet forces the slightly inclined magnetic field to wobble up and down as the planet rotates. Earth's magnetosphere and radiation belts are shown to scale.

You will recall that Earth's magnetosphere interacts with the solar wind to produce auroras, and the same thing occurs on

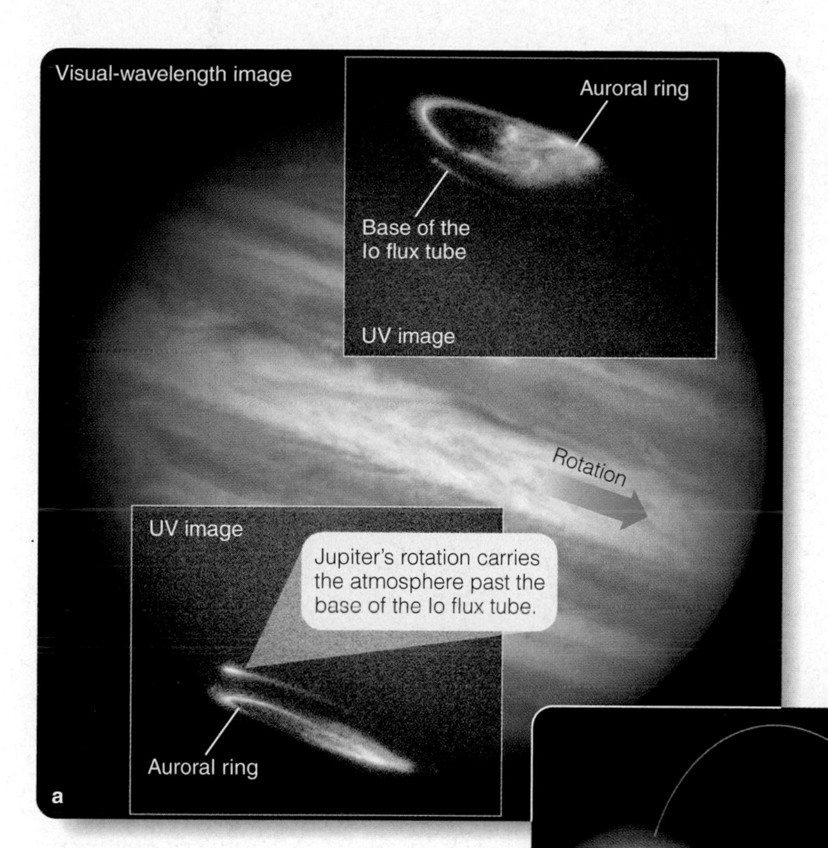
Visual-wavelength image
Auroral ring
Base of the
Io flux tube
UV image
Rotation
UV image
Jupiter's rotation carries the atmosphere past the base of the Io flux tube.
Auroral ring
a
The Io flux tube is an electric circuit connecting Io to Jupiter's magnetic field.
Io
Visual
UV
b

■ Active Figure 23-4

Jupiter's huge magnetic field funnels energy from the solar wind down to form rings of auroras around its magnetic poles, which are tipped some distance from the rotational poles. The same thing happens on Earth. The Io flux tube connects the small moon Io to the planet and carries a powerful electric current that creates spots of auroras where it touches the planet's atmosphere. (Jupiter: John Clarke, University of Michigan, NASA; Flux tube: NASA)

Jupiter. Charged particles in the magnetosphere leak downward along the magnetic field, and, where they enter the atmosphere, they produce auroras 1000 times more powerful than those on Earth. The auroras on Jupiter, like those on Earth, occur in rings around the magnetic poles (■ Figure 23-4).

The four Galilean moons of Jupiter orbit inside the magnetosphere, and some of the heavier ions in the radiation belts come from the innermost moon, Io. As you will see later in this chapter, Io has active volcanoes that spew gas and ash. Because Io orbits with a period of 1.8 days, and Jupiter's magnetic field rotates in only 10 hours, the wobbling magnetic field rushes past Io at high speed, sweeping up stray particles, accelerating them to high energy, and spreading them around Io's orbit in a doughnut of ionized gas called the **Io plasma torus.**

Jupiter's magnetic field interacts with Io to produce a powerful electric current (about a million amperes) that flows through a curving path called the **Io flux tube** from Jupiter out to Io and back to Jupiter. Small spots of bright auroras lie at the two points where the Io flux tube enters Jupiter's atmosphere (Figure 23-4).

Fluctuations in the auroras reveal that the solar wind buffets Jupiter's magnetosphere, but some of the fluctuations seem to be caused by changes in the magnetic dynamo deep inside the planet. In this way, studies of the auroras on Jupiter may help astronomers learn more about its liquid depths.

Jupiter's powerful magnetic field is invisible to your eyes, but the swirling cloud belts are beautiful in their complexity.

ThomsonNOW™ Sign in at www.thomsonedu.com and go to ThomsonNOW to see Astronomy Exercises "Auroras" and "Convection and Magnetic Fields."

Jupiter's Atmosphere

As you saw earlier, Jupiter is a liquid world that has no surface. The gaseous atmosphere blends gradually with the liquid hydrogen interior. Below the clouds of Jupiter lies the largest ocean in the solar system—and it has no surface and no waves.

When you look at Jupiter, all you see are clouds. These cloud layers lie deep inside a nearly transparent atmosphere of hydrogen and helium. You can detect this hydrogen atmosphere by noticing that Jupiter has limb darkening, just as the sun does (Chapter 8). When you look near the limb of Jupiter (the edge of its disk), the clouds are much dimmer (see Figure 23-1) because it is nearly sunset or sunrise along the limb. If you were on

Jupiter at that location, you would see the sun just above the horizon, and it would be dimmed by the atmosphere. In addition, light reflected from clouds must travel out at a steep angle through the atmosphere to reach Earth, dimming the light further. Jupiter is brighter near the center of the disk because the sunlight shines nearly straight down on the clouds.

Study **Jupiter's Atmosphere** on pages 518–519 and notice four important ideas:

1. The atmosphere is hydrogen rich, and the clouds are confined to a shallow layer.
2. The cloud layers lie at certain temperatures within the atmosphere where ammonia (NH_3), ammonium hydrosulfide (NH_4SH), and water (H_2O) can condense to form ice particles.
3. The belt–zone circulation is related to circulation driven by high- and low-pressure areas related to those on Earth.
4. Finally, the major spots on Jupiter, although they are only circulating storms, can remain stable for decades or even centuries.

Circulation in Jupiter's atmosphere is not totally understood. Observations made by the Cassini spacecraft as it raced past Jupiter on its way to Saturn revealed that the dark belts contain small rising storm systems too small to see in previous images. Evidently, the general circulation usually attributed to the belts and zones is much more complex when it is observed in more detail. Further understanding of the small-scale motions in Jupiter's atmosphere may have to await future planetary probes.

The highly complex spacecraft that have visited Jupiter are examples of how technology can give scientists the raw data they need to form their understanding of nature (**How Do We Know? 23-1**). Science is about understanding nature, and Jupiter is an entirely new kind of planet in your study. In fact, Jupiter has a feature that you did not find anywhere among the terrestrial planets. Jupiter has a ring.

ThomsonNOW Sign in at www.thomsonedu.com and go to ThomsonNOW to see Astronomy Exercise "Convection and Turbulence."

Jupiter's Ring

Astronomers have known for centuries that Saturn has rings, but it was not until 1979, when the Voyager 1 spacecraft sent back photos, that Jupiter's ring was discovered. Less than one percent as bright as Saturn's rings, the ghostly ring around Jupiter is a puzzle. What is it made of? Why is it there? A few simple observations will help you solve some of these puzzles.

Saturn's rings are made of bright ice chunks, but the particles in Jupiter's ring are very dark and reddish. You can conclude that its ring is rocky rather than icy.

You can also conclude that the ring particles are mostly microscopic. Photos show that the ring is very bright when illuminated from behind (■ Figure 23-5)—it is scattering light forward. Efficient **forward scattering** occurs when particles have diame-

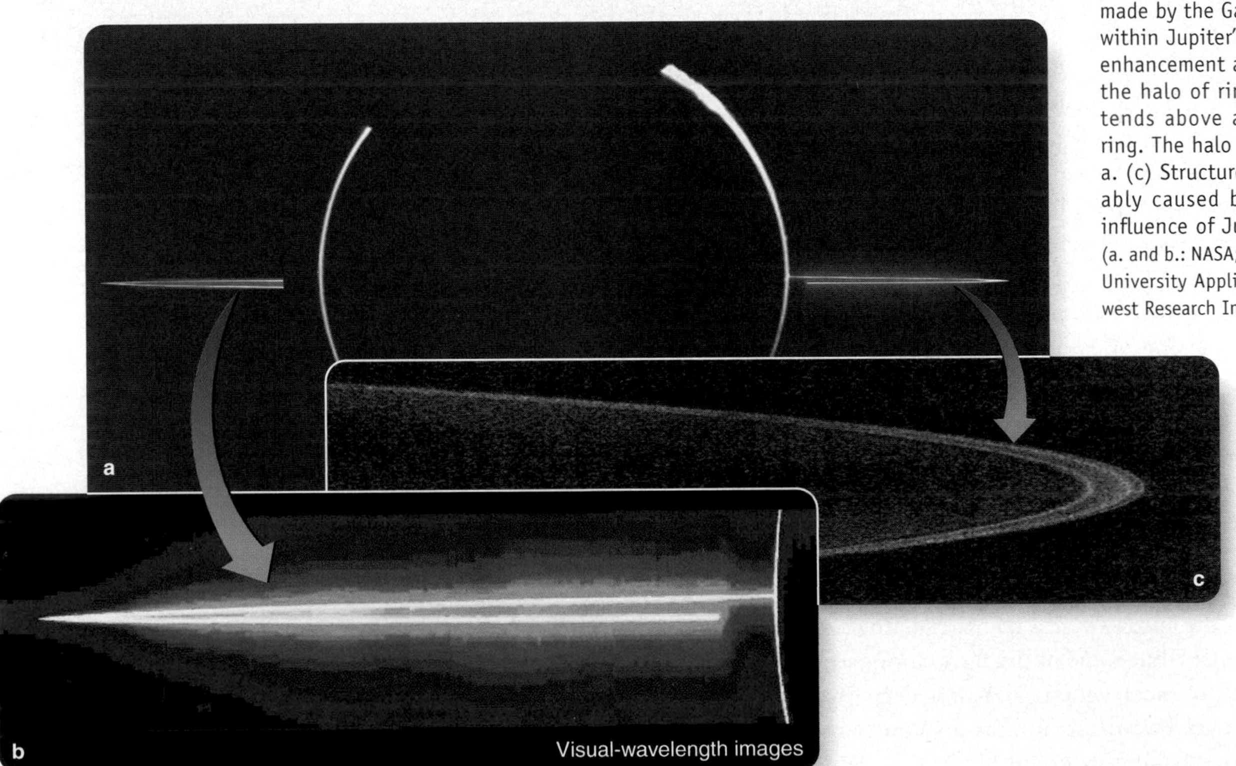

■ **Figure 23-5**

(a) The main ring of Jupiter, illuminated from behind, glows brightly in this visual-wavelength image made by the Galileo spacecraft from within Jupiter's shadow. (b) Digital enhancement and false color reveal the halo of ring particles that extends above and below the main ring. The halo is just visible in part a. (c) Structure in the ring is probably caused by the gravitational influence of Jupiter's inner moons. (a. and b.: NASA; c. NASA/Johns Hopkins University Applied Physics Lab/ Southwest Research Institute)

How Do We Know? 23-1

Science, Technology, and Engineering

Why would a robot need a brain to be a scientist? The Cassini space probe orbiting Saturn is a robot. It doesn't walk and talk like the Tin Man from *The Wizard of Oz,* but it is an electromechanical device with a stored program following its instructions to perform certain tasks. Some robots have artificial intelligence and can make decisions and take actions on their own. Robots can collect data that contribute to scientific understanding, but they can't actually do science themselves.

Science is nothing more than the logical study of nature, and the goal of science is understanding. Although much scientific knowledge proves to have tremendous practical value, the only goal of science is a better understanding of how nature works. Technology, in contrast, is the practical application of scientific knowledge to solve a specific problem. Engineering is the most practical form of technology. An engineer is likely to use well-understood technology to find a practical solution to a problem.

Scientists use robotic technology to explore the Saturn system. The robotic space probe can make delicate measurements, record images, and communicate with scientists back on Earth, but the ultimate goal is to understand Saturn, its rings, and its moons. Because the goal is to understand how nature works, it is science. Engineers may use the same technology to design a robot that can crawl into a nuclear reactor, search for a cracked pipe, and repair it. Their goal is not to understand nature better but to solve a specific problem, so they are properly described as engineers using technology.

Because scientists use technology in their work, it is easy to mistake the complicated instruments for the science. In a sense, science is what goes on in the heads of scientists as they try to understand natural processes.

Exploring other worlds is of cultural value. It helps us as humans understand ourselves. (NASA/JPL/Space Science Institute)

As you read about science, look for the ways scientists use technology but keep in mind the ultimate goal of science—understanding how nature works.

ters roughly the same as the wavelength of light, a few millionths of a meter. Large particles do not scatter light forward, so a ring filled with basketball-size particles would look dark when illuminated from behind. The forward scattering tells you that the ring is made of particles about the size of those in cigarette smoke.

Larger particles are not entirely ruled out. A sparse distribution of rocky objects ranging from pieces of gravel to boulders is possible, but objects larger than 4 km would have been detected in photos. The vast majority of the ring particles are microscopic dust.

The size of the ring particles is a clue to their origin, and so is their location. They orbit inside the **Roche limit,** the distance from a planet within which a moon cannot hold itself together by its own gravity. If a moon orbits relatively far from its planet, then the moon's gravity will be much greater than the tidal forces caused by the planet, and the moon will be able to hold itself together. If, however, a planet's moon comes inside the Roche limit, the tidal forces can overcome its gravity and pull the moon apart. The International Space Station can orbit inside Earth's Roche limit because it is held together by bolts and welds, and a single large rock can survive inside the Roche limit if it is strong enough not to break. However, a moon composed of separate rocks and particles held together by their mutual gravity could not survive inside a planet's Roche limit. Tidal forces would pull such a moon to pieces. If a planet and its moon have similar densities, the Roche limit is 2.44 planetary radii. Jupiter's main ring has an outer radius of 130,000 km (1.8 planetary radii) and lies inside the Roche limit, as do the rings of Saturn, Uranus, and Neptune.

Now you can understand the dust in Jupiter's ring. If a dust speck gets knocked loose from a larger rock orbiting inside the Roche limit, the rock's gravity cannot hold the dust speck. And the billions of dust specks in the ring can't pull themselves together to make a larger body—a moon—because of the tidal forces inside the Roche limit.

You can also be sure that the ring particles are not old. The pressure of sunlight and Jupiter's powerful magnetic field alter the orbits of the particles, and they gradually spiral into the planet. Images show faint ring material extending down toward the cloud tops, and this is evidently dust specks spiraling inward. Dust is also lost from the ring as electromagnetic effects force it out of the plane of the ring to form a low-density halo above and below the ring (Figure 23-5b). Yet another reason the ring particles can't be old is that the intense radiation around Jupiter grinds the dust specks down to nothing in a century or so. For all these reasons, the rings seen today can't be made up of material left over from the formation of Jupiter.

Obviously, the rings of Jupiter must be continuously resupplied with new material. Dust particles can be chipped off rocks ranging from gravel to boulders within the ring, and small moons that orbit near the outer edge of the rings lose particles as they are hit by meteorite impacts. Observations made by the

Galileo spacecraft show that the main ring is densest at its outer edge, where the small moon Adrastea orbits, and that another small moon, Metis, orbits inside the ring. Clearly these moons must be structurally strong to withstand Jupiter's tidal forces. Galileo images also reveal much fainter rings, called the **gossamer rings,** extending twice as far from the planet as the main ring. These gossamer rings are most dense at the orbits of two small moons, Amalthea and Thebe, more evidence that ring particles are being blasted into space by impacts on the moons.

Besides supplying the rings with particles, the moons help confine the ring particles and keep them from spreading outward. You will find that this is an important process in planetary rings when you study the rings of Saturn later in this chapter.

Your exploration of Jupiter reveals that it is much more than just a big planet. It is the gravitational and magnetic center of an entire community of objects. Occasionally the community suffers an intruder.

Comet Impact on Jupiter

Comets are very common in the solar system, and Jupiter, because of its strong gravity, probably gets hit by comets more often than most planets. But no one had ever seen it happen until 1994, when fragments from a comet disrupted by Jupiter's tidal forces a few years before looped back and smashed into the planet.

The comet, named Shoemaker–Levy 9 after its discoverers, Eugene and Carolyn Shoemaker and David H. Levy, had been captured into orbit around Jupiter sometime in the past. In 1992, it passed inside Jupiter's Roche limit and was pulled into at least 21 pieces that looped out away from Jupiter in long elliptical orbits. It was only then that the objects were discovered on a photograph. Drawn out into a long chain of small comets, the pieces fell back and slammed into Jupiter over a period of six days in July 1994 (■ Figure 23-6).

■ Figure 23-6

In 1994, fragments of Comet Shoemaker–Levy 9 struck Jupiter. Although these were the first comet impacts ever seen by Earth's inhabitants, such events are probably common occurrences in planetary systems. (Composite and visual images: NASA; IR sequence: Mike Skrutskie; IR image: University of Hawaii)

Impact L occurred out of sight beyond Jupiter's horizon.
Impacts were visible from the Galileo spacecraft.
Minutes after impact, the rising fireball was visible above the horizon.
Visual
Impact site just out of sight as seen from Earth
Only 9 minutes after impact L, the fireball was brilliant in the infrared.
Fragments of comet falling toward Jupiter
At visual wavelengths, impact sites were dark smudges that lasted for many days.
Infrared images
Visual image composite
Impact sites remained bright in the infrared as the rotation of Jupiter carried them into sight from Earth.
Larger than Earth
Visual
Infrared

The fragments were no bigger than a kilometer or so in diameter and contained a fluffy mixture of rock and ices. They hit Jupiter at speeds of about 60 km/s and released energy equivalent to a few million megatons of TNT. (At the height of the Cold War, all the nuclear weapons on Earth totaled about 80,000 megatons. The Hiroshima bomb was only 0.15 megaton.)

The impacts occurred just over Jupiter's horizon as seen from Earth, but they produced fireballs almost 3000 km high, and the rapid rotation of Jupiter brought the impact points within sight of Earth after only 15 minutes. Infrared telescopes easily spotted the glowing scars where the comets hit. As they cooled, the impact sites became dusty, dark spots shown at visual wavelengths in Figure 23-6. Visible through even small telescopes, some of these dark smudges were larger than Earth.

The comet impact was an astonishing spectacle, but what can it tell you about Jupiter? In fact, the impact was revealing in two ways. First, astronomers used the impacts as probes of Jupiter's atmosphere. By making assumptions about the nature of Jupiter's atmosphere and by using the most powerful computers, astronomers created models of a high-velocity projectile penetrating into Jupiter's upper atmosphere. By comparing the observed impacts with the models, astronomers were able to fine-tune the models to better represent Jupiter's atmosphere. This method of comparing models with reality is a critical part of science.

Second, the spectacle is a reminder that planets are sometimes hit by large objects such as asteroids and the heads of comets. Jupiter probably is hit every century or so. In 1690, the Italian astronomer Cassini observed a dark spot that appeared on Jupiter and changed over a period of days. Based on his description of the spot, modern astronomers recognize it as the scar of a comet impact.

Comets hit planets all the time. The impact in 1994 was just the first such event to occur since the rise of modern astronomy. You have seen evidence for large impacts on Earth, the moon, Venus, Mercury, and Mars, and you will see in Chapter 25 that a comet or asteroid impact on Earth may have changed the climate and killed the dinosaurs. A solar system is a dangerous place to put an inhabited planet.

The History of Jupiter

Your goal in studying any planet is to be able to tell its story—to describe how it got to be the way it is. While you can understand part of the story of Jupiter, there is still much to learn.

If the solar nebula theory for the origin of the solar system is correct, then Jupiter formed from the colder gases of the outer solar nebula, where ices were able to condense. Thus, Jupiter grew rapidly and became massive enough to capture hydrogen and helium gas from the solar nebula and form a deep liquid hydrogen envelope. Models are uncertain as to whether a heavy element core survives; it may have been mixed in with the convecting liquid hydrogen envelope, and one astronomer estimates the mass of Jupiter's heavy element core as "between 0 and 10 Earth masses."

In the interior of Jupiter, hydrogen exists as liquid metallic hydrogen, a very good electrical conductor. The planet's rapid rotation, coupled with the outward flow of heat from its hot interior, drives a dynamo effect that produces a powerful magnetic field. That vast magnetic field traps high-energy particles from the solar wind to form intense radiation belts and auroras.

The rapid rotation and large size of Jupiter cause belt–zone circulation in its atmosphere. Heat flowing upward from the interior causes rising currents in the bright zones, and cooler gas sinks in the dark belts. As on Earth, winds blow at the margins of these regions, and large spots appear to be cyclonic disturbances.

Although the age of planet building is long past, debris in the form of meteorites and occasional comets continues to hit Jupiter, as it does all the planets. Any debris left over from the formation of Jupiter would have been blown away long ago by the solar wind, so the dust trapped in Jupiter's thin ring must be young. It probably comes from meteorites hitting the innermost moons.

Your study of Jupiter has been challenging because the planet lacks a surface—it is difficult to imagine being there. Most of the surface features and processes you found on the terrestrial planets are missing on Jupiter, but, as the prototype of the Jovian worlds, it earns its place as the ruler of the solar system.

◄ SCIENTIFIC ARGUMENT ►

How do astronomers know Jupiter is hot inside?

A scientific argument is a way to test ideas, and sometimes it is helpful to test even the most basic ideas. You know that something is hot if you touch it and it burns your fingers, but you can't touch Jupiter. You also know something is hot if it is glowing bright red—it is red hot. But Jupiter is not glowing red hot. You can tell that something is hot if you can feel heat when you hold your hand near it. That is, you can detect infrared radiation with your skin. In the case of Jupiter, you would need greater sensitivity than the back of your hand, but infrared telescopes reveal that Jupiter is a source of infrared radiation; it is glowing in the infrared. Sunlight would warm Jupiter a little bit, but it is emitting 70 percent more infrared than it should. That means it must be hot inside. From models of the interior, astronomers conclude that the center must be five or six times hotter than the surface of the sun to make the planet glow in the infrared.

Astronomical understanding is usually based on simple observations, so build an argument to answer the following simple question. **How do astronomers know that Jupiter has a low density?**

◄ ►

23-3 Jupiter's Family of Moons

HOW MANY MOONS DOES JUPITER HAVE? Astronomers are finding more and more small moons, and the count is now over 60. (You will have to check the Internet to get the latest figure because new moons are frequently discovered.) Most of these moons are small and rocky, and many are probably captured asteroids. Four of the moons, those discovered by Galileo and now called the Galilean moons, are large and have interesting geologies (■ Figure 23-7).

Your study of the moons of Jupiter will illustrate three important principles in comparative planetology. First, a body's composition depends on the temperature of the material from which it formed. This is illustrated by the prevalence of ice as a building material in the outer solar system, where sunlight is weak. You are already familiar with the second principle: that cratering can reveal the age of a surface. Also, as you have seen in your study of the terrestrial planets, internal heat has a powerful influence over the geology of these larger moons.

Callisto: The Ancient Face

The outermost of Jupiter's four large moons, Callisto is half again as large in diameter as Earth's moon. Like all of Jupiter's larger satellites, Callisto is tidally locked to its planet, keeping the same side forever facing Jupiter. From its gravitational influence on passing spacecraft, astronomers can calculate Callisto's mass, and dividing that mass by its volume shows that its density is 1.79 g/cm^3. Ice has a density of about 1 and rock 2.5 to 4 g/cm^3, so Callisto must be a mixture of rock and ice.

Images from the Voyager and Galileo spacecraft show that the surface of Callisto is dark, dirty ice heavily pocked with craters (■ Figure 23-8). Old, icy surfaces in the solar system become dark because of dust added by meteorites and because meteorite impacts vaporize water, leaving any dust and rock in the ice behind to form a dirty crust. You may have seen the same thing happen to a city snowbank. As the snow evaporates over a few days, the crud in the snow is left behind to form a dirty rind. Break through that dirty surface, and the snow is much cleaner underneath.

Spectra of the surface show that in most places it is a 50/50 mix of ice and rock, but some areas are ice free. Nevertheless, the slumped shapes of craters suggests that the outer 10 km is mostly frozen water; ice isn't very strong, so big piles of it tend to slump under their own weight. The disagreement between the spectra and the shapes of craters can be understood when you recall that the spectra contain information about only the outer 1 mm of the surface, which can be quite dirty, while the shapes of craters tell you about the outermost 10 km, which appear to be rich in ice.

Delicate measurements of the shape of Callisto's gravitational field were made by the Galileo spacecraft, and they show that Callisto has never fully differentiated to form a dense core and a lower-density mantle. Its interior is a mixture of rock and ice. This is consistent with the observation that it has only a weak magnetic field of its own. A strong magnetic field could be generated by the dynamo effect in a liquid convecting core, and Callisto has no core. It does, however, interact with Jupiter's magnetic field in a way that suggests it has a layer of liquid water roughly 10 km thick about 100 km below its icy surface. Slow radioactive decay in Callisto's interior may produce enough heat to keep this layer of water from freezing.

Ganymede: A Hidden Past

The next Galilean moon inward is Ganymede, larger than Earth's moon (see Figure 23-7), larger than Mercury, and over three-quarters the diameter of Mars. Its density is 1.9 g/cm^3, and its influence on the Galileo spacecraft reveals that it has a rocky core, an ice-rich mantle, and a crust of ice 500 km thick. It may even have a small iron core at its center. It is large enough for radioactive decay to have melted its interior when it formed, allowing iron to sink to its center.

Ganymede's surface hints at an active past. Although a third of the surface is old, dark, and cratered, the rest is marked by bright parallel grooves. Because this bright **grooved terrain** (■ Figure 23-9a) contains fewer craters, it must be younger.

Observations show that the bright terrain was produced when the icy crust broke and water flooded up from below and froze. As the surface broke over and over, sets of parallel groves were formed. Some low-lying regions are smooth and appear to

■ **Figure 23-7**

The Gallilean moons of Jupiter from left to right are Io, Europa, Ganymede, and Callisto. The circle shows the size of Earth's moon. (NASA)

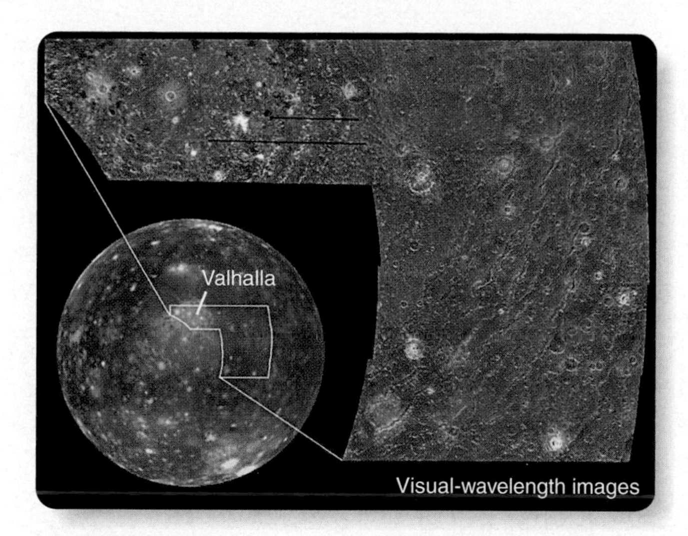

■ **Figure 23-8**

The dark surface of Callisto is dirty ice marked by craters in these visual-wavelength images. The youngest craters look bright because they have dug down to cleaner ice. Valhalla is the 4000-km-diameter scar of a giant impact feature, one of the largest in the solar system. Valhalla is so large and old that the icy crust has flowed back to partially heal itself, and the outer rings of Valhalla are shallow troughs marking fractures in the crust. (NASA)

have been flooded by water. Spectra reveal concentrations of salts such as those that would be left behind by the evaporation of mineral-rich water. Also, some features in or near the bright terrain appear to be caldera formed when subsurface water drained away and the surface collapsed (Figure 23-9b).

The Galileo spacecraft found that Ganymede has a magnetic field about 10 percent as strong as Earth's. It even has its own magnetosphere inside the larger magnetosphere of Jupiter. Mathematical models find it difficult to produce a magnetic field in a water-rich mantle, and there does not appear to be enough heat in Ganymede for it to have a molten metallic core. Astronomers wonder if its magnetic field is left over and frozen into the rock from a time when it was hotter and more active.

Ganymede's magnetic field fluctuates with the 10-hour period of Jupiter's rotation. The rotation of the planet sweeps its tilted magnetic field past the moon, and the two fields interact. That interaction reveals that the moon has a layer of liquid water about 170 km (100 mi) below its surface. The water layer may be about 5 km (3 mi) thick. It is possible that the water layer was thicker and closer to the surface long ago when the interior of the moon was warmer. That might explain the flooding that appears to have formed the bright grooved terrain.

Ganymede orbits rather close to a massive planet, and that exposes it to two unusual processes that many worlds never experience. **Tidal heating,** the frictional heating of a body by changing tides (■ Figure 23-10a) could have heated Ganymede's interior and added to the heat generated by radioactive decay. In a circular orbit, a moon experiences no tidal heating; but, at some point in the past, interactions with the other moons could have pushed Ganymede into a more elliptical orbit. Jupiter's gravity would have deformed the moon, and as Ganymede followed its elliptical orbit, tides would have flexed it, and friction would have heated it. Such an episode of tidal heating might have been enough to drive a dynamo to produce a magnetic field and break the crust to make the bright terrain.

The second process that affects Ganymede is the inward focusing of meteorites. Because massive planets like Jupiter draw

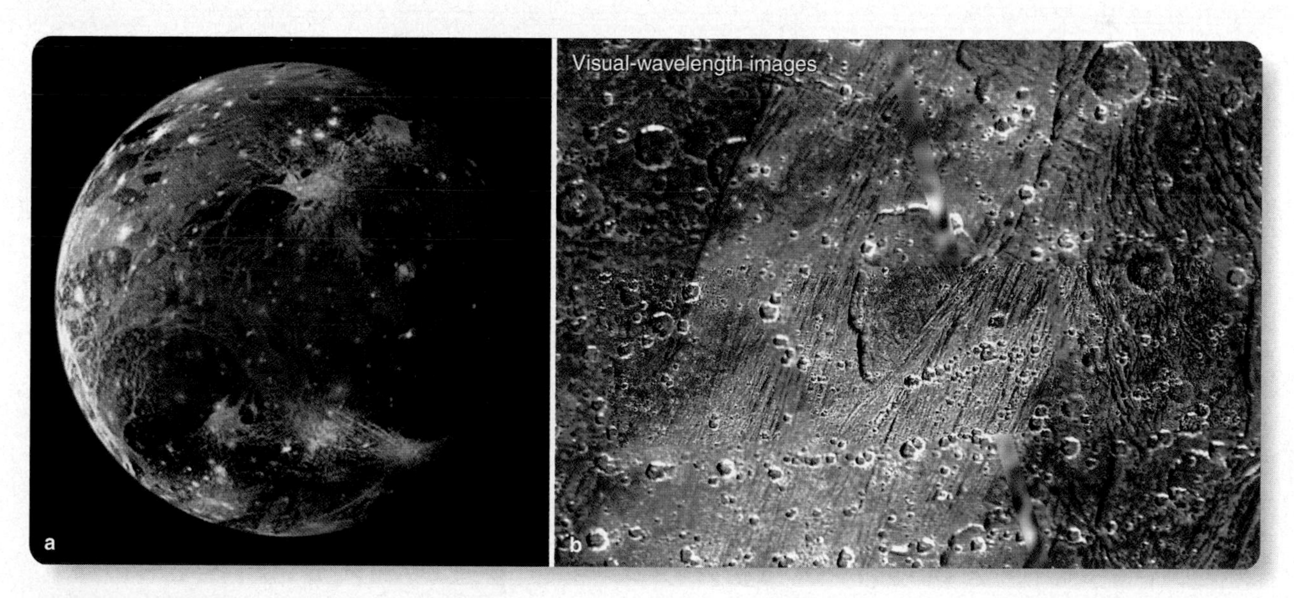

■ **Figure 23-9**

(a) This color-enhanced image of Ganymede shows the frosty poles at top and bottom, the old dark terrain, and the brighter grooved terrain. (b) A band of bright terrain runs from lower left to upper right, and a collapsed area, a possible caldera, lies at the center in this visual-wavelength image. Calderas form where subsurface liquid has drained away, and the bright areas do contain other features associated with flooding by water. (NASA)

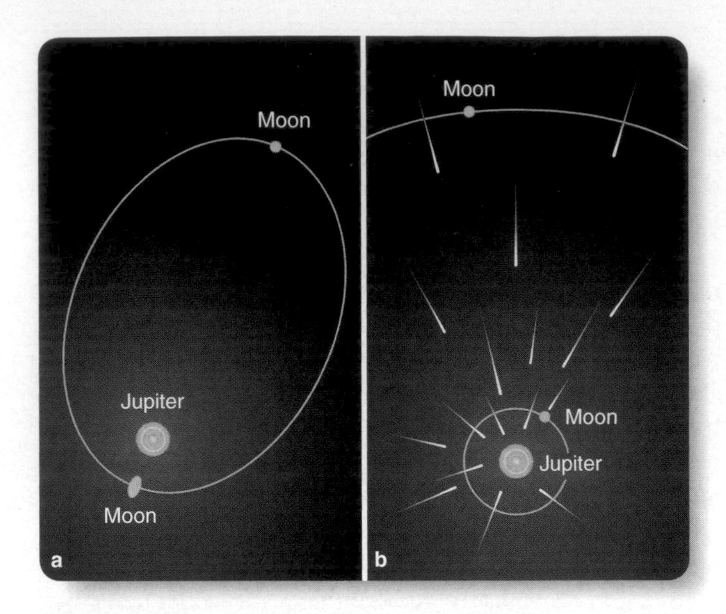

■ **Figure 23-10**

Two effects on planetary satellites. (a) Tidal heating occurs when changing tides cause friction within a moon. (b) The focusing of meteoroids exposes satellites in small orbits to more impacts than satellites in larger orbits receive.

debris inward, a moon orbiting near a massive planet is struck by many meteorites (Figure 23-10b). You should expect such a moon to have lots of craters, but the bright terrain on Ganymede has few craters. That part of Ganymede's surface must be only about 1 billion years old, and that should alert you that the Galilean moons are not just dead lumps of rock and ice. The closer you get to Jupiter, the more active the moons are.

Europa: A Hidden Ocean

The next Galilean moon inward is Europa, a bit smaller than Earth's moon. Europa has a density of about 3 g/cm^3, so it must be mostly rock and metal. Yet its surface is ice.

Europa lies closer to Jupiter than Ganymede, so it should be exposed to more meteorite impacts than Callisto or Ganymede, yet the icy crust of Europa is almost free of craters. Recent craters such as Pwyll are bright, but most are hardly more than blemishes in the ice (■ Figures 23-11a and b). Evidently the surface of Europa is active and erases craters almost as fast as they form. The number of impact scars on Europa suggests that the average age of its surface is only 10 million years. Other signs of activity include long cracks in the icy crust and regions where the crust has broken into sections that have moved apart as if they were icebergs floating on water (Figure 23-11c).

Europa's clean, bright face tells you its surface is young. The albedo of the surface is 0.69, meaning that it reflects 69 percent of the light that hits it. This high albedo is produced by clean ice. You have discovered that old, icy surfaces tend to be very dark, so Europa's high albedo means the surface is active, covering older surfaces with fresh ice.

Europa is too small to have retained much heat from its formation or from radioactive decay, and the Galileo spacecraft found that Europa has no magnetic field of its own. It cannot have a molten conducting core. Tidal heating, however, is important for Europa and apparently provides enough heat to keep the little moon active. In fact, the curving cracks in its crust reveal the shape of the tidal forces that flex it as it orbits Jupiter.

If you hiked on Europa with a compass in your hand, you would be in big trouble. Europa has no magnetic field of its own, but Jupiter rotates rapidly and drags its field past the little moon. That induces a fluctuating magnetic field in Europa that would make your compass wander uselessly. Europa's interaction with

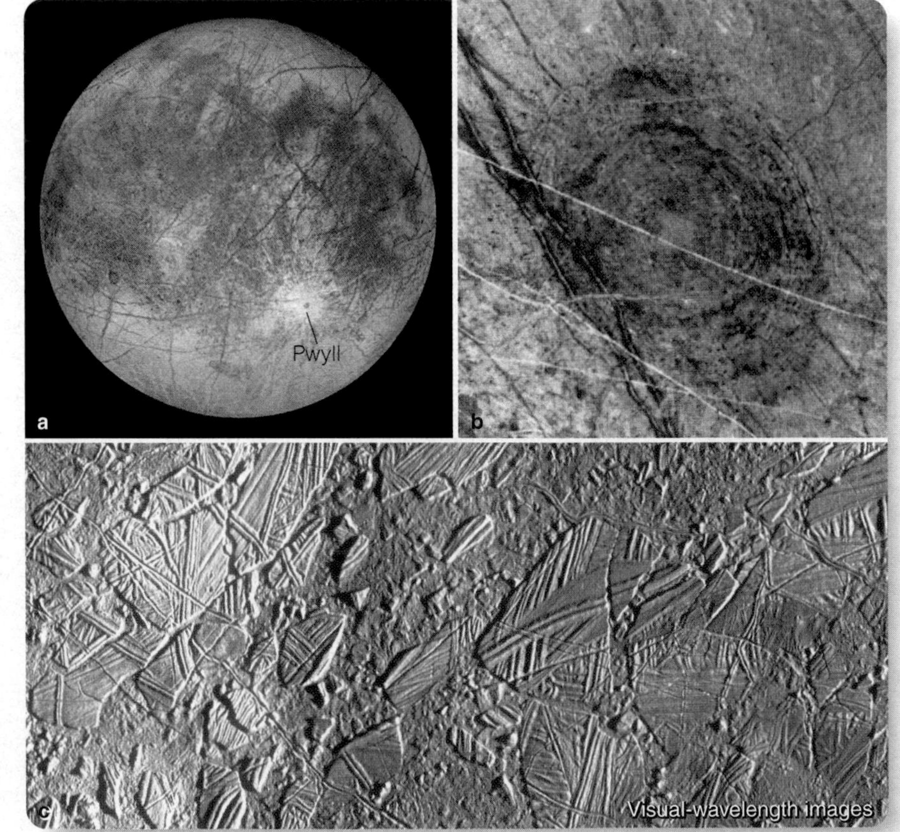

■ **Figure 23-11**

(a) The icy surface of Europa is shown here in natural color. Many faults are visible on its surface, but very few craters. The bright crater is Pwyll, a young impact feature. (b) This circular bull's-eye is 140 km in diameter. It is the remains of an impact by an object about the size of a mountain. Notice the younger cracks and faults that cross the older impact feature. (c) Like icebergs on the Arctic Ocean, blocks of crust on Europa appear to have floated apart. Spectra show that the blue ice is stained by salts such as those that would be left behind by mineral-rich water welling up from below and evaporating. White areas are ejecta from the impact that formed Pwyll. (NASA)

Jupiter's magnetic field reveals the presence of a liquid-water ocean lying about 15 km (10 mi) below the icy surface. The ocean may be as deep as 100 miles (■ Figure 23-12) and could contain twice as much water as all the oceans on Earth. It is surely rich in dissolved minerals, which would probably make it taste really bad. But those minerals make the water a good electrical conductor and allow it to interact with Jupiter's magnetic field. No one knows what might be swimming through such an ocean, and many scientists hope for a future mission to Europa to drill through the ice crust and sample the ocean below for signs of life.

Tidal heating makes Europa geologically active. Apparently, rising currents of water can break through the icy crust or melt surface patches. Many of the cracks show evidence that they have spread apart and that fresh water has welled up and frozen between. In other regions, compression in the crust is revealed by networks of faults and low ridges. Compression on Earth pushes up mountain ranges, but no such ranges appear on Europa. The icy crust isn't strong enough to support ridges higher than a kilometer or so.

Orbiting deep inside Jupiter's radiation belts, Europa is bombarded by high-energy particles that damage the icy surface. Water molecules are freed and broken up, then dispersed into a doughnut-shaped cloud spread round Jupiter and enclosing Europa's orbit. Flying past Jupiter in 2002, the Cassini spacecraft was able to image this cloud of excited gas. Europa's gas cloud should alert you that moons orbiting deep inside a massive planet's radiation belts are exposed to a form of erosion that is entirely lacking on Earth's moon.

Io: Bursting Energy

Geological activity is driven by heat flowing out of a planet's interior, and nothing could illustrate this principle better than Io, the innermost of Jupiter's Galilean moons. Photographs from the Voyager and Galileo spacecraft show no impact craters at all—surprising considering Jupiter's power to focus meteoroids inward (Figure 23-10b). No subtlety is needed to explain the missing craters. Over 150 active volcanoes are visible on Io's surface, blasting enough ash out over the surface to bury any newly formed craters (■ Figure 23-13). Unlike the dead or dormant outer Galilean moons, Io is bursting with energy.

Spectra reveal that Io has a tenuous atmosphere of gaseous sulfur and oxygen, but those gases can't be permanent. Even though the erupting volcanoes pour out about one ton of gases per second, the gases leak into space easily because of Io's low escape velocity. Also, any gas atoms that become ionized are swept away by Jupiter's rapidly rotating magnetic field. The ions produce a cloud of sulfur and sodium ions in a torus (a doughnut shape) enclosing Io's orbit (■ Figure 23-14).

You would need a very good spacesuit to visit the surface of Io. The temperature at the surface averages 130 K (−225°F) and the atmospheric pressure is very low. Because of the continuous volcanism and the sulfurous gases, Io's thin atmosphere is dirty and probably smelly with sulfur. In fact, the reddish color of Jupiter's small inner moon Amalthea may be caused by sulfur pollution escaping from Io. Your real problem as you bounced across the surface of Io under its low gravity would be radiation. Io is deep inside Jupiter's magnetosphere and radiation belts. Unless your spacesuit had astonishingly impressive shielding, the radiation would be lethal. Io, like Venus, may be a place that humans will never visit with any ease.

You can use basic observations to deduce the nature of Io's interior. From its density, 3.53 g/cm^3, you can conclude that it is rocky. Spectra reveal no trace of water at all, so there is no ice on Io. It is the driest world in our solar system. The oblateness of Io caused by its rotation and by the slight distortion produced by

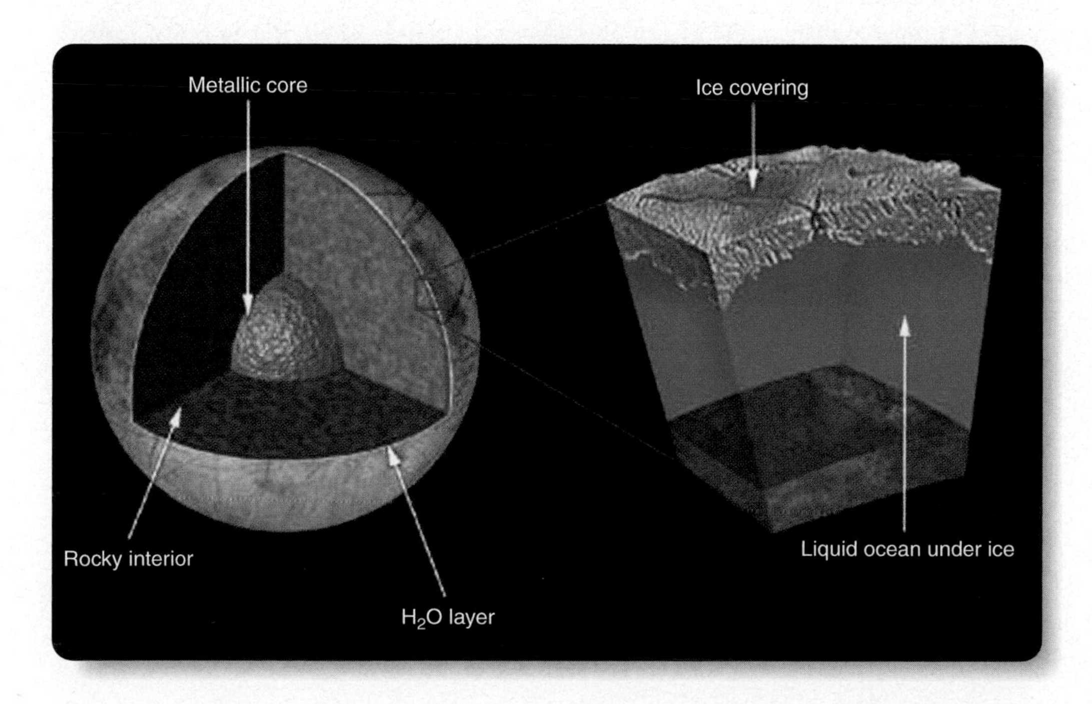

■ Figure 23-12

The gravitational influence of Europa on the passing Galileo spacecraft shows that the little moon has differentiated into a dense core and rocky mantle. Magnetic interactions with Jupiter show that it has a liquid-water ocean below its icy crust. Heat produced by tidal heating could flow outward as convection in such an ocean and drive geological activity in the icy crust. (NASA)

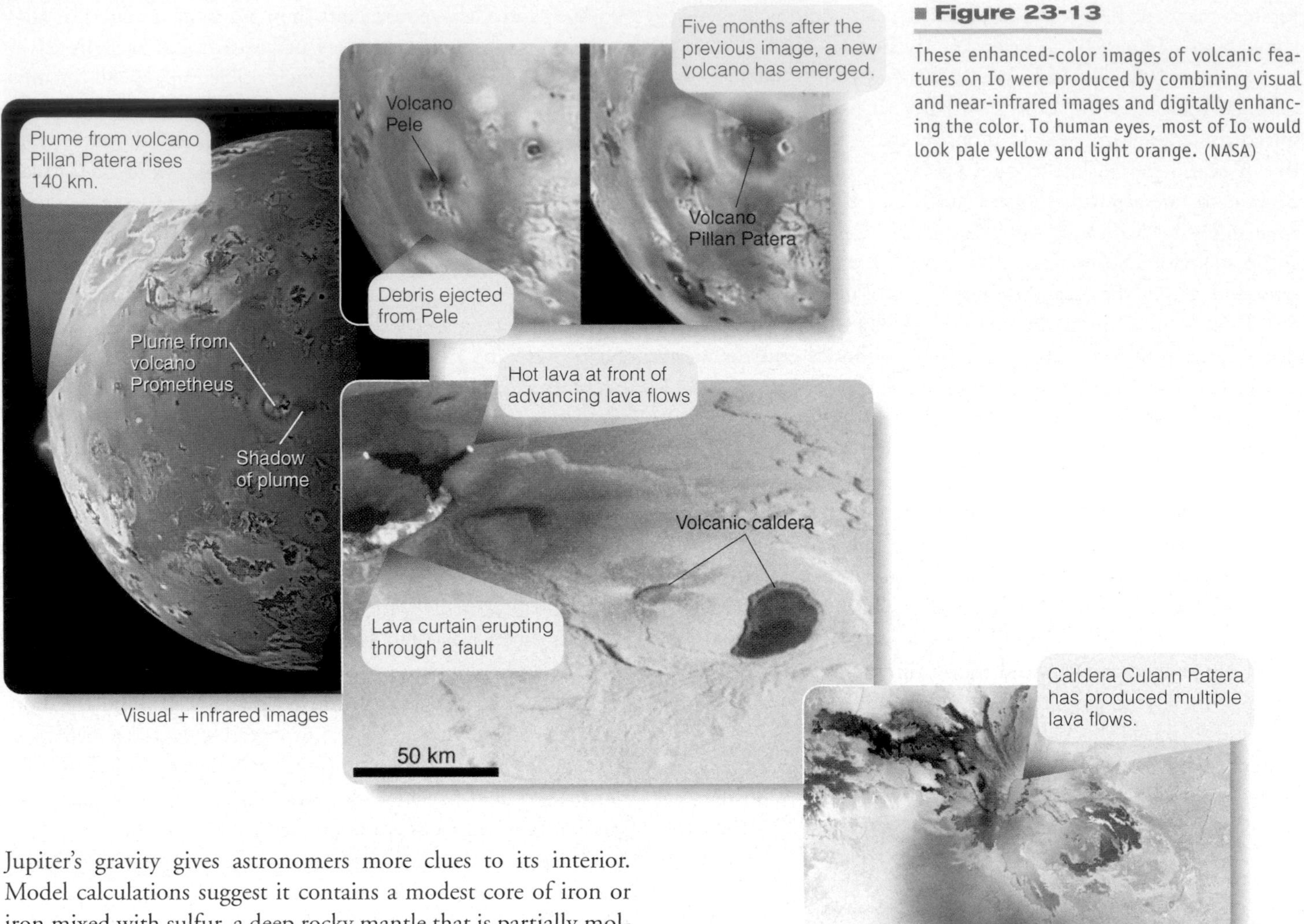

■ Figure 23-13

These enhanced-color images of volcanic features on Io were produced by combining visual and near-infrared images and digitally enhancing the color. To human eyes, most of Io would look pale yellow and light orange. (NASA)

Jupiter's gravity gives astronomers more clues to its interior. Model calculations suggest it contains a modest core of iron or iron mixed with sulfur, a deep rocky mantle that is partially molten, and a thin, rocky crust.

The colors of Io have been compared to those of a badly made pizza. The reds, oranges, and browns of Io are caused by sulfur and sulfur compounds, and an early theory proposed that the crust is mostly sulfur. The evidence says otherwise. Infrared measurements show that volcanoes on Io erupt lava with a temperature over 1500°C (2700°F), about 300°C hotter than lavas on Earth. Sulfur on Io would boil at only 550°C, so the volcanoes must be erupting molten rock and not just liquid sulfur. Also, a few isolated mountains exist that are as high as 18 km, twice the height of Mount Everest. Sulfur is not strong enough to support such high mountains. This evidence shows that the crust must be silicate rock.

Volcanism is continuous on Io. Plumes come and go over periods of months, but some volcanic vents, such as Pele, have been active since the Voyager spacecraft first visited Io in 1979 (Figure 23-13). Earth's explosive volcanoes eject lava and ash because of water dissolved in the lava. As rising lava reaches Earth's surface, the sudden decrease in pressure allows the water to come out of solution in the lava. It is like popping the cork on a bottle of champagne. The water flashes into vapor and blasts material out of the volcano, the process that was responsible for the Mount St. Helens explosion in 1980. But Io is dry. Instead, its volcanoes appear to be powered by sulfur dioxide dissolved in the magma. When the pressure on the magma is released, the sulfur dioxide boils out of solution and blasts gas and ash high above the surface in plumes up to 500 km high. Ash falling back to the surface produces debris layers around the volcanoes, such as that around Pele in Figure 23-13. Whitish areas on the surface are frosts of sulfur dioxide.

Great lava flows can be detected carrying molten material downhill, burying the surface under layer after layer. Sometimes lava bursts upward through faults to form long lava curtains, a form of eruption seen in Hawaii. Both of these processes are shown in Figure 23-13.

What powers Io? It is bursting with energy, but it is only 5 percent bigger than Earth's moon, which is cold and dead. Io is too small to have retained heat from its formation or to remain hot from radioactive decay. In fact, the energy blasting out of its volcanoes adds up to about three times more energy than it could make by radioactive decay in its interior.

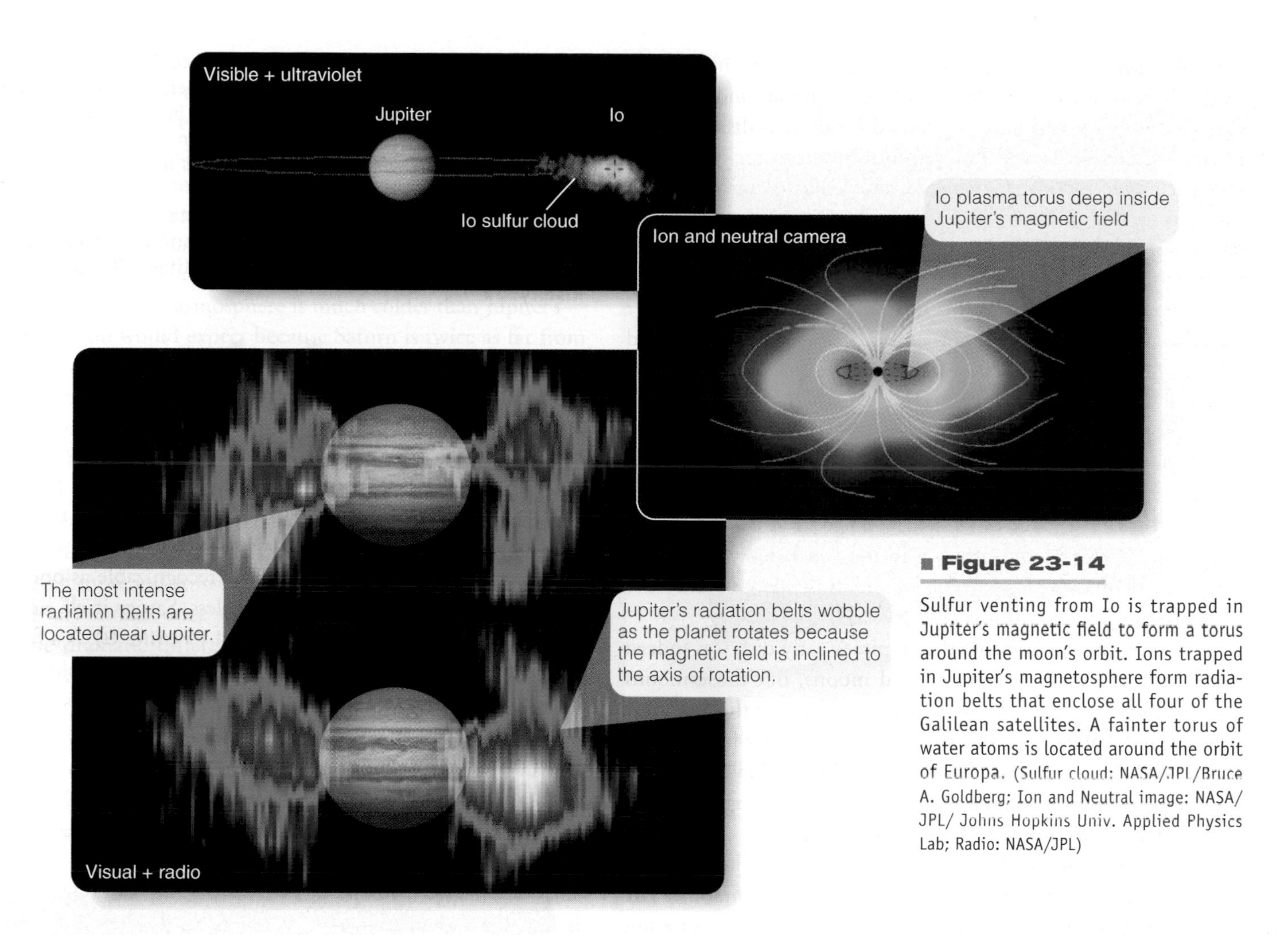
Visible + ultraviolet
Jupiter
Io
Io sulfur cloud
Ion and neutral camera
Io plasma torus deep inside Jupiter's magnetic field
The most intense radiation belts are located near Jupiter.
Jupiter's radiation belts wobble as the planet rotates because the magnetic field is inclined to the axis of rotation.
Visual + radio

■ Figure 23-14

Sulfur venting from Io is trapped in Jupiter's magnetic field to form a torus around the moon's orbit. Ions trapped in Jupiter's magnetosphere form radiation belts that enclose all four of the Galilean satellites. A fainter torus of water atoms is located around the orbit of Europa. (Sulfur cloud: NASA/JPL/Bruce A. Goldberg; Ion and Neutral image: NASA/JPL/ Johns Hopkins Univ. Applied Physics Lab; Radio: NASA/JPL)

Io is heated by the same kind of tidal heating that has affected Ganymede and Europa. Because Io is so close to Jupiter, the tides it experiences are powerful and should have forced Io's orbit to become circular long ago. Io, however, has fallen in with a bad crowd. Io, Europa, and Ganymede are locked in an orbital resonance; in the time it takes Ganymede to orbit once, Europa orbits twice and Io four times. This gravitational interaction keeps the orbits slightly elliptical; and Io, being closest to Jupiter, suffers dramatic tides, with its surface rising and falling by about 100 m. Tides on Earth move the solid ground by only a few centimeters. The resulting friction in Io is enough to melt the interior and drive volcanism. In fact, the energy flowing outward is continually recycling Io's crust. Deep layers melt, are spewed out through the volcanoes to cover the surface and are later covered themselves until they are buried so deeply that they are again melted.

Io is the most active world in our solar system because it orbits so close to Jupiter. The other Galilean moons are more distant and have had less dramatic histories.

The History of the Galilean Moons

Each time you have finished studying a world, you have tried to summarize its history. Now you have studied a system of four small worlds. Can you tell their story? To do that you need to draw on what you have learned about the moons and also on what you have learned about Jupiter and the origin of the solar system (Chapter 19).

The minor moons of Jupiter are probably captured asteroids, but the Galilean moons seem to be primordial. That is, they formed with Jupiter. Also, they seem to be a family of bodies in that their densities are related to their distance from Jupiter (■ Table 23-2).

From all the evidence, astronomers propose that the four moons formed in a disk-shaped nebula around Jupiter—a mini-solar nebula—in much the same way the planets formed from

■ Table 23-2 | The Galilean Satellites*

Name	Radius (km)	Density (g/cm³)	Orbital Period (days)
Io	1821	3.528	1.769
Europa	1561	3.014	3.551
Ganymede	2631	1.942	7.155
Callisto	2410	1.8344	16.689

*For comparison, the radius of Earth's moon is 1738 km, and its density is 3.36 g/cm³.

■ Active Figure 23-16

(a) Saturn's belt-zone circulation is not very distinct at visible wavelengths. These images were recorded when Saturn's southern hemisphere was tipped toward Earth. (NASA and E. Karkoschka) (b) Because Saturn is colder than Jupiter, the clouds form deeper in the hazy atmosphere. Notice that the three cloud layers on Saturn form at about the same temperature as do the three cloud layers on Jupiter.

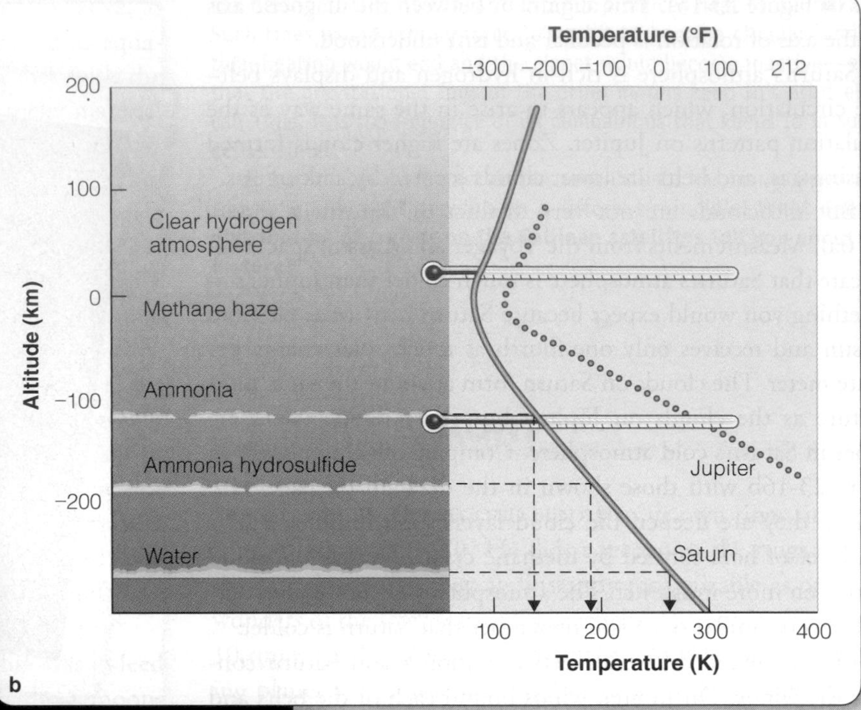

Saturn's Rings

Looking at the beauty and complexity of Saturn's rings, an astronomer once said, "The rings are made of beautiful physics." You could add that the physics is actually rather simple, but the result is one of the most astonishing sights in our solar system.

In 1609, Galileo became the first to see the rings of Saturn; but, perhaps because of the poor optics in his telescopes, he did not recognize the rings as a disk. He drew Saturn as three objects—a central body and two smaller ones on either side. In 1659, Christian Huygens realized that the rings were a disk surrounding but not touching the planet.

Understanding what the rings are has demanded over a century of human ingenuity. In 1859, James Clerk Maxwell (for whom the large mountain on Venus is named) proved mathematically that solid rings would be unstable. Saturn's rings, he concluded, had to be made of particles. In 1867, Daniel Kirkwood demonstrated that gaps in the rings were caused by resonances with some of Saturn's moons. Spectra of the rings eventually showed that the particles were mostly water ice.

Study **The Ice Rings of Saturn** on pages 532–533 and notice three points and a new term:

1. The rings are made up of billions of ice particles, each in its own orbit around the planet. But the rings can't be as old as Saturn. The rings must be replenished now and then by impacts on Saturn's icy moons or by the disruption of a small moon that wanders too close to the planet.

2. The gravitational effects of small moons called *shepherd satellites* can confine some rings in narrow strands or keep the edges of rings sharp. Moons can also produce waves in the rings that are visible as tightly wound ringlets.

3. The ring particles are confined to a thin layer in Saturn's equatorial plane. They are confined by small moons that are controlled by gravitational interactions with larger more distant moons. The rings of Saturn, and the rings of the other Jovian worlds, are created by and controlled by the planet's moons. Without the moons, there would be no rings.

Modern astronomers find simple gravitational interactions producing even more complex processes in the rings. Where particles orbit in resonance with a moon, the moon's gravity triggers spiral density waves in much the same way that spiral arms are produced in galaxies. The spiral density waves spread outward through the rings. If the moon follows an orbit that is inclined to the ring plane, the moon's gravity causes a different kind of waves—spiral bending waves—ripples extending above and below the ring plane, which spread inward. Both of these kinds of rings are shown in the inset ring image on page 533.

Many other processes occur in the rings. Specks of dust become electrically charged by sunlight, and Saturn's magnetic field lifts them out of the ring plane. Small moonlets imbedded in the rings produce gaps, waves, and scallops in the rings. The Cassini spacecraft has discovered faint rings lying far beyond those visible from Earth, and these appear to be related to moons (■ Figure 23-17).

Particle size ranges from snowlike powder to large particles and aggregates of particles. The subtle colors of the rings arise from contamination in the ice, and some areas have unusual compositions. Cassini's Division, for instance, is made up of particles that are richer in rock than most of the ring. No one knows how these differences in composition arise, but they must be related to the way the rings are formed and replenished.

Like a beautiful flower, the rings of Saturn are controlled by many different natural processes. Observations from spacecraft such as Voyager and Cassini will continue to reveal even more about the rings. Such missions are expensive, of course, but they are helping us understand what we are **(How Do We Know? 23-2)**.

The History of Saturn

The farther you journey from the sun, the more difficult it is to understand the history of the planets. Any fully successful history of Saturn should explain its low density, its peculiar magnetic field, and its beautiful rings. Planetary scientists can't tell a complete story yet, but you can understand a few of the principles that affected the formation of Saturn and its rings.

Saturn formed in the outer solar nebula, where ice particles were stable. It grew rapidly, becoming massive enough to capture hydrogen and helium gas directly from the nebula. The heavier elements probably form a denser core, and the hydrogen forms a liquid mantle containing liquid metallic hydrogen. The outward flow of heat from the core drives convection currents in this mantle that, coupled with the rapid rotation of the planet, produces its magnetic field. Because Saturn is smaller than Jupiter, it has less liquid metallic hydrogen, and its magnetic field is weaker.

Clearly the rings of Saturn are not primordial, made of material left over from the formation of the planet. That seems unlikely. Saturn, like Jupiter, would have been very hot when it

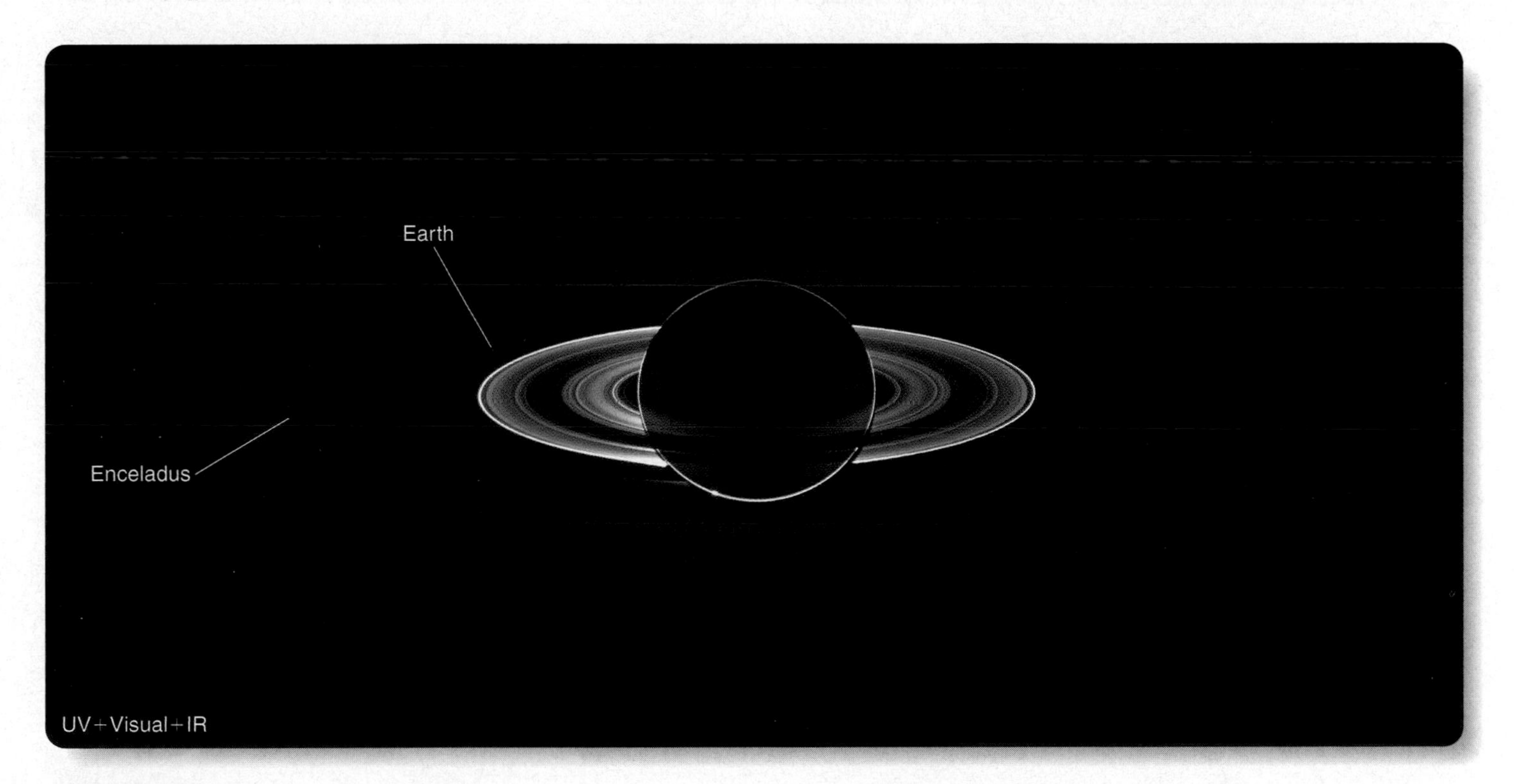

■ Figure 23-17

As its orbit carried the Cassini spacecraft through Saturn's shadow, it recorded this image, which has been adjusted to represent visual colors. Earth is visible as a faint blue dot just inside the G ring, and jets of ice particles vented from the moon Enceladus are visible at the left extreme of the larger E ring. Two faint rings were discovered in this image associated with small moons. (NASA/JPL/Space Science Institute)

The Ice Rings of Saturn

1 The brilliant rings of Saturn are made up of billions of ice particles ranging from microscopic specks to chunks bigger than a house. Each particle orbits Saturn in its own circular orbit. Much of what astronomers know about the rings was learned when the Voyager 1 spacecraft flew past Saturn in 1980, followed by the Voyager 2 spacecraft in 1981. The Cassini Spacecraft reached orbit around Saturn in 2004. From Earth, astronomers see three rings labeled A, B, and C. Voyager and Cassini images reveal over a thousand ringlets within the rings.

Saturn's rings can't be leftover material from the formation of Saturn. The rings are made of ice particles, and the planet would have been so hot when it formed that it would have vaporized and driven away any icy material. Rather, the rings must be debris from collisions between passing comets and Saturn's icy moons. Such impacts should occur every 10 million years or so, and they would scatter ice throughout Saturn's system of moons. The ice would quickly settle into the equatorial plane, and some would become trapped in rings. Although the ice may waste away due to meteorite impacts and damage from radiation in Saturn's magnetosphere, new impacts could replenish the rings with fresh ice. The bright, beautiful rings you see today may be only a temporary enhancement caused by an impact that occurred since the extinction of the dinosaurs.

Encke's division

Cassini's division

A ring

As in the case of Jupiter's ring, Saturn's rings lie inside the planet's Roché limit where the ring particles cannot pull themselves together to form a moon.

B ring

C ring
The Crepe ring

Earth to scale

Because it is so dark, the C ring has been called the Crepe ring, referring to the black, semitransparent cloth associated with funerals.

Visual-wavelength image

1a An astronaut could swim through the rings. Although the particles orbit Saturn at high velocity, all particles at the same distance from the planet orbit at about the same speed, so they collide gently at low velocities. If you could visit the rings, you could push your way from one icy particle to the next. This artwork is based on a model of particle sizes in the A ring.

The C ring contains boulder-size chunks of ice, whereas most particles in the A and B rings are more like golf balls, down to dust-size ice crystals. Further, C ring particles are less than half as bright as particles in the A and B rings. Cassini observations show that the C ring particles contain less ice and more minerals.

NASA

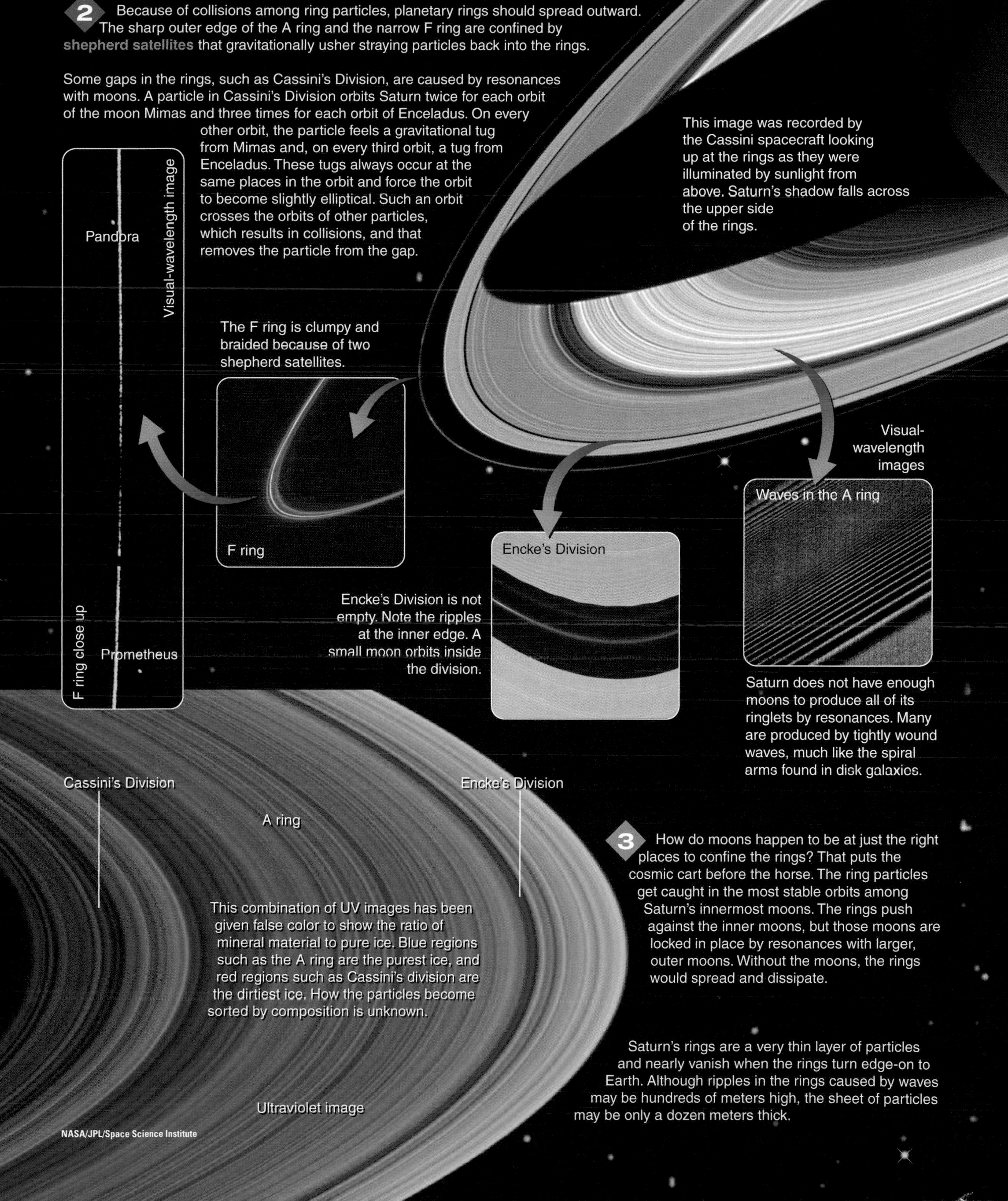
2 Because of collisions among ring particles, planetary rings should spread outward. The sharp outer edge of the A ring and the narrow F ring are confined by **shepherd satellites** that gravitationally usher straying particles back into the rings.
Some gaps in the rings, such as Cassini's Division, are caused by resonances with moons. A particle in Cassini's Division orbits Saturn twice for each orbit of the moon Mimas and three times for each orbit of Enceladus. On every other orbit, the particle feels a gravitational tug from Mimas and, on every third orbit, a tug from Enceladus. These tugs always occur at the same places in the orbit and force the orbit to become slightly elliptical. Such an orbit crosses the orbits of other particles, which results in collisions, and that removes the particle from the gap.
This image was recorded by the Cassini spacecraft looking up at the rings as they were illuminated by sunlight from above. Saturn's shadow falls across the upper side of the rings.
Visual-wavelength image
Pandora
Prometheus
F ring close up
The F ring is clumpy and braided because of two shepherd satellites.
F ring
Visual-wavelength images
Waves in the A ring
Encke's Division
Encke's Division is not empty. Note the ripples at the inner edge. A small moon orbits inside the division.
Saturn does not have enough moons to produce all of its ringlets by resonances. Many are produced by tightly wound waves, much like the spiral arms found in disk galaxies.
Cassini's Division
A ring
Encke's Division
This combination of UV images has been given false color to show the ratio of mineral material to pure ice. Blue regions such as the A ring are the purest ice, and red regions such as Cassini's division are the dirtiest ice. How the particles become sorted by composition is unknown.
Ultraviolet image
NASA/JPL/Space Science Institute
3 How do moons happen to be at just the right places to confine the rings? That puts the cosmic cart before the horse. The ring particles get caught in the most stable orbits among Saturn's innermost moons. The rings push against the inner moons, but those moons are locked in place by resonances with larger, outer moons. Without the moons, the rings would spread and dissipate.
Saturn's rings are a very thin layer of particles and nearly vanish when the rings turn edge-on to Earth. Although ripples in the rings caused by waves may be hundreds of meters high, the sheet of particles may be only a dozen meters thick.

How Do We Know? | 23-2

Who Pays for Science?

Why shouldn't you plan for a career as an industrial paleontologist? Searching out scientific knowledge can be expensive, and that raises the question of funding. Some science has direct applications, and industry supports such research. For example, pharmaceutical companies have large budgets for scientific research leading to the creation of new drugs. But some basic science is of no immediate practical value. Who pays the bill?

A paleontologist is a scientist who studies ancient life forms by examining fossils of plant and animal remains, and such research does not have commercial applications. Except for the rare Hollywood producer about to release a dinosaur movie, corporations can't make a profit from the discovery of a new dinosaur. The practical-minded stockholders of a company will not approve major investments in such research. Consequently, digging up dinosaurs, like astronomy, is poorly funded by industry.

It falls to government institutions and private foundations to pay the bill for this kind of research. The Keck Foundation has built two giant telescopes with no expectation of financial return, and the National Science Foundation has funded thousands of astronomy research projects for the benefit of society.

The discovery of a new dinosaur or a new galaxy is of no great financial value, but such scientific knowledge is not worthless. Its value lies in what it tells us about the world we live in. Such scientific research enriches our lives by helping us understand what we are. Ultimately, funding basic scientific research is a public responsibility that society must balance against other needs. There isn't anyone else to pick up the tab.

Sending the Cassini spacecraft to Saturn cost each American 56¢ per year over the life of the project.

formed, and that heat would have vaporized and driven off any leftover material. Also, such a hot Saturn would have had a very distended atmosphere, which would have slowed ring particles by friction and caused the particles to fall into the planet.

Planetary rings do not seem to be stable over 4.6 billion years, so the ring material must be more recent. Saturn's beautiful rings may have been produced within the last 100,000 years. One suggestion is that a small moon or an icy planetesimal came within Saturn's Roche limit, and tides pulled it apart. At least some of the resulting debris would have settled into the ring plane. Another possibility is that a comet struck one of Saturn's moons. Because both comets and moons in the outer solar system are icy, such a collision would produce icy debris. Bright planetary rings such as Saturn's may be temporary phenomena, forming when violent events produce fresh ice debris and then wasting away as the ice is gradually lost.

◄ SCIENTIFIC ARGUMENT ►

Why do the belts and zones on Saturn look so bland?

One of the most powerful tools of critical thought is simple comparing and contrasting. You can make that the theme of this scientific argument by comparing and contrasting Saturn with Jupiter. In the atmosphere of Jupiter, the dark belts form in regions where gas sinks, and zones form where gas rises. The rising gas cools and condenses to form icy crystals of ammonia, which are visible as bright clouds. Clouds of ammonia hydrosulfide and water form deeper, below the ammonia clouds, and are not as visible. Saturn is twice as far from the sun as Jupiter, so sunlight is four times dimmer. (Remember the inverse square law from Chapter 5.) The atmosphere is colder, and gas currents do not have to rise as far to reach cold levels and form clouds. That means the clouds are deeper in Saturn's atmosphere than in Jupiter's atmosphere. Because the clouds are deeper, they are not as brightly illuminated by sunlight and look dimmer. Also, a layer of methane-ice-crystal haze high above the ammonia clouds makes the clouds even less distinct.

Now build a new argument comparing the ring systems. **How is Saturn's ring system similar to and different from Jupiter's ring system?**

◄ ►

23-5 Saturn's Moons

SATURN HAS OVER 60 KNOWN SATELLITES—far too many to examine individually—but these moons share characteristics common to icy worlds. Most of them are small and dead, but one is big enough to have an atmosphere and perhaps even oceans and lakes.

Titan

Saturn's largest satellite is a giant ice moon with a thick atmosphere and a mysterious surface. From Earth it is only a dot of light, with no visible detail. Nevertheless, a few basic observations made from Earth can tell you a great deal about this strange world.

Titan's mass can be estimated from its influence on other moons, and its mass divided by its volume reveals that its density is 1.9 g/cm^3. Its uncompressed density (Chapter 19) is only 1.2 g/cm^3. Although it must have a rocky core, it must also contain a large amount of ices.

Titan is a bit larger than the planet Mercury and almost as large as Jupiter's moon Ganymede. Unlike these worlds, Titan has a thick atmosphere. Its escape velocity is low, but it is so far from the sun that it is very cold, and most gas atoms don't move fast enough to escape. (See Figure 22-13.) Methane was detected spectroscopically in 1944, and various hydrocarbons were found beginning in about 1970 (■ Table 23-3), but most of Titan's air is nitrogen with only 1.6 percent methane.

When the Voyager 1 and Voyager 2 spacecraft flew past Saturn in the early 1980s, their cameras could not penetrate Titan's hazy atmosphere (■ Figure 23-18). Measurements showed that

■ Table 23-3 | Some Organic Compounds Detected on Titan

C_2H_6	Ethane
C_2H_2	Acetylene
C_2H_4	Ethylene
C_3H_4	Methylacetylene
C_3H_8	Propane
C_4H_2	Diacetylene
HCN	Hydrogen cyanide
HC_3N	Cyanocetylene
C_2N_2	Cyanogen

Visual-wavelength images
Drainage channels were cut by flowing liquid methane.
Image recorded from 8 km above surface
Lakes of liquid methane look dark because they do not reflect radar waves
Icy grapefruit-size "rocks" on Titan are bathed in orange light from its hazy atmosphere.
False color Radar map
At visual wavelengths, Titan's heavy atmosphere hides its surface.
Visual

■ Figure 23-18

As the Huygens probe descended by parachute through Titan's smoggy atmosphere, it photographed the surface from an altitude of 8 km (26,000 ft). Although no liquid was present, dark drainage channels lead into the lowlands. Radar images reveal lakes of liquid methane and ethane around the poles. Once the Huygens probe landed on the surface, it radioed back photos showing a level plain and chunks of ice rounded by a moving liquid. (ESA/NASA/JPL/USGS/University of Arizona; NASA/JPL)

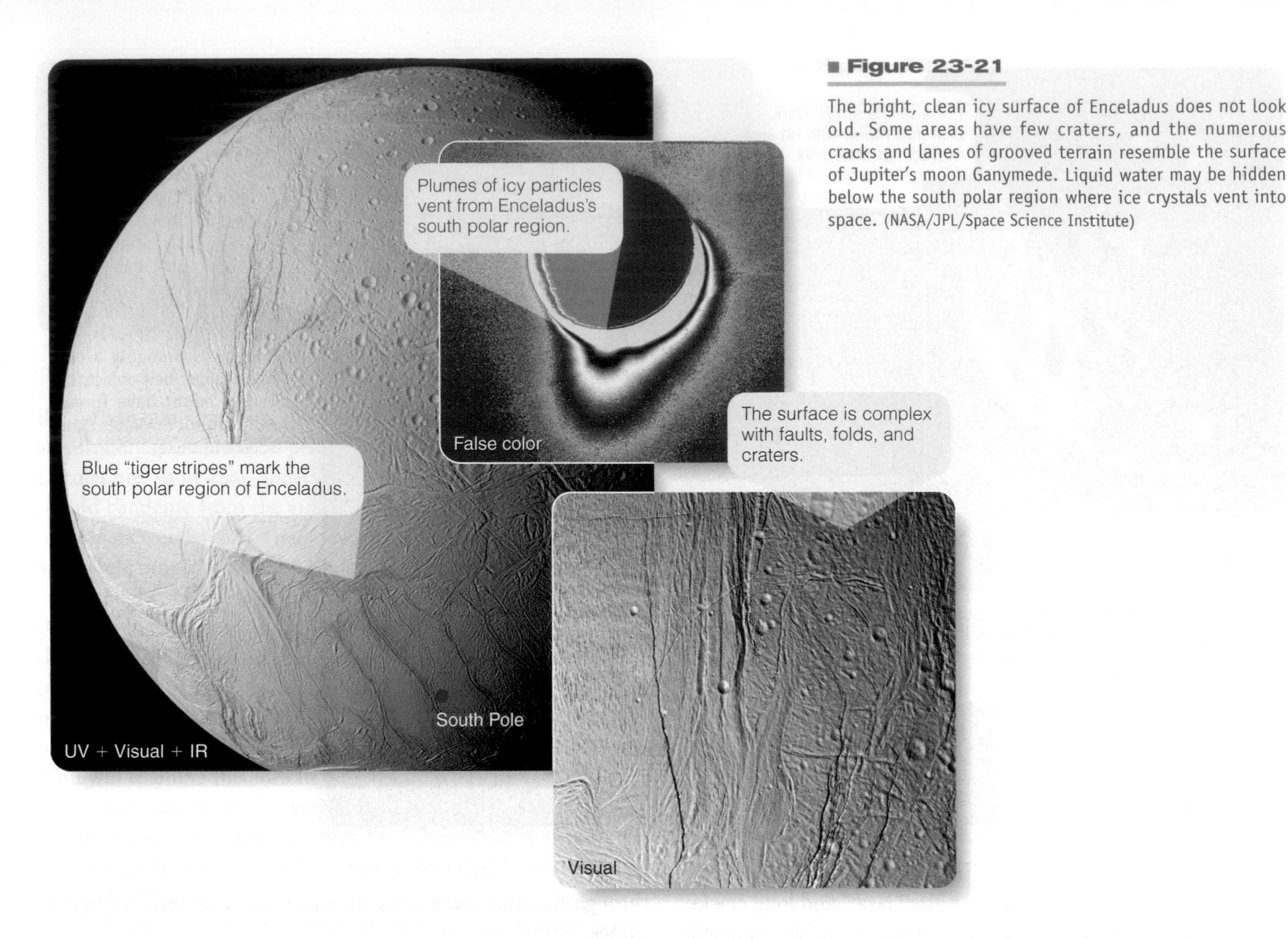
Plumes of icy particles vent from Enceladus's south polar region.
False color
The surface is complex with faults, folds, and craters.
Blue "tiger stripes" mark the south polar region of Enceladus.
South Pole
UV + Visual + IR
Visual

■ **Figure 23-21**

The bright, clean icy surface of Enceladus does not look old. Some areas have few craters, and the numerous cracks and lanes of grooved terrain resemble the surface of Jupiter's moon Ganymede. Liquid water may be hidden below the south polar region where ice crystals vent into space. (NASA/JPL/Space Science Institute)

material on its leading side. This dust could be produced by meteorites striking the outermost moon, Phoebe.

Another odd feature on Iapetus shows up in Cassini images—an equatorial ridge that stands as high as 13 km (8 mi) in some places. You can see the ridge clearly in Figure 23-22. The origin of this ridge is unknown, but it is not a minor feature. At 8 km high, it is over 50 percent higher than Mount Everest, and it extends for a long distance across the surface. That is a big pile of rock and ice. The ridge sits atop an equatorial bulge, and both ridge and bulge may have formed when Iapetus was young, spun rapidly, and was still mostly molten.

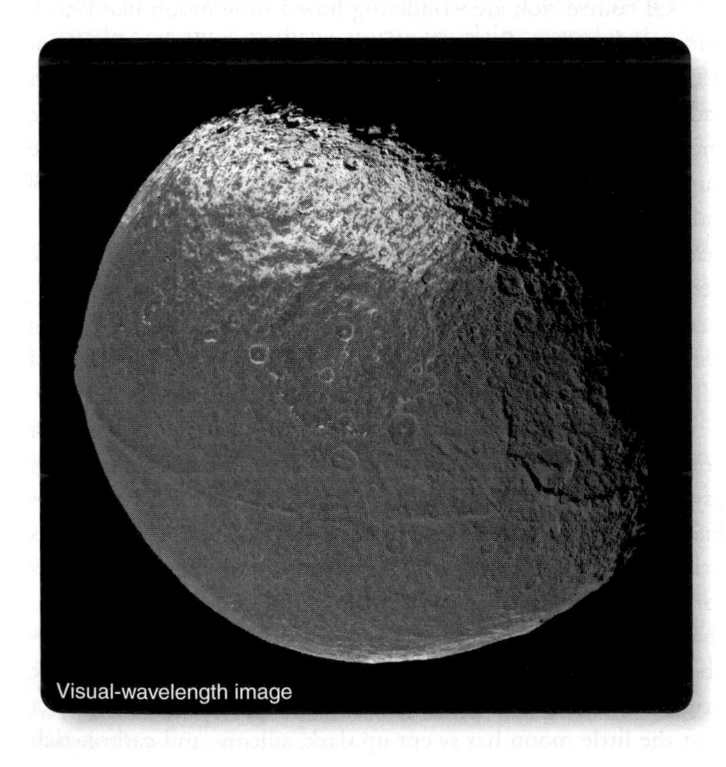
Visual-wavelength image

■ **Figure 23-22**

Like the windshield of a speeding car, the leading side of Saturn's moon Iapetus seems to have accumulated a coating of dark material. The poles and trailing side of the moon are much cleaner ice. The equatorial ridge is 20 km (12 mi) wide and up to 13 km (8 mi) high. It stretches roughly 1,300 km (800 mi) along the moon's equator. (NASA/JPL/Space Science Institute)

Saturn's moons illustrate a number of principles of comparative planetology. Small moons are irregular in shape, and old surfaces are dark and cratered. Resonances can trigger tidal heating, and that can in turn resurface moons and outgas atmospheres. Small moons can't keep atmospheres, but big, cold moons can. You are an expert in all of this, so you are ready to wonder where the moons came from.

The Origin of Saturn's Moons

Jupiter's four Galilean satellites seem clearly related to one another, and you can safely conclude that they formed with Jupiter. No such systematic relationship links Saturn's satellites. Planetary scientists suspect that many of the moons are captured icy planetesimals left over from the solar nebula and that the impacts of comets have so badly fractured them that they no longer show evidence of their common origin.

Understanding the origin of Saturn's moons is also difficult because the moons interact gravitationally in complicated ways, and the orbits they now occupy may differ from their earlier orbits. These interactions are dramatically illustrated by the two small moons that shepherd the F ring. An even more peculiar pair of moons is known as the coorbital satellites. These two irregularly shaped moonlets have orbits separated by only 100 km. Because one moon is about 200 km in diameter and the other about 100 km, they cannot pass in their orbits. Instead, the innermost moon gradually catches up with the outer moon. As the moons draw closer together, the gravity of the trailing moon slows the leading moon and makes it fall into a lower orbit. Simultaneously, the gravity of the leading moon pulls the trailing moon forward, and it rises into a higher orbit (■ Figure 23-23). The higher orbit has a longer period, so the trailing moon begins to fall behind the leading moon, which is now in a smaller, faster orbit. In this elegant dance, the moons exchange orbits and draw apart only to meet again and again. It seems very likely that these two moons are fragments of a larger moon destroyed by a major impact.

In addition to waltzing coorbital moons, the Voyager spacecraft discovered small moonlets trapped at the L4 and L5 Lagrangian points (see Figure 13-5) in the orbits of Dione and Tethys. These points of stability lie 60° ahead of and 60° behind the two moons, and small moonlets can become trapped in these regions. (You will see in Chapter 25 that asteroids are trapped in the Lagrangian points of Jupiter's orbit around the sun.) This gravitational curiosity is surely not unique to the Saturn system, and it warns that interactions between moons can dramatically alter their orbits.

Some of Saturn's larger moons, especially Titan, may have formed with the planet, but the orbital relationships and intense cratering suggest that the moons have interacted and may have collided with large planetesimals and the heads of comets in the

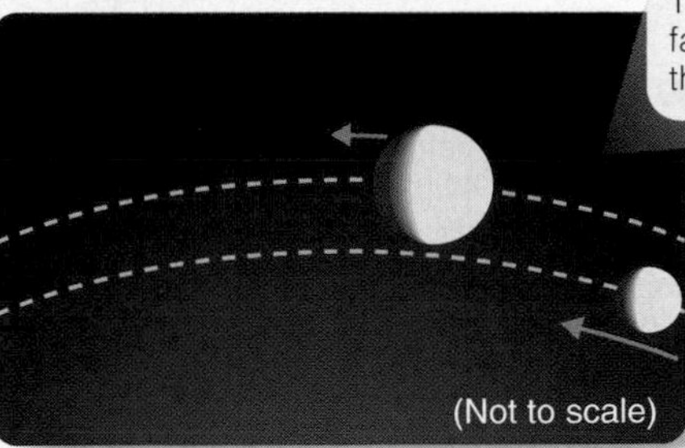

Coorbital Moons
The inner moon orbits faster and overtakes the outer moon.
(Not to scale)

The gravitational interaction pulls the outer moon backward and the inner moon forward.

As they approach, the inner moon moves to a higher orbit, and the outer moon sinks to a lower orbit.

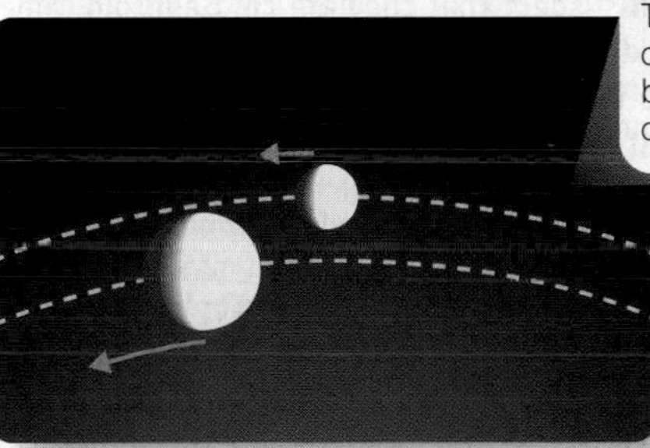

The moons have changed orbits, and the inner moon begins gaining on the outer moon.

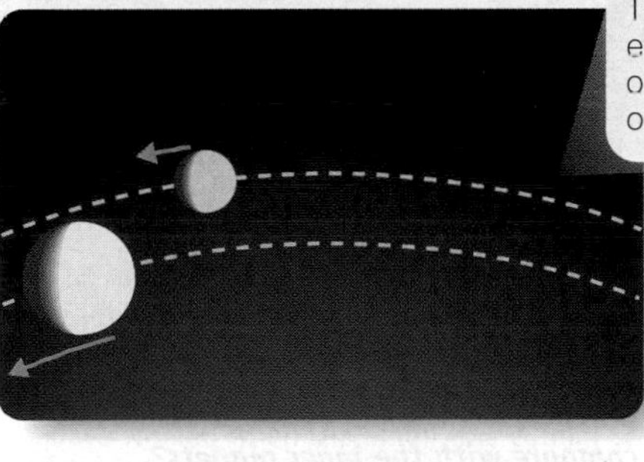

The inner moon will eventually overtake the outer moon from behind once again.

■ **Active Figure 23-23**

Saturn's coorbital moons follow nearly identical orbits. The moon in the lower orbit travels faster and always overtakes the other moon from behind. The moons interact and change orbits over and over again.

Review Questions

ThomsonNOW™ Assess your understanding of this chapter's topics with additional quizzing and animations at www.thomsonedu.com.

WebAssign The problems from this chapter may be assigned online in WebAssign.

1. Why is Jupiter more oblate than Earth? Do you expect all Jovian planets to be oblate? Why or why not?
2. How do the interiors of Jupiter and Saturn differ? How does this affect their magnetic fields?
3. What is the difference between a belt and a zone?
4. How can you be certain that Jupiter's ring does not date from the formation of the planet?
5. If Jupiter had a satellite the size of our own moon orbiting outside the orbit of Callisto, what would you predict for its density and surface features?
6. Why are there no craters on Io and few on Europa? Why should you expect Io to suffer more impacts per square kilometer than Callisto?
7. Why are the belts and zones on Saturn less distinct than those on Jupiter?
8. If Saturn had no moons, what do you suppose its rings would look like?
9. Where did the particles in Saturn's rings come from?
10. How can Titan keep an atmosphere when it is smaller than airless Ganymede?
11. If you piloted a spacecraft to visit Saturn's moons and wanted to land on a geologically old surface, what features would you look for? What features would you avoid?
12. Why does the leading side of some satellites differ from the trailing side?
13. **How Do We Know?** How would you know whether a person building a telescope was a scientist or an engineer?
14. **How Do We Know?** Why would you expect research in archaeology to be less well funded than research in chemistry?

Discussion Questions

1. Some astronomers argue that Jupiter and Saturn are unusual, while other astronomers argue that all solar systems should contain one or two such giant planets. What do you think? Support your argument with evidence.
2. Why don't the terrestrial planets have rings?

Problems

1. What is the maximum angular diameter of Jupiter as seen from Earth? What is the minimum? Repeat this calculation for Saturn and Titan. (*Hint:* Use the small-angle formula.)
2. The highest-speed winds on Jupiter are in the equatorial jet stream, which has a velocity of 150 m/s. How long does it take for these winds to circle Jupiter?
3. What are the orbital velocity and period of a ring particle at the outer edge of Jupiter's ring? At the outer edge of Saturn's A ring? (*Hints:* The radius of the edge of the A ring is 136,500 km. See Chapter 5.)
4. What is the angular diameter of Jupiter as seen from the surface of Callisto? (*Hint:* Use the small-angle formula.)
5. What is the escape velocity from the surface of Ganymede if its mass is 1.5×10^{26} g and its radius is 2628 km? (*Hint:* See Chapter 5.)
6. If you were to record the spectrum of Saturn and its rings, you would find light from one edge of the rings redshifted and light from the other edge blueshifted. If you observed at a wavelength of 500 nm, what difference in wavelength should you expect between the two edges of the rings? (*Hints:* See Problem 3 and Chapter 7.)
7. What is the difference in orbital velocity between particles at the outer edge of Saturn's B ring and particles at the inner edge of the B ring? (*Hint:* The outer edge of the B ring has a radius of 117,500 km, and the inner edge has a radius of 92,000 km.)
8. What is the difference in orbital velocity between the two coorbital satellites if the semimajor axes of their orbits are 151,400 km and 151,500 km? The mass of Saturn is 5.7×10^{26} kg. (*Hint:* See Chapter 5.)

Learning to Look

1. This image shows a segment of the surface of Jupiter's moon Callisto. Why is the surface dark? Why are some craters dark and some white? What does this image tell you about the history of Callisto?

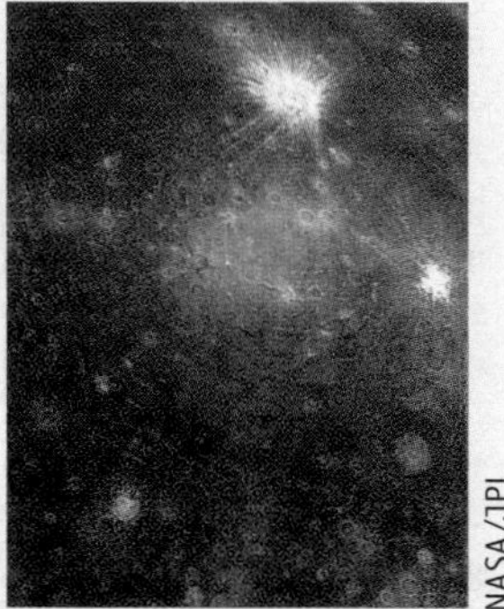

NASA/JPL

2. The Cassini spacecraft recorded this image of Saturn's A ring and Encke's division. What do you see in this photo that tells you about processes that confined and shape planetary rings?

NASA/JPL/Space Science Institute

Virtual Astronomy Labs

Lab 7: Planetary Atmospheres and Their Retention
This lab investigates the retention of atmospheres. You will explore the factors that govern the loss of atmospheric gases, and you will see why certain bodies can retain some gases but not others.

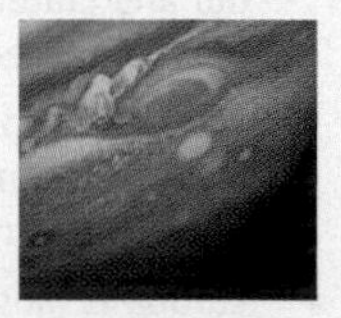

24 Uranus, Neptune, and the Dwarf Planets

Uranus is a cloudy, Jovian world far from the sun. It is orbited by dark, rocky particles that make up narrow rings much enhanced in this artist's impression.
(Don Dixon)

Guidepost

Two planets circle the sun in the twilight beyond Saturn. You will find Uranus and Neptune strangely different from Jupiter and Saturn but recognizable as planets. As you explore you will also discover a family of dwarf planets, which includes Pluto, which will give you important clues to the origin of our solar system. This chapter will help you answer five essential questions:

- **How is Uranus different from Jupiter?**
- **How did Uranus, its rings, and its moons form and evolve?**
- **How is Neptune different from Uranus?**
- **How did Neptune, its rings, and its moons form and evolve?**
- **How are Pluto and the dwarf planets related to the origin of the solar system?**

As you think about the search for worlds in the outer solar system, you will be able to answer a fundamental question about how science works:

- How Do We Know? **Why are nearly all truly important scientific discoveries made by accident?**

As you finish this chapter, you will have visited all of the major worlds in our solar system. But there is more to see. Vast numbers of small rocky and icy bodies orbit among the planets, and the next chapter will introduce you to these fragments from the age of planet building.

A good many things go around in the dark besides Santa Claus.

HERBERT HOOVER

OUT IN THE DARKNESS BEYOND SATURN, out where sunlight is 1000 times fainter than on Earth, there are things going around that Aristotle, Galileo, and Newton never imagined. They knew about Mercury, Venus, Mars, Jupiter, and Saturn, but our solar system includes worlds that were not discovered until after the invention of the telescope. The stories of these discoveries highlight the process of scientific discovery, and the characteristics of these dimly lit worlds will reveal more of nature's secrets from the birth of the solar system.

24-1 Uranus

IN MARCH 1781, Benjamin Franklin was in France raising money, troops, and arms for the American Revolution. George Washington and his colonial army were only six months away from the defeat of Cornwallis at Yorktown and the end of the war. In England, King George III was beginning to show signs of madness. And a German-speaking music teacher in the English resort city of Bath was about to discover the planet Uranus.

The Discovery of Uranus

William Herschel (■ Figure 24-1) came from a musical family in Hanover, Germany, but immigrated to England as a young man and eventually obtained a prestigious job as the organist at the Octagon Chapel in Bath.

To compose exercises for his students and choral and organ works for the chapel, Herschel studied the mathematical principles of musical harmony from a book by Professor Robert Smith of Cambridge. The mathematics in the book was so interesting that Herschel searched out other works by Smith, including a book on optics. Of course, it is not surprising that an 18th-century book on optics written by a professor in England relied heavily on Isaac Newton's discoveries and described the principles of Newton's reflecting telescope (see Chapter 6). Herschel and his brother Alexander began building telescopes, and William went on to study astronomy in his spare time.

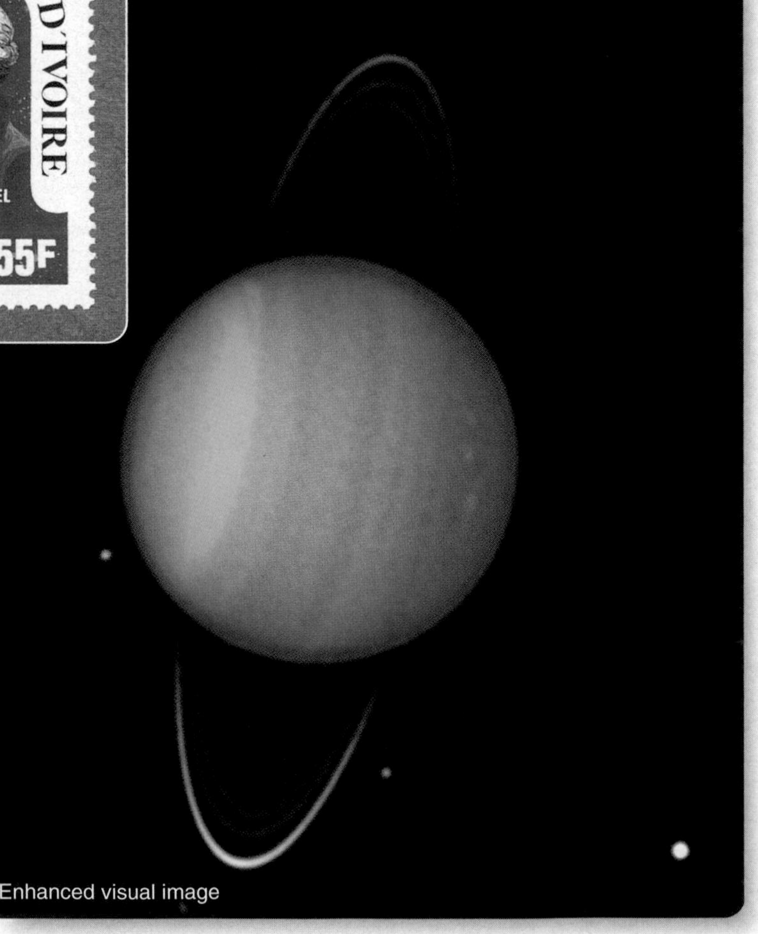

■ **Figure 24-1**

When William Herschel discovered Uranus in 1781, he saw only a tiny blue dot. He never knew how interesting the planet is. Even the designer of this stamp marking the passage of Comet Halley in 1986 had to guess at the planet's appearance. Voyager 2 visited Uranus in 1986, and now the Hubble Space Telescope and Earth-based giant telescopes can image the planet's clouds and thin rings. (NASA)

Herschel's telescopes were similar to Newton's in that they had metal mirrors, but Herschel's were much larger. Newton's telescope had a mirror about 1 in. in diameter, but Herschel developed ways of making much larger mirrors, and he soon had telescopes as long as 20 ft. One of his favorite telescopes was 7 ft long and had a mirror 6.2 in. in diameter. Using this telescope, he began the research project that led to the discovery of Uranus.

Herschel did not set out to search for a planet; he was trying to detect stellar parallax produced by Earth's motion around the sun. No one had yet detected this effect, although by the 1700s all astronomers accepted that Earth moved. Galileo had pointed out that parallax might be detected if a nearby star and a very distant star lay so nearly along the same line of sight that they looked like a very close double star through a telescope. In such a case, Earth's orbital motion would produce a parallactic shift in the position of the nearby star with respect to the more distant star (■ Figure 24-2). Herschel began to examine all stars brighter than eighth magnitude to search for double stars that might show parallax. That project alone took over two years.

On the night of March 13, 1781, Herschel set up his 7-ft telescope in his back garden and continued his work. He later wrote, "In examining the small stars in the neighborhood of H Geminorum, I perceived one that appeared visibly larger than the rest." As seen from Earth, Uranus is never larger in angular diameter than 3.7 seconds of arc, so Herschel's detection of the disk illustrates the quality of his telescope and his eye. At first he suspected that the object was a comet, but other astronomers quickly realized that it was a planet orbiting the sun beyond Saturn.

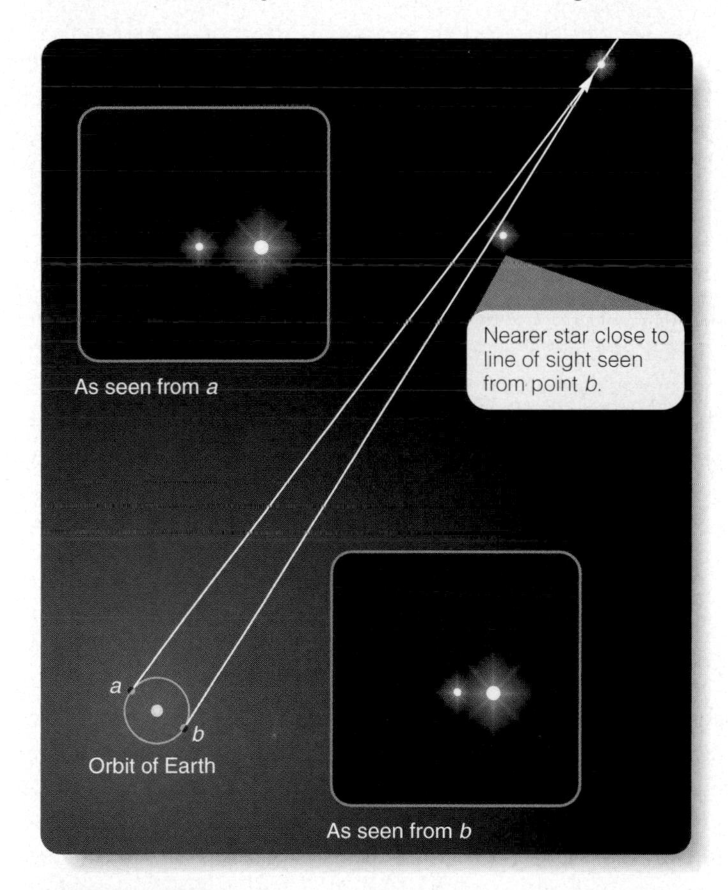

■ Figure 24-2

A double star consisting of a nearby star and a more distant star could be used to detect stellar parallax. Seen from point *b* in Earth's orbit, the two stars appear closer together than from point *a*. The effect is actually much too small to detect by eye.

The discovery of Uranus made Herschel world famous. Since antiquity, astronomers had known of five planets—Mercury, Venus, Mars, Jupiter, and Saturn—and they had supposed that the list was complete. Herschel's discovery extended the classical universe by adding a new planet. The English public accepted Herschel as their astronomer-hero, and, having named the new planet *Georgium Sidus* (George's Star) after King George III, Herschel received a royal pension. The former music teacher was welcomed into court society, where he eventually met and married a wealthy widow and took his place as one of the great English astronomers. His new financial position allowed him to build large telescopes on his estate and, with his sister Caroline, a talented astronomer herself, he attempted to map the extent of the universe. You read about their research in Chapter 15.

Continental astronomers were less than thrilled that an Englishman had made such a great discovery, and even some professional English astronomers thought Herschel a mere amateur. They called his discovery a lucky accident. Herschel defended himself by making three points. First, he had built some of the finest-quality telescopes then in existence. Second, he had been conducting a systematic research project and would have found Uranus eventually because he was inspecting all of the brighter stars visible with his telescope. And third, he had great experience seeing fine detail with his telescopes. As a musician, he knew the value of practice and applied it to the business of astronomical observing. In fact, records show that other astronomers had seen Uranus at least 17 times before Herschel, but each time they failed to notice that it was not a star. They plotted Uranus on their charts as if it was just another faint star.

This illustrates one of the ways in which scientific discoveries are made. Often, discoveries seem accidental, but on closer examination you find that the scientist has earned the right to the discovery through many years of study and preparation **(How Do We Know? 24-1)**. To quote a common saying, "Luck is what happens to people who work hard."

Continental astronomers, especially the French, insisted that the new planet not be named after an English king. They, along with many other non-English astronomers, stubbornly called the planet Herschel. Some years later, German astronomer J. E. Bode suggested the name Uranus, one of the oldest of the Greek gods.

Over the half-century following the discovery of Uranus, astronomers noted that Newton's laws did not exactly predict the observed position of the planet. Tiny variations in the orbital motion of Uranus eventually led to the discovery of Neptune, a controversial story you will read later in this chapter.

How Do We Know? 24-1

Scientific Discoveries

Why didn't Galileo expect to discover Jupiter's moons? In 1928, Alexander Fleming noticed that bacteria in a culture dish were avoiding a spot of mold. He went on to discover penicillin. In 1895, Conrad Roentgen noticed a fluorescent screen glowing in his laboratory when he experimented with other equipment. He discovered X rays. In 1896, Henri Becquerel stored a uranium mineral on a photographic plate safely wrapped in black paper. The plate was fogged, and Becquerel discovered natural radioactivity. Like many discoveries in science, these seem to be accidental; but, as you have seen in this chapter, "accidental" doesn't quite describe what happened.

The most important discoveries in science are those that totally change the way people think about nature, and it is very unlikely that anyone would predict such discoveries. For the most part, scientists work within a paradigm (How Do We Know? 4-1), a set of models, theories, and expectations about nature, and it is very difficult to imagine natural events that lie beyond that paradigm. Ptolemy, for example, could not have imagined galaxies because they were not part of his geocentric paradigm. That means that the most important discoveries in science are almost always unexpected.

An unexpected discovery, however, is not the same as an accidental discovery. Fleming discovered penicillin in his culture dish not because he was the first to see it, but because he had studied bacterial growth for many years; so, when he saw what many others must have seen before, he recognized it as important. Roentgen realized that the glowing screen in his lab was important, and Becquerel didn't discard that fogged photographic plate. Long years of experience prepared them to recognize the significance of what they saw.

A historical study has shown that each time astronomers build a telescope that significantly surpasses existing telescopes, their most important discoveries are unexpected. Herschel didn't expect to discover Uranus with his 7-foot telescope, and modern astronomers didn't expect to discover dark energy with the Hubble Space Telescope.

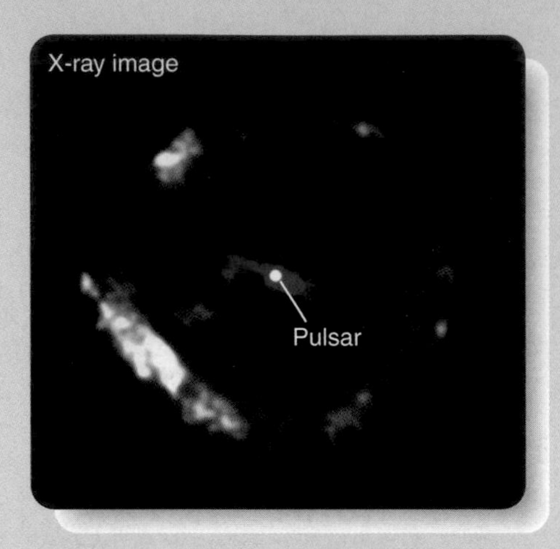

The discovery of pulsars, spinning neutron stars, was totally unexpected, as important scientific discoveries often are. (NASA/McGill/V. Kaspi et al.)

Scientists pursuing basic research are rarely able to explain the potential value of their work, but that doesn't make their discoveries accidental. They earn their right to those lucky accidents.

The Motion of Uranus

Uranus orbits nearly 20 AU from the sun and takes 84 years to go around once (**Celestial Profile 9**). The ancients thought of Saturn as the slowest of the planets, but Saturn orbits in a bit over 29 years. Uranus, being further from the sun, moves even slower than Saturn and has a longer orbital period.

The rotation of Uranus is peculiar. Earth rotates approximately upright in its orbit. That is, its axis of rotation is inclined only 23.5° from the perpendicular to its orbit. Uranus, in contrast, rotates on an axis that is inclined 97.9° from the perpendicular to its orbit. It rotates on its side (■ Figure 24-3).

The seasons on Uranus are extreme. The first good photographs of Uranus were taken in 1986, when the Voyager 2 spacecraft flew past. At that time, Uranus was in the segment of its orbit in which its south pole faced the sun. Consequently, its southern hemisphere was bathed in continuous sunlight, and a creature living on Uranus (an unlikely possibility, as you will discover later) would have seen the sun near the planet's south celestial pole. The sun was at winter solstice on Uranus in 1986, and you can see this at lower left in Figure 24-3. Over the next two decades, Uranus moved about a quarter of the way around its orbit, and, with the sun shining down from above the planet's equator, a citizen of Uranus would see the sun rise and set with the rotation of the planet. The sun reached the vernal equinox on Uranus in December 2007, and you can see that geometry by looking at the lower right in Figure 24-3. As Uranus continues along its orbit, the sun approaches the planet's north celestial pole, and the southern hemisphere of the planet experiences a lightless winter lasting 21 Earth years. In other words, the ecliptic on Uranus passes very near the planet's celestial poles, and the resulting seasons are extreme.

The Atmosphere of Uranus

Like Jupiter and Saturn, Uranus has no surface. The gases of its atmosphere—mostly hydrogen, 15 percent helium, and a few percent methane, ammonia, and water vapor—blend gradually into a fluid interior.

Seen through Earth-based telescopes, Uranus is a small, featureless blue disk. The blue color arises because the atmosphere contains methane, a good absorber of longer-wavelength photons. As sunlight penetrates into the atmosphere and is scattered back out, the longer-wavelength (red) photons are more likely to be absorbed. That means that the light entering your eye is richer in blue photons, giving the planet a blue color.

As Voyager 2 drew closer to the planet in late 1985, astronomers studied the images radioed back to Earth. Uranus was a pale blue ball with no obvious clouds, and only when the images were carefully computer enhanced was any banded structure detected (■ Figure 24-4). A few very high clouds of methane ice particles were detected, and their motions allowed astronomers to measure the rotation period of the planet.

You can understand the nearly featureless appearance of the atmosphere by studying the temperature profile of Uranus shown in ■ Figure 24-5. The atmosphere of Uranus is much colder than that of Saturn or Jupiter. Consequently, the three cloud layers of ammonia, ammonia hydrosulfide, and water that form the belts and zones in the atmospheres of Jupiter and Saturn lie very deep in the atmosphere of Uranus. These cloud layers, if they exist at all in Uranus, are not visible because of the thick atmosphere of hydrogen through which an observer has to look. The clouds that are visible on Uranus are clouds of methane ice crystals, which form at such a low temperature that they occur high in the atmosphere of Uranus. Figure 24-5 shows that there can be no methane clouds on Jupiter because that planet is too warm. The coldest part of Saturn's atmosphere is just cold enough to form a thin methane haze high above its more visible cloud layers. (See Figure 23-16b.)

The clouds and atmospheric banding faintly visible on Uranus appear to be the result of belt–zone circulation, which is a bit surprising. Uranus rotates on its side, so solar energy strikes its surface in a geometry quite different than that on Jupiter and Saturn. Evidently belt–zone circulation is dominated by the rotation of the planet and not by the direction of sunlight.

The Voyager 2 images from 1986 made some astronomers expect that Uranus was always a nearly featureless planet, but later observations have revealed that Uranus has seasons. Since 1986, Uranus has moved along its orbit, and spring has come to its northern hemisphere. The Hubble Space Telescope and giant Earth-based telescopes have detected changing clouds on Uranus including a dark cloud that may be a vortex resembling the spots on Jupiter (■ Figure 24-6). The clouds appear to be part of a seasonal cycle on Uranus, but its year lasts 84 Earth years, so you will have to be patient to see summer.

The Interior of Uranus

Astronomers cannot describe the interiors of Uranus and Neptune as accurately as they can the interiors of Jupiter and Saturn. Observational data are sparse, and the materials inside these planets are not as easy to model as simple liquid hydrogen.

The average density of Uranus, 1.3 g/cm^3, tells you that the planet must contain a larger share of dense materials than Jupiter or Saturn. Nearly all models of the interior of Uranus contain three layers. The uppermost layer, the atmosphere, is rich in hydrogen and helium. Below the atmosphere, a deep mantle must contain large amounts of water, methane, and ammonia ices mixed with

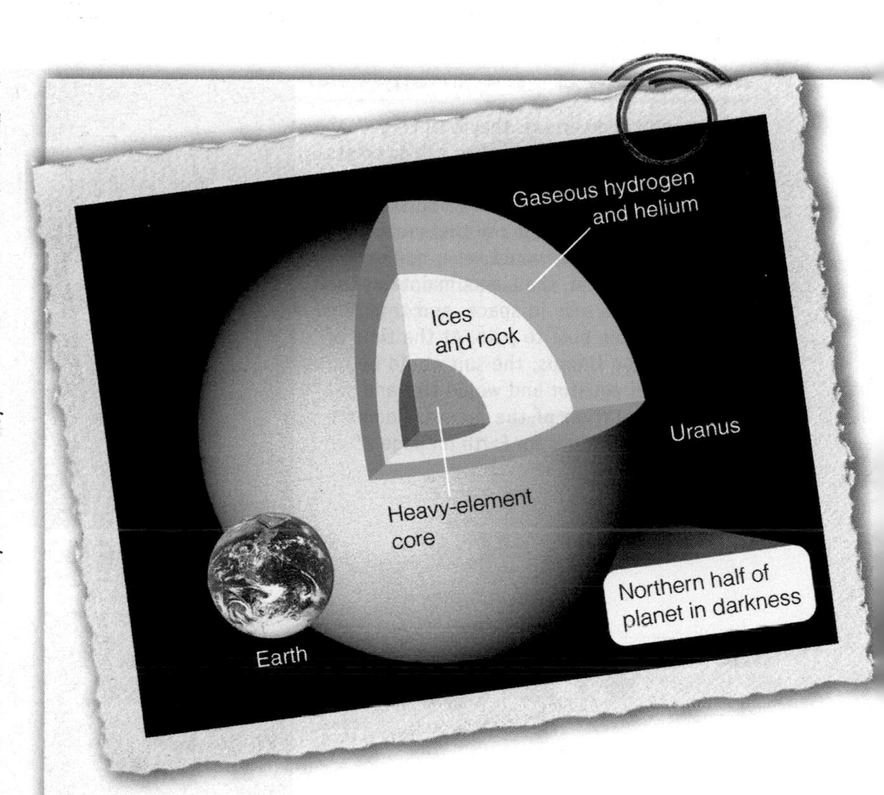

Uranus rotates on its side; when Voyager 2 flew past in 1986, the planet's south pole was pointed almost directly at the sun. (NASA)

Celestial Profile 9: Uranus

Motion:

Average distance from the sun	19.18 AU (28.69 × 10^8 km)
Eccentricity of orbit	0.0461
Maximum distance from the sun	20.1 AU (30.0 × 10^8 km)
Minimum distance from the sun	18.3 AU (27.4 × 10^8 km)
Inclination of orbit to ecliptic	0°46′23″
Average orbital velocity	6.81 km/s
Orbital period	84.013 y (30,685 days)
Period of rotation	17^h14^m
Inclination of equator to orbit	97°55′

Characteristics:

Equatorial diameter	51,118 km (4.01 $D_{\oplus}$)
Mass	8.69 × 10^{25} kg (14.54 $M_{\oplus}$)
Average density	1.318 g/cm^3
Gravity	0.919 Earth gravity
Escape velocity	22 km/s (1.96 $V_{\oplus}$)
Temperature above cloud tops	−220°C (−364°F)
Albedo	0.35

Personality Point:

Most creation stories begin with a separation of opposites, and Greek mythology is no different. Uranus (the sky) separated from Gaia (Earth) who was born from the void, Chaos. They gave birth to the giant Cyclops, Cronos (Saturn, father of Zeus) and his fellow Titans. Uranus is sometimes called the starry sky, but the sun (Helius), moon (Selene), and the stars were born later, so Uranus, one of the most ancient gods, began as the empty, dark sky.

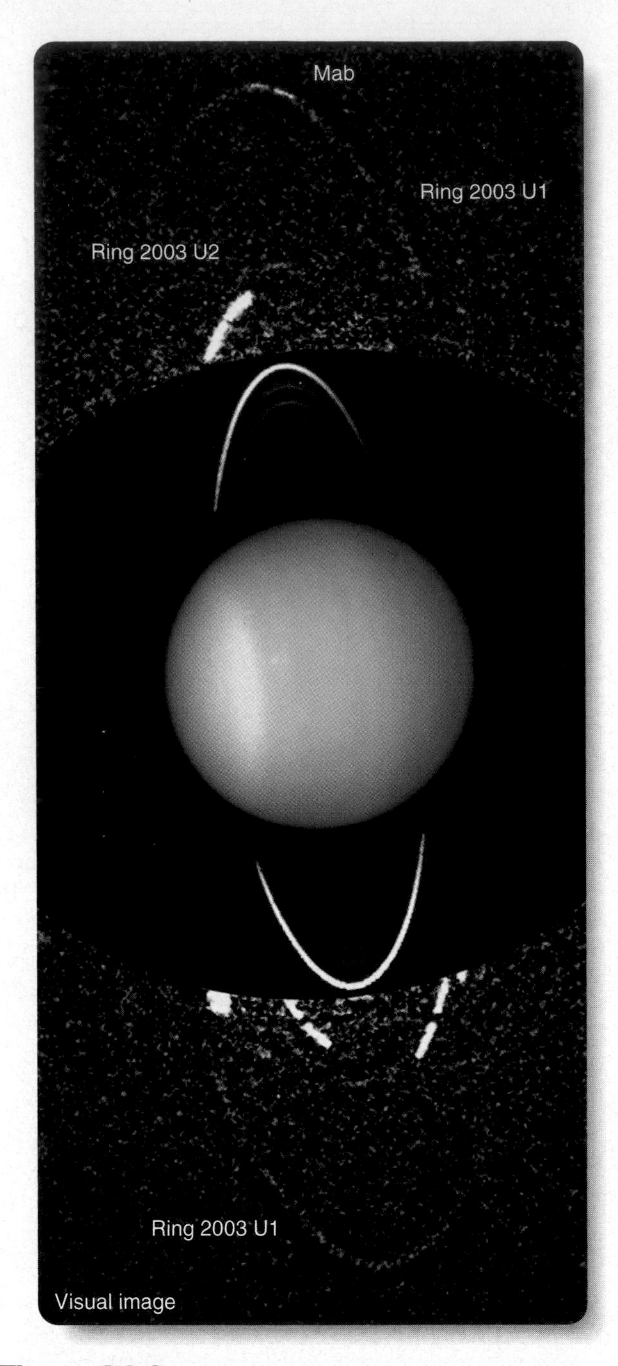

■ **Figure 24-9**

Two newly discovered rings orbit Uranus far outside the previously known rings. The outermost ring follows the orbit of the small moon Mab, only 7.5 km (12 mi) in diameter. Bright arcs in this photo were caused by moons moving along their orbits during the long time exposure. (NASA, ESA, M. Showalter, STEI Institute)

Miranda. The names Umbriel and Ariel are names from Alexander Pope's *The Rape of the Lock,* and the rest are from Shakespeare's *A Midsummer Night's Dream* and *The Tempest* (an Ariel also appears in *The Tempest*). Spectra show that the moons contain frozen water although their surfaces are dark. Planetary scientists assumed they were made of dirty ices, but little more was known of the moons before Voyager 2 flew through the system.

In addition to imaging the known moons, the Voyager 2 cameras discovered ten more moons too small to have been seen from Earth. Since then, the construction of new-generation telescopes and the development of new imaging techniques have allowed astronomers to find even more small moons orbiting Uranus. Roughly 30 are currently known, but there are almost certainly more to be found.

The smaller moons are all as dark as coal. They are icy worlds with surfaces that have been darkened by impacts. In addition, they orbit inside the radiation belts, and the radiation can convert methane trapped in their ices into dark carbon deposits and further darken their surfaces.

More is known about the five larger moons. They are all tidally locked to Uranus, and that means their south poles were pointed toward the sun in 1986. Voyager could not photograph their northern hemispheres, so the analysis of their geology must depend on images of only half their surfaces. The densities of the moons suggest that they contain relatively large rock cores surrounded by icy mantles, as shown in Figure 24-10.

Oberon, the outermost of the large moons, has a cratered surface, but evidence that it was once an active moon is visible in ■ Figure 24-11. A large fault crosses the sunlit hemisphere, and dark material, perhaps dirty water "lava," appears to have flooded the floors of some craters.

Titania is the largest of the five moons and has a heavily cratered surface, but it has no very large craters (Figure 24-11). This suggests that after the end of the heavy bombardment, when most of the large debris had been swept up, the young Titania underwent an active phase in which its surface was flooded with water that covered early craters with fresh ice. Since then, the craters that have formed are not as large as the largest of those that were erased. The network of faults that crosses Titania's surface is another sign of past activity.

Umbriel, the next moon inward, is a dark, cratered world with no sign of faults or surface activity (Figure 24-11). It is the darkest of the moons, with an albedo of only 0.16 compared with 0.25 to 0.45 for the other moons. Its crust is a mixture of rock and ice. A bright crater floor in one region suggests that some clean ice may lie at shallow depths in some regions.

In contrast to Umbriel, Ariel has the brightest surface of the five major satellites and shows clear signs of geological activity. It is crossed by faults over 10 km deep, and some regions appear to have been smoothed by resurfacing, as you can see in Figure 24-11. Crater counts show that the smoothed regions are younger than the other regions. Ariel may have been subject to tidal heating caused by an orbital resonance with Miranda and Umbriel.

Miranda is a mysterious moon. As you can see in Figure 24-11, it is the smallest of the five large moons, but it appears to have been the most active. In fact, its active past appears to have been quite unusual. Miranda is marked by oval patterns of grooves known as **ovoids** (■ Figure 24-12). These features were originally thought to have been produced when mutual gravita-

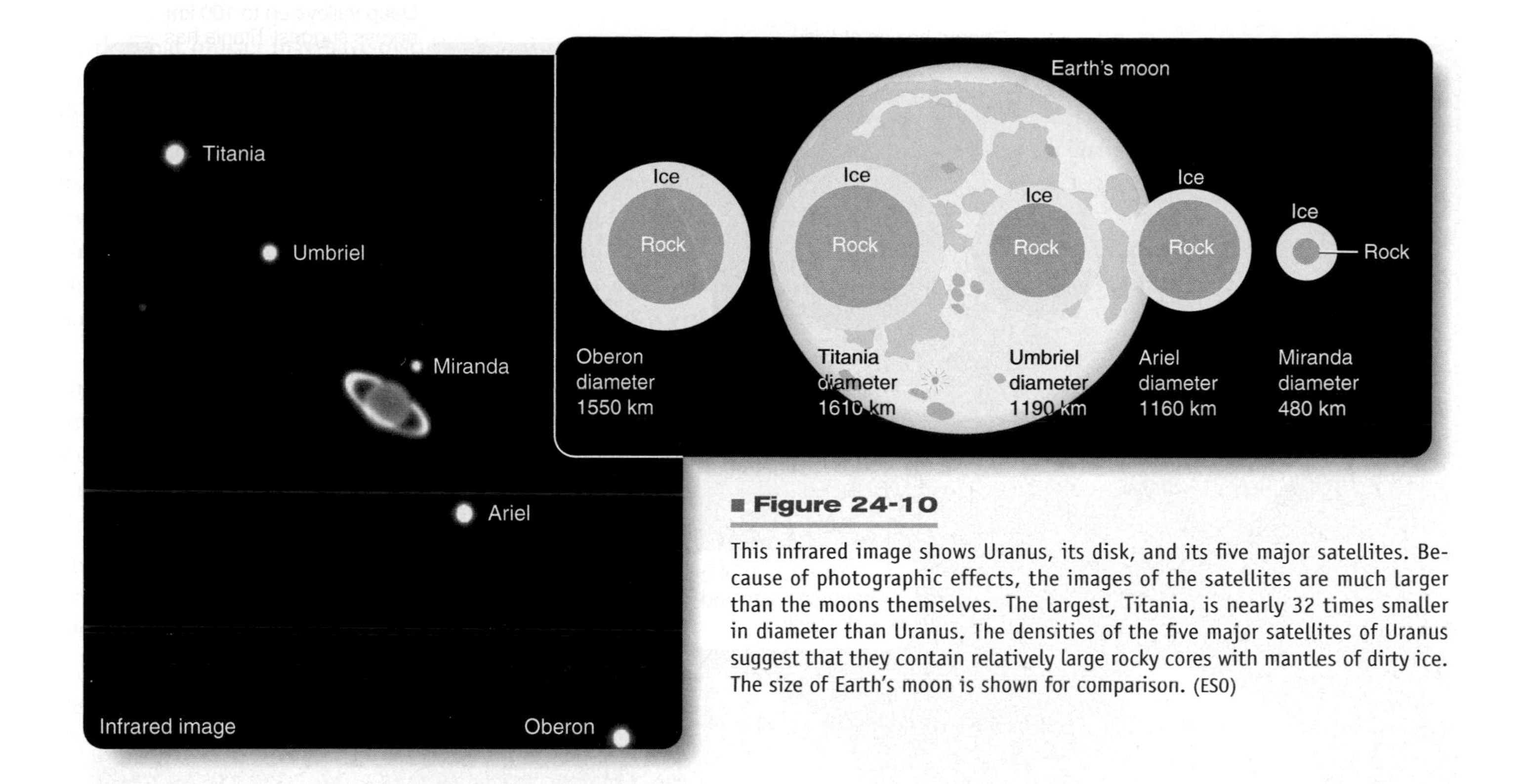

■ Figure 24-10

This infrared image shows Uranus, its disk, and its five major satellites. Because of photographic effects, the images of the satellites are much larger than the moons themselves. The largest, Titania, is nearly 32 times smaller in diameter than Uranus. The densities of the five major satellites of Uranus suggest that they contain relatively large rocky cores with mantles of dirty ice. The size of Earth's moon is shown for comparison. (ESO)

tion pulled together fragments of an earlier moon shattered by a major impact. More recent studies of the ovoids show that they are associated with faults, ice-lava flows, and rotated blocks of crust. These features suggest that a major impact was not involved. Rather, internal heat may have driven large, but very slow convection currents in Miranda's icy mantle that created the ovoids.

Certainly, Miranda has had a violent past. Near the equator, a huge cliff rises 20 km. If you stood in your spacesuit at the top of the cliff and dropped a rock over the edge, it would fall for 10 minutes before hitting the bottom. Nevertheless, the cliff and the ovoids are old. Miranda is no longer geologically active, and you can read hints of its history on its disturbed surface.

The peculiar geology of Miranda illustrates the problem astronomers face in trying to understand these icy moons. Has Miranda been dominated by interior or exterior factors? If the ovoids were caused by convection in Miranda's icy mantle, then heat rising from its interior must have been the dominant factor. Miranda is so small that its heat of formation must have been lost quickly, and it would not have been severely heated by radioactive decay. Tidal heating could have occurred if Miranda's orbit was made slightly elliptical by a resonance with other moons. Your knowledge of tidal heating in other moons can help you understand Miranda.

But external factors are also a possibility. Comets sweep through the solar system, and they occasionally strike planets and moons. The Jovian planets have no surfaces and would retain no permanent scars of such impacts, as Earth's astronomers saw in the summer of 1994 when a comet dramatically struck Jupiter. Their moons, however, could be battered or even disrupted by such impacts, and a major impact could have broken Miranda up and allowed it to re-form.

Although planetary scientists now prefer internal heat as an explanation of Miranda's geology, you should note that external factors such as major impacts do occur. Miranda is a reminder that modern astronomy recognizes the importance of catastrophic events in the history of the solar system. (See How Do We Know? 19-1.) You will see more hints of catastrophic events when you visit the moons of Neptune.

A History of Uranus

The challenge of comparative astronomy is to tell the story of a world, and Uranus may present the biggest challenge of all the objects in our solar system. Not only is it so distant that it is difficult to study, but it is also peculiar in many ways.

Uranus formed from the solar nebula, as did the other Jovian planets, but calculations show that Uranus and Neptune could not have grown to their present size in the slow-moving orbits they now occupy so far from the sun. Sophisticated mathematical models of planet formation in the solar nebula suggest that Uranus and Neptune formed closer to the sun, in the neighborhood of Jupiter and Saturn. Gravitational interactions among the Jovian planets could have gradually moved Uranus and Neptune outward to their present locations. This is an exciting hypothesis, and it illustrates how uncertain the histories of the outer planets really are.

The highly inclined axis of Uranus may have originated late in its formation when it was struck by a planetesimal as large as

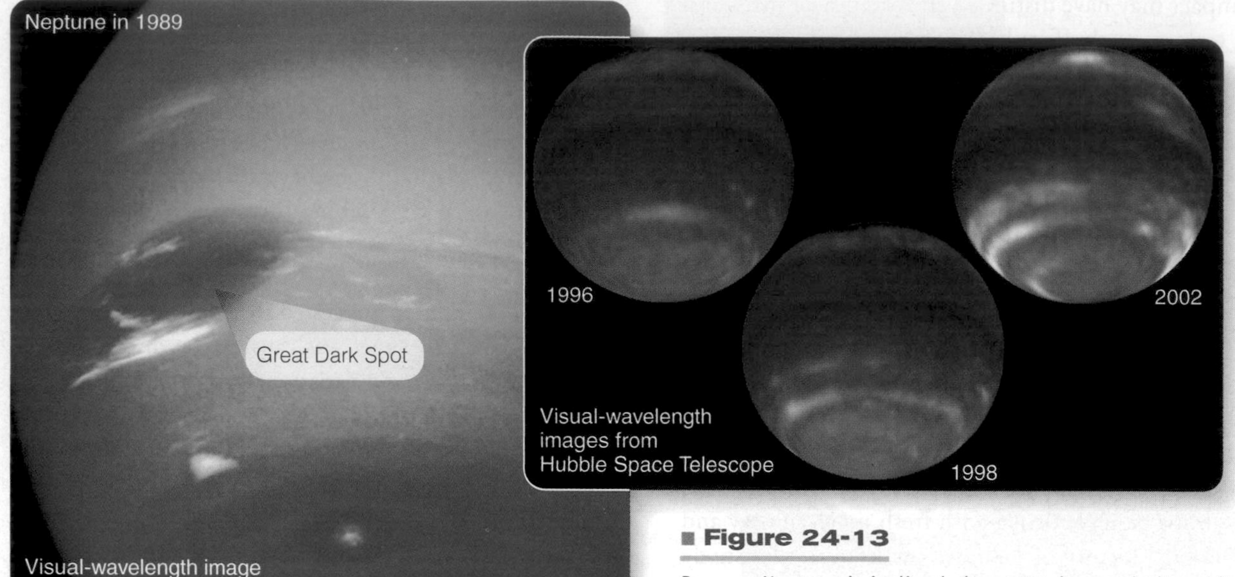

■ **Figure 24-13**

Because Neptune is inclined almost 29 degrees in its orbit, it experiences seasons that each last about 40 years. Since 1989, spring has come to the southern hemisphere, and the weather has clearly changed, which is surprising because sunlight on Neptune is 900 times dimmer than on Earth. (NASA, L. Sromovsky, and P. Fry, University of Wisconsin-Madison.)

more attention to what he saw in the background. Modern studies of Galileo's notebooks show that he saw Neptune on December 24, 1612, and again on January 28, 1613, but he plotted it as a star in the background of drawings of Jupiter. It is interesting to speculate about the response of the Inquisition had Galileo proposed that a planet existed beyond Saturn. Unfortunately for history, but perhaps fortunately for Galileo, he did not recognize Neptune as a planet, and its discovery had to wait another 234 years.

The discovery of Neptune has been recounted by historians as a triumph for Newtonian physics: The three laws of motion and the law of gravity had proved sufficient to predict the position and orbit of an unseen planet. Thus, the discovery of Neptune was fundamentally different from the discovery of Uranus—the existence of Neptune was predicted using basic laws. But a modern analysis shows that both Leverrier and Adams made unwarranted assumptions about the undiscovered planet's distance from the sun. By good fortune their assumptions made no difference in the 19th century, and the planet was close to the positions they predicted. Do you think they deserve less credit? They tried, and the other astronomers of the world didn't.

The Atmosphere and Interior of Neptune

Little was known about Neptune before the Voyager 2 spacecraft swept past it in August 1989. Seen from Earth, Neptune is a tiny, blue-green dot never more than 2.3 seconds of arc in diameter. Astronomers knew it was almost four times the diameter of Earth and about 4 percent smaller in diameter than Uranus **(Celestial Profile 10)**. Spectra revealed that, as in the case of Uranus, its blue-green color was caused by methane in its hydrogen-rich atmosphere. Methane absorbs red light, making Neptune look blue-green. Neptune's density showed that it was a Jovian planet rich in hydrogen, but almost no detail was visible from Earth, so even its period of rotation was uncertain.

Voyager 2 passed only 4905 km above Neptune's cloud tops, closer than any spacecraft had ever come to one of the Jovian planets. The images it captured revealed that Neptune is marked by dramatic belt–zone circulation parallel to the planet's equator. Voyager 2 also saw at least four cyclonic disturbances. The largest, dubbed the Great Dark Spot, looked similar to the Great Red Spot on Jupiter (■ Figure 24-13). The Great Dark Spot was located in the southern hemisphere and rotated counterclockwise, with a period of about 16 days. It appeared to be caused by gas rising from the planet's interior. When the Hubble Space Telescope began imaging Neptune in 1994, the Great Dark Spot was gone, and new cloud features were seen appearing and disappearing in Neptune's atmosphere (Figure 24-13). Evidently, the cyclonic disturbances on Neptune are not nearly as long lived as Jupiter's Great Red Spot.

The Voyager 2 images reveal other cloud features standing out against the deep blue of the methane-rich atmosphere. The white clouds are made up of crystals of frozen methane and range up to 50 km above the deeper layers, just where the temperature in Neptune's atmosphere is low enough for rising methane to freeze into crystals. (See Figure 24-5.) Presumably these features are related to rising convection currents that form clouds high in Neptune's atmosphere, where they catch sunlight and look bright. Special filters can reveal these bands in visual-wavelength images and at infrared wavelengths (■ Figure 24-14). Observations made by the Hubble Space Telescope suggest that atmospheric activity on Neptune may be related in some way to flares and other eruptions on the sun, but further observations are needed to explore this connection.

As on the other Jovian worlds, high winds circle Neptune parallel to its equator, but Neptune's winds blow at very high velocities and tend to blow backward—against the rotation of the planet. Why Neptune should have such high winds is not understood, and it is part of the larger problem of the belt–zone circulation.

Now that you have seen at least traces of belts and zones on all four of the Jovian planets (assuming that the faint clouds on Uranus are traces of belt–zone circulation), you can ask what drives this circulation. Because belts and zones remain parallel to a planet's equator even when the planet rotates at a high inclination, as in the case of Uranus, it seems reasonable to believe that the atmospheric circulation is dominated by the rotation of the planet and perhaps by circulation currents in the liquid interior and not by solar heating.

The same observations that helped define the interior of Uranus can be applied to Neptune. Models suggest that the interior contains a small heavy-element core, surrounded by a deep mantle of slushy ice mixed with heavier material with a chemical composition resembling rock. Neptune's magnetic field is a bit less than half as strong as Earth's and is tipped 47° from the axis of rotation. It is also offset 55 percent of the way to the surface (Figure 24-7). As in the case of Uranus, the field is probably generated by the dynamo effect acting in the conducting fluid mantle.

Neptune has more internal heat than Uranus, and part of that heat may be generated by radioactive decay in the minerals in its interior. Some of the energy may be released by denser material falling inward, including, as in the case of Uranus, diamond crystals formed by the disruption of methane.

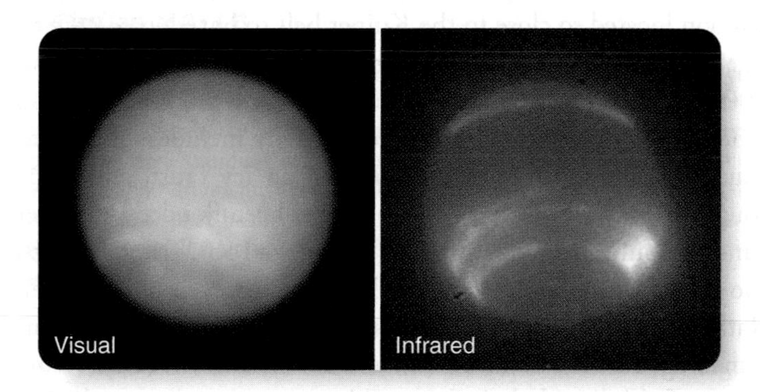

■ **Figure 24-14**

(a) This visual-wavelength image was made through filters that make belts of methane stand out. (NASA and Erich Karkoschka) (b) The color-corrected infrared image at right was recorded using one of the Keck 10-meter telescopes and adaptive objects. It shows the location of methane cloud belts around the planet. (U. C. Berkeley/W. M. Keck Observatory)

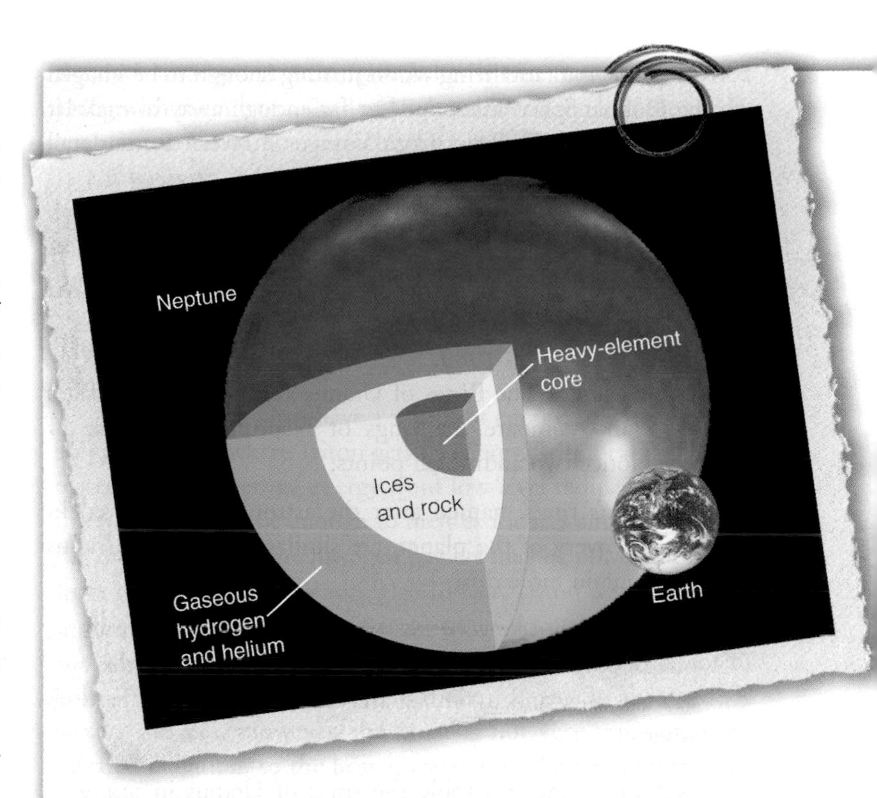

Neptune was tipped slightly away from the sun when the Hubble Space Telescope recorded this image. The interior is much like Uranus's, but there is more heat. (NASA)

Celestial Profile 10: Neptune

Motion:

Average distance from the sun	30.0611 AU (44.971 × 10^8 km)
Eccentricity of orbit	0.0100
Maximum distance from the sun	30.4 AU (45.4 × 10^8 km)
Minimum distance from the sun	29.8 AU (44.52 × 10^8 km)
Inclination of orbit to ecliptic	1°46′27″
Average orbital velocity	5.43 km/s
Orbital period	164.793 y (60,189 days)
Period of rotation	16^h3^m
Inclination of equator to orbit	28°48″

Characteristics:

Equatorial diameter	49,500 km (3.93 $D_{\oplus}$)
Mass	1.030 × 10^{26} kg (17.23 $M_{\oplus}$)
Average density	1.66 g/cm^3
Gravity	1.19 Earth gravities
Escape velocity	25 km/s (2.2 $V_{\oplus}$)
Temperature at cloud tops	−216°C (−357°F)
Albedo	0.35
Oblateness	0.027

Personality Point:

Because the planet Neptune looked so blue, astronomers named it after the Roman god of the sea, Neptune (Poseidon to the Greeks). His wife was Amphitrite, granddaughter of Ocean, one of the Titans. Neptune controlled the storms and waves and was a powerful god not to be trifled with. His three-pronged trident became the symbol for the planet Neptune.

observed arcs. Such delicate rings could not have survived since the origin of the planets, so the rings must be occasionally supplied with fresh particles generated by the impacts of meteorites and comets and Neptune's moons. Once again, the evidence of major impacts in the solar system's history assures you that such impacts do occur.

◄ SCIENTIFIC ARGUMENT ►

Why is Neptune blue but its clouds white?

To solve this problem you must build a scientific argument that follows a process step by step. When you look at something, you really turn your eyes toward it and receive light from the object. When you look at Neptune, the light you receive is sunlight that is reflected from various layers of Neptune and journeys to your eyes. Because sunlight contains a distribution of photons of all visible wavelengths, it looks white to human eyes, but sunlight entering Neptune's atmosphere must pass through hydrogen gas that contains a small amount of methane. While hydrogen is transparent, methane is a good absorber of longer wavelengths, so red photons are more likely to be absorbed than blue photons. Once the light is scattered from deeper layers, it must run this methane gauntlet again to emerge from the atmosphere, and again red photons are more likely to be absorbed. The light that finally emerges from Neptune and eventually reaches your eyes is poor in longer wavelengths and thus looks blue.

The methane-ice-crystal clouds lie at high altitudes, so sunlight does not have to penetrate very far into Neptune's atmosphere to reflect off the clouds, and consequently it loses many fewer of its red photons. The clouds look white.

This discussion shows how a careful, step-by-step analysis of a natural process can help you better understand how nature works. For example, build a step-by-step argument to answer the following: **Where does the energy come from to power Triton's surface geysers?**

◄ ►

24-3 The Dwarf Planets

IN 1930, A WORLD WAS FOUND orbiting beyond Neptune. Although a dense object smaller than Earth's moon rather than a low-density Jovian planet, the public welcomed it as the ninth planet and it was named Pluto. At the end of the 20th century, with much improved telescopes, astronomers found more such small worlds, and it became clear that Pluto was part of a large family of objects.

In 2006, the International Astronomical Union voted to toss Pluto out of the family of planets and make it part of a larger family of small worlds. To understand this highly controversial subject, you can start with Pluto and its discovery.

The Discovery of Pluto

Percival Lowell (1855–1916) was fascinated with the idea that an intelligent race built the canals he thought he could see on Mars (Chapter 22). Lowell founded Lowell Observatory in Flagstaff, Arizona, primarily for the study of Mars. Later, some say to improve the reputation of his observatory, he began to search for a planet beyond Neptune.

Lowell used the same method that Adams and Leverrier had used to predict the position of Neptune. Working from the observed irregularities in the motion of Neptune, Lowell predicted the location of an undiscovered planet beyond Neptune. He concluded it would contain about 7 Earth masses and would look like a 13th-magnitude object in eastern Taurus. Lowell searched for the planet photographically until his death in 1916.

In the late 1920s, 22-year-old amateur astronomer Clyde Tombaugh began using a homemade 9-in. telescope to sketch Jupiter and Mars from his family's wheat farm in western Kansas. He sent his drawings to Lowell Observatory, and the observatory director hired him without an interview. The young Tombaugh bought a one-way train ticket for Flagstaff not knowing what his new job would be like.

The observatory director set Tombaugh to work photographing the sky along the ecliptic around the predicted position of the planet. The search technique was a classic method in astronomy. Tombaugh obtained pairs of 14 × 17-inch glass plates exposed two or three days apart. To search a pair of plates, he mounted them in a blink comparator, a machine that allowed him to look through a microscope at a small spot on one plate and then at the flip of a lever see the same spot on the other plate. As he blinked back and forth, the star images did not move, but a planet would have moved along its orbit during the two or three days that elapsed before the second plate was exposed. So Tombaugh searched the giant plates, star image by star image, looking for an image that moved. A single pair of plates could contain 400,000 star images. He searched pair after pair and found nothing.

The observatory director turned to other projects, and Tombaugh, working alone, expanded his search to cover the entire ecliptic. For almost a year, Tombaugh exposed plates by night and blinked plates by day. Then on February 18, 1930, nearly a year after he had left Kansas, a quarter of the way through a pair of plates, he found a 15th magnitude image that moved (■ Figure 24-17). He later remembered, "'Oh,' I thought, 'I had better look at my watch; this could be a historic moment. It was within about 2 minutes of 4 PM [MST].'" The discovery was announced on March 13, the 149th anniversary of the discovery of Uranus and the 75th anniversary of the birth of Percival Lowell. The object was named Pluto after the god of the underworld and, in a way, after Lowell, because the first two letters in Pluto are the initials of Percival Lowell.

The discovery of Pluto seemed a triumph of discovery by prediction, but Tombaugh sensed something was wrong from the first moment he saw the image. It was moving in the right direction by the right amount, but it was 2.5 magnitudes too faint. Clearly, Pluto was not the 7-Earth-mass planet that Lowell had predicted. The faint image implied that Pluto was a small world with a mass too low to seriously alter the motion of Neptune.

■ **Figure 24-17**

Pluto is small and far away, so its image is indistinguishable from that of a star on most photographs. Clyde Tombaugh discovered the planet in 1930 by looking for an object that moved relative to the stars on a pair of photographs taken a few days apart. (Lowell Observatory photographs)

Later analysis has shown that the variations in the motion of Neptune, which Lowell used to predict the location of Pluto, were random uncertainties of observation and could not have led to a trustworthy prediction. The discovery of the new planet only 6° from Lowell's predicted position was apparently an accident, which proves that if you search long enough and know what to expect, you are likely to find something.

Pluto as a World

Pluto is very difficult to observe from Earth. Only a bit larger than 0.1 second of arc in diameter, it is only 65 percent the diameter of Earth's moon and shows little surface detail. No spacecraft has ever visited it, so there are no images of its surface. The New Horizons probe is now on its way to Pluto and will arrive in 2015.

Most planetary orbits in our solar system are nearly circular, but Pluto's is significantly elliptical. In fact, from January 21, 1979, to March 14, 1999, Pluto was closer to the sun than Neptune. The two worlds will never collide, however, because Pluto's orbit is inclined 17° and because the worlds orbit in resonance with each other and never come close together.

If you land on the surface of Pluto, your spacesuit will have to work hard to keep you warm. Orbiting so far from the sun, Pluto is cold enough to freeze most compounds that you think of as gases, and spectroscopic observations have found evidence of solid nitrogen ice on its surface with traces of frozen methane and carbon monoxide. The daytime temperature of about 50 K (-370°F) is enough to vaporize some of the nitrogen and carbon monoxide and a little of the methane to form a thin atmosphere around Pluto. This atmosphere was detected in 1988 when Pluto occulted a distant star, and the starlight faded gradually rather than winking out suddenly.

Pluto's largest moon was discovered on photographs in 1978. It is very faint and about half the diameter of Pluto (■ Figure 24-18). The moon was named Charon after the mythological ferryman who transports souls across the river Styx into the underworld. Two smaller moons were found in 2005.

The discovery of Charon was important for a number of reasons. Charon orbits Pluto in a nearly circular orbit in the plane of Pluto's equator. Observations show that the moon and Pluto are tidally locked to each other, and that Pluto's axis of rotation is highly inclined (■ Figure 24-19). Furthermore, the orbital motion of Charon allows the calculation of the mass of

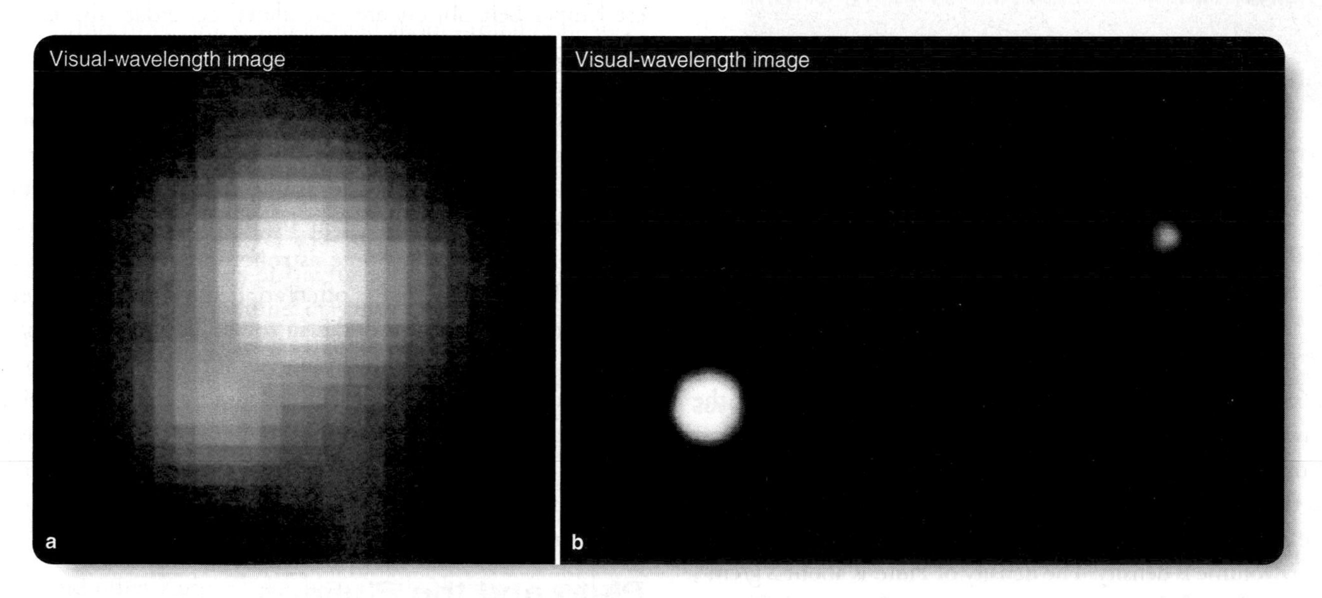

■ **Figure 24-18**

(a) A high-quality ground-based photo shows Pluto and its moon, Charon, badly blurred by seeing. (NASA) (b) The Hubble Space Telescope image clearly separates the planet and its moon and allows more accurate measurements of the position of the moon. (R. Albrecht, ESA/ESO Space Telescope European Coordinating Facility, NASA)

25 Meteorites, Asteroids, and Comets

Comet McNaught was bright in the sky as seen from the southern hemisphere in January and February 2007. Streamers in the tail are produced by variations in the release of gas and dust from the nucleus. (Peter Daalder)

Guidepost

Compared with planets, the comets and asteroids are unevolved objects much as they were when they formed 4.6 billion years ago. The fragments of these objects that fall to Earth, the meteors and meteorites, will give you a close look at these ancient planetesimals. As you explore, you will find answers to four essential questions:

- **Where do meteors and meteorites come from?**
- **What are the asteroids?**
- **Where do comets come from?**
- **What happens when an asteroid or comet hits Earth?**

The subjects of this chapter are often faint and hard to find. As you study them, you will answer an important question concerning the design of scientific experiments:

- **How Do We Know? How can what scientists notice bias their conclusions?**

When you finish this chapter, you will have an astronomer's insight into your place in nature. You live on the surface of a planet. There are other planets. Are they inhabited too? That is the subject of the next chapter.

When they shall cry "PEACE, PEACE" then cometh sudden destruction!

COMET'S CHAOS?—What Terrible events will the Comet bring?

FROM A RELIGIOUS PAMPHLET PREDICTING THE END OF THE WORLD BECAUSE OF THE APPEARANCE OF COMET KOHOUTEK, 1973

YOU ARE NOT AFRAID of comets, of course; but not long ago, people viewed them with terror. A few centuries ago, comets were thought to predict the deaths of kings or the arrival of plague. Even in 1910, when Earth passed through the tail of Comet Halley, millions of people panicked, thinking the world would be destroyed. Householders in Chicago stuffed rags around doors and windows to keep out poisonous gas, and con artists in Texas sold comet pills and inhalers to ward off the fumes.

Should we snicker at our silly ancestors? Today we see comets as graceful and beautiful visitors to our skies (■ Figure 25-1). Astronomers understand that comets and their rocky cousins the asteroids are ancient bodies that carry precious clues to the birth of the planets 4.6 billion years ago. But the evidence also shows that comets and asteroids do hit Earth now and then, and such an impact could cause a civilization-ending catastrophe. Perhaps comets and asteroids deserve our attention and cautious respect.

Unfortunately, comets and asteroids are far beyond your reach, so they are a challenge to study. You can begin with the fragments of these bodies that fall to Earth—the meteorites.

25-1 Meteorites

IN THE AFTERNOON OF NOVEMBER 30, 1954, Mrs. E. Hulitt Hodges of Sylacauga, Alabama, lay napping on her living room couch. An explosion and a sharp pain jolted her awake, and she found that a meteorite had smashed through the ceiling and

■ Figure 25-1

Comet Hyakutake swept through the inner solar system in 1996 and was dramatic in the northern sky. Seen here from Kitt Peak National Observatory, the comet passed close to the north celestial pole (behind the observatory dome in this photo). Notice the Big Dipper below and to the left of the head of the comet. (Courtesy Tod Lauer)

bruised her left leg. Mrs. Hodges is the only person known to have been injured by a meteorite. Coincidentally, Mrs. Hodges lived right across the street from the Comet Drive-In Theater.

Meteorite impacts on homes are not common, but they do happen. Statistical calculations show that a meteorite should damage a building somewhere in the world about once every 16 months. About two meteorites large enough to produce visible impacts strike somewhere on Earth each day, but most meteors are small particles ranging from a few centimeters down to microscopic dust. Earth gains about 40,000 tons of mass per year from meteorites of all sizes. That seems like a lot, but it is less than a thousandth of a trillionth of Earth's total mass.

Recall from Chapter 19 that astronomers distinguish between the words *meteoroid, meteor,* and *meteorite.* A small body in space is a meteoroid, but once it begins to vaporize in Earth's atmosphere it is called a meteor. If it survives to strike the ground, it is called a meteorite.

You should have two main questions concerning meteorites: Where in the solar system do these objects come from, and what can meteorites tell you about the origin of the solar system? To answer these questions, you must consider the orbits of meteoroids and the minerals found in meteorites.

Meteoroid Orbits

Meteoroids are much too small to be visible through even the largest telescope. They are visible only when they fall into Earth's atmosphere at 10 to 30 km/s, roughly 30 times faster than a rifle bullet, and are vaporized by friction with the air. The average meteoroid is about the mass of a paper clip and vaporizes at an altitude of about 80 km above Earth's surface. In doing so it produces a bright streak of fire that you see as a meteor. The trail of a meteor points back along the path of the meteoroid, so if you could study the direction and speed of meteors, you could get clues to their orbits.

One way to backtrack meteor trails is to observe meteor showers. On any clear night, you can see 3 to 15 meteors an hour, but on some nights you can see a shower of dozens of meteors an hour that are obviously related to each other. To confirm this, try observing a meteor shower. Pick a shower from ■ Table 25-1 and on the appropriate night stretch out in a lawn chair and watch a large area of the sky. When you see a meteor, sketch its path on the appropriate sky chart from the back of this book. In just an hour or so you will discover that most of the meteors you see seem to come from a single area of the sky, the **radiant** of the shower (■ Figure 25-2a). In fact, meteor showers are named after the constellation from which they seem to radiate. The Perseid shower radiates from the constellation Perseus in mid-August.

Observing a meteor shower is a natural fireworks show, but it is even more exciting when you understand what a meteor shower tells you. The fact that the meteors in a shower appear to come from a single point in the sky, the radiant, means that the meteoroids were traveling through space along parallel paths. When they encounter Earth and are vaporized in the upper atmosphere, you see their fiery tracks in perspective; so they appear

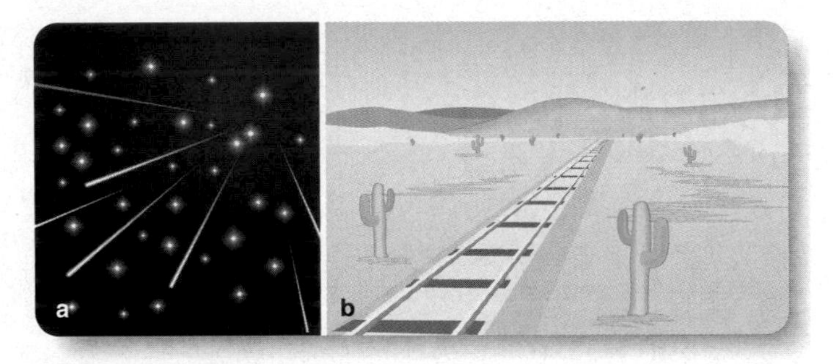

■ **Figure 25-2**

(a) Meteors in a meteor shower enter Earth's atmosphere along parallel paths, but perspective makes them appear to diverge from a radiant point. (b) Similarly, parallel railroad tracks appear to diverge from a point on the horizon.

■ **Table 25-1 | Meteor Showers**

Shower	Dates	Hourly Rate	Radiant* R. A.	Radiant* Dec.	Associated Comet
Quadrantids	Jan. 2–4	30	15^h24^m	50°	
Lyrids	April 20–22	8	18^h4^m	33°	1861 I
η Aquarids	May 2–7	10	22^h24^m	0°	Halley?
δ Aquarids	July 26–31	15	22^h36^m	−10°	
Perseids	Aug. 10–14	40	3^h4^m	58°	1982 III
Orionids	Oct. 18–23	15	6^h20^m	15°	Halley?
Taurids	Nov. 1–7	8	3^h40^m	17°	Encke
Leonids	Nov. 14–19	6	10^h12^m	22°	1866 I Temp
Geminids	Dec. 10–13	50	7^h28^m	32°	

*R. A. and Dec. give the celestial coordinates (right ascension and declination) of the radiant of each shower.

to come from a single radiant point, just as railroad tracks seem to come from a single point on the horizon (Figure 25-2b).

Studies of meteor-shower radiants reveal that these meteors are produced by bits of matter orbiting the sun along the paths of comets. The vaporizing head of the comet releases bits of rock that become spread along its entire orbit (■ Figure 25-3). When Earth passes through this stream of material, you see a meteor shower. In some cases the comet has wasted away and is no longer visible, but in other cases the comet is still prominent though somewhere else along its orbit. For example, in May Earth comes near the orbit of Comet Halley, and Earthlings see the Eta Aquarids shower. Around October 20, Earth passes near the other side of the orbit of Comet Halley, and you can see the Orionids shower. Evidently at least some meteoroids come from comets.

Even when there is no shower, you will still see meteors, which are called **sporadic meteors** because they are not part of specific showers. Many of these are produced by stray bits of matter that were released long ago by comets. Such comet debris gets spread throughout the inner solar system, and bits fall into Earth's atmosphere even when there is no shower.

Another way to backtrack meteor trails is to photograph the same meteor from two locations on Earth a few miles apart. Then astronomers can use triangulation to find the altitude, speed, and direction of the meteor as it moves through the atmosphere and work backward to find out what its orbit looked like before it entered Earth's atmosphere. Not surprisingly, these studies confirm that meteors belonging to showers and some sporadic meteors have orbits that are similar to the orbits of comets. A few sporadic meteors, however, have orbits that lead back to the asteroid belt between Mars and Jupiter. From this you can conclude that meteors have a dual source: Many come from comets, but some come from the asteroid belt. To learn more about meteors, you need to examine those that make it to Earth's surface, the meteorites. You can begin with their dramatic arrivals.

Meteorite Impacts on Earth

When a meteor is massive enough and strong enough to survive its plunge through Earth's atmosphere and reach Earth's surface, it is called a meteorite. A large meteorite hitting Earth might dig an impact crater much like those on the moon (see page 464).

Over 150 impact craters have been found on Earth. The Barringer Meteorite Crater near Flagstaff, Arizona, is a good example (■ Figure 25-4). It was created about 50,000 years ago by a meteorite about as large as a big building (40 m) that hit at a

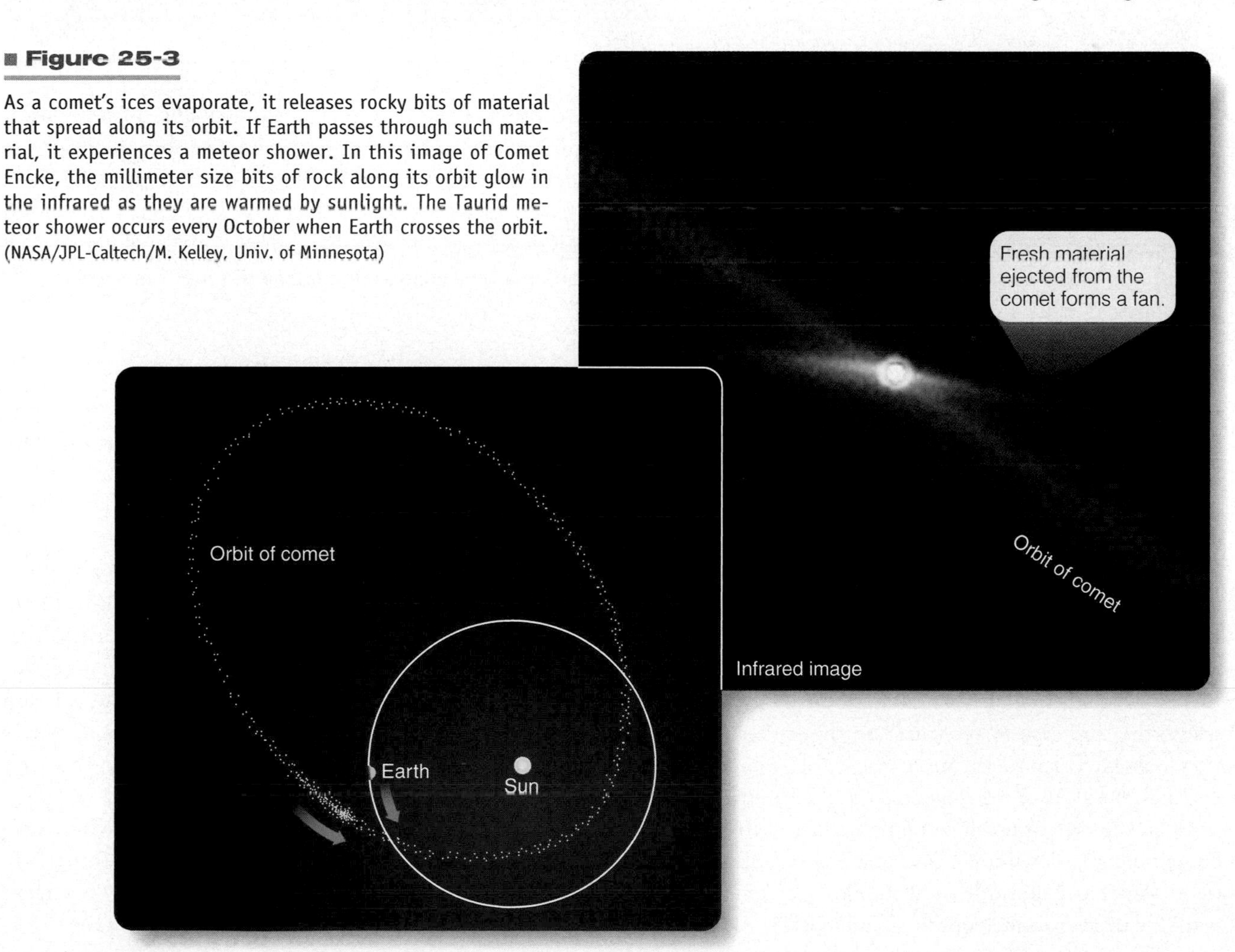

■ Figure 25-3

As a comet's ices evaporate, it releases rocky bits of material that spread along its orbit. If Earth passes through such material, it experiences a meteor shower. In this image of Comet Encke, the millimeter size bits of rock along its orbit glow in the infrared as they are warmed by sunlight. The Taurid meteor shower occurs every October when Earth crosses the orbit. (NASA/JPL-Caltech/M. Kelley, Univ. of Minnesota)

■ **Active Figure 25-4**

(a) The Barringer Meteorite Crater (near Flagstaff, Arizona) is nearly a mile in diameter and was formed about 50,000 years ago by the impact of an iron meteorite roughly 40 m in diameter. It hit with energy equivalent to that of a 3-megaton hydrogen bomb. Notice the raised and deformed rock strata all around the crater. For scale, locate the brick building on the far rim at right. (M. Seeds) (b) Like all larger-impact features, the Barringer Meteorite Crater has a raised rim and scattered ejecta. (USGS)

speed of 11 km/s and released as much energy as a large thermonuclear bomb. Debris at the site shows that the meteorite was composed of iron.

The Barringer crater is 1.2 km in diameter and 200 m deep. It seems large when you stand on the edge, and the hike around it, though beautiful, is long and dry. Nevertheless, the crater is actually small compared with some other impact features on Earth. About 65 million years ago, an impact in the northern Yucatán of Central America produced a crater, now covered by sediment, that was between 180 and 300 km in diameter, almost as big as the state of Ohio. This impact has been blamed for changing Earth's climate and causing the extinction of the dinosaurs. Later in this chapter, you can come back to this Earth-shaking event.

Large meteorites can produce large craters, but most meteorites are small and don't create significant craters. Also, craters erode rapidly on Earth, so most meteorites are found without associated craters. The best place to look for meteorites turns out to be certain areas of Antarctica—not because more meteorites fall there but because they are easy to recognize on the icy terrain. Most meteorites look like Earth rocks, but on the ice of Antarctica there are no natural rocks. Also, the slow creep of the ice toward the sea concentrates the meteorites in certain areas where the ice runs into mountain ranges, slows down, and evaporates. Teams of scientists travel to Antarctica and ride snowmobiles in systematic sweeps across the ice to recover meteorites (■ Figure 25-5).

Meteorites that are seen to fall are called **falls;** a fall is known to have occurred at a given time and place, and thus the meteorite is well documented. A meteorite that is discovered but was not seen to fall is called a **find.** Such a meteorite could have fallen thousands of years ago. The distinction between falls and finds will be important as you analyze the different kinds of meteorites.

An Analysis of Meteorites

Meteorites can be divided into three broad categories: *Iron* meteorites are solid chunks of iron and nickel, *stony* meteorites are silicate masses that resemble Earth rocks, and *stony-iron* meteorites are mixtures of iron and stone. These types are illustrated in ■ Figure 25-6.

Iron meteorites are easy to recognize because they are heavy, dense lumps of iron-nickel steel—a magnet will stick to them. That explains an important bit of statistics. Iron meteorites make up 66 percent of finds (■ Table 25-2) but only 6 percent of falls. Why? Because an iron meteorite does not look like a rock. If you trip over one on a hike, you are more likely to recognize it as something odd, carry it home, and show it to your local museum. Also, some stony meteorites deteriorate rapidly when exposed to weather; irons survive longer. That means there is a **selection effect** that makes it more likely that iron meteorites will be found (**How Do We Know? 25-1**). That only 6 percent of falls are irons shows that iron meteorites are fairly rare.

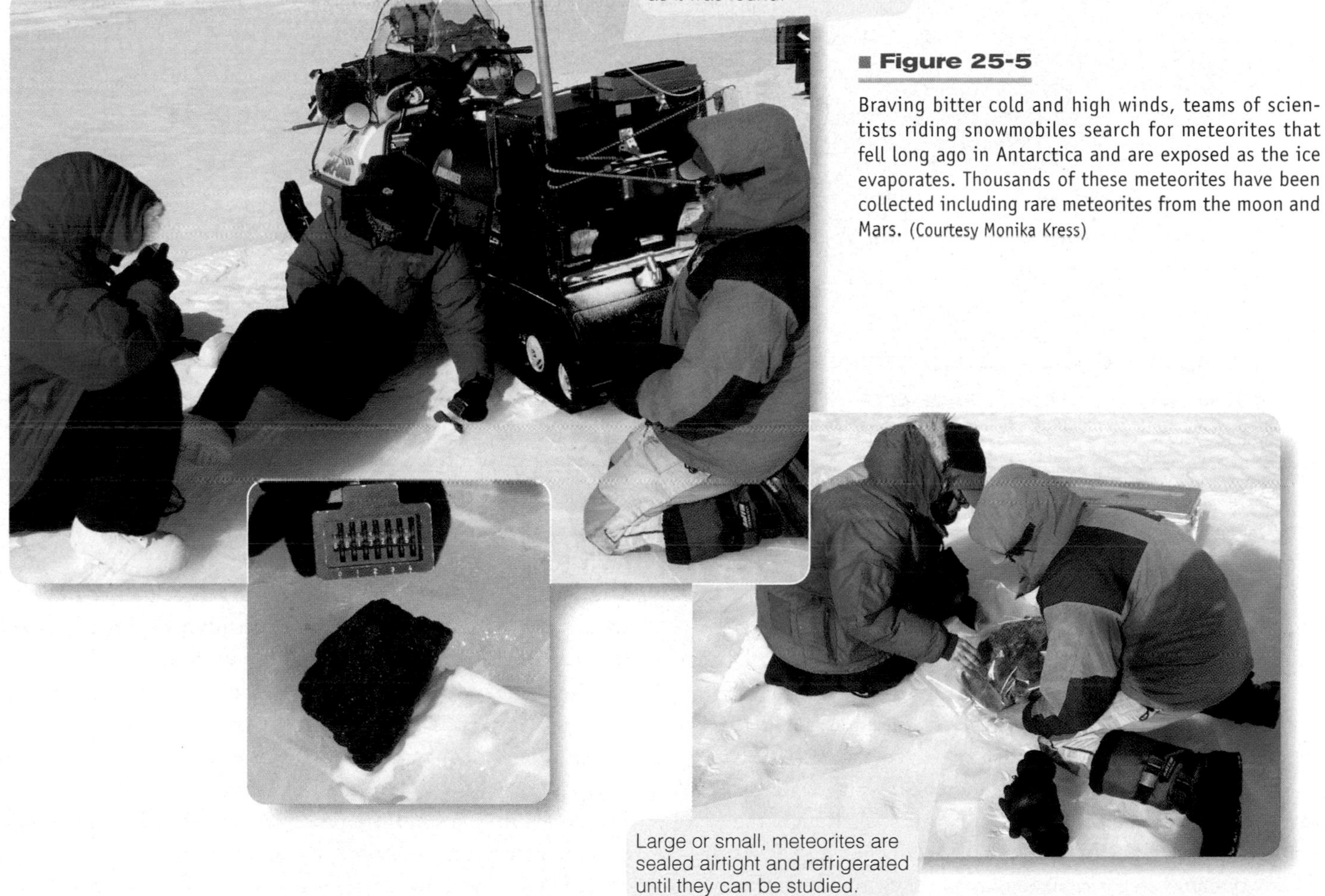

Each meteorite is assigned a number and photographed as it was found.
Large or small, meteorites are sealed airtight and refrigerated until they can be studied.

■ Figure 25-5

Braving bitter cold and high winds, teams of scientists riding snowmobiles search for meteorites that fell long ago in Antarctica and are exposed as the ice evaporates. Thousands of these meteorites have been collected including rare meteorites from the moon and Mars. (Courtesy Monika Kress)

When iron meteorites are sliced open, polished, and etched with nitric acid, they reveal regular bands called **Widmanstätten patterns** (Figure 25-6). These patterns are caused by alloys of iron and nickel called kamacite and taenite that formed crystals billions of years ago as the molten iron cooled and solidified. The size and shape of the bands indicate that the molten metal cooled very slowly, no faster than 20 K per million years.

A lump of molten metal floating in space would cool very quickly. The Widmanstätten pattern tells you that the molten metal must have been well insulated to have cooled so slowly. Such slow cooling is typical of the interiors of bodies at least 30 to 50 km in diameter. On the other hand, the iron meteorites show no effects of the very high pressures that would exist inside larger bodies. Evidently, the iron meteorites

Iron meteorites are very heavy for their size and have a dark, irregular surface.
Stony meteorites tend to have a fusion crust caused by melting in Earth's atmosphere.
A stony-iron meteorite cut and polished reveals a mixture of iron and rock.
Chondrules are small, glassy spheres found in chondrites.
Cut, polished, and etched with acid, iron meteorites show a Widmanstätten pattern.
This carbonaceous chondrite contains chondrules and volatiles, including carbon, that make the rock very dark.

■ Figure 25-6

The three main types of meteorites, irons, stones, and stony-iron, have distinctive characteristics. (Lab photos courtesy of Russell Kempton, New England Meteoritical Services)

■ Table 25-2 | Proportions of Meteorites

Type	Falls (%)	Finds (%)
Stony	92	26
Iron	6	66
Stony-iron	2	8

formed from the cooling interiors of planetesimal-sized objects.

In contrast to irons, **stony meteorites** are relatively common. Among falls, 92 percent are stones. Although there are many different types of stony meteorites, you can classify them into two main categories depending on their physical and chemical content—chondrites and achondrites.

Roughly 80 percent of all meteorite falls are stony meteorites called **chondrites,** which look like dark gray, granular rocks. The chemical composition of chondrites is the same as that of a sample of matter from the sun with the most volatile gases removed. The classification of meteorites has become quite sophisticated, and there are many types of chondrites, but in general they appear to be samples of the original material that condensed from the solar nebula. Some have slight mineral differences, showing that the solar nebula was not totally uniform when this material condensed.

Most types of chondrites contain **chondrules,** round bits of glassy rock no larger than 5 mm across (Figure 25-6). To be glassy rather than crystalline, the chondrules must have cooled from a molten state quickly, within a few hours, but their origin is not clear. One theory is that they are bits of matter that were suddenly melted by shock waves spreading through the solar nebula. Whatever their origin, they are very old.

All of the different types of chondrites contain some volatiles such as water, and this shows that the meteoroids were never heated to high temperatures. Differences among the many types of chondrites are a result of their condensation in different parts of the solar nebula and from processes that altered their composition after they formed. Some, for example, appear to have been altered by the presence of water released by the melting of ice.

Among the chondrites, the **carbonaceous chondrites** are rare, only about 5.7 percent of falls. These dark gray, rocky meteorites contain volatiles including water and carbon compounds that would have been driven off if the meteoroid had been heated much above room temperature. It has been common for astronomers to think of the carbonaceous chondrites as the least altered of the chondrites and therefore the most likely objects to provide clues to the nature of the solar nebula. But many types of chon-

How Do We Know? | 25-1

Selection Effects

How is a red insect like a red car? Scientists must plan ahead and design their research projects with great care. Biologists studying insects in the rain forest, for example, must choose which ones to catch. They can't catch every insect they see, so they might decide to catch and study any insect that is red. If they are not careful, a selection effect could bias their data and lead them to incorrect conclusions without their ever knowing it.

For example, suppose you needed to measure the speed of cars on a highway. There are too many cars to measure every one, so you might reduce the workload and measure only red cars. It is quite possible that this selection criterion will mislead you because people who buy red cars may be more likely to be younger and drive faster. Should you measure only brown cars? No, because older, more sedate people might tend to buy brown cars. Only by very carefully designing your experiment can you be certain that the cars you measure are traveling at representative speeds.

Astronomers understand that what you see through a telescope depends on what you notice, and that is powerfully influenced by selection effects. The biologists in the rain forest, for example, should not catch and study only red insects. Often, the most brightly colored insects are poisonous or at least taste bad to predators. Catching only red insects could produce a result highly biased by a selection effect.

Things that are bright and beautiful, such as spiral galaxies, may attract a disproportionate amount of attention. Scientists must be aware of such selection effects. (Hubble Heritage Team/STScI/AURA/NASA)

drites are also essentially unaltered samples of the solar nebula. A carbonaceous chondrite is shown in Figure 25-6.

One of the most important meteorites ever recovered was a carbonaceous chondrite that was seen falling on the night of February 8, 1969, near the Mexican village of Pueblito de Allende. The brilliant fireball was accompanied by tremendous sonic booms and showered an area about 50 km by 10 km with over 4 tons of fragments. About 2 tons were recovered.

Studies of the Allende meteorite disclosed that it contained—besides volatiles and chondrules— small, irregular inclusions rich in calcium, aluminum, and titanium. Now called **CAIs,** for calcium–aluminum-rich inclusions, these bits of matter are highly refractory (■ Figure 25-7); that is, they can survive very high temperatures. If you could scoop out a ton of the sun's surface matter and cool it, the CAIs would be the first particles to form. As the temperature fell, other materials would condense in accord with the condensation sequence described in Chapter 19. When the material finally reached room temperature, you would find that all of the hydrogen, helium, and a few other gases like argon and neon had escaped and that the remaining lump, weighing about 18 kg (40 lb), had almost the same composition as the Allende meteorite, including CAIs. The Allende meteorite seems to be a very old sample of the solar nebula.

Unlike the chondrites, stony meteorites called **achondrites** (7.1 percent of falls) are highly modified. They contain no chondrules and no volatiles. This suggests that they have been hot enough to melt chondrules and drive off volatiles, leaving behind rock with compositions similar to Earth's basalts.

■ Figure 25-7

A sliced portion of the Allende meteorite showing round chondrules and irregularly shaped white inclusions called CAIs. (NASA)

Iron meteorites and stony meteorites make up most falls, but 2 percent of falls are meteorites that are made up of mixed iron and stone. These **stony-iron meteorites** appear to have solidified from a region of molten iron and rock—the kind of environment you might expect to find deep inside a planetesimal with a molten iron core and a rock mantle.

The Origin of Meteorites

Where do meteorites come from? Even though the iron, stony, and stony-iron meteorites seem very different from each other, their properties show that they all formed from the solar nebula. Some appear not to have been modified since they formed, but others have been heated slightly or even melted sometime after formation.

Meteorites almost certainly do not come from comets. Most cometary particles are very small specks of low-density, almost fluffy, material. When these specks enter Earth's atmosphere, you see them incinerated as meteors. Most meteors are produced by this cometary debris, but such meteors are small and weak and do not survive to reach the ground. Although most meteors come from comets, most meteorites are stronger, denser chunks of matter—more like fragments of asteroids.

Meteorites must have come not from cometlike planetesimals rich in ices but instead from asteroid-like planetesimals rich in metals and rock, which have evolved in complicated ways and eventually were broken during collisions. You already know that the solar nebula was full of rocky planetesimals, but you must be wondering about two things: How did these planetesimals evolve to produce both iron and stony meteorites? And when did these planetesimals break up?

At least some of the planetesimals must have melted and differentiated to produce the iron and achondritic meteorites, but what produced this heat? Planets the size of Earth can accumulate a great deal of heat from the slow decay of radioactive atoms such as uranium, thorium, and the radioactive isotope of potassium. The heat is trapped deep underground by thousands of kilometers of insulating rock. But in a small planetesimal, the insulating layers are not as thick, and the heat leaks out into space as fast as the slowly decaying atoms can produce it. If a small planetesimal is to melt, it must have a more rapidly decaying heat source.

Modern studies have shown that some meteorites must have contained the radioactive element aluminum-26. Aluminum-26 decays to form magnesium-26 with a half-life of only 715,000 years, so all of the aluminum-26 is now gone. But the magnesium-26 can be detected in the laboratory, and that shows that aluminum-26 was once present. Aluminum-26 decays so rapidly the heat could have melted the center of a planetesimal as small as 20 km in diameter.

The origin of the aluminum-26 has interesting implications for the solar system. Supernova explosions can manufacture aluminum-26. If the solar nebula was enriched in aluminum-26 from a nearby supernova explosion, the explosion must have occurred just before the formation of our solar system. In fact, some astronomers wonder if it was shock waves from the supernova explosion that compressed gas clouds and triggered the formation of the sun and planets.

The melting of a planetesimal's interior would allow differentiation as heavy metals sank to the center to form a molten metal core and the less-dense silicates floated upward to form a stony mantle. Once the aluminum-26 had decayed away, the planetesimal would slowly cool and solidify. If such a planetesimal were broken up (■ Figure 25-8), fragments from the center would look like iron meteorites with their Widmanstätten patterns.

The achondrites, which have been strongly heated or melted, appear to come from the mantles and surfaces of differentiated planetesimals. The stony-iron meteorites probably come from the core–mantle boundaries where iron and stone were mixed. But the chondrites are probably fragments of smaller bodies that never melted. Volatile-rich meteorites such as the carbonaceous chondrites may have been part of smaller bodies that formed farther from the sun where temperatures were lower.

There is even more evidence that the meteorites are the result of the breaking up of larger bodies. For example, some meteorites are breccias—collections of stony fragments cemented together. Breccias are found on Earth and are very common on the moon, but studies of the meteoric breccias show that they were produced by impacts. A collision between planetesimals would produce fragments, and the slower-moving particles would fall back to the surface of the planetesimal to form a regolith, a soil of broken rock fragments. Later impacts may add to this regolith and stir it. Still later, an impact may be violent enough that the fragments are pressed together so hard they momentarily melt where they touch. Almost instantly, the material cools, and the fragments weld themselves together to form a breccia. Much later, as the planetesimals were broken up by collisions, the layers of breccia were exposed, shattered, and became brecciated meteoroids.

When did these planetesimals break up? Recall from Chapter 19 that radioactive dating shows that meteorites formed about 4.6 billion years ago. But that age tells you only when the parent bodies formed, melted, differentiated, and cooled. The collisions that broke up the planetesimals could not have happened that long ago, because small meteoroids would have been swept up by the planets in a billion years or less. Also, cosmic rays that strike meteoroids in space produce isotopes such as helium-3, neon-20, and argon-38. Studies of these atoms in meteorites show that most meteorites have not been exposed to cosmic rays for more than a few tens of millions of years. That means that the thousands of meteorites now in museums around the world must have been broken from planetesimals somewhere in our solar system within the last billion years or less. Where are those planetesimals?

To answer that question, you need only recall that many meteoroid orbits lead back to the asteroid belt. The asteroids are evidently the planetesimals from which meteorites are born.

The Origin of Meteorites

A large planetesimal can keep its internal heat long enough to differentiate.
Collisions break up the layers of different composition.
Silicates
Cratering collisions
Iron

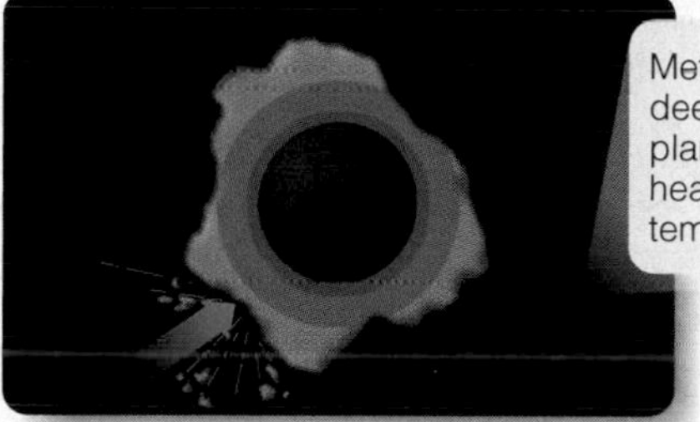

Meteorites from deeper in the planetesimal were heated to higher temperatures.

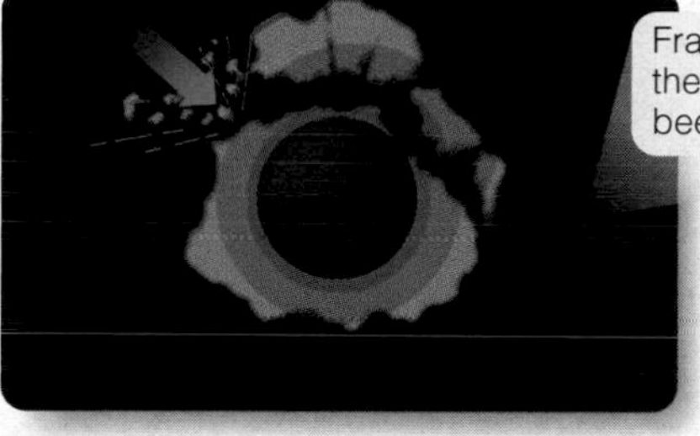

Fragments from near the core might have been melted entirely.

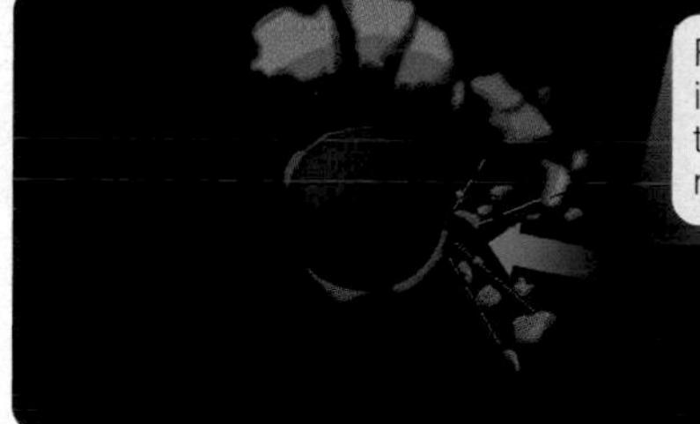

Fragments of the iron core would fall to Earth as iron meteorites.

■ **Figure 25-8**

Planetesimals formed when the solar system was forming may have melted and separated into layers of different density and composition. The fragmentation of such a body could produce many types of meteorites. (Adapted from a diagram by C. R. Chapman)

◄ **SCIENTIFIC ARGUMENT** ►

How can meteors come from comets, but meteorites come from asteroids?

This is a revealing argument because it contains a warning that seeing is not enough in science; thinking about seeing is critical. A selection effect can determine what you notice when you observe nature, and a very strong selection effect prevents people from finding meteorites that originated in comets. Cometary particles are physically weak, and they vaporize in Earth's atmosphere easily. Very few ever reach the ground, and people are unlikely to find them. Furthermore, even if a cometary particle reached the ground, it would be so fragile that it would weather away rapidly, and, again, people would be unlikely to find it. Asteroidal particles, however, are made from rock and metal and so are stronger. They are more likely to survive their plunge through the atmosphere and more likely to survive erosion on the ground. Meteors from the asteroid belt are rare. Almost all of the meteors you see come from comets, but not a single meteorite is known to be cometary.

The meteorites are valuable because they provide hints about the process of planet building in the solar nebula. Build a new argument but, as always, think carefully about what you see. **Why are most falls stony, but most finds are irons?**

◄ ►

25-2 Asteroids

UNTIL RECENTLY, FEW ASTRONOMERS KNEW or cared much about asteroids. They were small chunks of rock drifting between the orbits of Mars and Jupiter that occasionally marred long-exposure photographs by drifting past more interesting objects. Asteroids were more irritation than fascination.

Now you know differently. The evidence from meteorites shows that the asteroids are the last remains of the rocky planetesimals that built the planets 4.6 billion years ago. The study of the asteroids gives you a way to explore the ancient past of our planetary system.

Properties of Asteroids

In Chapter 19, you learned that most of the asteroids orbit between Mars and Jupiter and that images recorded by passing spacecraft reveal them to be small, complex worlds with irregular shapes and heavily cratered surfaces (Figure 19-3).

Study **Observations of Asteroids** on pages 578–579 and notice four important points:

1. Most asteroids are irregular in shape and battered by impact cratering. In fact, some appear to be rubble piles of broken fragments.
2. Some asteroids are double objects or have small moons in orbit around them. This is further evidence of collisions among the asteroids.
3. A few larger asteroids show signs of geological activity on their surfaces that may have been caused by volcanic activity when the asteroid was young.
4. Asteroids can be classified by their albedo and color to reveal clues to their compositions.

Observations of Asteroids

1 Seen from Earth, asteroids look like faint points of light moving in front of distant stars. Not many years ago they were known mostly for drifting slowly through the field of view and spoiling long time exposures. Some astronomers referred to them as "the vermin of the sky." Spacecraft have now visited asteroids, and the images radioed back to Earth show that the asteroids are mostly small, gray, irregular worlds heavily cratered by impacts.

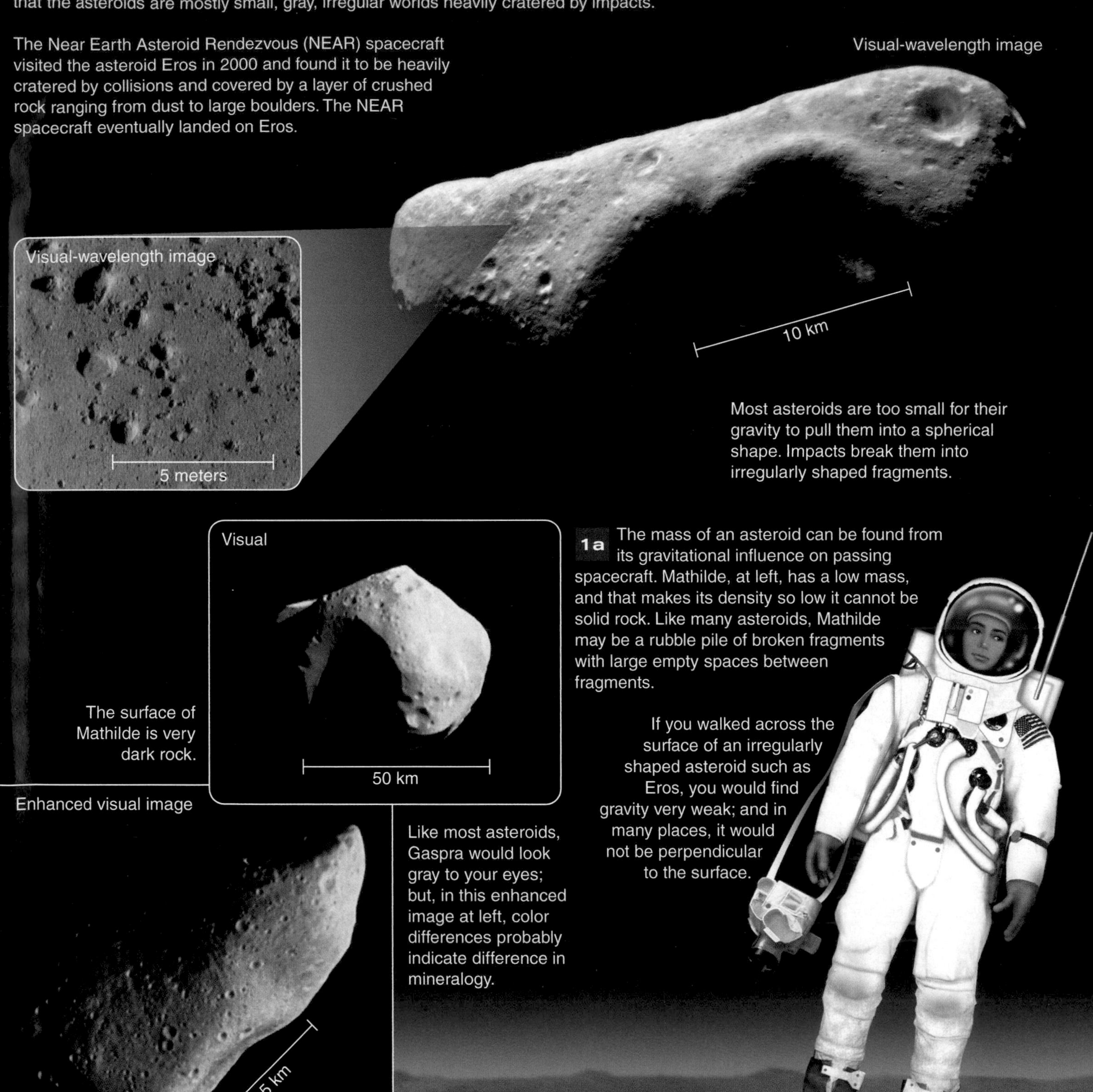

The Near Earth Asteroid Rendezvous (NEAR) spacecraft visited the asteroid Eros in 2000 and found it to be heavily cratered by collisions and covered by a layer of crushed rock ranging from dust to large boulders. The NEAR spacecraft eventually landed on Eros.

Most asteroids are too small for their gravity to pull them into a spherical shape. Impacts break them into irregularly shaped fragments.

1a The mass of an asteroid can be found from its gravitational influence on passing spacecraft. Mathilde, at left, has a low mass, and that makes its density so low it cannot be solid rock. Like many asteroids, Mathilde may be a rubble pile of broken fragments with large empty spaces between fragments.

The surface of Mathilde is very dark rock.

If you walked across the surface of an irregularly shaped asteroid such as Eros, you would find gravity very weak; and in many places, it would not be perpendicular to the surface.

Like most asteroids, Gaspra would look gray to your eyes; but, in this enhanced image at left, color differences probably indicate difference in mineralogy.

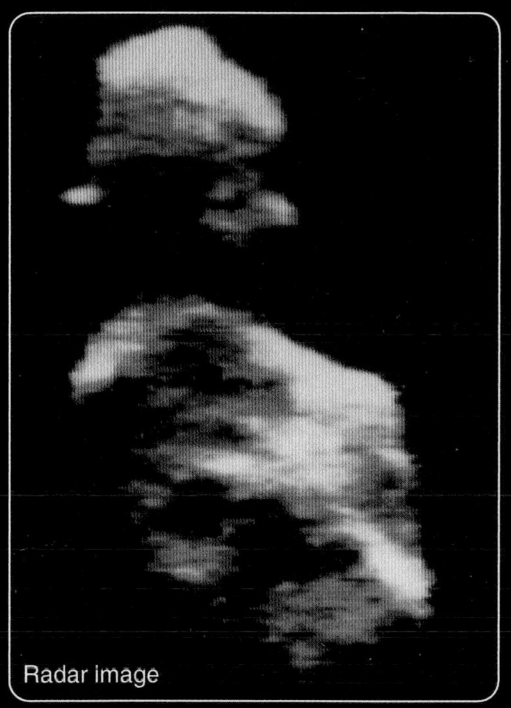

2 Asteroids that pass near Earth can be imaged by radar. The asteroid Toutatis is revealed to be a double object—two objects orbiting close to each other or actually in contact.

Ida

Double asteroids are more common than was once thought, reflecting a history of collisions and fragmentation. The asteroid Ida is orbited by a moon Dactyl only about 1.5 km in diameter.

Dactyl

30 km

Enhanced visual + infrared

Occasional collisions among the asteroids release fragments, and Jupiter's gravity scatters them into the inner solar system as a continuous supply of meteorites.

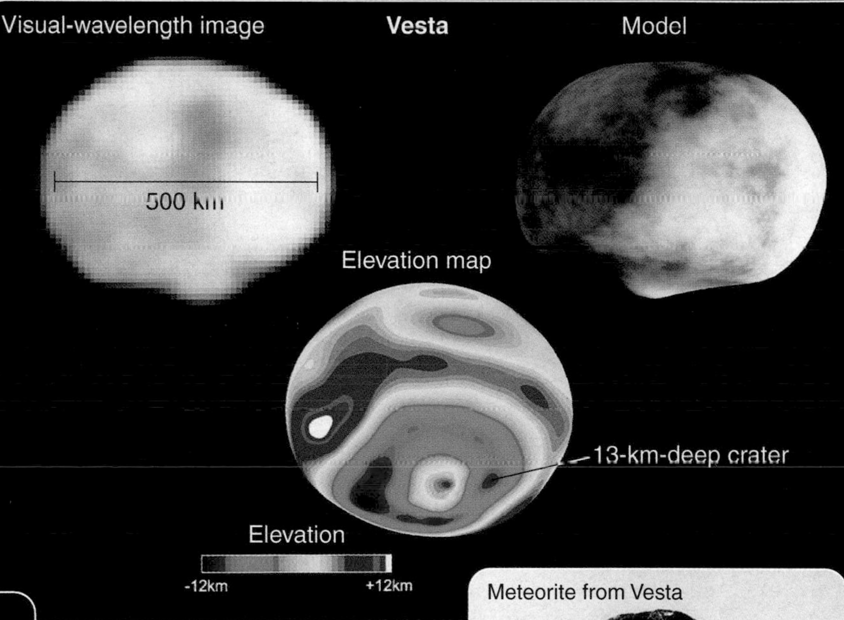

3 The large asteroid Vesta, as shown at right, provides evidence that some have suffered geological activity. No spacecraft has visited it, but its spectrum resembles that of solidified lava. Images made by the Hubble Space Telescope allow the creation of a model of its shape. It has a huge crater at its south pole. A family of small asteroids is evidently composed of fragments from Vesta, and a certain class of meteorites, spectroscopically identical to Vesta, are believed to be fragments from the asteroid. The meteorites appear to be solidified basalt.

Meteorite from Vesta

5 cm
2 in.

3a Vesta appears to have had internal heat at some point in its history, perhaps due to the decay of radioactive minerals. Lava flows have covered at least some of its surface.

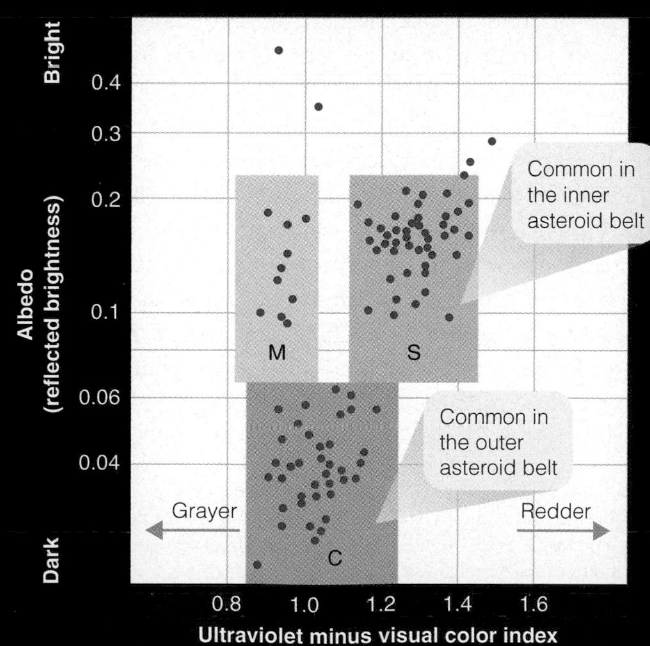

4 Although asteroids would look gray to your eyes, they can be classified according to their albedos (reflected brightness) and spectroscopic colors. As shown at left, S-types are brighter and tend to be reddish. They are the most common kind of asteroid and appear to be the source of the most common chondrites.

M-type asteroids are not too dark but are also not very red. They may be mostly iron-nickel alloys.

C-type asteroids are as dark as lumps of sooty coal and appear to be carbonaceous.

Collisions among asteroids must have been occurring since the formation of the solar system, and astronomers have found evidence of catastrophic impacts powerful enough to shatter an asteroid. Early in the 20th century, Japanese astronomer Kiyotsugu Hirayama discovered that some groups of asteroids share similar orbits. Each group is distinct from other groups, but asteroids within a group have the same average distance from the sun, the same eccentricity, and the same inclination. Up to 20 of these **Hirayama families** are known, and modern observations show that the asteroids in a family typically share similar spectroscopic characteristics. Evidently, a family is produced by a catastrophic collision that breaks a single asteroid into a family of fragments that continue traveling along similar orbits around the sun. Evidence shows that one family was produced only 5.8 million years ago in a collision between asteroids 3 km and 16 km in diameter traveling at about 5 km/s (11,000 mph), a typical speed for asteroid collisions. Evidently the fragmentation of asteroids is a continuing process.

In 1983, the Infrared Astronomy Satellite detected the infrared glow of sun-warmed dust scattered in bands throughout the asteroid belt. These dust bands appear to be the products of past collisions. The dust will eventually be blown away, but because collisions occur constantly in the asteroid belt, new dust bands will presumably be produced as the present bands dissipate.

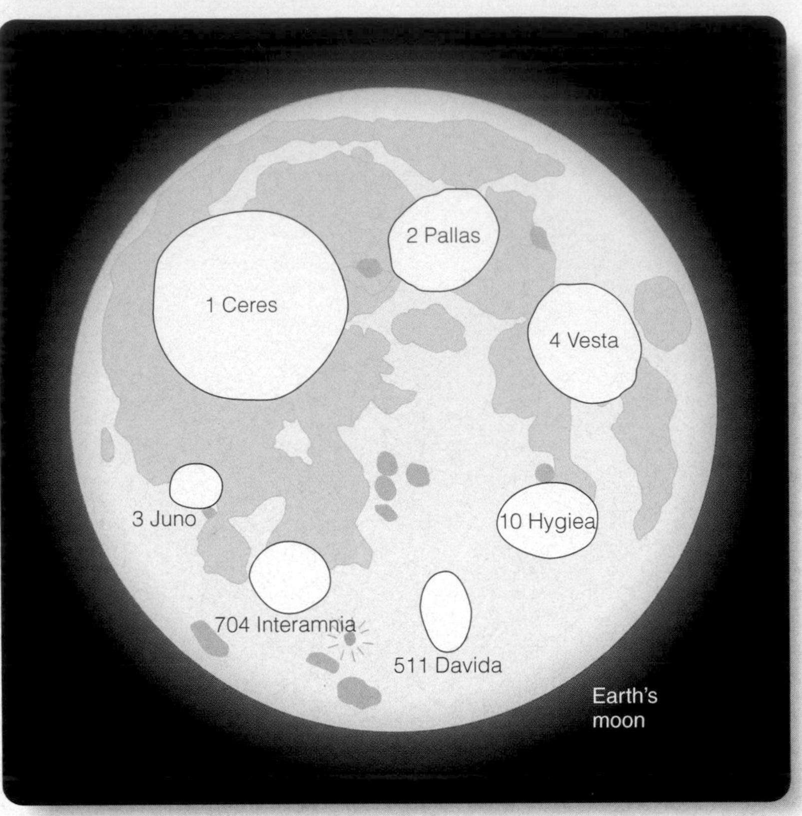

■ **Figure 25-9**

The relative size and approximate shape of the larger asteroids are shown here compared with the size of Earth's moon. Smaller asteroids can be highly irregular in shape.

What are the asteroids? Where did they come from? There are clues hidden among their orbits, and you can begin that story at its beginning.

The Asteroid Belt

The first asteroid was discovered on January 1, 1801 (the first night of the 19th century), by the Sicilian monk Giuseppe Piazzi. It was later named Ceres after the Roman goddess of the harvest (thus our word *cereal*).

Astronomers were excited by Piazzi's discovery because there seemed to be a gap where a planet might exist between Mars and Jupiter at an average distance from the sun of 2.8 AU. Ceres fit right in; its average distance from the sun is 2.766 AU. But it was a bit small to be a planet, and three more objects—Pallas, Juno, and Vesta—were discovered in the following years, so astronomers realized that Ceres and the other asteroids were not true planets.

Today over 100,000 asteroids have well-charted orbits. Many more are as yet undiscovered, but they are all small bodies. All of the larger asteroids in the asteroid belt have been found. Only three are larger than 400 km in diameter, and most are much smaller (■ Figure 25-9).

If you discovered an asteroid, you would be allowed to choose a name for it, and asteroids have been named for spouses, lovers, dogs, Greek gods, politicians, and others.* Once an orbit has been calculated, the asteroid is assigned a number listing its position in the catalog known as the *Ephemerides of Minor Planets.* Thus, Ceres is known as 1 Ceres, Pallas as 2 Pallas, and so on.

Although a few asteroids follow orbits that bring them into the inner solar system or outward among the Jovian planets, most orbit in the asteroid belt between Mars and Jupiter, and you might suspect that massive Jupiter was responsible for their origin. Certainly the distribution of asteroids in the belt is strongly affected by Jupiter's gravitation. Certain regions of the belt, called **Kirkwood's gaps** after their discoverer, Daniel Kirkwood (1814–1895), are almost free of asteroids (■ Figure 25-10). These gaps lie at certain distances from the sun where an asteroid would find itself in resonance with Jupiter. For example, if an asteroid lay 3.28 AU from the sun, it would orbit twice around the sun in the time it took Jupiter to orbit once. On alternate orbits, the asteroid would find Jupiter at the same place in space tugging outward. The cumulative perturbations would rapidly change the asteroid's orbit until it was no longer in resonance with Jupiter. This ex-

*Some sample asteroid names: Olga, Chicago, Vaticana, Noel, Ohio, Tea, Gaby, Fidelio, Hagar, Geisha, Dudu, Tata, Mimi, Dulu, Tito, Zulu, Beer, and Zappafrank (after the late musician Frank Zappa).

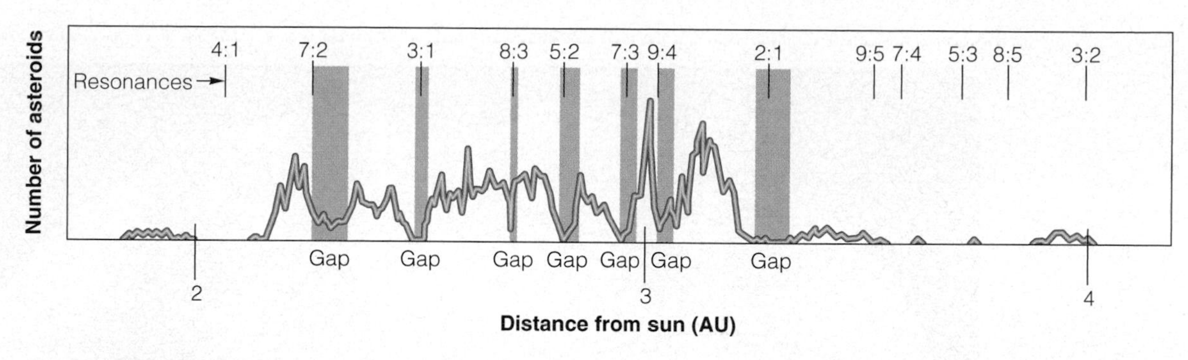

■ **Figure 25-10**

Here the red curve shows the number of asteroids at different distances from the sun. Purple bars mark Kirkwood's gaps, where there are few asteroids. Note that these gaps match resonances with the orbital motion of Jupiter.

ample is a 2:1 resonance, but gaps occur in the asteroid belt at many resonances, including 3:1, 5:2, and 7:3. You will recognize that Kirkwood's gaps in the asteroid belt are produced in the same way as some of the gaps in Saturn's rings. Both sets of gaps were discovered by Daniel Kirkwood (Chapter 23).

Modern research shows that the motion of asteroids in Kirkwood's gaps is described by a theory in mathematics that deals with chaotic behavior. As an example, consider how the smooth motion of water sliding over the edge of a waterfall decays rapidly into a chaotic jumble. The same theory of chaos that describes the motion of the water shows how the slowly changing orbit of an asteroid within one of Kirkwood's gaps can suddenly become a long, elliptical orbit that carries the asteroid into the inner solar system, where it is likely to be removed by a collision with Mars, Earth, or Venus. In this way, Jupiter's gravity can throw many meteoroids from the asteroid belt into the inner solar system.

Nonbelt Asteroids

You don't have to go all the way to the asteroid belt if you want to visit an asteroid. Some of the most interesting follow orbits that cross the orbits of the terrestrial planets or wander among the Jovian worlds. In fact, some asteroids even share orbits with the larger planets.

The **Apollo–Amor objects** are asteroids whose orbits carry them into the inner solar system. The Amor objects follow orbits that cross the orbit of Mars but don't reach the orbit of Earth. The Apollo objects have Earth-crossing orbits. These Apollo–Amor objects are dangerous. Jupiter's influence makes their orbits precess. About one-third will be thrown into the sun, and a few will be ejected from the solar system, but many of these objects are doomed to collide with a planet—perhaps ours. Earth is hit by an Apollo object once every 250,000 years, on average. With a diameter of up to 2 km, they hit with the power of a 100,000-megaton bomb and can dig craters 20 km in diameter.

Over 2300 Apollo objects are known, and none of those will hit Earth in the foreseeable future. The bad news is that there are about 1000 of these near-Earth Objects (NEOs) larger than 1 km in diameter, the minimum size of an impactor that could cause global effects on Earth. More than half a dozen teams are now searching for these NEOs. For example, LONEOS (Lowell Observatory Near Earth Object Search) is searching the entire sky visible from Lowell Observatory once a month. The LINEAR (Lincoln Near-Earth Asteroid Research) telescope in New Mexico has been very successful in finding NEOs and in finding new main-belt asteroids (■ Figure 25-11). The combined searches should be able to locate all of the largest NEOs by 2010.

This is a serious issue because even a small asteroid could do serious damage. For example, in late December 2004, an asteroid was discovered that was predicted to have a 2.6 percent chance of striking Earth on April 13, 2029. The object is large enough to do significant damage over a wide area but is not large enough to alter Earth's climate. Fortunately, further observations revealed that the object will not hit Earth in 2029.

Objects a few tens of meters or less in diameter fragment and explode in Earth's atmosphere, but the shock waves from their explosions could still cause serious damage on Earth's surface. Declassified data from military satellites show that Earth is hit about once a week by meter-size asteroids. Larger impacts produce more damage but are much less common.

It is easy to assume that the Apollo–Amor objects are rocky asteroids that have been thrown into their extreme orbits by events in the main asteroid belt. At least some of these objects, however, may be comets that have exhausted their ices and become trapped in short orbits that keep them in the inner solar system. You will see later in this chapter that the distinction between comets and asteroids is not totally clear.

There are also nonbelt asteroids in the outer solar system. These objects, being farther from the sun, move more slowly. The object Chiron, found in 1977, appears to be about 170 km in diameter. Its orbit carries it from just inside the orbit of Uranus to just inside the orbit of Saturn. Although it was first classified as an asteroid, its status is now less certain. Ten years after its discovery, Chiron surprised astronomers by suddenly brightening as it releasing jets of vapor and dust much like a comet.

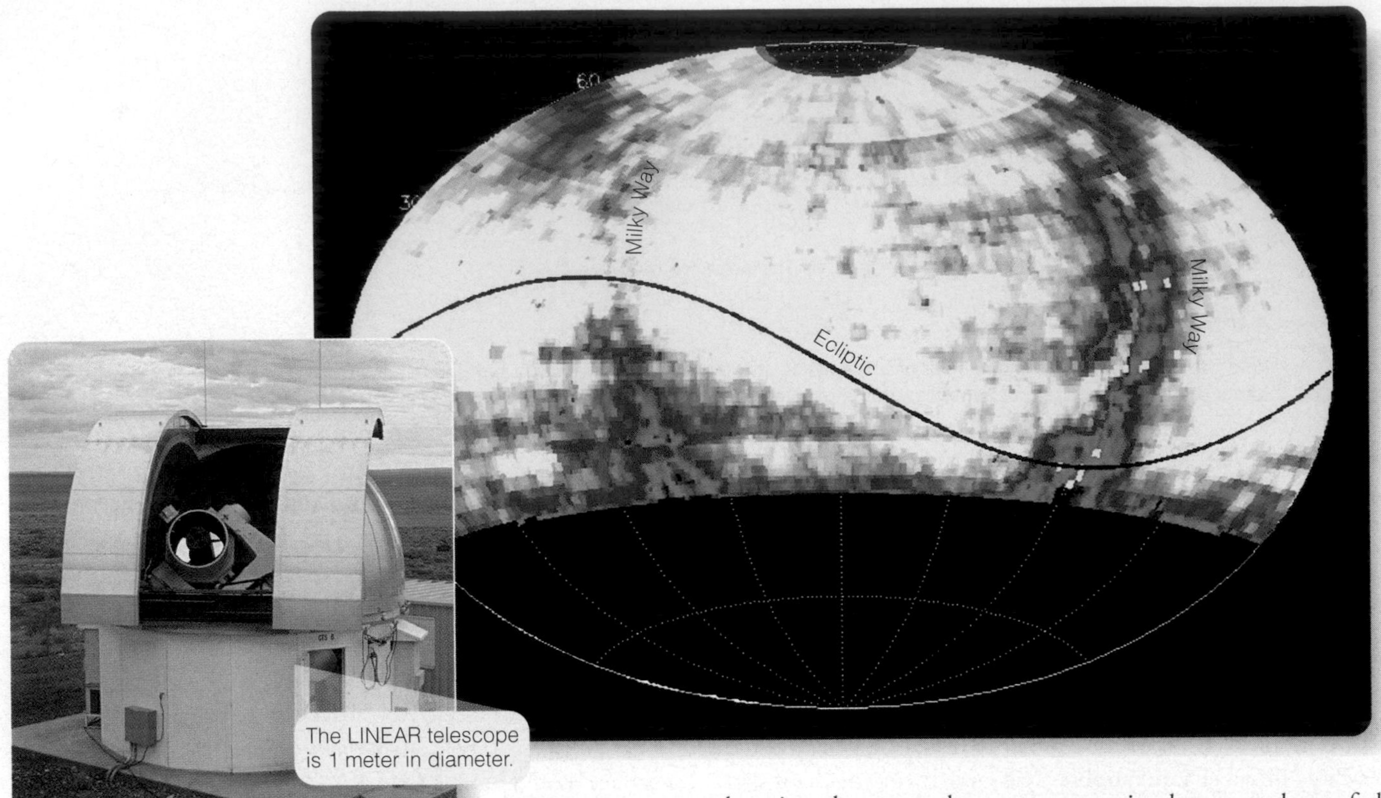

Figure 25-11

The LINEAR telescope searches for asteroids every clear night when the moon is not bright. The diagram shows the thoroughness of its search over the entire sky for one year. Asteroids are hard to discover in front of the starry Milky Way. (MIT/Lincoln Labs)

Studies of older photographs showed that Chiron had done this before sometimes when it was even farther from the sun. Astronomers now suspect that it may have a rocky crust covering deposits of ices such as solid nitrogen, methane, and carbon monoxide. Thus, Chiron may be more comet than asteroid, and it serves as a warning that the distinction is not clear-cut.

Jupiter ushers two groups of asteroids around its own orbit. These nonbelt asteroids have become trapped in the Lagrangian points along Jupiter's orbit. (See Figure 13-5.) These points lie 60° ahead of and 60° behind the planet and are regions where the gravitation of the sun and Jupiter combine to trap small bodies (■ Figure 25-12). Like cosmic sinkholes, the Lagrangian points have trapped chunks of debris now called **Trojan asteroids.** Individual asteroids are named after the heroes of the Trojan War (588 Achilles, 624 Hektor, 659 Nestor, and 1143 Odysseus, for example). Slightly over 1000 Trojan asteroids are known, but only the brightest have been given names.

Astronomers have also found a few objects in the Lagrangian points of the orbits of Mars and Neptune. Other planets, including Earth, may have Trojan asteroids trapped in their orbits. As technology allows astronomers to detect smaller objects, they are learning that our solar system contains large numbers of these small bodies. The challenge is to explain their origin.

The Origin of the Asteroids

You have concluded your study of each planet by trying to summarize its history. Can you tell the story of the asteroids? Begin with an idea that didn't work out.

An old theory held that the asteroids are the remains of a planet that broke up. Modern astronomers discount that idea, but it survives as a **Common Misconception.** Once formed, a planet is very difficult to tear apart. Rather, astronomers now understand that the asteroids are the remains of planetesimals that were unable to form a planet. Jupiter's gravity stirred the planetesimals just inside its orbit and caused collisions at unusually high velocities. These impacts tended to break up the planetesimals rather than assemble them into a planet. Orbital resonances helped to eject material, and by now most of the planetesimal objects have been lost—swept up by planets, consumed by the sun, captured as satellites, or ejected from the solar system. The asteroid belt today contains hardly 4 percent the mass of Earth's moon.

Even though most of the planetesimals have been lost, the objects left behind carry clues to their origin. The C-type asteroids have albedos smaller than 0.06 and would look very dark to your eyes. They are probably made of carbon-rich material similar to that in carbonaceous chondrites. The S-type asteroids have albedos of 0.1 to 0.2, so they would look brighter and spectroscopically redder. The M-type asteroids are bright but not as red.

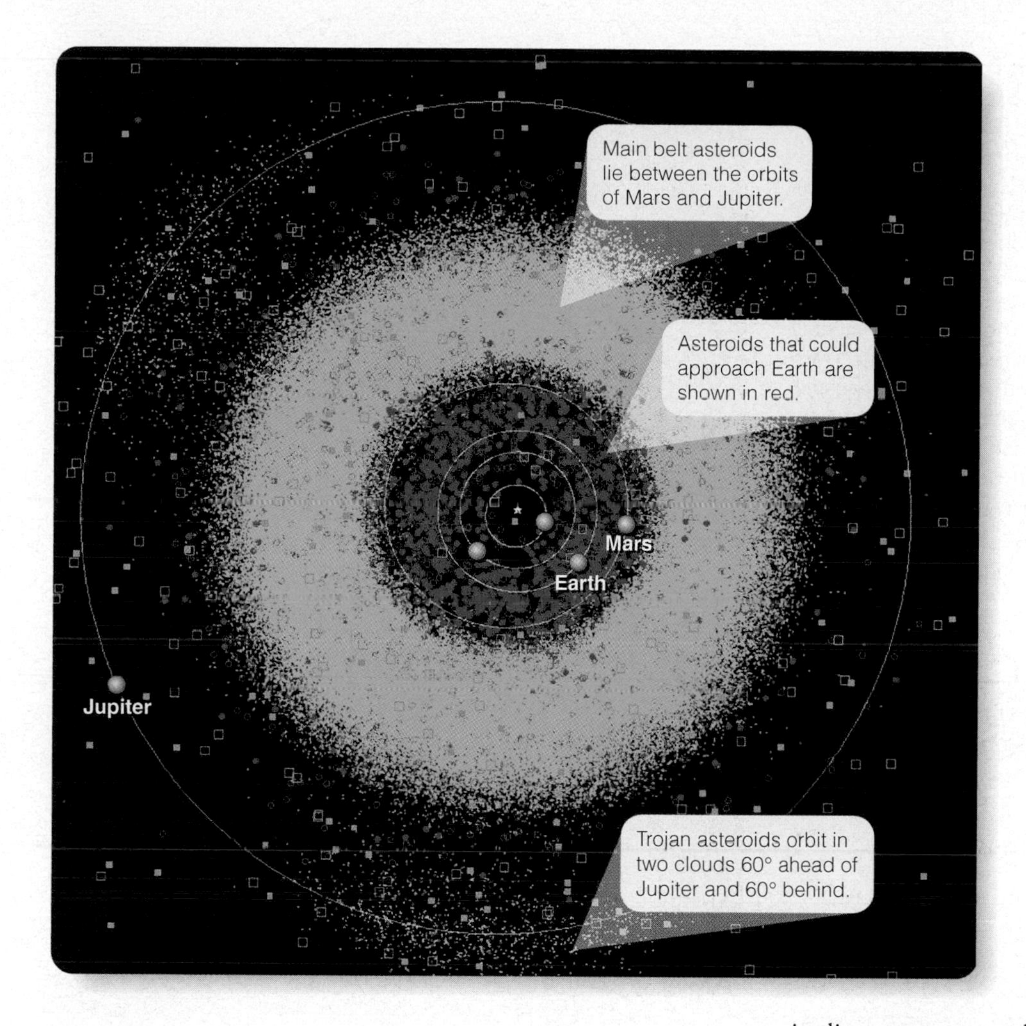

Main belt asteroids lie between the orbits of Mars and Jupiter.
Asteroids that could approach Earth are shown in red.
Mars
Earth
Jupiter
Trojan asteroids orbit in two clouds 60° ahead of Jupiter and 60° behind.

Figure 25-12

This diagram plots the position of known asteroids inside or near the orbit of Jupiter on a specific day. Squares, filled or empty, show the location of known comets. Although asteroids and comets are small bodies and lie far apart, there are a great many of them in the inner solar system. (Minor Planet Center)

S-type asteroids are believed to be rocky, but M-type asteroids appear to be metal rich and may be the iron cores of fragmented asteroids.

Although S-type asteroids are very common in the inner asteroid belt, their spectroscopic colors are different from the chondrites—the most common kind of meteorite. New evidence from the analysis of moon rocks and from observations of Eros, an S-type asteroid, shows that bombardment by micrometeorites can redden and darken S-type asteroids. Therefore, it seems that the most common kind of meteorites comes from the most common kind of asteroid.

A few other types of asteroids are known, and a number of individual asteroids have been found that are unique, but these three classes contain a majority of the known asteroids.

How did these three types originate? A clue lies in their distribution in the asteroid belt. The S-type asteroids are much more common in the inner belt, but there is almost none beyond a distance of about 3.45 AU. In contrast, the C types are rather rare in the inner belt but are very common in the outer belt. This distribution reflects differences in the temperature of the solar nebula during the formation of the planetesimals. It was cooler in the outer belt, so the planetesimals that formed there tended to be volatile-rich carbonaceous chondrites. In the warmer regions of the inner belt, the composition of the growing planetesimals was more like that of the chondrites.

Earlier you saw evidence that Vesta has at some time in the past been at least partly resurfaced by lava flowing up from its interior. How could a small asteroid be heated enough to produce lava flows? One source of heat in newly formed asteroids could be the decay of short-lived radioactive elements such as aluminum-26. The smallest asteroids lose their heat too fast to melt, but aluminum-26 decays fast enough to melt the interiors of bodies larger than a few dozen kilometers in diameter. Thus it is not so surprising that Vesta and some other asteroids larger than about 100 km in diameter were modified by lava flows.

In contrast, the largest main-belt asteroid, Ceres, is now recognized as a dwarf planet (Chapter 24). It is spherical, about 900 km in diameter, but does not seem to have been modified by internal heating. The light reflected from its surface suggests a claylike material related to carbonaceous chondrites. This is surprising, because clays form when minerals are exposed to water. In fact, spectroscopic observations reveal water ice on Ceres. Evidently some asteroids may have had significant amounts of water bound into their crusts when they were young.

All of this evidence suggests that the asteroids are the broken remains of planetesimals that formed in the solar nebula as planet building began. The largest remaining asteroids, such as Ceres and Vesta, may be largely unbroken planetesimals, and some of these may have experienced some surface evolution due to internal heat or the presence of water. Nevertheless, the vast majority of the asteroids are fragments, and many may consist of bodies that were shattered and then re-formed as gravity pulled the fragments back together. Compositional differences between asteroids seem to be due to temperature differences in the ancient solar nebula from which the planetesimals formed. The presence of massive Jupiter orbiting nearby prevented the original planetesimals from accreting to build a planet. Instead, collisions fragmented them, and nearly all of the material has been lost.

◄ SCIENTIFIC ARGUMENT ►

What evidence makes you think that the asteroids have been fragmented?

Some of the best scientific arguments test the interpretation of evidence. If you understand the evidence, you hold the key to the science. To begin, you might note that the solar nebula theory of the formation of the solar system predicts that planetesimals collided and either stuck together or fragmented. This is suggestive, but it is not evidence. A theory can never be used as evidence to support some other theory or hypothesis. Evidence takes the form of observations or experimental results, so you need to turn to observations of asteroids. Spacecraft photographs of asteroids such as Ida, Gaspra, and Eros show irregularly shaped little worlds heavily scarred by impact craters. In fact, observations of some asteroids show what may be pairs of bodies in contact, and the Galileo image of Ida reveals its small satellite, Dactyl. Furthermore, some meteorites appear to come from the asteroid belt, and a few have been linked to specific asteroids such as Vesta. There are even families of asteroids that seem to be fragments from a single collision. All of this evidence suggests that the asteroids have been broken up by violent impacts.

The impact fragmentation of asteroids has been important, but it has not erased all traces of the original planetesimals from which the asteroids formed. Build another argument based on evidence. **What evidence can you cite that reveals what those planetesimals were like?**

◄ ►

25-3 Comets

FEW THINGS IN ASTRONOMY ARE MORE BEAUTIFUL than a bright comet hanging in the night sky (■ Figure 25-13). It is a **Common Misconception** that comets whiz rapidly across the sky. Meteors shoot across the sky like demented fireflies, but a comet moves with the stately grace of a great ship at sea, its motion hardly apparent. Night by night it shifts slightly against the stars and may remain visible for weeks. Faint comets are common; a number are discovered every year. But a truly bright comet appears about once a decade. Comet Hyakutake in 1996 (Figure 25-1) and Comet Hale–Bopp in 1997 were both dramatic, but the later comet was so bright that you might class it with the great comets such as Comet Halley in 1910. A patient person might see half a dozen or more bright comets in a lifetime.

While everyone enjoys the beauty of comets, astronomers study them for their cargo of clues to the origin of our solar system.

Properties of Comets

As always, you should begin your study of a new kind of object by summarizing its observational properties. What do comets look like, and how do they behave? The observations are the evidence that reveals the secrets of the comets.

Study **Comet Observations** on pages 586–587 and notice three important properties of comets and three new terms:

■ **Figure 25-13**

Comet Hale–Bopp was very bright in the sky in 1997. A comet can remain bright in the sky for weeks as it sweeps along its orbit through the inner solar system. (Dean Ketelsen)

1. Comets have two kinds of tails shaped by the solar wind and solar radiation. Gas and dust released by a comet's icy nucleus produces a head or *coma* and is then blown outward. The gas produces a *type I,* or *gas, tail,* and the dust produces a *type II,* or *dust, tail.*
2. Notice the importance of dust in comets. It not only produces dust tails but spreads throughout the solar system.
3. Evidence shows that comet nuclei are fragile and can break into pieces.

Astronomers can put these and other observations together to study the structure of comet nuclei.

ThomsonNOW Sign in at www.thomsonedu.com and go to ThomsonNOW to see Astronomy Exercise "Comets."

The Geology of Comet Nuclei

The nuclei of comets are quite small and cannot be studied in detail using Earth-based telescopes. Nevertheless, astronomers are beginning to understand the geology of these peculiar objects.

Comet nuclei contain ices of water and other volatile compounds such as carbon dioxide, carbon monoxide, methane, ammonia, and so on. These are the kinds of compounds that should have condensed from the outer solar nebula, and that makes astronomers think that comets are ancient samples of the gases and dust from which the planets formed.

When comet nuclei approach the sun, the ices absorb energy from sunlight and sublime—change from a solid directly into a gas—producing the observed tails. As the gases break down and combine chemically, they produce the many compounds found in comet tails. Vast clouds of hydrogen gas observed around the heads of comets are derived from the breakup of molecules from the ices.

Five spacecraft flew past the nucleus of Comet Halley when it visited the inner solar system in 1985 and 1986. Other spacecraft flew past the nuclei of Comet Borrelly in 2001 and Comet Wild 2 in 2004. The Deep Impact probe hit Comet Tempel 1 in 2005. Photos show that these comet nuclei are irregular in shape and very dark, with jets of gas and dust spewing from active regions (■ Figure 25-14). In general, these nuclei are darker than a lump of coal, which suggests the composition of the carbon-rich meteorites called carbonaceous chondrites.

From the gravitational influence of a nucleus on a passing spacecraft, astronomers can calculate the mass and density of the nucleus. Comet nuclei appear to have densities of 0.1 to 0.25 g/cm^3, much less than the density of ice. Comet nuclei are evidently not solid balls of ice but must be fluffy mixtures of ices and rocky dust with significant amounts of empty space.

Photographs of the comae (plural of *coma*) of comets often show jets springing from the nucleus and being swept back by the pressure of sunlight and by the solar wind to form the tail (Figure 25-14). Studies of the motions of these jets as the nucleus rotates reveal that the jets originate from active regions that may be faults or vents. As the rotation of a cometary nucleus carries an active region into sunlight, it begins venting gas and dust, and as it rotates into darkness it shuts down. The nuclei of comets appear to have a crust of rocky dust left behind when the ices vaporize. Breaks in that crust can expose ices to sunlight, and vents can occur in those regions. It also seems that some comets have pockets of volatiles buried below the crust. When one of those pockets is exposed and begins to vaporize, the comet can suffer a dramatic outburst.

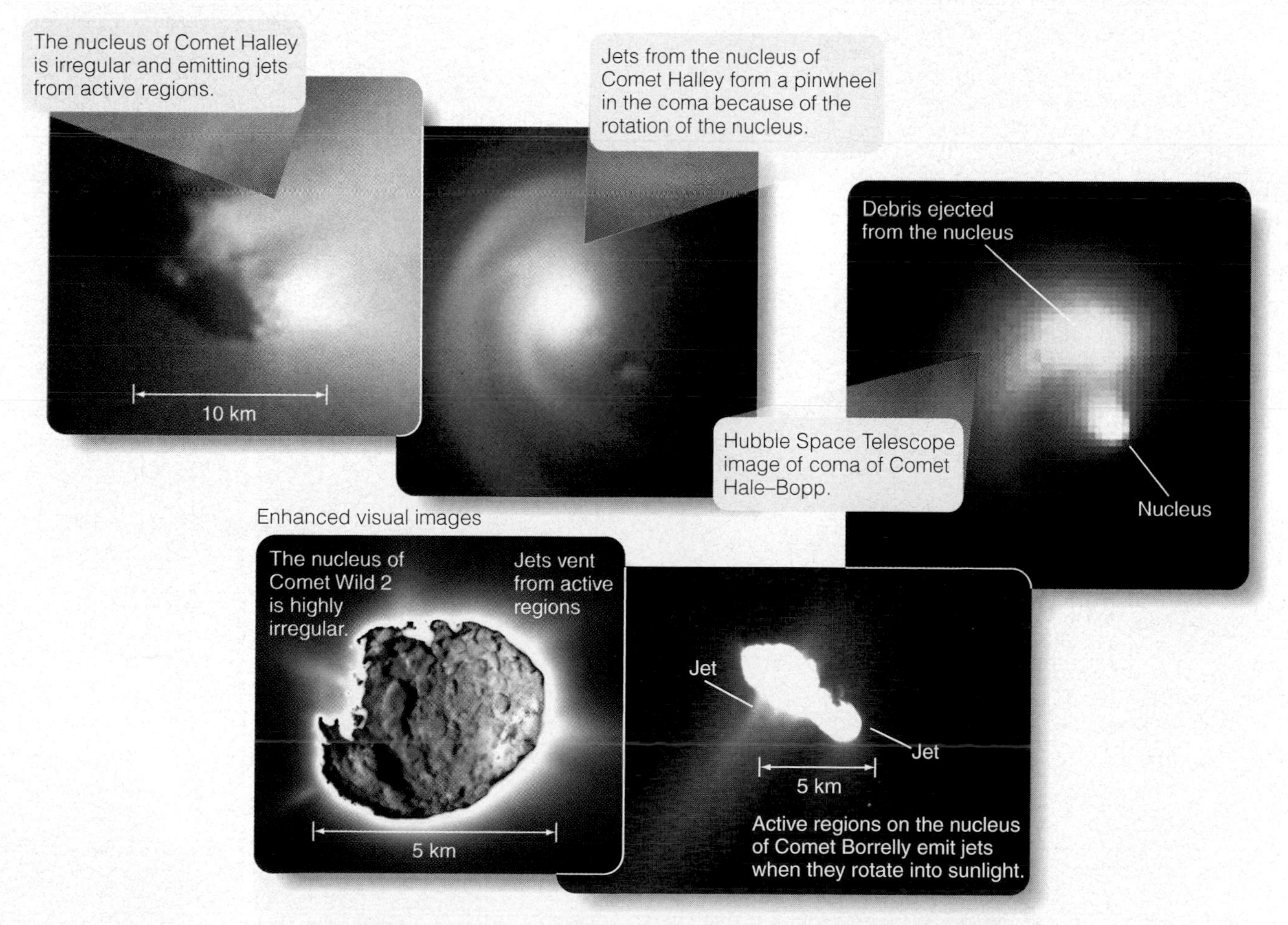

The nucleus of Comet Halley is irregular and emitting jets from active regions.
10 km
Jets from the nucleus of Comet Halley form a pinwheel in the coma because of the rotation of the nucleus.
Debris ejected from the nucleus
Hubble Space Telescope image of coma of Comet Hale–Bopp.
Nucleus
Enhanced visual images
The nucleus of Comet Wild 2 is highly irregular.
Jets vent from active regions
5 km
Jet
Jet
5 km
Active regions on the nucleus of Comet Borrelly emit jets when they rotate into sunlight.

■ Figure 25-14

Visual-wavelength images made by spacecraft and by the Hubble Space Telescope show how the nucleus of a comet produces jets of gases from regions where sunlight vaporizes ices. (Halley nucleus: © 1986 Max-Planck Institute; Halley coma: Steven Larson; Comets Borrelly, Hale–Bopp and Wild 2: NASA)

In 2005, the Deep Impact spacecraft released an instrumented impactor into the path of comet Tempel 1. Exactly as planned, the nucleus of the comet ran into the impactor at almost 10 km/s (23,000 mph). The impactor penetrated the crust of the nucleus and blasted material out into space where the mother ship could analyze it (■ Figure 25-15). The burst of vapor and dust was not only detected from the mother ship but was also seen by Earth-based telescopes.

The nuclei of comets are not strong. A nucleus can be ripped apart by the violence of gases bursting through the crust or by the tidal forces produced if the comet passes near a massive body. A number of comets have been observed to break into two or more parts while near Jupiter or near the sun. On July 8, 1992, Comet Shoemaker–Levy 9 passed within 1.29 planetary radii of Jupiter's center, well within its Roche limit, and tidal forces ripped the nucleus into at least 21 pieces (■ Figure 25-16a). The fragmented pieces were as large as a few kilometers in diameter and spread into a long string of objects that looped away from Jupiter and then fell back to strike the planet and produce massive impacts over a period of six days in July 1994 (Figure 23-6). The fragmentation of comet nuclei may explain long chains of craters like those shown in Figure 25-16b and c found on the Earth's moon and on some other planetary surfaces in the solar system.

The Solar and Heliospheric Observatory (SOHO) was put into space to observe the sun, but it has also discovered over a thousand comets called "sun grazers" because they come very close to the sun, in some cases 70 times closer to the sun's surface than the planet Mercury (■ Figure 25-17). In fact, as many as three comets a week plunge into the sun and are destroyed. Most sun grazers belong to one of four groups, where the comets in each group have very similar orbits. Like the Hirayama families of asteroids, these comet groups appear to be made up of fragments of larger cometary nuclei.

Sun grazers can be destroyed by the sun, but even normal comets that don't come so close can suffer sun damage. Each passage around the sun vaporizes many millions of tons of ices, so the nucleus slowly loses its ices until there is nothing left but dust and rock falling along an orbit around the sun. The fate of a comet is clear. The mystery is its origin.

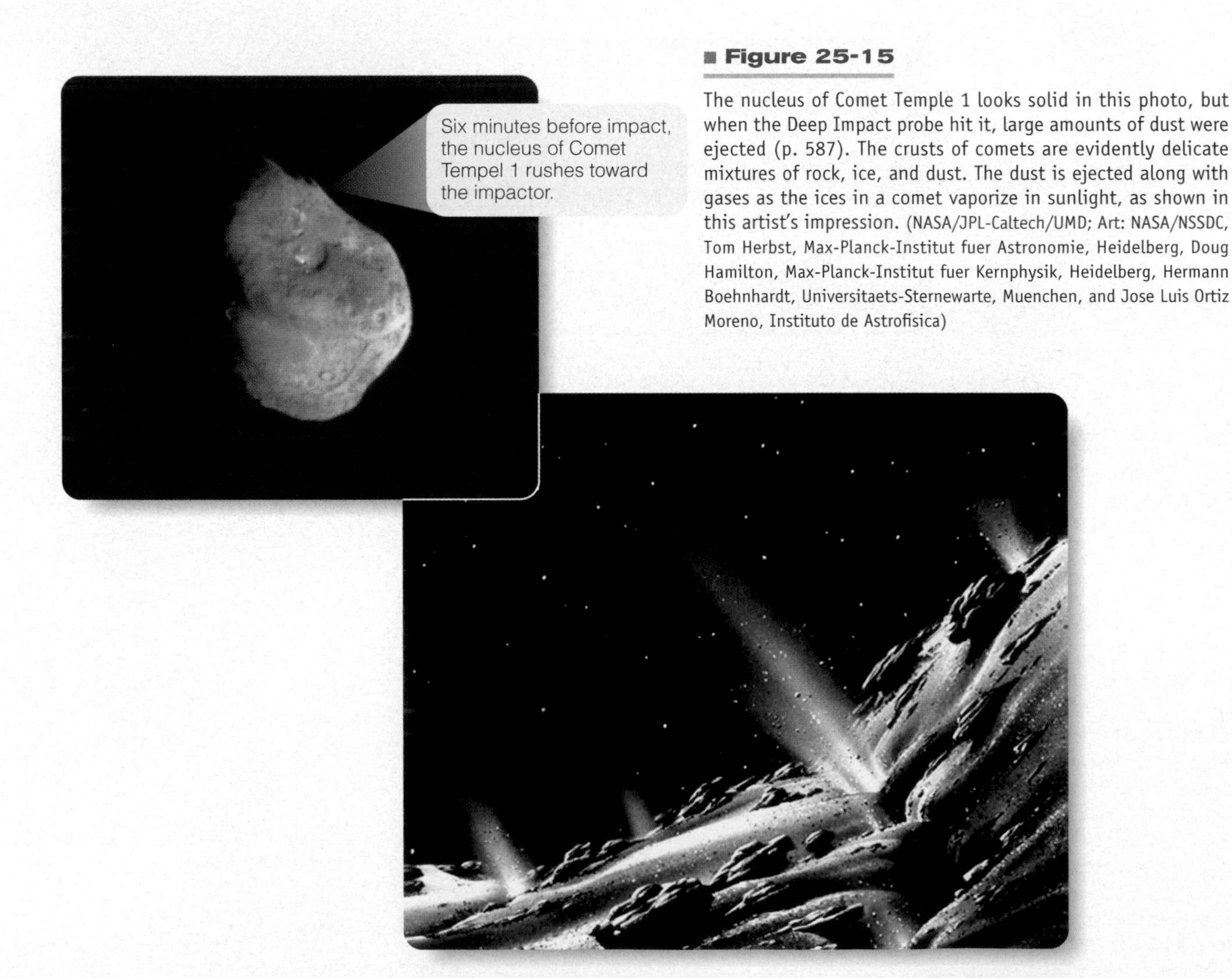

Six minutes before impact, the nucleus of Comet Tempel 1 rushes toward the impactor.

■ Figure 25-15

The nucleus of Comet Temple 1 looks solid in this photo, but when the Deep Impact probe hit it, large amounts of dust were ejected (p. 587). The crusts of comets are evidently delicate mixtures of rock, ice, and dust. The dust is ejected along with gases as the ices in a comet vaporize in sunlight, as shown in this artist's impression. (NASA/JPL-Caltech/UMD; Art: NASA/NSSDC, Tom Herbst, Max-Planck-Institut fuer Astronomie, Heidelberg, Doug Hamilton, Max-Planck-Institut fuer Kernphysik, Heidelberg, Hermann Boehnhardt, Universitaets-Sternewarte, Muenchen, and Jose Luis Ortiz Moreno, Instituto de Astrofisica)

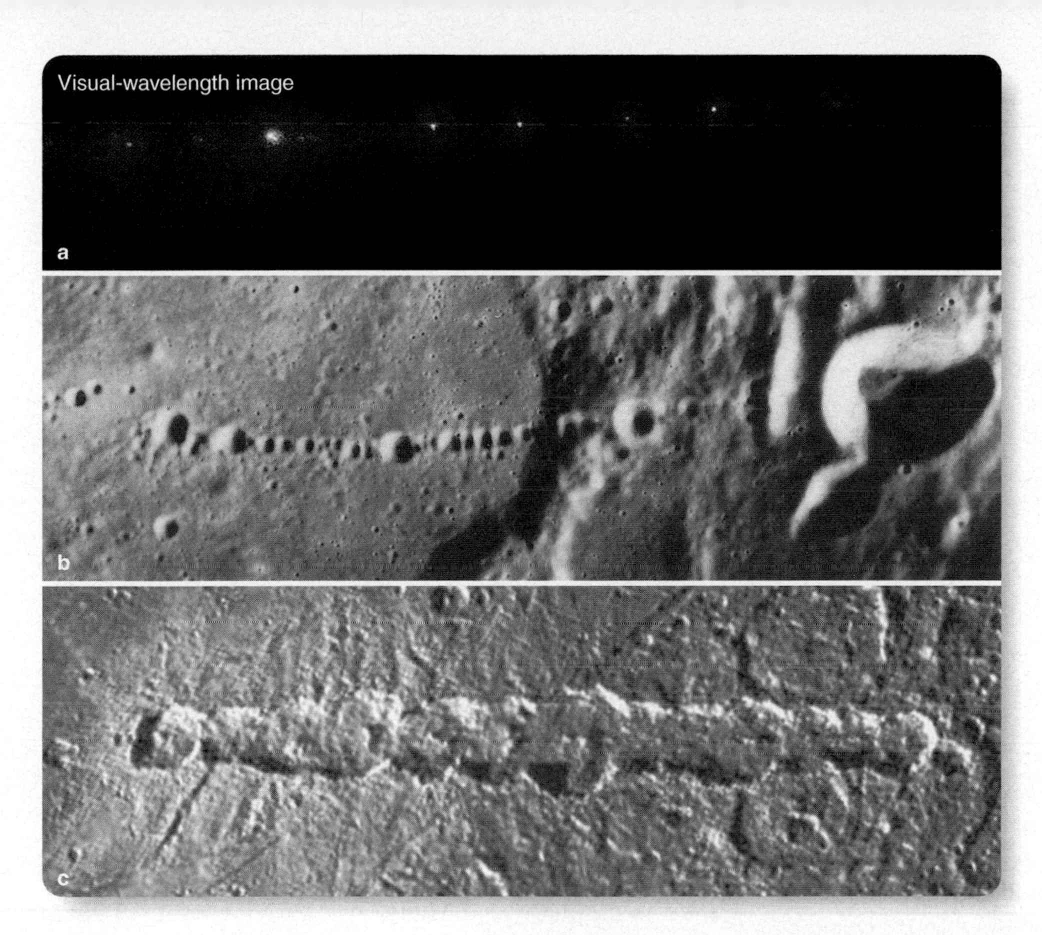

■ **Figure 25-16**

(a) Tides from Jupiter pulled apart the nucleus of Comet Shoemaker–Levy 9 to form a long strand of icy bodies and dust that fell back to strike Jupiter two years later. (b) A 40-km-long crater chain on the moon and (c) another 140-km-long crater chain on Jupiter's moon Callisto were apparently formed by the impact of fragmented comet nuclei. The impacts that form such chains probably occur within a span of seconds. (NASA)

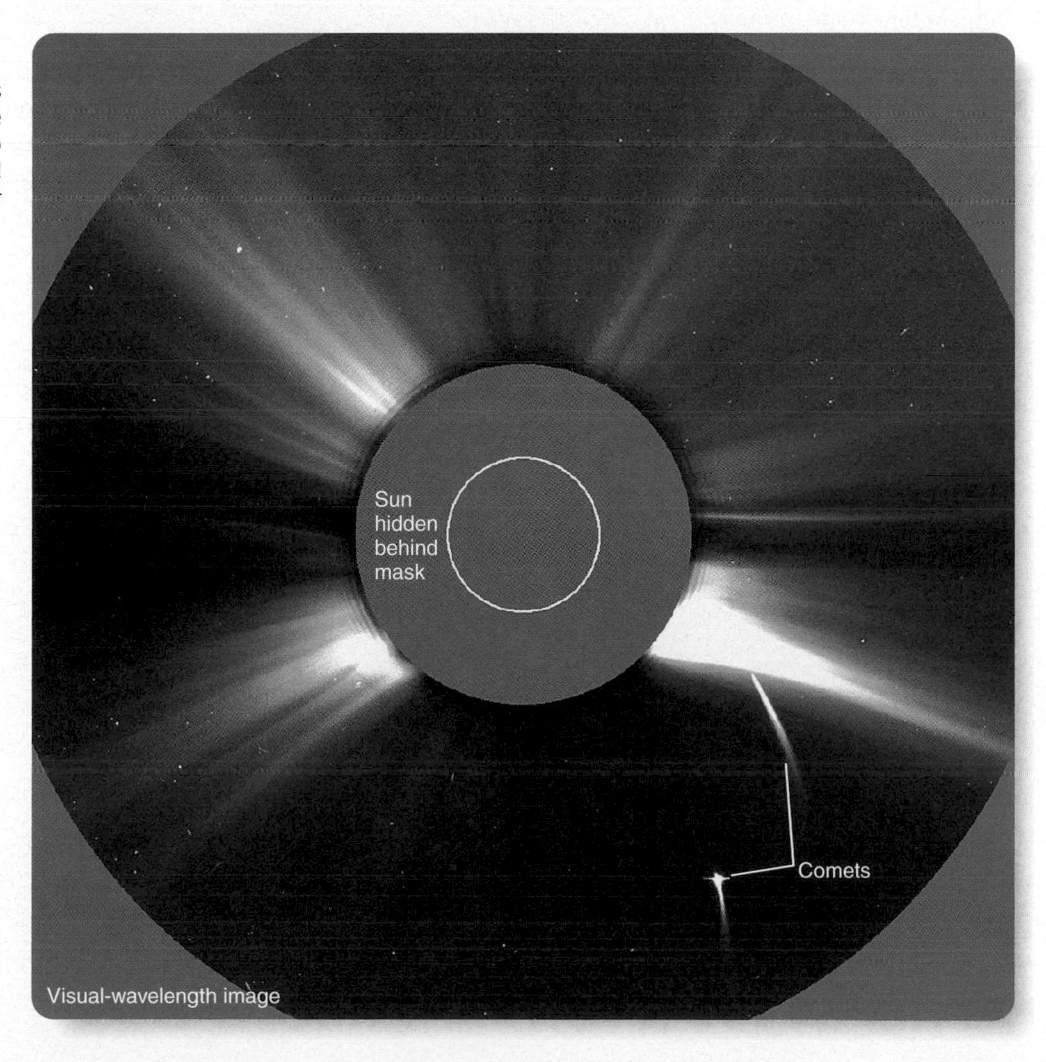

■ **Figure 25-17**

The SOHO observatory can see comets rounding the sun on very tight orbits. Some of these sun-grazing comets, like the two shown here, are destroyed by the heat and are not detected emerging on the other side of the sun. (SOHO)

The Origin of Comets

Family relationships among the comets can give you clues to their origin. Most comets have long, elliptical orbits with periods greater than 200 years and are known as long-period comets. Their orbits are randomly inclined, with comets falling into the inner solar system from all directions. As many circle the sun clockwise as counterclockwise.

In contrast, about 100 or so of the 600 well-studied comets have orbits with periods less than 200 years. These short-period comets follow orbits that lie within 30° of the plane of the solar system, and most revolve around the sun counterclockwise—the same direction the planets orbit. Comet Halley, with a period of 76 years, is a short-period comet.

A comet cannot survive long in an orbit that brings it into the inner solar system. The heat of the sun vaporizes ices and reduces comets to inactive bodies of rock and dust. A comet may last only 100 to 1000 orbits around the sun. Also, encounters with planets, especially Jupiter, can fling a comet into the sun or out of the solar system. And as you have seen, comets hit planets, and that's the end of those unlucky comets. Even if it didn't vaporize in the sun's heat, a comet couldn't survive more than about half a million years before being swept up by a planet. The comets visible in our skies can't have survived in their present orbits for 4.6 billion years since the formation of the solar system, and that means there must be a continuous supply of new comets. Where do they come from?

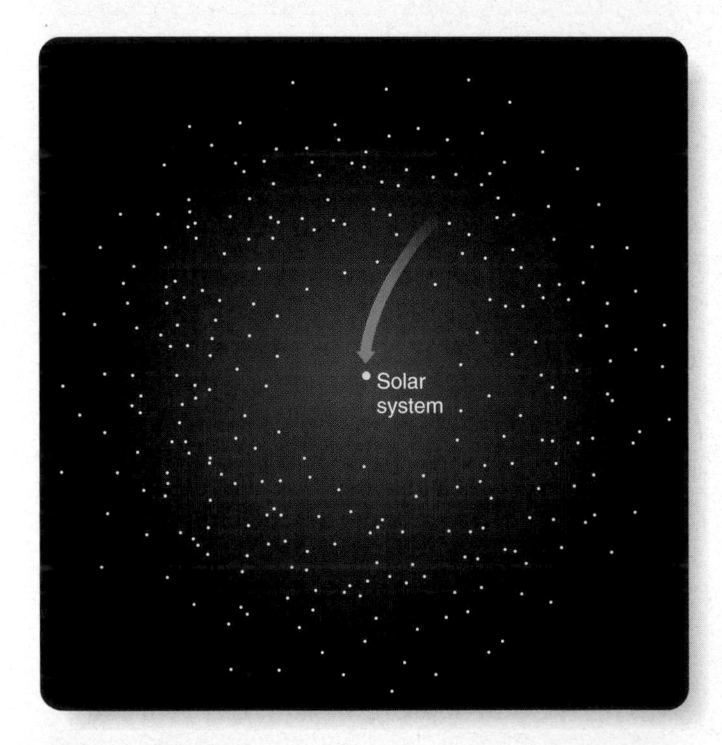

■ **Figure 25-18**

The long-period comets appear to originate in the Oort cloud. Objects that fall into the solar system from this cloud arrive from all directions.

In the 1950s, the Dutch astronomer Jan Oort proposed that the long-period comets are objects that fall in from what became known as the **Oort cloud,** a spherical cloud of icy bodies that extends from about 10,000 to 100,000 AU from the sun (■ Figure 25-18). Astronomers estimate that the cloud contains several trillion icy bodies. Far from the sun, they are very cold, lack comae and tails, and are invisible from Earth. The gravitational influence of occasional passing stars perturbs a few of these objects and causes them to fall into the inner solar system, where the heat of the sun warms their ices and transforms them into comets. Because the Oort cloud is spherical, long-period comets can fall inward from any direction.

It is not outlandish that stars pass close enough to affect the Oort cloud. Data from the Hipparcos satellite show that the star Gliese 710 will pass within 1 ly (about 63,000 AU) of our solar system in about a million years. That may shower the inner solar system with Oort cloud objects, which, warmed by the sun, will become comets.

But saying that comets come from the Oort cloud only pushes the mystery back one step. Where did those icy bodies come from? In preceding chapters, you have studied the origin and evolution of our solar system so carefully that the answer leaps out at you. Those are some of the icy planetesimals that formed in the outer solar nebula.

The bodies in the Oort cloud, however, could not have formed at their present location, because the solar nebula would not have been dense enough so far from the sun. And if they had formed from the solar nebula, they would be distributed in a disk and not in a sphere. Astronomers think the Oort cloud planetesimals formed in the outer solar system near the present orbits of Uranus and Neptune. As those planets grew more massive and were pushed outward by gravitational interactions with Jupiter and Saturn, they swept up many of these planetesimals, but they also ejected some out of the solar system. Most of those objects vanished into space, but perhaps 10 percent had their orbits modified by the gravity of stars passing nearby and became part of the Oort cloud. Some of them later became the long-period comets.

The long-period comets originate in the Oort cloud, and some of the short-period comets do too. Some short-period comets, including Comet Halley, appear to have begun as long period comets from the Oort cloud and had their orbits altered by a close encounter with Jupiter. But that process can't explain all of the short-period comets. Some follow orbits that could not have been produced by Oort cloud objects interacting with Jupiter. There must be another source of icy bodies in our solar system. To find the answer, you need only look beyond the orbit of Neptune and study the icy Kuiper belt objects.

Comets from the Kuiper Belt

You first met the Kuiper belt in Chapter 19 when you studied the origin of the solar system. In Chapter 24, you discussed the largest Kuiper belt objects as examples of dwarf planets. In this chapter, it is important to study the smaller bodies of the Kuiper belt because they are one of the sources of comets.

In 1951, Dutch-American astronomer Gerard P. Kuiper proposed that the formation of the solar system should have left behind a belt of small, icy planetesimals beyond the Jovian planets and in the plane of the solar system. Such objects were first discovered in 1992 and are now known as Kuiper belt objects. You should know, however, that in 1943 and 1949, astronomer Kenneth Edgeworth published papers that included a paragraph speculating about objects beyond Pluto. Consequently, you may see the Kuiper belt referred to as the Edgeworth–Kuiper belt. Actually, lots of astronomers have speculated about one or more bodies beyond Pluto, and most astronomers refer to the band of objects as the Kuiper belt.

The Kuiper belt objects are small, icy bodies (■ Figure 25-19) that orbit in the plane of the solar system extending from the orbit of Neptune out to about 50 AU from the sun. Some objects loop out as far as 150 AU, but they seem to have been scattered by gravitational interactions with passing stars. The entire Kuiper belt, containing as many as 70,000 objects as big as 100 km in diameter, would be hidden behind the yellow dot representing the solar system in Figure 25-18. Although some Kuiper belt objects are surprisingly large, most are quite small.

Can this belt of ancient, icy worlds generate short-period comets? Because Kuiper belt objects orbit in the same direction as the planets and in the plane of the solar system, it is possible for an object thrown inward to be perturbed by Jupiter into an orbit resembling those of the short-period comets. Collisions and interactions among the KBOs must be a rare but continuing process, so there must be a continuous supply of small, icy bodies from the Kuiper belt falling into the inner solar system.

Comets vary in brightness and orbit, but they look very similar. Nevertheless, there are at least two kinds of comets in our solar system. Some originate in the Oort cloud far from the sun. Others come from the Kuiper belt not too far beyond Neptune. But they all share one characteristic—they are ancient icy bodies that were born when the solar system was young.

◄ SCIENTIFIC ARGUMENT ►

How do comets help explain the formation of the planets?

This scientific argument pulls together a number of ideas. According to the solar nebula hypothesis, the planets formed from planetesimals that accreted in a disk-shaped nebula around the forming sun. In the outer solar nebula, it was cold, and the planetesimals would have contained large amounts of ices. Many of these planetesimals were destroyed when they fell together to make the planets, but some survived. The icy bodies of the Oort cloud and the Kuiper belt are the last surviving icy planetesimals in our solar system. When these bodies fall into

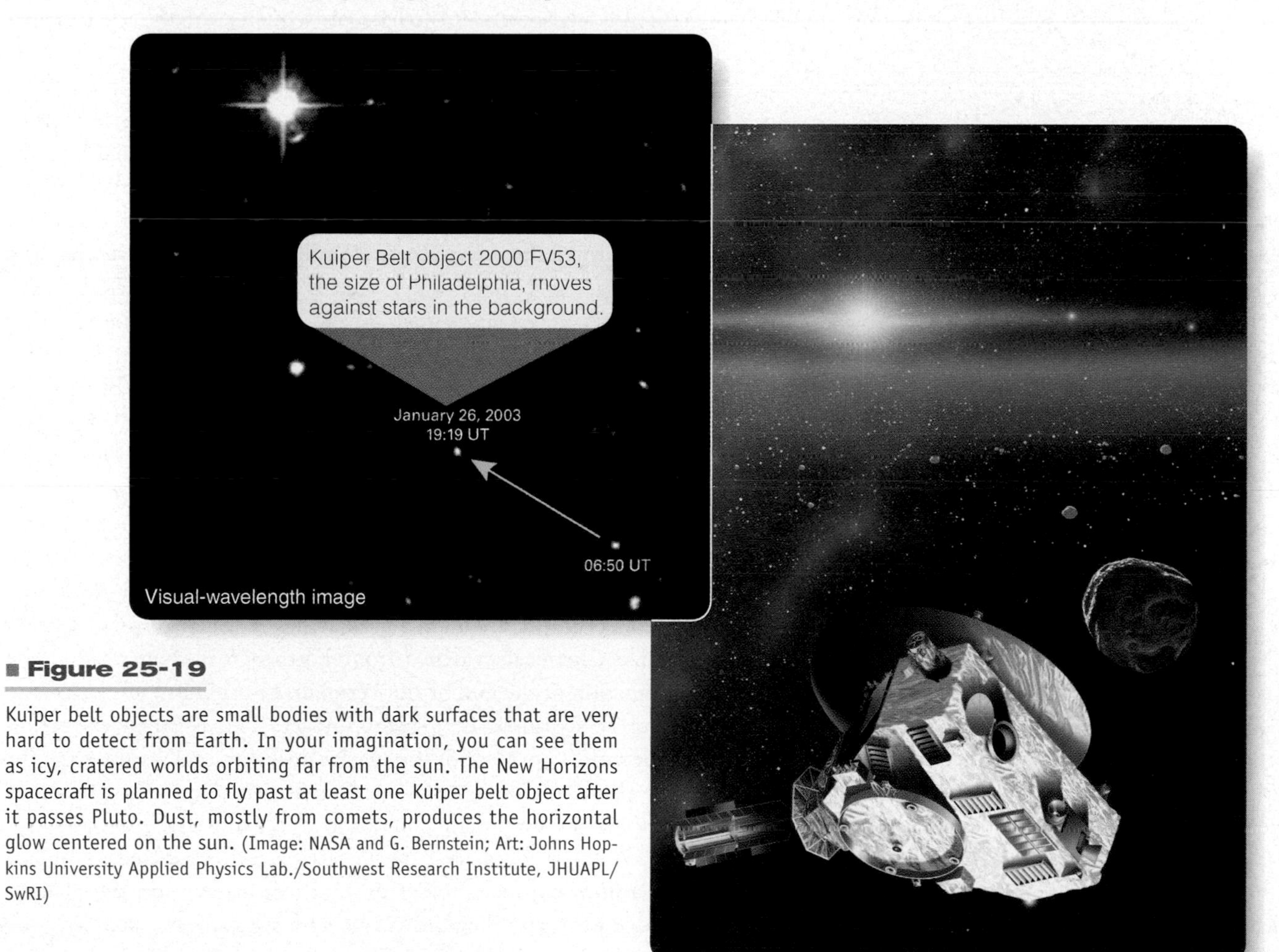
Kuiper Belt object 2000 FV53, the size of Philadelphia, moves against stars in the background.
January 26, 2003
19:19 UT
06:50 UT
Visual-wavelength image

■ Figure 25-19

Kuiper belt objects are small bodies with dark surfaces that are very hard to detect from Earth. In your imagination, you can see them as icy, cratered worlds orbiting far from the sun. The New Horizons spacecraft is planned to fly past at least one Kuiper belt object after it passes Pluto. Dust, mostly from comets, produces the horizontal glow centered on the sun. (Image: NASA and G. Bernstein; Art: Johns Hopkins University Applied Physics Lab./Southwest Research Institute, JHUAPL/SwRI)

the inner solar system, they become comets, and the gases they release reveal that they are rich in volatile materials such as water, carbon dioxide, carbon monoxide, methane, and ammonia. These are the ices you would expect to find in icy planetesimals. Furthermore, comets are rich in dust, and the planetesimals must have included large amounts of dust frozen into the ices when they formed. The nuclei of comets are frozen samples of the ancient solar nebula.

Nearly all of the mass of a comet is in the nucleus, but the light you see comes from the coma and the tail. Build an argument pulling together ideas about cometary nuclei. **What kind of spectra do comets produce, and what does that tell you about their nuclei?**

◄ ►

25-4 Impacts on Earth

ASTEROIDS AND COMETS FALL through our solar system like runaway trucks on a busy highway. These objects must hit planets now and then. In fact, it is these collisions that built the planets, and most of the original planetesimals have now been either incorporated into the planets or ejected from the solar system. Nevertheless, planets still get hit, as was dramatically apparent in 1994 when Comet Shoemaker–Levy hit Jupiter. Like any other planet, Earth must get hit now and then, but where is the evidence?

Impacts and Dinosaurs

Comet nuclei and asteroids may seem small compared to Earth, but they hit with tremendous energy. Calculations show that such impacts throw vast quantities of dust into the atmosphere and cause widespread changes in climate. The extinction of entire species, including the dinosaurs, may have been caused by comet impacts. One of the most important clues is an element called iridium.

In 1980, Luis and Walter Alvarez announced the discovery of unusual amounts of the element iridium in sediments laid down at the end of the Cretaceous period 65 million years ago—just when the dinosaurs and many other species became extinct. Iridium is rare in Earth's crust but common in meteorites, leading the Alvarezes to suggest that a major meteorite impact at the end of the Cretaceous threw vast amounts of iridium-rich dust into the atmosphere. This dust might have plunged Earth into a winter that lasted many years, killing off many species of plants and animals, including all of the dinosaurs.

This theory was met with skepticism at first, but soon scientists found the iridium anomaly in sediments of the same age all over the world. Others found related elements and mineral forms typical of meteorite impacts. At the same time, theorists studying nuclear weapons predicted that a nuclear war would throw so much dust into the atmosphere that our planet would be plunged into a "nuclear winter" that could last a number of years. A major impact would do the same thing. Within a few years, most scientists agreed that the Cretaceous extinctions were caused by one or more major impacts.

Mathematical models combined with data from the comet impacts on Jupiter in 1994 have yielded a plausible scenario of events likely to follow a major impact on Earth. Of course, creatures living near the site of the impact would probably die in the initial shock, and an impact at sea would create tsunami (tidal waves) many hundreds of meters high that would devastate coastal regions for many kilometers inland even halfway around the world.

But the worst effects would begin after the initial explosion. Whether it occurred on land or sea, a major impact would excavate large amounts of pulverized rock and loft it high above the atmosphere. As this material fell back, Earth's atmosphere would be turned into a glowing oven as red-hot rock streamed through the air in a rain of meteors, and the heat would trigger massive forest fires around the world. Soot from such fires has been detected in layers of clay laid down at the end of the Cretaceous.

Once the firestorms cooled, the remaining dust in the atmosphere would block sunlight and produce deep darkness for a year or more. And no matter where on Earth an impact occurred, it would almost certainly vaporize large amounts of limestone. The carbon dioxide released from the limestone would produce intense acid rain. All of these consequences make it surprising that any life could have survived such an impact.

Geologists have located a crater at least 150 km in diameter centered near the village of Chicxulub in the northern Yucatán (■ Figure 25-20). Although the crater is totally covered by sediments, mineral samples show that it contains shocked quartz typical of impact sites and that it is the right age. A gigantic impact formed the crater about 65 million years ago, just when the dinosaurs and many other species died out, and many Earth scientists now believe that this is the scar of the impact that ended the Cretaceous.

At first, these climate-changing impacts were blamed on large meteorites, but the nucleus of a comet would be just as damaging. In fact, you have seen in this chapter that the distinction between an asteroid and a comet is not clear-cut.

The theory that an impact caused the extinction at the end of the Cretaceous period was once highly controversial, but the accumulated evidence has made it a widely accepted idea. Now Earth scientists are applying what they learned to other extinctions. Chemical evidence from Earth's crust suggests that the extinction at the end of the Permian 250 million years ago may have been caused by a giant impact. The Permian extinction killed off 95 percent of the species on Earth and set the stage for the rise of the dinosaurs. The dinosaurs may owe their origin as well as their eventual extinction to giant impacts.

A solar system is a dangerous place for a planet. Comets, asteroids, and meteoroids constantly rain down on the planets, and Earth gets hit quite often. The Chicxulub crater isn't the only large impact scar on Earth. A giant crater has been found

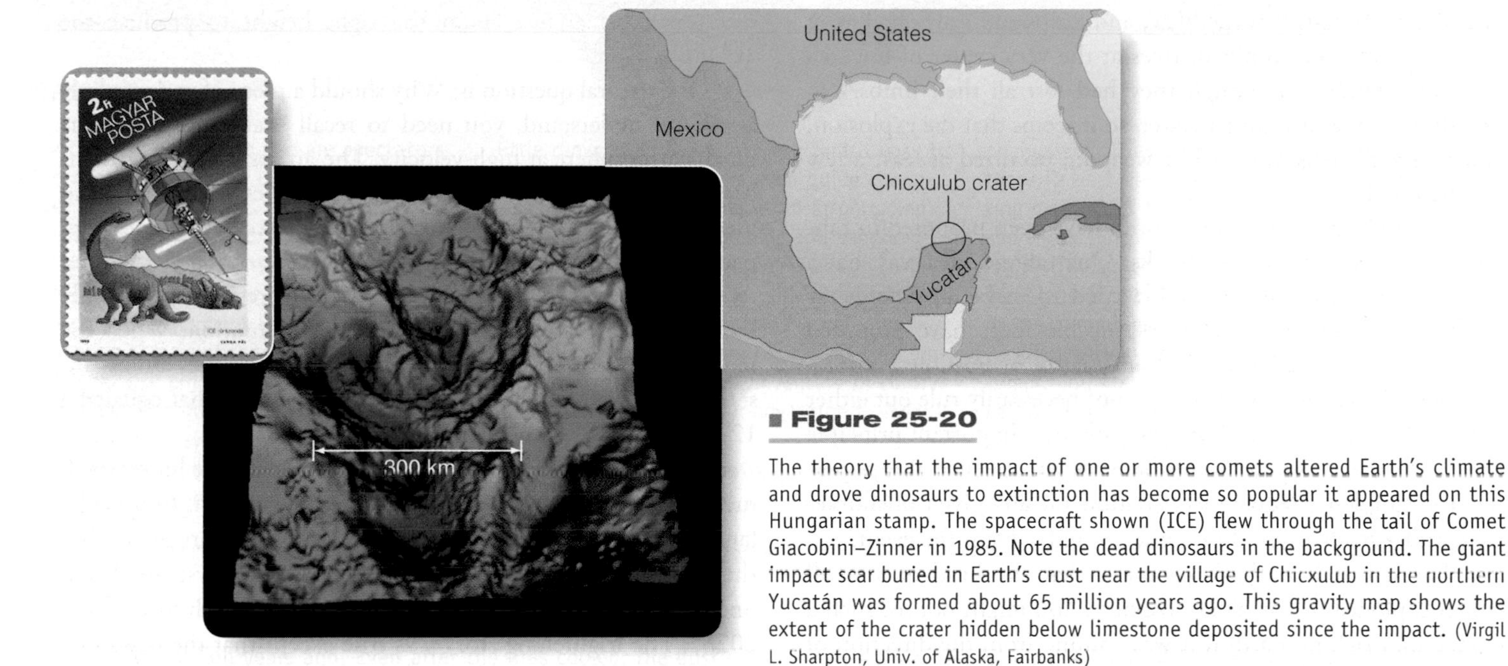

■ Figure 25-20

The theory that the impact of one or more comets altered Earth's climate and drove dinosaurs to extinction has become so popular it appeared on this Hungarian stamp. The spacecraft shown (ICE) flew through the tail of Comet Giacobini–Zinner in 1985. Note the dead dinosaurs in the background. The giant impact scar buried in Earth's crust near the village of Chicxulub in the northern Yucatán was formed about 65 million years ago. This gravity map shows the extent of the crater hidden below limestone deposited since the impact. (Virgil L. Sharpton, Univ. of Alaska, Fairbanks)

buried under sediment in Iowa, and another giant impact crater may underlie most of Chesapeake Bay. What would it have been like to live on Earth when such an impact occurred? Humanity got a hint in 1908 when something hit Siberia.

The Tunguska Event

On the morning of June 30, 1908, scattered reindeer herders and homesteaders in central Siberia were startled to see a brilliant blue-white fireball brighter than the sun streak across the sky. Still descending, it exploded with a blinding flash and an intense pulse of heat. One eyewitness account states:

> *The whole northern part of the sky appeared to be covered with fire. . . . I felt great heat as if my shirt had caught fire . . . there was a . . . mighty crash. . . . I was thrown on the ground about [7 m] from the porch. . . . A hot wind, as from a cannon, blew past the huts from the north.*

The blast was heard up to 1000 km away, and the resulting pulse of air pressure circled Earth twice. For a number of nights following the blast, European astronomers, who knew nothing of the explosion, observed a glowing reddish haze high in the atmosphere.

Travel in the wilderness of Siberia was difficult early in the 20th century; moreover, World War I, the Bolshevik Revolution, and the Russian Civil War prevented expeditions from reaching the site before 1927. When at last an expedition arrived, it found that the blast had occurred above the Stony Tunguska River valley and had flattened trees in an irregular pattern extending out 30 km (■ Figure 25-21). The trees were knocked down radially

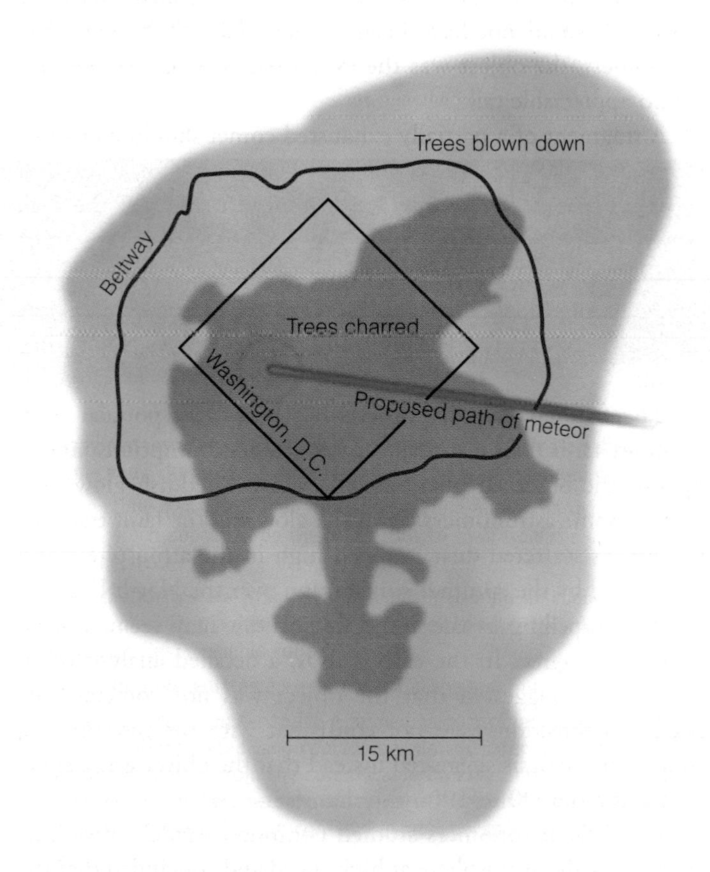

■ Active Figure 25-21

The 1908 Tunguska event in Siberia destroyed an area the size of a large city. Here the area of destruction is superimposed on a map of Washington, D.C., and its surrounding beltway. In the central area, trees were burned; and, in the outer area, trees were blown down in a pattern away from the path of the meteor.

26 Life on Other Worlds

Every life form we know of has evolved to live somewhere on Earth. The Wekiu bug lives with the astronomers at 13,800 feet atop Hawaiian volcano Mauna Kea. It inhabits the icy cinders and eats insects carried up by ocean breezes. (Kris Koenig/Coast Learning Systems)

Guidepost

This chapter is either unnecessary or critical, depending on your point of view. If you believe that astronomy is the study of the physical universe above the clouds, then you are done; the last 25 chapters completed your study of astronomy. But, if you believe that astronomy is the study of your role in the evolution of the universe, then everything you have done so far was just preparation for this chapter.

This chapter focuses on four questions about life on Earth and on other worlds:

What is life?

How did life originate on Earth?

Could life begin on other worlds?

Could Earthlings communicate with civilizations on other worlds?

These are difficult questions, but often in science asking a question is more important than getting an answer.

The origin of life is a difficult scientific issue, and it will help you to consider an important question about how science works:

How Do We Know? How do scientists evaluate the sources of evidence?

Just as you must judge the worth of facts and opinions every day, scientists must choose carefully to avoid being misled.

Did I solicit thee from darkness to promote me?

JOHN MILTON, *PARADISE LOST*

AS LIVING THINGS, WE HAVE BEEN PROMOTED from darkness. The atoms heavier than helium that are necessary components of our bodies did not exist at the beginning of the universe but were created by successive generations of stars. The elements from which we are made are not unique to our solar system, so it is possible that life began on other worlds and evolved to intelligence.

Your goal in this chapter is to try to understand the origin and evolution of life on Earth and other worlds. Although unknowns remain, the evidence is illuminating.

26-1 The Nature of Life

WHAT IS LIFE? Philosophers have struggled with that question for thousands of years, so it is not possible to answer it in a single chapter. But you might think of life as the process by which a living thing extracts energy from its surroundings, maintains itself, and modifies the surroundings to promote its survival. All living things, no matter how different, share certain characteristics.

The Physical Basis of Life

The physical basis of life on Earth is the element carbon (■ Figure 26-1). Because of the way carbon atoms bond to each other and to other atoms, they can form long, complex, stable chains that are capable of storing and transmitting information.

It is possible that life on other worlds could use silicon instead of carbon, but this seems unlikely because silicon chains are much less stable than their carbon counterparts. Science fiction has proposed even stranger life forms based on electromagnetic fields and ionized gas, and none of these possibilities can be ruled out. These hypothetical life forms make for fascinating speculation, but they can't be studied systematically as life on Earth can be. Consequently this chapter is concerned with the origin and evolution of carbon-based life.

What makes a lump of carbon-based molecules a living thing? The answer lies in the transmission of information from one molecule to another.

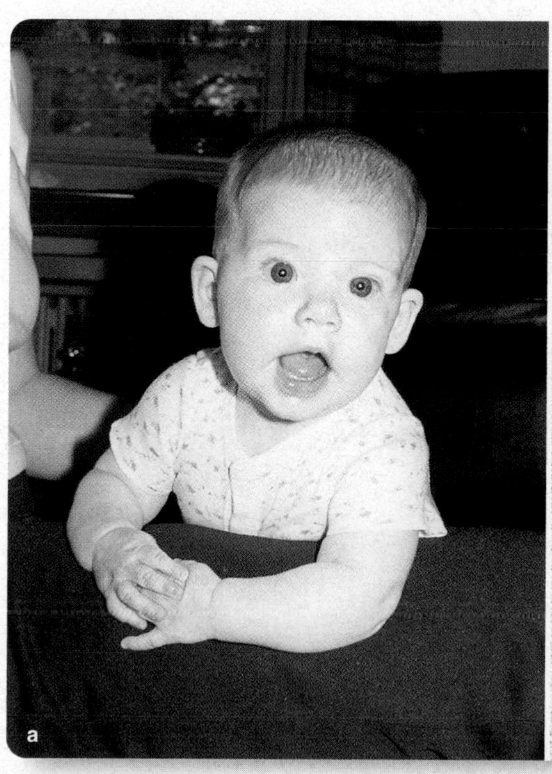

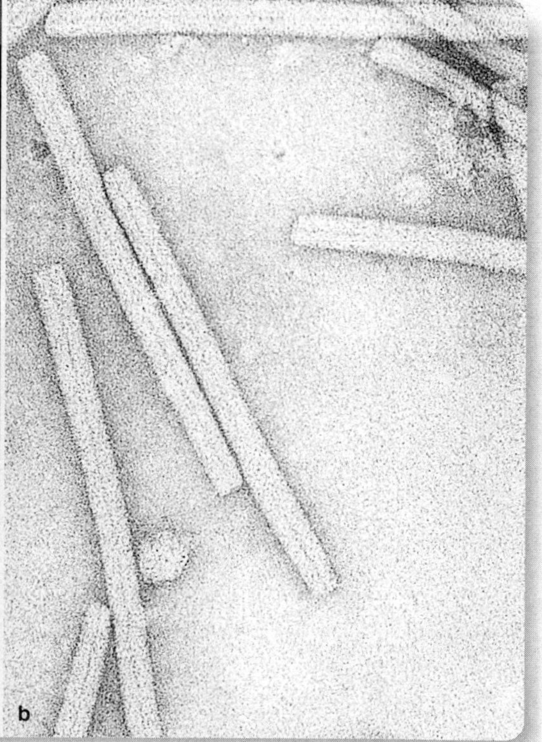

■ Figure 26-1

All living things on Earth are based on carbon chemistry. Even the long molecules that carry genetic information, DNA and RNA, have a framework defined by chains of carbon atoms. (a) Katie, a complex mammal, contains about 30 AU of DNA. (Michael Seeds) (b) Each rod of the tobacco mosaic virus contains a single spiral strand of RNA about 0.01 mm long. (L. D. Simon)

Information Storage and Duplication

Every task a living cell performs is carried out by chemicals that it manufactures. Cells must store recipes for all these chemicals, use the recipes when they need them, and pass them on to their offspring.

Read **DNA: The Code of Life** on pages 600–601 and notice three important points and seven new terms:

1. The chemical recipes of life are stored as templates on *DNA (deoxyribonucleic acid)* molecules. The templates automatically guide specific chemical reactions within the cell.
2. A DNA molecule looks like a ladder with chemical bases as rungs, and the order of the rungs provide recipes that combine *amino acids* to make molecules such as *proteins* and *enzymes* that govern the structure and operation of the cell. A related molecule called *RNA (ribonucleic acid)* copies the recipes for use.
3. The recipes stored in DNA are the genetic information handed down as *genes* in *chromosomes* to offspring. The DNA molecule reproduces itself when a cell divides so that each new cell contains a copy of the original information.

DNA: The Code of Life

1 The key to understanding life is information — the information that guides all of the processes in an organism. In most living things on Earth, that information is stored on a long spiral molecule called **DNA (deoxyribonucleic acid)**.

1a The DNA molecule looks like a spiral ladder with rails made of phosphates and sugars. The rungs of the ladder are made of four chemical bases arranged in pairs. The bases always pair the same way. That is, base A always pairs with base T, and base G always pairs with base C.

1b Information is coded on the DNA molecule by the order in which the base pairs occur. To read that code, molecular biologists have to "sequence the DNA." That is, they must determine the order in which the base pairs occur along the DNA ladder.

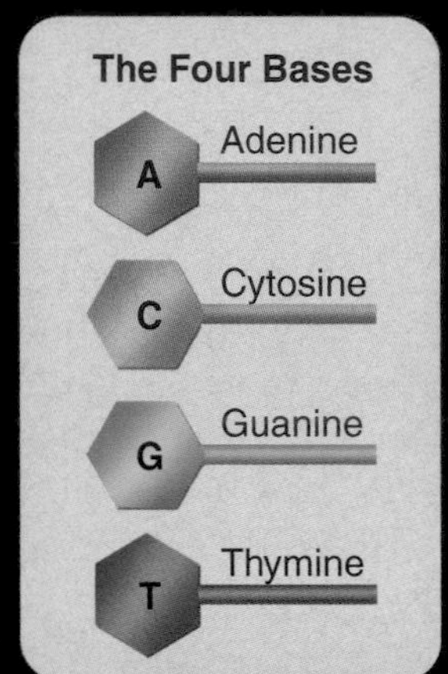

2 DNA automatically combines raw materials to form important chemical compounds. The building blocks of these compounds are relatively simple **amino acids.** Segments of DNA act as templates that guide the amino acids to join together in the correct order to build specific **proteins,** chemical compounds important to the structure and function of organisms. Some proteins called **enzymes** regulate other processes. In this way, DNA recipes regulate the production of the compounds of life.

The traits you inherit from your parents, the chemical processes that animate you, and the structure of your body are all encoded in your DNA. When people say "you have your mother's eyes," they are talking about DNA codes.

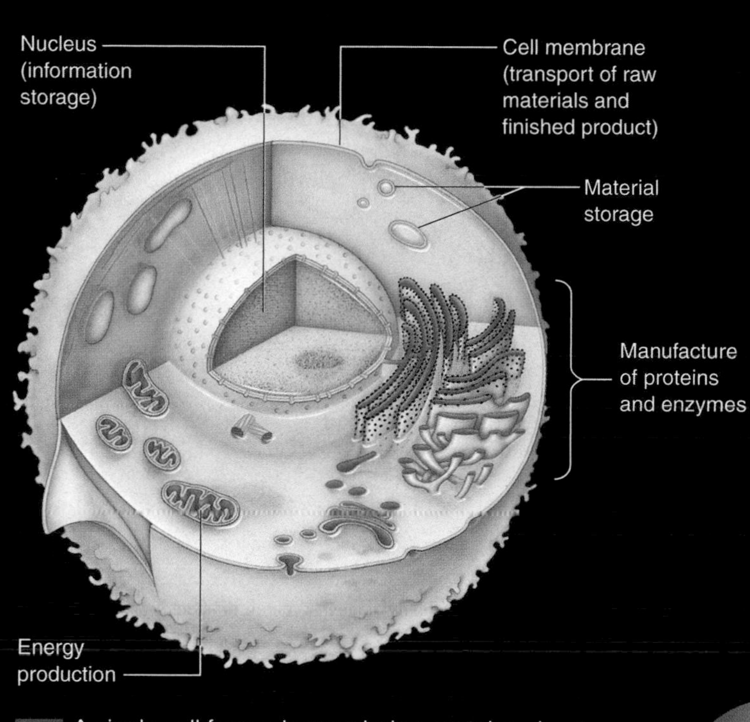

2a A cell is a tiny factory that uses the DNA code to manufacture chemicals. Most of the DNA remains safe in the nucleus of a cell, and the code is copied to create a molecule of **RNA (ribonucleic acid).** Like a messenger carrying blueprints, the RNA carries the code out of the nucleus to the work site where the proteins and enzymes are made.

2b A single cell from a human being contains about 1.5 meters of DNA containing about 4.5 billion base pairs — enough to record the entire works of Shakespeare 200 times. A typical human contains a total of about 600 AU of DNA. Yet the DNA in each cell, only 1.5 meters in length, contains all of the information to create a new human. A clone is a new creature created from the DNA code found in a single cell.

Original DNA

Copy DNA

Copy DNA

ThomsonNOW™

Sign in at www.thomsonedu.com and go to ThomsonNOW to see the Active Figure called "DNA." Explore the structure of DNA.

3 DNA, coiled into a tight spiral, makes up the **chromosomes** that are the genetic material in a cell. A **gene** is a segment of a chromosome that controls a certain function. When a cell divides, each of the new cells receives a copy of the chromosomes, as genetic information is handed down to new generations.

3a To divide, a cell must duplicate its DNA. The DNA ladder splits, and new bases match to the exposed bases of the ladder to build two copies of the original DNA code. Because the base pairs almost always match correctly, errors in copying are rare. One set of the DNA code goes to each of the two new cells.

Cell Reproduction by Division

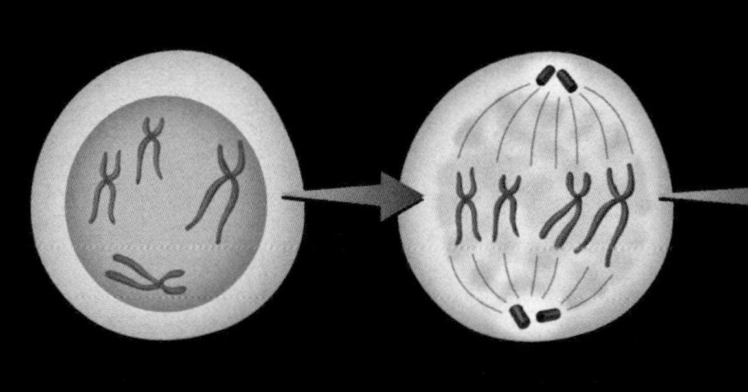

As a cell begins to divide, its DNA duplicates itself.

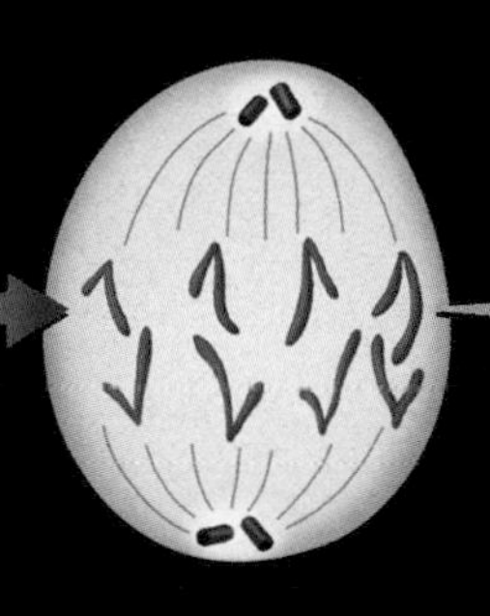

The duplicated chromosomes move to the middle.

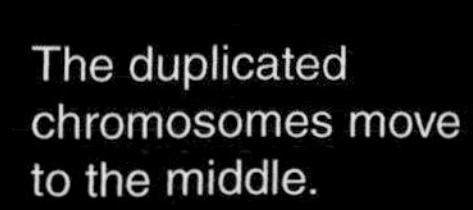

The two sets of chromosomes separate, and . . .

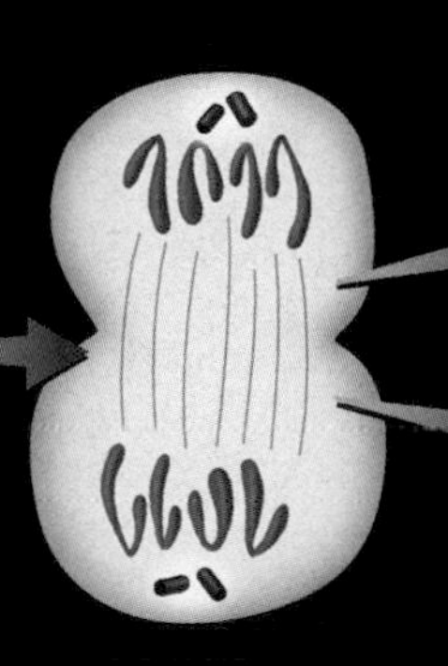

the cell divides to produce . . .

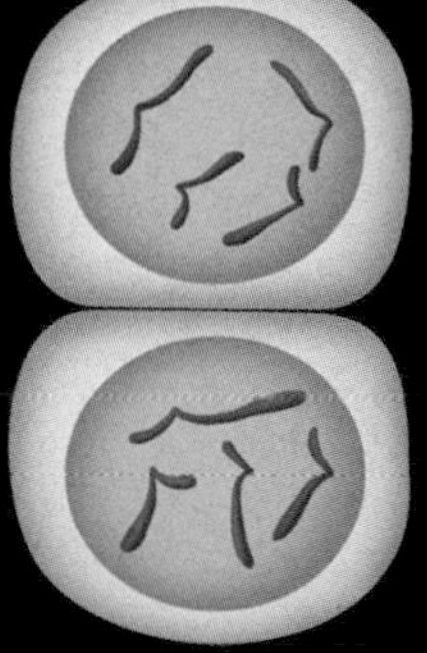

two cells, each containing a full set of the DNA code.

To produce successful offspring, a cell must be able to make accurate copies of its DNA. But it is also important for the continued existence of the species that not all the DNA copies be exact duplicates.

Modifying the Information

Earth's environment changes continuously. To survive, species must change as their food supply, climate, or home terrain changes. If the information stored in DNA could never change, then life would quickly go extinct. The process by which life adjusts itself to its changing environment is called **evolution.**

When an organism reproduces, its offspring each receive a copy of its DNA. But sometimes mistakes are made in the copying process, and a copy is slightly different from the original. Offspring born with random alterations to their DNA are called **mutants.** Most mutations make no difference, but some mutations are fatal, killing the afflicted organisms before they can reproduce. In rare but vitally important cases, a mutation can actually help an organism survive.

Mutations produce variation among the members of a species. All of the squirrels in the park may look the same, but they carry a wide range of genetic variation. Some may have slightly longer fur or faster-growing claws. These variations make almost no difference until the environment changes. If the environment becomes colder, for example, a squirrel with a heavier coat of fur will, on average, survive longer and produce more offspring than its short-furred contemporaries. Likewise, the offspring that inherit this beneficial variation will also live longer and have more offspring of their own. As the years pass, a larger and larger proportion of the squirrels in the park will have heavier fur. Over time, the species can evolve until nearly the entire population shares the trait. These differing rates of survival and reproduction are examples of **natural selection**—the process that adapts species to their changing environments by selecting from the huge array of random variations those that would most benefit the survival of the species. As natural selection increases the frequency of advantageous genes and reduces the frequency of disadvantageous genes, the species evolves to better fit its environment.

It is a **Common Misconception** that evolution is random, but that is not true. Variation is random, but natural selection is not random because progressive changes in a species are shaped by changes in the environment.

◄ SCIENTIFIC ARGUMENT ►

Why can't the information in DNA be permanent?

Sometimes the most valuable scientific arguments are those that challenge common misconceptions. It seems obvious that mistakes shouldn't be made in copying DNA, but variation is necessary for long-term survival. For example, the DNA in a starfish contains all the information the starfish needs to grow, develop, survive, and reproduce. The information must be passed on to the starfish's offspring for them to survive. But that information must change if the environment changes. A change in the ocean's temperature may kill the specific shellfish that the starfish eat. If none of the starfish is able to digest any other kind of food—if all the starfish have the same DNA—they will all die. But if from each mass of starfish eggs a few are born with the ability to digest a different kind of shellfish, the species may be able to carry on.

The survival of life depends on this delicate balance between reliable reproduction and the introduction of small variations in DNA. Now build a new argument. **How does DNA make copies of itself?**

26-2 The Origin of Life

IT IS OBVIOUS that the 4.5 billion chemical bases that make up human DNA did not just come together by chance. The key to understanding the origin of life lies in the processes of evolution. The complex interplay of environmental factors with the DNA of generation after generation of organisms drove some life forms to become more sophisticated over time, until they became the unique and specialized creatures on Earth today.

This means that life on Earth could have begun very simply, even as simple carbon-chain molecules that automatically made copies of themselves. Of course, this hypothesis requires evidence. What evidence exists regarding the origin of life on Earth?

The Origin of Life on Earth

The oldest fossils are all the remains of sea creatures, and that shows that life began in the sea. Identifying the oldest fossils is not easy, however. Rock from western Australia that is nearly 3.5 billion years old contains features that some experts identify as microscopic bacteria (■ Figure 26-2). Fossils this old are difficult to recognize because the earliest living things contained no easily preserved hard parts like bones or shells and because they were

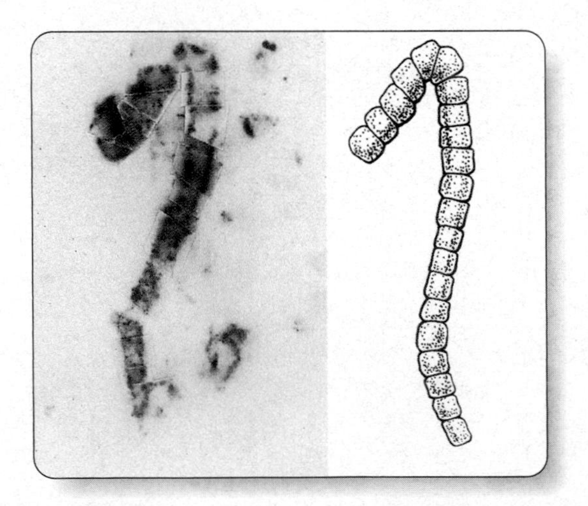

■ Figure 26-2

This microscopic filament resembles modern bacterial forms (artist's reconstruction at right) and may be one of the oldest fossils known. It was found in the 3.5-billion-year-old chert of the Pilbara Block in northwestern Australia. (Courtesy J. William Schopf)

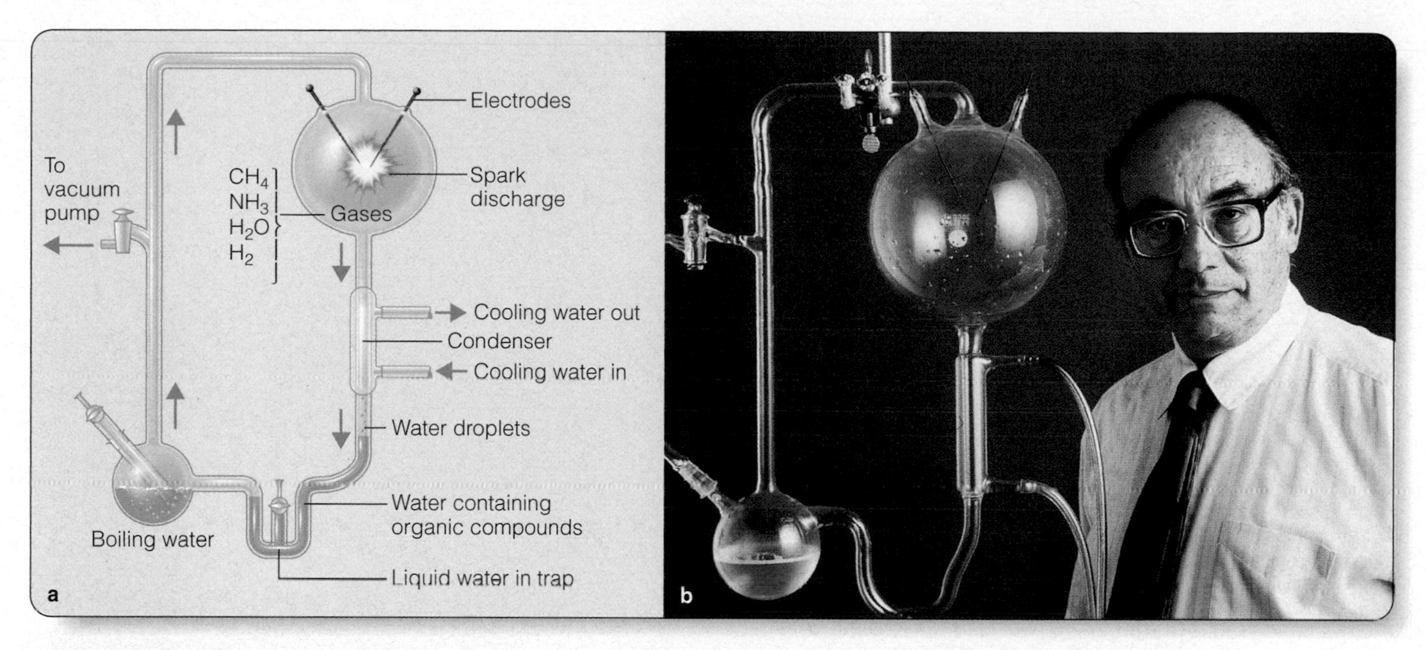

■ Figure 26-3

(a) The Miller experiment circulated gases through water in the presence of an electric arc. This simulation of primitive conditions on Earth produced amino acids, the building blocks of proteins. (b) Stanley Miller with a Miller apparatus. (Courtesy Stanley Miller)

microscopic. Nevertheless, the Strelley Pool Chert, a rock formation in western Australia, contains signs of life that are roughly 3.4 billion years old.

The evidence, though scarce, is clear: Simple organisms lived in Earth's oceans as early as 3.4 billion years ago. Where did these simple organisms come from?

An important experiment performed by Stanley Miller and Harold Urey in 1952 sought to recreate the conditions under which life on Earth began. The **Miller experiment** consisted of a sterile, sealed glass container holding water, hydrogen, ammonia, and methane. An electric arc inside the apparatus created sparks to simulate the effects of lightning in Earth's early atmosphere (■ Figure 26-3).

Miller and Urey let the apparatus run for a week and then analyzed the material inside. They found that the arc had produced many organic molecules from the raw material of the experiment, including such important building blocks of life as amino acids. (Remember that an organic molecule is just a molecule with a carbon-chain structure and need not be derived from a living thing.) When the experiment was run again using different energy sources, such as hot silica to represent molten lava spilling into the ocean, similar molecules were produced. Even the ultraviolet radiation present in sunlight was sufficient to produce complex organic molecules.

According to updated models of the formation of the solar system, Earth's early atmosphere probably consisted of carbon dioxide, nitrogen, and water vapor instead of hydrogen, ammonia, and methane. When these gases are processed in a Miller apparatus, lesser but still significant numbers of organic molecules are produced.

The Miller experiment is important because it shows that complex organic molecules form naturally in a wide variety of circumstances. Lightning, sunlight, and hot lava pouring into the oceans are just some of the processes that naturally produce the complex molecules that make life possible. If you could travel back in time, you would find Earth's first oceans filled with a rich mixture of organic compounds called the **primordial soup.**

Many of these organic compounds would have linked up to form larger molecules. Amino acids, for example, can link together to form proteins by joining ends and releasing a water molecule (■ Figure 26-4). It was initially thought that this must have occurred in sun-warmed tidal pools where organic molecules were concentrated by evaporation. But violent episodes of volcanism and catastrophic meteorite impacts on Earth would probably have destroyed any evolving life forms at the surface, so it is now thought that the early linkage of complex molecules took place on the ocean floor near the hot springs at midocean ridges.

These complex organic molecules were still not living things. Even though some proteins may have contained hundreds of amino acids, they did not reproduce but rather linked and broke apart at random. But because some molecules are more stable than others, and some bond together more easily than others, this **chemical evolution** eventually concentrated the various smaller molecules into the most stable larger forms. Eventually, somewhere in the oceans, a molecule formed that automatically

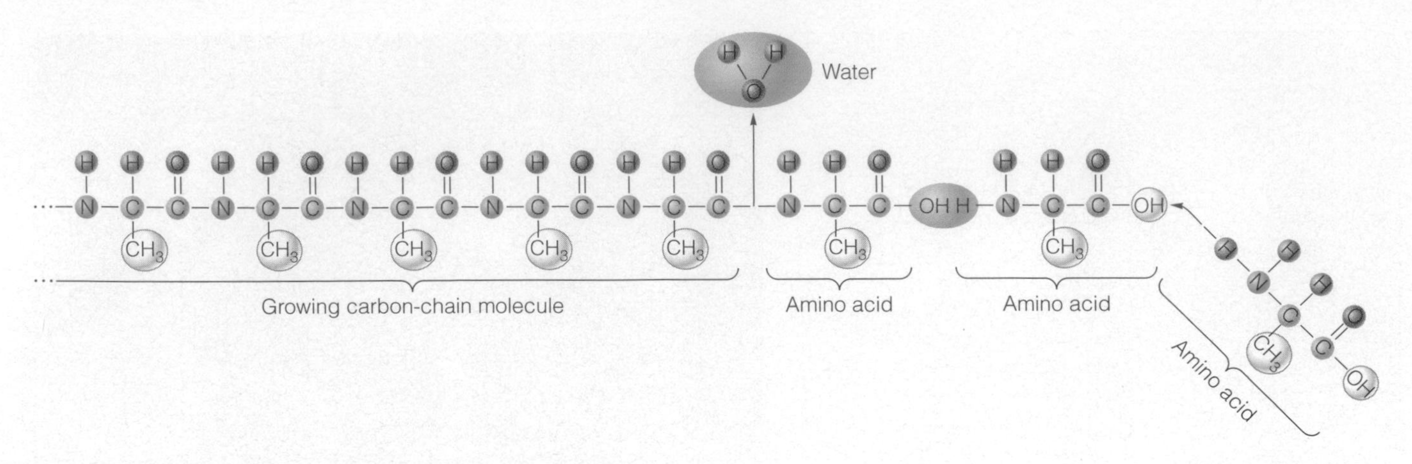

■ **Figure 26-4**

Amino acids can link together through the release of a water molecule to form long carbon-chain molecules. The amino acid in this hypothetical example is alanine, one of the simplest.

made copies of itself. At that point, the chemical evolution of molecules became the biological evolution of living things.

An alternative theory for the origin of life holds that reproducing molecules may have arrived here from space. Radio astronomers have found a wide variety of organic molecules in the interstellar medium, and similar compounds have been found inside meteorites (■ Figure 26-5). The Miller experiment shows how easy it is for organic molecules to form, so it is not surprising to find them in space. But the hypothesis that life arrived on Earth from space is presently untestable, so, although it is fun to speculate, the hypothesis is of little practical value.

Whether the first reproducing molecules formed here on Earth or in space, the important thing is that they could have formed by natural processes. Scientists know enough about these processes to feel confident about them, even though some of the steps remain unknown.

Even the structure of a cell may have arisen automatically because of the way molecules interact during chemical evolution. If a dry mixture of amino acids is heated, the acids form long, proteinlike molecules that, when poured into water, collect to form microscopic spheres that function in ways similar to cells (■ Figure 26-6). They have a thin membrane surface, they absorb material from their surroundings, they grow in size, and they divide and bud just as cells do. They contain no large molecule that copies itself, however, so they are not alive.

An alternative theory proposes that the replicating molecule developed first. Such a molecule would have been exposed to damage if it had been bare, so the first to manufacture or attract a protective coating of protein would have had a significant survival advantage. If this was the case, the protective cell membrane was a later development of biological evolution.

The oldest cells must have been single-celled organisms much like modern bacteria. These kinds of cells are preserved in the rocks from Australia described earlier. Another example of early life are the **stromatolites,** mineral formations deposited layer upon layer by growing mats of photosynthetic bacteria (■ Figure 26-7). If photosynthetic bacteria were common long ago, they may have begun adding oxygen, a product of photosynthesis, to Earth's early atmosphere. An oxygen abundance of only 0.1 percent would have created an ozone screen, protecting organisms from the sun's ultraviolet radiation.

■ **Figure 26-5**

A sample of the Murchison meteorite, a carbonaceous chondrite that fell in 1969 near Murchison, Australia. Analysis of the interior of the meteorite revealed evidence of amino acids. Whether the first building blocks of life originated in space is unknown, but the amino acids found in meteorites illustrate how commonly amino acids and other complex molecules occur even in the absence of living things. (Courtesy Chip Clark, National Museum of Natural History)

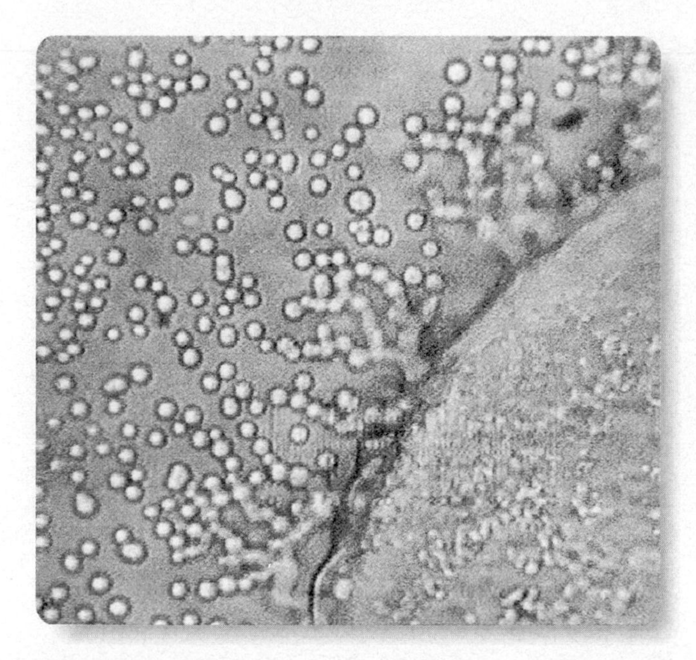

■ **Figure 26-6**

Single amino acids can be assembled into long proteinlike molecules. When such material cools in water, it can form microspheres, microscopic spheres with double-layered boundaries similar to cell membranes. Microspheres may have been an intermediate stage in the evolution of life between complex molecules and cells holding molecules reproducing genetic information. (Courtesy Sidney Fox and Randall Grubbs)

Over the course of eons, the natural processes of evolution gave rise to stunningly complex multicellular life forms with their own widely differing ways of life. It is a **Common Misconception** to imagine that life is too complex to have evolved from such simple beginnings. It is possible because small variations accumulate and are handed down, but it took huge amounts of time.

Geologic Time

Life has existed on Earth for roughly 3.4 billion years, but there is no evidence of anything more than simple organisms until about 600 million years ago, when life suddenly branched into a wide variety of complex forms like the trilobites (■ Figure 26-8). This sudden increase in complexity is known as the **Cambrian explosion,** and it marks the beginning of the Cambrian period. If you represented the entire history of Earth on a scale diagram,

■ **Active Figure 26-7**

A fossil stromatolite from western Australia is one of the oldest known fossils (left). It is believed to be 3.5 billion years old. Stromatolites were formed, layer by layer, by mats of bacteria living in shallow water. Such life may have been common in shallow seas when Earth was young (right). (Mural by Peter Sawyer; photo courtesy Chip Clark, National Museum of Natural History)

■ **Figure 26-8**

Trilobites, such as the fossil shown here, made their first appearance in the Cambrian period, when life became complex and specialized. Anomalocaris (rear at right center and looming at upper right) was about the size of your hand and had specialized organs including eyes, coordinated fins, gripping mandibles, and a powerful, toothed maw. Notice Opabinia at center right with its long snout. (Fossil: Grundy Observatory photograph; Art: Smithsonian and D. W. Miller)

the Cambrian explosion would be near the top of the column shown at the left of ■ Figure 26-9. The emergence of most animals familiar to you today, including fishes, amphibians, reptiles, birds, and mammals, would be crammed into the topmost part of the chart, above the Cambrian explosion.

If you magnify this portion of the diagram, as shown on the right side of Figure 26-9, you can get a better idea of when these events occurred in the history of life. Creatures like us have walked the Earth for about 3 million years, a long time by normal standards, but it makes only a narrow red line at the top of the diagram. All of recorded history would be a microscopically thin line at the very top.

To understand just how thin this line is, imagine that the entire 4.5-billion-year history of the Earth is compressed onto a yearlong video. Imagine that you began watching this video on January 1. You would not see any signs of life until March or early April, and the slow evolution of the first simple forms would take up the rest of the spring and summer and most of the fall. Suddenly, in mid-November, you would see the trilobites and other complex organisms of the Cambrian explosion.

You would see no life of any kind on land until November 28, but once it appeared it would diversify quickly, and by December 12 you would see dinosaurs walking the continents. By the evening of Christmas Day, they would be gone, and mammals and birds would be on the rise.

If you watched closely, you might see the first humanoid forms by suppertime on New Year's Eve, and by late evening you could see humans making the first stone tools. The Stone Age would last until 11:45 PM, and the first towns and cities would appear at about 11:54. Suddenly things would begin to happen at lighting speed. Babylon would flourish, the Pyramids would rise, and Troy would fall. The Christian era would begin 14 seconds before the New Year. Rome would fall; the Middle Ages and the Renaissance would flicker past. The Declaration of Independence would be signed one second before the end of the video. (Put your videotape of the history of life on Earth into perspective by comparing it to the entire history of the universe as shown on the inside cover of this book.)

By imagining the history of Earth as a yearlong video, you have gained some perspective on the rise of life. Tremendous amounts of time were needed for the first simple living things to evolve in the oceans, but as life became more complex, new forms arose more and more quickly as the hardest problems—how to reproduce, how to take in energy from the environment, how to move around—were solved. The easier problems, like what to eat, where to live, and how to raise young, were solved in different ways by different organisms, leading to the diversity that you see today.

Even human intelligence—that which appears to set us apart from other animals—may be the unique solution to an evolutionary problem posed to our ancient ancestors. A smart animal is better able to escape predators, outwit its prey, and feed and shelter itself and its offspring, so under certain conditions evolution is likely to favor the rise of intelligence. Could intelligent life arise on other worlds? To try to answer this question, you will need to estimate the chances of any type of life arising on other worlds, then assess the likelihood of that life developing intelligence.

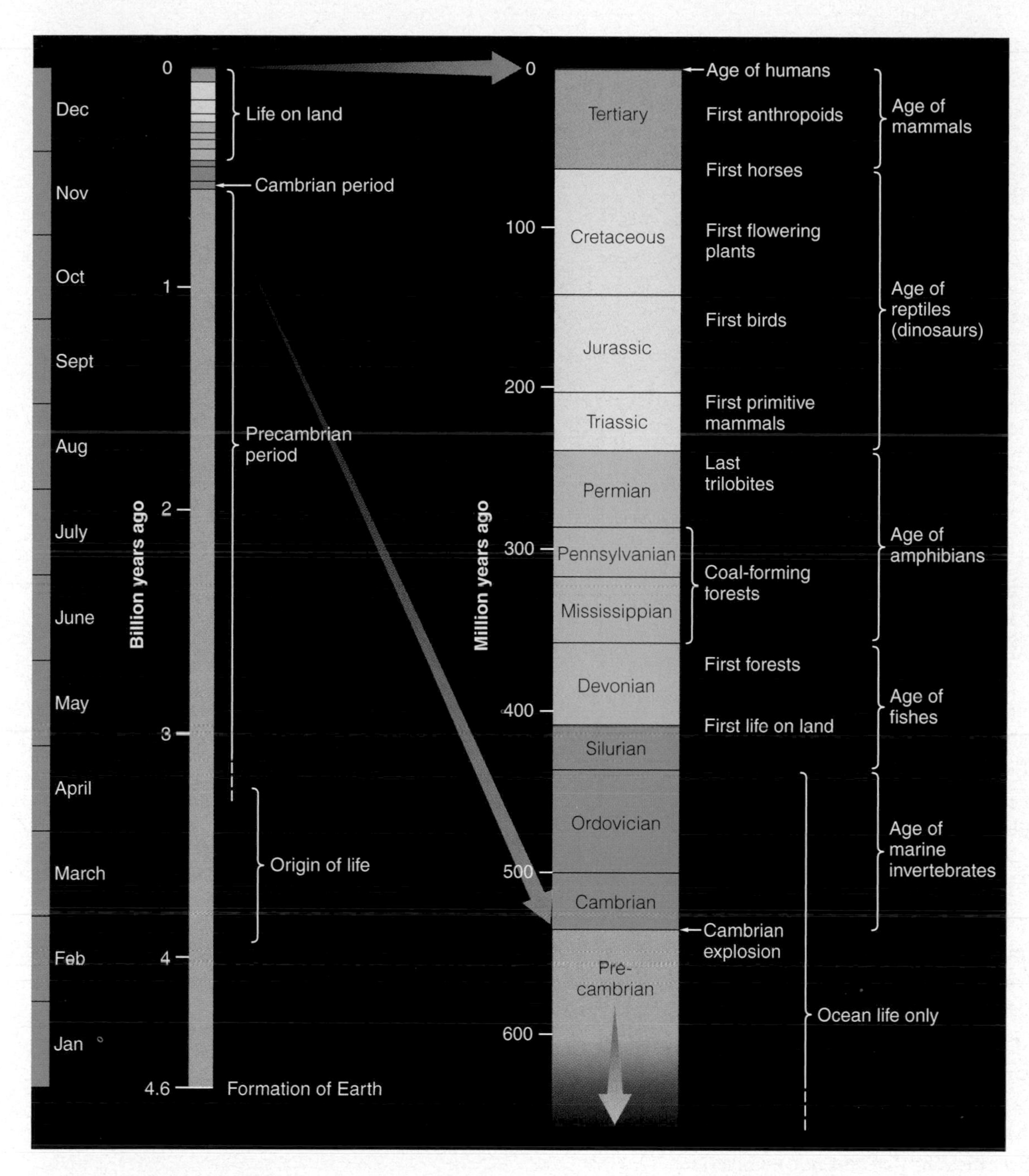

Active Figure 26-9

Complex life has developed on Earth only recently. If the entire history of Earth were represented in a time line (left), you would have to magnify the end of the line to see details such as life leaving the oceans and dinosaurs appearing. The age of humans would still be only a thin line at the top of your diagram. If the history of Earth were a yearlong videotape, humans would not appear until the last hours of December 31.

Life in Our Solar System

Could there be carbon-based life elsewhere in our solar system? Liquid water seems to be a requirement of carbon-based life, necessary both for vital chemical reactions and as a medium to transport nutrients and wastes. It is not surprising that life developed in Earth's oceans and stayed there for billions of years before it was able to colonize the land. Any world harboring living things must have significant quantities of liquid water.

Many worlds in the solar system can be eliminated immediately. The moon and Mercury are airless, and liquid water would boil away into space immediately. Venus has traces of water vapor in its atmosphere, but it is too hot for liquid water to survive on the surface. The Jovian planets have deep atmospheres, and at a certain level it is likely that water condenses into liquid droplets. However, it seems unlikely that life could have originated there. Isolated water droplets could not mingle to mimic the rich primordial oceans of Earth, where complex molecules interacted and grew. Additionally, powerful currents in the gas giants' atmospheres would quickly carry any reproducing molecules that did form into inhospitable regions of the atmosphere.

Three of the Jovian satellites, however, could potentially support life. Jupiter's moon Europa appears to have a liquid-water ocean below its icy crust, and minerals dissolved in the water could provide a rich source of raw material for chemical evolution. But Europa's ocean is kept warm and liquid by tidal heating, and that can change as the orbits of the moons interact. Europa may have been frozen solid at times in its history, which would probably have destroyed any living organisms that had developed there.

Saturn's moon Titan is rich in organic molecules. You learned in Chapter 23 how sunlight converts the methane in Titan's atmosphere into organic smog particles that settle to the surface. The chemistry of any life that could evolve from these molecules and survive in Titan's lakes of liquid methane is unknown. It is fascinating to consider possibilities, but Titan's extremely low temperature of −179°C (−290°F) could slow chemical reactions to the point where life is impossible.

Observations of water venting from the south polar region of Saturn's moon Enceladus show that it has liquid water below its crust. It is possible that life could have begun in that water, but the moon is very small and is warmed by tidal heating. It is unlikely to have had liquid water for the extended time necessary for the rise of life.

Mars is a possible home for life because, as you learned in Chapter 17, there is a great deal of evidence that liquid water once flowed on its surface. Even so, the evidence for living organisms on the surface is not encouraging. The robotic spacecraft Viking 1 and Viking 2 landed on Mars in 1976 and tested soil samples for living organisms. No evidence clearly indicated the presence of life in the soil. If life survives on Mars, it may be hidden below ground where there is water and where ultraviolet radiation from the sun cannot penetrate.

In 1996, news media published exciting stories regarding chemical and physical traces of life on Mars discovered inside a Martian meteorite found in Antarctica (■ Figure 26-10). Scientists were excited by the announcement, but they immediately began testing the evidence. Their results suggested that the unusual chemicals may have been the result of Earthly contamination or may have formed by processes that did not involve life. Features that were originally taken to be fossils of ancient Martian organisms could be nonorganic mineral formations. This is the only direct evidence yet found regarding potential life on Mars, but it is highly controversial and not generally accepted. Conclusive evidence of life on Mars may have to wait until a geologist from Earth can scramble down dry Martian streambeds and crack open rocks looking for fossils.

■ Figure 26-10

(a) Meteorite ALH84001 is one of a dozen meteorites known to have originated on Mars. It was claimed that the meteorite contained chemical and physical traces of ancient life on Mars, including what appear to be fossils of microscopic organisms shown in part b. The evidence has not been confirmed, and the validity of the claim is highly questionable. (NASA)

You have found no strong evidence for the existence of other life in the solar system. Now your search will take you to distant planetary systems.

Life in Other Planetary Systems

Could life exist on other planets? You already know that there are many different kinds of stars and that many of these stars have planets. As a first step toward answering this question, you can try to identify the kinds of stars that seem most likely to have stable planetary systems where life could evolve.

If a planet is to be a suitable home for living things, it must be in a stable orbit around its sun. This is simple in a planetary system like our own, but most planetary orbits in binary systems are unstable. Planets in such systems are usually swallowed up by one of the stars or ejected from the system. Half the stars in the galaxy are members of binary systems and are unlikely to have planets and support life.

But just because a star is single does not necessarily make it a good candidate for sustaining life. Earth required 1 or 2 billion years to produce the first cells and 4.6 billion years for intelligence to emerge. Massive stars that live only a few million years will not do. If Earth's example is at all representative, then stars hotter than about F5 are too short lived for life to develop. Main-sequence G, K, and possibly the faint M stars are candidates.

The luminosity of a star is also important. Astronomers have defined a **life zone** (or **ecosphere**) around a star as a region within which planets have temperatures that could permit the existence of liquid water. A low-luminosity star has a small life zone, and a high-luminosity star has a large one. The sun's life zone extends from around the orbit of Venus to the orbit of Mars. Obviously, other factors are important; Venus lies in the sun's life zone but has no liquid water because of its intense greenhouse effect. A life zone is only a rough guide to a star's suitability for life.

Recent discoveries make the idea of a life zone seem even less useful. Scientists on Earth are finding life in places previously judged inhospitable, such as the bottoms of icy lakes in Antarctica and far underground inside solid rock. Life has been found in boiling hot springs with highly acidic water. It is difficult for scientists to pin down a range of conditions and state with certainty, "These conditions are necessary for life." You should also note that three of the environments listed as possible havens for life, Titan, Europa, and Enceladus, are in the outer solar system and lie far outside the sun's traditional life zone. Stable planets inside the life zones of long-lived stars are the places where life seems most likely, but, given the tenacity and resilience of Earth's life forms, there are almost certainly other, seemingly inhospitable, places in the universe where life exists.

◄ SCIENTIFIC ARGUMENT ►

What evidence indicates that life is at least possible on other worlds?

A good scientific argument involves careful analysis of evidence. Fossils on Earth show that life originated in the oceans at least 3.4 billion years ago, and biologists have proposed relatively simple chemical processes that could have created the first reproducing molecules. Fossils show that life developed slowly at first. The pace of evolution quickened about half a million years ago, when life took on complex forms. Later, when life emerged onto the land, it evolved rapidly into diverse forms. Intelligence is a relatively recent development; it is only a few million years old.

If this process occurred on Earth, it seems reasonable that it could have occurred on other worlds as well. Life may begin and eventually evolve to intelligence on any world where conditions are right. **What are the conditions you should expect on other worlds that host life?**

◄ ►

26-3 Communication with Distant Civilizations

VISITING OTHER WORLDS is, for now, impossible. But if other civilizations exist, perhaps we can communicate with them. Nature places restrictions on such conversations, but the main problem lies in the life expectancy of civilizations.

Travel between the Stars

The distances between stars are almost beyond comprehension. The fastest commercial jet would take about 4 million years to reach the nearest star. The obvious way to overcome these huge distances is with tremendously fast spaceships, but no ship could travel faster than the speed of light, and even the closest stars are many light-years away.

Nothing can exceed the speed of light, and accelerating a spaceship close to the speed of light takes huge amounts of energy. Even if you travel more slowly, your rocket would require massive amounts of fuel. If you were piloting a spaceship the size of a yacht to a star 5 light-years away, and you wanted to arrive in 10 years, you would use 40,000 times as much energy to get there as the entire United States consumes in a year.

These limitations not only make it difficult for humans to leave the solar system, but they would also make it difficult for aliens to visit Earth. Reputable scientists have studied "unidentified flying objects" (UFOs) and have never found any evidence that Earth is being visited or has ever been visited by aliens (see **How Do We Know? 26-1**). Humans are unlikely to ever meet aliens face-to-face. The only way to communicate is by radio.

Radio Communication

Nature places restrictions on travel through space, and it also restricts astronomers' ability to communicate with distant civilizations by radio. One restriction is based on simple physics. Radio signals are electromagnetic waves and travel at the speed of light. Due to the distances between the stars, the speed of radio waves would severely limit astronomers' ability to carry on normal conversations with distant civilizations. Decades could elapse between asking a question and getting an answer.

So, rather than try to begin a conversation, one group of astronomers decided to broadcast a simple message of friendship. In 1974, astronomers at the Arecibo radio telescope transmitted a signal toward the globular cluster M13, 26,000 light-years away. When the signal arrives, 26,000 years in the future, alien astronomers may be able to decode it.

The Arecibo beacon is an anticoded message, meaning that it is specifically designed to be easily decoded. The message is a string of 1679 pulses and gaps. Pulses can be represented as 1s, and gaps as 0s. The string can be arranged in only two possible ways, as 23 rows of 73 or as 73 rows of 23. The second arrangement forms a picture that describes life on Earth (■ Figure 26-11).

Earth is sending out many other signals. Short-wave radio signals, such as TV and FM, have been leaking into space for the last 60 years or more. Any civilization within 60 light-years could already have detected us.

But this works both ways. Alien signals, whether intentional messages of friendship or the blather of their daytime TV, could be arriving at Earth now. Astronomers all over the world are

How Do We Know? 26-1

Judging Evidence

Why don't scientists take UFOs seriously? Scientists deal with evidence, and much of that evidence is produced by other people. How does a scientist know what evidence to respect and what to dismiss? The answer is reputation; scientists depend on the reputation of a source of information.

UFOs have visited Earth, some people claim, and you might wonder why that evidence isn't considered in discussions of life in the universe. There have been plenty of reports of flying saucers and alien abductions, but scientists don't take them seriously because of the reputation of the sources of these stories. Most are reported in tabloid newspapers and sensational magazines that also report sightings of Bigfoot and babies with bat wings. TV specials on viewer-hungry cable networks are not reliable sources of scientific data.

Scientists look at the source of information and consider its reputation. Papers in respected scientific journals have been peer reviewed—checked by experts. Reports from well-known research centers are taken seriously. Scientists also consider the personal reputation of other experts. Fraud in science is quite rare, but it does happen, and a researcher who has knowingly published a fraudulent paper has a ruined reputation and will probably never be trusted again. Scientists protect their reputations and depend on the reputations of others.

Respected scientists have studied UFOs and found no evidence that they represent real visits by aliens from other worlds. Consequently, scientists do not take such tabloid reports seriously and instead focus their attention on more reliable evidence from sources they trust.

UFOs from space are fun to think about, but there is no credible evidence that they are real.

pointing radio telescopes at the most likely stars and listening for alien civilizations.

Which wavelengths should astronomers monitor? Wavelengths longer than 30 cm would get lost in the background noise of the Milky Way Galaxy, while wavelengths shorter than about 1 cm are absorbed in Earth's atmosphere. This is the radio window that is open for communication.

Even this restricted window contains millions of possible radio-frequency bands and is too wide to monitor easily, but astronomers may have found a way to narrow the search. Within this communications window lie the 21-cm line of neutral hydrogen and the 18-cm line of OH (■ Figure 26-12). The interval between the lines is named the **"water hole"** because H plus OH yields water. Any civilizations sophisticated enough to do radio astronomy must know of these lines and appreciate their significance, and it is hoped that they would choose a wavelength between these lines to broadcast a message of their own.

A number of searches for extraterrestrial radio signals have been made, and some are now under way. This field of study is known as **SETI,** Search for Extra-Terrestrial Intelligence, and it has generated heated debate among astronomers, philosophers, theologians, and politicians. Congress funded a NASA search for a short time but ended support in the early 1990s because it feared negative public reaction. In fact, the annual cost of a major search is only about as much as a single Air Force attack helicopter, but much of the reluctance to fund searches probably stems from issues other than cost. The discovery of alien intelligence would cause a huge change in our worldview, akin to Galileo's discovery of the moons of Jupiter, and some turmoil would inevitably result.

In spite of the controversy, the search continues. The NASA SETI project canceled by Congress was completed using private funds and renamed Project Phoenix. The SETI Institute, founded in 1984, has pursued a number of important searches and is currently building a new radio telescope with the University of California, Berkeley (■ Figure 26-13).

There is even a way for you to help with the search. The Berkeley SETI team, with the support of the Planetary Society, has recruited about 4 million owners of personal computers that are connected to the Internet. Participants download a screen saver that searches data files from the Arecibo radio telescope for signals whenever the owner is not using the computer. For information, locate the seti@home project at seti//setiathome.ssl.berkeley.edu/

The search continues, but radio astronomers struggle to hear anything against the worsening babble of noise from Earth. Wider and wider sections of the electromagnetic spectrum are being used for Earthly communication, and this, combined with stray radio noise from electronic devices including everything from computers to refrigerators, makes hearing faint signals difficult. It would be ironic if we fail to detect signals from another world because our own world has become too noisy. Ultimately, the chances of success depend on the number of inhabited worlds in the galaxy.

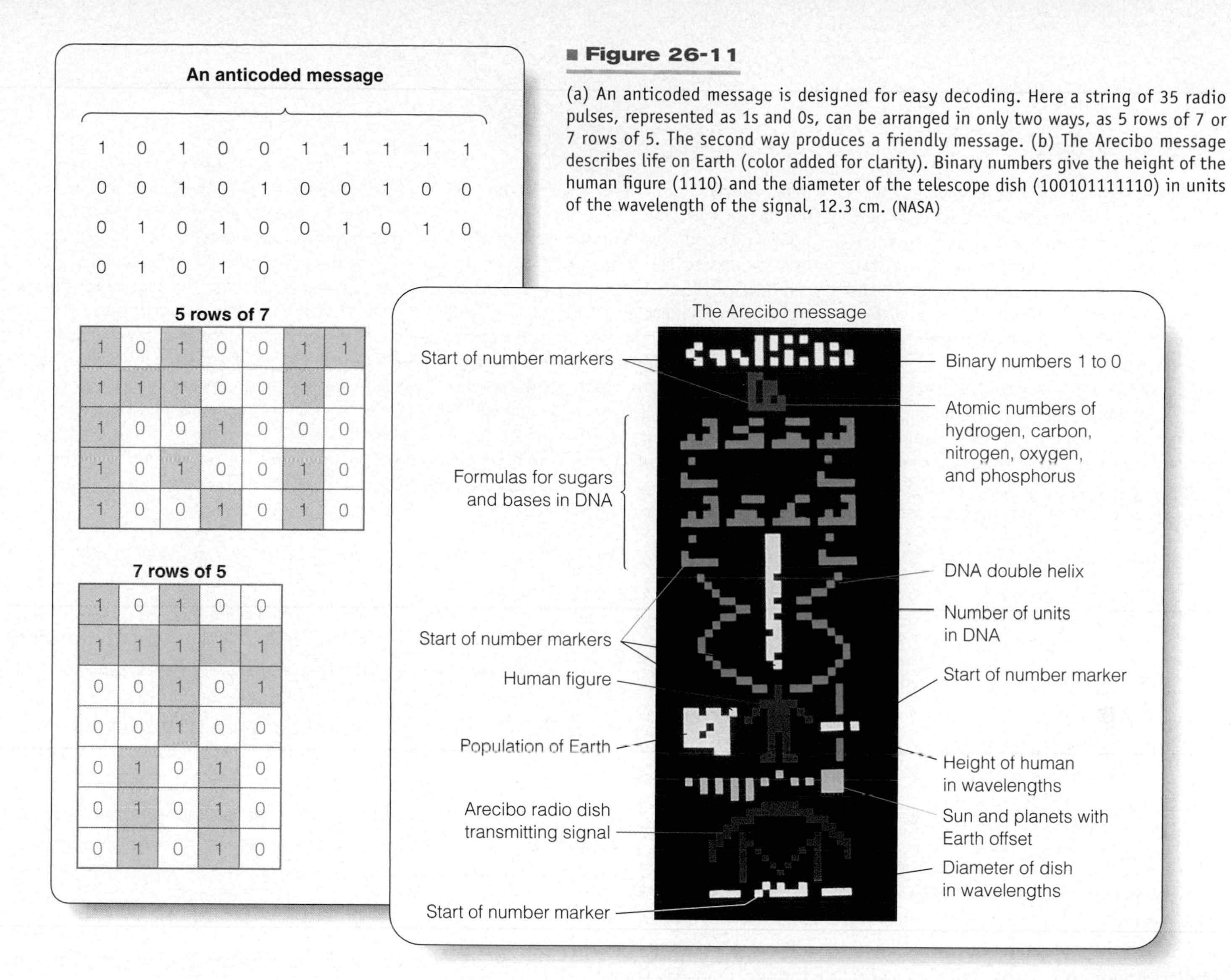

■ Figure 26-11

(a) An anticoded message is designed for easy decoding. Here a string of 35 radio pulses, represented as 1s and 0s, can be arranged in only two ways, as 5 rows of 7 or 7 rows of 5. The second way produces a friendly message. (b) The Arecibo message describes life on Earth (color added for clarity). Binary numbers give the height of the human figure (1110) and the diameter of the telescope dish (100101111110) in units of the wavelength of the signal, 12.3 cm. (NASA)

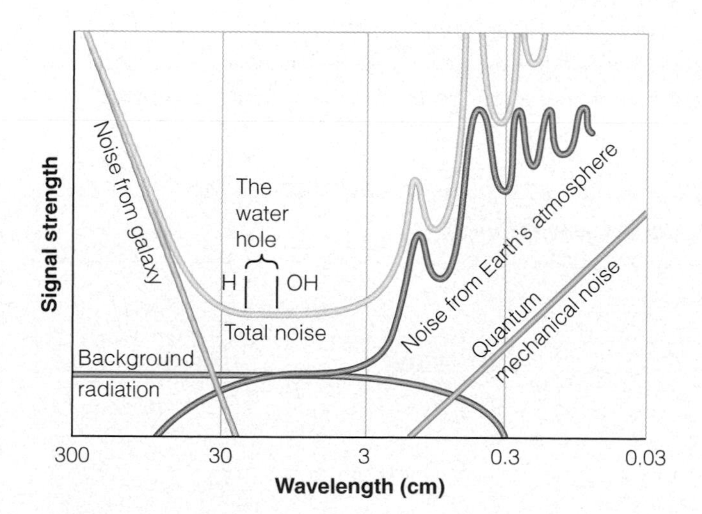

■ Active Figure 26-12

Radio noise from various sources makes it difficult to detect distant signals at wavelengths longer than 30 cm or shorter than 1 cm. In this range, radio emission from H atoms and from OH marks a small wavelength range dubbed the water hole, which may be a likely place for communication.

■ Figure 26-13

The Allen Telescope Array, now being built in California, will eventually grow to include 350 radio dishes, each 6 meters in diameter, in an arrangement precisely designed to maximize resolution. As radio astronomers aim the telescope at galaxies and nebulae of interest, state-of-the-art computer systems will analyze stars in the field of view searching for signals from distant civilizations. (SETI Institute)

What Are We?

Participants

The matter you are made of appeared in the big bang and was cooked into a wide range of atoms inside stars. Your atoms have been inside at least two or three generations of stars. Eventually, your atoms became part of a nebula that contracted to form the sun and the planets of the solar system. You are made of very old atoms.

Your atoms have been part of Earth for the last 4.6 billion years. They have been recycled over and over through dinosaurs, stromatolites, fish, bacteria, grass, birds, worms, and many other living things. Now you are using your atoms, but when you are done with them, they will go back to the Earth and be used again and again.

When the sun swells into a giant star and dies in a few billion years, Earth's atmosphere and oceans will be driven away, and at least the outer few kilometers of Earth's crust will be vaporized and blown outward to become part of the nebula around the white-dwarf remains of the sun. Your atoms are destined to return to the interstellar medium and will become part of future generations of stars.

The message of astronomy is that we are not observers; we are participants. We are part of the universe. Among all of the galaxies, stars, planets, planetesimals, and bits of matter, humans are objects that can think, and that means we can understand what we are.

Is the human race the only thinking species? If we are, we bear the sole responsibility to understand and admire the universe. The detection of signals from another civilization would demonstrate that we are not alone, and such communication would end the self-centered isolation of humanity and stimulate a reevaluation of the meaning of our existence. We may never realize our full potential as humans until we communicate with nonhuman intelligent life.

How Many Inhabited Worlds?

Given enough time, the searches will find other inhabited worlds, assuming that there are at least a few out there. If intelligence is common, scientists should find signals relatively soon—within the next few decades—but, if intelligence is rare, it may take much longer.

Simple arithmetic can give you an estimate of the number of technological civilizations with which you might communicate, N_c. The first formula to be proposed for N_c is named the **Drake equation** after the radio astronomer Frank Drake, a pioneer in the search for extraterrestrial intelligence. The version of the Drake equation presented here is modified slightly from its original form:

$$N_c = N^* \cdot f_P \cdot n_{LZ} \cdot f_L \cdot f_I \cdot F_S$$

N^* is the number of stars in our galaxy, and f_P represents the fraction of stars that have planets. If all single stars have planets, f_P is about 0.5. The factor n_{LZ} is the average number of planets in a solar system suitably placed in the life zone, f_L is the fraction of suitable planets on which life begins, and f_I is the fraction of those planets where life evolves to intelligence. These factors can be roughly estimated, but the remaining factor is much more uncertain.

F_S is the fraction of a star's life during which the life form is communicative. If a society survives at a technological level for only 100 years, the chances of communicating with it are small. But a society that stabilizes and remains technological for a long time is much more likely to be detected. For a star with a life span of 10 billion years, F_S can range from 10^{-8} for extremely short-lived societies to 10^{-4} for societies that survive for a million years. ■ Table 26-1 summarizes the likely range of values for F_S and the other factors.

If the optimistic estimates are true, there could be a communicative civilization within a few dozen light-years of Earth. On the other hand, if the pessimistic estimates are true, Earth may be the only planet in our galaxy capable of communication.

■ Table 26-1 | The Number of Technological Civilizations per Galaxy

Variables		Estimates	
		Pessimistic	Optimistic
N^*	Number of stars per galaxy	2×10^{11}	2×10^{11}
f_P	Fraction of stars with planets	0.01	0.5
n_{LZ}	Planets per star in life zone for over 4 billion years	0.01	1
f_L	Fraction of suitable planets on which life begins	0.01	1
f_I	Fraction of planets with life where intelligence evolves	0.01	1
F_S	Fraction of star's life during which a technological society survives	10^{-8}	10^{-4}
N_C	Number of communicative civilizations per galaxy	<1	10^7

◄ SCIENTIFIC ARGUMENT ►

Why does the number of civilizations that could be detected depend on how long civilizations survive at a technological level?

This scientific argument depends on the timing of events. If you turned a radio telescope to the sky and scanned millions of frequency bands for many stars, you would be taking a snapshot of the universe at a particular time. Other civilizations must be broadcasting at this time if they are to be detected. If most civilizations survive for a long time, there is a much greater chance that you will detect one of them in your snapshot than if civilizations tend to fall quickly due to nuclear war or environmental collapse. If most civilizations last only a short time, there may be none capable of transmitting during the short interval when Earthlings are capable of building radio telescopes to look for them.

The speed at which astronomers can search for signals is limited because computers must search many frequency intervals, but not all frequencies inside Earth's radio window are subject to intensive search. Build a new argument to explain: **Why is the water hole an especially good place to listen?**

◄ ►

Study and Review

Summary

26-1 I The Nature of Life

What is life?

- The process of life extracts energy from the surroundings, maintains the organism, and modifies the surroundings to promote the organism's survival.
- Living things have a physical basis—the arrangement of matter and energy that makes life possible. Life on Earth is based on carbon chemistry.
- Living things must also have a controlling unit of information, which you can recognize as genetic information passed to each new generation.
- Genetic information for life on Earth is stored in long carbon-chain molecules such as **DNA (deoxyribonucleic acid).**
- The DNA molecule stores information in the form of chemical bases linked together like the rungs of a ladder. Copied by the **RNA (ribonucleic acid)** molecule, the patterns of bases act as recipes that combine **amino acids** to manufacture **proteins** and **enzymes** that determine the structure and operation of the cell.
- The recipes are **genes** that are arranged along segments of DNA called **chromosomes.** When a cell divides, the DNA molecules split lengthwise and duplicate themselves so that each of the new cells can receive a copy of the genetic information.
- Errors in duplication can produce **mutants,** organisms that contain new DNA information and have new properties. This causes variation among individuals in a population, and **natural selection** determines which of these variations are best suited for survival. **Evolution** is the process by which life adjusts itself to its changing environment. Variation in genetic codes can become widespread among individuals in a species.
- Evolution is not random. Variation is random, but natural selection is controlled by the environment.

26-2 I The Origin of Life

How did life originate on Earth?

- The oldest fossils on Earth are about 3.4 billion years old. The fossils show that life began in the oceans.
- Life began on Earth as very simple organisms like bacteria, and it evolved into more complex creatures.
- The **Miller experiment** shows that the building blocks of life form easily and naturally under a wide range of circumstances. The early ocean, filled with these organic molecules, was a **primordial soup** in which **chemical evolution** concentrated simple molecules into larger, stronger molecules.
- Biological evolution begins when a molecule begins producing copies of itself. Then advantageous variations are preserved, and disadvantageous variations do not survive.
- **Stromatolites,** among the oldest fossils, are formed as mats of simple bacteria that trap mineral grains and build columns of layered sediments.
- Not until about half a billion years ago did life become complex in what is called the **Cambrian explosion.**
- Life emerged from the oceans only about 0.4 billion years ago, and human intelligence developed over the last 3 million years.

Could life begin on other worlds?

- Life as we know it on Earth requires liquid water and moderate temperatures.
- No other planet in our solar system is known to harbor life at present. Most are too hot or too cold.
- Life seems at least possible on the Jovian moons Europa, Titan, and Enceladus, where liquid water may exist. Also life might have begun on Mars before it became too cold and dry and might still survive underground where water is available and the ultraviolet radiation from the sun cannot penetrate.
- Because the origin of life and its evolution into intelligent creatures took so long on Earth, scientists eliminate short-lived stars such as middle- and upper-main-sequence stars as homes for life.
- Main-sequence G and K stars are thought to be likely candidates for searches for life. The faint M stars are also possibilities.
- The **life zone** or **ecosphere** around a star is the region in which a planet may have conditions suitable for life. It may be larger than scientists had expected, given the wide variety of living things being found in extreme environments on Earth.

OUR JOURNEY TOGETHER IS OVER, but before we part company, there is one last thing to discuss—the place of humanity in the universe. Astronomy gives us some comprehension of the workings of stars, galaxies, and planets, but its greatest value lies in what it teaches us about ourselves. Now that you have surveyed astronomical knowledge, you can better understand your own position in nature.

To some, the word *nature* conjures up visions of furry rabbits hopping about in a forest glade. Others think of blue-green ocean depths or windswept mountaintops. As diverse as these images are, they are all Earthbound. Having studied astronomy, you see nature as a beautiful mechanism composed of matter and energy interacting according to simple rules to form galaxies, stars, planets, mountaintops, ocean depths, and forest glades.

Perhaps the most important astronomical lesson is that humanity is a small but important part of the universe. Most of the universe is the lifeless cold of deep space or the intense heat inside stars. Only on the surfaces of a few planets can the chemical bonds of living things survive.

If life is special, then intelligence is precious. The universe must contain many planets devoid of life, planets where the wind has blown unfelt for billions of years. There may exist planets where life has developed but has not become complex, planets where the wind stirs dark forests and insects, fish, birds, and animals watch the passing days unaware of their own existence. It is intelligence, human or alien, that gives meaning to the landscape.

Science is the process by which intelligence tries to understand the universe. Science is not the invention of new devices or processes. It does not create home computers, cure the mumps, or manufacture plastic spoons—that is engineering and technology, the adaptation of scientific understanding for practical purposes. Science is understanding nature, and astronomy is understanding on the grandest scale. Astronomy is the science by which the universe, through its intelligent lumps of matter, tries to understand its own existence.

As the primary intelligent species on this planet, we are the custodians of a priceless gift—a planet filled with living things. This is especially true if life is rare in the universe. In fact, if Earth is the only inhabited planet, our responsibility is overwhelming. We are the only creatures who can take action to preserve the existence of life on Earth, and, ironically, our own actions are the most serious hazards.

The future of humanity is not secure. We are trapped on a tiny planet with limited resources and a population growing faster than our ability to produce food. We have already driven some creatures to extinction and now threaten others. We are changing the climate of our planet in ways we do not fully understand. Even if we reshape our civilization to preserve our world, the evolution of the sun will eventually destroy Earth, and if humanity is unable to leave the solar system, our history will end.

This is a depressing prospect, but a few factors are comforting. First, everything in the universe is temporary. Stars die, galaxies die, perhaps the entire universe will someday end. That our distant future is limited only assures us that we are a part of a much larger whole. Second, we have a few billion years to prepare, and a billion years is a very long time. Only 10,000 years ago, our ancestors were building the first cities; a few million years ago, they were learning to walk erect and communicate; and a billion years ago, our ancestors were microscopic organisms living in the primeval oceans. It may be unlikely that in a billion years we humans will still be the dominant species on Earth.

Our responsibility is not to save our race for all eternity but to behave as dependable custodians of our planet, preserving it, admiring it, and trying to understand it. That calls for drastic changes in our behavior toward other living things and a revolution in our attitude toward our planet's resources. Whether we can change our ways is debatable—humanity is far from perfect in its understanding, abilities, or intentions. However, you must not imagine that we and our civilization are less than precious. We have the gift of intelligence, and that is the finest thing this planet has ever produced.

We shall not cease from exploration
And the end of all our exploring
Will be to arrive where we started
And know the place for the first time.

—*T. S. Eliot, "Little Gidding"*

Excerpt from "Little Gidding" in Four Quartets, Copyright 1942 by T. S. Eliot and renewed 1970 by Esme Valerie Eliot, reprinted by permission of Harcourt, Inc. and Faber & Faber, Ltd.

Appendix A
Units and Astronomical Data

Introduction

THE METRIC SYSTEM IS USED WORLDWIDE as the system of units, not only in science but also in engineering, business, sports, and daily life. Developed in 18th-century France, the metric system has gained acceptance in almost every country in the world because it simplifies computations.

A system of units is based on the three fundamental units for length, mass, and time. Other quantities, such as density and force, are derived from these fundamental units. In the English (or British) system of units (commonly used only in the United States, Tonga, and Southern Yemen, but not in Great Britain) the fundamental unit of length is the foot, composed of 12 inches. The metric system is based on the decimal system of numbers, and the fundamental unit of length is the meter, composed of 100 centimeters.

Because the metric system is a decimal system, it is easy to express quantities in larger or smaller units as is convenient. You can express distances in centimeters, meters, kilometers, and so on. The prefixes specify the relation of the unit to the meter. Just as a cent is 1/100 of a dollar, so a centimeter is 1/100 of a meter. A kilometer is 1000 m, and a kilogram is 1000 g. The meanings of the commonly used prefixes are given in ■ Table A-1.

■ Table A-1 | Metric Prefixes

Prefix	Symbol	Factor
Mega	M	10^6
Kilo	k	10^3
Centi	c	10^{-2}
Milli	m	10^{-3}
Micro	μ	10^{-6}
Nano	n	10^{-9}

The SI Units

ANY SYSTEM OF UNITS based on the decimal system would be easy to use, but by international agreement, the preferred set of units, known as the *Système International d'Unités* (SI units) is based on the meter, kilogram, and second. These three fundamental units define the rest of the units, as given in ■ Table A-2.

The SI unit of force is the newton (N), named after Isaac Newton. It is the force needed to accelerate a 1 kg mass by 1 m/s^2, or the force roughly equivalent to the weight of an apple at Earth's surface. The SI unit of energy is the joule (J), the energy produced by a force of 1 N acting through a distance of 1 m. A joule is roughly the energy in the impact of an apple falling off a table.

■ Table A-2 | SI Metric Units

Quantity	SI Unit	English Unit
Length	Meter (m)	Foot
Mass	Kilogram (kg)	Slug (sl)
Time	Second (s)	Second (s)
Force	Newton (N)	Pound (lb)
Energy	Joule (J)	Foot-pound (fp)

Exceptions

Units can help you in two ways. They make it possible to make calculations, and they can help you to conceive of certain quantities. For calculations, the metric system is far superior, and it is used for calculations throughout this book.

But Americans commonly use the English system of units, so for conceptual purposes this book also expresses quantities in English units. Instead of saying the average person would weigh 133 N on the moon, it might be more helpful to express the weight as 30 lb. Consequently, this text commonly gives quantities in metric form followed by the English form in parentheses: The radius of the moon is 1738 km (1080 mi).

In SI units, density should be expressed as kilograms per cubic meter, but no human hand can enclose a cubic meter, so this unit does not help you grasp the significance of a given density. This book refers to density in grams per cubic centimeter. A gram is roughly the mass of a paperclip, and a cubic centimeter is the size of a small sugar cube, so you can conceive of a density of 1 g/cm^3, roughly the density of water. This is not a bothersome departure from SI units because you will not have to make complex calculations using density.

Conversions

TO CONVERT FROM ONE METRIC UNIT to another (from meters to kilometers, for example), you have only to look at the prefix. However, converting from metric to English or English to metric is more complicated. The conversion factors are given in ■ Table A-3.

■ Table A-3 | Conversion Factors

1 inch = 2.54 centimeters	1 centimeter = 0.394 inch
1 foot = 0.3048 meter	1 meter = 39.36 inches= 3.28 feet
1 mile = 1.6093 kilometers	1 kilometer = 0.6214 mile
1 slug = 14.594 kilograms	1 kilogram = 0.0685 slug
1 pound = 4.4482 newtons	1 newton = 0.2248 pound
1 foot-pound = 1.35582 joules	1 joule = 0.7376 foot-pound
1 horsepower = 745.7 joules/s	1 joule/s = 1 watt

Example: The radius of the moon is 1738 km. What is this in miles? Table A-3 indicates that 1 mile equals 1.609 km, so

$$1738 \text{ km} \times \frac{(1 \text{ mile})}{(1.609 \text{ km})} = 1080 \text{ miles}$$

Temperature Scales

IN ASTRONOMY, as in most other sciences, temperatures are expressed on the Kelvin scale, although the centigrade (or Celsius) scale is also used. The Fahrenheit scale commonly used in the United States is not used in scientific work.

Temperatures on the Kelvin scale are measured from absolute zero, the temperature of an object that contains no extractable heat. In practice, no object can be as cold as absolute zero, although laboratory apparatuses have reached temperatures lower than 10^{-6} K. The scale is named after the Scottish mathematical physicist William Thomson, Lord Kelvin (1824–1907).

The centigrade scale refers temperatures to the freezing point of water (0°C) and to the boiling point of water (100°C). One degree centigrade is $\frac{1}{100}$ the temperature difference between the freezing and boiling points of water, thus the prefix *centi.* The centigrade scale is also called the Celsius scale after its inventor, the Swedish astronomer Anders Celsius (1701–1744).

The Fahrenheit scale fixes the freezing point of water at 32°F and the boiling point at 212°F. Named after the German physicist Gabriel Daniel Fahrenheit (1686–1736), who made the first successful mercury thermometer in 1720, the Fahrenheit scale is used only in the United States.

It is easy to convert temperatures from one scale to another using the information given in ■ Table A-4.

■ Table A-4 | Temperature Scales

	Kelvin (K)	Centigrade (°C)	Fahrenheit (°F)
Absolute zero	0 K	−273°C	−459°F
Freezing point of water	273 K	0°C	32°F
Boiling point of water	373 K	100°C	212°F

Conversions:
$K = °C + 273$
$°C = (\frac{5}{9})(°F - 32)$
$°F = (\frac{9}{5})\,C + 32$

Powers of 10 Notation

POWERS OF 10 make writing very large numbers much simpler. For example, the nearest star is about 43,000,000,000,000 km from the sun. Writing this number as 4.3×10^{13} km is much easier.

Very small numbers can also be written with powers of 10. For example, the wavelength of visible light is about 0.0000005 m. In powers of 10 this becomes 5×10^{-7} m.

The powers of 10 used in this notation appear below. The exponent tells you how to move the decimal point. If the exponent is positive, move the decimal point to the right. If the exponent is negative, move the decimal point to the left. For example, 2×10^3 equals 2000.0, and 2×10^{-3} equals 0.002.

$$10^5 = 100{,}000$$
$$10^4 = 10{,}000$$
$$10^3 = 1{,}000$$
$$10^2 = 100$$
$$10^1 = 10$$
$$10^0 = 1$$
$$10^{-1} = 0.1$$
$$10^{-2} = 0.01$$
$$10^{-3} = 0.001$$
$$10^{-4} = 0.0001$$

If you use scientific notation in calculations, be sure you correctly enter the numbers into your calculator. Not all calculators accept scientific notation, but those that can have a key labeled EXP, EEX, or perhaps EE that allows you to enter the exponent of ten. To enter a number such as 3×10^8, press the keys 3 EXP 8. To enter a number with a negative exponent, you must use the change-sign key, usually labeled +/− or CHS. To enter the number 5.2×10^{-3}, press the keys 5.2 EXP +/−3. Try a few examples.

To read a number in scientific notation from a calculator you must read the exponent separately. The number 3.1×10^{25} may appear in a calculator display as 3.1 25 or on some calculators as 3.1 10^{25}. Examine your calculator to determine how such numbers are displayed.

ASTRONOMY, AND SCIENCE IN GENERAL, is a way of learning about nature and understanding the universe. To test hypotheses about how nature works, scientists use observations of nature. The tables that follow contain some of the basic observations that support science's best understanding of the astronomical universe. Of course, these data are expressed in the form of numbers, not because science reduces all understanding to mere numbers, but because the struggle to understand nature is so demanding that science must use every tool available. Quantitative thinking—reasoning mathematically—is one of the most powerful tools ever invented by the human brain. Thus these tables are not nature reduced to mere numbers but numbers supporting humanity's growing understanding of the natural world around us.

Table A-5 | Constants

Astronomical unit (AU)	$= 1.495979 \times 10^{11}$ m
Parsec (pc)	= 206,265 AU
	$= 3.085678 \times 10^{16}$m
	= 3.261633 ly
Light-year (ly)	$= 9.46053 \times 10^{15}$ m
Velocity of light (c)	$= 2.997925 \times 10^{8}$ m/s
Gravitational constant (G)	$= 6.67 \times 10^{-11}$ m^3/s^{2}kg
Mass of Earth ($M_{\oplus}$)	$= 5.976 \times 10^{24}$ kg
Earth equatorial radius ($R_{\oplus}$)	= 6378.164 km
Mass of sun ($M_{\odot}$)	$= 1.989 \times 10^{30}$ kg
Radius of sun ($R_{\odot}$)	$= 6.9599 \times 10^{8}$ m
Solar luminosity ($L_{\odot}$)	$= 3.826 \times 10^{26}$ J/s
Mass of moon	$= 7.350 \times 10^{22}$ kg
Radius of moon	= 1738 km
Mass of H atom	$= 1.67352 \times 10^{-27}$ kg

Table A-6 | Units Used in Astronomy

1 angstrom (Å)	$= 10^{-8}$ cm
	$= 10^{-10}$ m
1 astronomical unit (AU)	$= 1.495979 \times 10^{11}$ m
	$= 92.95582 \times 10^{6}$ miles
1 light-year (ly)	$= 6.3240 \times 10^{4}$ AU
	$= 9.46053 \times 10^{15}$ m
	$= 5.9 \times 10^{12}$ miles
1 parsec (pc)	= 206,265 AU
	$= 3.085678 \times 10^{16}$ m
	= 3.261633 ly
1 kiloparsec (kpc)	= 1000 pc
1 megaparsec (Mpc)	= 1,000,000 pc

Table A-7 | Properties of Main-Sequence Stars

Spectral Type	Absolute Visual Magnitude (M_V)	Luminosity*	Temp. (K)	λ_{max} (nm)	Mass*	Radius*	Average Density (g/cm^3)
O5	−5.8	501,000	40,000	72.4	40	17.8	0.01
B0	−4.1	20,000	28,000	100	18	7.4	0.1
B5	−1.1	790	15,000	190	6.4	3.8	0.2
A0	+0.7	79	9900	290	3.2	2.5	0.3
A5	+2.0	20	8500	340	2.1	1.7	0.6
F0	+2.6	6.3	7400	390	1.7	1.4	1.0
F5	+3.4	2.5	6600	440	1.3	1.2	1.1
G0	+4.4	1.3	6000	480	1.1	1.0	1.4
G5	+5.1	0.8	5500	520	0.9	0.9	1.6
K0	+5.9	0.4	4900	590	0.8	0.8	1.8
K5	+7.3	0.2	4100	700	0.7	0.7	2.4
M0	+9.0	0.1	3500	830	0.5	0.6	2.5
M5	+11.8	0.01	2800	1000	0.2	0.3	10.0
M8	+16	0.001	2400	1200	0.1	0.1	63

*Luminosity, mass, and radius are given in terms of the sun's luminosity, mass, and radius.

Table A-8 | The Brightest Stars

Star	Name	Apparent Visual Magnitude (m_V)	Spectral Type	Absolute Visual Magnitude (M_V)	Distance (ly)
α CMa A	Sirius	−1.47	A1	1.4	8.7
α Car	Canopus	−0.72	F0	−3.1	98
α Cen	Rigil Kentaurus	−0.01	G2	4.4	4.3
α Boo	Arcturus	−0.06	K2	−0.3	36
α Lyr	Vega	0.04	A0	0.5	26.5
α Aur	Capella	0.05	G8	−0.6	45
β Ori A	Rigel	0.14	B8	−7.1	900
α CMi A	Procyon	0.37	F5	2.7	11.3
α Ori	Betelgeuse	0.41	M2	−5.6	520
α Eri	Achernar	0.51	B3	−2.3	118
β Cen AB	Hadar	0.63	B1	−5.2	490
α Aql	Altair	0.77	A7	2.2	16.5
α Tau A	Aldebaran	0.86	K5	−0.7	68
α Cru	Acrux	0.90	B2	−3.5	260
α Vir	Spica	0.91	B1	−3.3	220
α Sco A	Antares	0.92	M1	−5.1	520
α PsA	Fomalhaut	1.15	A3	2.0	22.6
α Gem	Pollux	1.16	K0	1.0	35
α Cyg	Deneb	1.26	A2	−7.1	1600
β Cru	Beta Crucis	1.28	B0.5	−4.6	490

Table A-9 | The Nearest Stars

Name	Absolute Magnitude (M_v)	Distance (ly)	Spectral Type	Apparent Visual Magnitude (m_v)
Sun	4.83		G2	−26.8
Proxima Cen	15.45	4.28	M5	11.05
α Cen A	4.38	4.3	G2	0.1
α Cen B	5.76	4.3	K5	1.5
Barnard's Star	13.21	5.9	M5	9.5
Wolf 359	16.80	7.6	M6	13.5
Lalande 21185	10.42	8.1	M2	7.5
Sirius A	1.41	8.6	A1	−1.5
Sirius B	11.54	8.6	white dwarf	7.2
Luyten 726-8A	15.27	8.9	M5	12.5
Luyten 726-8B (UV Cet)	15.8	8.9	M6	13.0
Ross 154	13.3	9.4	M5	10.6
Ross 248	14.8	10.3	M6	12.2
ϵ Eri	6.13	10.7	K2	3.7
Luyten 789-6	14.6	10.8	M7	12.2
Ross 128	13.5	10.8	M5	11.1
61 CYG A	7.58	11.2	K5	5.2
61 CYG B	8.39	11.2	K7	6.0
ϵ Ind	7.0	11.2	K5	4.7
Procyon A	2.64	11.4	F5	0.3
Procyon B	13.1	11.4	white dwarf	10.8
Σ2398 A	11.15	11.5	M4	8.9
Σ2398 B	11.94	11.5	M5	9.7
Groombridge 34 A	10.32	11.6	M1	8.1
Groombridge 34 B	13.29	11.6	M6	11.0
Lacaille 9352	9.59	11.7	M2	7.4
τ Ceti	5.72	11.9	G8	3.5
BD + 5° 1668	11.98	12.2	M5	9.8
L 725-32	15.27	12.4	M5	11.5
Lacaille 8760	8.75	12.5	M0	6.7
Kapteyn's Star	10.85	12.7	M0	8.8
Kruger 60 A	11.87	12.8	M3	9.7
Kruger 60 B	13.3	12.8	M4	11.2

Table A-10 | Properties of the Planets

Physical Properties (Earth = ⊕)

Planet	Equatorial Radius (km)	Equatorial Radius (⊕ = 1)	Mass (⊕ = 1)	Average Density (g/cm³)	Surface Gravity (⊕ = 1)	Escape Velocity (km/s)	Sidereal Period of Rotation	Inclination of Equator to Orbit
Mercury	2439	0.382	0.0558	5.44	0.378	4.3	58.646^{d}	0°
Venus	6052	0.95	0.815	5.24	0.903	10.3	243.01^{d}	177°
Earth	6378	1.00	1.00	5.497	1.00	11.2	$23^{h}56^{m}04.1^{s}$	23°27′
Mars	3396	0.53	0.1075	3.94	0.379	5.0	$24^{h}37^{m}22.6^{s}$	25°19′
Jupiter	71,494	11.20	317.83	1.34	2.54	61	$9^{h}55^{m}30^{s}$	3°5′
Saturn	60,330	9.42	95.147	0.69	1.16	35.6	$10^{h}13^{m}59^{s}$	26°24′
Uranus	25,559	4.01	14.54	1.19	0.919	22	$17^{h}14^{m}$	97°55′
Neptune	24,750	3.93	17.23	1.66	1.19	25	$16^{h}3^{m}$	28°48′

Orbital Properties

Planet	Semimajor Axis (*a*) (AU)	Semimajor Axis (*a*) (10^6 km)	Orbital Period (*P*) (y)	Orbital Period (*P*) (days)	Average Orbital Velocity (km/s)	Orbital Eccentricity	Inclination to Ecliptic
Mercury	0.3871	57.9	0.24084	87.969	47.89	0.2056	7°0′16″
Venus	0.7233	108.2	0.61515	224.68	35.03	0.0068	3°23′40″
Earth	1	149.6	1	365.26	29.79	0.0167	0°
Mars	1.5237	227.9	1.8808	686.95	24.13	0.0934	1°51′09″
Jupiter	5.2028	778.3	11.867	4334.3	13.06	0.0484	1°18′29″
Saturn	9.5388	1427.0	29.461	10,760	9.64	0.0560	2°29′17″
Uranus	19.18	2869.0	84.013	30,685	6.81	0.0461	0°46′23″
Neptune	30.0611	4497.1	164.793	60,189	5.43	0.0100	1°46′27″

■ Table A-11 | Principal Satellites of the Solar System

Planet	Satellite	Radius (km)	Distance from Planet (10^3 km)	Orbital Period (days)	Orbital Eccentricity	Orbital Inclination
Earth	Moon	1738	384.4	27.322	0.055	5°8′43″
Mars	Phobos	14 × 12 × 10	9.38	0.3189	0.018	1°.0
	Deimos	8 × 6 × 5	23.5	1.262	0.002	2°.8
Jupiter	Metis	20	126	0.29	0.0	0°.0
	Adrastea	12 × 8 × 10	128	0.294	0.0	0°.0
	Amalthea	135 × 100 × 78	182	0.4982	0.003	0°.45
	Thebe	50	223	0.674	0.0	1°.3
	Io	1820	422	1.769	0.000	0°.3
	Europa	1565	671	3.551	0.000	0°.46
	Ganymede	2640	1071	7.155	0.002	0°.18
	Callisto	2420	1884	16.689	0.008	0°.25
	Leda	~8	11,110	240	0.146	26°.7
	Himalia	~85	11,470	250.6	0.158	27°.6
	Lysithea	~20	11,710	260	0.12	29°
	Elara	~30	11,740	260.1	0.207	24°.8
	Ananke	15	21,200	631	0.169	147°
	Carme	22	22,350	692	0.207	163°
	Pasiphae	35	23,300	735	0.40	147°
	Sinope	20	23,700	758	0.275	156°
Saturn	Pan	10	133.570	0.574	0.000	0°
	Atlas	20 × 15 × 15	137.7	0.601	0.002	0°.3
	Prometheus	70 × 40 × 50	139.4	0.613	0.003	0°.0
	Pandora	55 × 35 × 50	141.7	0.629	0.004	0°.05
	Epimetheus	70 × 50 × 50	151.42	0.694	0.009	0°.34
	Janus	110 × 80 × 100	151.47	0.695	0.007	0°.14
	Mimas	196	185.54	0.942	0.020	1°.5
	Enceladus	250	238.04	1.370	0.004	0°.0
	Tethys	530	294.67	1.888	0.000	1°.1
	Calypso	17 × 11 × 12	294.67	1.888	0.0	~1°?
	Telesto	12	294.67	1.888	0.0	~1°?
	Dione	560	377	2.737	0.002	0°.0
	Helene	20 × 15 × 15	377	2.74	0.005	0°.15
	Rhea	765	527	4.518	0.001	0°.4
	Titan	2575	1222	15.94	0.029	0°.3
	Hyperion	205 × 130 × 110	1484	21.28	0.104	~0°.5
	Iapetus	720	3562	79.33	0.028	14°.72
	Phoebe	110	12,930	550.4	0.163	150°
Uranus	Cordelia	20	49.8	0.3333	~0	~0°
	Ophelia	15	53.8	0.375	~0	~0°
	Bianca	25	59.1	0.433	~0	~0°
	Cressida	30	61.8	0.462	~0	~0°
	Desdemona	30	62.7	0.475	~0	~0°
	Juliet	40	64.4	0.492	~0	~0°
	Portia	55	66.1	0.512	~0	~0°
	Rosalind	30	69.9	0.558	~0	~0°
	Belinda	30	75.2	0.621	~0	~0°
	Puck	85 ± 5	85.9	0.762	~0	~0°
	Miranda	242 ± 5	129.9	1.414	0.017	3°.4

Continued

■ Table A-11 | Principal Satellites of the Solar System *(continued)*

	Ariel	580 ± 5	190.9	2.520	0.003	0°
	Umbriel	595 ± 10	266.0	4.144	0.003	0°
	Titania	805 ± 5	436.3	8.706	0.002	0°
	Oberon	775 ± 10	583.4	13.463	0.001	0°
	Caliban	40	7164	579	0.082	139.2°
	Stephano	~20	7900	676		
	Sycorax	80	12,174	1284	0.509	152.7°
	Prospero	~20	16,100	1950		
	Setebos	~20	17,600	2240	0.539	
Neptune	Naiad	30	48.2	0.296	~0	~0°
	Thalassa	40	50.0	0.312	~0	~0°
	Despina	90	52.5	0.333	~0	~0°
	Galatea	75	62.0	0.396	~0	~0°
	Larissa	95	73.6	0.554	~0	~0°
	Proteus	205	117.6	1.121	~0	~0°
	Triton	1352	354.59	5.875	0.00	160°
	Nereid	170	5588.6	360.125	0.76	27.7°

■ Table A-12 | Meteor Showers

			Radiant		
Shower	**Dates**	**Hourly Rate**	**R.A.**	**Dec.**	**Associated Comet**
Quadrantids	Jan. 2–4	30	15^h24^m	50°	
Lyrids	April 20–22	8	18^h4^m	33°	1861 I
η Aquarids	May 2–7	10	22^h24^m	0°	Halley?
δ Aquarids	July 26–31	15	22^h36^m	−10°	
Perseids	Aug. 10–14	40	3^h4^m	58°	1982 III
Orionids	Oct. 18–23	15	6^h20^m	15°	Halley?
Taurids	Nov. 1–7	8	3^h40^m	17°	Encke
Leonids	Nov. 14–19	6	10^h12^m	22°	1866 I Temp
Geminids	Dec. 10–13	50	7^h28^m	32°	

Table A-13 | Greatest Elongations of Mercury

Evening Sky	Morning Sky
Jan. 22, 2008	March 3, 2008
May 14, 2008	July 1, 2008
Sept. 11, 2008	Oct. 22, 2008*
Jan. 4, 2009	Feb. 13, 2009
Apr. 26, 2009*	June 13, 2009
Aug. 24, 2009	Oct. 6, 2009*
Dec. 18, 2009	Jan. 27, 2010
Apr. 8, 2010*	May 26, 2010
Aug. 7, 2010	Sept. 19, 2010*
Dec. 1, 2010	Jan. 9, 2011
March 23, 2011*	May 7, 2011
July 20, 2011	Sept. 3, 2011*
Nov. 14, 2011	Dec. 23, 2011
March 5, 2012*	April 18, 2012
July 1, 2012	Aug. 16, 2012
Oct. 26, 2012	Dec. 4, 2012

Elongation is the angular distance from the sun to a planet.
*Most favorable elongations.

Table A-14 | Greatest Elongations of Venus

Evening Sky	Morning Sky
Jan. 14, 2009	June 5, 2009
Aug. 20, 2010	Jan. 8, 2011
March 27, 2012	Aug. 15, 2012
Nov. 1, 2013	March 22, 2014
June 6, 2015	Oct. 26, 2015
Jan. 12, 2017	June 3, 2017
Aug. 17, 2018	Jan. 6, 2019

Venus does not reach greatest elongation during 2008.

Table A-15 | The Greek Alphabet

Α, α alpha	Η, η eta	Ν, ν nu	Τ, τ tau
Β, β beta	Θ, θ theta	Ξ, ξ xi	Υ, υ upsilon
Γ, γ gamma	Ι, ι iota	Ο, ο omicron	Φ, φ phi
Δ, δ delta	Κ, κ kappa	Π, π pi	Χ, χ chi
Ε, ϵ epsilon	Λ, λ lambda	Ρ, ρ rho	Ψ, ψ psi
Ζ, ζ zeta	Μ, μ mu	Σ, σ sigma	Ω, ω omega

■ Table A-16 | Periodic Table of the Elements

Atomic number → 11
Symbol → Na
Atomic mass → 22.99

Atomic masses are based on carbon-12. Numbers in parentheses are mass numbers of most stable or best-known isotopes of radioactive elements.

← Transition Elements →

Period	Group IA(1)	IIA(2)	IIIB(3)	IVB(4)	VB(5)	VIB(6)	VIIB(7)	VIII (8)	VIII (9)	VIII (10)	IB(11)	IIB(12)	IIIA(13)	IVA(14)	VA(15)	VIA(16)	VIIA(17)	Noble Gases (18)
1	1 H 1.008																	2 He 4.003
2	3 Li 6.941	4 Be 9.012											5 B 10.81	6 C 12.01	7 N 14.01	8 O 16.00	9 F 19.00	10 Ne 20.18
3	11 Na 22.99	12 Mg 24.31											13 Al 26.98	14 Si 28.09	15 P 30.97	16 S 32.06	17 Cl 35.45	18 Ar 39.95
4	19 K 39.10	20 Ca 40.08	21 Sc 44.96	22 Ti 47.90	23 V 50.94	24 Cr 52.00	25 Mn 54.94	26 Fe 55.85	27 Co 58.93	28 Ni 58.7	29 Cu 63.55	30 Zn 65.38	31 Ga 69.72	32 Ge 72.59	33 As 74.92	34 Se 78.96	35 Br 79.90	36 Kr 83.80
5	37 Rb 85.47	38 Sr 87.62	39 Y 88.91	40 Zr 91.22	41 Nb 92.91	42 Mo 95.94	43 Tc 98.91	44 Ru 101.1	45 Rh 102.9	46 Pd 106.4	47 Ag 107.9	48 Cd 112.4	49 In 114.8	50 Sn 118.7	51 Sb 121.8	52 Te 127.6	53 I 126.9	54 Xe 131.3
6	55 Cs 132.9	56 Ba 137.3	57* La 138.9	72 Hf 178.5	73 Ta 180.9	74 W 183.9	75 Re 186.2	76 Os 190.2	77 Ir 192.2	78 Pt 195.1	79 Au 197.0	80 Hg 200.6	81 Tl 204.4	82 Pb 207.2	83 Bi 209.0	84 Po (210)	85 At (210)	86 Rn (222)
7	87 Fr (223)	88 Ra 226.0	89** Ac (227)	104 Rf (261)	105 Db (262)	106 Sg (263)	107 Bh (262)	108 Hs (265)	109 Mt (266)	110 Ds (269)	111 Uuu (272)	112 Uub (277)	113 Uub (284)	114 Uuq (285)	115 Uub (288)	116 Uuh (289)		

Inner Transition Elements

Lanthanide Series 6 *	58 Ce 140.1	59 Pr 140.9	60 Nd 144.2	61 Pm (145)	62 Sm 150.4	63 Eu 152.0	64 Gd 157.3	65 Tb 158.9	66 Dy 162.5	67 Ho 164.9	68 Er 167.3	69 Tm 168.9	70 Yb 173.0	71 Lu 175.0
Actinide Series 7 **	90 Th 232.0	91 Pa 231.0	92 U 238.0	93 Np 237.0	94 Pu (244)	95 Am (243)	96 Cm (247)	97 Bk (247)	98 Cf (251)	99 Es (252)	100 Fm (257)	101 Md (258)	102 No (259)	103 Lr (260)

The Elements and Their Symbols

Actinium	Ac	Cesium	Cs	Hafnium	Hf	Mercury	Hg	Protactinium	Pa	Tellurium	Te
Aluminum	Al	Chlorine	Cl	Hassium	Hs	Molybdenum	Mo	Radium	Ra	Terbium	Tb
Americium	Am	Chromium	Cr	Helium	He	Neodymium	Nd	Radon	Rn	Thallium	Tl
Antimony	Sb	Cobalt	Co	Holmium	Ho	Neon	Ne	Rhenium	Re	Thorium	Th
Argon	Ar	Copper	Cu	Hydrogen	H	Neptunium	Np	Rhodium	Rh	Thulium	Tm
Arsenic	As	Curium	Cm	Indium	In	Nickel	Ni	Rubidium	Rb	Tin	Sn
Astatine	At	Darmstadtium	Ds	Iodine	I	Niobium	Nb	Ruthenium	Ru	Titanium	Ti
Barium	Ba	Dubnium	Db	Iridium	Ir	Nitrogen	N	Rutherfordium	Rf	Tungsten	W
Berkelium	Bk	Dysprosium	Dy	Iron	Fe	Nobelium	No	Samarium	Sm	Uranium	U
Beryllium	Be	Einsteinium	Es	Krypton	Kr	Osmium	Os	Scandium	Sc	Vanadium	V
Bismuth	Bi	Erbium	Er	Lanthanum	La	Oxygen	O	Seaborgium	Sg	Xenon	Xe
Bohrium	Bh	Europium	Eu	Lawrencium	Lr	Palladium	Pd	Selenium	Se	Ytterbium	Yb
Boron	B	Fermium	Fm	Lead	Pb	Phosphorous	P	Silicon	Si	Yttrium	Y
Bromine	Br	Fluorine	F	Lithium	Li	Platinum	Pt	Silver	Ag	Zinc	Zn
Cadmium	Cd	Francium	Fr	Lutetium	Lu	Plutonium	Pu	Sodium	Na	Zirconium	Zr
Calcium	Ca	Gadolinium	Gd	Magnesium	Mg	Polonium	Po	Strontium	Sr		
Californium	Cf	Gallium	Ga	Manganese	Mn	Potassium	K	Sulfur	S		
Carbon	C	Germanium	Ge	Meitnerium	Mt	Praseodymium	Pr	Tantalum	Ta		
Cerium	Ce	Gold	Au	Mendelevium	Md	Promethium	Pm	Technetium	Tc		

Appendix B
Observing the Sky

OBSERVING THE SKY WITH THE NAKED EYE is of no more importance to modern astronomy than picking up pretty pebbles is to modern geology. But the sky is a natural wonder unimaginably bigger than the Grand Canyon, the Rocky Mountains, or any other natural wonder that tourists visit every year. To neglect the beauty of the sky is equivalent to geologists neglecting the beauty of the minerals they study. This supplement is meant to act as a tourist's guide to the sky. You analyzed the universe in the regular chapters, but here you will admire it.

The brighter stars in the sky are visible even from the centers of cities with their air and light pollution. But in the countryside, only a few miles beyond the cities, the night sky is a velvety blackness strewn with thousands of glittering stars. From a wilderness location, far from the city's glare, and especially from high mountains, the night sky is spectacular.

Using Star Charts

THE CONSTELLATIONS ARE A FASCINATING CULTURAL HERITAGE of our planet, but they are sometimes a bit difficult to learn because of Earth's motion. The constellations above the horizon change with the time of night and the seasons.

Because Earth rotates eastward, the sky appears to rotate westward around Earth. A constellation visible in the southern sky soon after sunset will appear to move westward, and in a few hours it will disappear below the horizon. Other constellations will rise in the east, so the sky changes gradually through the night.

In addition, Earth's orbital motion makes the sun appear to move eastward among the stars. Each day the sun moves about twice its own diameter, about one degree, eastward along the ecliptic, and consequently, each night at sunset the constellations are about one degree farther toward the west.

Orion, for instance, is visible in the evening sky in January; but, as the days pass, the sun moves closer to Orion. By March, Orion is difficult to see in the western sky soon after sunset. By June, the sun is so close to Orion it sets with the sun and is invisible. Not until late July is the sun far enough past Orion for the constellation to become visible rising in the eastern sky just before dawn.

Because of the rotation and orbital motion of Earth, you need more than one star chart to map the sky. Which chart you select depends on the month and the time of night. The charts given in this appendix show the evening sky for each month.

Two sets of charts are included for two typical locations on Earth. The Northern Hemisphere charts show the sky as seen from a northern latitude typical of the United States and central Europe. The Southern Hemisphere star charts are appropriate for readers in Earth's Southern Hemisphere, including Australia, southern South America, and southern Africa.

To use the charts, select the appropriate chart and hold it overhead as shown in Figure B-1. If you face south, turn the chart until the words *southern horizon* are at the bottom of the chart. If you face other directions, turn the chart appropriately.

Figure B-1

To use the star charts in this book, select the appropriate chart for the date and time. Hold it overhead and turn it until the direction at the bottom of the chart is the same as the direction you are facing.

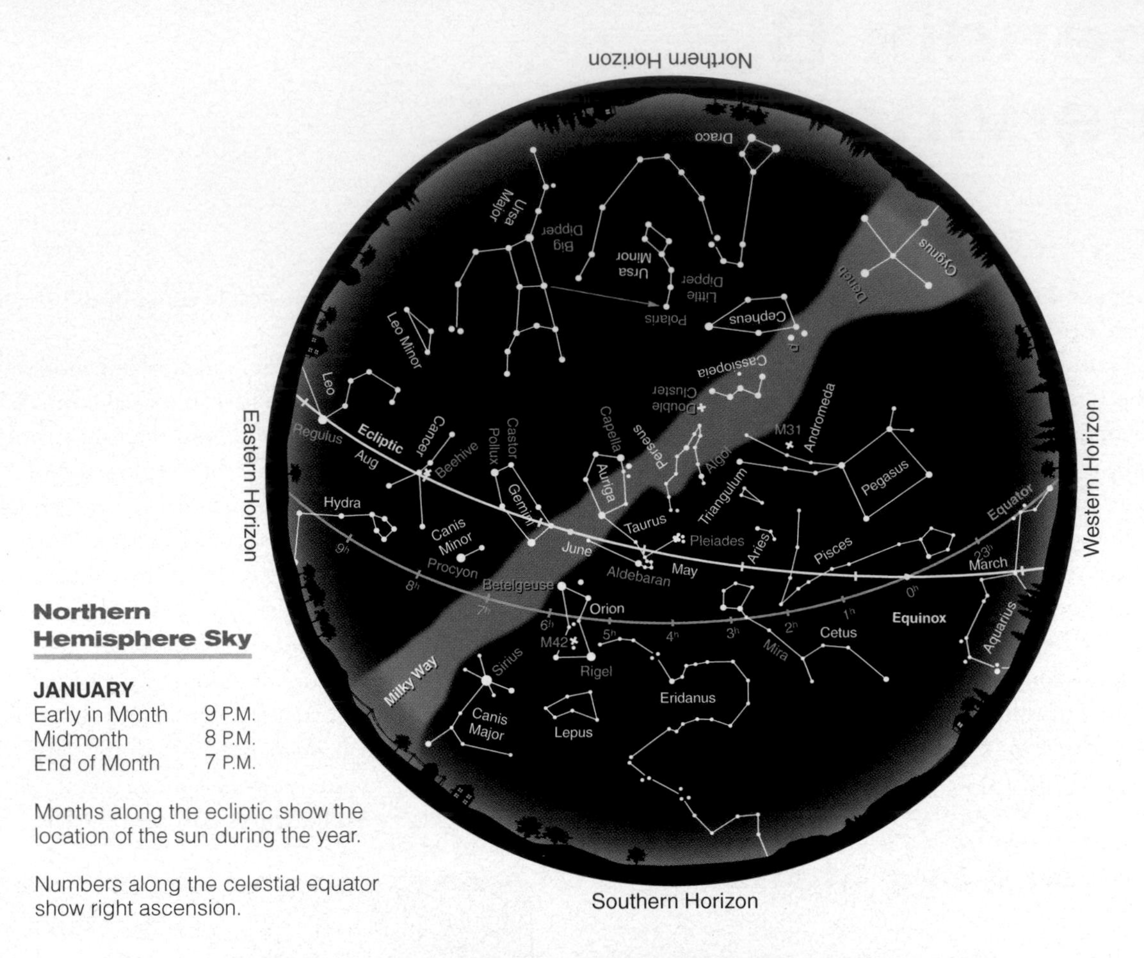

Northern Hemisphere Sky

JANUARY

Early in Month	9 P.M.
Midmonth	8 P.M.
End of Month	7 P.M.

Months along the ecliptic show the location of the sun during the year.

Numbers along the celestial equator show right ascension.

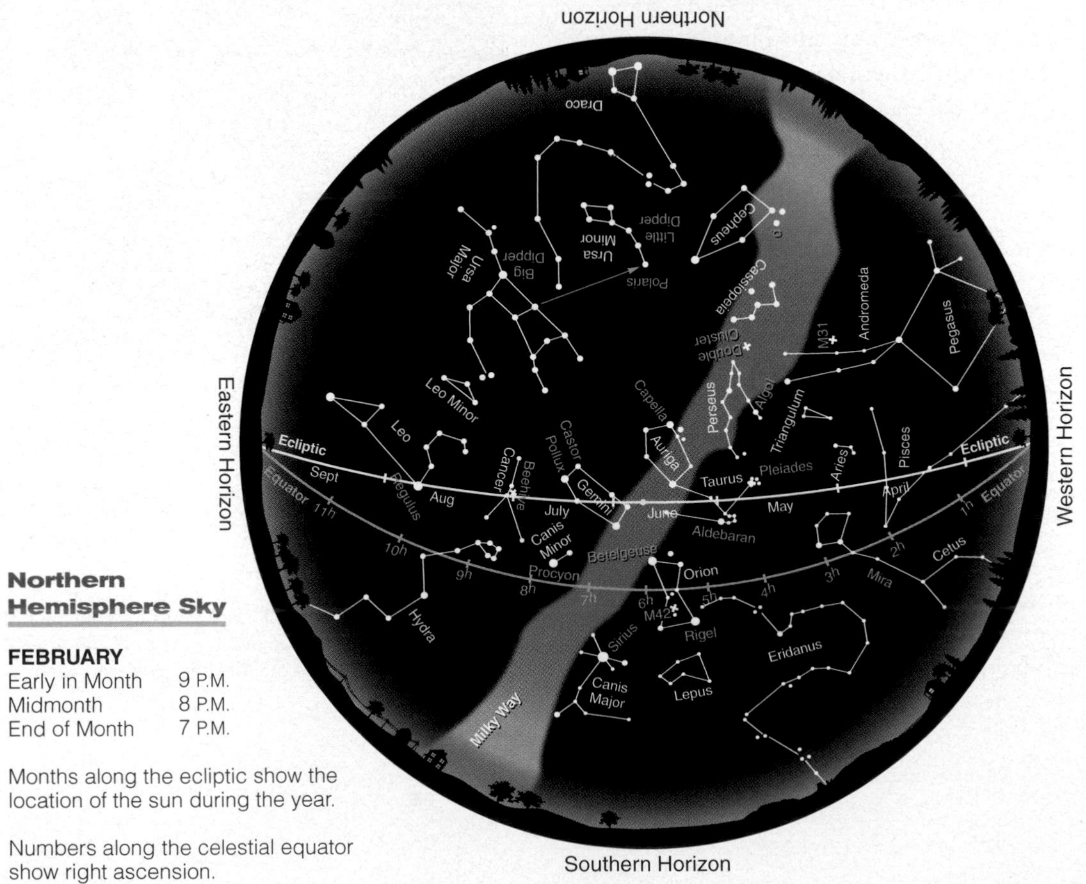

Northern Hemisphere Sky

FEBRUARY

Early in Month	9 P.M.
Midmonth	8 P.M.
End of Month	7 P.M.

Months along the ecliptic show the location of the sun during the year.

Numbers along the celestial equator show right ascension.

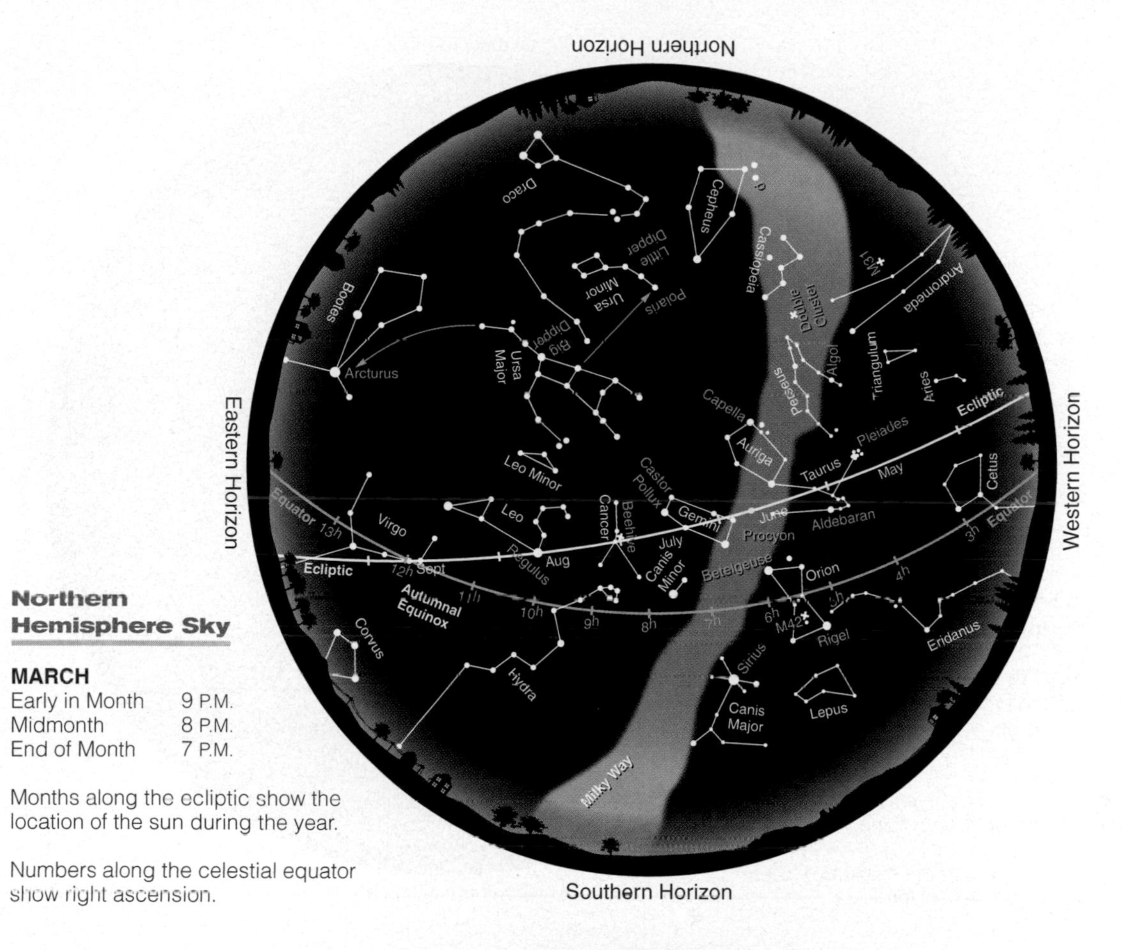

Northern Hemisphere Sky

MARCH

Early in Month	9 P.M.
Midmonth	8 P.M.
End of Month	7 P.M.

Months along the ecliptic show the location of the sun during the year.

Numbers along the celestial equator show right ascension.

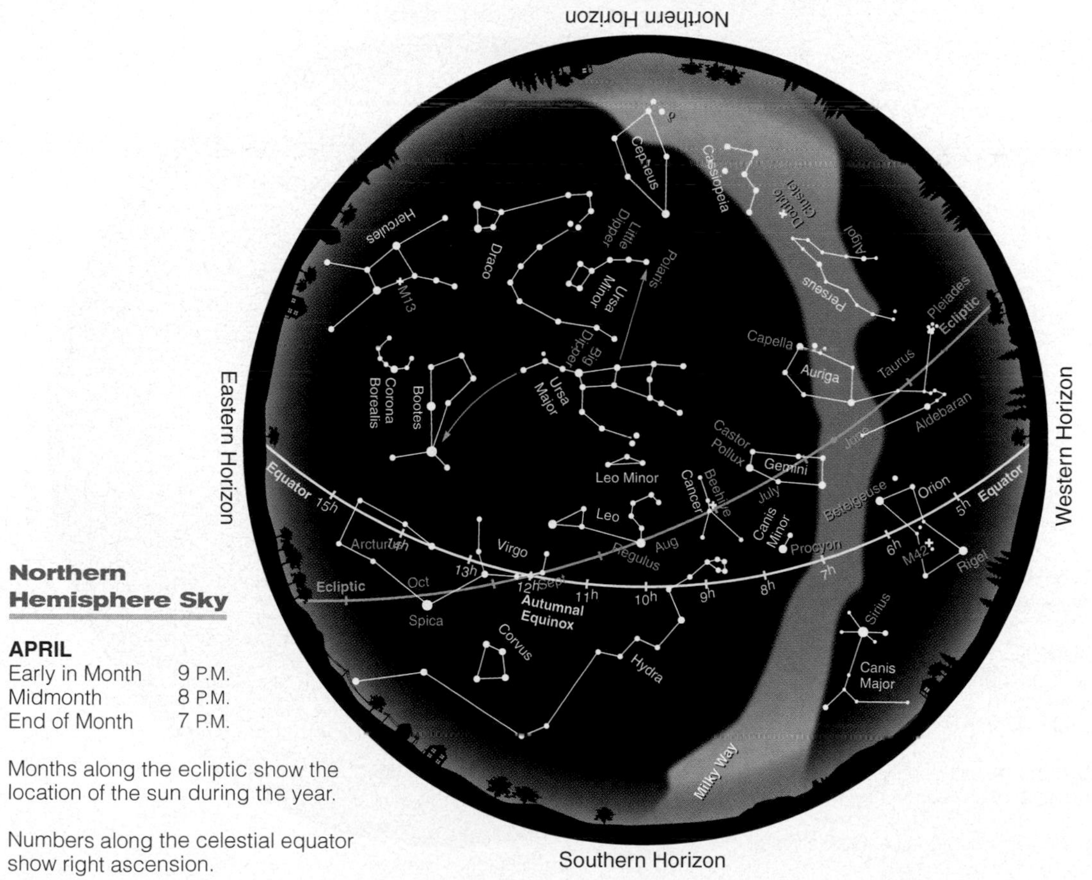

Northern Hemisphere Sky

APRIL

Early in Month	9 P.M.
Midmonth	8 P.M.
End of Month	7 P.M.

Months along the ecliptic show the location of the sun during the year.

Numbers along the celestial equator show right ascension.

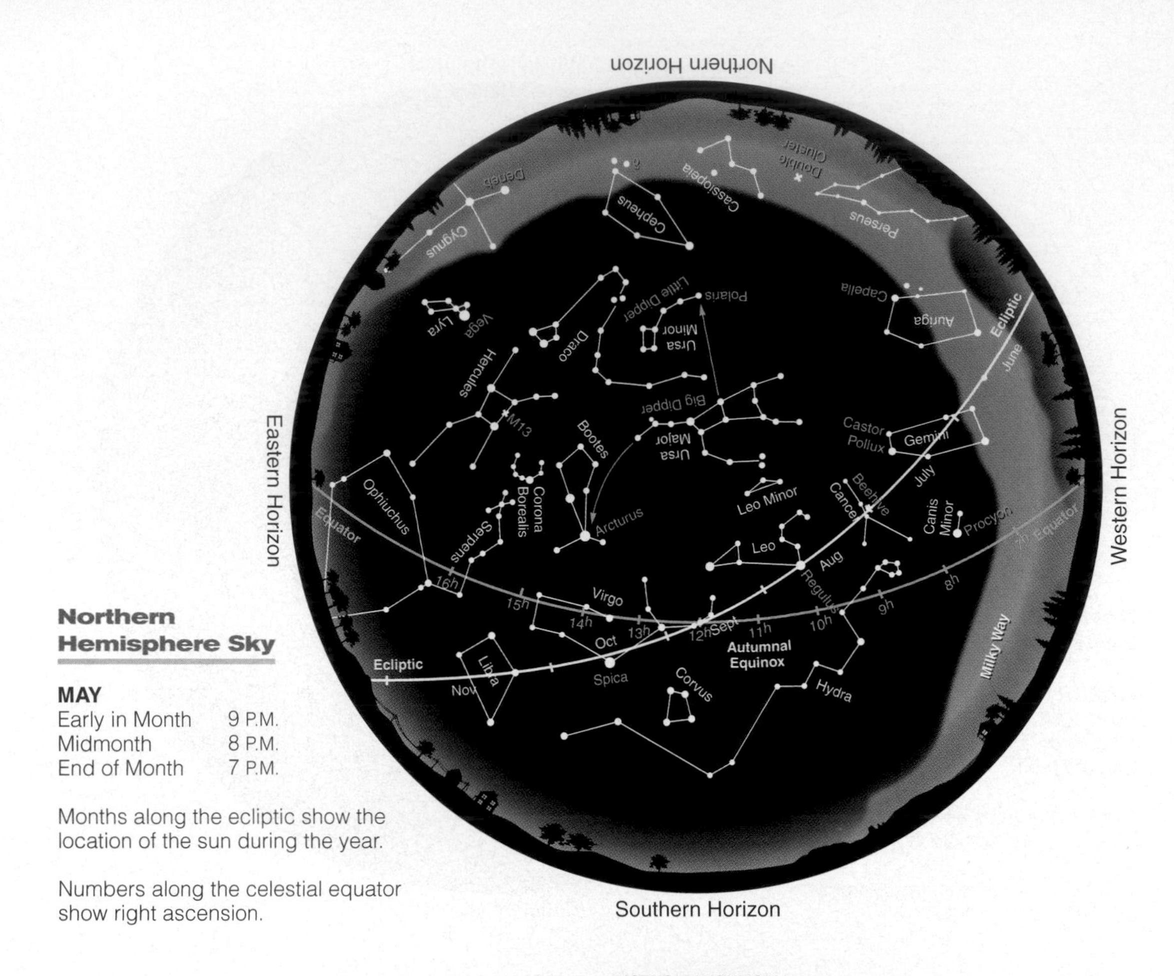

Northern Hemisphere Sky

MAY

Early in Month	9 P.M.
Midmonth	8 P.M.
End of Month	7 P.M.

Months along the ecliptic show the location of the sun during the year.

Numbers along the celestial equator show right ascension.

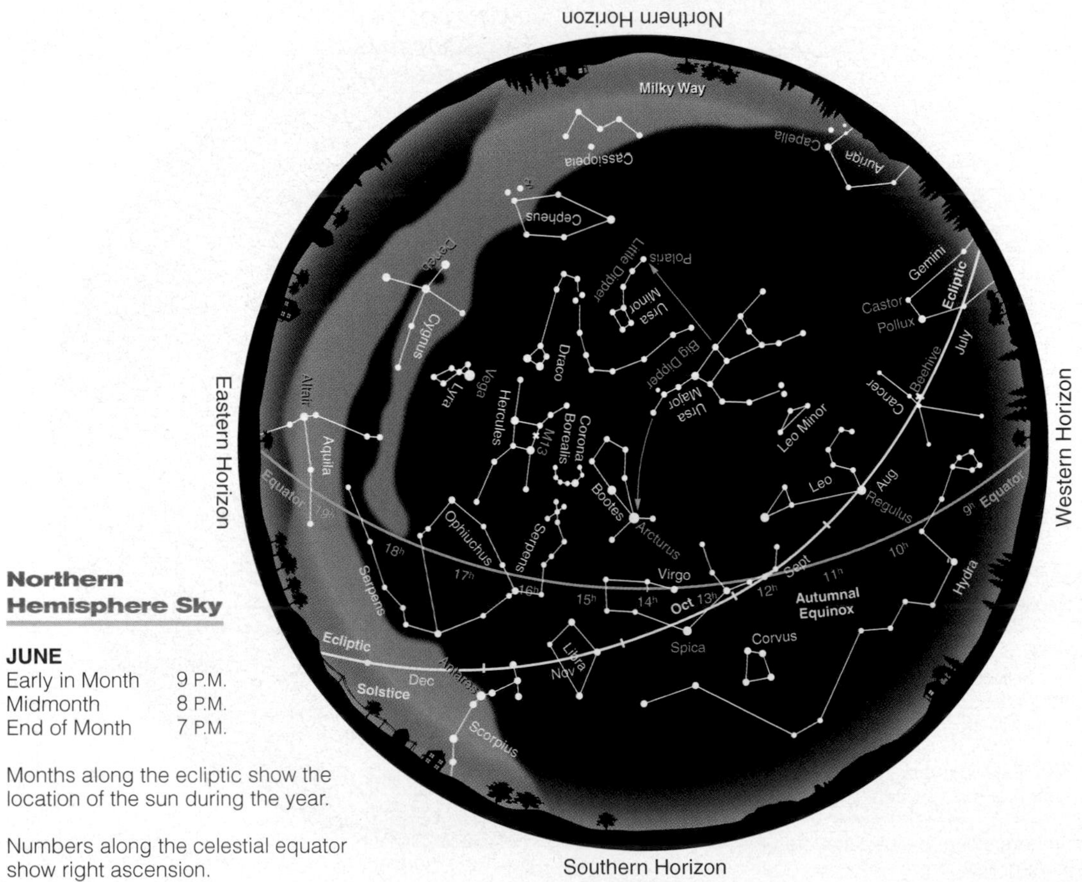

Northern Hemisphere Sky

JUNE

Early in Month	9 P.M.
Midmonth	8 P.M.
End of Month	7 P.M.

Months along the ecliptic show the location of the sun during the year.

Numbers along the celestial equator show right ascension.

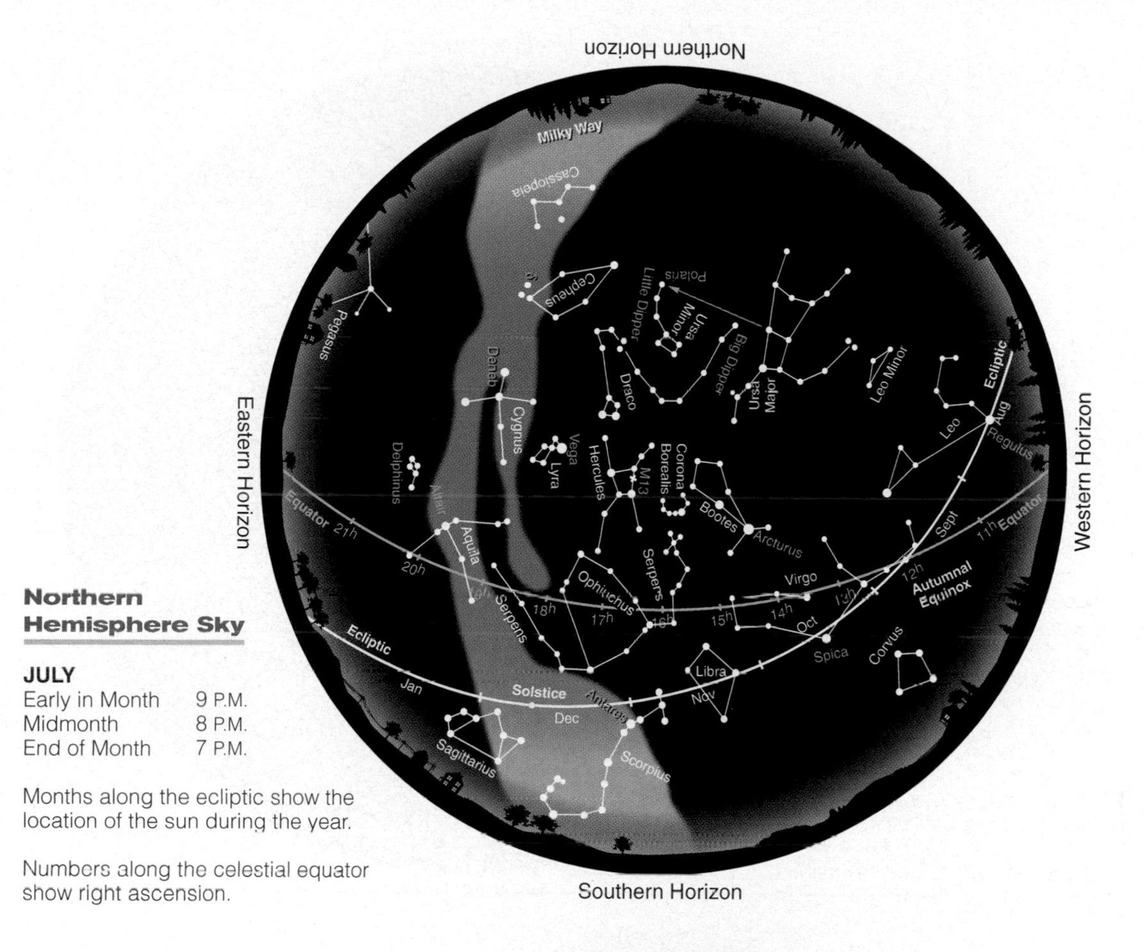

Northern Hemisphere Sky

JULY

Early in Month	9 P.M.
Midmonth	8 P.M.
End of Month	7 P.M.

Months along the ecliptic show the location of the sun during the year.

Numbers along the celestial equator show right ascension.

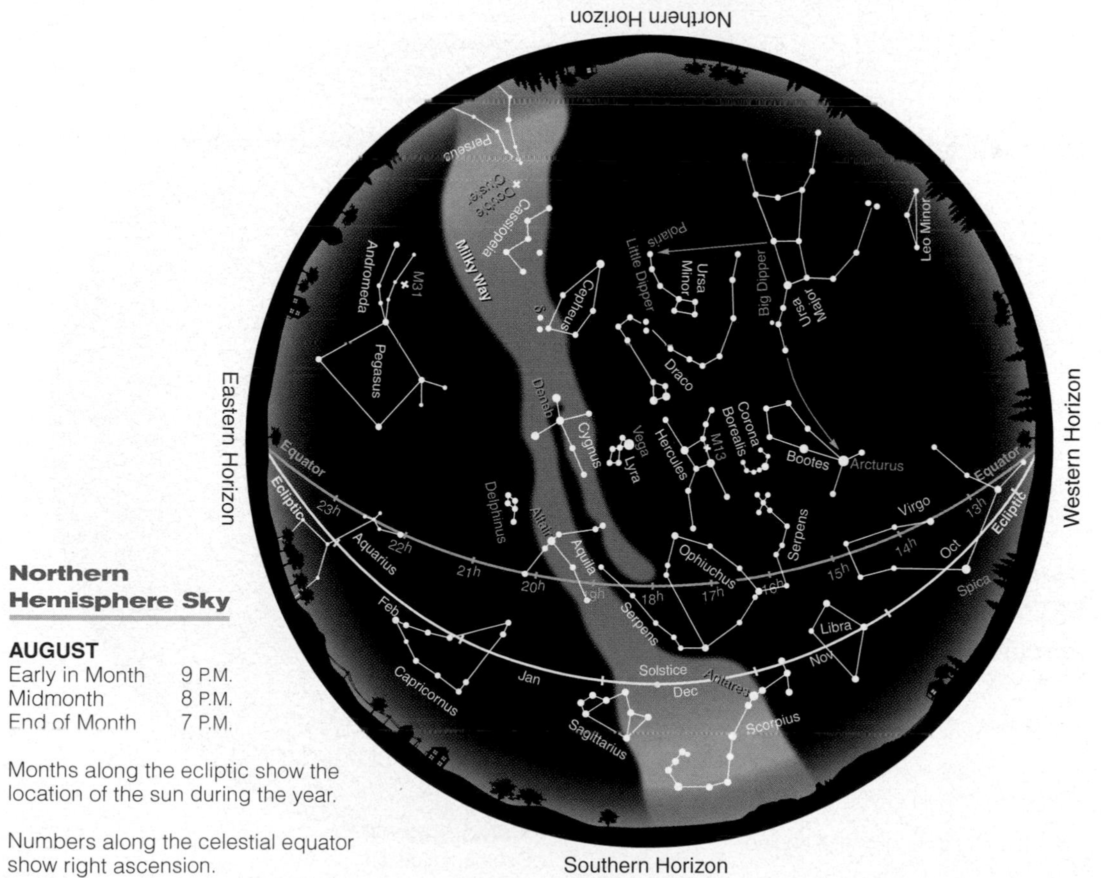

Northern Hemisphere Sky

AUGUST

Early in Month	9 P.M.
Midmonth	8 P.M.
End of Month	7 P.M.

Months along the ecliptic show the location of the sun during the year.

Numbers along the celestial equator show right ascension.

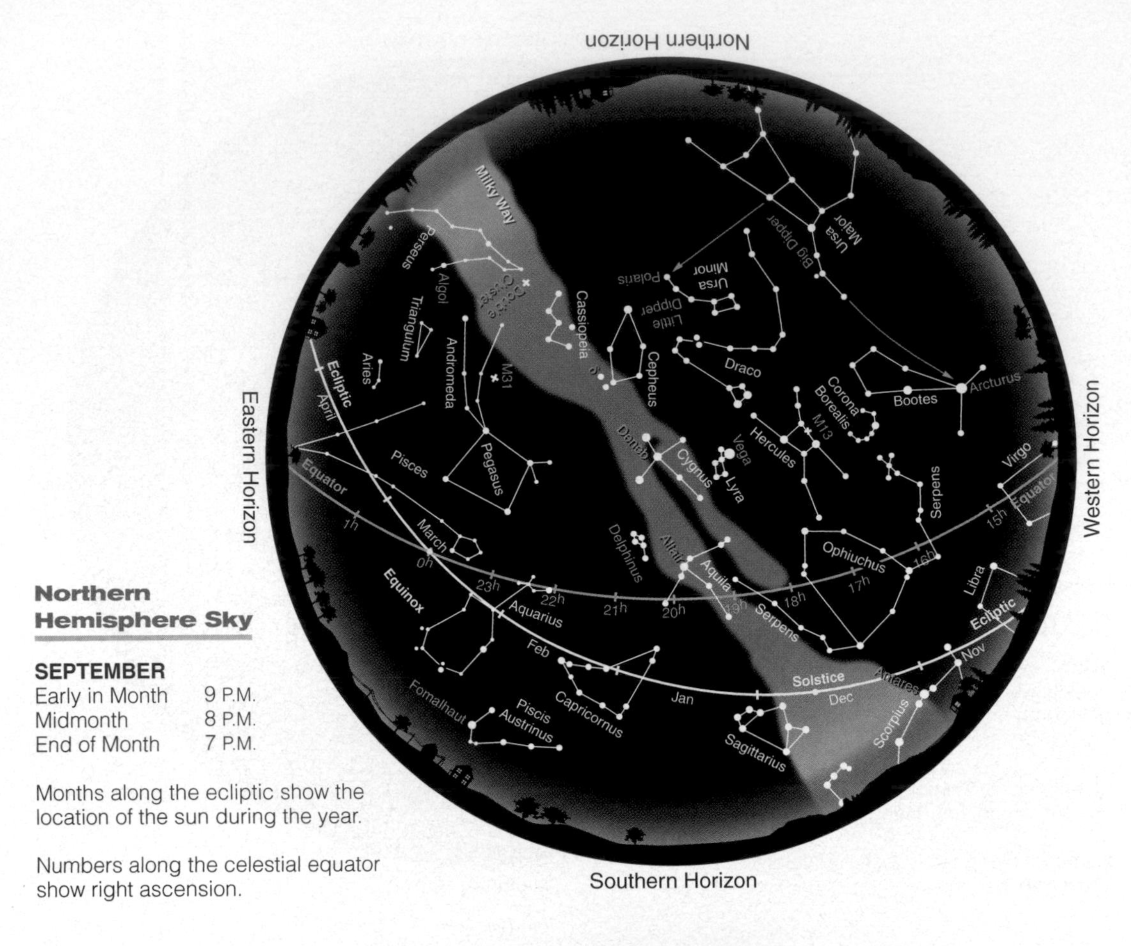

Northern Hemisphere Sky

SEPTEMBER

Early in Month	9 P.M.
Midmonth	8 P.M.
End of Month	7 P.M.

Months along the ecliptic show the location of the sun during the year.

Numbers along the celestial equator show right ascension.

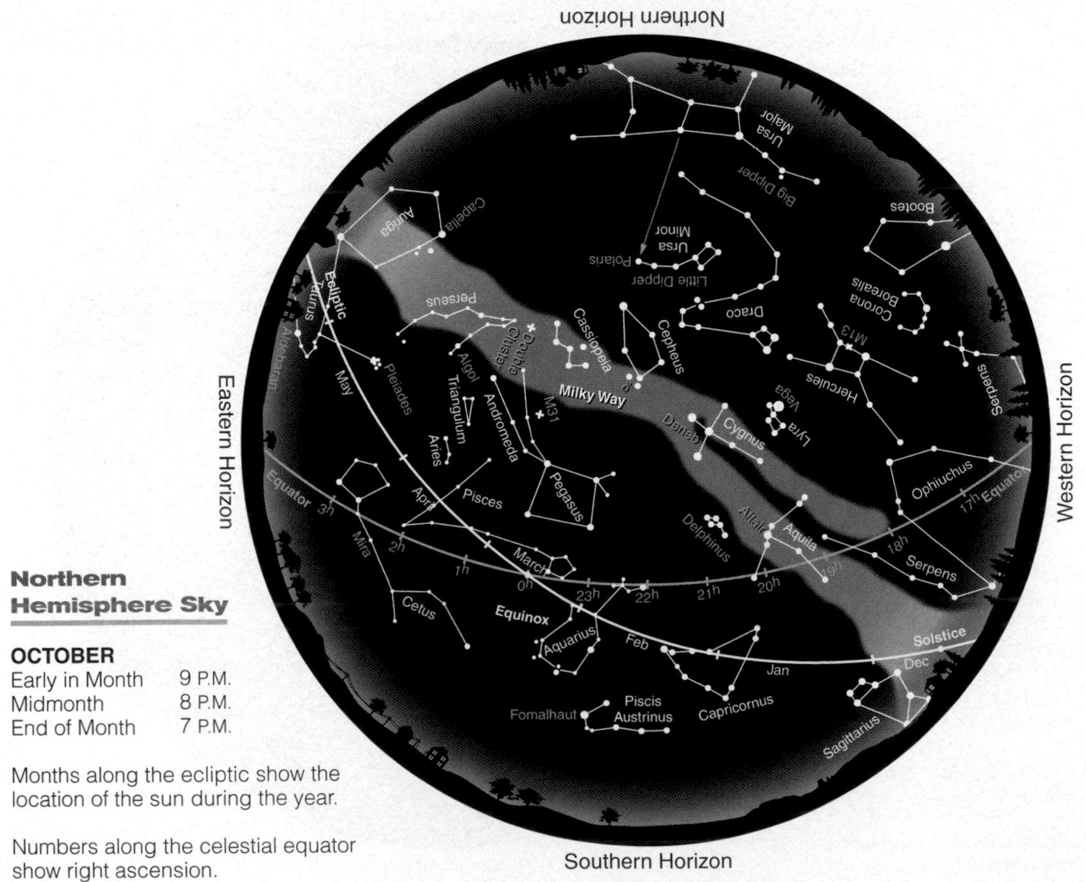

Northern Hemisphere Sky

OCTOBER

Early in Month	9 P.M.
Midmonth	8 P.M.
End of Month	7 P.M.

Months along the ecliptic show the location of the sun during the year.

Numbers along the celestial equator show right ascension.

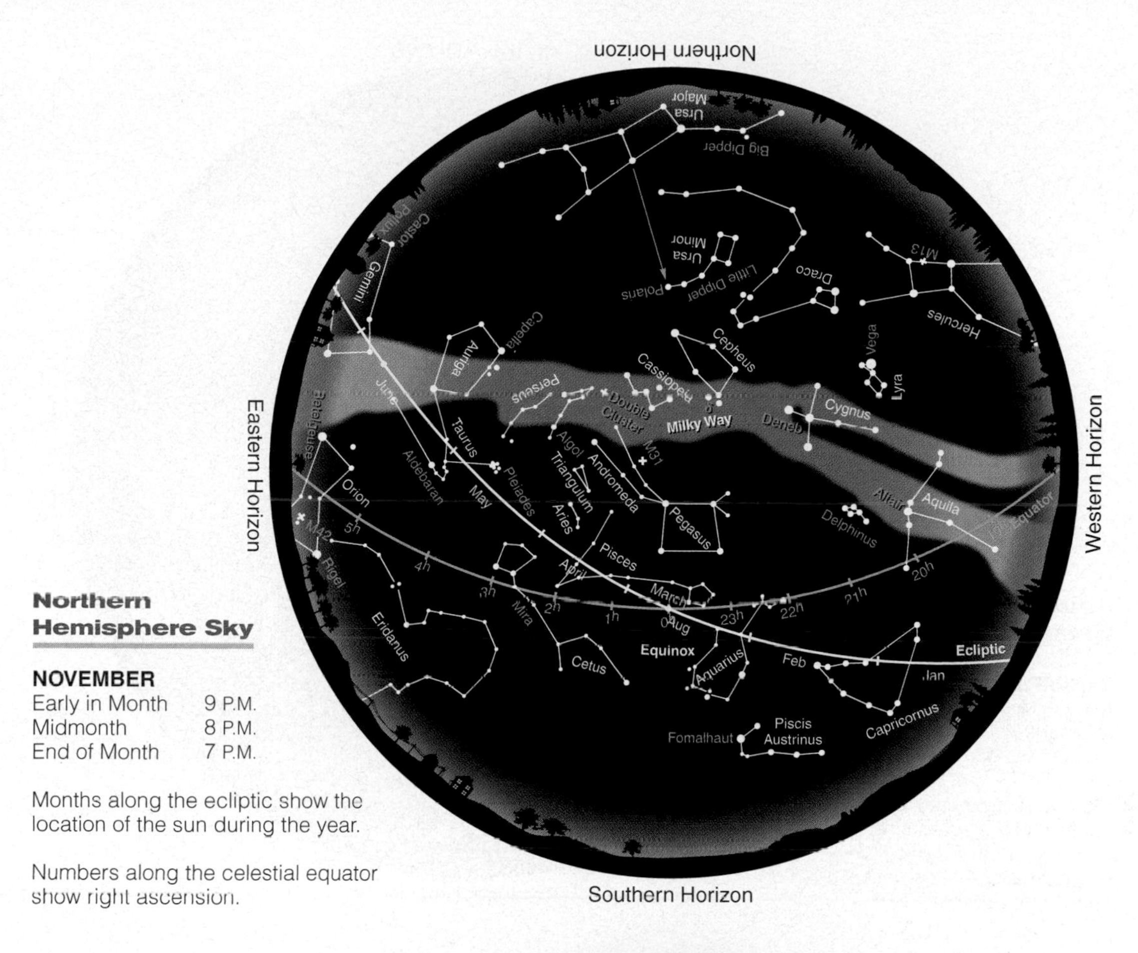

Northern Hemisphere Sky

NOVEMBER

Early in Month	9 P.M.
Midmonth	8 P.M.
End of Month	7 P.M.

Months along the ecliptic show the location of the sun during the year.

Numbers along the celestial equator show right ascension.

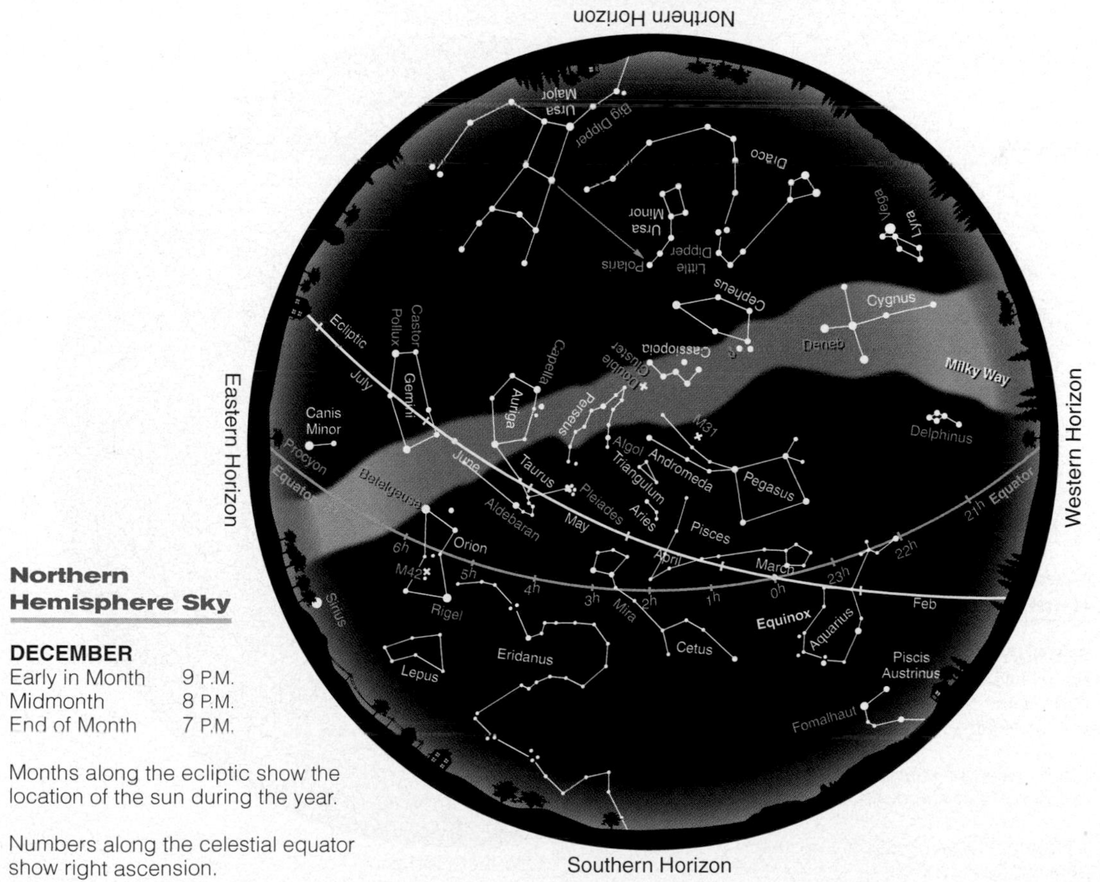

Northern Hemisphere Sky

DECEMBER

Early in Month	9 P.M.
Midmonth	8 P.M.
End of Month	7 P.M.

Months along the ecliptic show the location of the sun during the year.

Numbers along the celestial equator show right ascension.

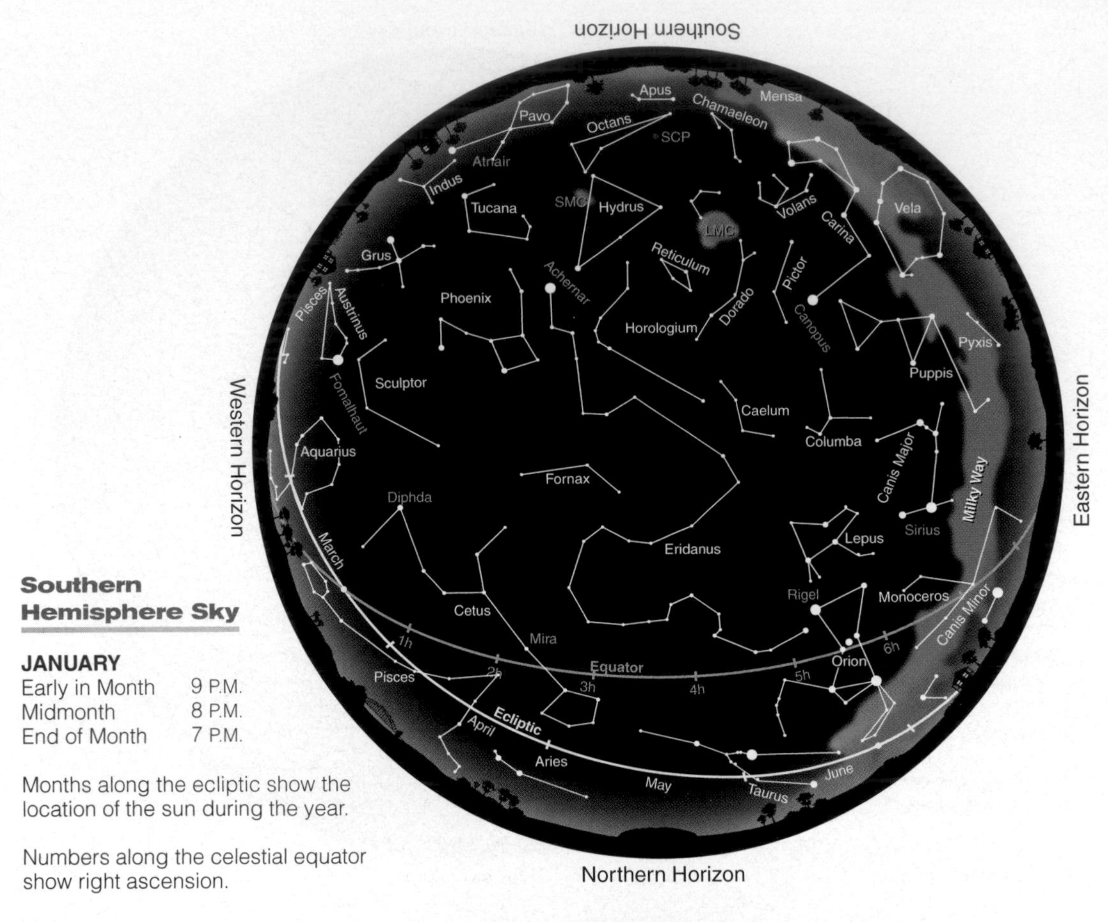

Southern Hemisphere Sky

JANUARY

Early in Month	9 P.M.
Midmonth	8 P.M.
End of Month	7 P.M.

Months along the ecliptic show the location of the sun during the year.

Numbers along the celestial equator show right ascension.

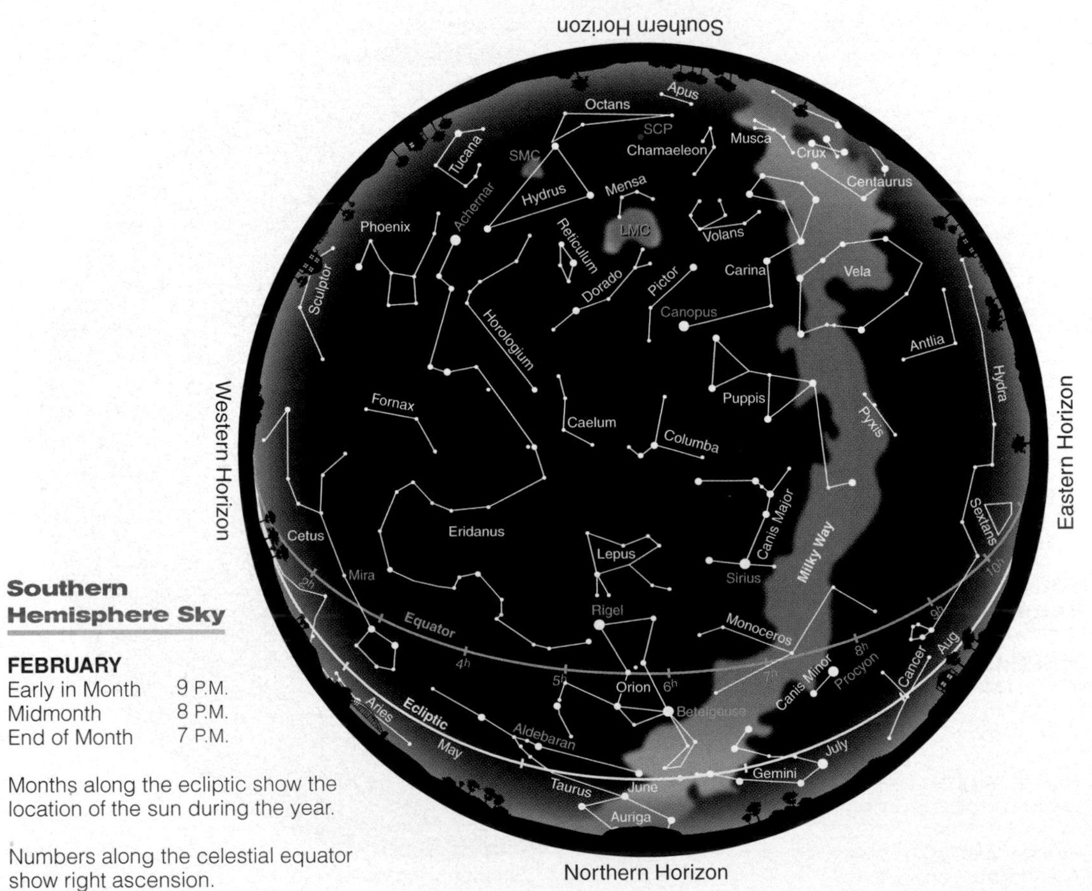

Southern Hemisphere Sky

FEBRUARY

Early in Month	9 P.M.
Midmonth	8 P.M.
End of Month	7 P.M.

Months along the ecliptic show the location of the sun during the year.

Numbers along the celestial equator show right ascension.

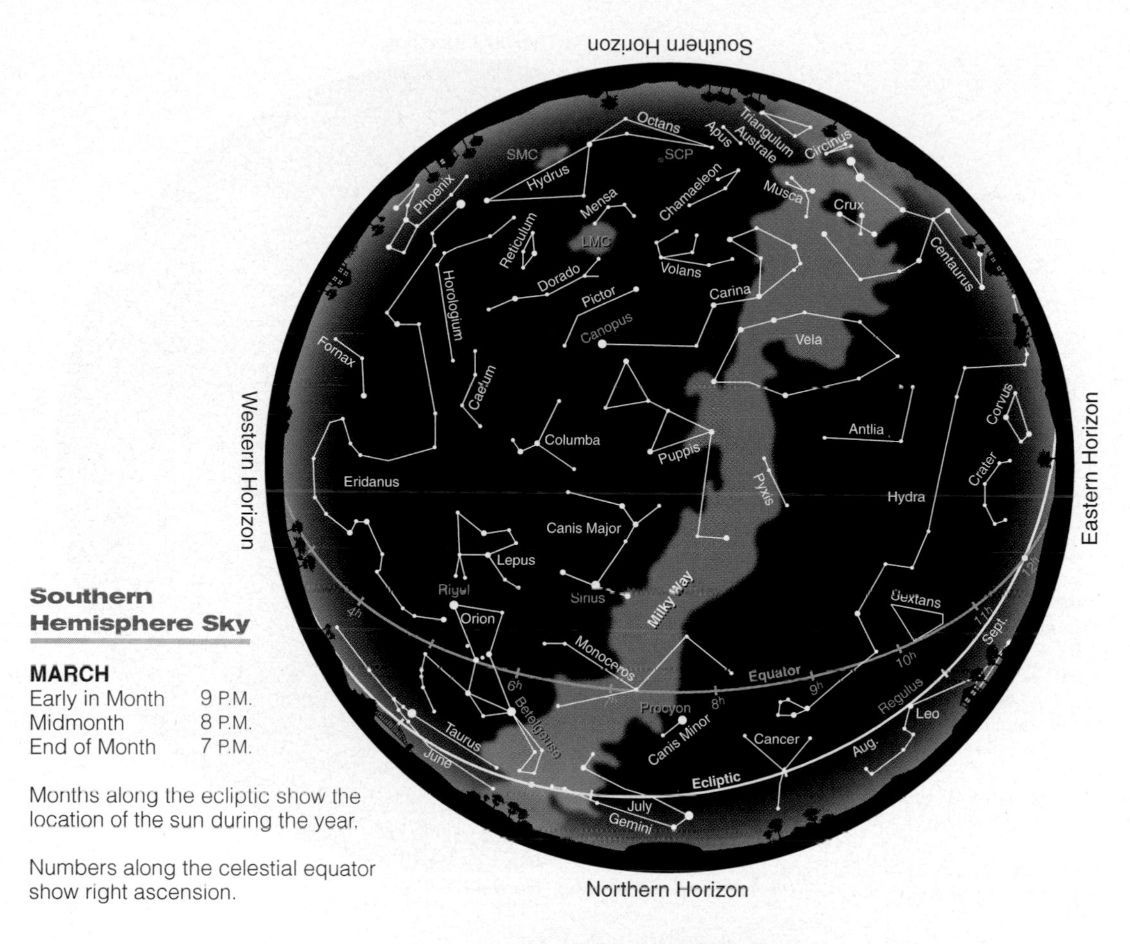

Southern Hemisphere Sky

MARCH

Early in Month	9 P.M.
Midmonth	8 P.M.
End of Month	7 P.M.

Months along the ecliptic show the location of the sun during the year.

Numbers along the celestial equator show right ascension.

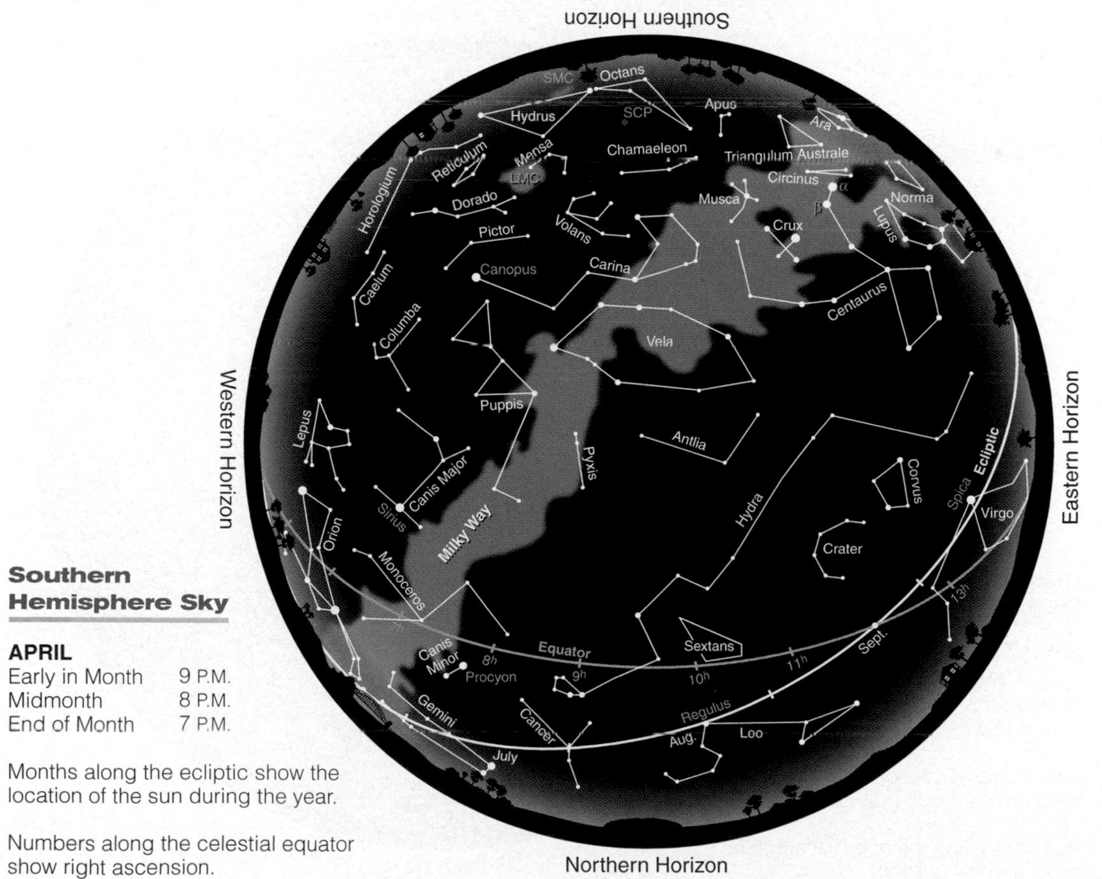

Southern Hemisphere Sky

APRIL

Early in Month	9 P.M.
Midmonth	8 P.M.
End of Month	7 P.M.

Months along the ecliptic show the location of the sun during the year.

Numbers along the celestial equator show right ascension.

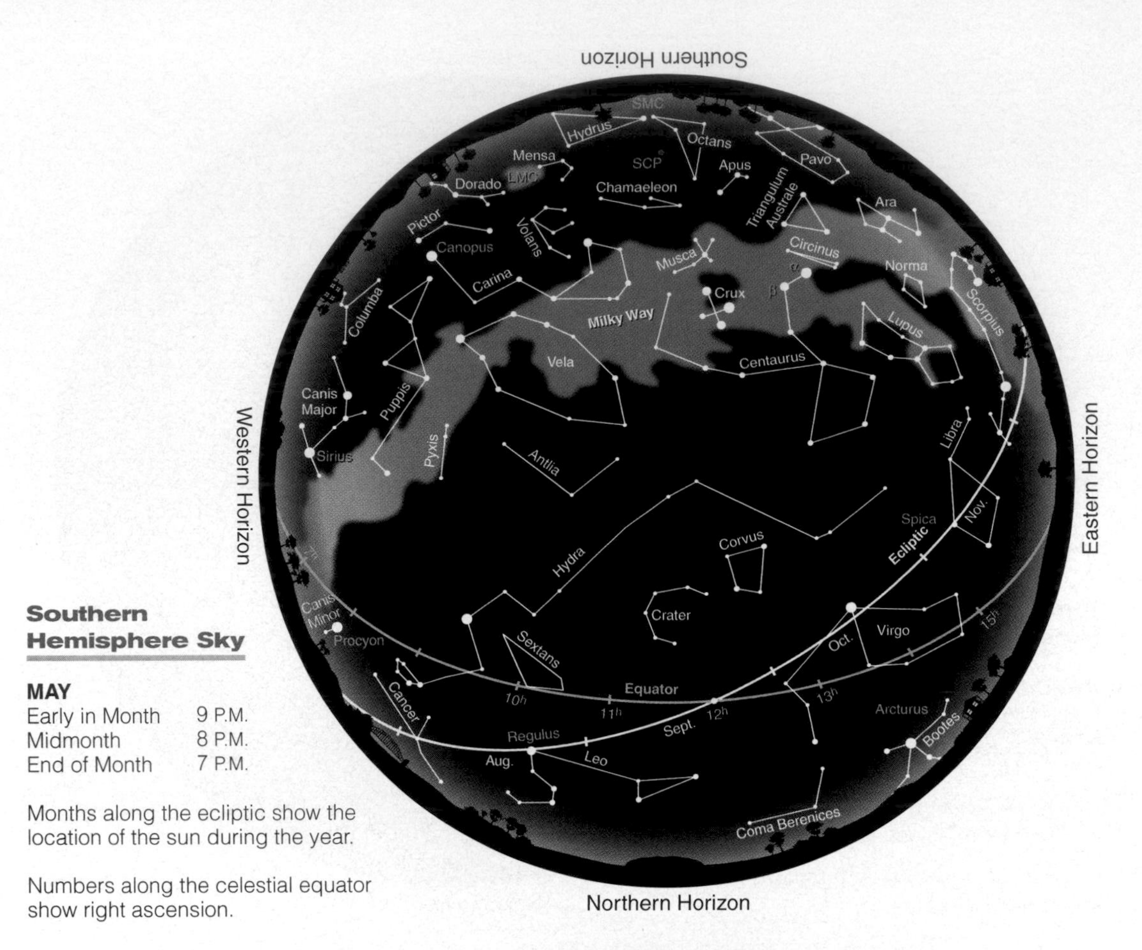

Southern Hemisphere Sky

MAY

Early in Month 9 P.M.
Midmonth 8 P.M.
End of Month 7 P.M.

Months along the ecliptic show the location of the sun during the year.

Numbers along the celestial equator show right ascension.

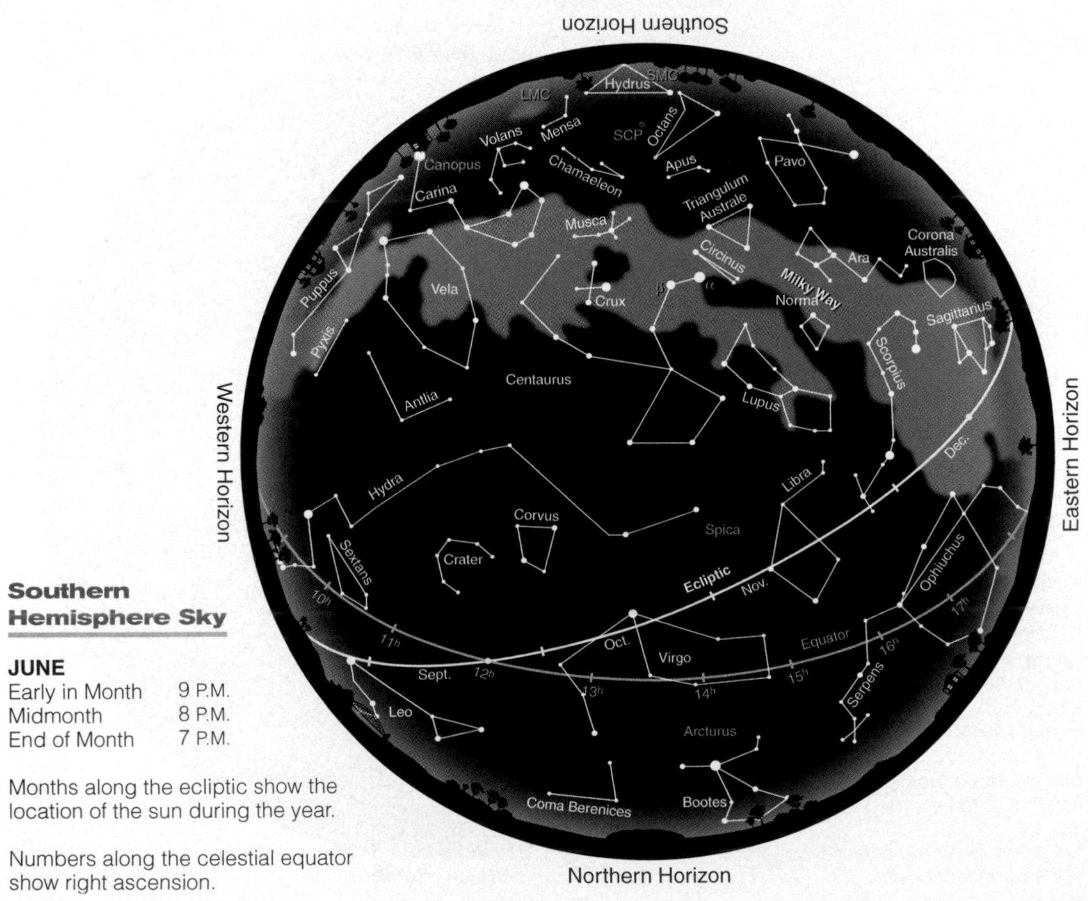

Southern Hemisphere Sky

JUNE

Early in Month 9 P.M.
Midmonth 8 P.M.
End of Month 7 P.M.

Months along the ecliptic show the location of the sun during the year.

Numbers along the celestial equator show right ascension.

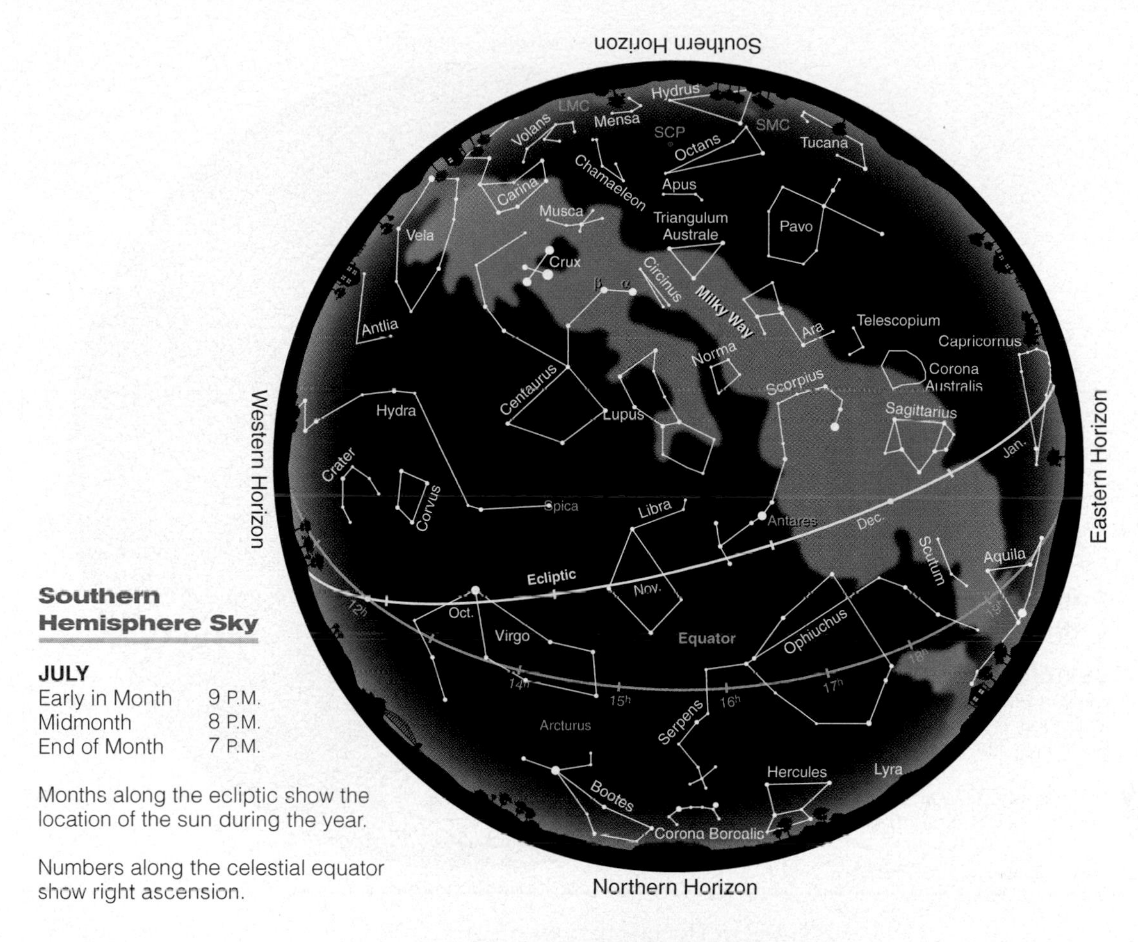

Southern Hemisphere Sky

JULY

Early in Month	9 P.M.
Midmonth	8 P.M.
End of Month	7 P.M.

Months along the ecliptic show the location of the sun during the year.

Numbers along the celestial equator show right ascension.

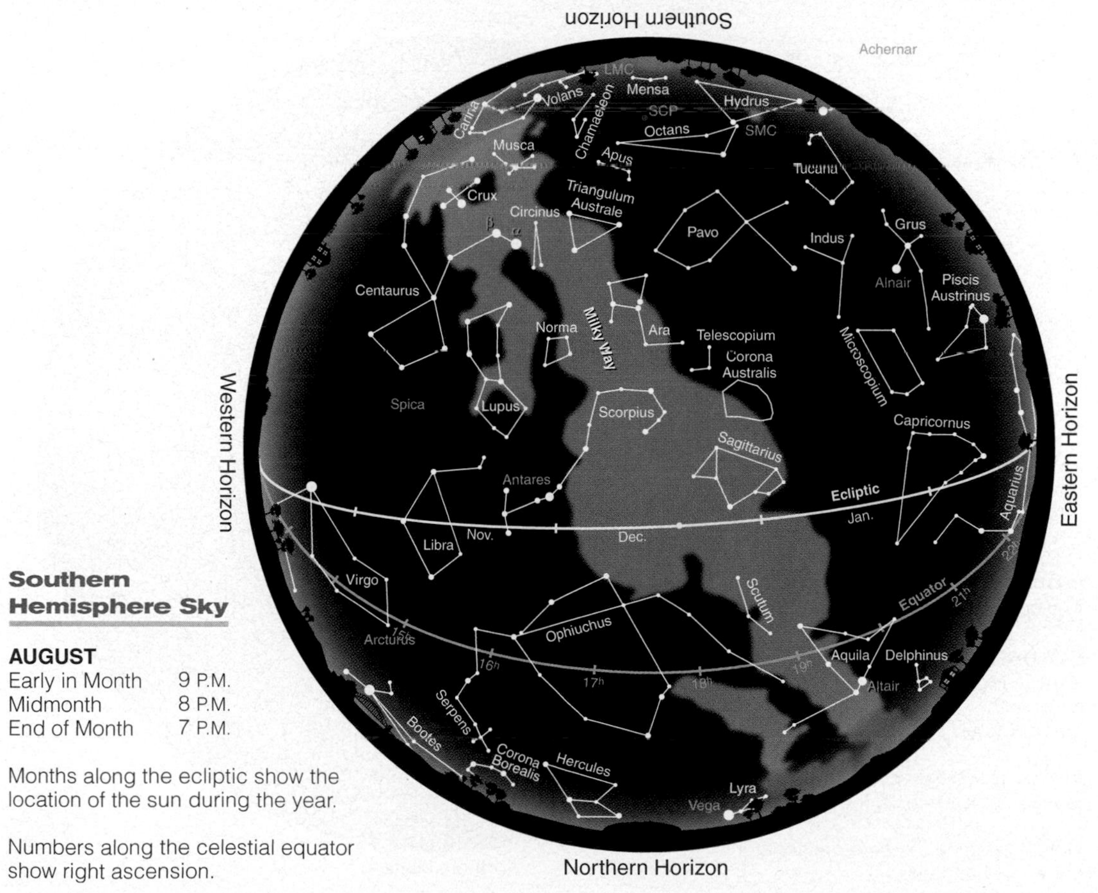

Southern Hemisphere Sky

AUGUST

Early in Month	9 P.M.
Midmonth	8 P.M.
End of Month	7 P.M.

Months along the ecliptic show the location of the sun during the year.

Numbers along the celestial equator show right ascension.

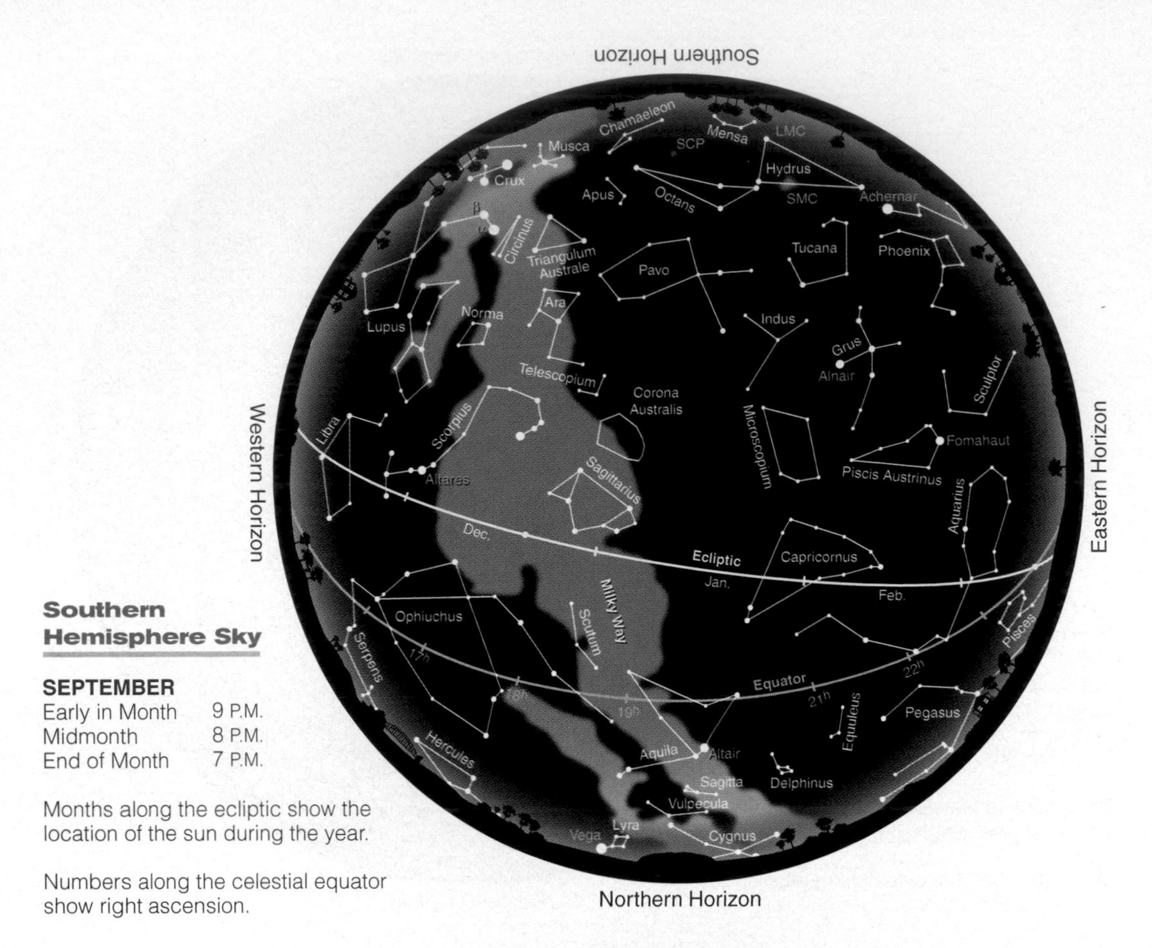

Southern Hemisphere Sky

SEPTEMBER

Early in Month	9 P.M.
Midmonth	8 P.M.
End of Month	7 P.M.

Months along the ecliptic show the location of the sun during the year.

Numbers along the celestial equator show right ascension.

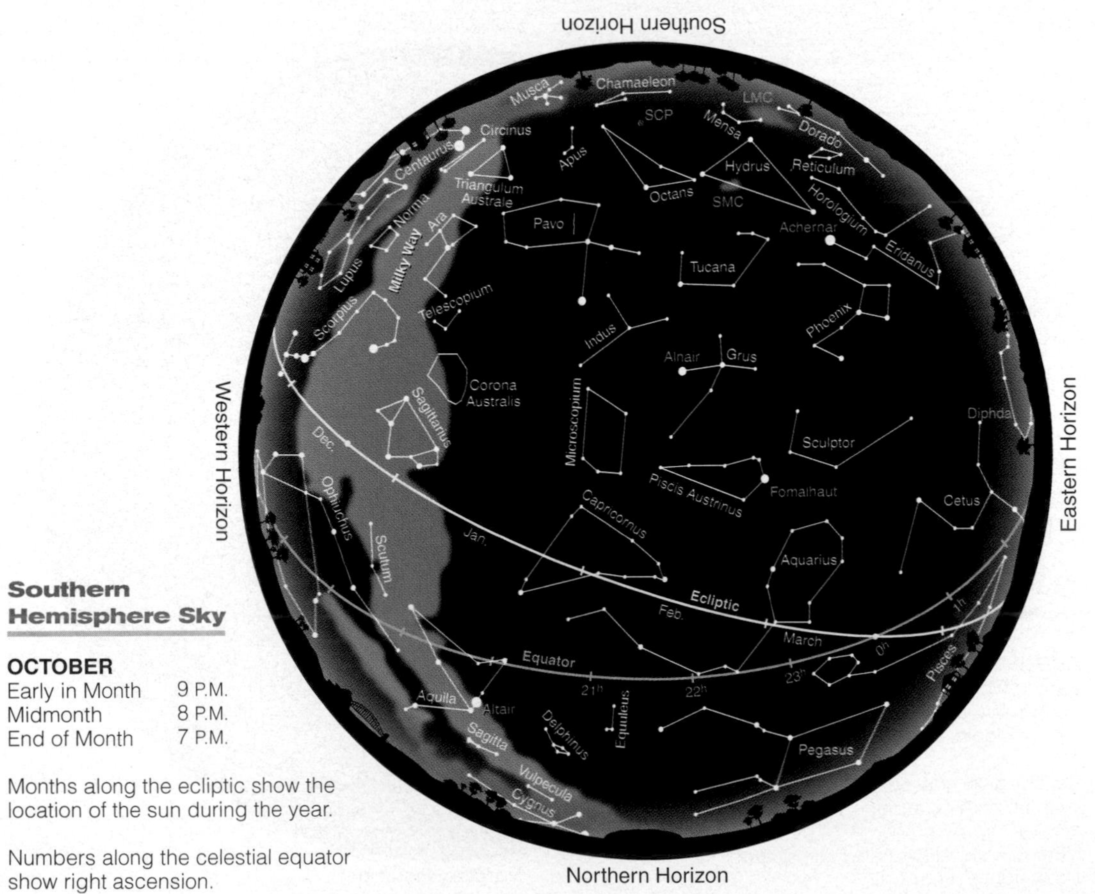

Southern Hemisphere Sky

OCTOBER

Early in Month	9 P.M.
Midmonth	8 P.M.
End of Month	7 P.M.

Months along the ecliptic show the location of the sun during the year.

Numbers along the celestial equator show right ascension.

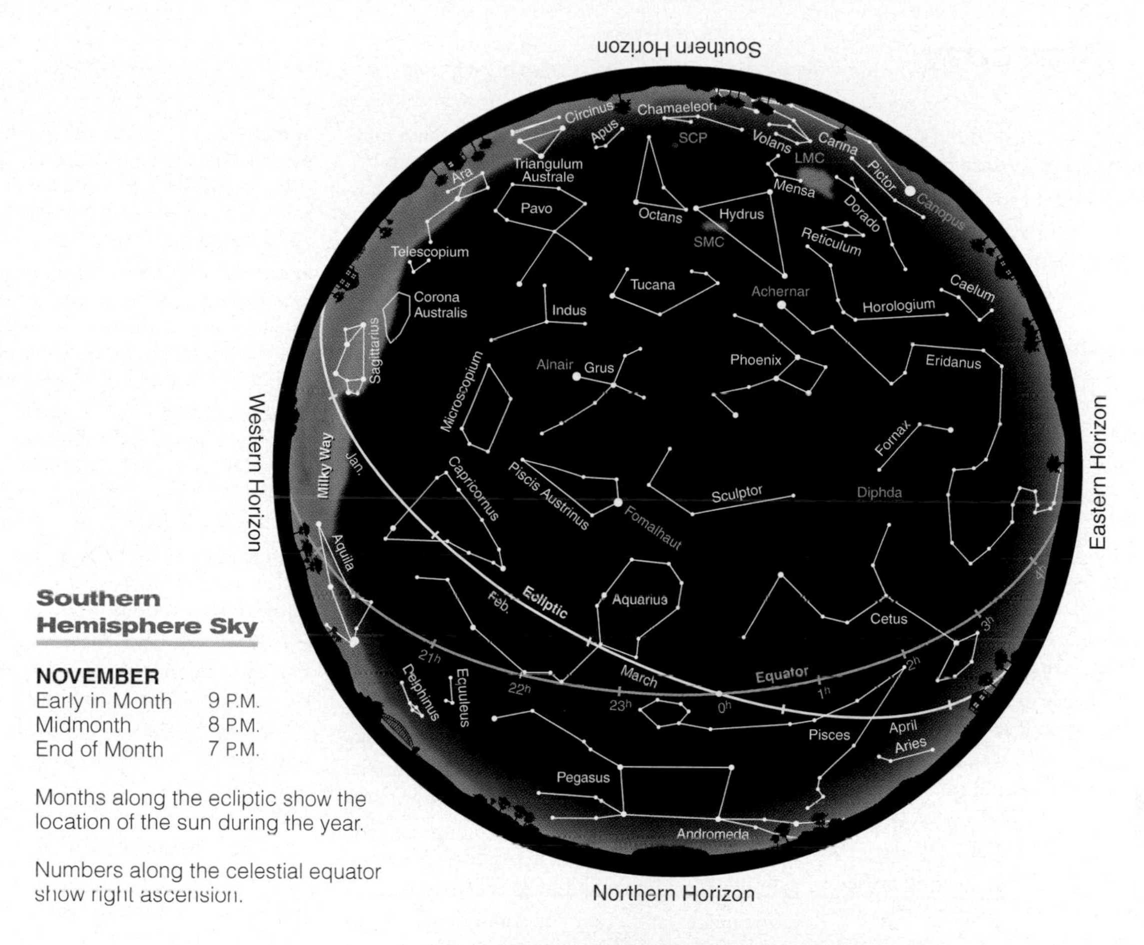

Southern Hemisphere Sky

NOVEMBER

Early in Month 9 P.M.
Midmonth 8 P.M.
End of Month 7 P.M.

Months along the ecliptic show the location of the sun during the year.

Numbers along the celestial equator show right ascension.

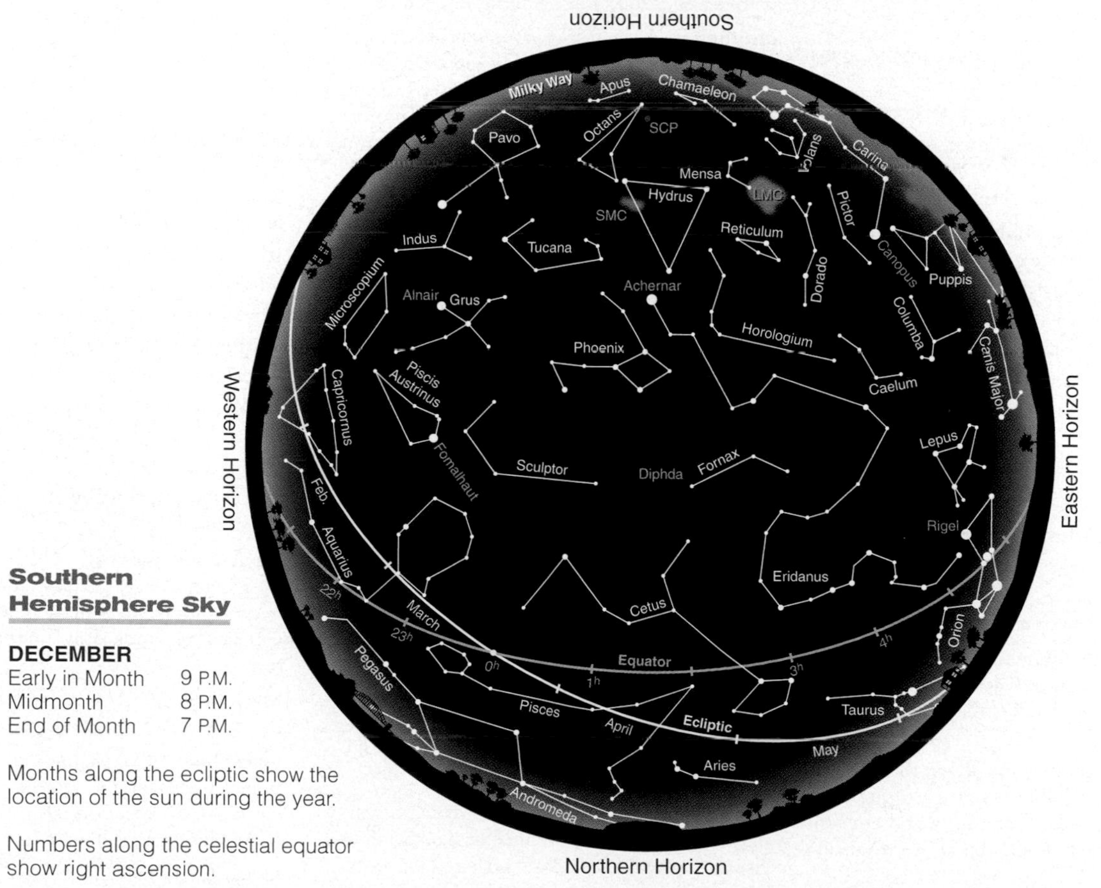

Southern Hemisphere Sky

DECEMBER

Early in Month 9 P.M.
Midmonth 8 P.M.
End of Month 7 P.M.

Months along the ecliptic show the location of the sun during the year.

Numbers along the celestial equator show right ascension.

Glossary

Numbers in parentheses refer to the page where the term is first discussed in the text.

absolute age (462) An age determined in years, as from radioactive dating (see also *relative age*).

absolute bolometric magnitude (180) The absolute magnitude you would observe if you could detect all wavelengths.

absolute visual magnitude (M_V) (179) Intrinsic brightness of a star; the apparent visual magnitude the star would have if it were 10 pc away.

absolute zero (132) The lowest possible temperature; the temperature at which the particles in a material, atoms or molecules, contain no energy of motion that can be extracted from the body.

absorption line (136) A dark line in a spectrum; produced by the absence of photons absorbed by atoms or molecules.

absorption spectrum (dark-line spectrum) (136) A spectrum that contains absorption lines.

acceleration (82) A change in a velocity; a change in either speed or direction. (See *velocity*.)

acceleration of gravity (80) A measure of the strength of gravity at a planet's surface.

accretion (428) The sticking together of solid particles to produce a larger particle.

accretion disk (273) The whirling disk of gas that forms around a compact object such as a white dwarf, neutron star, or black hole as matter is drawn in.

achondrite (575) Stony meteorite containing no chondrules or volatiles.

achromatic lens (107) A telescope lens composed of two lenses ground from different kinds of glass and designed to bring two selected colors to the same focus and correct for chromatic aberration.

active galactic nucleus (AGN) (368) The central energy source of an active galaxy.

active galaxy (368) A galaxy that is a source of excess radiation, usually radio waves, X rays, gamma rays, or some combination.

active optics (113) Optical elements whose position or shape is continuously controlled by computers.

active region (163) An area on the sun where sunspots, prominences, flares, and the like occur.

adaptive optics (113) Computer-controlled telescope mirrors that can at least partially compensate for seeing.

albedo (455) The fraction of the light hitting an object that is reflected.

alt-azimuth mounting (113) A telescope mounting capable of motion parallel to and perpendicular to the horizon.

Amazonian period (505) On Mars, the geological era from about 3 billion years ago to the present marked by low-level cratering, wind erosion, and small amounts of water seeping from subsurface ice.

amino acid (600) One of the carbon-chain molecules that are the building blocks of protein.

Angstrom (Å) (104) A unit of distance; 1 Å = 10^{-10} m; often used to measure the wavelength of light.

angular diameter (21) A measure of the size of an object in the sky; numerically equal to the angle in degrees between two lines extending from the observer's eye to opposite edges of the object.

angular distance (21) A measure of the separation between two objects in the sky; numerically equal to the angle in degrees between two lines extending from the observer's eye to the two objects.

angular momentum (87) The tendency of a rotating body to continue rotating; mathematically, the product of mass, velocity, and radius.

angular momentum problem (418) An objection to Laplace's nebular hypothesis that cited the slow rotation of the sun.

annular eclipse (42) A solar eclipse in which the solar photosphere appears around the edge of the moon in a bright ring, or annulus. The corona, chromosphere, and prominences cannot be seen.

anomalous X-ray pulsars (295) Highly magnetic neutron stars (magnetars) that emit X rays but spin slowly with periods of 5 to 10 seconds.

anorthosite (468) Rock of aluminum and calcium silicates found in the lunar highlands.

antimatter (397) Matter composed of antiparticles, which on colliding with a matching particle of normal matter annihilate and convert the mass of both particles into energy. The antiproton is the antiparticle of the proton, and the positron is the antiparticle of the electron.

aphelion (25) The orbital point of greatest distance from the sun.

apogee (42) The orbital point of greatest distance from Earth.

Apollo–Amor object (581) Asteroid whose orbit crosses that of Earth (Apollo) and Mars (Amor).

apparent visual magnitude (m_v) (16) The brightness of a star as seen by human eyes on Earth.

archaeoastronomy (53) The study of the astronomy of ancient cultures.

association (226) Group of widely scattered stars (10 to 1000) moving together through space; not gravitationally bound into a cluster.

asterism (14) A named group of stars not identified as a constellation, e.g., the Big Dipper.

asteroid (420) Small rocky world; most asteroids lie between Mars and Jupiter in the asteroid belt.

astronomical unit (AU) (4) Average distance from Earth to the sun; 1.5×10^8 km, or 93×10^6 miles.

atmospheric window (105) Wavelength regions in which Earth's atmosphere is transparent—at visual, infrared, and radio wavelengths.

aurora (169) The glowing light display that results when a planet's magnetic field guides charged particles toward the north and south magnetic poles, where they strike the upper atmosphere and excite atoms to emit photons.

autumnal equinox (24) The point on the celestial sphere where the sun crosses the celestial equator going southward. Also, the time when the sun reaches this point and autumn begins in the Northern Hemisphere—about September 22.

Babcock model (164) A model of the sun's magnetic cycle in which the differential rotation of the sun winds up and tangles the solar magnetic field in a 22-year cycle. This is thought to be responsible for the 11-year sunspot cycle.

Balmer series (137) Spectral lines in the visible and near-ultraviolet spectrum of hydrogen produced by transitions whose lowest orbit is the second.

barred spiral galaxy (346) A spiral galaxy with an elongated nucleus resembling a bar from which the arms originate.

basalt (452) Dark, igneous rock characteristic of solidified lava.

belt (519) One of the dark bands of clouds that circle Jupiter parallel to its equator; generally red, brown, or blue-green.

belt–zone circulation (511) The atmospheric circulation typical of Jovian planets. Dark belts and bright zones encircle the planet parallel to its equator.

big bang (394) The theory that the universe began with a violent explosion from which the expanding universe of galaxies eventually formed.

big rip (408) The possible fate of the universe if dark energy increases rapidly and the expansion of space-time pulls galaxies, stars, and ultimately atoms apart.

binary star (187) One of a pair of stars that orbit around their common center of mass.

binding energy (129) The energy needed to pull an electron away from its atom.

bipolar flow (229) Oppositely directed jets of gas ejected by some protostellar objects.

birth line (225) In the H–R diagram, the line above the main sequence where protostars first become visible.

black body radiation (132) Radiation emitted by a hypothetical perfect radiator; the spectrum is continuous, and the wavelength of maximum emission depends only on the body's temperature.

black dwarf (270) The end state of a white dwarf that has cooled to low temperature.

black hole (300) A mass that has collapsed to such a small volume that its gravity prevents the escape of all radiation; also, the volume of space from which radiation may not escape.

blazar (374) See *BL Lac object.*

BL Lac object (374) Object that resembles a quasar; thought to be the highly luminous core of a distant galaxy emitting a jet almost directly toward Earth.

blueshift (142) The shortening of the wavelengths of light observed when the source and observer are approaching each other.

Bok globule (228) Small, dark cloud only about 1 ly in diameter that contains 10 to 1000 $M_\odot$ of gas and dust; thought to be related to star formation.

bow shock (450) The boundary between the undisturbed solar wind and the region being deflected around a planet or comet.

breccia (468) A rock composed of fragments of earlier rocks bonded together.

bright-line spectrum (136) See *emission spectrum.*

brown dwarf (245) A very cool, low-luminosity star whose mass is not sufficient to ignite nuclear fusion.

butterfly diagram (162) See *Maunder butterfly diagram.*

CAI (575) Calcium–aluminum-rich inclusions found in some meteorites.

Cambrian explosion (605) The sudden appearance of complex life forms at the beginning of the Cambrian period 0.6 to 0.5 billion years ago. Cambrian rocks contain the oldest easily identifiable fossils.

capture hypothesis (472) The theory that the moon formed elsewhere in the solar system and was later captured by Earth.

carbonaceous chondrite (574) Stony meteorite that contains both chondrules and volatiles. These may be the least altered remains of the solar nebula still present in the solar system.

carbon deflagration (280) The process in which the carbon in a white dwarf is completely consumed by nuclear fusion, producing a type Ia supernova explosion.

carbon–nitrogen–oxygen (CNO) cycle (231) A series of nuclear reactions that use carbon as a catalyst to combine four hydrogen atoms to make one helium atom plus energy; effective in stars more massive than the sun.

Cassegrain focus (112) The optical design of a reflecting telescope in which the secondary mirror reflects light back down the tube through a hole in the center of the objective mirror.

catastrophic hypothesis (417) Explanation for natural processes that depends on dramatic and unlikely events, such as the collision of two stars to produce our solar system.

celestial equator (20) The imaginary line around the sky directly above Earth's equator.

celestial sphere (20) An imaginary sphere of very large radius surrounding Earth to which the planets, stars, sun, and moon seem to be attached.

center of mass (89) The balance point of a body or system of bodies.

Cepheid variable star (258) Variable star with a period of 1 to 60 days; the period of variation is related to luminosity.

Chandrasekhar limit (270) The maximum mass of a white dwarf, about 1.4 $M_\odot$; a white dwarf of greater mass cannot support itself and will collapse.

charge-coupled device (CCD) (115) An electronic device consisting of a large array of light-sensitive elements used to record very faint images.

chemical evolution (603) The chemical process that led to the growth of complex molecules on the primitive Earth. This did not involve the reproduction of molecules.

chondrite (574) A stony meteorite that contains chondrules.

chondrule (574) Round, glassy body in some stony meteorites; thought to have solidified very quickly from molten drops of silicate material.

chromatic aberration (107) A distortion found in refracting telescopes because lenses focus different colors at slightly different distances. Images are consequently surrounded by color fringes.

chromosome (601) One of the bodies in a cell that contains the DNA carrying genetic information.

chromosphere (44) Bright gases just above the photosphere of the sun.

circular velocity (88) The velocity required to remain in a circular orbit about a body.

circumpolar constellation (21) Any of the constellations so close to the celestial pole that they never set (or never rise) as seen from a given latitude.

closed orbit (89) An orbit that returns to its starting point; a circular or elliptical orbit. (See *open orbit.*)

closed universe (401) A model universe in which the average density is great enough to stop the expansion and make the universe contract.

cluster method (353) The method of determining the masses of galaxies based on the motions of galaxies in a cluster.

CNO cycle (231) See *carbon–nitrogen–oxygen cycle.*

cocoon (224) The cloud of gas and dust around a contracting protostar that conceals it at visible wavelengths.

cold dark matter (403) Invisible matter in the universe composed by heavy, slow-moving particles such as WIMPs.

collapsar (309) See *hypernova.*

collisional broadening (144) The smearing out of a spectral line because of collisions among the atoms of the gas.

coma (586) The glowing head of a comet.

comet (421) One of the small, icy bodies that orbit the sun and produce tails of gas and dust when they near the sun.

compact object (270) A star that has collapsed to form a white dwarf, neutron star, or black hole.

comparative planetology (444) The study of planets by comparing the characteristics of different examples.

comparison spectrum (116) A spectrum of known spectral lines used to identify unknown wavelengths in an object's spectrum.

composite volcano (490) A volcano built up of layers of lava flows and ash falls. These are steep sided and typically associated with subduction zones.

condensation (428) The growth of a particle by addition of material from surrounding gas, one atom or molecule at a time.

condensation hypothesis (472) The hypothesis that the moon and Earth formed by condensing together from the solar nebula.

condensation sequence (427) The sequence in which different materials condense from the solar nebula at increasing distances from the sun.

conservation of energy law (242) One of the basic laws of stellar structure. The amount of energy flowing out of the top of a shell must equal the amount coming in at the bottom plus whatever energy is generated within the shell.

conservation of mass law (242) One of the basic laws of stellar structure. The total mass of the star must equal the sum of the masses of the shells, and the mass must be distributed smoothly throughout the star.

constellation (13) One of the stellar patterns identified by name, usually of mythological gods, people, animals, or objects; also, the region of the sky containing that star pattern.

continuous spectrum (136) A spectrum in which there are no absorption or emission lines.

convection (152) Circulation in a fluid driven by heat; hot material rises, and cool material sinks.

convective zone (159) The region inside a star where energy is carried outward as rising hot gas and sinking cool gas.

corona (44, 492) The faint outer atmosphere of the sun; composed of low-density, very hot, ionized gas. On Venus, round networks of fractures and ridges up to 1000 km in diameter.

coronagraph (153) A telescope designed to photograph the inner corona of the sun.

coronal gas (213) Extremely high-temperature, low-density gas in the interstellar medium.

coronal hole (169) An area of the solar surface that is dark at X-ray wavelengths; thought to be associated with divergent magnetic fields and the source of the solar wind.

coronal mass ejection (CME) (169) Gas trapped in the sun's magnetic field.

cosmic microwave background radiation (395) Radiation from the hot clouds of the big bang explosion. Because of its large redshift, it appears to come from a body whose temperature is only 2.7 K.

cosmic ray (124) A subatomic particle traveling at tremendous velocity that strikes Earth's atmosphere from space.

cosmological constant (Λ) (407) Einstein's constant that represents a repulsion in space to oppose gravity.

cosmological principle (400) The assumption that any observer in any galaxy sees the same general features of the universe.

cosmology (390) The study of the nature, origin, and evolution of the universe.

Coulomb barrier (158) The electrostatic force of repulsion between bodies of like charge; commonly applied to atomic nuclei.

Coulomb force (129) The repulsive force between particles with like electrostatic charge.

critical density (402) The average density of the universe needed to make its curvature flat.

critical point (513) The temperature and pressure at which the vapor and liquid phases of a material have the same density.

dark age (398) The period of a few hundred million years during which the universe expanded in darkness. Extends from soon after the big bang glow faded into the infrared to the formation of the first stars.

dark energy (407) The energy of empty space that drives the acceleration of the expanding universe.

dark-line spectrum (136) See *absorption spectrum.*

dark matter (323) Nonluminous material that is detected only by its gravitational influence.

dark nebula (205) A nonluminous cloud of gas and dust visible because it blocks light from more distant stars and nebulae.

debris disk (435) A disk of dust found by infrared observations around some stars. The dust is debris from collisions among asteroids, comets, and Kuiper belt objects.

deferent (61) In the Ptolemaic theory, the large circle around Earth along which the center of the epicycle moved.

degenerate matter (251) Extremely high-density matter in which pressure no longer depends on temperature, due to quantum mechanical effects.

density (144) The amount of matter per unit volume in a material; measured in grams per cubic centimeter, for example.

density wave theory (331) Theory proposed to account for spiral arms as compressions of the interstellar medium in the disk of the galaxy.

deuterium (158) An isotope of hydrogen in which the nucleus contains a proton and a neutron.

diamond ring effect (44) A momentary phenomenon seen during some total solar eclipses when the ring of the corona and a bright spot of photosphere resemble a large diamond set in a silvery ring.

differential rotation (164) The rotation of a body in which different parts of the body have different periods of rotation; this is true of the sun, the Jovian planets, and the disk of the galaxy.

differentiation (429) The separation of planetary material according to density.

diffraction fringe (108) Blurred fringe surrounding any image caused by the wave properties of light. Because of this, no image detail smaller than the fringe can be seen.

direct gravitational collapse (431) The proposed process by which a Jovian planet might skip the accretion of a solid core and form quickly from the gases of the solar nebula.

disk component (318) All material confined to the plane of the galaxy.

distance indicator (348) Object whose luminosity or diameter is known; used to find the distance to a star cluster or galaxy.

distance modulus ($m_V - M_V$) (179) The difference between the apparent and absolute magnitude of a star; a measure of how far away the star is.

distance scale (350) The combined calibration of distance indicators used by astronomers to find the distances to remote galaxies.

DNA (deoxyribonucleic acid) (600) The long carbon-chain molecule that records information to govern the biological activity of the organism. DNA carries the genetic data passed to offspring.

Doppler broadening (144) The smearing of spectral lines because of the motion of the atoms in the gas.

Doppler effect (142) The change in the wavelength of radiation due to relative radial motion of source and observer.

double-exhaust model (372) The theory that double radio lobes are produced by pairs of jets emitted in opposite directions from the centers of active galaxies.

double-lobed radio source (370) A galaxy that emits radio energy from two regions (lobes) located on opposite sides of the galaxy.

Drake equation (612) A formula for the number of communicative civilizations in our galaxy.

dust (type II) tail (586) The tail of a comet formed of dust blown outward by the pressure of sunlight. (See *gas tail.*)

dwarf planet (564) An object that orbits the sun and has pulled itself into a spherical shape but has not cleared its orbital lane of other objects. Pluto is a dwarf planet.

dynamo effect (164) The process by which a rotating, convecting body of conducting matter, such as Earth's core, can generate a magnetic field.

east point (20) The point on the eastern horizon exactly halfway between the north point and the south point; exactly east.

eccentric (58) In astronomy, an off-center circular path.

eccentricity, *e* (72) A measure of the flattening of an ellipse. An ellipse of $e = 0$ is circular. The closer to 1 e becomes, the more flattened the ellipse.

eclipse season (47) That period when the sun is near a node of the moon's orbit and eclipses are possible.

eclipse year (48) The time the sun takes to circle the sky and return to a node of the moon's orbit; 346.62 days.

eclipsing binary system (191) A binary star system in which the stars eclipse each other.

ecliptic (23) The apparent path of the sun around the sky.

ecosphere (609) A region around a star within which a planet can have temperatures that permit the existence of liquid water.

ejecta (464) Pulverized rock scattered by meteorite impacts on a planetary surface.

electromagnetic radiation (103) Changing electric and magnetic fields that travel through space and transfer energy from one place to another—for example, light, radio waves, and the like.

electron (128) Low-mass atomic particle carrying a negative charge.

ellipse (72) A closed curve enclosing two points (foci) such that the total distance from one focus to any point on the curve back to the other focus equals a constant.

elliptical galaxy (346) A galaxy that is round or elliptical in outline; it contains little gas and dust, no disk or spiral arms, and few hot, bright stars.

emission line (136) A bright line in a spectrum caused by the emission of photons from atoms.

emission nebula (204) A cloud of glowing gas excited by ultraviolet radiation from hot stars.

emission spectrum (bright-line spectrum) (136) A spectrum containing emission lines.

energy (91) The capacity of a natural system to perform work—for example, thermal energy.

energy level (131) One of a number of states an electron may occupy in an atom, depending on its binding energy.

enzyme (600) Special protein that controls processes in an organism.

epicycle (61) The small circle followed by a planet in the Ptolemaic theory. The center of the epicycle follows a larger circle (deferent) around Earth.

equant (61) The point off-center in the deferent from which the center of the epicycle appears to move uniformly.

equatorial mounting (113) A telescope mounting that allows motion parallel to and perpendicular to the celestial equator.

ergosphere (302) The region surrounding a rotating black hole within which one could not resist being dragged around the black hole. It is possible for a particle to escape from the ergosphere and extract energy from the black hole.

escape velocity (89) The initial velocity an object needs to escape from the surface of a celestial body.

evening star (23) Any planet visible in the sky just after sunset.

event horizon (300) The boundary of the region of a black hole from which no radiation may escape. No event that occurs within the event horizon is visible to a distant observer.

evolution (602) The process by which life adjusts itself to fit its changing environment.

evolutionary hypothesis (417) Explanation for natural events that involves gradual changes as opposed to sudden catastrophic changes—for example, the formation of the planets in the gas cloud around the forming sun.

excited atom (131) An atom in which an electron has moved from a lower to a higher orbit.

extrasolar planet (435) A planet orbiting a star other than the sun.

eyepiece (107) A short-focal-length lens used to enlarge the image in a telescope; the lens nearest the eye.

fall (572) A meteorite seen to fall. (See *find*.)

false-color image (115) A representation of graphical data in which the colors are altered or added to reveal details.

field (85) A way of explaining action at a distance; a particle produces a field of influence (gravitational, electric, or magnetic) to which another particle in the field responds.

field of view (3) The area visible in an image; usually given as the diameter of the region.

filament (153) On the sun, a prominence seen silhouetted against the solar surface.

filtergram (153) An image (usually of the sun) taken in the light of a specific region of the spectrum—e.g., an H-alpha filtergram.

find (572) A meteorite that is found but was not seen to fall. (See *fall*.)

fission hypothesis (472) The theory that the moon formed by breaking away from Earth.

flare (169) A violent eruption on the sun's surface.

flatness problem (404) In cosmology, the circumstance that the early universe must have contained almost exactly the right amount of matter to close space-time (to make space-time flat).

flat universe (402) A model of the universe in which space-time is not curved.

flocculent (333) Woolly, fluffy; used to refer to certain galaxies that have a woolly appearance.

flux (16) A measure of the flow of energy onto or through a surface. Usually applied to light.

focal length (107) The distance from a lens to the point where it focuses parallel rays of light.

folded mountain range (452) A long range of mountains formed by the compression of a planet's crust—for example, the Andes on Earth.

forbidden line (203) A spectral line that does not occur in the laboratory because it depends on an atomic transition that is highly unlikely.

forward scattering (516) The optical property of finely divided particles to preferentially direct light in the original direction of the light's travel.

free-fall contraction (223) The early contraction of a gas cloud to form a star during which internal pressure is too low to resist contraction.

frequency (104) The number of times a given event occurs in a given time; for a wave, the number of cycles that pass the observer in 1 second.

galactic cannibalism (360) The theory that large galaxies absorb smaller galaxies.

galactic corona (323) The low-density extension of the halo of a galaxy; now suspected to extend many times the visible diameter of the galaxy.

galactic fountain (326) A region of the galaxy's disk in which gas heated by supernova explosions throws gas out of the disk where it can fall back and spread metals through the disk.

galaxy (6) A very large collection of gas, dust, and stars orbiting a common center of mass. The sun and Earth are located in the Milky Way Galaxy.

Galilean satellites (423) The four largest satellites of Jupiter, named after their discoverer, Galileo.

gamma-ray burster (308) An object that produces a sudden burst of gamma rays; thought to be associated with neutron stars and black holes.

gas (type I) tail (586) The tail of a comet produced by gas blown outward by the solar wind. (See *dust tail*.)

gene (601) A unit of DNA containing genetic information that influences a particular inherited trait.

general theory of relativity (96) Einstein's more sophisticated theory of space and time, which describes gravity as a curvature of space-time.

geocentric universe (57) A model universe with Earth at the center, such as the Ptolemaic universe.

geosynchronous satellite (88) An Earth satellite in an eastward orbit whose period is 24 hours. A satellite in such an orbit remains above the same spot on Earth's surface.

giant molecular cloud (212) Very large, cool cloud of dense gas in which stars form.

giant (183) Large, cool, highly luminous star in the upper right of the H–R diagram, typically 10 to 100 times the diameter of the sun.

glitch (291) A sudden change in the period of a pulsar.

global warming (456) The gradual increase in the surface temperature of Earth caused by human modifications to Earth's atmosphere.

globular cluster (256) A star cluster containing 50,000 to 1 million stars in a sphere about 75 ly in diameter; generally old, metal-poor, and found in the spherical component of the galaxy.

gossamer ring (520) The dimmest part of Jupiter's ring produced by dust particles orbiting near small moons.

grand unified theory (GUT) (405) Theory that attempts to unify (describe in a similar way) the electromagnetic, weak, and strong forces of nature.

granulation (151) The fine structure visible on the solar surface caused by rising currents of hot gas and sinking currents of cool gas below the surface.

grating (116) A piece of material in which numerous microscopic parallel lines are scribed; light encountering a grating is dispersed to form a spectrum.

gravitational collapse (429) The stage in the formation of a massive planet when it grows massive enough to begin capturing gas directly from the nebula around it.

gravitational lensing (355) The effect of the focusing of light from a distant galaxy or quasar by an intervening galaxy to produce multiple images of the distant body.

gravitational radiation (296) As predicted by general relativity, expanding waves in a gravitational field that transport energy through space.

gravitational redshift (303) The lengthening of the wavelength of a photon due to its escape from a gravitational field.

greenhouse effect (455) The process by which a carbon dioxide atmosphere traps heat and raises the temperature of a planetary surface.

grooved terrain (522) Region of the surface of Ganymede consisting of bright, parallel grooves.

ground state (131) The lowest permitted electron orbit in an atom.

half-life (424) The time required for half of the atoms in a radioactive sample to decay.

halo (320) The spherical region of a spiral galaxy containing a thin scattering of stars, star clusters, and small amounts of gas.

heat (132) Energy flowing from a warm body to a cool body by the agitation of particles such as atoms or molecules.

heat of formation (430) In planetology, the heat released by the infall of matter during the formation of a planetary body.

heavy bombardment (420) The period of intense meteorite impacts early in the formation of the planets, when the solar system was filled with debris.

heliocentric universe (59) A model of the universe with the sun at the center, such as the Copernican universe.

helioseismology (154) The study of the interior of the sun by the analysis of its modes of vibration.

helium flash (251) The explosive ignition of helium burning that takes place in some giant stars.

Herbig–Haro object (229) A small nebula associated with star formation that varies irregularly in brightness.

Hertzsprung–Russell diagram (181) A plot of the intrinsic brightness versus the surface temperature of stars; it separates the effects of temperature and surface area on stellar luminosity; commonly absolute magnitude versus spectral type but also luminosity versus surface temperature or color.

Hesperian period (505) On Mars, the geological era from the decline of heavy cratering and lava flows and the melting of subsurface ice to form the outflow channels.

HI cloud (208) An interstellar cloud of neutral hydrogen.

HII region (204) A region of ionized hydrogen around a hot star.

Hirayama family (580) Family of asteroids with orbits of similar size, shape, and orientation; thought to be fragments of larger bodies.

homogeneous (400) The property of being uniform. In cosmology, the characteristic of the universe in which, on the large scale, matter is uniformly spread through the universe.

horizon (20) The line that marks the apparent intersection of Earth and the sky.

horizon problem (405) In cosmology, the circumstance that the primordial background radiation seems much more isotropic than could be explained by the standard big bang theory.

horizontal branch (257) In the H–R diagram of a globular cluster, the sequence of stars extending from the red giants toward the blue side of the diagram; includes RR Lyrae stars.

horoscope (26) A chart showing the positions of the sun, moon, planets, and constellations at the time of a person's birth; used in astrology to attempt to read character or foretell the future.

hot dark matter (403) Invisible matter in the universe composed of low-mass, high-velocity particles such as neutrinos.

hot Jupiter (438) A massive and presumably Jovian planet that orbits close to its star and consequently has a high temperature.

hot spot (372) In radio astronomy, a bright spot in a radio lobe.

H–R diagram (181) See *Hertzsprung–Russell diagram.*

Hubble constant (H) (351) A measure of the rate of expansion of the universe; the average value of velocity of recession divided by distance; about 70 km/s/megaparsec.

Hubble law (351) The linear relation between the distance to a galaxy and its radial velocity.

Hubble time (394) An upper limit on the age of the universe derived from the Hubble constant.

hydrostatic equilibrium (233) The balance between the weight of the material pressing downward on a layer in a star and the pressure in that layer.

hypernova (309) The explosion produced as a very massive star collapses into a black hole; thought to be responsible for at least some gamma-ray bursts.

hypothesis (83) A conjecture, subject to further tests, that accounts for a set of facts.

ice line (427) In the solar nebula, the boundary beyond which water vapor and other compounds could form ice particles.

inflationary universe (405) A version of the big bang theory that includes a rapid expansion when the universe was very young.

infrared radiation (104) Electromagnetic radiation with wavelengths intermediate between visible light and radio waves.

instability strip (258) The region of the H–R diagram in which stars are unstable to pulsation; a star passing through this strip becomes a variable star.

intercloud medium (208) The hot, low-density gas between cooler clouds in the interstellar medium.

intercrater plain (477) The relatively smooth terrain on Mercury.

interferometry (114) The observing technique in which separated telescopes combine to produce a virtual telescope with the resolution of a much-larger-diameter telescope.

interstellar absorption lines (206) Dark lines in some stellar spectra that are formed by interstellar gas.

interstellar dust (206) Microscopic solid grains in the interstellar medium.

interstellar extinction (206) The dimming of starlight by gas and dust in the interstellar medium.

interstellar medium (203) The gas and dust distributed between the stars.

interstellar reddening (206) The process in which dust scatters blue light out of starlight and makes the stars look redder.

intrinsic variable star (258) A variable star driven to pulsate by processes in its interior.

inverse square law (84) The rule that the strength of an effect (such as gravity) decreases in proportion as the distance squared increases.

Io flux tube (515) A tube of magnetic lines and electric currents connecting Io and Jupiter.

ion (129) An atom that has lost or gained one or more electrons.

ionization (129) The process in which atoms lose or gain electrons.

Io plasma torus (515) The doughnut-shaped cloud of ionized gas that encloses the orbit of Jupiter's moon Io.

iron meteorite (572) A meteorite composed mainly of iron–nickel alloy.

irregular galaxy (347) A galaxy with a chaotic appearance, large clouds of gas and dust, and both population I and population II stars, but without spiral arms.

island universe (343) An older term for a galaxy.

isotopes (129) Atoms that have the same number of protons but a different number of neutrons.

isotropic (400) The conditions of being uniform in all directions. In cosmology, the characteristic of the universe in which, in its general properties, it looks the same in every direction.

joule (J) (91) A unit of energy equivalent to a force of 1 newton acting over a distance of 1 meter; 1 joule per second equals 1 watt of power.

Jovian planet (422) Jupiter-like planet with large diameter and low density.

jumbled terrain (471) Strangely disturbed regions of the moon opposite the locations of the Imbrium Basin and Mare Orientale.

Kelvin temperature scale (132) The temperature, in Celsius (centigrade) degrees, measured above absolute zero.

Keplerian motion (322) Orbital motion in accord with Kepler's laws of planetary motion.

Kerr solution (300) A solution to the equations of general relativity that describes the properties of a rotating black hole.

kiloparsec (kpc) (316) A unit of distance equal to 1000 pc, or 3260 ly.

kinetic energy (91) Energy of motion. Depends on mass and velocity of a moving body.

Kirchhoff's laws (136) A set of laws that describe the absorption and emission of light by matter.

Kirkwood's gaps (580) Regions in the asteroid belt in which there are very few asteroids; caused by orbital resonances with Jupiter.

Kuiper belt (421) The collection of icy planetesimals that orbit in a region from just beyond Neptune out to about 50 AU.

Lagrangian point (272) Point of stability in the orbital plane of a binary star system, planet, or moon. One is located 60° ahead and one 60° behind the orbiting bodies; another is located between the orbiting bodies.

large-impact theory (472) The theory that the moon formed from debris ejected during a collision between Earth and a large planetesimal.

large-scale structure (409) The distribution of galaxy clusters and superclusters in walls and filaments surrounding voids mostly empty of galaxies.

late heavy bombardment (420) The surge in cratering impacts in the solar system that occurred about 3.9 billion years ago.

L dwarf (140) A type of star that is even cooler than the M stars.

life zone (609) A region around a star within which a planet can have temperatures that permit the existence of liquid water.

light curve (191) A graph of brightness versus time commonly used in analyzing variable stars and eclipsing binaries.

light-gathering power (108) The ability of a telescope to collect light; proportional to the area of the telescope objective lens or mirror.

lighthouse model (290) The explanation of a pulsar as a spinning neutron star sweeping beams of radio radiation around the sky.

light pollution (110) The illumination of the night sky by waste light from cities and outdoor lighting, which prevents the observation of faint objects.

light-year (ly) (5) The distance light travels in one year.

limb (152, 462) The edge of the apparent disk of a body, as in "the limb of the moon."

limb darkening (152) The decrease in brightness of the sun or other body from its center to its limb.

line of nodes (47) The line across an orbit connecting the nodes; commonly applied to the orbit of the moon.

liquid metallic hydrogen (513) A form of hydrogen under high pressure that is a good electrical conductor.

lobate scarp (476) A curved cliff such as those found on Mercury.

local bubble or void (214) A region of high-temperature, low-density gas in the interstellar medium in which the sun happens to be located.

look-back time (350) The amount by which you look into the past when you look at a distant galaxy; a time equal to the distance to the galaxy in light-years.

luminosity (L) (179) The total amount of energy a star radiates in 1 second.

luminosity class (184) A category of stars of similar luminosity; determined by the widths of lines in their spectra.

lunar eclipse (35) The darkening of the moon when it moves through Earth's shadow.

Lyman series (137) Spectral lines in the ultraviolet spectrum of hydrogen produced by transitions whose lowest orbit is the ground state.

magnetar (295) A class of neutron stars that have exceedingly strong magnetic fields; thought to be responsible for soft gamma-ray repeaters.

magnetic carpet (154) The widely distributed, low-level magnetic field extending up through the sun's visible surface.

magnetosphere (450) The volume of space around a planet within which the motion of charged particles is dominated by the planetary magnetic field rather than the solar wind.

magnifying power (110) The ability of a telescope to make an image larger.

magnitude–distance formula (179) The mathematical formula that relates the apparent magnitude and absolute magnitude of a star to its distance.

magnitude scale (15) The astronomical brightness scale; the larger the number, the fainter the star.

main sequence (182) The region of the H–R diagram running from upper left to lower right, which includes roughly 90 percent of all stars.

mantle (445) The layer of dense rock and metal oxides that lies between the molten core and Earth's surface; also, similar layers in other planets.

mare (462) (Plural: *maria*) One of the lunar lowlands filled by successive flows of dark lava; from the Latin word for *sea*.

mass (84) A measure of the amount of matter making up an object.

mass–luminosity relation (194) The more massive a star is, the more luminous it is.

Maunder butterfly diagram (162) A graph showing the latitude of sunspots versus time; first plotted by W. W. Maunder in 1904.

Maunder minimum (163) A period of less numerous sunspots and other solar activity from 1645 to 1715.

megaparsec (Mpc) (348) A unit of distance equal to 1 million pc.

metals (323) In astronomical usage, all atoms heavier than helium.

metastable level (203) An atomic energy level from which an electron takes a long time to decay; responsible for producing forbidden lines.

meteor (421) A small bit of matter heated by friction to incandescent vapor as it falls into Earth's atmosphere.

meteorite (424) A meteor that has survived its passage through the atmosphere and strikes the ground.

meteoroid (424) A meteor in space before it enters Earth's atmosphere.

micrometeorite (465) Meteorite of microscopic size.

midocean rise (452) One of the undersea mountain ranges that push up from the seafloor in the center of the oceans.

Milankovitch hypothesis (27) The hypothesis that small changes in Earth's orbital and rotational motions cause the ice ages.

Milky Way (6) The hazy band of light that circles the sky, produced by the combined light of billions of stars in our Milky Way Galaxy.

Milky Way Galaxy (6) The spiral galaxy containing the sun; visible at night as the Milky Way.

Miller experiment (603) An experiment that reproduced the conditions under which life began on Earth and amino acids and other organic compounds were manufactured.

millisecond pulsar (298) A pulsar with a period of approximately a millisecond, a thousandth of a second.

minute of arc (21) An angular measure; each degree is divided into 60 minutes of arc.

molecular cloud (212) An interstellar gas cloud that is dense enough for the formation of molecules; discovered and studied through the radio emissions of such molecules.

molecule (129) Two or more atoms bonded together.

momentum (82) The tendency of a moving object to continue moving; mathematically, the product of mass and velocity.

morning star (23) Any planet visible in the sky just before sunrise.

multiringed basin (465) Very large impact basin in which there are concentric rings of mountains.

mutant (602) Offspring born with altered DNA.

nadir (20) The point on the bottom of the sky directly under your feet.

nanometer (nm) (104) A unit of length equal to 10^{-9} m.

natural law (83) A conjecture about how nature works in which scientists have overwhelming confidence.

natural motion (80) In Aristotelian physics, the motion of objects toward their natural places—fire and air upward and earth and water downward.

natural selection (602) The process by which the best traits are passed on, allowing the most able to survive.

neap tide (91) Ocean tide of low amplitude occurring at first- and third-quarter moon.

nebula (203) A cloud of gas and dust in space.

nebular hypothesis (417) The proposal that the solar system formed from a rotating cloud of gas.

neutrino (158) A neutral, massless atomic particle that travels at or nearly at the speed of light.

neutron (128) An atomic particle with no charge and about the same mass as a proton.

neutron star (288) A small, highly dense star composed almost entirely of tightly packed neutrons; radius about 10 km.

Newtonian focus (112) The focal arrangement of a reflecting telescope in which a diagonal mirror reflects light out the side of the telescope tube for easier access.

Noachian period (504) On Mars, the era that extends from the formation of the crust to the end of heavy cratering and includes the formation of the valley networks.

node (46) A point where an object's orbit passes through the plane of Earth's orbit.

nonbaryonic matter (403) In cosmology, a suspected component of the dark matter composed of matter that does not contain protons and neutrons.

north celestial pole (20) The point on the celestial sphere directly above Earth's North Pole.

north point (20) The point on the horizon directly below the north celestial pole; exactly north.

nova (273) From the Latin "new," a sudden brightening of a star, making it appear as a "new" star in the sky; thought to be associated with eruptions on white dwarfs in binary systems.

nuclear bulge (320) The spherical cloud of stars that lies at the center of spiral galaxies.

nuclear fission (157) Reaction that splits nuclei into less massive fragments.

nuclear fusion (157) Reaction that joins the nuclei of atoms to form more massive nuclei.

nucleosynthesis (325) The production of elements heavier than helium by the fusion of atomic nuclei in stars and during supernovae explosions.

nucleus (of an atom) (128) The central core of an atom containing protons and neutrons; carries a net positive charge.

O association (226) A large, loosely bound cluster of very young stars.

objective lens or mirror (107) The main optical element in an astronomical telescope. The large lens at the top of the telescope or large mirror at the bottom.

oblateness (514) The flattening of a spherical body, usually caused by rotation.

observable universe (392) The part of the universe that is visible from Earth's location in space and time.

occultation (552) The passage of a larger body in front of a smaller body.

Olbers's paradox (391) The conflict between observation and theory as to why the night sky should or should not be dark.

Oort cloud (390) The cloud of icy bodies—extending from the outer part of our solar system out to roughly 100,000 AU from the sun—that acts as the source of most comets.

opacity (232) The resistance of a gas to the passage of radiation.

open cluster (256) A cluster of 10 to 10,000 stars with an open, transparent appearance and stars not tightly grouped; usually relatively young and located in the disk of the galaxy.

open orbit (89) An orbit that does not return to its starting point; an escape orbit. (See *closed orbit.*)

open universe (402) A model universe in which the average density is less than the critical density needed to halt the expansion.

outflow channel (501) Geological feature on Mars that appears to have been caused by sudden flooding.

outgassing (430) The release of gases from a planet's interior.

ovoid (554) Geological feature on Uranus's moon Miranda thought to be produced by circulation in the solid icy mantle and crust.

ozone layer (454) In Earth's atmosphere, a layer of oxygen ions (O_3) lying 15 to 30 km high that protects the surface by absorbing ultraviolet radiation.

paradigm (64) A commonly accepted set of scientific ideas and assumptions.

parallax (60) The apparent change in the position of an object due to a change in the location of the observer. Astronomical parallax is measured in seconds of arc.

parsec (pc) (116) The distance to a hypothetical star whose parallax is one second of arc; 1 pc = 206,265 AU = 3.26 ly.

partial lunar eclipse (39) A lunar eclipse in which the moon does not completely enter Earth's shadow.

partial solar eclipse (40) A solar eclipse in which the moon does not completely cover the sun.

Paschen series (137) Spectral lines in the infrared spectrum of hydrogen produced by transitions whose lowest orbit is the third.

passing star hypothesis (417) The proposal that our solar system formed when two stars passed near each other and material was pulled out of one to form the planets.

path of totality (42) The track of the moon's umbral shadow over Earth's surface. The sun is totally eclipsed as seen from within this path.

penumbra (35) The portion of a shadow that is only partially shaded.

penumbral lunar eclipse (39) A lunar eclipse in which the moon enters the penumbra of Earth's shadow but does not reach the umbra.

perigee (42) The orbital point of closest approach to Earth.

perihelion (25) The orbital point of closest approach to the sun.

period–luminosity relation (258) The relation between period of pulsation and intrinsic brightness among Cepheid variable stars.

permitted orbit (130) One of the energy levels in an atom that an electron may occupy.

photon (104) A quantum of electromagnetic energy; carries an amount of energy that depends inversely on its wavelength.

photosphere (44) The bright visible surface of the sun.

planet (4) A nonluminous object, larger than a comet or asteroid, that orbits a star.

planetary nebula (266) An expanding shell of gas ejected from a star during the latter stages of its evolution.

planetesimal (428) One of the small bodies that formed from the solar nebula and eventually grew into protoplanets.

plastic (449) A material with the properties of a solid but capable of flowing under pressure.

plate tectonics (452) The constant destruction and renewal of Earth's surface by the motion of sections of crust.

plutino (565) One of the icy Kuiper belt objects that, like Pluto, are caught in a 3:2 orbital resonance with Neptune.

polar axis (113) The axis around which a celestial body rotates.

poor cluster (348) An irregularly shaped cluster that contains fewer than 1000 galaxies, many spiral, and no giant ellipticals.

population I star (323) Star rich in atoms heavier than helium; nearly always a relatively young star found in the disk of the galaxy.

population II star (323) Star poor in atoms heavier than helium; nearly always a relatively old star found in the halo, globular clusters, or the nuclear bulge.

positron (158) The antiparticle of the electron.

potential energy (91) The energy a body has by virtue of its position. A weight on a high shelf has more potential energy than a weight on a low shelf.

precession (19) The slow change in the direction of Earth's axis of rotation; one cycle takes nearly 26,000 years.

pressure (209) A force exerted over a surface; expressed as force per unit area.

pressure *(P)* wave (447) In geophysics, a mechanical wave of compression and rarefaction that travels through Earth's interior.

primary lens or mirror (107) The main optical element in an astronomical telescope. The large lens at the top of the telescope tube or the large mirror at the bottom.

prime focus (112) The point at which the objective mirror forms an image in a reflecting telescope.

primeval atmosphere (454) Earth's first air, composed of gases from the solar nebula.

primordial soup (603) The rich solution of organic molecules in Earth's first oceans.

prominence (44, 168) Eruption on the solar surface; visible during total solar eclipses.

proper motion (177) The rate at which a star moves across the sky; measured in seconds of arc per year.

protein (600) Complex molecule composed of amino acid units.

proton (128) A positively charged atomic particle contained in the nucleus of atoms; the nucleus of a hydrogen atom.

proton–proton chain (158) A series of three nuclear reactions that build a helium atom by adding together protons; the main energy source in the sun.

protoplanet (429) Massive object resulting from the coalescence of planetesimals in the solar nebula and destined to become a planet.

protostar (224) A collapsing cloud of gas and dust destined to become a star.

protostellar disk (225) A gas cloud around a forming star flattened by its rotation.

pseudoscience (26) A subject that claims to obey the rules of scientific reasoning but does not. Examples include astrology, crystal power, and pyramid power.

pulsar (290) A source of short, precisely timed radio bursts; thought to be a spinning neutron star.

pulsar wind (291) The flow of high-energy particles that carries most of the energy away from a spinning neutron star.

quantum mechanics (130) The study of the behavior of atoms and atomic particles.

quasar (quasi-stellar object, or QSO) (378) Small, powerful source of energy thought to be the active core of a very distant galaxy.

quasi-periodic oscillation (QPO) (306) A high-speed flickering in the radiation from an accretion disk evidently caused by material spiraling inward.

quintessence (407) The proposed energy of empty space that causes the acceleration of the expanding universe.

radial velocity *(V_r)* (143) That component of an object's velocity directed away from or toward Earth.

radiant (570) The point in the sky from which meteors in a shower seem to come.

radiation pressure (225) The force exerted on the surface of a body by its absorption of light.

Small particles floating in the solar system can be blown outward by the pressure of the sunlight.

radiative zone (159) The region inside a star where energy is carried outward as photons.

radio galaxy (368) A galaxy that is a strong source of radio signals.

radio interferometer (118) Two or more radio telescopes that combine their signals to achieve the resolving power of a larger telescope.

ray (464) Ejecta from a meteorite impact, forming white streamers radiating from some lunar craters.

recombination (398) The stage within a million years of the big bang when the gas became transparent to radiation.

reconnection (169) The process in the sun's atmosphere by which opposing magnetic fields combine and release energy to power solar flares.

red dwarf (183) Cool, low-mass star on the lower main sequence.

redshift (142) The lengthening of the wavelengths of light seen when the source and observer are receding from each other.

reflecting telescope (106) A telescope that uses a concave mirror to focus light into an image.

reflection nebula (204) A nebula produced by starlight reflecting off dust particles in the interstellar medium.

refracting telescope (106) A telescope that forms images by bending (refracting) light with a lens.

regolith (469) A soil made up of crushed rock fragments.

reionization (399) The stage in the early history of the universe when ultraviolet photons from the first stars ionized the gas filling space.

relative age (462) The age of a geological feature referred to other features. For example, relative ages reveal that the lunar maria are younger than the highlands.

resolving power (108) The ability of a telescope to reveal fine detail; depends on the diameter of the telescope objective.

resonance (474) The coincidental agreement between two periodic phenomena; commonly applied to agreements between orbital periods, which can make orbits more stable or less stable.

retrograde motion (60) The apparent backward (westward) motion of planets as seen against the background of stars.

revolution (22) The motion of an object in a closed path about a point outside its volume; Earth revolves around the sun.

rich cluster (348) A cluster containing over 1000 galaxies, mostly elliptical, scattered over a volume about 3 Mpc in diameter.

rift valley (452) A long, straight, deep valley produced by the separation of crustal plates.

ring galaxy (261) A galaxy that resembles a ring around a bright nucleus; thought to be the result of a head-on collision of two galaxies.

RNA (ribonucleic acid) (601) A long carbon-chain molecule that uses the information stored in DNA to manufacture complex molecules necessary to the organism.

Roche limit (517) The minimum distance between a planet and a satellite that holds itself together by its own gravity. If a satellite's orbit brings it within its planet's Roche limit, tidal forces will pull the satellite apart.

Roche lobe (272) In a system with two bodies orbiting each other, the volume of space dominated by the gravitation of one of the bodies.

Roche surface (272) In a system with two bodies orbiting each other, the outer boundary of the volume of space dominated by the gravitation of one of the bodies.

rotation (22) The turning of a body about an axis that passes through its volume; Earth rotates on its axis.

rotation curve (322, 352) A graph of orbital velocity versus radius in the disk of a galaxy.

rotation curve method (353) The procedure for finding the mass of a galaxy from its rotation curve.

RR Lyrae variable star (258) Variable star with a period of 12 to 24 hours; common in some globular clusters.

Sagittarius A* (338) The powerful radio source located at the core of the Milky Way Galaxy.

saros cycle (48) An 18-year $11\frac{1}{3}$-day period after which the pattern of lunar and solar eclipses repeats.

Schmidt-Cassegrain focus (112) The optical design of a reflecting telescope in which a thin correcting lens is placed at the top of a Cassegrain telescope.

Schwarzschild radius (R_s) (301) The radius of the event horizon around a black hole.

scientific method (2) The reasoning style by which scientists test theories against evidence to understand how nature works.

scientific model (17) An intellectual concept designed to help you think about a natural process without necessarily being a conjecture of truth.

scientific notation (4) The system of recording very large or very small numbers by using powers of 10.

secondary atmosphere (454) The gases outgassed from a planet's interior; rich in carbon dioxide.

secondary crater (464) A crater formed by the impact of debris ejected from a larger crater.

secondary mirror (112) In a reflecting telescope, the mirror that reflects the light to a point of easy observation.

second of arc (21) An angular measure; each minute of arc is divided into 60 seconds of arc.

seeing (109) Atmospheric conditions on a given night. When the atmosphere is unsteady, producing blurred images, the seeing is said to be poor.

seismic wave (447) A mechanical vibration that travels through Earth; usually caused by an earthquake.

seismograph (447) An instrument that records seismic waves.

selection effect (572) An influence on the probability that certain phenomena will be detected or selected, which can alter the outcome of a survey.

self-sustaining star formation (333) The process by which the birth of stars compresses the surrounding gas clouds and triggers the formation of more stars; proposed to explain spiral arms.

semimajor axis, *a* (72) Half of the longest axis of an ellipse.

SETI (610) Search for Extra-Terrestrial Intelligence.

Seyfert galaxy (368) An otherwise normal spiral galaxy with an unusually bright, small core that fluctuates in brightness; thought to indicate the core is erupting.

Shapley–Curtis Debate (344) The 1920 debate between Harlow Shapley and Heber Curtis over the nature of the spiral nebulae.

shear *(S)* wave (447) A mechanical wave that travels through Earth's interior by the vibration of particles perpendicular to the direction of wave travel.

shepherd satellite (533) A satellite that, by its gravitational field, confines particles to a planetary ring.

shield volcano (490) Wide, low-profile volcanic cone produced by highly liquid lava.

shock wave (222) A sudden change in pressure that travels as an intense sound wave.

sidereal drive (113) The motor and gears on a telescope that turn it westward to keep it pointed at a star.

sidereal period (37) The period of rotation or revolution of an astronomical body relative to the stars.

singularity (300) The object of zero radius into which the matter in a black hole is thought to fall.

sinuous rille (462) A narrow, winding valley on the moon caused by ancient lava flows along narrow channels.

small-angle formula (40) The mathematical formula that relates an object's linear diameter and distance to its angular diameter.

smooth plain (477) Apparently young plain on Mercury formed by lava flows at or soon after the formation of the Caloris Basin.

soft gamma-ray repeater (SGR) (295) An object that produces repeated bursts of lower-energy gamma rays; thought to be produced by magnetars.

solar constant (167) A measure of the energy output of the sun; the total solar energy striking 1 m^2 just above Earth's atmosphere in 1 second.

solar eclipse (40) The event that occurs when the moon passes directly between Earth and the sun, blocking your view of the sun.

solar nebula theory (418) The proposal that the planets formed from the same cloud of gas and dust that formed the sun.

solar system (4) The sun and the nonluminous objects that orbit it, including the planets, comets, and asteroids.

solar wind (154) Rapidly moving atoms and ions that escape from the solar corona and blow outward through the solar system.

south celestial pole (20) The point of the celestial sphere directly above Earth's South Pole.

south point (20) The point on the horizon directly above the south celestial pole; exactly south.

special relativity (95) The first of Einstein's theories of relativity, which dealt with uniform motion.

spectral class or type (138) A star's position in the temperature classification system O, B, A F, G, K, and M. Based on the appearance of the star's spectrum.

spectral line (116) A dark or bright line that crosses a spectrum at a specific wavelength.

spectral sequence (138) The arrangement of spectral classes (O, B, A, F, G, K, M) ranging from hot to cool.

spectrograph (116) A device that separates light by wavelength to produce a spectrum.

spectroscopic binary system (189) A star system in which the stars are too close together to be visible separately. You see a single point of light, and only by taking a spectrum can you determine that there are two stars.

spectroscopic parallax (186) The method of determining a star's distance by comparing its apparent magnitude with its absolute magnitude, as estimated from its spectrum.

spherical component (320) The part of the galaxy including all matter in a spherical distribution around the center (the halo and nuclear bulge).

spicule (153) Small, flamelike projection in the chromosphere of the sun.

spiral arm (6) Long, spiral pattern of bright stars, star clusters, gas, and dust that extends from the center to the edge of the disk of spiral galaxies.

spiral galaxy (346) A galaxy with an obvious disk component containing gas; dust; hot, bright stars; and spiral arms.

spiral nebula (343) Nebulous object with a spiral appearance observed in early telescopes; later recognized as a spiral galaxy.

spiral tracer (329) Object used to map the spiral arms (e.g., O and B associations, open clusters, clouds of ionized hydrogen, and some types of variable stars).

sporadic meteor (571) A meteor not part of a meteor shower.

spring tide (91) Ocean tide of high amplitude that occurs at full and new moon.

standard candle (348) Object of known brightness that astronomers use to find distance—for example, Cepheid variable stars and supernovae.

star (4) A celestial object composed of gas held together by its own gravity and supported by nuclear fusion occurring in its interior.

starburst galaxy (359) A bright blue galaxy in which many new stars are forming, thought to be caused by collisions between galaxies.

star-formation pillar (227) The column of gas produced when a dense core of gas protects the nebula behind it from the energy of a nearby hot star that is evaporating and driving away a star-forming nebula.

stellar model (242) A table of numbers representing the conditions in various layers within a star.

stellar parallax (p) (176) A measure of stellar distance. (*See* parallax.)

stellar wind (225) Hot gases blowing outward from the surface of a star. The equivalent for another star of the solar wind.

stony-iron meteorite (575) A meteorite that is a mixture of stone and iron.

stony meteorite (574) A meteorite composed of silicate (rocky) material.

stromatolite (604) A layered fossil formation caused by ancient mats of algae or bacteria that build up mineral deposits season after season.

strong force (157) One of the four forces of nature; the strong force binds protons and neutrons together in atomic nuclei.

subduction zone (452) A region of a planetary crust where a tectonic plate slides downward.

subsolar point (484) The point on a planet that is directly below the sun.

summer solstice (24) The point on the celestial sphere where the sun is at its most northerly point; also, the time when the sun passes this point, about June 22, and summer begins in the Northern Hemisphere.

sunspot (150) Relatively dark spot on the sun that contains intense magnetic fields.

supercluster (409) A cluster of galaxy clusters.

supergiant (183) Exceptionally luminous star, 10 to 1000 times the sun's diameter.

supergranule (152) A large granule on the sun's surface including many smaller granules.

superluminal expansion (382) The apparent expansion of parts of a quasar at speeds greater than the speed of light.

supernova (277) A "new" star appearing in Earth's sky and lasting for a year or so before fading. Caused by the violent explosion of a star.

supernova remnant (281) The expanding shell of gas marking the site of a supernova explosion.

supernova (type Ia) (280) The explosion of a star, thought to be caused by the transfer of matter to a white dwarf.

supernova (type I) (280) The explosion of a white dwarf that has gained matter from a companion star and exceeded the Chandrasekhar limit (type Ia) or the explosion of a massive star that has lost its outer layers of hydrogen to a companion star (type Ib).

supernova (type II) (280) The explosion of a star, thought to be caused by the collapse of a massive star.

synchrotron radiation (281) Radiation emitted when high-speed electrons move through a magnetic field.

synodic period (37) The period of rotation or revolution of a celestial body with respect to the sun.

T association (226) A large, loosely bound group of T Tauri stars.

T dwarf (140) A very-low-mass star at the bottom end of the main sequence with a cool surface and a low luminosity.

temperature (132) A measure of the velocity of random motions among the atoms or molecules in a material.

terminator (461) The dividing line between daylight and darkness on a planet or moon.

terrestrial planet (442) Earthlike planet—small, dense, rocky.

theory (83) A system of assumptions and principles applicable to a wide range of phenomena that have been repeatedly verified.

thermal energy (132) The energy stored in an object as agitation among its atoms and molecules.

thermal pulse (266) Periodic eruptions in the helium fusion shell in an aging giant star; thought to aid in ejecting the surface layers of the stars to form planetary nebulae.

tidal coupling (461) The locking of the rotation of a body to its revolution around another body.

tidal heating (523) The heating of a planet or satellite because of friction caused by tides.

tidal tail (360) A long strand of gas, dust, and stars drawn out of a galaxy interacting gravitationally with another galaxy.

time dilation (303) The slowing of moving clocks or clocks in strong gravitational fields.

total lunar eclipse (35) A lunar eclipse in which the moon completely enters Earth's dark shadow.

total solar eclipse (40) A solar eclipse in which the moon completely covers the bright surface of the sun.

totality (38) The period during a solar eclipse when the sun's photosphere is completely hidden by the moon, or the period during a lunar eclipse when the moon is completely inside the umbra of Earth's shadow.

transition (137) The movement of an electron from one atomic orbit to another.

transition region (152) The layer in the solar atmosphere between the chromosphere and the corona.

triple alpha process (251) The nuclear fusion process that combines three helium nuclei (alpha particles) to make one carbon nucleus.

Trojan asteroid (582) Small, rocky body caught in Jupiter's orbit at the Lagrangian points, 60° ahead of and behind the planet.

T Tauri star (228) Young star surrounded by gas and dust contracting toward the main sequence.

turnoff point (256) The point in an H–R diagram where a cluster's stars turn off the main sequence and move toward the red giant region, revealing the approximate age of the cluster.

21-cm radiation (209) Radio emission produced by cold, low-density hydrogen in interstellar space.

ultraluminous infrared galaxy (362) A highly luminous galaxy so filled with dust that most of its energy escapes as infrared photons emitted by warmed dust.

ultraviolet radiation (105) Electromagnetic radiation with wavelengths shorter than visible light but longer than X rays.

umbra (35) The region of a shadow that is totally shaded.

uncompressed density (427) The density a planet would have if its gravity did not compress it.

unified model (374) The attempt to explain the different kinds of active galaxies and quasars by a single model.

uniform circular motion (56) The classical belief that the perfect heavens could move only by the combination of constant motion along circular orbits.

valley networks (501) Dry drainage channels resembling streambeds found on Mars.

Van Allen belt (451) One of the radiation belts of high-energy particles trapped in Earth's magnetosphere.

variable star (258) A star whose brightness changes periodically.

velocity (82) A rate of travel that specifies both speed and direction.

velocity dispersion method (353) A method of finding a galaxy's mass by observing the range of velocities within the galaxy.

vernal equinox (24) The place on the celestial sphere where the sun crosses the celestial equator moving northward; also, the time of year when the sun crosses this point, about March 21, and spring begins in the Northern Hemisphere.

vesicular basalt (468) A porous rock formed by solidified lava with trapped bubbles.

violent motion (80) In Aristotelian physics, motion other than natural motion. (See *natural motion.*)

visual binary system (188) A binary star system in which the two stars are separately visible in the telescope.

water hole (610) The interval of the radio spectrum between the 21-cm hydrogen radiation and the 18-cm OH radiation, likely wavelengths to use in the search for extraterrestrial life.

wavelength (103) The distance between successive peaks or troughs of a wave; usually represented by λ.

wavelength of maximum intensity (λ_{max}) (133) The wavelength at which a perfect radiator emits the maximum amount of energy; depends only on the object's temperature.

weak force (157) One of the four forces of nature; the weak force is responsible for some forms of radioactive decay.

west point (20) The point on the western horizon exactly halfway between the north point and the south point; exactly west.

white dwarf (183) The remains of a dying star that has collapsed to the size of Earth and is slowly cooling off; at the lower left of the H–R diagram.

Widmanstätten pattern (573) Bands in iron meteorites due to large crystals of nickel–iron alloys.

winter solstice (24) The point on the celestial sphere where the sun is farthest south; also, the time of year when the sun passes this point, about December 22, and winter begins in the Northern Hemisphere.

X-ray burster (306) An object that produces occasional X-ray flares. Thought to be caused by mass transfer in a closed binary star system.

Zeeman effect (163) The splitting of spectral lines into multiple components when the atoms are in a magnetic field.

zenith (20) The point directly overhead on the sky.

zero-age main sequence (ZAMS) (246) The locus in the H–R diagram where stars first reach stability as hydrogen-burning stars.

zodiac (26) The band around the sky centered on the ecliptic within which the planets move.

zone (519) One of the yellow-white regions that circle Jupiter parallel to its equator.

Answers to Even-Numbered Problems

Chapter 1:
2. 3475 km; **4.** 1.05×10^8 km; **6.** about 1.2 seconds; **8.** 75,000 years; **10.** about 27

Chapter 2:
2. 4; **4.** 2800; **6.** A is brighter than B by a factor of 170; **8.** 66.5°; 113.5°

Chapter 3:
2. a) full; b) first quarter; c) waxing gibbous; d) waxing crescent; **4.** 29.5 days later on about March 30th; 27.3 days later on about March 24th; **6.** 6850 arc seconds or about 1.9°; **8.** a) The moon won't be full until Oct. 17; b) The moon will no longer be near the node of its orbit; **10.** August 12, 2026 [July 10, 1972 + 3 × (6585 ⅓ days)]. In order to get Aug. 12 instead of Aug. 11, you must take into account the number of leap days in the interval.

Chapter 4:
2. Retrograde motion: Jupiter, Saturn, Uranus, Neptune, and Pluto; Never seen as crescents: Jupiter, Saturn, Uranus, Neptune, and Pluto; **4.** Mars, about 18 seconds of arc; the maximum angular diameter of Jupiter is 50 seconds of arc; **6.** $\sqrt{27} = 5.2$ years

Chapter 5:
2. The force of gravity on the moon is about ⅙ the force of gravity on Earth; **4.** 7350 m/s; **6.** 5070 s (1 hr and 25 min); **8.** The cannonball would move in an elliptical orbit with Earth's center at one focus of the ellipse; **10.** 6320 s (1 hr and 45 min)

Chapter 6:
2. 3m; **4.** Either Keck telescope has a light-gathering power that is 1.56 million times greater than the human eye; **6.** No, his resolving power should have been about 5.8 seconds of arc at best; **8.** 0.013 m (1.3 cm or about 0.5 inches); **10.** about 50 cm (From 400 km above, a human is about 0.25 seconds of arc from shoulder to shoulder.)

Chapter 7:
2. 150 nm; **4.** by a factor of 16; **6.** 250 nm; **8.** a) B; b) F; c) M; d) K; **10.** about 0.58 nm

Chapter 8:
2. 730 km; **4.** 9×10^{16} J; **6.** 0.222 kg; **8.** about 3.5 times; **10.** 400,000 years

Chapter 9:
2. 63 pc; absolute magnitude is 2; **4.** about B7; **6.** about 1580 solar luminosities; **8.** 160 pc; **10.** a, c, c, d (use Figure 9-14 to determine the absolute magnitudes); **12.** 3.69 days or 0.010 years; about 1.2 solar masses; about 0.67 and 0.53 solar masses; **14.** 1.38×10^6 km, about the size of the sun

Chapter 10:
2. 60,000 nm; **4.** 0.0001; **6.** 1.5×10^6 years; **8.** 4.2×10^{35} kg or 210,000 solar masses (*Note:* Each hydrogen molecule contains two H atoms.)

Chapter 11:
2. 24.5 km/s; **4.** 2.98 km/s; **6.** 9.46×10^{10} s (about 3000 years); **8.** There are four ^{1}H nuclei in the figure. They are used in a series of reactions to build up nuclei from ^{12}C to ^{15}N. In the last reaction (with the ^{15}N) a ^{12}C and a ^{4}He are produced. Because the ^{12}C can be used in the next cycle, the net is four ^{1}H nuclei in with one ^{4}He nucleus out; **10.** 7.9×10^{33} kg

Chapter 12:
2. about 9.8×10^6 years for a 16-solar-mass star; about 5.7×10^5 years for a 50-solar-mass star; **4.** about 1×10^6 times less than present or about 1.4×10^{-6} g/cm^3; **6.** 2.4×10^{-9} or 1/420,000,000; **8.** about 3 pc; **10.** 3.04 minutes early after 1 year; 30.4 minutes early after 10 years

Chapter 13:
2. about 1.8 ly; **4.** about 16,000 years old; **6.** about 940 years ago (approximately 1060 AD); **8.** 2400 pc

Chapter 14:
2. 7.1×10^{25} J/s or about 0.19 solar luminosity; **4.** 820 km/s (assuming mass is one solar mass); **6.** about 11 seconds of arc; **8.** about 490 seconds

Chapter 15:
2. about 16 percent; **4.** 3.8×10^6 years; **6.** overestimate by a factor of 1.58; **8.** about 21 kpc; **10.** 7.8×10^{10} solar masses; **12.** 1500 K

Chapter 16:
2. 2.58 Mpc; **4.** 131 km/s; **6.** 28.6 Mpc; **8.** 1.64×10^8 yr; **10.** 4.49×10^{41} kg

Chapter 17:
2. 7.8×10^6 years; **4.** 0.024 pc; **6.** −28.5; **8.** about 29,800 km/s; **10.** 0.16

Chapter 18:
2. 57; 17.5 billion years; 11.7 billion years; **4.** 1.6×10^{-30} gm/cm^3; **6.** 76 km/Mpc; **8.** 16.6 billion years; 11 billion years; the universe could be older

Chapter 19:
2. It will look $206{,}265^2 = 4.3 \times 10^{10}$ times fainter, which is about 26.6 magnitudes fainter; 22.6 mag; **4.** about 3.3 times the half-life, or 4.3 billion years; **6.** large amounts of methane and water ices; **8.** about 1300

Chapter 20:
2. about 17 percent; **4.** 81×10^6 yr; they have been subducted; **6.** 0.22 percent

Chapter 21:
2. The rate at which an object radiates energy is proportional to its surface area, which is proportional to its radius squared (r^2). However, the energy an object has stored as heat is proportional to its mass and hence to its volume, and that is proportional to its radius cubed (r^3). So the cooling time will be proportional to the amount of stored energy divided by the rate of cooling, which is the same as the radius cubed divided by the radius squared (r^3/r^2). That shows that the cooling time is proportional to the radius (r); and that means that the bigger an object is, the longer it takes to cool; **4.** No. Their angular diameter would be only 0.5 second of arc. They would be visible in photos taken from orbit around the moon; **6.** 0.5 second of arc (assuming an astronaut seen from above is 0.5 meter in diameter); no; **8.** 10.0016 cm; **10.** Mercury, $V_e = 4250$ m/s; moon, $V_e = 2380$ m/s; Earth, $V_e = 11{,}200$ m/s

Chapter 22:
2. 33,400 km (39,500 km from the center of Venus); **4.** 61 seconds of arc; **6.** 260 km; **8.** 82 seconds of arc

Chapter 23:
2. 35 Earth days; **4.** 4.4°; **6.** about 0.056 nm; **8.** 5.2 m/s

Chapter 24:
2. 42 seconds of arc; **4.** 256 m/s; **6.** 8.8 s; yes; **8.** 12.3 km/s; **10.** 1.04×10^{26} kg (17.2 Earth masses)

Chapter 25:
2. one billion; **4.** 79 km; **6.** 0.18 km/s; **8.** 3.28 AU and 2.5 AU; **10.** 9 million km (0.06 AU); too small; **12.** 33 Earth masses

Chapter 26:
2. 8.9 cm; 0.67 mm; **4.** about 1.3 solar masses; **6.** 380 km; **8.** pessimistic, 2×10^{-5}; optimistic, 10^7

Index

Boldface page numbers indicate definitions of key terms.

A

absolute age, **462**
absolute bolometric magnitude, **180**
absolute visual magnitude, 178–179
absolute zero, **132**
absorption spectra, **136**
accelerated motion, 95
acceleration, **82**
  dark energy, 407–408
  gravity, 80
accretion, **428**
accretion disks, **273**
  fluctuations from, 310
  observations, 306–307
achondrites, 575, 576
active galactic nuclei (AGN), **368**
  elliptical galaxies, 375
  quasars and, 386
  supermassive blackholes and, 371–372
  visible spectrum, 375
active galaxies, **368**
  double-lobed radio sources, 370
  Seyfert, 368–370
  supermassive black holes in, 371–374, 376–378
  unified models, 374–376
  violence in, 377
active optics, 111, 113
Adams, John Couch, 557–558, 562
Adrastea, 520
Airy, George, 557
Aitkin Basin, 471
Al Magisti (The Greatest), 61
albedo, **455**
Alcor, 191, 193
Aldebaran, 15
  absolute bolometric magnitude, 180
  characterization, 249
  expansion of, 249–250
Aldrin, Edwin Jr. "Buzz," 461, 467
Alfonsine Tables, The, 58, 63, 69–70
Alfonso X, 58
Algol
  eclipses, 192
  mass transfer, 272–273
  paradox, 275
ALH84001 meteorite, 608
Allen Telescope Array, 611
Allende meteorite, 575
Almagest, 58
Alpha Centauri, 15, 18
Alpha Lyrae. *See* Vega
Alpha Persei. *See* Algol
Alpha Regio, 487–488
Alpher, Ralph, 395
Alpheratz, 13
Alt-azimuth mounting, 113
aluminum-26, 576
Alvarez, Luis, 592
Alvarez, Walter, 592
Amalthea, 520
amino acids
  assembly, 605
  DNA and, 599–600
  meteorites and, 604
Anaximander, 56
Andromeda Galaxy
  characterization, 14
  dust clouds in, 334
  dwarf galaxies in, 358
  look-back time, 350
Angstrom (Ä), **104**
angular diameter, 42–43
angular distance, **21**
angular momentum, **87**
  conservation of, 273
  problem, **418**
annular eclipses, **42,** 43
anorthosite, **468**
Antares, **14**
antimatter, **397**
aphelion, 23, 25
Aphrodite Terra, 488
apogee, **42**
Apollo Missions, 461, 463–467
Apollo-Amor objects, **581**
apparent visual magnitudes, **16**
arc, minutes of, **21**
archaeoastronomy, **53,** 55
Arecibo dish, 474
Arecibo message, 611
Arecibo Observatory, 120
Ariel, 554
Aristarchus, 58
Aristotle, 56–58, 80, 85
Armstrong, Neil, 461, 467
Ascraeus Mons, 496
associations, **226**
  Milky Way, 320
  O and B, 329–330, 332
  T Tauri stars, 226–227
asterisms, **14**
asteroid belt, 421, 580–581
asteroids, **420**
  collisions among, 437, 580
  evidence of, 420–421
  fragmentation, 583–584
  impact on Earth, 592–595
  mass of, 578
  nonbelt, 581–582
  observations of, 578–579
  origin of, 582–584
  properties of, 577, 580
  types of, 582–583
Astroid 1997XF$_{11}$, 594
astrology, 23–24, 26–27
Astronomia Nova, 71
astronomical image, 109
astronomical unit (AU), 5
astronomy
  constants, 619
  Copernicus revolution, 59, 62–63
  Galileo's contributions, 64–68
  Greek astronomy, 55–56
  history of, 15–17
  Kepler's analysis, 70–74
  measurement in, 2–10, 16–17, 176–177, 617–619
  modern, 74–75
  Newton's contributions, 79–86
  perspective, 9
  planetary motion puzzle, 68–74
  post-Newton, 92–94
  Ptolemaic universe, 58–59
  roots of, 53–59
  Tycho's observations, 68–70
  units used in, 619
atlas, 14
atmospheres
  Earth's, 454–457
  Jovian planets, 511
  Jupiter, 515–516, 518–519
  Mars, 496–497
  Neptune, 558–560
  solar, 150–156
  terrestrial planets, 445–446
  Uranus, 546–547, 549
  Venus, 484–485
atmospheric windows, 105–106
atomic spectra. *See* spectra
atoms
  behavior, 129–130
  excitation, 131–132
  model, 128
  neutral, 222
  solar system comparison to, 41
  types, 128–129

Augustine of Hippo, 67
auroras
  Earth, 167
  Jupiter, 515
  Saturn, 529

B

B associations, 329–330, 332
B0 stars, 304
Baade, Walter, 288
Babcock model, **164,** 166–167
Balmer lines, 135, 138
Balmer, Lyman, 135
Barberini, Cardinal Antonio, 65
Barnard 86, 207
Baronius, Cardinal Cesare, 67
barred spiral galaxies, 345–346, 362
Barringer Meteorite Crater, 571–572
basalt, 452
Becklin-Neugebauer (BN) object, 237
Becquerel, Henri, 546
Bell, Jocelyn, 290
Bellarmine, Cardinal Roberto, 66, 67
BeppoSAX, 309
Bethe, Hans, 226
Beta Pictoris, 435
Beta Regio, 488
Betelgeuse
  brightness, 15
  characterization, 250
  expansion of, 250
  parallax, 186
  size, 183
big bang
  argument for, 400
  background radiation from, 395–396
  deuterium produced in, 403
  elements produced during, 327
  impact of, 8–9
  lithium produced in, 403
  look-back-time, 394
  model, problems, 404–405
  necessity of, **394**–395
  nuclear reactions in, 397–398
  star formation, 399
  story of, 396–400
  understanding, 400
Big Dipper
  ancient people's view of, 13
  observing, 15
  quasars in, 381
  spectra, 191
big rip, **408**
binary systems
  accretion disks, 273
  black holes in, 304
  calculating masses, 188
  eclipsing, 190–193
  evolution of, 271–275
  general aspects, 187–188
  mass transfer, 272–273
  neutron stars in, 295–297
  nova explosions, 273–274
  scuba diving comparison, 195
  spectroscopic, **189**–190
  visual, **188**–189
binding energy, **129**–130
bipolar flows, 226
birth line, **225**
BL Lac objects, **374**
black body radiation, **132,** 134–135
Black Cloud, 207
black dwarfs, **270**
Black Eye Galaxy. *See* M64 galaxy
black holes, **300**
  accretion disk observations, 306–307
  candidates, 305
  charged, 302
  escape velocity, 299–300
  formation, 301
  gravitational pull, 302–303
  light leaving, 303
  Milky Way, 336
  relativistic effects, 302–303
  Schwarzschild, 300–302
  search for, 303–306
  supermassive, 338
  X-ray emission, 304, 306
Black Widow, 298
blazars, **374**
blue photons, 204
blueshift, **142**
Bode, J. E., 545
Bok globules, 226, 228
bow shock, **450**
Brahe, Tycho, 73
breccias, **468**
Brecht, Bertolt, 67
brightness
  comets, 421
  distance and, 178
  Greek letter indicators, 14
  intrinsic, 178–180
  Neptune, 553
  Sirius, 197
  stars, 15–16
  supergiants, 187
brown dwarfs, **245**–247
Bruno, Giordano, 67–68
Burroughs, Edgar Rice, 495

C

C rings, 532
C-type asteroids, 582–583
C153 galaxy, 362
calcium emissions, 166
calcium, once-ionized, 206, 208
calibrations, 318
Callisto, 522
Caloris Basin, 476–477
Cambrian explosion, **605**
Cambrian periods, 606
Canis major, 196–198
Canis Major Dwarf Galaxy, 356
Canis Majoris, 271
Cannon, Annie J., 138
Capella, 14
capture hypothesis, **472**
carbon deflagration, **280**
carbon dioxide, 455–457, 486
carbon fusion, 253, 254
carbon monoxide (CO)
  abundance of, 214
  importance of, 212
  in radio maps, 331
carbon-oxygen core, 280
carbonaceous chondrites, **574**
Cartwheel Galaxy, 361
Cassegrain focus, 111
Cassini spacecraft, 516–517, 531
Cassini's Divisions, 533
Cat's Eye Nebula, 269
catastrophic hypothesis, **417**
cause and effect principle, 85, 130
celestial equator, 24
celestial pole, 19–20
celestial sphere, 17–18
cell reproduction, 601
Centarus A, 373
Center for High Angular Resolution Astronomy (CHARA), 183
center of mass, 86, 89
Cepheids
  calibration, 326
  discovery, 259
  distances, 317–318, 348
  identification of, 344
  locating, 349
Ceres, 564, 580, 583
chains of inference, 189
Chandra X-ray Observatory, 123, 124, 408
Chandrasekhar, Subrahmanyan, 270
Chandrasekhar limit, **270**
CHARA array, 115
charge-coupled devices (CCDs), 115
Charon, 563–564

chemical evolution, **603**
Chicxulub crater, 592–593
Chinese eclipse symbol, 40
Chiron, 581–582
chlorofluorocarbons (CFCs), **456**
chondrites, **574**
chondrules, **574**
chromatic aberration, **107**
chromosomes
  DNA and, 599
  genes and, 601
  recognition of, 3
chromosphere, **44**
  magnetic fields, 166–170
  solar, 152
circular velocity, 86, 88–90
circumpolar constellations, 18
climate change
  greenhouse effect, **455**–456
  Milankovitch hypothesis, 27–28
closed clusters, 255
closed orbits, 86, 89
closed universe, **401**
clouds
  black, 207
  classification, 207–208
  Jupiter's, 519
  Magellanic, 328
  molecular (*See* molecular clouds)
  Oort, 590
  Saturn, 529
  Venusian, 484
cluster method, **353**
clusters
  aging of, 256
  Coma, 354, 362
  evolution of, 255–258
  galaxies', 356–358
  globular (*See* globular clusters)
  grouping of, 7–8
  H-R diagrams, 256–257
  Local Group, 356–358
  Milky Way, 320
  observing, 255
  types of, 255–256
CNO cycle, 231–232, 234
cobalt atoms, 282–283
cocoons, **224**
cold dark matter, **403**
collapsars, **309**
colliding galaxies
  evidence of, 358–359
  frequency of collisions, 358
  process, 360–361
Collins, Michael, 467
collisional broadening, **144**
colors
  galaxies', 348
  intensity measurement, 115–116
  sky, 3
  total lunar eclipse, 38
  true, production, 115
Columbus, Christopher, 13, 58
Coma cluster, 354, 362
Comet 73P/Schwassmann-Wachmann, 587
Comet Borrelly, 585
Comet Encke, 571
Comet Giacobini-Zinner, 593
Comet Hale-Bopp, 121, 584
Comet Halley, 584–585, 590
  discovery, 544
  nuclei, 421
  orbit, Earth nearing, 571
  panic over, 569
Comet Hyakutake, 569, 584
Comet Linear, 454, 587
Comet McNaught, 568
Comet Mrkos, 586
Comet Shoemaker-Levy 9, 520, 588
Comet Tempel 1, 585, 588
Comet Wild 2, 585, 587
comets, **421**
  coma, 586
  history, 569
  impact on Earth, 592–595
  impact on Jupiter, 520–521
  Kuiper Belt, 590–592
  nuclei, 584–585
  orbits, 424
  origins of, 590
  planet formation and, 591
  properties, 584
  tails, 586
compact objects, **270**
  accretion disks, 306–307, 310
  gamma-ray bursters, 308–310
  jet energy from, 307–308
  X-ray busters, 306
comparative planetology, **444**
composite volcanos, 499
Compton Gamma Ray Observatory, 122, 124, 308
condensation, **428, 472**
condensation sequence, **427**–428
conservation of energy law, 242
conservation of mass law, **242**
constellations, **13**
  ancient, 13
  astrology and, 23–24
  circumpolar, 18
  northern hemisphere sky, 628–634
  precession, **19,** 26
  southern hemisphere sky, 634–639
  star charts, 627–639
  stars in, 14
contour maps, 119
contraction, 223–224
convection, **152**
convective zone, **159**
Copernicus, Nicolaus
  Earth's origins hypothesis, 417
  education of, 59
  finite universe belief of, 392
  heliocentric theory, 62–63
  influence of, 73
Copernicus 200, 464
Cordelia, 552
corona
  activity, 153–154
  density, 153
  holes, 169
  magnetic fields, 168–170
  spectra, 153
  temperature, 154
coronae, **492**
coronagraphs, **153**
coronal gas, **213**–215
coronal mass ejections (CMEs), 166, 169
cosmic background radiation, 395–396
cosmic element building, 397
cosmic expansion
  acceleration of, 406–407
  big bang and, 397
  discovery of, 393
  relativity theory and, 401
cosmic jets, 370, 372–373
cosmic rays, 124
cosmological constant, **407**
cosmological principle, **400–401**
cosmology, **390**
  dark matter in, 402–404
  twenty-first century, 404–413
Coulomb barrier, **158,** 231
Coulomb force, **129**
Crab Nebula
  age of, 281
  discovery of, 280
  energy, 291
  filaments of, 280–281
  pulsar in, 290, 294–295
craters
  commonness, 422
  formation, 446
  history of, 469
  impact, 462, 464–465
  Jovian, 511
  lunar, 462

craters *(continued)*
Phobos, 506
scarring by, 433
Credo ut intelligame, 67
Cretaceous period, 592
Crick, Francis, 18
critical density, **402**–404
critical point, **513**
Curtis, Heber, 344
curvature
space-time, 97–99
universe, 406, 409–413
Cuvier, Georges, 420
Cygnus A, 372, 374
Cygnus superbubble, 213–214
Cygnus X-1, 304, 307

D

dark age, **398**
dark energy, **407**–408
dark matter
detection of, 355–356
galaxical, 354–356
gravitational lensing, 355–356
peculiarity of, 356
universe, 402–404
dark nebulae, 203, 205
Darwin, Charles, 345
data manipulation, 487
Davis, Raymond, Jr., 159–160
de Buffon, Georges-Louis, 417
de Coulomb, Charles-Augustin, 129
de Laplace, Pierre-Simon, 417
De Revolutionibus Orbium Coelestrium (On the Revolutions of the Celestial Spheres), 59, 62, 66
De Stella Nova (The New Star), 69
death of stars. *See* stellar death
debris disks, **435**–436
Deep Impact spacecraft, 585, 587–588
deferent, 58, 61
deflagration, 280
degenerate matter, 250–**251,** 270
Deimos, 505–507
densities, **144**
measuring, 145
pressure *versus,* 209
stars, 194–195
uncompressed, **427**
density wave theory, **331**–333
Descartes, René, 417
deuterium, **158**
fusion, 244–445
produced in big bang, 403
Devil's Hole, 28–29
Dialogo Dei Due Massimi Sistemi (Dialogue Concerning the Two Chief World Systems), 66, 67
diameters
galaxies', 351–352
H-R diagram, 181–183
interferometric observations, 183
diamond ring effect, **44**
Dicke, Robert, 395
differential rotation, **164,** 322
differentiation, **429,** 446
diffraction fringe, **108**
Diggers, Thomas, 391
dinosaurs' extinction, 592–593
Dione, 537, 539
direct gravitational collapse, **431**
dish reflector, 118
disk component, **318**–321
disks
debris, **435**–436, 440
planet-forming, 434–435
distance indicators, **348**
distance modules, **179**
distance scale, **350**
distances
angular, **21**
brightness and, 178
Cepheids, 317–318, 348
galaxies, measuring, 348–350
magnitude formula, **179**
planets from sun, 5
quasars, 379–382
stars, measuring, 175–178
distant civilizations, 608–612
distributary fan, 502
DNA, 18, 599–602
Doppler broadening, **144**
Doppler effect, **142**
calculating velocity, 143–144
measuring, 142–143
redshift interpretation, 393, 401
spectroscopic binary systems, 189
double pulsars, 295–297
double-exhaust model, 370, 372
double-lobed radio sources, **370,** 377
Drake equation, **612**
dust
comet, 586
disks, 434 (*See* Debris disks)
galaxies rich in, 345, 347
interstellar, 212–213, 216
dwarf elliptical, 362–363
dwarf galaxies, 358
dwarf planets, 562–565
dwarf stars
black, **270**
brown, **245**–247
density, 194–195
red, **183,** 245, 265–266
white, 194–195, 267–271, 280
dynamo effect, **164**

E

Eagle Nebula, 174, 227, 229–230
Earth
active crusts, 451–454
Apollo objects hits, 581
asteroid's impact on, 592–595
atmosphere, 105, 454–457
auroras on, 167
climate change, 27–28
comet's impact on, 592–595
cosmic rays striking, 124
death of, 274–275
dwarf planet *versus,* 565
early history, 446–447
eclipsing binary from, 191–192
geologic time, 605–607
interior, 447–450
leaving, 565
life origins, 602–605
local supernovas and, 283–284
magnetic field, 450–451
meteorite impact, 569–572, 594
moon's orbit, 35–38
nature of, 56–58
ocean floors, 453
orbit, 27, 41, 88–89
origins, 417–419
rotation, 4, 18–23, 30
rotation around sun, 30
shadow, 35, 38
size of, 419
stellar parallax, 545
summer solstice, 24
sun's death and, 274–275
surface, 445
tidal forces, 92
Venus *versus,* 493
winter solstice, 25
earthquakes, 448
eccentrics, 58
eclipse season, **47**
eclipse year, **49**
eclipses
conditions for, 46–47
cycle pattern, 48
lunar, 35–39

predicting, 46–49
saros cycle, 48–49
solar, **33,** 40–46
space perspective, 47–49
eclipsing binary systems, 190–193
ecliptic, **23,** 24
ecosphere, **609**
Eddington, Arthur, 226
Edgeworth, Kenneth, 591
Egg Nebula, 269
Egyptian mythology, 53
Einstein, Albert, 78, 397
about, 94
dark matter equivalency, 408
gravity description, 355
relativity theory, 94–98, 401
space-time, 300
electromagnetic radiation, **103**
spectrum, 104–106, 116–117
wavelength, 103–104
electron microscope, 448
electron shells, 129–131
electrons, **128**
energy levels, 251
orbits, 137
element-building process, 325
elements
periodic table, 626
solar, 141
Elephant Trunk, 228
Eliot, T. S., 616
ellipse, **72**
elliptical galaxies
AGN in, 374
characterization, 345–346, 348
collisions and, 363
distance indicators, 351
interactions, 359, 362
origins, 359
elongations, 625
Elysium region, 500
emission nebulae, 203–204
emission spectra, **136**
Enceladus, 537–538, 540
Encke's Divisions, 533
energy, **91**
binding, 129–130
electron, 251
flow, 267
jets of, 307–308
kinetic, 91
levels, **131**
nuclear fusion, 156–161
photon, 104
potential, 91
thermal, **132**
transport, 159, 232
enzymes, 599
Ephemerides of Minor Planets, 580
epicycle, 58
equant, 58
Equatorial mounting, 113
equinoxes, 23
equivalence principle, **97,** 98
Eratosthenes, 57, 58
ergosphere, **302**
Eris, 564
Eros, 420–421, 578
escape velocity, 299–300
Eta Aquarids shower, 571
Eta Carenae
light and gas flowing from, 223
mass, 245
nebula, 175
ethane, 536
Eudoxus, 56
Euler, 464
Europa, 524–525, 607
European Extremely Large telescope (E-ELT), 113
evaporating gaseous globules (EGGS), 229
evening star, **23**
event horizon, 300
evolution, **602**
biological, 602
chemical, 603
galaxies, 356–364
Martian moons, 505–507
slow surface, 447
stellar (*See* stellar evolution)
evolutionary hypothesis, **417**
excited atom, **131**
extinction
dinosaurs', 592–593
interstellar, **206**–207
extrasolar planets, **435**–440
Extreme Ultraviolet Explorer (EUVE), 214
eyepiece, **107**

F

falling universe, 99
false-color images, 115
far-infrared range, 121
Faraday, Michael, 540
Feynman, Richard, 368
field, 85
field of view, **3**
filtergram, **153**
firestorms, 361
fission. *See* nuclear fission
fission hypothesis, **472**
Fitzgerald, Edward, 128
flares, 166, 169
flat universe, **402**
flatness problem, **404**
Fleming, Alexander, 546
flocculent, **333**
flooding, 446–447
floppy mirrors, 113
flux, 16
focal length, **107**
focus, 111–113
forbidden lines, **203,** 206
forward scattering, 516
Fra Mauro, 467
fraud in science, 305
Frederick II, King, 69, 70
free-fall contraction, **223**
frequency, **104**
Frost, Robert, 103
full moon, 35
fusion. *See* nuclear fusion

G

galactic cannibalism, 358, 360
galactic corona, 323
galactic disk, 327
galactic fountains, 325–**326**
galaxies, **6**. *See also* Milky Way Galaxy
classification of, 345
colliding, 358–363
color, 348
Coma cluster, 354
dark matter in, 354–356
diameter, 351–352
discovery of, 343–344
distances, 348–350, 356
farthest, 363
grand-design spiral pattern, 338
gravitational lensing, 402
grouping of, 7
Hubble Law, 350–351
luminosity, 351–352
mass, 352–353
numbers of, 344–345
origins, 359, 362–363
properties, 353
starburst, **359**
supermassive black holes in, 353
types of, 345
ultraluminous infrared, 362
Galaxy 3C31, 374
Galaxy 3C75, 374
Galilean moons. *See also* specific moons
discovery, 423
history of, 527–528

Galilean moons *(continued0*
orbits, 512, 515
size, 420
Galilei, Galileo, 52, 392
about, 64
Copernican model defense, 64–66
Earth's origins hypothesis, 417
gravity experiment, 80–81
influence of, 79–82
Neptune observations, 557–558
observations, 65
Saturn's rings observations, 530
solar observations, 161
tidal forces, 91
trial of, 52, 67–68
Galileo moons, 522
Galileo spacecraft, 513, 518, 520
gamma rays, 105, 397
gamma-ray bursters, **308**–310
Gamow, George, 395
Ganymede, 522–524
Gaposchkin, Sergei, 142
gas
big bang and, 396–400
coronal, **213**–215
galaxies rich in, 345
interstellar, 212–213, 216
low-density, 207
opacity of, **232**
Gaspra, 578
gathering light, 108
general theory of relativity, **96**
big bang and, 401
planetary obits and, 97–98
space-time curvature, 97–99
genes, 599
genetics, 3
geocentric universe, 57, 58
geologic time, 605–607
George III, King, 545
Georgium Sidus (George's star), 545
geosynchronous satellites, 88
Giant Magellan Telescope, 111
giant molecular clouds, 221–223, 332
giants, **183**
density, 194–195
expansion of, 249–250
lifespan, 254
Gliese 710, 590
glitch, **291,** 294
Global Oscillation Network Group (GONG), 155
global warming, **456**
globular clusters
distance measures and, 348
distribution, 316–317, 319
evolution of, 255–257
halos and, 320, 327
locating, 349–350
metal-poor, 324
turnoff points, 326
Goodricke, John, 258
gossamer rings, **520**
Grand Maximum, 170
grand unified theories (GUTs), **405**
grand-design spiral pattern, 338
granulation, **151**–152
grating, 116
gravitation
collapse, **429**
lensing, **355**–356, 380, 402
mutual, 84–86
radiation, 296
redshift, 303
gravity
acceleration of, 80
Einstein's description, 355
field, 85
law of, 80, 85
moon, 90
motion and, 80–83
relativity and, 94–98
solar, 98
star formation and, 284
tides and, 90–92
Great Dark Spot, 558
Great Observatories Origins Deep Survey (GOODS) program, 345
Greek alphabet, 625
Greek astronomy, 55–56
greenhouse effect, **455**–456, 485
Gregory XV, Pope, 66
grindstone model, 316
grooved terrain, **522**
ground state, **131**

H

half-life, **424**
Halley, Edmund, 391
halo, **320,** 327–328
Harmonice Mundi (Harmony of the World), 71–72
HD80715, 190
heat. *See* thermal energy
heat of formation, **430**
heavy bombardment, **420**
heliocentric universe, 59, 63
helioseismology, **154**–155
helium
core expansion, 249–253
elements heavier than, 325
flash, **251**–252
fusion, 251–253
in big bang, 397–398
Helix Nebula, 268
Herbig-Haro objects, 226, 229
Hercules galaxy cluster, 357
Hercules X-1, 297
Herman, Roger, 395
Herschel, Alexander, 544
Herschel, Caroline, 315
Herschel, William
about, 315
galaxy discoveries by, 343
Uranus discovery by, 544–545, 557
Hertzsprung, Ejnar, 181, 316
Hertzsprung-Russell (H-R) diagram, **181**
function of, 182–183
star clusters, 256–257
Hesperian period, **505**
Hewish, Anthony, 290
HH30, 227
HI clouds, **208,** 214–215
Hickson Compact Group 87, 351
high-speed computers, 114
HII clouds, 214
HII gas, 208
HII regions, 203–204
Hipparchus, 15–17, 19, 58
Hipparcos satellite, 177, 186, 326
Hirayama, Kiyotsugu, 580
horizon, **20**
horizon problem, **405**
horoscopes, **26**
hot dark matter, **403**
hot Jupiters, **438**
hot spots
cosmic jets and, 372
presence of, 370
rising magma and, 452
Hubble constant, **351,** 394
Hubble Deep Field, 344–345
Hubble Law, 350–**351,** 379
Hubble Space Telescope
acceleration detection, 408
imaging limits, 420
launch of, 122
observation parameters, 124
star detection, 16
universe curvature detection, 406
young star detection, 435
Hubble time, 394
Hubble, Edwin P., 348
about, 122
cosmological constant, 407
galaxy measurement law, 350–351, 393
galaxy shape system, 345
photographs, 344
Hulse, Russell, 296
Humason, Milton, 350–351
Huygens probe, 535

Huygens, Christian, 530
Hydra cluster, 393
hydrogen
heavy (*See* deuterium)
ionized, 208
liquid metallic, **513**
molecules, 211
molecules in water, 270
neutral, 209–211
hydrogen atoms, 128–129
energy levels, 210
photon absorption, 131
transitions in, 137
hydrogen fusion, 157–159, 245
hydrogen-fusing shell, 249–250
hydrostatic equilibrium, **233,** 244–245
hypernovae, **309**
hypothesis, **83,** 466, 472

I

Iapetus, 537–538
ice ages, 27, 30
ice cores, 284
ice line, **427**
ice rings, 532–533
icebergs, 270
Ida, 579
imaging systems, 115–116
Imbrium Basin, 470, 476
impact cratering, 462, 464–465
Indian eclipse symbol, 40
inference, chains of, 189
inflationary theory, 404–406
inflationary universe, **405**
Infrared Astronomy Satellite, 212
infrared radiation, **104**
dust, 212–213, 216
long wavelengths, 105
measurement, 121
Infrared Telescope Facility (IRAF), 121
instability strip, **258,** 259
interacting galaxies, 360–361
intercrater plains, **477**
interferometric observations, 183
interferometry, 114–115
International Astronomical Union, 13
International Ultraviolet Explorer (IUF), 124
interstellar absorption lines, **206**–209
interstellar cycle, 215–218
interstellar extinction, **206**–207
interstellar medium, 203
21-cm radiation, 209–211
absorption lines, 206–209
clouds, 207–209
components of, 214–215
dust in, 212–213
evidence for, 209
extinction, 206
gas and dust in, 212–213, 216
magnetic fields, 222
mass, 217
model of, 214–218
molecules, 211–212
nebulae, 203–206
pressure, 209
reddening, 206
stars from, 221–230
turbulence in, 222
ultraviolet spectrum, 214
visible-wavelength observations, 203–209
X rays from, 213–214
interstellar reddening, **206**–207
intrinsic brightness, 178–180
intrinsic variables, **258**
cepheid, 258
evolution, 258–262
period changes in, 261
pulsating, 258–261
RR Lyrae, 258
inverse square law, **84,** 178
Io
atmosphere, 525
distance from Jupiter, 514
interior, 525–526
internal heat, 528
orbit, 512
tidal heating, 527
volcanos, 526
Io flux tube, **515**
ion, **129**
ionization, **129**
ionized hydrogen, 208
ions, 206
iron
core, 276–278
in massive stars, 426
meteorites, **572**–574
irregular galaxies, 345, 347–348, 351
Ishtar Terra, 488, 492–493, 494
island universe, **343**
isotopes, **129**
isotropic, **400**

J

J1550-564, 306
James Webb Space Telescope, 123, 124
John Paul II, Pope, 67–68
joules, **91,** 134
Jovian planets, 420. *See also* specific planets
atmospheres, 511
characteristics, 422–423
list of, 511
possibility of life on, 607–608
problems with, 430–431
rotation, 431–432
satellite systems, 511
temperature escape velocities, 497
jumbled terrain, **471**
Juno, 580
Jupiter
atmosphere, 515–516, 518–519
celestial profile, 513
comets impact on, 520–521
elements on, 513
history, 521
interior of, 423
magnetic field, 514–515
mass of, 245
moons, 65, 512, 520–528, 539
orbit, 439
rings, 516–520
Saturn *versus,* 528–529
size of, 419
surveying, 512–514
tides, 589

K

Kant, Immanuel, 343
Kapteyn, Jacobus C., 316
Keck Foundation, 534
Keck telescopes, 113, 115
Kelvin temperature scale, **132**
Kennedy, John, 463
Kepler, Johannes, 52, 64, 391
about, 70
analysis of, 70–72
Earth's origin hypothesis, 417
influence of, 79–80
Mercury's orbit, 97
motion laws of, 72–74, 87, 90–94, 188
supernova, 281
Keplerian motion, **322**
Kerr, Roy P., 302
Kerr black hole, **302**
kilometer, **3**
kiloparsec (kpc), **316**
kinetic energy, **91**
Kirchhoff, Gustav, 136
Kirchhoff's laws, 135, 136
Kirkwood, Daniel, 580–581
Kirkwood gaps, 580
Kitt Peak National Observatory, 112, 114
Koch, Robert, 85
Kuhn, Thomas, 64
Kuiper, Gerard P., 421, 591
Kuiper belt, **421**
comets in, 590–592
ice objects, 432
Neptune and, 565

Kuiper belt *(continued)*
objects in, 564–565
orbits, 431

L

L dwarfs, **140**
Lagrange, Joseph Louis, 272
Lagrangian points, **272**–273
Lakshmi Planum, 488, 492–493
Large Binocular Telescope (LBT), 111, 114–115
large dark nebulae, 205
Large Magellanic Cloud, 281
large-impact theory, **472**
large-scale structure, **409**
late heavy bombardment, **420**
law of expansion, 393
laws of motion, 82–83, 88–89
Leavitt, Henrietta S., 316
lens, 107
Leverrier, Urbain Jean, 557–558, 562
Levy, David H., 520
life
evolution of, 602
in distant civilizations, 609–613
in solar system, 607–609
origins of, 602–604
physical basis of, 599
time line, 607
life zone, **609**
light
analyzing, 116–117
colors, 3, 38
gathering, 108
intensity, 84
leaving black holes, 303
matter interaction, 131–135
particles, 104
stars (*See* starlight)
velocity of, 95–96
visible, 103, 104
waves, 103–104
light curve, **191**
light pollution, **110**
light-year (ly), **6**
light-gathering power, **108**
light-house model, **290,** 292–293
limb darkening, **152**
limb of, **152**
Lincoln Near-Earth Asteroid Research (LINEAR), 581
line of nodes, **47**
linear diameter, 42–43
liquid metallic hydrogen, **513**
Lithium, 403
Little Ice Age, 167
lobate scarps, **476,** 479
local bubble, **214**
Local Group, 356–358, 358
local supernovas, 283–284
local void, **214**
long-period comets, 590
look-back time, **350**
big bag, 394
quasars, 385–386
Lowell, Percival, 494–495, 562–563
Lowell Observatory, 562
Lowell Observatory Near Earth Object Search (LONEOS), 581
lower main-sequence stars, 265–266, 266
luminosity, **179**
calculating, 180
classification, 184–186
diameters and, 180–181
galaxies', 351–352
H-R diagram, 181–183
mass relationship, **194**–195
luminosity classes, **184**–186
luminous stars, 197
lunar eclipses, 35–39
lunar landing module (LM), 463, 467
lunatic, 34
Luther, Martin, 59, 74
Lynx Arc, 1

M

M stars, 196, 225
M-type asteroids, 582–583
M2-9 Galaxy, 269
M32 Galaxy, 359
M33 Galaxy, 354
M64 Galaxy, 359, 361
M87 Galaxy, 371, 385
M101 Galaxy, 342
Maat Mons, 490, 499
Magellan spacecraft, 486
Magellanic clouds, 328, 347
magnetars, **295**
magnetic carpet, **154**
magnetic cycle
Babcock model, 164–165
extension of, 166–167
Maunder butterfly and, 165
powering, 164
magnetic fields
active regions, 163–164
Earth, 450–451
energy stored in, 166–167
Europa, 523
Ganymede, 523
interstellar medium, 222
Jupiter's, 514–515
neutron stars, 295
spinning, 292
Uranus, 550
magnetic resonance imaging (MRI), 487
magnetosphere, **450**
magnifying power, **110**
magnitude scale, **15**–17
magnitude-distance formula, **179**
main-sequence stars
evolution of, 249–255
globular clusters, 257
H-R diagram, 182
life of, 245–248
lower, 265–271
properties of, 620
stellar models, 242–244
mantle, 445
maps
radar, 486
radio, 329–331
radio energy, 119
spiral arms, 329–331
Mare Crisium, 470
Mare Humorum, 470
Mare Imbrium, 467, 470–471
Mare Orientale, 465
Mare Serenitatis, 470
maria, **462**
Mariner 10 spacecraft, 473, 477
Mars
atmosphere, 496–497
canals on, 494–496
celestial profile, 495
geology of, 499–501
history of, 504–505
meteorites from, 425
moons, 505–507
polar caps, 496, 498
possibility of life on, 608
temperature, 483, 497
volcanoes, 499–500
water on, 501–504
Mars Express, 502
Mars Global Surveyor, 501
Mars Odyssey probe, 498, 502
Mars Orbiter, 501
mass, **84**
asteroid, 578
center of, 86, 89
Eta Carinae, 245
galaxies', 352–353
gravity and, 85
interstellar medium, 217
Jupiter, 245
Milky Way, 321–323
neutron stars, 288

sun, 91
sun-like stars loss of, 266
supermassive blackholes, 376–377
Titan, 534–535
transfer, 272–273
Mass-luminosity relation, **194,** 244–245
massive stars
core, sudden collapse of, 289
death of, 275–284
iron core, 276–278
iron production in, 426
nuclear fusion in, 275–276
supernova deaths, 277–279
Mathematical Syntaxis, 61
Mathilde, 578
matter. *See also* dark matter
light interaction, 131–135
origin of, 426–427
Mauna Kea, 598
Mauna Loa, 499
Maunder minimum, 161, 165, 167
Maxwell Montes, 487–488
Maxwell, James Clark, 405, 487, 530
measurement
astronomical, 2–10, 16–17, 619
color intensity, 115–116
constants, 619
Hubble Law, 350–351, 393
infrared radiation, 121
methods, 176–177
metric conversions, 617–618
powers of 10 notation, 618
SI units, 617
temperature scales, 618
megaparsec (Mpc), **348**
Mendel, Gregor, 3
Mercury
celestial profile, 475
distance from sun, 423
elongations, 625
history of, 478–479
interior of, 478
movement, 23
orbit, 97, 419, 473–475
plains of, 477–478
rotation, 474–475
size of, 422
surface of, 476–478
Merope, 205
MESSENGER space craft, 474, 477–478
metals, **323**
abundance of, 326, 339
orbits and, 329
metastable levels, 203
meteor showers, 570, 624
meteorites, **424**
amino acids in, 604
analysis of, 572–576
characterization, 425–426
impact on Earth, 594
orbits, 570–571
origins of, 576–577
proportions of, 574
terrestrial impact, 569–572
types of, 574
meteoroid, **424**
meteors, **421,** 424
methane, 536
Metis, 520
metric conversions, 617–618
metric system, 617
microwaves, 104–105, 396
Milankaiteh, Milutin, 27
Milankovitch hypothesis, **27**–29
Milky Way Galaxy, **6**
age of, 326
black hole in, 336
characterization, 7
discovery, 316–318
disk of, 318–321
dwarf galaxies in, 358
element-building process, 325
first studies, 315–316
galactic fountains, 325–326
galaxies in, 356, 427
gas and dust in, 323
history of, 326–329
mass of, 321–323
neutral hydrogen in, 211
nucleus, 338–339
origin of, 323–329
rifts and holes, 205
rotation, 322–323
size of, 351
spiral arms, 329–338
stellar populations, 323–325
structure, 318–321
Miller experiment, 603
Miller, Stanley, 603
millisecond pulsars, **298**
Mimas, 533
Miranda, 552, 554–556
mirrors, 112–113
Mitchell, John, 300
Mizar, 191
molecular clouds
density, **214**
star birth in, 221–223
molecules, **129,** 211–212
momentum, **82**
momentum angular, **87**
moon (of Earth)
angular diameter, 40–42
Apollo Missions, 461, 463–467
atmosphere, 461–462
celestial profile, 461
craters, 462, 464–465, 473
diameters, 42
eclipses, 35–39
gravity, 90
history of, 469–472
human connection to, 49
light pollution and, 110
motion of, 33–34, 42
orbit, 4, 46
origin, 472–473
phases, 34–35
rocks, 467–469
rotation, 461
shadows, 42–44
surface, 462
moons
coorbital, 539
geosynchronous, 88
Jovian, 511
Jupiter, 65, 512, 520–528, 539
Martian, 505–507
Neptune, 560–561
Pluto, 563–564
principal, 623–624
Saturn, 534–546
small, 530
Uranus, 551–555, 557
morning star, **23**
motion. *See* planetary motion
Mount St. Helens, 490
mountains, 420
Murchison meteorite, 604
mutants, **602**
mutual gravitation, 84–86
Myan eclipse symbol, 40
Mysterium Cosmographicum, 70

N

N44, 221
N44C, 216
nanometers (nm), **104**
National Astronomy and Ionosphere Foundation, 120
National Radio Astronomy Observatory, 119
National Science Foundation, 534
natural law, **83**
natural motions, 80–82
natural selection, **602**
NGC2264, 228
NCG7251 galaxy, 361
neap tides, **91**

Near Earth Asteroid Rendezvous (NEAR), 577
NEAR spacecraft, 420–421
near-Earth Objects (NEOs), 581
nebulae, **203**
- forbidden lines, 203, 206
- NGC 6751, 128
- planetary, **266**–270
- solar, 427–430, 433–434
- spiral (*See* spiral nebulae)
- types, 203–205
- young stars in, 228

nebular hypothesis, **417**–418
Neptune
- atmosphere, 558–560
- brightness, 553
- celestial profile, 559
- colors, 562
- discovery of, 557–559
- Great Dark Spot, 558
- history of, 561–562
- interior, 558–560
- interior of, 423
- Kuiper belt and, 565
- moons, 560–561
- rings, 552–553, 560
- size of, 419

Nereid, 560
neutral atoms, 222
neutral hydrogen, 209–211
neutrinos, **158**
- in neuron star formation, 288
- solar, 159–161

neutron stars, **288**
- accretion disk observations, 306–307
- complexity of, 290–291
- mass, 288
- pulsars, 290–299
- recognizing, 294–295
- starquakes, 294
- theoretical prediction of, 288–290
- X-ray bursters, 306
- X-ray detection, 299

neutrons, **128**
New grange, 53–54
Newton, Isaac, 71, 74, 78
- about, 79
- influences on, 79–82
- laws of motion, 82–83, 85, 88–89
- light, views of, 104
- reflecting telescope of, 544
- tidal forces, views of, 91

Newtonian focus, 112
GC383 Galaxy, 374
602 cluster, 238
NGC1068 Galaxy, 375
NGC1068 galaxy, 371
NGC1300 Galaxy, 338
NGC1309 Galaxy, 350
NGC1316 Galaxy, 367
NGC1569 Galaxy, 359
NGC1705 Galaxy, 359
NGC2998 Galaxy, 352
NGC3603 Galaxy, 221
NGC4038 Galaxy, 361
NGC4039 Galaxy, 361
NGC4258 Galaxy, 369
NGC4261 Galaxy, 371
NGC5128 Galaxy, 360, 373–374
NGC6751 Galaxy, 128
NGC7052 Galaxy, 371
NGC7674 Galaxy, 369
Noachian period, **504**
nodes, **46, 47**
nonbaryonic matter, **403**–404
nonvisible wavelengths, 163
north celestial pole, 19–20, 24
north point, **20**
North Pole, 21
North Star. *See* Polaris
Northern Gemini telescope, 103
Northern Hubble Deep Field, 344
nova, **273**–274
Nova T Pyxidis, 275
nuclear bulge, **320**
nuclear fission, **157**
nuclear fusion, **157**
- aging stars, 269
- carbon, 254
- deuterium, 245
- helium, 251–253
- hydrogen, 245
- massive stars and, 254, 275–276
- solar, 156–161

nuclei, **128**
nucleosynthesis, 325

O

O associations, **226**–227
- brightness, 332
- spiral arms and, 330
- star formation, 227

Oberon, 554
object lens, **107**
oblateness, **514**
observable universe, 392
occulation, **552**
Olbers, Heinrich, 391, 392
Olbers's paradox, **391**
oligarch, **564**
Olympus Mons, 491, 499, 500
once-ionized calcium, 206, 208
Oort, Jan, 590
Oort cloud, 590
opacity of gas, **232**
open clusters, 255–256, 326
open orbits, 86
open universe, **402**
Ophelia, 552
Ophiuchus, 26
Ophiuchus dark cloud, 332
Opportunity spacecraft, 499, 503
optical telescopes, 106–107
optics, adaptive, 113
orbits
- closed, 86, 89
- comets, 424
- Earth, 5, 27, 88–89
- electron, 137
- Jupiter, 439
- lunar, 4, 46
- Mercury, 97, 419, 474–475
- metal and, 329
- Milky Way, 320–321
- open, 86
- sun, 322

Orcus, 564
Orion
- brightness, 15
- mythology, 53
- names for, 13–14
- visibility, 627

Orion Nebula
- disks surrounding, 434
- stars in, 224, 235–238, 432
- type of, 204

Orionids shower, 571
oscillations, 306–307
outflow channels, **501**
outgassing, **430**
ovoids, **554**–555
oxygen
- atmospheric, 454–455, 604
- ions, 206

ozone layer, **454**–457

P

Pagel, Bernard, 329
Pallas, 580
paradigm, **64**
parallax
- defined, 176
- discovery of, **58**
- measuring, 177
- spectroscopic, **186**

Paranal Observatory, 110, 114
parsec (pc), **176**
Parsons, William, 343
partial lunar eclipses, 39
particle accelerators, 93
passing star hypothesis, **417**
Pasteur, Louis, 83
path of totality, **42**
Pathfinder spacecraft, 499
Paul V, Pope, 66
Pauli Exclusion Principle, 250
Payne-Gaposchkin, Cecilia, 141–142
Pegasus, 13–14
penumbra, **35,** 39, 162
penumbral lunar eclipses, 39
Penzias, Arno, 395
perigee, **42**
perihelion, 25, 97
period-luminosity diagram, 315–316
period-luminosity relation, **258,** 259–260
periodic table, 626
Permian, 592
permitted orbits, 130
perpetual motion machines, 160
Perseus galaxy cluster, 377–378
Philolaus, 56
Philosophiae Naturalis Principia Mathematica (Mathematical Principles of Natural Philosophy), 92–94
Phobos, 505–507
Phoebe, 536–537
photons, **104**
  atom absorption of, 131
  big bang and, 397–398
  blue, 204
  energy of, 105
  wavelengths, 132–133
photosphere, **44**
  stars and, 132
  sun, 150–152, 165
Piazzi, Giuseppe, 580
Pioneer spacecraft, 514
Pioneer Venus probe, 485, 486
Planck's constant, 104
planetary motion, **5**
  accelerated, 95
  Aristotle's ideas, 80
  Galileo's observation, 80–82
  gravity and, 80–83
  Kepler's analysis, 70–74
  Kepler's laws, 72–74, 87, 90
  laws of, 82–83, 88–89
  lunar, 34
  momentum, 82
  moon, 42
  natural, 80–82
  nature of, 68
  Newton's laws, 88–90
  orbital, 86–94
  planetary (*See* planetary motion)
  Ptolemaic view, 58
  relativity and, 97–98
  Rudolphine Tables, 73–74
  simple uniform circular, 58
  stars, 177–178
  Tycho's observations, 68–70
  uniform circular, 56
  Uranus, 546
  velocity and, 86
  violent, 80–81
planetary nebulae, **266**–270
planetesimals, **428**
  disappearance, 582
  evolution, 576
  formation, 428, 590
  fragmentation, 583
  icy, 430
  largest, 429
planets. *See also* Jovian planets; terrestrial planets; specific planets
  around young stars, 434–435
  characterization, 4–5
  comparisons, 444
  development stages, 446–447
  distance from sun, 5
  dwarf, 562–565
  extra solar, 435–440
  formation, 417–419, 591
  Jovian, 420, 422–423
  movement, 23–26
  orbits, 5
  outermost, 511
  properties of, 622
  pulsar, 298–299
  terrestrial, 420, 422–423
  types, 419–420, 434
  visibility, 23
plastic, **449**
plate tectonics, 451, 453
Plato, 56, 58, 67
Pleiades, 205, 215–216
plutinos, 564–**565**
Pluto
  discovery of, 562–563
  moons, 563–564
  plutinos and, 564–565
  profile, 563
Poe, Edgar Allan, 392
polar caps, 496, 498
Polaris, 15
  north celestial pole, 19–20
  spectral class, 140
  telescopes fixed on, 111
pollution, **110**
poor clusters, **356**–358
Pope, Alexander, 79, 554
Population I stars, **323**–324
Population II stars, **323**–325
potassium, 424
potential energy, **91**
powers of 10 notation, 618
precession, **19,** 26, 27
predictions, 93
pressure, 209
pressure (P) waves, **447**
pressure-temperature thermostat, 235
Prima Narratio (First Narrative), 62
primary lens, **107**
primary mirror, **107**
primeval atmosphere, **454**
primordial soup, **603**
Principia, 92–94
processes, 325
prominences, **44,** 168
proper motion, **177**–178
proteins, 599–600
proton-proton chain, **158,** 231
protoplantes, 429–430
protostar, **224**
  contracting, 248
  disks in, 225
protostelar disks, **225**
Prutenic Tables, The, 63, 69
pseudoscience, **26,** 27
Ptolemaic universe, 58–60
Ptolemy, Claudius, 15, 58–63, 59
pulsar wind, **291**
pulsars, **290**
  3C58, 293
  B1509, 293
  binary, 295–297
  causes, 258–259
  Cepheids, 259–261
  discovery of, 290
  fastest, 297–298
  model, 290–294
  planets, 298–299
  PSR 1257 + 12, 298–299
  PSR J1740-5340, 298
  PSR J1748-2446ad, 298
  recognizing, 295
Puppis A supernova, 293
pyramids, 27
Pythagoras, 56

Q

- quanta, 120
- Quantitative thinking, 619
- quantum mechanics, 130, 158–159
- Quaoar, 564
- Quasar 351+026, 382
- Quasar 3C175, 383
- Quasar 3C273, 378–379, 382, 384
- Quasar 3C48, 378–379
- quasars
  - discovery of, 378–379
  - distance, 379–382
  - distant galaxies, 382–385
  - fuzz, 380
  - look-back time, 385–386
  - model, 385
  - spectra, 378–380
  - superluminal expansion, 382–385
- quasi-periodic oscillations (QPOs), 306–307
- quasi-stellar objects (QSOs). *See* Quasars
- quiescent prominence, 168
- quintessence, **407**

R

- radar, 142–143
- radar maps, 486
- radial velocity, **143**
- radiant, **570**
- radiation. *See also* electromagnetic radiation; infrared radiation
  - 21-cm, **209**–211
  - black body, **132**
  - cosmic background, 395–396, 398
  - gravitational, 296
  - heat, 132–134
  - irregularities in, 411–413
  - laws of, 134–135
  - wavelength, 142
- radiation pressure, **225**
- radiative zone, **159**
- radio communication, 609–611
- radio energy maps, 118, 119
- radio interferometers, 118
- radio lobes, 370
- radio maps, 329–331
- radio telescopes
  - advantages, 120
  - function of, 118
  - largest, 119
  - limitations, 118
  - parts, 117–118
  - radiation detection by, 211
- io waves, 103–105
- ctive dating, 424–425
- radius, stars, 180–181
- recombination, **398**
- red dwarfs, **183**
  - death of, 265–266
  - main sequence, 245
- red light, 38
- red shifted galaxies, 363
- reddening, **206**–207
- redshifts, **142**
  - expression of, 401
  - quasars, 378–383
  - relativity theory and, 401
  - size of, 393
  - survey, 410
- reflecting telescope, **106**
- reflection nebulae, 203–205
- refracting telescope, **106**–107
- regolith, **469**
- reionization, **399**
- relative ages, **462**
- relativity
  - general theory of, 96–99
  - principle of, 94
  - special, 94–96
- Renaissance, 74
- resolving power, **108,** 114–115
- resonance, **474**
- revolution, **22**
- Rheticus, Joachim, 59, 62
- rich clusters, **356,** 358
- rift valleys, 451
- Rigel, 15
  - characterization, 249
  - expansion of, 250
  - spectral class, 140
- ring galaxies, 358, 361
- rings
  - Jovian planets, 423
  - Neptune, 552–553, 560
  - Saturn, 530–533
  - Uranus, 551
- RNA, 599, 601
- robotics, 517
- Roché limit, **517,** 532
- Roché lobes, **272**
- Roché surface, **272**
- Roentgen, Conrad, 546
- Rossi X-ray Timing Explorer, 307
- rotation curve, **322, 352**–353
- rotations, **22**
  - differential, **164,** 322
  - Earth, 18–19, 30
  - ecliptic, **23**
  - Jovian planets, 431–432
  - M64 galaxy, 361
  - Mercury, 474–475
  - Milky Way, 322–323
  - solar, 165
  - Uranus, 419, 548
  - Venus, 419, 484
- Rover Opportunity, 496
- Rover Spirit, 499, 503
- RR Lyrae stars
  - calibration, 326
  - discovery, 259
  - visibility, 317
- Rubidium, 424
- Rudolph II, 71
- *Rudolphine Tables, The,* 70–71, 73–74
- Russell, Henry Norris, 181
- Rutherford, Ernest, 41

S

- S-type asteroids, 582
- S0 galaxies, 347, 362
- Sagan, Carl, 615
- Sagittarius
  - globular clusters in, 316, 318
  - location, 13
  - radiation from, 335
- Sagittarius A (Sgr A), 336, 338
- Sagittarius Dwarf Galaxy, 356–357
- Saha, Meghnad, 141
- Sapas Mons, 489–490, 492
- saros cycle, 48–49
- satellites. *See* moons
- Saturn
  - auroras, 529
  - celestial profile, 529
  - characteristics, 528–530
  - composition, 513
  - gravitational collapse, 430
  - history, 531–534
  - interior of, 423
  - Jupiter *versus,* 528–529
  - moons, 534–546
  - rings, 530–533
  - size of, 419
- Schiaparelli, Virginio, 495
- Schmidt, Maarten, 378
- Schwarzschild, Karl, 300
- Schwarzschild black holes, 300–302
- Schwarzschild radius, **301**
- science
  - analogy use by, 393
  - arguments in, 8
  - as system of knowledge, 400
  - calibrations in, 318
  - cause and effect, 85
  - classification in, 345

confirmation and, 166
data manipulation in, 487
discoveries, 546
evidence, 29
fraud in, 305
funding, 534
hypotheses in, 83, 466
models, 18
paradigm, **64**
process, 325, 616
scientific method, **3**
skepticism in, 438
technology, 517
scientific notation, **4**
scientific revolution, 64, 75
Scorpius, 13, 321
Scott, David, 81
Search for Extraterrestrial Intelligence (SETI), **610**
seasons, 23–25
second law of motion (Kepler), 89
secondary atmosphere, **454**
Sedna, 564
seeing, **109**
Segmented mirrors, 113
seismic waves, **447**
seismographs, **447**–449
selection effects, **572,** 575
self-sustaining star formation, **333**
semimajor axis, **72**
SETI project, **610**
Seyfert, Carl K., 368
Seyfert galaxies, **368**
energy in, 377
shapes of, 369–370
types of, 369
shadows
Earth, 35, 38, 46
lunar, 42–44
Shapley, Harlow, 316–318, 343, 344
Shapley-Curtis Debate, **344**
shear (S) waves, **447**–449
shepherd satellites, 530
shield volcanos, 490, 499
shock wave, **222**
Shoemaker, Carolyn, 520
Shoemaker, Eugene, 520
shooting stars. *See* meteors
short-period comets, 590
sidereal drive, 113
sidereal period, 34, 38
Sidereus Nuncius (Sidereal Messenger), 65
singularity, **300**
sinuous rilles, **462**
Sirius
brightness, 197–198
discovery of companion, 267
location, 15
mythology, 53
naming, 14
orbital motions, 190
spectral class, 140
sky
atlas of, 14
celestial sphere, 17–18
galaxies in, 390
stars in, 13–14
sun's location, 23
Slipher, V. M., 350
Sloan Digital Sky Survey, 114
Sloan Great Wall, 409
slow surface evolution, 447
small-angle formula, 42
Smith, Robert, 544
smooth plains, **477**
SN1987A, 281–288
Snow, John, 85
soft gamma-ray repeaters (SGRs), **295**
Sojourner spacecraft, 499
Solar and Heliospheric Observatory (SOHO), 588–589
solar constant, 167–168, **167**–168
solar eclipses
features, 44
lunar diameters and, 40–42
observing, 44–46
symbols, 40
total, **33**
solar flares, 166, 169
solar nebula
chemical composition, 427
clearing, 433–434
planetesimals in, 428–430
temperature changes, 429
theory, **418**–419
solar system, 7
age of, 424–426
atom comparison to, 41
characteristics, 431–432
density variation in, 427–428
Earth's origins, 417–419
general view, 419
life in, 607–608
planet types, 419–420
principle satellites in, 623–624
properties, 426
shape of, 426
space debris, 420–424
solar winds, **154,** 169
solids, 427–428
solstices, 23–25, 53
sound, 103
sound, Doppler effect, 142–143
south celestial pole, 20, 24
south point, **20**
space
debris, 420–424
molecules, 211–212
telescopes in, 120–125
Space Interferometry Mission, 115
space-time
curvature, 97–99
shape of, 400–404
special relativity, 94–96
spectra, **104**
AGN, 375
analysis, 136–137
classification, 138–140
comparison, 116–117
digital, 140
Doppler effect, **142**–144
ends of, 121, 124
formation, 135
granulation, 151–152
Jupiter, 512–513
line shape, 144–145
quasar, 378–380
solar, 151
spectral lines, 116
visible, 104–106
spectral classes, **138**–140
spectral sequence, **138**
spectrograph, 116–117
spectroscopic binary systems, **189**–190
spectroscopic parallax, **186**
speed, 82
sphere 62, 196
spherical component, **320**
Spica, 15, 221
spicules, **153**
spinning magnetic fields, 292
spiral arms, **7**
density wave theory, 331–333
radio maps, 329–331
star formation in, 333–338
sun's location and, 321
tracing, 329
spiral galaxies
characterization, 345–346, 348
collisions and, 363
interactions, 362
measuring, 349
spiral nebulae, **343**–344
Spitzer Space Telescope, 123
sporadic meteors, 571

spring tides, **91**
standard candles, **348**
Standard Model, 93
star 51 Pegasi, 436–437
star charts
  Canis major, 196–198
  northern hemisphere sky, 628–634
  southern hemisphere sky, 628–634
star formation. *See also* stellar evolution
  big bang and, 399
  CNO cycle, 231–232, 234
  contracting heating and, 223–224
  evidence of, 225–230
  giant molecular clouds and, 221–223
  gravity and, 284
  Milky Way and, 339
  Orion Nebula and, 235–238
  pressure-temperature, 235
  proton-proton chain, 231
  protostars, 224–225
  rapid, 359
  self-sustaining, 333
  spiral arms and, 333–338
  stellar energy and, 230–232
  stellar structure and, 232–235
star-formation pillars, **227**
starburst galaxies, **359**
stardust, 284
starlight, 98, 132
starquakes, 294
stars
  absolute visual magnitude, 178–179
  ancient views of, 13–14
  asterisms, 14
  binary system, 187–194, 271–275
  brightest, 620
  brightness, 6, 15–16, 178
  clusters (*See* clusters)
  composition, 140–142
  constellations, 13–14
  death (*See* stellar death)
  density, 194–195
  diameters, 180–187
  diffraction fringes, 108
  disk component and, 318–321
  distance modulus, **179**
  distance, measuring, 175–178
  energy produced by, 157
  favorite, 14–15
  first catalog, 15–16
  giant, 182–183
  in Milky Way, 323–325
  intensity, 16–17
  interferometric observations, 183
  intrinsic brightness, 178–180
  life expectancy, 248–249
  luminosity, 179–182, 194–195
  luminosity classification, 184–185
  luminous, 197
  magnetic cycle, 166
  magnitude, 16–17
  main-sequence, 242–255, 620
  mathematical models of, 244, 246
  metal-poor, 329
  most common, 196
  names of, 14
  nearest, 621
  neutron (*See* neutron stars)
  nuclear reactions in, 254
  parallax, 176
  photographing, 109
  photosphere and, 132
  proper motion, 177–178
  radius, 180–181
  spectra, 134–144
  spectroscopic parallax, 186
  supergiant, 182–183
  survey of, 194–199
  temperature, 138
  variable (*See* intrinsic variables)
Stefan-Boltzmann law, 134
stellar death
  binary evolution, 271–275
  Earth's future and, 274–275
  lower-main-sequence stars, 265–271
  mass loss and, 266
  massive stars, 275–284
  planetary nebulae and, 266–269
  red dwarfs, 265–266
  supernovas, 277–284
stellar energy, 230–232
stellar evolution
  binary stars, 271–275
  clusters, 255–258
  evidence in clusters, 255–258
  evidence in variable stars, 258–262
  main-sequence stars, 249–255
  mass transfer with, 272–273
  post-main-sequence, 249–255
  uncertainties, 255
stellar models
  equations, 244
  laws of, 242–243
  main sequences, 244–247
stellar parallax, **176,** 545
stellar populations, 323–325
stellar structure
  energy transport, 232
  inside stars, 234–235
  sun, 233–234
stellar winds, **225,** 267–270, 519
Stonehenge, 53–54
stony meteorites, 572
Stony Tunguska River valley, 593
stony-iron meteorites, 572, **575**–576
Stratospheric Observatory for Infrared Astronomy (SOFIA), 121
Strelley Pool Chert, 603
stromatolites, **604**–605
strong force, **157**
Stukely, W., 53
subatomic particles, 130
subduction zones, 452–453
subsolar point, **484**
summer solstice
  ancient myths, 53, 55
  Earth's rotation and, 24–25
sun
  activity, 161–171
  age of, 426
  angular diameter, 40–43
  annual motion, 22–23
  atmosphere, 150–156
  calcium emissions, 166
  characterization, 5–6
  chromosphere, 152
  corona, 153–154
  cycles of, 22–30
  death of, 271
  Earth's orbit around, 22–23, 30
  eclipses, **33,** 40–45
  elements in, 141
  energy constant, 167–171
  energy generation, 157
  energy transport from, 159
  fusion, 156–161
  gravity, 98
  helioseismology, 154–155
  limb of, **152**
  linear diameter, 42–43
  magnetic cycle, 164–166
  main-sequence, 246–247, 254
  mass, 91
  mass loss, 266
  Mercury distance from, 423
  Milky Way location, 321
  model of, 243
  neutrino problem, 159–161
  observing, 161
  orbital period, 72, 322
  photosphere, 150–152
  planets distance from, 5–6
  power of, 171
  profile, 151
  seasons, 23–25

spectrum, 151
structure, 233–234
Sun Dagger, 55
sun-like stars, 266
sunlight, 91
sunspot cycle, **161**–162
sunspots, **150**
characterization, 161
grouping, 164
measuring, 163
occurrence, 165
visibility, 162
superclusters, **409**
supergiants, **183**
brightness, 187
density, 194–195
lifespan, 254
supergranules, **152**
superluminal expansion, 382–385
supermassive blackholes
active galaxies and, 369, 374–378
formation, 353
mass of, 376–377
origins, 376–378
supernova remnants (SNR), 335
supernovas, **277**
dark energy and, 407
distance measures, 348–349
explosions, 277–278, 284
Kepler's, 281
local, 283–284
observation of, 280–281
remnant, 282
SN1987A, 281–283
type I, 280
type Ia, 348–349, 407
type II, 280
types of, 279–280
surface
Earth, 445
Mercury, 476–478
moon, 462
Roche, **272**
slow, evolution, 447
Venus, 485–489
surveyor's method, 175–176
Swicky, Fritz, 288
synchrotron radiation, **281**
synodic period, 34, 38
Systèm International d'Unités (SI units), 617

T

T associations, **226**
T dwarfs, **140**
T Tauri stars
gas disk age, 430
location, 228
mass, 226–227
Tarantula Nebula, 347
Taurid meteor shower, 571
Taurus-Littrow, 468
Taylor, Joseph, 296
telescopes, 16, **103**. *See also* specific telescopes
adaptive optics, 113
buying, 111
costs of, 107
Galileo's use of, 65, 67
high resolution, 114–115
high-speed computers, 114
infrared, 121
lens, 107
mountings, 113
new generation, 111–114
optical, 106–111, 115
photosphere blocking, 153
powers of, 108–111
radio, 117–120
size, 103
space, 120–125
special instruments on, 115–117
traditional, 111–113
temperatures, **132**
corona, 154
diameters and, 180–181
escape velocities, 497
heat *versus,* 133
Mars, 483, 497
measuring, 132
Mercury, 476–477
prediction, 28
scales, 618
solar nebula, 429
star, 138
transition region, 152
Uranus, 547, 550
terminator, **461**
terrestrial planets, 420
atmospheres, 445–446
characteristics, 422–423
development states, 446–447
formation, 445
interiors, 445
lists of, 444
rotation, 431–432
size comparison, 444
surfaces, 445
Tethys, 536–537, 539
Thales of Miletus, 48, 56
Tharsis rise, 500, 504
Thebe, 520
theories, **83**
facts and, 215
hypotheses and, 466
observations *versus,* 391
predictions, 93
star formation, 226
thermal energy, **132**
star formation and, 222–224
temperature *versus,* 133
thermal pulses, **266**
Third law of motion (Kepler), 90–94, 188
Thomson, J. J., 41
Thuban, 19
tides
causes, 90
coupled, 461
distortion, 360
forces, 90–92, 91–92
heating, 523
moon's effect on, 90–91
tails, 358
time dilation, **303**
Titan, 534–536, 539–540, 607
Titania, 554, 555
Tombaugh, Clyde, 562
torus, 527
total eclipses, 194
total lunar eclipse, 35–39
total solar eclipse, **33,** 44–45
totality, **38**
length of, 44
path of, **42**
transition region, **152**
Trapezium, 236
Trapezium stars, 238
Trifid nebula, 217
trilobites, 83, 606
triple alpha process, **251**
Triton, 560–561
Trojan asteroids, 582
Tunguska event, 592–595
turbulence, 222
21-cm radiation, **209**–211
Two-Degree-Field Redshift Survey, 410
Two-Micron All Sky Survey (2 MASS), 114
Tycho Brahe, 68–70, 464

U

ultraluminous infrared galaxies, 362
ultraviolet radiation, 105, 124, 214
umbra, **35,** 38, 162
Umbriel, 554
uncompressed densities, **427**
unidentified flying objects (UFOs), 609–610

unified models, 374
uniform circular motion, 56
universe, 7
acceleration, 406–408
age of, 394, 408–409, 412
big bang and, 327, 396–400
closed, 401
colors in, 3
curvature of, 409–413
dark matter in, 402–404
edge to, 390–391
elements in, 326
expansion of, 393, 397
falling, 99
fate of, 408–409
flat, 402, 412–413
galaxies in, 318
geocentric, 57, 58
heliocentric, 59, 63
history, 8–9
homogeneity, 400–401
inflationary, 404–406
introduction to, 390–400
irregularities in, 411–413
island, **343**
isotropy, 400–401
life in, 608–613
model, 401–402
necessity of a beginning, 391–393
observable, 392
open, 402
origins, 350
Ptolemaic, 58–60
reionization of, 399
space-time curvature, 97–99
structure of, 409–413
Upsilon Andromedae, 436
Uranus
atmosphere, 546–547
celestial profile, 547
discovery, 544–546
history of, 555, 557
interior of, 423, 547, 549–551, 557
magnetic fields, 550
moons, 551–555, 557
motion of, 546
rings, 551
rotation, 419, 548
size of, 419
temperatures, 547, 550
Urban VIII, Pope, 66
Urey, Harold, 603
Ursa Major, 14

V

Valles Marineries, 500
valley networks, **501**
Van Allen belts, **451**
van de Hulst, H. C., 210
variable stars. *See* intrinsic variables
Vega
brightness, 16
dust surrounding, 435
north celestial pole, 19
visibility, 15
Vela pulsar, 295
Vela satellites, 308
velocity, **82**
circular, 86, 90
Doppler effect, 143–144
escape, 299–300
escape, calculating, 86–87
light, 95–96
orbital, 86
radial, **143**
velocity dispersion method, **353**
Venus
celestial profile, 483
clouds over, 484
Earth *versus,* 493
elongations, 625
Galileo's observation of, 65
greenhouse, 485
history, 493–494
movement, 23
names of, 487–488
radar maps of, 486–488
rotation, 419, 484
surface, 485–489
volcanism on, 489–493
water on, 485
Vernal equinox, 24
Very Large Array (VLA), 118
Very Large Telescope (VLT), 111, 115
Very Long Baseline Array (VLBA), 118
Very Long Baseline Interferometry (VLBI), 382
vesicular, **468**
Vesta, 579–580, 583
Victoria, Queen, 540
Viking spacecraft, 499, 501
Violent motion, 80–81
Virgo cluster
galaxies in, 356, 409
redshift, 393
spiral galaxies in, 362
virus, 325, 448
visible light
ends of, 121, 124
waves in, 103
visible wavelengths
clouds, 207–209
energy and, 104
extinction, 206
interstellar absorption lines, 206–207
nebulae, 203, 206, 269
reddening, 206
visual binary system, **188**–189
Volatile-rich meteorites, 576
volcanos
Deimos, 507
Io, 526
Martian, 499–500
Phobos, 507
Venusian, 489–493
von Fraunhofer, Joseph, 128
Voyager spacecrafts, 514, 516, 532

W

waning gibbous moon, 35–37
War of the Worlds, The (Wells), 495
warm rooms, 114
water
carbon dioxide solubility, 486
hydrogen molecules in, 270
Martian, 501–504
Venus and, 485
water hole, **610,** 613
Watson, James, 18
wavelength of maximum intensity, 133
wavelengths, **103**
far-infrared range, 121
frequency and, 104
longer, 105
microwaves, 104–105
nonvisible, 163
photon, 132–133
regions, 105–106
ultraviolet, 105
visible light, 104, 203–209
waxing gibbous moon, 35–37
weak force, **157**
weakly interacting massive particles (WIMPS), 403–404
Wekiu bug, 598
Wells, H. G., 495
Wells, Orson, 495
Whirlpool galaxy, 360
white dwarfs
core of, 280
density, 194–195

discovery, 267
energy flow, 267
future of, 270
interior of, 267, 270
mass, 194
surface gravity of, 270
Widmanstätten patterns, **573**
Wien's law, 134
Wilczyk, Frank, 405
Wilkinson Microwave Anisotropy Probe (WMAP), 410–412
Wilson, E. O., 8
Wilson, Robert, 395
winter solstice, 24–25, 53
WR124, 271

X

X Cygni, 261
X rays
absorbtion of, 105
binaries, 304, 307
black holes emitting, 304, 306
bursters, **306**
from interstellar medium, 213–214
in visible light, 103
neutron star detection, 299
XTEJ1751-305, 298

Z

Zeeman effect, 161, 163
zenith, **20**
zero-age main sequence (ZAMS), **245**–248
zodiac, **23**